Lecture Notes in Networks and Systems

Volume 735

The series "Lecture Notes in Networks and Systems" publishes the latest developments in Networks and Systems—quickly, informally and with high quality. Original research reported in proceedings and post-proceedings represents the core of LNNS.

Volumes published in LNNS embrace all aspects and subfields of, as well as new challenges in, Networks and Systems.

The series contains proceedings and edited volumes in systems and networks, spanning the areas of Cyber-Physical Systems, Autonomous Systems, Sensor Networks, Control Systems, Energy Systems, Automotive Systems, Biological Systems, Vehicular Networking and Connected Vehicles, Aerospace Systems, Automation, Manufacturing, Smart Grids, Nonlinear Systems, Power Systems, Robotics, Social Systems, Economic Systems and other. Of particular value to both the contributors and the readership are the short publication timeframe and the world-wide distribution and exposure which enable both a wide and rapid dissemination of research output.

The series covers the theory, applications, and perspectives on the state of the art and future developments relevant to systems and networks, decision making, control, complex processes and related areas, as embedded in the fields of interdisciplinary and applied sciences, engineering, computer science, physics, economics, social, and life sciences, as well as the paradigms and methodologies behind them.

Indexed by SCOPUS, INSPEC, WTI Frankfurt eG, zbMATH, SCImago.

All books published in the series are submitted for consideration in Web of Science.

For proposals from Asia please contact Aninda Bose (aninda.bose@springer.com).

Celestine Iwendi · Zakaria Boulouard ·
Natalia Kryvinska
Editors

Proceedings of ICACTCE'23—The International Conference on Advances in Communication Technology and Computer Engineering

New Artificial Intelligence and the Internet of Things Based Perspective and Solutions

Springer

Editors
Celestine Iwendi
University of Bolton
Bolton, UK

Zakaria Boulouard
Hassan II University Casablanca
El Mansouria, Morocco

Natalia Kryvinska
Department of Information Systems
Comenius University
Bratislava, Slovakia

ISSN 2367-3370 ISSN 2367-3389 (electronic)
Lecture Notes in Networks and Systems
ISBN 978-3-031-37163-9 ISBN 978-3-031-37164-6 (eBook)
https://doi.org/10.1007/978-3-031-37164-6

This Springer imprint is published by the registered company Springer Nature Switzerland AG
The registered company address is: Gewerbestrasse 11, 6330 Cham, Switzerland

Preface

Over the past few years, there has been a significant acceleration in the development of new technologies, particularly in urban areas due to the growth of the population. This has led to the emergence of the concept of smart cities. With the exponential growth of urbanization and the digital revolution, cities are being equipped with advanced information and communication technologies to aid decision-making and improve urban functionality across a range of sectors.

Artificial Intelligence and the Internet of Things have emerged as must-have technologies across many industries, including agriculture, healthcare, and industry. The scope of applications for AI and IoT is vast, and these technologies are now extensively used in developed countries. However, emerging countries are still in the early stages of adopting these technologies.

The COVID-19 pandemic has underscored the need for sustainable development solutions, particularly in emerging countries, which have been disproportionately affected by the pandemic. The pandemic has also highlighted the role of advanced technologies in facilitating remote work, education, and healthcare.

In this context, the International Conference on Advances in Communication Technology and Computer Engineering (ICACTCE'23) was organized in Bolton, UK from 24–25 February 2023. The conference aimed to bring together innovative scientists, professors, research scholars, students, and industrial experts in the field of Computing and Communication to a common forum.

The scope of the conference was to provide a common platform for researchers, engineers, academicians, and industrial professionals from all over the world to present their research results and development activities in various topics of Engineering and Technology. The conference covered a broad range of topics, including Smart cities, Artificial Intelligence, the Internet of Things, Communication Networks, and Data Analytics.

The event was a comprehensive two-day program that featured a range of activities designed to explore the latest developments and applications of communication networks, artificial intelligence, and related fields. The program included keynote speeches from some of the most influential leaders in both academia and industry, oral communications by experts, and special sessions that provided attendees with the opportunity to delve deeper into specific topics of interest, such as AI for Mental Health and AI for the Digital Transformation of Business Practices.

One of the highlights of the event was the lineup of featured speakers, which included four distinguished professionals who have made significant contributions to the fields of communication networks, artificial intelligence, and related areas. These plenary speakers included:

- **Prof. Dr. Jyotika Athavale**, a Senior Technical Leader at Nvidia Corporation and President-Elect of the IEEE Computer Society. With a large experience in the industry, Prof. Dr. Athavale has become a leading figure in the field of computer science, and her work has been widely recognized for its contributions to the development of cutting-edge technologies.
- **Prof. Dr. Sudha Ramasamy**, a Senior Lecturer at the Division of Production Systems, University of West, Sweden, and Vice President of "IEEE Women in Engineering" Sweden Chapter. Prof. Dr. Ramasamy is an accomplished researcher and educator with a deep understanding of the complex interplay between artificial intelligence, robotics, and automation.
- **Prof. Dr. Imed Romdhani**, an Associate Professor in Networking at Edinburgh Napier University, UK. Prof. Dr. Romdhani is a highly respected expert in the field of networking, and his work has been instrumental in the development of new technologies that have helped to advance the field of communication networks.
- **Prof. Dr. Mohammed M. Alani**, a Cybersecurity Professor at Seneca College and Research Fellow at Ryerson University, Toronto, Canada. Prof. Dr. Alani is a renowned expert in the field of cybersecurity, and his work has been widely recognized for its contributions to the development of new strategies and technologies for protecting communication networks from malicious attacks.

We were extremely gratified to have such distinguished speakers share their insights and expertise with us at the event. Their presentations covered a wide range of topics, from the latest developments in artificial intelligence and communication networks to the ethical and social implications of these technologies.

In addition to the keynote speeches, the event featured a number of oral presentations by experts in the field, as well as special sessions that provided attendees with the opportunity to engage in deeper discussions on specific topics. The oral presentation sessions covered a wide range of topics, including the latest developments in communication networks, the use of artificial intelligence in healthcare and other industries.

Overall, the event was a resounding success, and we are proud to have brought together so many experts and professionals from around the world to explore the latest developments and applications of communication networks, artificial intelligence, and related fields. We look forward to hosting similar events in the future and continuing to support the advancement of these important technologies.

Bolton, UK — Celestine Iwendi
El Mansouria, Morocco — Zakaria Boulouard
Bratislava, Slovakia — Natalia Kryvinska

Contents

Communication Networks and Cybersecurity

Enhancing the Performance of Cloud Environment by a Novel Three-Stage Task Scheduling Policy

R. Thilagavathy, R. Deebalakshmi, S. Jayasankari, and V. Nivedita

Abstract Cloud computing is a method of granting remote access to assets that have been dispersed. In distributed decentralized systems such as clouds, one of the most important performance problems is how to ensure that activities are finished within the expected durations while fairly and efficiently utilizing available bandwidth and compensating for variations in resource performance. The cloud client and specialist co-op look at cloud administration from different perspectives. The cloud customer is concerned with reducing response time and administration expenses, while the cloud provider is concerned with making effective use of cloud assets, ensuring administration reliability, and lowering maintenance costs. Previously, numerous models such as replication heuristics, migration, and reservation were used to address these types of issues using a number of different scheduling strategies; nevertheless, still they have certain unsolved issues. The proposed work presents a dynamic three-stage job scheduling model in the cloud to improve task scheduling performance. The proposed model employs the job attributes to bind them with the resource which has the shortest possible completion time, while simultaneously using the providers' resources efficiently and satisfying the quality of service (QoS) standards of cloud users. This method of scheduling works by calculating the expected value for every task that is proportionate to how the features of that job relate to the characteristics of all other jobs combined. As a result, their virtual machine selection should be more

R. Thilagavathy (✉)
Department of Computing Technologies, School of Computing, College of Engineering and Technology, SRM Institute of Science and Technology (Deemed to Be University), Kattankulathur, Chennai, India
e-mail: thilagar2@srmist.edu.in

R. Deebalakshmi · S. Jayasankari · V. Nivedita
Department of Computer Science and Engineering, SRM Institute of Science and Technology (Deemed to Be University), Tiruchirapalli, India

S. Jayasankari
School of Computing, SASTRA Deemed to be University India, Thanjavur, Tamilnadu, India

V. Nivedita
CSE Department, SRM Institute of Science and Technology, Ramapuram, Chennai, Tamilnadu, India

C. Iwendi et al. (eds.), *Proceedings of ICACTCE'23—The International Conference on Advances in Communication Technology and Computer Engineering*, Lecture Notes in Networks and Systems 735, https://doi.org/10.1007/978-3-031-37164-6_1

in line with their expectations, and hence more equitable. The testing results show a significant reduction in task completion time as well as energy usage.

Keywords Cloud computing · Energy efficiency · Job classification · Job scheduling · Virtual machine management · QoS

1 Introduction

Cloud computing has recently established itself as one of the most powerful frameworks for distributed computing [1, 2]. A cloud is a distributed and parallel framework composed of a number of networked and virtualized PCs which are powerfully provisioned and seen as one or more computing assets. Virtualization is a key to the success of distributed computing. Virtual Machines (VMs) are deployed to hosts in Cloud systems based on their host resource use, rather than their long-term and overall utilization. The scheduling and distribution procedures, like many others, are computationally expensive and have an impact on the performance of the deployed VMs. Job scheduling is amongst the most critical issues in cloud computing., especially when the task's deadline and cost are taken into account.

In cloud work scheduling, virtual machines (VMs) are essential as a crucial actuator. To fulfill task deadlines, we need to cut down on the time it takes to create virtual machines, wait for tasks, and execute them. We must schedule jobs to execute on the most relevant VMs to decrease task execution expenses. Clouds are on-demand computing platforms that often have many "centralized" resources, like storage and CPUs. Cloud computing systems have basic benefits over conventional distributed computing systems-notably grids, in terms of extensibility, durability, flexibility, and long-term viability [3, 4]. Also, users can use a collection of cloud resources as needed in a resource-as-a-service model.

Virtualization is one of the most used methods for recognizing clouds which enables cloud service providers to deliver practically any form of computing system, eliminating the complexity and cost of supporting a broad variety of software and hardware to fulfill customers' demands [5, 6]. Clouds, unlike most grids, are owned and operated through major IT service providers (e.g., Google, Amazon, and GoGrid), and are priced on the basis of usage or subscription. Amazon's Elastic Compute Cloud (EC2), for example, leased out cloud processors on hourly basis to the consumers [7, 8]. In cloud computing the traditional task scheduling methods, only evaluate how to ensure the required QoS for users or how to improve job scheduling efficiency; they rarely propose a scheme that focuses on combining all these qualities at the same time. At this moment, numerous task scheduling strategies are utilized as indicators in cloud computing sectors [9–11]. By emphasizing efficiency, most of them may be classified as cloud-based [12, 13]. Some of them spend a lot of time thinking about user fairness, which has an impact on competence.

Cloud resources, or at least those allotted to a given users are stable, uniform, and closely connected regularly. In addition, the characteristics, amount of resources, and categories of resources to be reserved can all be customized to meet the user's needs. These characteristics distinguish cloud resources as an appropriate environment for the systems that deal with uncertainty. However, some of these systems are not convenient due to financial expenditures and the use of cloud resources. Clouds contain several characteristics, including dynamicity and large scale. Physical hosts may be allowed to gradually enter or exit clouds for cloud service providers, while cloud administration clients' jobs are allowed to submit and be planned effectively. As a rule, task dispersion isn't uniform, necessitating greater oversight and planning of tasks [14, 15]. The tasks required to build cloud architecture are usually diverse and heterogeneous, with varying Quality of Service (QoS) requirements.

Advanced reservation, in particular, is not appropriate for dealing with cloud resource performance changes. Migration may be a better option, although it cannot be simply incorporated into diverse computing systems such as clouds, and it requires time-consuming and sophisticated operations such as check-pointing. Because replication tends to spend resources redundantly for its numerous clones, it is an untenable solution. As a result, we've decided to pay attention to reallocation as an additional support strategy for cloud computing scheduling techniques. Dynamic scheduling algorithms are frequently cited as one of the best current options for problem-solving, as they offer ideal solution but only locally.

A dynamic algorithm may not be the ideal option for all issues; nevertheless, it may offer an optimal or near-optimal solution for wide-range of situations. So, in this paper, we propose a dynamic method along with an energy-efficient allocation scheme in a cloud context to fulfill user's QoS needs also for effective utilization of resources. The suggested dynamic model is divided into three stages, the first of which is the classification stage based on QoS (see Fig. 1). In the first stage, we divide jobs into two categories depending on their bandwidth (BW) and completion time (CT). Then, for each work, a general expectation value is assigned based on its features (for example expected bandwidth, expected execution time, and job length) in comparison to all other jobs in the same category.

After that, each job at each type is assigned a priority value, and by using the scheduling scheme jobs are assigned to a certain resource [16]. Finally, in the third stage of the dynamic model, a simple method for dealing with tasks that do not fulfill the mapping criterion is used, which in this case adjusts the job's general expectation value based on the job's attributes and available resources and then it will be moved further for the reallocation stage. This method effectively utilizes the VMs and reduces the failure rate.

2 Related Works

In the IT industry, cloud computing has evolved into a more powerful technology that delivers resources for improved business profit growth. When the number of users attempting to use a resource in a cloud environment grows, job scheduling becomes increasingly important in reducing the datacenter load's complexity. The Berger

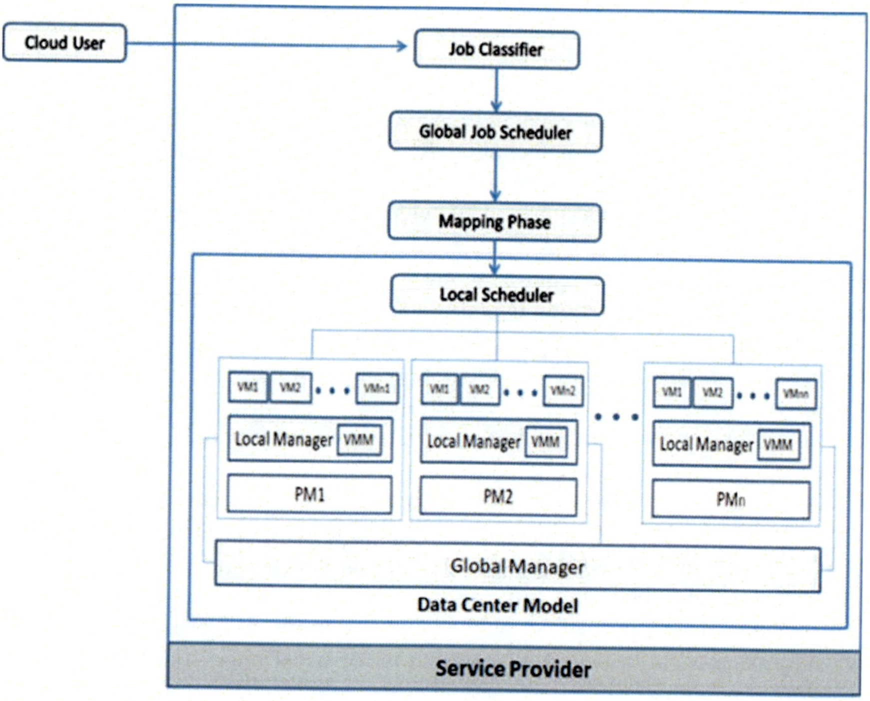

Fig. 1 System model of the service provider which consists of two models known as a three-stage global task scheduler and data center (s)

model [9], Bags-of-Tasks [10], Multi-QoS Cost-Based Scheduling [11], Priority Scheduling [12, 13], Utility-Based Scheduling [15], Optimal Workflow Scheduling [17] are just a few of the algorithms, approaches, and strategies that have been presented to help with this.

Zhang et al. [4] proposed a two-stage solution for reducing irrational task allocation and improving work scheduling performance in clouds. Using historical scheduling data, a job classifier prompted by a Bayes classifier's design concept is utilized to categorize the jobs in the first step. As a result, a variety of virtual computers are created, it saves time on job scheduling by eliminating the need to create virtual machines. The jobs are dynamically assigned to multiple virtual computers during the second step. Srichandan et al. [5] offered a work scheduling technique based on a hybrid algorithm, which combines the key features of two of the most extensively utilized bio-inspired heuristic techniques in cloud computing: genetic algorithms and bacteria for aging algorithms. This research makes a two-fold contribution. Both ecologically and economically, the first goal is to lower the make span, and the second goal is to reduce energy usage.

Thilagavathy et al. [6] show how to employ a meta-heuristic approach known as "Grey-Wolf Optimization" to make most of a cloud environs. The results of the simulation demonstrate that, the work completion time has decreased significantly.

Aldulaimy et al. [7] the authors have introduced a novel Mixed Type Placement technique that places virtual machines of multiple types of jobs in the same physical machine. The amount of physically active machines in the data center is reduced as a result of this method, resulting in a significant reduction in total energy usage. Xu et al. [9] illustrate a job organization technique based on the "Berger model of distributive justice social theories". This paradigm generates an approach for task management that considers both justice and efficiency. It compares the projected state of the jobs to the actual state." Likewise, to obtain a solution that meets the user's QoS requirements, an Optimal Workflow-based Scheduling (OWS) technique [17] was carried out. This project focuses on cloud scheduling of many sorts of operations. The utilization of CPU resources was significantly improved by using this method.

Huang et al. [18] proposed an alternative task scheduling model based on QoS, in which two elements (completion time and cost) have been considered from the user's perspective to fulfill user's needs on time and at a fair cost. However, the cost of the waiting job queue is not factored into this scheduling mechanism. Different scheduling policies in the fair scheduler (like FIFO or delay scheduling) were employed, where the challenges of work scheduling were examined because of its importance in cloud computing [19].

Deore et al. [20] describe an energy-efficient task planning and distribution technique that claims to reduce the number of physically active machines. Feller [21] the IaaS cloud service concept was used to provide on-demand VM provisioning as a compute infrastructure. In virtualized cloud data centers, the bin packing approach is used to provide distributed dynamic consolidation of virtual computers [22]. Wang et al. [23] present a task scheduling procedure based on the catastrophic genetic algorithm (CGA) to reach global optimal for jobs that need to be migrated to the cloud. The technique uses a fitness function to calculate the overall job completion time and penalty factor. CGA's simulation findings demonstrate that it effectively schedules jobs to the VM, but it fails to manage dynamic and real-time circumstances.

For scheduling work in the cloud, Attiya et al. [24] offer simulated annealing (SA) based Harris hawk's optimization (HHO) method known as HHOSA. The usage of SA as a local search algorithm in HHOSA method increases the pace of convergence and accuracy of results provided via the normal HHO algorithm. In cloud computing, scheduling virtual machines for user requests is an NP-complete problem.

Sundaravadivazhagan et al. [31] presents a scheduling policy which deconstructs the conventional methods that depend on client requirement and job length. According to this model, People who are too preoccupied with errands are not given any extra consideration once they arrive. A Tree structured Deep Convolutional Neural Network (T-CNN) with sheep flock optimization method was used by Karthikeyan et al. [32] to predict the workload for sustainable cloud data centers.

The majority of existing job scheduling algorithms use assumptions or historical values to justify jobs in a static manner. However, if such assumptions are made statically without the consideration of job characteristics, the mapping of jobs to VMs may be affected and the fairness judgment may be impacted. As a result, the suggested scheduler gathers user tasks and distributes them to available resources by considering job qualities and demands such as waiting time, deadline, and duration, as well as resource parameters and properties. It also categorizes tasks based on QoS requirements and uses a Mixed Type Placement method to assign VMs to PMs, ensuing in a reduction in the amount of active physical machines and a substantial reduction in overall data center energy usage.

3 Proposed System

The suggested DTS (Dynamic Task Scheduling) is built on the cloud computing paradigm, in which users request services from a Cloud provider in order to complete tasks. As illustrated in Fig. 1, the system is made up of two primary components: User and the Provider. The cloud user sends the task (or collection of tasks) to the cloud provider for execution, usually with an implicit QoS requirement.

As illustrated in Fig. 1. Two models make up the Provider component: A three-stage Global Task Scheduler, and Data Center (s). The global scheduler serves as a link between users and cloud infrastructure. The Task Scheduler first categorizes jobs according to the provided QoS metric. After that, it analyses and evaluates the service needs of the tasks supplied before deciding whether to admit or refuse them depending on resource availability. Finally, it chooses the data center that will run the processes in order to decrease energy usage while maintaining QoS.

3.1 Three-Stage Task Scheduling

As illustrated in Fig. 2, the proposed scheduling method consists of three phases, one of which is the classification step based on QoS. In the early phases, we divide jobs into two categories: completion-time (CT) based and band-width (BW) based jobs. After that, we allot a general expectation value for each job based on its features (such as expected bandwidth, job length and expected execution time) in relation to all other jobs in the same category. The scheduling method will then be used to map each job to a specific resource in the second Phase. Finally, in the third phase we introduce a fundamental strategy for dealing with tasks that do not meet the mapping requirements. At this stage, the jobs are prepared for the reallocation step depending on their features and available resources.

Job Classification. The jobs-classifier assigns a classification to a task based on its characteristics and baseline QoS settings (CT and BW). Then, each accessible Virtual Machine (VM) is assigned a priority value. Finally, the jobs are sent to various scheduler sections, which will dispatch them to the most suitable available resource for execution (VMs). When users submit jobs in the task organization section, it dynamically calculates the characteristics of each job and QoS properties, followed by the initial expectation value for each job in each class will be calculated. Finally,

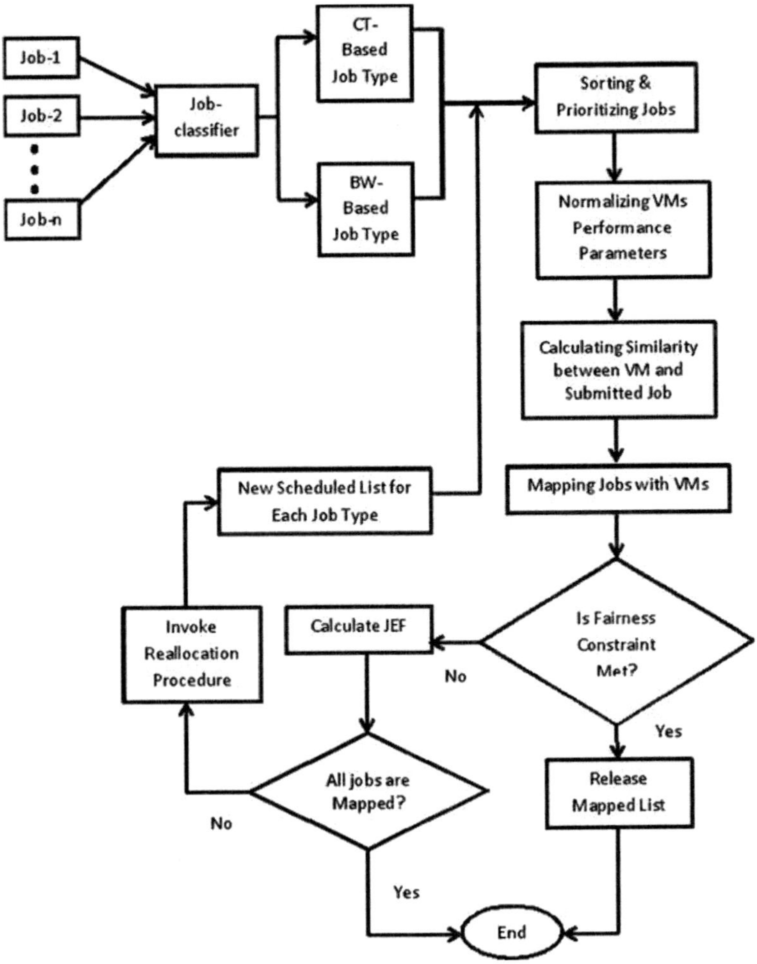

Fig. 2 Work-flow diagram of three-stage task scheduling

these values are stored in the job attributes vector which contains Job's Class-Type, Starting Time, Expected Finishing Time, Priority, Expected Bandwidth and Justice Evaluation Factor (JEF).

Virtualization is used in cloud computing to assign host resources to the VM layer. Assume the following properties of the virtual machine (VMi) resource set:

$$A_i = A_{i1}, A_{i2}, A_{i3} \tag{1}$$

where Ai1 denotes the CPU, Ai2 denotes memory, and Ai3 denotes bandwidth. VMi's performance is represented by the following vector:

$$P_i = P_{i1}, P_{i2}, P_{i3} \tag{2}$$

where Pi1, Pi2 and Pi3 are the related performance value of the attributes Ai1, Ai2 and Ai3 respectively. The job's general expectation function is created in order to establish a fair resource selection.

This function takes into account a range of user QoS work parameters and changes the virtual machine performance ratio accordingly. The following is an example of an expectation function:

$$ET_i = ET_{i1}, ET_{i2}, ET_{i3} \tag{3}$$

$$\sum_{n=1}^{3} ET_{in} = 1 \tag{4}$$

At this point, we calculate General Expected Value (GEV) for each job in each job-class (band-width based and Completion-time based jobs) that is proportionate to the correlation between a job's attributes with all other jobs; this approach is known as "Atomization," and it follows the Eq. (5):

$$GEV_A = ET_{Ai}/Sum(ET_{An}) \tag{5}$$

where GEVA reflects the work length in CT-type jobs and the bandwidth required for BW-type tasks.

Job Scheduling. The atomization approach was used to construct the initial expectation value for each task in each category, and the available resources were gathered and sorted in descending order in a VM list (VMi_List). Then the job's resource mapping began with the selection of a virtual machine. To fulfill the fairness requirements, the justice evaluation function (JEF) is utilized in the resource selection process. The JEF is calculated as follows:

$$\mathrm{JEF} = \theta(\log(\mathrm{R_{alloc}}) - \log(\mathrm{R_{exp}})) \tag{6}$$

where ϴ is a constant which lies between 0 and 1. Ralloc is the total amount of resources allocated to job Ji after mapping and R_{exp} refers to the total amount of resources that J_i expected to get as part of its service. Fairness is attained when R_{alloc} equals R_{exp} and JEF equals 0. Each job J_i can be mapped to any VM in the list of available VM_i as long as the required number of VM_i for execution does not go beyond the available number of VM_i. Then, normalization groups the virtual machines' performance parameters into a list named P[i] that ranges from [0, 1], allowing the comparison with the ET[i] vector of general expectations to proceed. The normalized value is calculated as:

$$\mathrm{NP} = (\mathrm{P_{cur}} - \mathrm{P_{min}})/(\mathrm{P_{max}} - \mathrm{P_{min}}) \tag{7}$$

where P_{cur} is the performance parameter's current value, P_{min} is the performance parameter's minimum value, and P_{max} is the performance parameter's maximum value. The resultant performance parameters list NP of VM_i may then be utilized to match the task ET expectations function (Eq. (4)). The following formula may be used to estimate the correlation between the performance vector and the initial general expectation vector:

$$S = 1 - (NP[i] * ET[i])/(NP[i] * ET[i])^2 \tag{8}$$

Finally, jobs are iteratively assigned to the suitable virtual machine until all jobs have been mapped. If one of these tasks invalidates the prior restrictions, the task will be remapped after changing the initial general expectation value as indicated in Eq. (5). The system will verify whether all tasks have been mapped in the last round, and if they haven't, it will use the reallocation technique.

Reallocation. Reallocation procedure for all failed bandwidth type jobs is given as follows,

```
for all failed bandwidth type jobs do
    Sort jobs according to their bandwidth in an ascending manner.
    Pick the job that has the largest length (hence needs the largest available VM's BW)
    Sort VMs according to their remaining available bandwidth in an ascending style.
    Select the 1st VM in the sorted BW VM list and mark it as preferred VM.
    if preferred VM's available remaining bandwidth matches the BW required to execute the largest job then
        Map that job to that VM
    else
        Move the job to the next VM in the preferred VM list
    end
end
```

Reallocation procedure for all completion type jobs is given as follows,

```
for all failed completion time type jobs do
    Sort jobs according to their length in a descending manner.
    Pick the job that has the smallest length
    Sort VMs according to their remaining length in an ascending style.
    Select the 1st VM in the sorted timeVM list and mark it as preferred VM.
    if preferred VMs available remaining capacity matches the required one to execute the largest job then
        Map that job to that Vm
    else
        Divide them in half and move to the next preferred Vm
    end
end
```

Finally, each job type's new scheduled list is returned by the reallocation method.

4 Performance Evaluation

A genuine workload trace in a cloud computing environment was simulated using the CloudSim emulator. The simulation is based on Google workload traces [25], which were gathered from big cloud platform (about 11,500 compute nodes over 20 days).

Different sorts of jobs are shown in the traces [26, 29, 30]. Table 1. shows the job parameter settings utilized in this simulation (expected bandwidth, expected time, length, file size, and output size). There are 50, 200, 400, 600, 800, and 1000 tasks in the test sets. Figures 3, 4, 5 show the outcomes of the simulation. The suggested model's (DTS) performance is compared to existing methods like Berger model [9], Fuzzy Logic [27], and Round Robin [28] techniques using execution time, bandwidth, and total used energy as metrics.

Figure 3 and Table 2 demonstrate that the DTS method has a faster task execution time than the other three techniques (the Berger model algorithm, fuzzy logic algorithm and Round Robin algorithm). The DTS algorithm's completion time is less to the other three methods in comparison because the suggested scheme lowered the job's needed execution time.

Figure 4 depicts a comparison of the allotted virtual machine bandwidth across tasks of type-2 class (BW-Type Jobs). DTS-MTP allows jobs 6–10 to utilize almost the same expected amount of BW, indicating more fairness because each job's bandwidth is equivalent to its expected bandwidth.

A comparative analysis of Class-1 (Completion Time) type jobs based on CPU utilization is depicted in Fig. 5. It shows that the DTS and fuzzy logic algorithms

Table 1 Job characteristics

Clouldlet_ Id	Class type	Length	File size	Output size	Expected time	Expected bandwidth
1	1	2000	1800	600	150	–
2	1	4850	3000	1200	500	–
3	1	600	350	200	100	–
4	1	1500	700	500	150	–
5	1	2500	2000	700	250	–
6	2	3000	2100	1500	–	2500
7	2	4000	2700	1800	–	3000
8	2	1800	1000	500	–	2000
9	2	750	300	300	–	1000
10	2	3200	2500	1000	–	2500

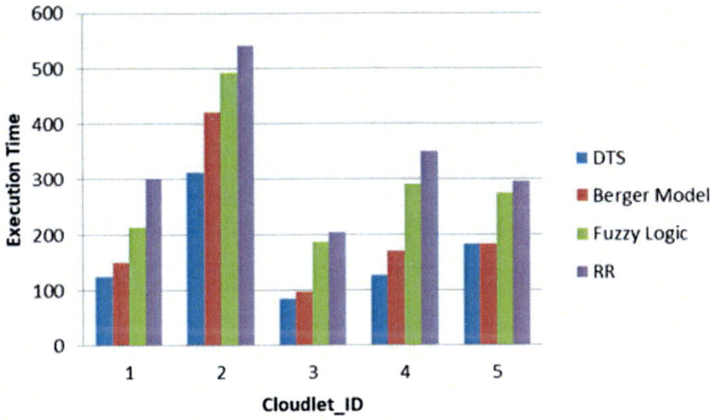

Fig. 3 Comparison of class-1 type jobs in-terms of execution time

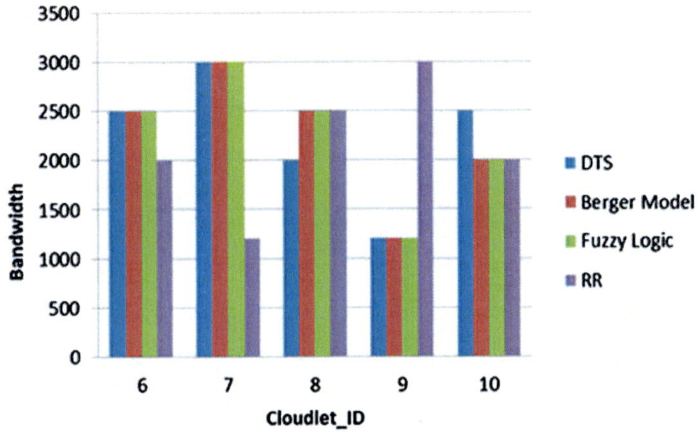

Fig. 4 Comparison of Class -2 type Jobs in-terms of Bandwidth

provide high processing power (CPU usage) and greater fairness; the Berger and RR algorithms produce poor outcomes.

5 Conclusion

The proposed DTS paradigm in this paper operates on two levels. A dynamic three-stage task scheduling scheme is used at the first level, which classifies jobs based on completion time (CT) and bandwidth (BW), and then map the jobs to the highly

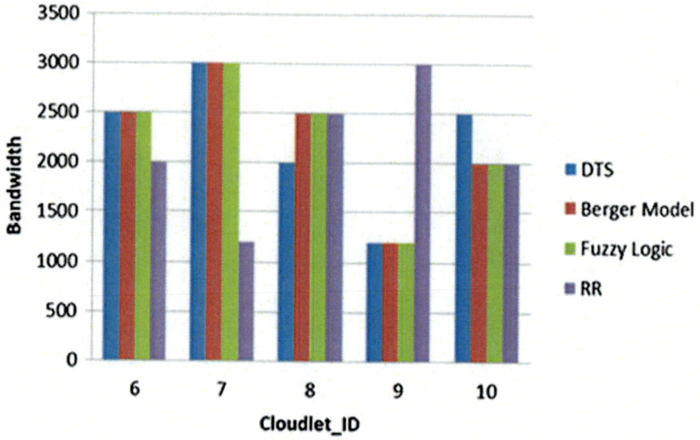

Fig. 5 Comparison of Class -1 type Jobs in-terms of CPU Utilization

Table 2 Execution time

Cloudlet ID	Execution time			
	DTS	Berger model	Fuzzy logic	RR
1	125	150	214	300
2	312	420	492	540
3	85	98	186	205
4	127	170	290	350
5	182	182	274	296
6	220	276	308	365
7	305	374	428	470
8	131	135	280	310
9	87	115	240	285
10	226	325	398	435

suitable resources (VMs) which are available in a way that fulfills users' QoS expectations while also achieving resource utilization by using scheduling and reallocation process.

The impact of the suggested technique on some QoS parameters such as availability, throughput, and budget will be investigated in future research. In addition, to make the proposed architecture more effective in the cloud computing paradigm, VM management techniques such as VM migration and consolidation can also be implemented on the proposed model.

References

1. Buyya, R., Ranjan, R., Calheiros, R.N.: Modeling and simulation of scalable cloud computing environments and the CloudSim toolkit: challenges and opportunities. In: 2009 International Conference on High Performance Computing & Simulation, pp. 1–11. IEEE (2009)
2. Hewitt, C.: ORGs for scalable, robust, privacy-friendly client cloud computing. IEEE Internet Comput. **12**(5), 96–99 (2008)
3. Radu, L.D.: Green cloud computing: a literature survey. Symmetry **9**(12), 295 (2017)
4. Zhang, P., Zhou, M.: Dynamic cloud task scheduling based on a two-stage strategy. IEEE Trans. Autom. Sci. Eng. **15**(2), 772–783 (2017)
5. Srichandan, S., Kumar, T.A., Bibhudatta, S.: Task scheduling for cloud computing using multi-objective hybrid bacteria foraging algorithm. Future Comput. Inf. J. **3**(2), 210–230 (2018)
6. Thilagavathy, R., Sabitha, R.: Using cloud effectively in concept based text mining using grey wolf self-organizing feature map. Clust. Comput. **22**(5), 10697–10707 (2019)
7. Aldulaimy, A., Zantout, R., Zekri, A., Itani, W.: Job classification in cloud computing: the classification effects on energy efficiency. In: 2015 IEEE/ACM 8th International Conference on Utility and Cloud Computing (UCC), pp. 547–552. IEEE (2015)
8. Wankhede, P., Talati, M., Chinchamalatpure, R.: Comparative study of cloud platforms-Microsoft Azure, Google Cloud Platform and Amazon EC2. J. Res. Eng. Appl. Sci. **5**(2), 60–64 (2020)
9. Xu, B., Zhao, C., Hu, E., Hu, B.: Job scheduling algorithm based on Berger model in cloud environment. Adv. Eng. Softw. **42**(7), 419–425 (2011)
10. Benoit, A., Marchal, L., Pineau, J.F., Robert, Y., Vivien, F.: Offline and online master-worker scheduling of concurrent bags-of-tasks on heterogeneous platforms. In: 2008 IEEE International Symposium on Parallel and Distributed Processing, pp. 1–8. IEEE (2008)
11. Abdullah, M., Othman, M.: Cost-based multi-QoS job scheduling using divisible load theory in cloud computing. Procedia Comput. Sci. **18**, 928–935 (2013)
12. Ghanbari, S., Othman, M.: A priority based job scheduling algorithm in cloud computing. Procedia Eng. **50**, 778–785 (2012)
13. Potluri, S., Rao, K.S.: Quality of service based task scheduling algorithms in cloud computing. Int. J. Electr. Comput. Eng. **7**(2), 1088–1095 (2017)
14. Wu, X., Deng, M., Zhang, R., Zeng, B., Zhou, S.: A task scheduling algorithm based on QoS-driven in cloud computing. Procedia Comput. Sci. **17**, 1162–1169 (2013)
15. Yang, B., Xu, X., Tan, F., Park, D.H.: An utility-based job scheduling algorithm for cloud computing considering reliability factor. In: 2011 international conference on cloud and service computing, pp. 95–102. IEEE (2011)
16. Chang, R.S., Lin, C.Y., Lin, C.F.: An adaptive scoring job scheduling algorithm for grid computing. Inf. Sci. **207**, 79–89 (2012)
17. Varalakshmi, P., Ramaswamy, A., Balasubramanian, A., Vijaykumar, P.: An optimal workflow based scheduling and resource allocation in cloud. In: International Conference on Advances in Computing and Communications, pp. 411–420. Springer, Berlin, Heidelberg (2011)
18. Huang, Q.Y., Huang, T.L.: An optimistic job scheduling strategy based on QoS for cloud computing. In: 2010 International Conference on Intelligent Computing and Integrated Systems, pp. 673–675. IEEE (2010)
19. Ibrahim, I.M.: Task scheduling algorithms in cloud computing: a review. Turkish J. Comput. Math. Educ. (TURCOMAT) **12**(4), 1041–1053 (2021)
20. Deore, S.S., Patil, A.N.: Energy-efficient job scheduling and allocation scheme for virtual machines in private clouds. Energy **5**(1), 56–60 (2013)
21. Feller, E.: Autonomic and energy-efficient management of large-scale virtualized data centers (Doctoral dissertation, Université Rennes 1) (2012)
22. Beloglazov, A.: Energy-efficient management of virtual machines in data centers for cloud computing (Doctoral dissertation) (2013)

23. Wang, S., Li, Y., Pang, S., Lu, Q., Wang, S., Zhao, J:. A task scheduling strategy in edge-cloud collaborative scenario based on deadline. Scientific Programming, 2020, Article ID 3967847(2020)
24. Attiya, I., Abd Elaziz, M., Xiong, S.: Job scheduling in cloud computing using a modified harris hawks optimization and simulated annealing algorithm. Computat. Intell. Neurosci. (2020)
25. Wankhede, P., Talati, M., Chinchamalatpure, R.: Comparative study of cloud platforms-microsoft azure, google cloud platform and amazon EC2. Int. J. Res. Eng. Appl. Sci. **5**(2), 60–645 (2020)
26. Reiss, C., Wilkes, J., Hellerstein, J.L.: Google cluster-usage traces: format+ schema. Google Inc., White Paper, 1–14(2011)
27. Pandey, P., Singh, S.: Fuzzy logic based job scheduling algorithm in cloud environment. Comput. Model New Technol. **21**, 25–30 (2017)
28. Eucalyptus, https://www.eucalyptus.com, [Online] (2015)
29. Chowdhury, M.: Time and energy-efficient hybrid job scheduling scheme for mobile cloud computing empowered wireless sensor networks. Int. J. Ad Hoc Ubiquitous Comput. **37**(1), 26–36 (2021)
30. Ru, J., Yang, Y., Grundy, J., Keung, J., Hao, L.: An efficient deadline constrained and data locality aware dynamic scheduling framework for multitenancy clouds. Concurr. Comput. Practice Exp. **33**(5), e6037 (2021)
31. Sundaravadivazhagan, B., Malathi, V., Kavitha, V.: A novel credit grounded job scheduling algorithm for the cloud computing environment. 2022 International Conference on Inventive Computation Technologies (ICICT), Nepal, pp. 912–919(2022)
32. Karthikeyan, R., Balamurugan, V., Cyriac, R., Sundaravadivazhagan, B.: COSCO2: AI-augmented evolutionary algorithm based workload prediction framework for sustainable cloud data centers. Trans. Emerging Telecommun. Technol. **34**(1), e4652 (2022)

A New Approach to Coupled IR Sensor-Based Real-Time System for Alerting Elderly People of Accidents in Rest Room

A. L. Sriram and B. Thamotharan

Abstract In this paper, an IR sensor-based Real-Time Rest Room system helps in detecting accidents or abrupt falling, or movements of elderly people when they are also in a restroom. It also helps in optimizing the consumption of power and also minimizes the memory utilization that has been proposed. The system detects the rapid change in infrared radiation that happens when people enter its detection range then this signal enables and alerts the smart software system integrated with it If there is a sudden rapid change in the radiation and motion detector output the smart system switches on the webcam and alarm system which is integrated through the relay. There will be two sensors with height levels of 4 feet and 1 foot respectively If both detect the movements of the person then the system will enable the alarm and capture the videos respectively. This concludes that the person has fallen and has met an accident or heart attack. Since it is a passive sensor the system gets activated only when the person enters the restroom and gets into sleep mode when there is no intruder. Thus it saves a lot of power consumption and unwanted recording thus saving memory space.

Keywords Infrared sensor · Electromagnetic radiation · Alarm system · Relay

1 Introduction

In this paper, we proposed a new approach to solving elderly senior people's accidents in the restroom. The elderly people when using the restroom, due to their age factor, unfortunately, they may fall inside the restroom. This may cause a very serious impact on their health. Sometimes this kind of circumstance creates hard injuries to elderly persons. Our proposed approach will alert the members of the family to save the elder person from accidents in the restroom (Fig. 1).

A. L. Sriram (✉) · B. Thamotharan
SASTRA Deemed University, Thanjavur, India
e-mail: als@it.sastra.edu

C. Iwendi et al. (eds.), *Proceedings of ICACTCE'23—The International Conference on Advances in Communication Technology and Computer Engineering*, Lecture Notes in Networks and Systems 735, https://doi.org/10.1007/978-3-031-37164-6_2

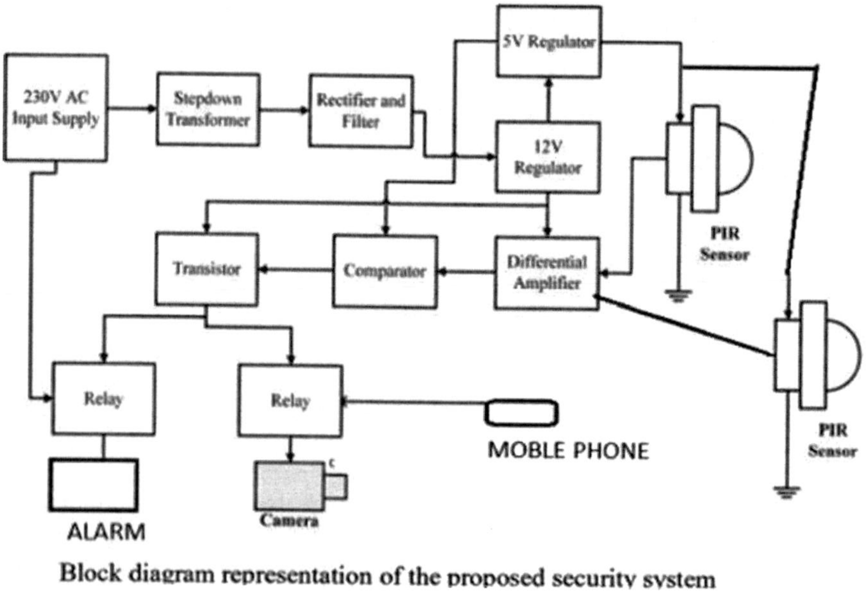

Fig. 1 Block Diagram for proposed security system

The proposed system works flow is as follows:

Step i. The software is written to check continuously whether both the sensors and the alarm system are active throughout.
Step ii. When somebody enters the restroom within the sensor range it checks the output of both the sensors and accordingly generates 3.2 V.
Step iii. This output is amplified to a desired level to activate the relay of the alarm and the web cam systems.
Step iv. Once activate the relay of the alarm and the webcam systems are the software compares the defined criteria and activates the mobile phone emergency call.
Step v. Next it captures the image and stores it in the storage system attached after raising the alarm and activating the mobile phone emergency call.
Step vi. Each time whoever enters the washroom the system repeats the process 24/7 and ensures that no untoward accidents happen with elderly people like falling or heart attacks etc.

2 System Analysis

The proposed coupled IR sensor-based system has two divisions, the first part is. hardware and the second one is. Software. According to the output response of the coupled sensors at the particular height these two hardware and software work in sync to ensure perfect working.

Hardware Parts

The hardware part consists of a couple of Infra-Red sensors, a power supply, an amplifier, a window detection circuit, an alarm system, a webcam, and a mobile phone.

Coupled Infra-Red Sensor

This forms the central part of the proposed system. The function of the proposed system is based on infrared radiation that body of a human is being emitted. The infrared sensor is mostly employed in a security systems to accurately find human motion [1].

The electromagnetic radiation of Infrared (IR) light and its wavelength is between 0.7 and 300 μm. Individuals are a source of IR radiations with body temperature radiating at a wavelength of 10 μm to 12 μm. The gate, drain, and the source is the three pins that it has [2]. Whenever there is an unsync. in the output from two sensors, it sends a signal intimating so (Fig. 2).

Power Supply Unit

The conversion of getting 16 V DC from 230 V,50 Hz AC power is taken care of by this unit. This conversion is done by employing a step-down 16-0-16 center-tapped transformer full wave rectifier. The capacitors help in eliminating the AC ripples by effectively using a passive filter. The following Figure depicts the circuit for the power supply (Fig. 3).

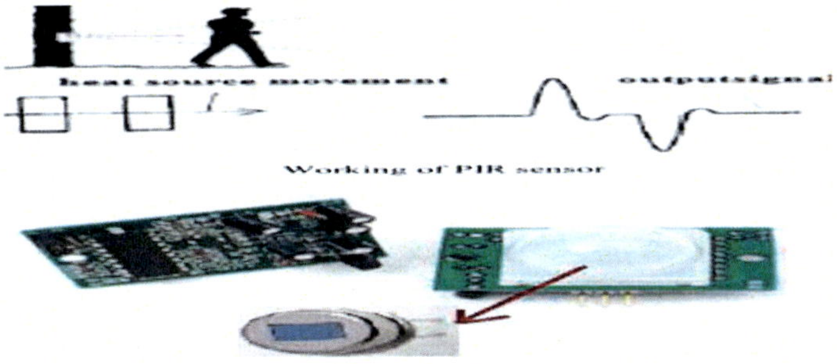

Fig. 2 System diagram

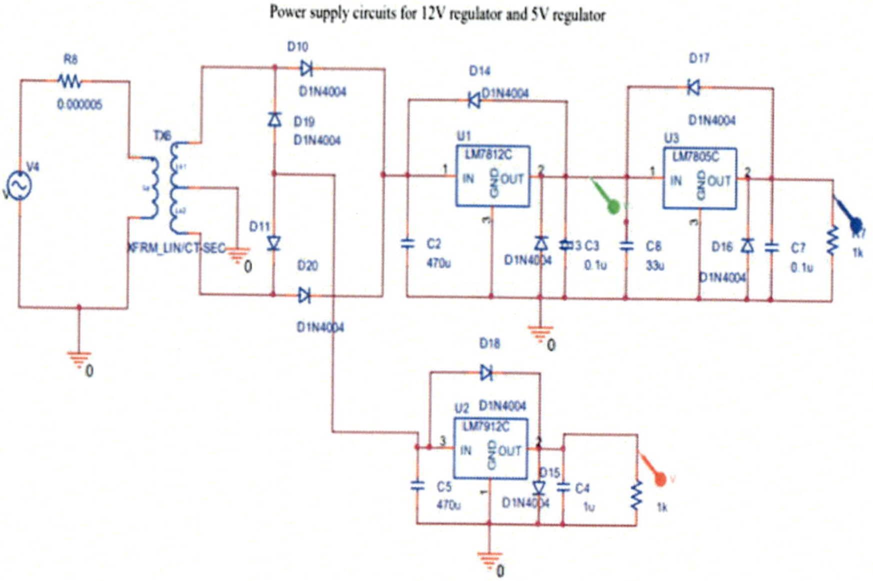

Fig. 3 Block diagram—power supply circuits

Differential Amplifier and Window Comparator

The output from both the coupled sensors is amplified using a differential amplifier. The sensor output is amplified to 5 V from 3.2 V. At the window comparator, if this voltage is inputted, then it is compared with the two reference voltage inputs [3]. Thus the diodes are forward biased or reverse biased is decided based on the above-said response of the coupled sensors shown in Figure thereby effectively controlling the transistor input. This output which comes from the window comparator controls the base current at the transistor which acts as a switch that regulates the flow of voltage to the relay of the alarm system and video system.

Webcam and Alarm System: The coupled IR sensors are used to detect any person entering the washroom and this information is analyzed the moment there is a default the video camera and alarm system are enabled and also the mobile phone emergency call is activated. The sensor data that is transmitted from a non-standard-attitude needs to be changed to the one from a standard attitude [4]. The relay after the transistor switch will act as a switch to ensure the supply of power to an Alarm system, mobile call activation, and the webcam.

Software Parts

The software side is developed with an integration of both Microsoft Visual C++ and OpenCV 2.4.4 library files like opencv_ml244.lib, and opencv_video244.lip. OpenCV library files help in image and video manipulations and analysis.

The object-oriented language C++ is a very flexible language that comprises both low-level and high-level features. The software runs and checks the hardware

Fig. 4 Flow chart diagram

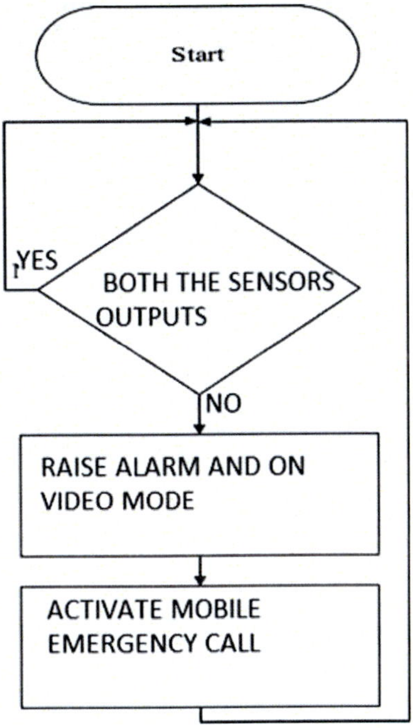

working 24/7 also it analyses the output from the sensors and ascertains the condition of the elderly person inside the wash room [5]. if any untoward incident happens the moment the software activates the hardware part to raise alarm and also trigger the emergency call from the mobile phone (Fig. 4).

Algorithm: The execution of the logic of the proposed software system integrated with this system is as follows.

1. Program starts
2. Check if all the hardware parts are active and function
3. If there is any entry into the washroom the system analyses the output of both sensors coupled together.
4. If only one output means the person has fallen because of any reason and the moment it activates the alarm and also makes an emergency call in the mobile phone connected
5. The process is repeated indefinitely forever.

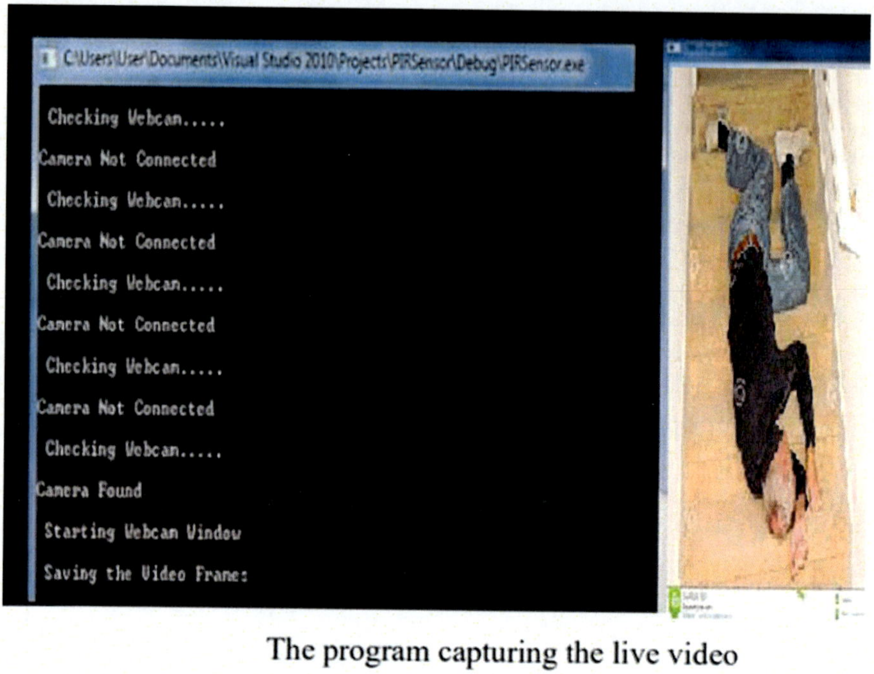

Fig. 5 Output image of the proposed system

3 System Testing

In this paper, the proposed system was tested for its functionality after unit and integration testing is done at normal room temperature [6]. Also, software simulation is done and ensured that the whole of the proposed functionality is achieved, with the sensor range set manually.

It is found in the final system testing that if any person enters the rest room the system gets into active mode and once both the sensors do not respond the alarm and exception management system get activated as expected. The output is captured and included in the image which is given below (Fig. 5).

4 Systttem Performance Analysssis

The proposed system not only promises a very high and accurate response but also helps a lot in saving power and unwanted memory usage [7]. The system gets into active mode from its passive mode as soon as a person enters the rest room and upon any eventuality, the system responds within a time duration of 4 to 9 s.

5 Conclusion and Future Enhancements

In this paper, a level-coupled Infrared sensor-based security system for alerting accidents with an elderly person inside a restroom has been proposed. It has been seen through simulation and prototype testing that the system performs very well and could play a major role in alerting against accidents and eventually helping elderly people get the required attention and medical aid in time [8].

The system is also very efficient in terms of energy and space optimization and utilization. The system is best when particularly utilized in places where very aged people live. Also, the system is very simple to maintain and get installed.

The system that has been proposed can be optimized in many ways and can be developed at a very competitive and affordable price. Also, the efficiency can be increased from 90 to 100%.

References

1. Di Paolo Emilio, M.: Data Acquisition Systems—From Fundamental to Applied Design, pp. 1–2. Springer Publication, New York (2013)
2. Data Acquisition Tutorial:: Radio-Electronics.com (http://www.radioelectronics.com/info/t_and_m/data-acquisition/data-acquisition.php), visited on 8th March 2018
3. Berman, H.L., Hills, L.A.: Infrared intrusion detector system 3,73,718 (1972)
4. Wang, R., Xie, F., Zhao, J., Zhang, B., Sun, R., Yang, J.: Smartphone sensors-basedabnormal driving behaviors detection: serial-feature network. IEEE Sens. J. (2020)
5. Herbert, L., Berman, Los, A.H., Calif: Infrared Intrusion Detector System 3,703,718 (1972)
6. Yuvaraj, S., Ramesh, S.: Improved response time on safety mechanism based on PIR sensor. International J. Emerging Technol. Adv. Eng. **2**(4) (2012)
7. Anonymous (2007) The Electromagnetic Spectrum [online] http://science.hq.nasa.gov/kids/imagers/ems/infrared.html
8. Li, Z., Liu, F., Yang, W., Peng, S., Zhou, J.: A survey of convolutional neural networks: analysis, applications, and prospects. IEEE Trans. Neural Netw. Learn. Syst. (2021)

Gesture Recognition Based Virtual Mouse and Keyboard

B. R. Sandhya, C. Amrutha, and S. Ashika

Abstract Today, computer vision has advanced to an extent where a machine can recognize its owner by running a straightforward picture processing application. In the modern day, individuals use this vision for various activities in their daily life, including automatic cars, face recognition, and color detection. Hand gesture recognition is important and common since it also plays a role in the understanding of sign language. In the proposed framework, Human Computer Interaction (HCI) approach has been used. An optical mouse and keyboard that employ hand motions are created using computer vision. The computer's camera will recognize various hand gestures made by a user, and the computer's mouse or pointer moves according to the movements of the motions. Users may even conduct right and left clicks using various gestures. Similar to this, several gestures might be used to operate the keyboard, such as the one-fingered gesture for selecting an alphabet and the two-fingered motion for swiping left and right. It won't require a cable or other external accessories to function as a virtual mouse and keyboard.

Keywords HCI · Gesture recognition · Cursor control · Feature extraction

B. R. Sandhya (✉)
Department of CSE, Nitte Meenakshi Institute of Technology, Bengaluru, India
e-mail: sandhya.br@nmit.ac.in

C. Amrutha
Department of ISE, Sri Venkateshwara College of Engineering, Bengaluru, India
e-mail: amrutha.c_ise@svcengg.edu.in

S. Ashika
Department of CSE, Sri Venkateshwara College of Engineering, Bengaluru, India
e-mail: ashika.s_cse@svcengg.edu.in

C. Iwendi et al. (eds.), *Proceedings of ICACTCE'23—The International Conference on Advances in Communication Technology and Computer Engineering*, Lecture Notes in Networks and Systems 735, https://doi.org/10.1007/978-3-031-37164-6_3

1 Introduction

With the growth in technology, gadgets are growing smaller in size. Some gadgets have gone wireless, while others have gone dormant. The proposed framework suggests a technology that might cause some gadgets to become dormant in the future, which is the future of HCI (Human–Computer Interaction). Today's state-of-the-art computer vision technology allows a computer to recognize its owner with the help of a straightforward picture processing application. People are employing this vision in many activities of various aspects of daily life like face recognition, color detection, automatic cars, and so on. Computer vision is utilized in this research to create an alternative method in the place of optical mouse and keyboard using hand motions. The aim is to create a virtual mouse and keyboard based on gesture recognition. The technique of identifying and understanding a stream of continuous sequential gestures from a particular set of input data is known as gesture recognition. The most intuitive and natural techniques to interface with a computer are through gestures. The technology for input and output is based on gesture interactions. The system doesn't need any additional hardware outside the laptop or computer. It uses built-in webcam that comes with every laptop.

The computer's camera will consider multiple hand gestures made by a user, and the computer's mouse or pointer will move in accordance with the movements of the actions. Users may even execute right and left clicks using various gestures. Similarly, the keyboard operations may be utilized with certain alternative gestures. The goal is to replace a conventional or standard device with a basic camera which can control keyboard and mouse cursor functionalities. A camera is the sole device the Virtual Mouse employs to establish communication between the user and the system. It enables the user to interact with a machine and perform mouse activities without the need for any mechanical or physical devices. A virtual Human–Computer Interaction device serves as an alternative to the touch screen that uses a camera or other picture capturing devices to replace the actual mouse or keyboard. A software program will continuously use the webcam to track the user's actions, interpret them, and convert them into the motion of a pointer, much like a real mouse.

A modern computer can now recognize its owner by employing a simple picture processing algorithm, thanks to advancements in computer vision. At this stage of development, people use this vision for many aspects of daily living including face identification, color detection, automated vehicles, and other things. Computer vision is employed in this method to create an optical mouse and keyboard that uses hand signaling. An image of users hands with varied motions will be scanned by the computers web camera. The user's hand's varied motions, and the mouse or pointer will move in reaction to the movement of the gestures, including performing right- and left-clicking with different gestures. Similarly, other gestures may be used to manipulate the keyboard such as a one-finger gesture to pick an alphabet and a two-figure gesture to swipe left and right. It will work as a virtual mouse and keyboard in the absence of any cables or other devices. The camera is the sole piece of hardware for the project, and Python IDLE is used for coding. An algorithm is then devised by mapping the mouse and keyboard capabilities to the weaknesses using defect calculations. The computer will comprehend the gesture and react appropriately if you map a handful of them using the mouse and keyboard.

When the computer camera captures the image of the person seated in front of the computer, a small green box will appear in the screen's core. The code will be parsed and compared with the items shown in the green box. If it does, a red border will turn up. It indicates that the object has been detected by the computer and that it is feasible to move the mouse pointer by moving the object. This will assist in creating a virtual computational experience as well as computer security.

Hand gestures will be utilized in place of various items to move the cursor, click the right or left mouse button, and similarly, a straightforward motion can perform keyboard activities remotely that would have been performed on a real keyboard. The box will only exhibit a green border if the gesture does not match; however, if the known gesture is recognized, the box will display a red border. A system that is efficient in minimising computing cost while retainng high accuracy in hand gesture recognition was proposed by Roman Golovanov and et al. [1].

1.1 The Work's Inspiration

This idea has been picked to make a direct connection between people and technology. This results in a new degree of interaction for the user reliance on gadgets like the mouse, keyboard, etc. would decrease. Thanks to gesture control technology, which would also simplify the process of controlling a system. Initially, this technology is primarily used in video games (like Xbox Kinect), but if applied motion/gesture control technology to commonly used electronics like computers, televisions, etc., for various gesture control like clicking, scrolling, etc., the application of this technology would be more diverse. Additionally, our own sign language can be developed, which would make communicating with deaf and hard of hearing.

2 Literature Review

This chapter provides information on a review of the literature on virtual keyboards and mice that identify gestures. Nikhil et al. [2], used the method of working with gesture comparison algorithms, skin pixel detection along with the hand segmentation from stored frames which needs to be done. The foundation subtraction technique is used to separate the hand area from the foundation. At that stage, fingers are divided into portions so that they can be recognized and perceived. Finally, a common classifier is used to predict the names of hand motions. This system's accuracy ranged from 78 to 90%. The system is not effective against a challenging or complicated background. It can be observed from the literature analysis that earlier systems either had a virtual keyboard or mouse. These devices can't totally replace the necessity for a mouse and keyboard. Through the use of basic computer vision and multimedia techniques, they created a machine-user interface that enables hand gesture recognition.

Kumari and et al. [3] used the methodology of the key by using the mouse where the vital quantity was of colors. But the system was managed by a number of colors. Instead of using distinct movements, the amount of colors is used to complete a function. Using a webcam as a sensor and coloured caps to track the user's hand in two dimensions, and able to identify up to five different hand movements that are translated into mouse functions. The user's face is first photographed, and the eye location is obtained. This information is then utilized as a reference point. The Integral Projection technique was utilized to determine the position of the eye movement, and the HaarCascade approach was employed to extract facial characteristics. The outcomes showed that the various colour bands of various colours execute various behaviors.

Ramakant et al. [4] have created a system that controls the cursor movement using two fingers. They have proposed a model which can control the movements of the mouse using the two fingers and the gestures made by them. In the paper, "Fusion of Information From Data Glove and a Camera for Hand Gesture Recognition" [5], the author proposed a model that recognizes hand gestures using an algorithm called the back propagation algorithm based on data from a data-glove and for all static hand gestures a camera. A system that uses nothing more than a low resolution webcam.

Mishra et al. [6] implemented the methods that used the IR sensors identify and capture the picture of the hand in the shot. Following pre-processing, hand tracking and segmentation are used to retrieve the characteristics. As an input to the following stage, the extracted features are clustered to create discrete vectors. Recognition methods based on a hand gesture dictionary (database) are used to classify the retrieved characteristics. This allows for the performance of the output gesture. Two Ultrasonic Sensors with Arduino are used, the hands are placed in front of them and calculation of the distance between the sensor and hand is done. By converting the data into keyboard click actions, a specific library called PyAutoGUI may use this information to carry out the necessary computer operations. The observation drawn was that five separate computer operations may be accomplished by calibrating the hand motions produced in front of the ultrasonic sensors only few simple hand motions conducted in front of the computer will execute some computer activities without the assistance of a mouse or keyboard.

Marker-based technique and marker-less approach are two categories that can be used to categorize vision-based hand motion recognition. The marker-based strategy calls for the user to use gloves or colour markers, but the marker-free method does not [7]. Mainly concentrates on creating a webcam-based virtual human–computer interaction. A camera was used to capture hand motions using a color-detection approach. But if the operational background is light bright coloured, objects will not be detected and the system might function more slowly on some PCs with weak computational power. A Sign Language recognition using American Sign Language is discussed in the paper [8]. American Sign Language was proposed for identification of signs by Mehreen Hurroo and et al.

shukla A and et al. proposed an AI virtual mouse in their work [9], however it has a number of drawbacks, including a little loss of precision when utilizing the right click mouse function and some trouble selecting text by clicking and dragging. Where there is a sample of light present, this system will operate effectively. If more than one face is recognized, mistakes could happen. Users may encounter slow text entry rates and a need to recalibrate regularly when using the virtual keyboards. There are two basic approaches in vision Technology: The model-based approach, which involves building a three-dimensional model of the hand before doing hand recognition and Image-Based Method where the image is first taken by the camera, and then detection is carried out. A new recognition method utilising the location, angle, and velocity features coupled was proposed by Ho-Sub Yoon et al. Another real-time gesture recognition system with support for 46 American sign languages, the letter spelling alphabet, and numbers was proposed in. The gesture recognition system also uses hidden Markov models, which have the advantages of expansibility, time-scale invariance, automatic segmentation, and classification. Since the suggested virtual mouse can access the palm node directly and without the usage of any additional hardware, it uses the hand motion recognition technique rather than the time-consuming hand tracking procedures. A method of categorizing dynamic hand motions using hand skeleton modeling was proposed by Quentin De Smed and et al. [10]. Three hand gesture datasets with a collection of fine and coarse diverse gestures recorded in various settings are used in the experiments. This method has been evaluated in terms of latency, and the results show benefits for low-latency hand gesture recognition systems. A method for controlling a computer using six static and eight dynamic hand gestures was proposed by Hussain and et al. [11]. Patil M and et al. [12] proposed Gesture Based Virtual Mouse And Keyboard to recognise hand motions. But if more than 1 face is detected, error may occur.

3 Design and Methodology

3.1 Design

The system consists of two sensors which has specific functionalities. The input is the video of the user hand which is captured lively using web camera of the computer. It extracts the required features of the hand, the image is segmented and filtered to reduce the noise in the image, performs the hand gestures and the hand gestures are detected i.e. the number of fingers raised are preprocessed, on the basis of this, feature are extracted. After Feature extraction is done, by using conditionals statements in the program, the feature are matched. As the features matches, it automatically controls the mediaplayer and gives us the required results.

The Use case diagram Fig. 1 shown below explains the different entities involved in making interaction with the System. As this System is based on Human Computer Interaction so it basically includes the user, computer and the medium to connect

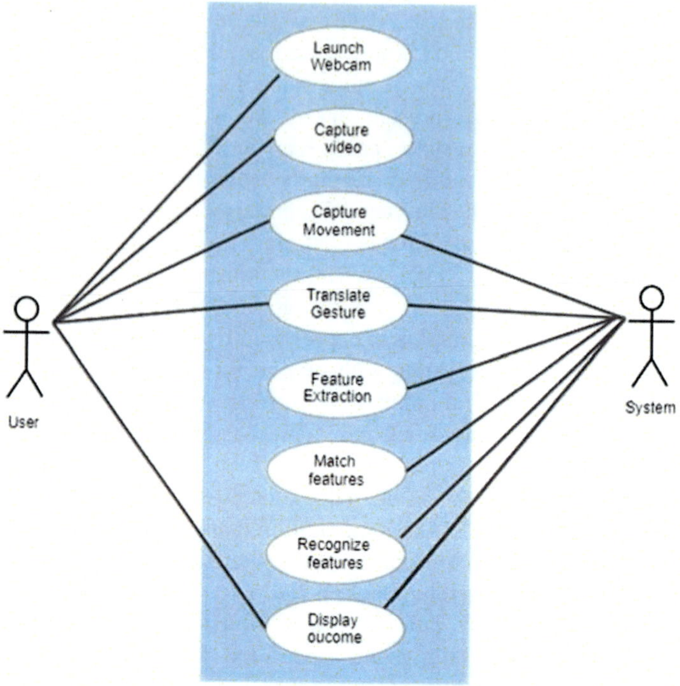

Fig. 1 Use case diagram

both digitally that is web camera which is connected to System. As the execution of the program starts, it firstly invokes the web camera to take RGB image of the hand. Then the image is segmented and filtered to reduce the noise in the image. After the removal of the noise, the hand gestures are detected. It involves the gesture detection which detects the gestures from the user. Then the next stage involved is the preprocessing of the gestures from the user which is later passed onto the feature extraction process. Then it recognizes the features extracted and execution occurs.

3.2 Methodology

Because the HAAR Cascade method is so effective at removing background noise, so utilized it. A machine learning technique called HAAR Cascade is used to recognize objects in images and real-time videos. A feature-based classifier offers good accuracy and speed. It has fewer instructions and requirements. The packages used in the proposed method is opencv-python, mediapipe, pynput and autopy. Figure 2 below displays the proposed system's block diagram.

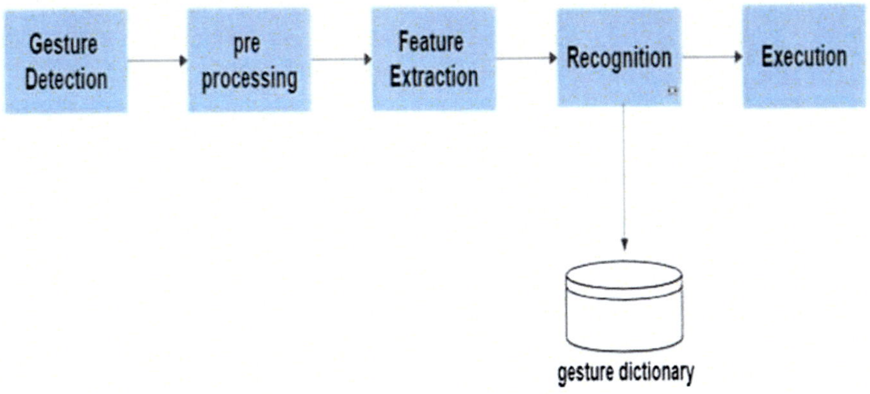

Fig. 2 Block Diagram

The steps/process involved in gesture recognition is:

- **Gesture detection**

It is a kind of user interface for perceptual computing that enables computers to record and understand human movements as commands. The capacity of a computer to comprehend gestures and carry out orders based on such movements is the broad definition of gesture detection.

- **Preprocessing**

Preliminary processed data that has been prepared for analysis or primary processing. When preparing data for a user requires a number of steps, the phrase can be used to represent any first or preparatory processing stage.

- **Feature extraction**

It is a kind of diminishing the dimensions in which a sizable portion of the picture's pixels are successfully represented so as to effectively capture interesting portions of the image.

- **Recognition**

Here Recognition is an act of identifying someone or something and also identifying hand gestures.

- **Gesture dictionary**

It contains different hand gesture images which we have stored.

- **Execution**

The word "executing process" refers to any particular stage of an activity or project during which the act of implementing occurs, as well as the predefined steps taken to reach that stage in order to successfully complete this action.

It Consists of 2 Modules

Virtual Keyboard, Virtual Mouse.

Virtual Keyboard

- Real-time video shooting with a computer's webcam
- Processing each picture frame from the recorded video
- For object tracking, It will cycle through all the contours and overlay a rectangle on each one.

Virtual Mouse

- Our system's Virtual Mouse module uses hand recognition and finger counting to carry out mouse operations.
- It is capable of various actions: left and right clicks, left and right movements, up and down movements, and left and right movement, selection of folders, web browsers and opening of that.

3.3 *Hardware and Software Requirements*

Hardware may employ a single web camera comes with laptop. Additionally, there must be some sort of tracking system in place to keep track of the palm's position, form, orientation, etc. Some magnetic or inertial sensors are used by tracking devices to carry out their function. Gesture recognition systems can be implemented in software using programming languages like C, C++, JAVA, and Python together with some actual tracking algorithms, opencv-python, mediapipe framework, pynput and autopy libraries installed in python.

3.4 *Benefits of Proposed Work*

- It is not safe to touch the devices in the post-COVID-19 environment because there are chances of virus spread, and we are not sure that spread of covid-19 virus came to end. So the proposed system help to control the system mouse function with out using physical mouse.
- A person who is too concerned of getting contaminated by germs can also make use of the proposed system.
- Gestures are useful and can be used as an alternative when getting a physical controller takes a lot of effort or money. Controlling the television or electronics in a car, like the radio or air conditioner, is an example of such usage.
- Robots can be used to help with patient rehabilitation. This can be accomplished by placing the appropriate seniors on a person's body and using a screen to read the values of the results the sensors provide.

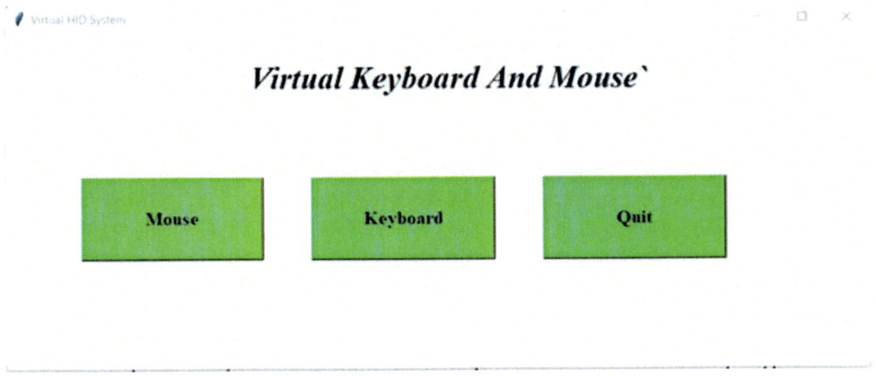

Fig. 3 User interface design

4 Results and Discussion

The outcomes of the suggested system has been discussed here.

The above Fig. 3 depicts how the user interface page looks like, it contains three options as shown in the snapshot. If we click on keyboard option the virtual keyboard will be displayed on the screen, similarly if we click on mouse option the virtual mouse will get opened and if we want to close this application we can click on quit option. This is the first depiction of the user interface in this project which can be further improvised in the future with more aspect of options.

The above Fig. 4 depicts the selection of virtual keyboard by the user among 3 options like Mouse, Keyboard, Quit.

The above Fig. 5 shows how the virtual keyboard looks like when it is displayed on the screen. Keys can be used for the typing characters, words, and sentences. These keys can be again modernized by updating the keyboard with features such as special characters, various emojis and additional symbols as well. The maximum distance the user operates is 5 feet. The quit option which is shown above in above Fig. 6 selected for quiting the application.

The above Fig. 7 shows how the virtual mouse looks like when it is displayed on the screen. The webcam captures the finger points and the points in the palm which would basically count as gestures to the system during the implementation of the process.

5 Conclusion and Future Enhancement

A fresh approach has been put forth to raise the System flexibility and speed of response. To date, a system for directing the mouse pointer and keyboard and use a live camera to carry out its role. Implementation of all mouse functions, including left-

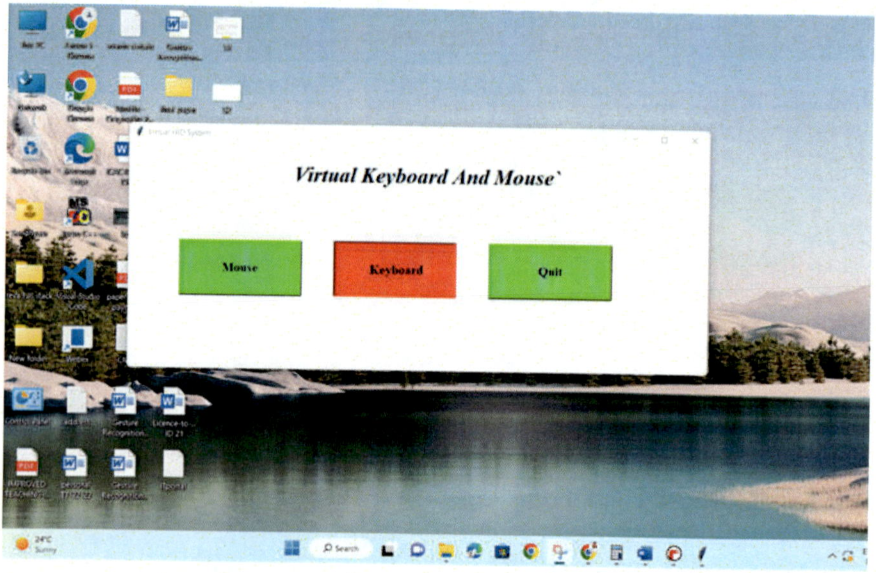

Fig. 4 Selection of virtual keyboard by user

Fig. 5 Virtual keyboard

and right-clicking, scrolling of the pages, selecting the objects and performing double click on that to open the folders, web browsers and applications like paint, notepad and many more that uses hand gestures. Keyboard implementations like typing of letters, words, sentences and many more. All these features are developed in python Language using Open CV library with media pipe framework, pynput library, NumPy

Fig. 6 Selecting quit

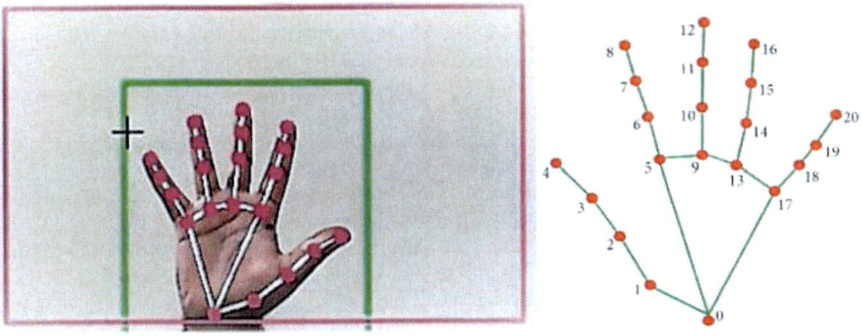

Fig. 7 Virtual mouse

and autopy library. Use of the skin segmentation process allows for the separation of the hand's color and image from its surrounding area. Having trouble getting your entire body in the frame is effectively solved by the procedure of eliminating an arm. In conclusion, the proposed technique can construct a realistic user interface and detect and recognize hand movements to control mouse and keyboard functionalities. Nurses and physicians may not be allowed to touch the display or trackpad for health and safety concerns, but they still need to manage the system. Medical apps offer extra gesture capabilities. Proper gestures, such as using your finger as a virtual mouse or swiping with your hand, are a faster and safer method to use your smartphone. In the future, we will be able to improve the webcam with the flash light. To totally replace the computer mouse with hand tracking and solely utilise gestures to operate the system, we would like to improve the software's performance. In order to totally eliminate the need for a mouse or keyboard, we would also like to expand the system's control to a general-purpose computer controller. Additionally, we can also include number keys, functional keys and few special characters including emojis, utilising multiple fingers to expand and contract windows when using multiple windows.

References

1. Golovanov, R., Vorotnev, D., Kalina, D.: Combining Hand Detection and Gesture Recognition Algorithms for Minimizing Computational Cost (2022)
2. Nikhil, C.D.S., Someswara Rao, C.U., Brumancia, E., Indira, K., Anandhi, T., Ajitha, P.: Finger Recognition & Gesture Based Virtual Keyboard. ICCES 2020, pp 1321–1324 (2020)
3. Kumari, P., Singh, S., Pasi, V.Kr.: Cursor control using hand gestures. Recent trends in future prospective in engineering & management technology 2016. Int. J. Comput. Appl., 0975–8887 (2019)
4. Ramakant, Sharma, V.K., Kumar, V., Iqbal, Md., Tawara, S., Jayaswal, V.: Real Time Static and Dynamic Hand Gesture Recognition. IJERT (2017)
5. Ramakant: Fusion of information from data-glove and a camera for hand gesture recognition. IJERT **3**(3), 1395–1400 (2014)
6. Mishra, S., Dhikhi, T.: Gesture control keyboard. Int. J. Res. Eng. Sci. Manag. **1**(10), 678–682 (2018)
7. Banerjee, A., Ghosh, A., Bharadwaj, K., Saikia, H.: Mouse control using a web camera based on colour detection. Int. J. Comput. Trends Technol. (IJCTT) **9**(1) (2014)
8. Hurroo, M., Elham, M.: Sign language recognition system using convolutional neural network and computer vision. IJERT **9**(12) (2020)
9. Shukla, A., Katiyar, D., Goel, G.: Gesture Recognition-based AI virtual mouse. Ijraset J. Res. Appl. Sci. Eng. Technol. (2013)
10. De Smedt, Q., Wannous, H., Vandeborre, J.-P.: Heterogeneous hand gesture recognition using 3D dynamic skeletal data. Comput. Vis. Image Understanding **181**, 60–72 (2019)
11. Hussain S., Saxena R., Han X., Khan J.A, Shin H.: Hand Gesture Recognition Using Deep Learning, pp. 48–49, ISOCC (2017)
12. Patil, M., Disale, S., Phalke, H., Ranade, U., Khot, S.: Survey on gesture based virtual mouse and keyboard. Int. J. Creative Res. Thoughts **10**(5) (2022)
13. Yoon, H.-S., Soh, J., Bae, Y.J., Yang, H.S.: Hand gesture recognition using combined features of location, angle and velocity **34**(7), 1491–1501 (2001)

DDoS Attack Prediction Using Decision Tree and Random Forest Algorithms

Tao Hai, Jincheng Zhou, Oluwabukola A. Adetiloye, Shirin Abolfath Zadeh, Yanli Yin, and Celestine Iwendi

Abstract The most common network attacks are Denial of Service (DoS) and Distributed Denial of Service (DDoS) attacks which causes packet loss by delaying the exchange of information, thereby altering the data packets sent through networks which affect the integrity and reliability of the data. Over time, various machine learning models have been identified and presented by researchers to predict and prevent DoS and DDoS attacks. Many researchers have proposed and used different machine learning techniques to predict DoS and DDoS attacks, however, there is still a need for improvement in the accuracy of prediction and more evaluation of these algorithms and a need for more algorithms to be explored. Hence, this paper improves on existing works by re-evaluating and comparing the accuracy between Decision Tree and Random Forest Algorithms in predicting DDoS attacks. The results of the paper show that Random Forest (RF) Regression model is the best-fit model for the cleaned DDoS SDN dataset used because it is more accurate as it has a lesser mean squared error of 0.21091041940417007 for the test data compared to the mean squared error value of Decision Tree Regression (DTR) Model. Hence, the paper concludes that the RF model is the best-fit model to be used in predicting DDoS attacks. However, the paper proposes that more machine learning algorithms should be explored, implemented, and re-evaluated in detecting DDoS attacks.

Keywords Distributed Denial of Service (DDoS) attack prediction · Decision tree algorithm · Random forest algorithm

T. Hai · J. Zhou
School of Computer and Information, Qiannan Normal University for Nationalities, Duyun, Guizhou 558000, China

T. Hai · Y. Yin
School of Information and Artificial Intelligence, Nanchang Institute of Science and Technology, Nanchang, China

O. A. Adetiloye (✉) · S. A. Zadeh · C. Iwendi
School of Creative Technologies, University of Bolton, Bolton BL3 5AB, UK
e-mail: scholar_bukola@ieee.org

C. Iwendi et al. (eds.), *Proceedings of ICACTCE'23—The International Conference on Advances in Communication Technology and Computer Engineering*, Lecture Notes in Networks and Systems 735, https://doi.org/10.1007/978-3-031-37164-6_4

1 Introduction

Over time, as acceptance of technology and the use of the internet increases, hackers have found it easy to develop advanced methods for attacking digital devices and networks [1]. One of the urgent issues that require an immediate solution in information security is the prediction of network attacks [2]. The most common network attacks are Denial of Service (DoS) and Distributed Denial of Service (DDoS) attacks which causes packet loss by delaying the exchange of information, thereby altering the data packets sent through networks which affect the integrity and reliability of the data [3]. The main goal of these attacks is to overwhelm application servers by sending constant and numerous requests to the server which leads to network traffic thereby making it hard for users to easily access internet services. DoS and DDoS attacks force businesses' services to stop which in turn ruins organizations' reputations, alters and slows down service delivery, communication breakdowns, increases the cost of providing services, energy consumption, and detrimental effects on virtual and co-virtual systems, among other negative impacts [4, 5]. This paper compares two supervised machine learning techniques specifically Decision Tree and Random Forest Regression to predict DDoS attacks. The paper also presents some machine learning models that have been identified by researchers to predict and detect DDoS attacks.

2 Related Works

Advancement in science and technology has made the security and privacy of many organisations more susceptible to hacking [4, 6]. Over time, various machine learning models have been identified and presented by researchers to predict and prevent DoS and DDoS attacks [5, 7]. A survey carried out by some researchers identified that users of every organisation do not have the knowledge on how to handle these attacks. The researchers presented that with the use of some machine learning models which are Random Forest, Naïve Bayes and Multilayer Perceptron procedures, DoS attacks were predicted and prevented with an accuracy of 98.5% [1, 8]. Wang Zidong investigated the security of networks while taking DoS attacks as a type of network attack on random systems in accordance with the analysis of cyberattack predictions [9]. However, Zhang Jin suggested a time-delay system solution for DoS attacks with upper bound [10]. The paper by Kaur et al. (2019) presents DoS Attack Prediction using machine learning techniques. Kaur et al. (2019) further discuss the disadvantages of one of the machine learning techniques explored which is k-nearest neighbors *algorithm* (*KNN*) and the benefits of adopting a hybrid approach with higher accuracy of 95.2–97.4% [11]. Paper [12] proposed the use of a Support Vector Machine and Self Organized Map as the machine learning models to detect real-time denial of service attacks. Their dataset was passed onto Support Vector Machine in which data is classified based on protocols such that if there is a little chance of attack, the packet data would be terminated based on termination policies which prevents the packet from any form of attack, and when output data is vulnerable to attacks even though not attacked, it is sent to Self-Organized Map which detects and applies termination policies. Paper [13] discusses the use of supervised machine learning algorithms in predicting DoS attacks and the algorithms used on a generated dataset are Support Vector Machine, Decision Tree, Logistic Regression and KNN with accuracy 94.4%, 94.3%, 94.2%, and 94.1% respectively. Sahingoz et al in paper [14] proposed Phishing URL detection system based on machine learning models. They created a dataset using 36,383 valid urls and 37,172 phishing urls. Data on valid and phishing urls is retrieved from Yandex Search API and PhishTank respectively. Random Forest algorithm was used in detecting phishing URLs with an accuracy of 97.98%. Paper [4] proposed the use of Logistic Regression Model in predicting DoS attacks by using SelectKBest to extract 15 features and used as input features. Url is tokenized and each token is divided into valid and phishing token and the dataset is trained using logistic regression model. Paper [5] carried out a comparative analysis between different machine learning models in predicting Distributed Denial of Service (DDoS) attacks. The specific machine learning models compared are decision tree, decision table, bayesian network, random forest, and support vector machine methods. The results of the comparative analysis shows decision tree as the best fit model for the DDoS prediction with higher accuracy. Through systematic literature review, it has been discovered that many researchers have proposed and used different machine learning techniques to predict DoS and DDoS attacks, however, there is still need for improvement in accuracy of prediction and more evaluation of these algorithms and more algorithms to be explored. Hence, this paper improves on existing works by re-evaluating and comparing accuracy between Decision Tree and Random Forest Algorithms in predicting DDoS attacks.

3 Materials and Methods

The tools and libraries used for the study are NumPy and pandas, Scikit Learn, and Google Colaboratory as a web Integrated Environment (IDE). Figure 1 shows the libraries imported.

Fig. 1 Imported libraries

3.1 Data Understanding

The dataset used for this study is an open access dataset gotten from https://www.kaggle.com/datasets/aikenkazin/ddos-sdn-dataset [15]. The dataset is named DDoS SDN dataset with 104345 rows and 23 columns. The dataset contains 3 categorical and 20 numeric features (including the label). The dataset was produced as part of research at Bennett University. Statistics on flow and port are generated in two separate CSV files which are merged to produce the final dataset. The dataset is specific to Software-defined Networking (SDN). Ten mininet topologies with switches connected to a single Ryu controller are created at the beginning of the project. Network simulation is performed for both legitimate traffic, such as Transmission Control Protocol (TCP), User Datagram Protocol (UDP), and Internet Control Message Protocol (ICMP), as well as malicious traffic, such as TCP Syn attack, UDP flood attack, and ICMP assault. The data collection contains a total of 23 features, some of which are calculated and some of which are retrieved from the switches. Switch-id, Packet count, Byte Count, Duration sec, Duration nsec, which is Duration in Nanoseconds, Source IP, Destination IP, and Total Duration are among the features that were extracted. The number of bytes transported from the switch port is denoted by tx bytes, while the number of bytes received on the switch port is denoted by rx bytes. The date and time are displayed in the dt field after being converted to numbers, and a flow is observed every 30 seconds. Among the calculated features are Packet per flow which is the number of packets in a single flow, and byte per flow is the number of bytes in a single flow. tx kbps and rx kbps are the data transfer and receiving rates, and port bandwidth is the sum of tx kbps and rx kbps. Packet Rate is the number of packets sent per second and is calculated by dividing the packet per flow by monitoring interval, the number of packet ins messages, and the total flow entries in the switch. The class name, which is displayed in the last column, shows whether the traffic type is malicious or benign. Malicious traffic has label 1, while benign traffic has label 0. A 250 min network simulation is run, and 1,04,345 rows of data are gathered. More data can be gathered by running the simulation again for a predetermined interval [16].

3.2 Data Cleaning

The DDoS SDN Dataset is cleaned using python codes on Google Colab. Figures 2, 3, 4, 5, 6, 7 and 8 shows the screenshots of the steps taken in cleaning the data by running statistics and removing missing rows for rx_kbps and tot_kbps. The dataset has 103839 rows and 23 columns after cleaning.

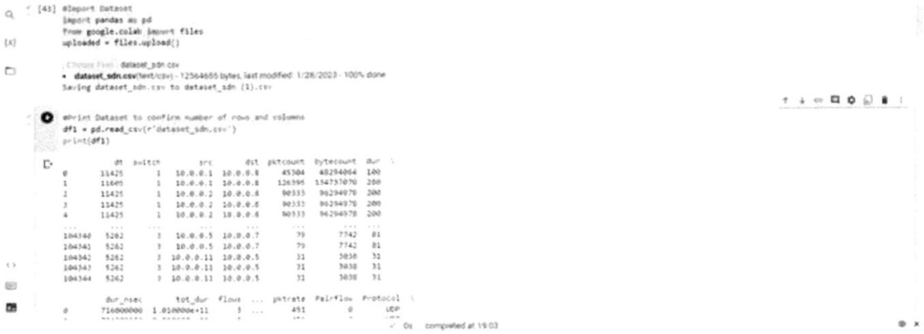

Fig. 2 Data import and reading

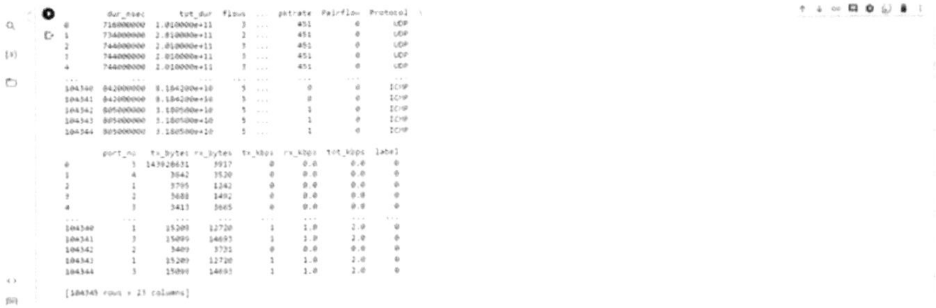

Fig. 3 Data reading continues with confirmation of rows and columns [104345 rows × 23 columns]

Fig. 4 Statistics on missing values per column

Fig. 5 Removing missing rows for rx_kbps and tot_kbps

Fig. 6 Final statistics on missing values per column

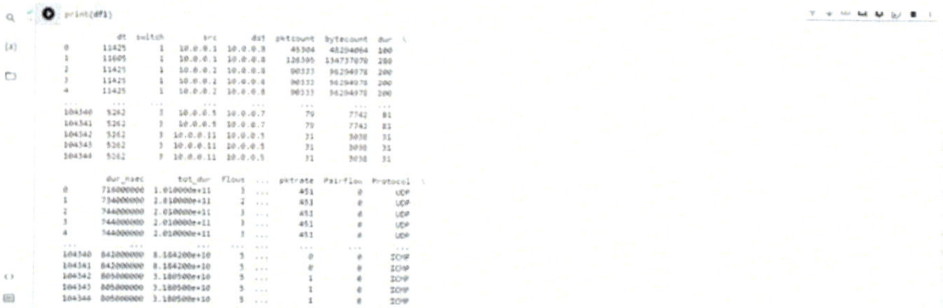

Fig. 7 Printing cleaned dataset

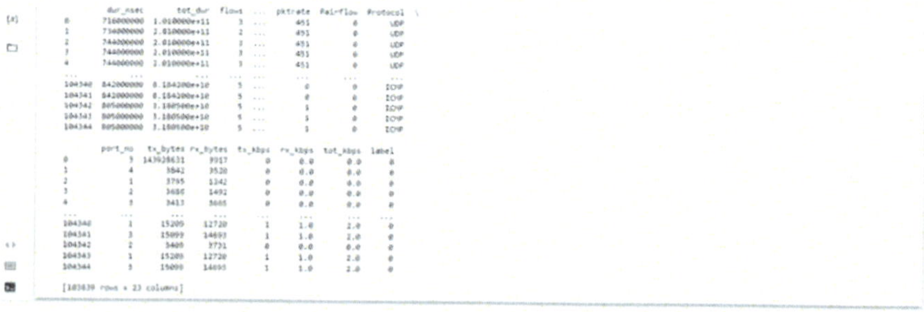

Fig. 8 Confirmation of rows and columns after data cleaning [103839 rows × 23 columns]

4 Results (Decision Tree and Random Forest Regression Models)

Decision Tree and Random Forest Regression models are created to predict DDoS using the cleaned DDoS SDN dataset. 80% of the dataset is for training and 20% for testing. The cleaned dataset is normalised before applying machine learning techniques. Figures 9, 10, 11, 12, 13, 14, 15 and 16 show the lines of codes written to create the models.

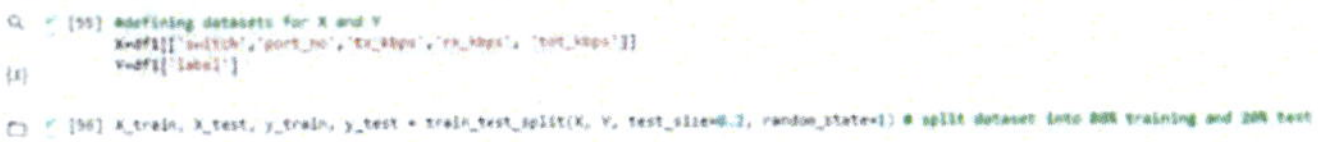

Fig. 9 Dataset definition and splitting (80% Training, 20% Test)

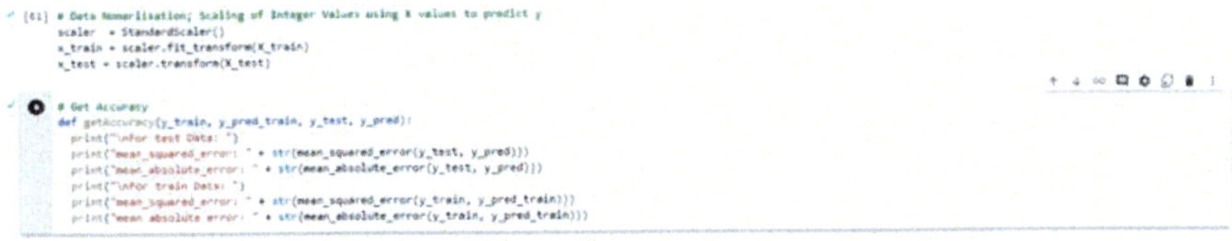

Fig. 10 Data normalisation and accuracy parameters (Performance Metrics)

Fig. 11 Importing decision tree regressor and training decision tree regression (DTR) Model

Fig. 12 Predicting test and train data using the created DTR model

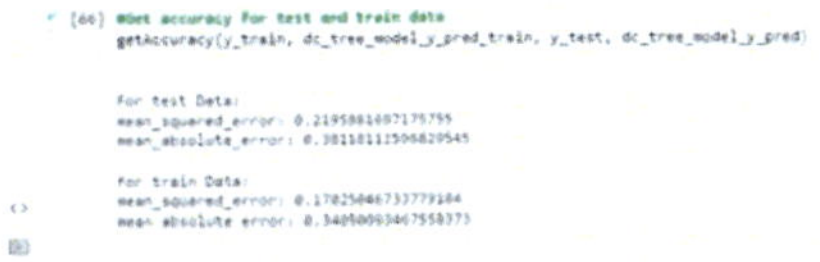

Fig. 13 Accuracy

Fig. 14 Importing random forest regressor and training random forest (RF) model

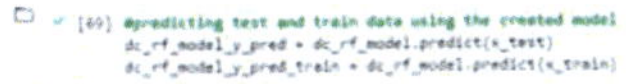

Fig. 15 Predicting test and train data using the created RF model

Fig. 16 Accuracy

5 Discussion

This paper adopts Supervised Machine Learning techniques in training and testing DDoS SDN dataset and specifically used regression because the aim of the paper is to predict DDoS attacks. Comparing the results gotten for Decision Tree Regression (DTR) Model with Random Forest (RF) Regression Model, it is evident that RF Regression Model is the best-fit model for the cleaned DDoS SDN dataset because it has a lower mean squared error of 0.21091041940417007 for test data compared to the mean squared error value of DTR Model for test data which is 0.2195881697175755. Hence, this paper concludes that RF Regression Model is the best-fit model to be used in predicting DDoS attacks [17–21].

6 Conclusion

Cybersecurity attacks are evolving and becoming increasingly difficult to detect. Hence, the prediction and detection of damaging security attacks in cyberspace are becoming a big concern. Therefore, a potent tool for overcoming these difficulties is machine learning. The use of machine learning techniques in detecting DDoS attacks has helped to predict and prevent the attacks in time. However, this work proposes that more machine learning algorithms should be explored, implemented, and re-evaluated in detecting DDoS attacks.

Acknowledgements This work was supported by Key Laboratory of Advanced Manufacturing Technology of the Ministry of Education of Guizhou University (GZUAMT2022KF[07]), the National Natural Science Foundation of China (No.61862051), the Science and Technology Foundation of Guizhou Province (No.[2019]1299, No.ZK[2022]449), the Top-notch Talent Program of Guizhou province (No.KY[2018]080), the Natural Science Foundation of Education of Guizhou province(No.[2019]203) and the Funds of Qiannan Normal University for Nationalities (No. qnsy2019rc09). The Educational Department of Guizhou under Grant NO. KY[2019]067.

References

1. Lakshminarasimman, S., Ruswin, S., Sundarakantham, K.: Detecting ddos attacks using decision tree algorithm. 2017 Fourth International Conference on Signal Processing, Communication and Networking (ICSCN) (2017)
2. Bakhareva, N., Shukhman, A., Matveev, A., Polezhaev, P., Ushakov, Y., Legashev, L.: Attack detection in enterprise networks by machine learning methods. 2019 International Russian Automation Conference (RusAutoCon) (2019)
3. Yang, T., Yue, D., Cheng, Z.: Data-driven based predictive control for networked nonlinear systems with energy-limited DOS attacks. 2017 IEEE Conference on Energy Internet and Energy System Integration (EI2) (2017)
4. Chavan, N., Kukreja, M., Jagwani, G., Nishad, N., Deb, N.: DDoS attack detection and botnet prevention using machine learning. 2022 8th International Conference on Advanced Computing and Communication Systems (ICACCS) (2022)
5. Gulia, N., Solanki, K., Dalal, S.: Comparative analysis to identify the effective machine learning method for prediction of DDOS attack. 2022 10th International Conference on Reliability, Infocom Technologies and Optimization (Trends and Future Directions) (ICRITO) (2022)
6. Iwendi, C., Jalil, Z., Javed, A.R., Reddy G. T., Kaluri, R., Srivastava, G., Jo, O.: KeySplitWatermark: Zero Watermarking Algorithm for Software Protection Against Cyber-Attacks. IEEE Access **8**, 72650–72660 (2020).https://doi.org/10.1109/access.2020.2988160
7. Manoharan, P., Walia, R., Iwendi, C., Ahanger, T.A., Suganthi, S.T., Kamruzzaman, M.M., Bourouis, S., Alhakami, W., Hamdi, M.: SVM-based generative adverserial networks for federated learning and edge computing attack model and outpoising. Expert. Syst. (2022). https://doi.org/10.1111/exsy.13072
8. Alkasassbeh, M., Al-Naymat, G., B.A, A., Almseidin, M. : Distributed denial of service attacks using data mining techniques. (IJACSA) Int. J. Adv. Comput. Sci. Appl. **7**(1) (2016)
9. Hu, L., Wang, Z., Naeem, W.: Security analysis of stochastic networked control systems under false data injection attacks. 2016 UKACC 11th International Conference on Control, Belfast, pp. 1–6 (2016)
10. Zhang, J., Peng, C., Masroor, S., Sun, H., Chai, L.: Stability analysis of networked control systems with denial-of-service attacks. 2016 UKACC 11th International Conference on Control, Belfast, pp. 1–6 (2016)
11. Kaur, G., Gupta, P.: Hybrid approach for detecting ddos attacks in software defined networks. In: 2019 Twelfth International Conference on Contemporary Computing (IC3) 2019 Aug 8, pp. 1–6. IEEE
12. Manoj, M., Keerthi, Mp,, Kiran Kumar, M., ViswaTeja, D., Narayan, S.: Detection of ddos attacks using hybrid machine learning algorithms. Int. J. Creative Res. Thoughts (IJCRT) **8**(2), 1178–1180 (2020), ISSN:2320–2882
13. Arpitha, K.S., Hema, M.K., Sona, G., Reddy, S.K., Gu, P.B.: DDoS attacks using machine learning. J. Xi'an Univ. Archit. Technol. **2**(4), 3380–3384 (2020)
14. Sahingoz, O.K., Buber, E., Demir, O., Diri, B.: Machine learning based phishing detection from URLs. Expert Syst. Appl. **117**, 345–357 (2019)
15. DDoS Dataset. (n.d.). www.kaggle.com. https://www.kaggle.com/datasets/devendra416/ddos-datasets
16. Ahuja, N., Singal, G., Mukhopadhyay, D.: DDOS attack SDN dataset. Data.mendeley.com, 1 (2020). https://doi.org/10.17632/jxpfjc64kr.1
17. Wang, C.: State prediction for smart grids under DOS attack using state correlations underoptimized PMU deployment. 2022 5th International Symposium on Autonomous Systems (ISAS) (2022)
18. Wang, T., Guo, Z., Chen, H., Liu, W.: BWManager: mitigating denial of service attacks in software-defined networks through bandwidth prediction. IEEE Trans. Netw. Serv. Manage. **15**(4), 1235–1248 (2018)

19. Rachmadi, S., Mandala, S., Oktaria, D.: Detection of DOS attack using AdaBoost algorithm on IOT system. 2021 International Conference on Data Science and Its Applications (ICoDSA) (2021)
20. More, K.K., Gosavi, P.B.: A real time system for denial of service attack detection based on multivariate correlation analysis approach. 2016 International Conference on Electrical, Electronics, and Optimization Techniques (ICEEOT) (2016)
21. Lucky, G., Jjunju, F., Marshall, A.: A lightweight decision-tree algorithm for detecting ddos flooding attacks. 2020 IEEE 20th International Conference on Software Quality, Reliability and Security Companion (QRS-C) (2020)

The Impact of Control Plane Attacks in Software Defined Networks Environments

M. Anand Kumar, Senthilkumar Muthusamy, Mohamed Ashik, R. Karthikeyan, and B. Sundaravadivazhagan

Abstract The Internet of Everything (IoE) and 5G are two initiatives that have embraced Software-Defined Networking (SDN), a model that is intensely changing communications systems. Even though SDN has been around for more than 20 years, it is still improving, and the technology industry is placing more and more demands on SDN networks to make them more creative, adaptable, and reliable. Unfortunately, there is an enormous security flaw in the SDN architecture's components that might be used by intruders to execute critical attacks which disrupt the components and their operations. SDN networks have several benefits for managing infrastructure and predicting development. The supply of services was made easier with centralized control, which offers a comprehensive and in-depth perspective of the network. Due to the enormous advantages, SDN is deployed in several applications instead of traditional network architecture. As far as large-scale applications are concerned, there is a huge demand to implement strong security strategies in information allotment and resources across different platforms. Even though several recent research works are recently concentrated on finding the limitations of SDN architecture components, types of security attacks, and possible solutions to overcome the limitations. SDN has encountered several security issues like man-in-the-middle attacks, host hijacking ARP poisoning, and saturation attacks. This research work mainly focuses on control

M. A. Kumar
School of Information Science, Presidency University, Bangalore, India

S. Muthusamy
Department of Engineering, University of Technology and Applied Sciences-Al Mussanah, Muladdah, Oman

M. Ashik
College of Computing and Information Sciences, University of Technology and Applied Sciences - Salalah, Salalah, Oman

R. Karthikeyan
Department of CSE (AI&ML), Vardhaman College of Engineering, Hyderabad, Telangana, India

B. Sundaravadivazhagan (✉)
Department of Information Technology, University of Technology and Applied Sciences-Al Mussanah, Muladdah, Oman
e-mail: sundaravadi@act.edu.om

C. Iwendi et al. (eds.), *Proceedings of ICACTCE'23—The International Conference on Advances in Communication Technology and Computer Engineering*, Lecture Notes in Networks and Systems 735, https://doi.org/10.1007/978-3-031-37164-6_5

plane attacks that affect the total operations of network devices, including switches and the SDN controller. It was also identified that these attacks have a huge impact on the network components of SDN environments. Experimental results provide a detailed analysis of control plane attacks and their impact on the SDN networks.

Keywords Attacks · ARP spoofing · Software- defined-networks · Controller · CPU · Control channel · And flow control

1 Introduction

The realm of communications has benefited greatly from traditional networks. Although currently used network infrastructures are less flexible due to the rapid demand for cloud services for business applications. Modern-day business applications need a more flexible environment to expand services and applications based on their needs. The network architecture of SDN separates the control plane from network devices and other components. The external software entity was used to implement the environment [1]. The external agency configures each device's forwarding tables with the complete information on the network topology. The external entity has complete control over the resources and the components of the network. Contrarily, the functionalities of the control plane and network control features like implementations of routing protocols are very complex in traditional networks. Network devices become substantially less complex and more affordable than traditional devices when the control plane is removed from them and implemented in the independent SDN controller [2].

A variety of network devices are under the administrative control of an SDN controller, which contains information about the network topology. Application programming interfaces (APIs) of the SDN enable the creation of networking applications like traffic engineering and fostering network innovation [3]. The development of novel network applications is difficult or impossible in traditional networks.

One or more SDN controllers operate various network components inside their domains using standard protocols (such as OpenFlow). To provide high scalability and availability, each SDN controller will run in numerous instances, each of which manages a specific set of network components and serves as a backup for other instances [4]. Additionally, SDN controller instances can federate with controllers in other domains or communicate with one another within the same domain. Different SDN controllers may also exist for multiple-layer control or scalability. Numerous innovative network applications can be deployed with the separation of data and control planes. The SDN architecture is presented in the following Fig. 1.

The data plane, the control plane, and the application plane are the three layers that makeup SDN architecture. Through an interface, the data between the components are transmitted. The controller is a key element that separates both the data and control planes [5]. Through a northbound interface, the SDN controller communicates with applications; through a southbound interface, it communicates with the data plane.

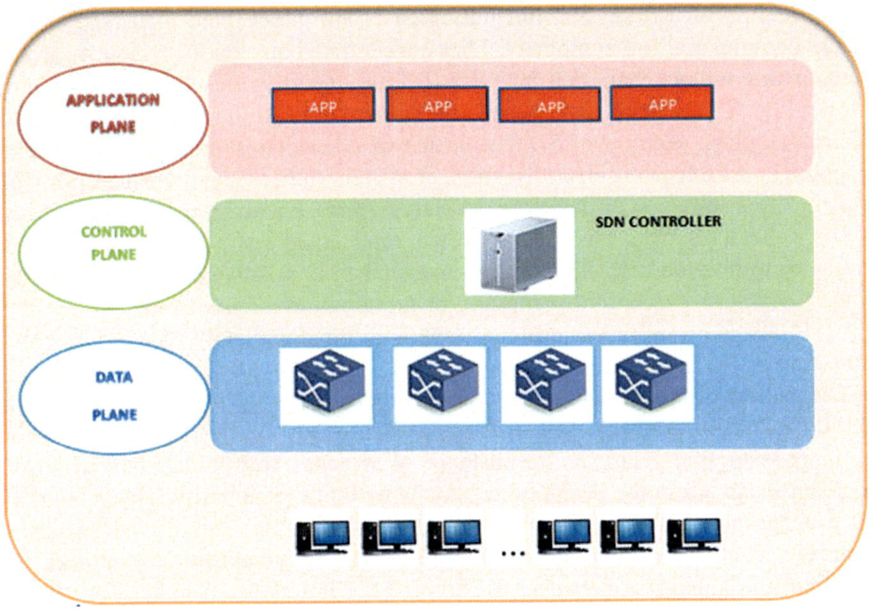

Fig. 1 SDN architecture

With the use of an east–west bound interface, distributed controllers can be linked together [6]. The controller offers a unified sight of the whole network and makes managing other devices easier.

2 Literature Review

A lot of research works on SDN security issues and remedies for privacy, integrity, and the availability of the components that make up the SDN design have been done recently. This section presents the recent review of control panel security issues related to SDN networks. The paper [7] presented a method for grouping SDN controllers into uniform units and raised their security concerns by retaining the SDN's strong security characteristics. To address the trade-off between communication performance and connection security, encryption techniques have been developed. For preventing malicious access to the network and creating an adaptation of the infrastructure components, the authors of [8] developed a Software-Defined Perimeter framework. By properly mitigating the denial-of-service attacks, it is possible to avoid unwanted bandwidth consumption in the network. It was done with the integration of SDP and SDN that enables their examination through virtual network test beds.

The authors of [9] suggest integrating SDN and blockchain to handle cyber-attacks on Industrial IoT equipment. They concentrated on creating a security architecture using a blockchain that includes IDS and integrity-checking techniques to prevent interference with SDN-based IoT systems. The authors of [10] compared the control plane security of SDN with that of other conventional networks. The proposed approach offers error detection, consistency checks, and the possibility of implementing strong security features in SDN systems. For active networks, securing dynamic elements from the attack-injected application was a big challenge. It is necessary to provide the authentication mechanism to guarantee the node security in the SDN networks before permitting or granting any access to the component. But still, active nodes face a difficult issue with the complexity of managing and protecting active nodes [11].

The authors of the work [12] presented methods for analyzing network traffic to identify attacks; in addition, it is necessary to identify attacks by taking into account the inner state, particularly in the instance of switches and virtualized end hosts. Such hosts can aid in the detection of threats with fine granularity. The authors of this paper proposed methods for protecting the network components such as switches, controllers, and trustworthy host machines in SDNs from harmful node attacks. To safeguard the enterprise network, the article [13] introduced SANE as a way to regulate the flow at the central server. SANE behaves similarly to SDN even if they don't employ an OpenFlow switch or an SDN controller. The study did not take into account every SDN threat channel.

In SDN-enabled cloud networking, the authors of [14] suggested a dynamic game-based security framework. The framework can be used for different SDN-based designs, although it does not take into account specific SDN attack vectors. Online learning was employed in the paper [15] to modify the flow dimension. This allowed for scalable dimension with the best placement of observation rules across several switches and nursing of the most revealing flows. To select the measurement interval for traffic monitoring, the authors of [16] developed a table to monitor the table entries in the proxy. Corresponding flow rules are sent to the open-flow switches. Rather than for all flows, just those observed for the chosen flows are exchanged for Flow-stats-requests/replies.

From the literature review, it was identified that Replication attacks in DNS, SMTP protocols, flow attacks of TCP/IP, SYN attacks, Denial-of-service; flooding attacks with UDP, incorporated controller vulnerabilities, Flow table attacks, MITM attacks, network topology attacks are more dangerous attacks that affect SDN components and devices seriously. These attacks have a huge impact on the SDN environments that are to be addressed and resolved.

3 Control-Plane Security

The ability to communicate between manually configured and fixed-function network devices is extremely limited in conventional network architecture [17]. A particular change can have a domain effect on the network's efficiency and even bring the whole thing to a halt. The exponential increase in data traffic, complex network architecture, and growing demands for higher network performance render the traditional approach to network management ineffective [18]. The implementation of the control plane comes next after the control and data planes have been separated. Running the program that implements the control plane is one option. By doing so, it is implied that each switch behaves autonomously and interacts with its network peers to build a personal routing table. BGP, OSPF, RIP, and other convenient protocols can already be used for this purpose [19]. The Internet has been using a distributed control plane exactly like this for more than 25 years.

4 Control-Plane Attacks

4.1 Host Location Hijacking Attacks

The controller's resources, especially the control of the data plane are vulnerable to the host location hijacking attacks [20]. By using resources designated for the controller, the intruder engaging in this operation frequently slows down the network, sometimes even rendering it unreachable to end users. This will bring down the entire network and components of the networks from the perspective of the opponents by overloading the SDN control plane. This is due to the centralized nature of the SDN controller [21]. The controller is responsible for the resource management of the entire network, which could occasionally result in a single point of failure. Using host location hijacking attacks might create distributed denial of service or simple denial-of-service vulnerability on the data layer of SDN [22]. Moreover, there aren't many security limits on host location updates in the present iterations of many SDN controllers. As a result, using the host profile index, adversaries can simply spoof any network identity. The following Fig. 2 presents the attack scenario in the SDN environment.

4.2 Man-in-the-Middle Attacks

When an attacker intervenes in an ongoing data transfer or communication, it is indicated as a "man-in-the-middle" attack [23]. Here intruder acts as a "middleman" during the data transfer and then pretends to be both authorized parties. This gives a hacker the ability to deliver harmful hyperlinks or other types of content to

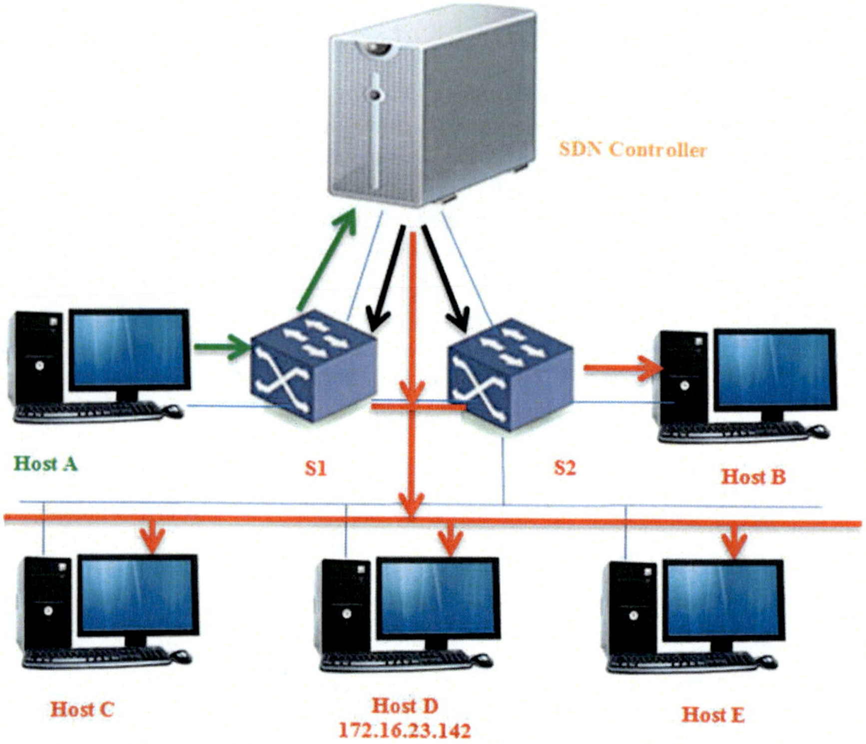

Fig. 2 Host Hijacking attack scenario

both valid persons. But it takes more time to detect such type of false information in the network. The intruder will have complete control to steal data and information coming from either party. The data plane acts as a host node when communication is initiated with the control plane in SDN networks. The control plane acts as a server node for communication purposes, which allows the MITM attack to occur at that point.

4.3 ARP Spoofing Attacks

Using the ARP spoofing technique, attackers can poison caches by sending incorrect IP packets to Access points that are routed into the recipient ARP cache. ARP Request and ARP Response are the two categories of ARP Spoofing assaults [24]. In a request attack, the source IP-MAC in the ARP header of an ARP request message is fabricated. When the victim receives this spoof ARP message, it updates its ARP cache database with the attacker's forged IP-MAC pair. The host's MAC will receive any further packets from the victim with the fake IP. The victim will have complete control over the connection. The traffic transmitted to the target network user can then be monitored or restricted by attackers. A response attack is the third kind of attack in which the attacker either sends forged ARP responses or forges genuine ARP responses in response to legitimate ARP requests to map the next hop IP to the adversary MAC address. The hackers use spoofing techniques such as Address Resolution Protocol (ARP) or ARP poisoning to intercept the data.. An ARP spoofing attack allows a hacker to trick a device into delivering communications to them instead of the intended recipient. The hacker can obtain vital information like passwords and credit card details through transmissions from your device.

5 Experiments and Results

The Mininet simulator, which is running on a virtual machine with the Windows operating system, is used to test security attacks. The Windows platform was able to run virtualized computer systems thanks to the adoption of Hyper-V virtualization technologies. Three primary methods frequently employed for evaluating communication network systems have been identified as analytical modeling, measurement, and evaluation. The evaluation makes use of 60 Core i7 CPUs running at 3.40 GHz, one IBM Intel server, 8 GB of RAM, and Windows 64-bit operating systems. The following Fig. 3 presents the experimental topology which is used to experiment with this research work.

Threshold values are assigned to the network elements with varying parameters before measuring the impact of assaults in an SDN environment. These variables are then compared to obtain reliable outcomes after assessing the attacks. CPU, memory, storage, and throughput utilization are the threshold variables that are used for the evaluation. The recorded parameter values before the attack scenario are shown in the following Table 1.

In this section, the impact of control plane attacks on the SDN environment is investigated. CPU performance, the packet delivery ratio, flow request analysis, and control channel bandwidth are the parameters that were used in this experiment. The impact of control plane attacks is estimated using the aforementioned parameter, as shown below.

5.1 CPU Utilization

One crucial factor used to assess the impact of attacks on the security of the control plane is the CPU utilization. The volume and type of attacks that are executed in the environment determine the actual CPU utilization. Due to non-CPU resource requirements, certain operations demand a lot of CPU time while others require less. The initial threshold value for CPU utilization is 16%. CPU utilization is calculated using the following formula

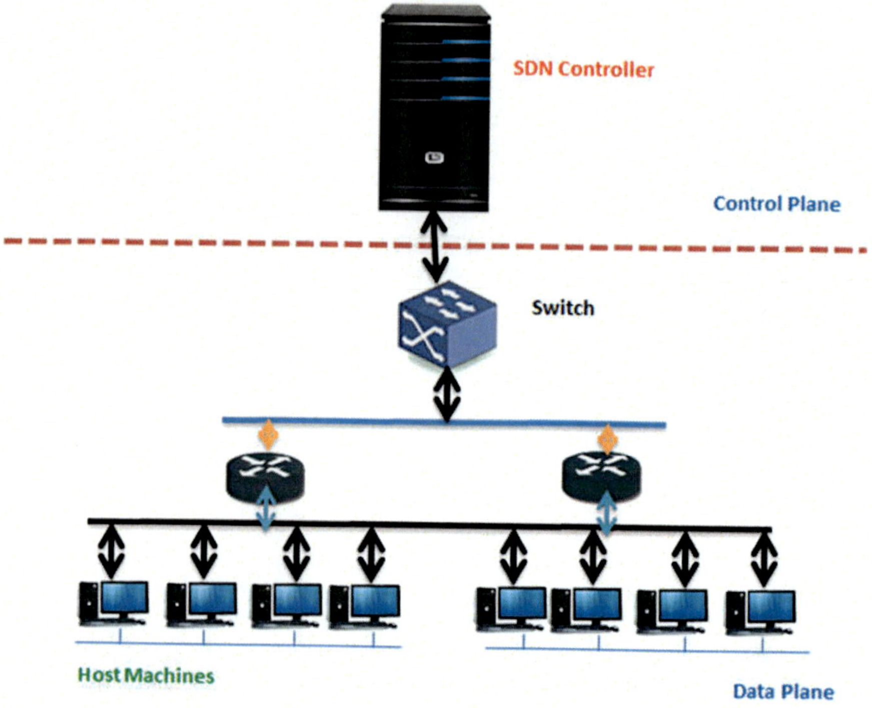

Fig. 3 Experimental topology of CPU utilization

Table 1 Threshold parameters

S. no	Threshold parameters	Initial values (%)
1	Memory utilization	21
2	CPU utilization	16
3	Throughput	93
4	Storage space	57

$$\text{CPUUtilization}(\%) = \frac{\text{Ut(UtilizedTime)}}{\text{Ct(Capacity)}} \quad (1)$$

When compared to MITM and ARP assaults, the investigation reveals that the host location hijacking attacks have very high CPU consumption. This is because 16 out of the network's 60 machines were the targets of hijacking attacks. For all three attacks, CPU usage is minimal at the beginning. However, it gradually increased over a very short time, reaching a point where hijacking attacks account for more than 94% of all CPU usage. The statistics for each of the three control plane attacks at various time intervals are presented in the Table 2. The comparison of CPU usage for control plane attacks is shown in Fig. 4.

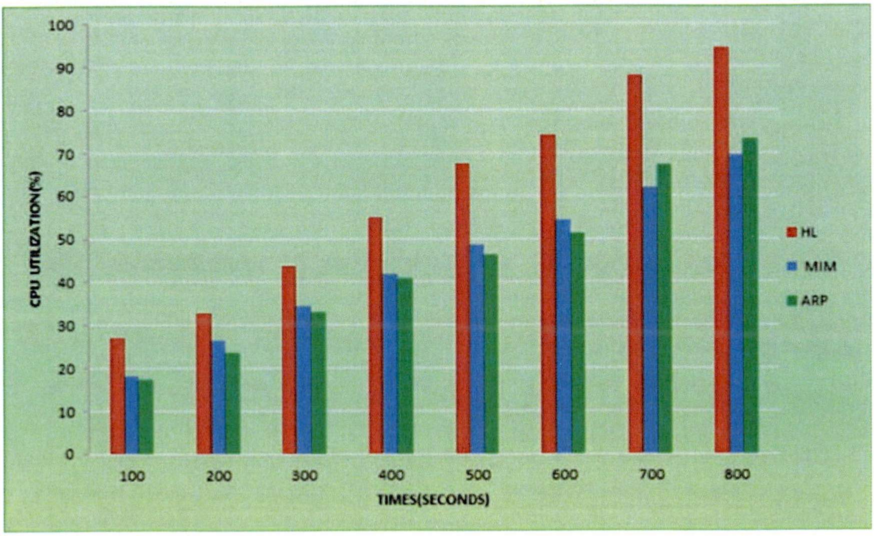

Fig. 4 CPU utilization

Table 2 Initial threshold parameters

Time in seconds	Model		
	HLH	MIM	RP
100	27.2	18.3	17.2
200	32.8	26.4	23.7
300	43.8	34.6	33.1
400	54.9	41.7	40.7
500	67.6	48.5	46.2
600	74.2	54.2	51.2
700	87.8	61.8	67.2
800	94.5	69.4	73.2

5.2 *Flow Control Analysis*

One of the essential elements of SDN traffic is the flow request. More flow restrictions will be installed by the end switch due to MITM assaults. The majority of assaults solely use the flow request functionalities to overwhelm the SDN Controller. According to the analysis, man-in-the-middle assaults generate more packet-in messages than the other two attacks combined. Additionally, it demonstrates unequivocally that more erroneous packet-in requests occur during the attack. In the first stages of an ARP attack, the typical message packet is sent at a rate of fewer than 1000 messages per minute as shown in Table 3 (Fig. 5).

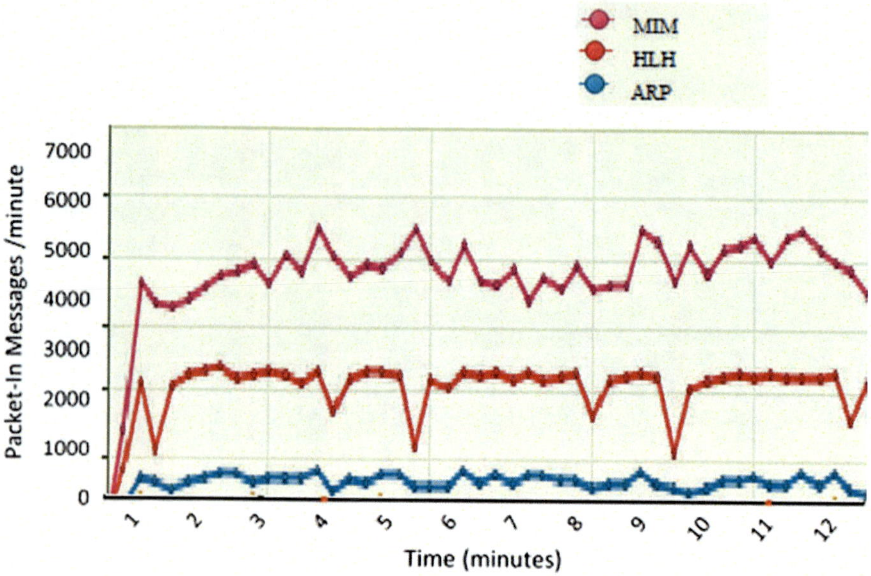

Fig. 5 Flow control

Table 3 Flow control data

Time in seconds	Packet in messages		
	HLH	MIM	ARP
60	3105	4525	643
120	3326	4658	421
180	3458	4129	634
240	3351	5984	643
300	1205	4872	654
360	3487	5435	423
420	3569	4987	754
480	3124	5894	354
540	3485	5124	486
600	1356	4985	234
660	2956	4756	458
720	1654	4084	198

5.3 *Packet Delivery Ratio*

The ratio of the total number of packets transmitted by the source computer to the total number of packets received by the destination machine is known as the packet delivery ratio. When evaluating the packet delivery ratio, the packet loss ratio is also an important factor to consider. TCP packets are transferred from the source machine (Host A) to the destination machine (Host E) in the experiment. The counter is then used to keep track of how many packets are successful and how many were not. The Packet delivery ratio is too low for all three attacks in the attack scenario. The table gives a detailed breakdown of a packet sent and delivered statistics for the SDN network throughout analysis. Figure 6 unequivocally demonstrates that in all three instances, the packet delivery rate is less than 60%. The formula that is used to calculate the packet delivery ratio is as follows (Table 4).

$$\text{Packet\_delivery\_ratio} = \frac{\text{Data\_Received}}{\text{Data\_received} + \text{Data\_Loss}} \quad (2)$$

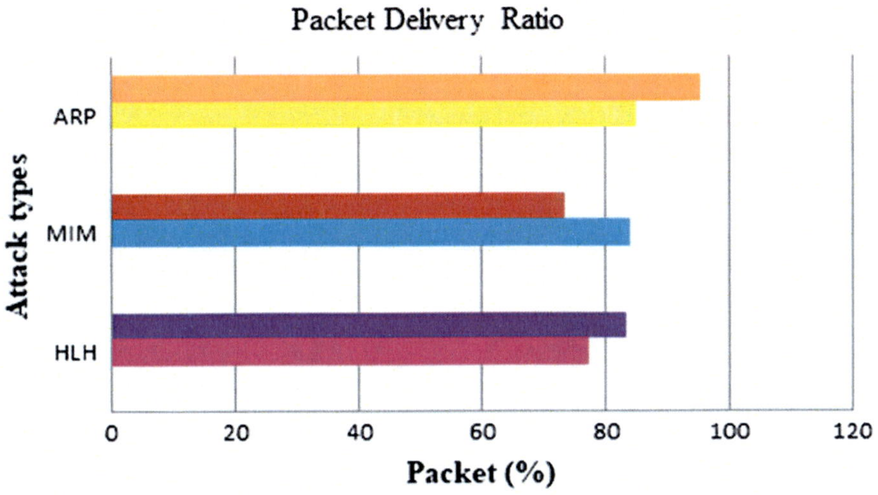

Fig. 6 Packet delivery ratio

Table 4 Packet delivery ratio

Model	Packet delivery (%)			
	Experiment 1		Experiment 2	
	Dropped	Percentage	Dropped	Percentage
HLH	1450	77.48	2978	83.25
MIM	1474	84.08	1401	73.67
ARP	684	85.02	5603	95.45

6 Conclusion

The most recent developments in a software-defined network (SDN) offer new tools for more straightforward network operations. By reducing the complexity of the network topology, this method offers network operators a straightforward abstraction mechanism. A network may be designed and programmed in a flexible fashion thanks to the division of hardware and software components. The devices are frequently aggressively attacked through high-rate Man-in-the-Middle attacks that flood them with numerous malicious packets to deplete their resources. Control channel-based MITM attacks are typically launched through a single host and last for a relatively short time. However, in an SDN context, several hosts or devices can execute simultaneous host location hijacking assaults. As control plane-based attacks employ only a little amount of network traffic while being very similar to the original traffic or host, network administrators are finding it challenging to identify them. The paper assessed three recent threats, including the host location hijacking, man-in-the-middle, and ARP poisoning, which have a significant impact on SDN networks. The final research has shown that these assaults have a significant impact on SDN networks' ability to deliver services to the network's end devices.

References

1. Jiménez, M.B., Fernández, D., Rivadeneira, J.E., Bellido, L., Cárdenas, A.: A survey of the main security issues and solutions for the SDN architecture. IEEE Access **9**, 122016–122038 (2021)
2. Feamster, N., Rexford, J., Zegura, E.: The road to SDN: an intellectual history of programmable networks. ACM Comput. Commun. Rev. **44**(2), 87–98 (2014)
3. Cox, J.H., Chung, J., Donovan, S., Ivey, J., Clark, R.J., Riley, G., et al.: Advancing software-defined networks: a survey. IEEE Access **5**, 25487–25526 (2017)
4. Thyagaturu, A.S., Mercian, A., McGarry, M.P., Reisslein, M., Kellerer, W.: Software defined optical networks (SDONs): a comprehensive survey. IEEE Commun. Surveys Tuts. **18**(4), 2738–2786 (2016)
5. Lv, Z., Kumar, N.: Software defined solutions for sensors in 6G/IoE. Comput. Commun. **153**, 42–47 (2020)
6. Ahmad, I., Namal, S., Ylianttila, M., Gurtov, A.: Security in software defined networks: a survey. IEEE Commun. Surv. Tuts. **17**(4), 2317–2346 (2015)
7. Sood, K., Karmakar, K.K., Yu, S., Varadharajan, V., Pokhrel, S.R., Xiang, Y.: Alleviating heterogeneity in SDN-IoT networks to maintain QoS and enhance security. IEEE Internet Things J. **7**(7), 5964–5975 (2020)
8. Sallam, A., Refaey, A., Shami, A.: On the security of SDN: a completed secure and scalable framework using the software-defined perimeter. IEEE Access **7**, 146577–146587 (2019)
9. Derhab, A., Guerroumi, M., Gumaei, A., Maglaras, L., Ferrag, M.A., Mukherjee, M., Khan, F.A.: Blockchain and random subspace learning-based IDS for SDN-enabled industrial IoT security. Sensors **19**(14), 3119 (2019)
10. Abdou, A., van Oorschot, P.C., Wan, T.: Comparative analysis of control plane security of SDN and conventional networks. IEEE Commun. Surv. Tuts. **20**(4), 3542–3559 (2018)
11. Liu, Z., Campbell, R., Mickunas, M.: Active security support for active networks. IEEE Trans. Syst. Man Cybern. Part C Appl. Rev. **33**(4), 432–445 (2003)

12. Paolucci, F., Civerchia, F., Sgambelluri, A., Giorgetti, A., Cugini, F., Castoldi, P.: P4 edge node enabling stateful traffic engineering and cyber security. IEEE/OSA J. Opt. Commun. Netw. **11**(1), A84–A95 (2019)
13. Scott-Hayward, S., Natarajan, S., Sezer, S.: A survey of security in software defined networks. IEEE Commun. Surveys Tuts. **18**(1), 623–654 (2016)
14. Giotis, K., Argyropoulos, C., Androulidakis, G., Kalogeras, D., Maglaris, V.: Combining OpenFlow and sFlow for an effective and scalable anomaly detection and mitigation mechanism on SDN environments. Comput. Netw. **62**(5), 122–136 (2014)
15. Su, Z., Wang, T., Xia, Y., Hamdi, M.: FlowCover: low-cost flow monitoring scheme in software defined networks. Proc. IEEE Global Commun. Conf. (GLOBECOM), 1956–1961 (2014)
16. Taniguchi, Y., Tsutsumi, H., Iguchi, N., Watanabe, K.: Design and evaluation of a proxy-based monitoring system for openflow networks. Sci. World J. (2016)
17. Hyder, M.F., Ismail, M.A.: Securing control and data planes from reconnaissance attacks using distributed shadow controllers, reactive and proactive approaches. IEEE Access **9**, 21881–21894 (2021)
18. Chang, S.-Y., Park, Y., Ashok Babu, B.B.: Fast IP hopping randomization to secure hop-by-hop access in SDN. IEEE Trans. Netw. Serv. Manage. **16**(1), 308–320 (2019)
19. Achleitner, S., Porta, T.F.L., McDaniel, P., Sugrim, S., Krishnamurthy, S.V., Chadha, R.: Deceiving network reconnaissance using SDN-based virtual topologies. IEEE Trans. Netw. Service Manage. **14**(4), 1098–1112 (2017)
20. Saraswat, S., Agarwal, V., Gupta, H.P., Mishra, R., Gupta, A., Dutta, T.: Challenges and solutions in software defined networking: a survey. J. Netw. Comput. Appl. **141**, 23–58 (2019)
21. Nagarathna, R., Shalinie, S.M.: SLAMHHA: a supervised learning approach to mitigate host location hijacking attack on SDN controllers. In: Proceedings 4th International Conference Signal Process, Communications Network (ICSCN), pp. 1–7 (2017)
22. Li, Y., Qin, Y., Zhang, P., Herzberg, A.: SDN-enabled cyber-physical security in networked microgrids. IEEE Trans. Sustain. Energy **10**(3), 1613–1622 (2019)
23. Brooks, M., Yang, B.: A man-in-the-middle attack against OpenDay- Light SDN controller. In: Proceedings 4th annual ACM conference resistance information technology (RIIT), pp. 45–49 (2019)
24. Pradhan, A., Mathew, R.: Solutions to vulnerabilities and threats in software de_ned networking (SDN). Proc. Comput. Sci. **171**, 2581–2589 (2020)

Compressed Data Aggregation Using Regulated Duty Cycle for WSN-IOT Networks

Anuradha Gupta, Tarun Gulati, Amit Kumar Bindal, Karan Kumar, A. C. Ramachandra, and Július Seleck

Abstract IoT sensors can send the data using different modes i.e. periodically/one time only/continues streaming using different application-types i.e. periodic/single shot/streaming etc. and resource consumption may also vary as per requirement. In order to utilize resources in optimal way, for each application-type, there is need to regulate the data aggregation. It may be affected by other parameters also i.e. transmission range, Signal-to-Noise-Ratio (SNR) etc. This paper introduces a compressed data aggregation approach (using Zlib library) that takes care of application behavior by enforcing a transmission duty cycle and its performance will be analyzed using different parameters (SNR/Throughput/Packet delivery ratio/Residual Energy etc.).

Keywords IoT · WSN · Data aggregation · Sensors · Compression

A. Gupta · T. Gulati · K. Kumar (✉)
ECE Department, M. M. Engineering College, M. M. Deemed to be University, Mullana, Ambala, India
e-mail: karan.170987@gmail.com

A. K. Bindal
Department of Computer Science and Engineering, M. M. Engineering College, M. M. Deemed to be University, Mullana, Ambala, India

A. C. Ramachandra
Department of Electronics and Communication Engineering, Nitte Meenakshi Institute of Technology, Bengalore, India

J. Seleck
Faculty of Management, Comenius University in Bratislava, Odbojárov 10, 82005 Bratislava 25, Slovakia

C. Iwendi et al. (eds.), *Proceedings of ICACTCE'23—The International Conference on Advances in Communication Technology and Computer Engineering*, Lecture Notes in Networks and Systems 735, https://doi.org/10.1007/978-3-031-37164-6_6

1 Introduction

Proficiencies of traditional wireless sensor networks (WSN) can be extended by integrating Internet of Things, called WSN-IoT networks as shown in Fig. 1.

Following are the barriers for optimal data aggregation over such networks:

(a) Application Types: WSN-IoT may use the different application types and resource consumption may vary w.r.t. each type as discussed below:
   (i) Periodic application: It sends the data periodically and may consume excessive resources as shown in Fig. 2.
   (ii) Non-Periodic application: It is used for one time data transmission only and it consumes fewer resources for small data volume but in case of bulk data, it may cause extra overhead as shown in Fig. 3.

(b) Efficiency of data aggregation: It is difficult to maintain the efficiency of data aggregation process as it may vary for each application type. It may be degraded for large scale periodic/non periodic transmission also.

(c) Resource Availability: WSN-IoT networks use limited resources. IoT enabled operations over WSN can consume excessive resources thus may reduce network life-span.

Fig. 1 WSN-IOT

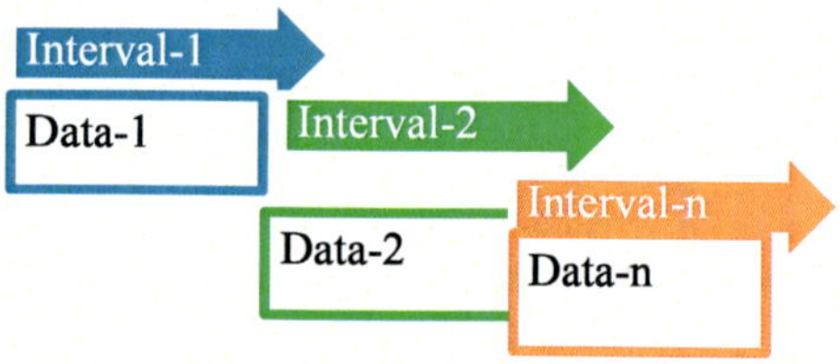

Fig. 2 Data send-periodic application

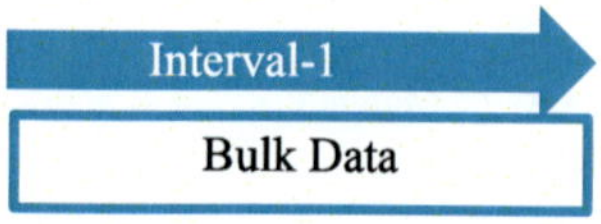

Fig. 3 Data send-non-periodic application

(d) Data complexity: Field data can be arranged in to simple or large volumes as well as it may have different structure/format etc. Accuracy of data aggregation may vary with heterogeneous data types.
(e) Error prone Transmission: Data may be corrupted due to various factors (i.e. SNR variations [1]/low transmission power/long distance transmission/ interference etc.). Aggregation over invalid data may degrade its accuracy.
(f) Quality of Services: It is hard to achieve in heterogeneous WSN-IoT environment due to different constraints (i.e. delay/span of coverage area/dynamic topology/limited computational resources etc.) [1–3].

2 Literature Survey

Following section highlights the contribution made by different researchers in the relevant domain:

Sirajudheen and Rekh [4] a proposed a fuzzy logic based data aggregation method for healthcare applications. Experimental results show that it can manage the non-linear and complex data under the constraints of huge data volume and extends the network life-span as compared to existing data aggregation scheme for healthcare scenario.

Nabil et al. [5] introduced an analytical model that can perform data aggregation over large scale IoT network. It builds ransom transmission schedule by analyzing different packet size/transmission rate/duty cycle frequencies/traffic density to ensure the reliable communication. Results show that it outperforms under the constraints of Signal-to-Interference-Plus-Noise-Ratio/transmission rate/delay etc.

Jiao et al. [6] developed a solution to optimize the delay factor for data aggregation process over IoT networks. It performs tree based aggregation along with graph based scheduling to minimize intermediate delay. Analysis shows its performance in terms of accuracy as compared to baseline methods.

Su et al. [7] developed a decentralized data aggregation method for IoT networks. It forms a aggregation group for nodes and each member can join this group at any time for aggregation and can also leave after process completion. Analysis shows that it can improve aggregation accuracy as well as overall computational cost and it can also perform under compromised network.

Ibrahim et al. [8] proposed a data aggregation scheme for IoT networks. It analyzes the traffic patterns and builds an abstract traffic stream by suppressing the duplicate traffic types in current transmission. Experiments show that it performs optimal and accurate aggregation as well as it also reduced the congestion level over IoT network.

Said et al. [9] introduced a node based aggregation approach for IoT healthcare applications. Aggregation is performed at node level and finally it is encrypted and forwarded to server that uses a message extractor to decrypt incoming data. Simulation shows that it offer lightweight and secure transmission over IoT network w.r.t. minimal computational/transmission/resource consumption etc.

Saha et al. [10] introduced a rule based aggregation scheme for the IoT based software defined networks. As per current traffic flows, different rules are built and finally, duplicate rules are suppressed and heuristic based data forwarding is performed. Simulation results show that it can reduce overall delay/duplicate traffic flows fulfills the QoS constraints w.r.t. IoT network.

Vo et al. [11] investigated the delay issue caused by aggregation process over IoT-WSN and introduced a time constrained data aggregation scheme to resolve this. It builds a tree based aggregation model that collects data through shortest paths and performs aggregation and finally, uses a multi-channel duty cycle for transmission.

Sharanappa and Kakkasageri [12] presented an efficient data aggregation scheme for IoT-WSN. It uses multiple agents to perform aggregation at node level thus condenses the unnecessarily traffic flow over network. Agents can traverse the entire network to perform data aggregation and results show performance in terms of optimal resource consumption/efficient transmission/extended network life-span etc.

Li et al. [13] introduced a data aggregation scheme for IoT network that collects the sensed data from the remote areas where human intervention is not feasible. It performs aggregation using Aerial data aggregators that collects data from different IoT devices shared by various drones and transmission schedule is prepared by drones using reinforcement learning algorithm. Analysis shows that overall aggregation efficiency can be enhanced using reinforcement technique as compared to tradition methods.

Yousefi et al. [14] explored the relation of data aggregation w.r.t. various parameters i.e. efficient transmission, minimal delay factor and optimal resource consumption etc. Study shows that heterogeneous network operations can affect the performance of data aggregation methods as well as it can also have impact over some other factors (i.e. reliability/network life-span etc.) and these are open issues.

Ullah et al. [15] explored the issues related to the secure data sharing, aggregation and transmission for healthcare services over IoT. Analysis shows that efficient data aggregation can optimize the size of data to be transmitted however, security provision may reduce the efficiency of data aggregation scheme and it can also increase the transmission cost/delay factor etc.

Zhang et al. [16] introduced a compressed sensing based data aggregation method for IoT networks. It performs data mapping to relate compressed and reconstructed data under the constraints of measuring matrix. Experimental results shows its performance in terms of optimal data transmission over network, energy consumption, delay and it also maintains acceptable level of various parameters as compared to traditional compressed sensing methods.

Ibrahim et al. [17] developed a aggregation scheme to handle the hybrid data at large scale over IoT platform. It performs cyclic data collection, aggregation and also performs compression before data forwarding to sink device. Simulation results show that it can efficiently reduce the amount of redundancy, energy consumption and transmission load.

Pappas et al. [18] developed the aggregator approach to collect the data from IoT network. Aggregators are deployed at node level to reduce the queuing delay for incoming traffic. Simulation analysis shows that it offers optimal value of SNR and its performance varies as per the channel conditions/traffic load etc. however, it can be further extended for delay aware network operations.

Du et al. [19] merged the IoT network with cloud/edge computing platform to reduce the overall energy consumption. First of all, data collection is done at IoT node level and after that data compressed and synchronized with cloud through edge nodes using shortest routes. Experimental results show that this scheme reduces the computational overhead for IoT network and extends its life-span also.

Ullah et al. [20] integrated the WSN-IoT platform with Fog computing to provide the efficient data aggregation support and secure communication for healthcare applications. It uses compression and encryption at node level before forwarding and at server side; it is decrypted and extracted further. Simulation results show that it outperforms in terms of optimal resource consumption/transmission rate etc.

Dehkordi et al. [21] investigated the various aggregation methods for WSN-IoT platform and these can be mainly categorized as structure less and structured methods. Study shows that data aggregation performance and energy consumption varies w.r.t. deployment environment (under water/wireless/terrestrial/gas/air).

Alfarraj [22] used a machine learning algorithm to assist the data aggregation process over IoT network. It analyzes the incoming flow and regulates the aggregation process thus reduces the response time/route cost error etc. and finally, aggregated data is synchronized with cloud for reliable communication. Experiments show that it enhances the overall efficiency of data aggregation with minimal transmission failure rate.

Wang et al. [23] introduced a deep learning based data aggregation method for IoT network. It uses support vector regression to predict rate of data collection and performs aggregation in secure way and finally, aggregated data is forward to the destination through energy efficient shortest routes. Simulation shows that it minimizes the overall resource consumption.

Gupta et al. [24] proposed a data aggregation scheme for IoT devices that collects the data using healthcare applications and wearable sensors. Data aggregation and encryption is perfrmed at sensor level and its authentication is verified at base station. Analysis shows that it offers dual security layer for commination with efficient data aggregation/validation.

Li et al. [25] introduced a data aggregation scheme for IoT networks that performs data sampling in a uniform manner using Bernoulli equation. Analytical study shows that it can accurately estimate the sample size w.r.t. traffic and can measure the probability factors for aggregation. Simulation results show its performance in terms of efficient aggregation/optimal energy consumption as compared with existing scheme.

Aziz et al. [26] a proposed a data aggregation scheme for WSN-IoT networks. It uses multiple clusters for data collection and compression sensing technique Grey Wolf based method forwards the aggregated data to base station through optimal paths. Simulation results indicates its performance in terms of efficient data collection rate/energy consumption/minimal reconstruction rate etc.

Du et al. [27] proposed a compressed sensing algorithm for data aggregation over WSN-IoT network. Data collection is done by field sensors and a prediction model is also used for its forecasting and after node level aggregation, data is synchronized with cloud platform. Experimental results show that it is an energy efficient scheme that outperforms under the constraints of network dynamics.

Amarlingam et al. [28] proposed a low complexity data aggregation scheme for WSN-IoT network. It performs compressed sensing and aggregation at cluster level and minimizes the computational overhead at field sensors and forms a sparse measurement matrix for entire network to reconstruct the aggregated data. Simulation analysis shows that it offers higher data recovery rate under the constraints of restricted resource consumption.

Gupta and Bindal [29] investigated the data aggregation schemes for WSN-IoT networks and found that performance of aggregation methods may vary due to different factors. Analysis shows that all above factors may increase the computational overhead and resource consumption also, if aggregation method is not compatible with the traffic payload.

Sankar et al. [30] developed a data aggregation method for Agriculture-IoT network. It forms grid for aggregation using rational calculation and grid head is elected to forward the data using an optimal transmission schedule. Simulation results show its efficiency in terms of minimum delay/resource consumption/congestion control etc.

Kumar and Champa [31] introduced a data aggregation scheme for IoT networks. It monitors the incoming data streams as well as also identifies the probability of data outliers in a given stream that can reduce the efficiency of aggregation process. After the detection of any outliers, it is filtered out. Simulation results show that it retains higher data accuracy for end users with optimal energy consumption and it can be further extended using machine learning methods.

Saleem et al. [32] integrated the Fog computing support with IoT platform. Edge nodes collect the data from IoT devices and perform aggregation at node level and output is finally encrypted and transmitted. At server side, only authenticated messages are treated and rest of all are ignored. Simulation results show that it offers reliable and secure communication with optimal consumption of resources.

Subbiah and Prema [33] proposed a data aggregation scheme for WSN-IoT that measures the data gathering status at each member at cluster level in a given coverage area and secure aggregation is performed w.r.t assigned time slots. Simulation results show that it offers optimal delay and consumes less energy during communication.

3 Compressed Data Aggregation for WSN-IOT Using Regulated Duty Cycle

Figure 4 shows the WSN-IoT communication in which IoT sensors collects different data (it may be heterogeneous/structure less/structured type) and transmit to the gateway simultaneously and finally, it is submitted to server. Performance of the WSN-IoT network may be degraded due to the simultaneously transmission, packet collision, retransmission etc. In order to resolve this issue, in this paper, a duty cycle

Fig. 4 WSN-IoT communication

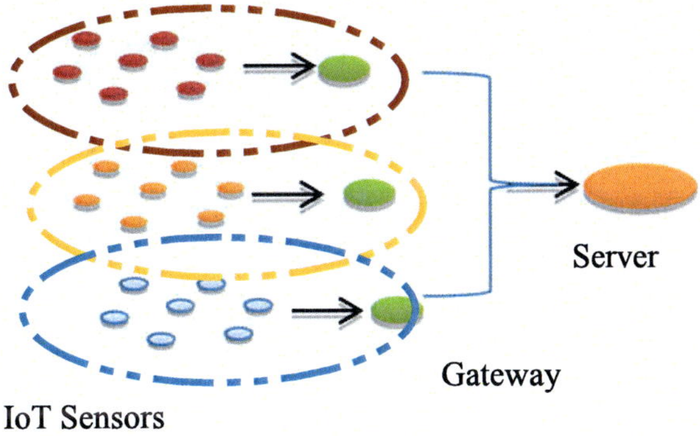

based aggregation approach is introduced that configured the application types as periodic and non-periodic. Life span of duty cycle is managed by gateway that allows the members to forward the data to gateway. In case of periodic transmission, it is regulated as desired however, in case of non-periodic transmission, it is initialized for a short interval. After each transmission, duty cycle is expired and no more transmission to gateway is allowed and after that gateway forwards the aggregated data to server.

Flowchart 1 shows the following steps:

Step 1: First of all WSN-IoT network is initialized.

Step 2: Application type is configured with sampling interval.

Step 3: Gateway is initialized that collects the field data and forwards it to server.

Step 4: Gateway initializes a duty cycle for member nodes for transmission.

Step 5: During active duty cycle, data transmission and forwarding is allowed and if there is no more transmission from members, duty cycle is terminated by gateway. Gateway performs zlib library based compression (DEFLATE algorithm) over received data during aggregation and forwards all data to server. As shown in Fig. 5.

Step 6: If there no more transmission from members, all nodes status is marked as idle. If yes than above steps are repeated w.r.t. sampling interval of periodic/ non-periodic application.

For the implementation of proposed scheme, network simulator version 3 (NS-3) was used with IoT sensor density 100/200/400/600, Long Range Radio communication standard for IoT, Wireless terrain is 1000 × 5000, initial energy is 350 J, Rx/Tx is 10, simulation time is 4000 s, simulation scenarios are: IoT-n: normal data aggregation over WSN-IoT and IoT-r: regulated duty cycle based compressed data aggregation. For data compression, Zlib data compression library (zlib 1.2.13) [34], Fig. 6 shows the simulation scenario for the above discussed parameters.Fig. 6Simulation scenario-Data aggregation- WSN-IoT

Figure 7 shows the Throughput of IoT-DA-n and IoT-DA-r scenarios using periodic application w.r.t. IoT sensor density. Outcomes indicate that IoT-DA-r delivers higher Throughput as compared to IoT-DA-n. In case of IoT-DA-n, with 100 IoT

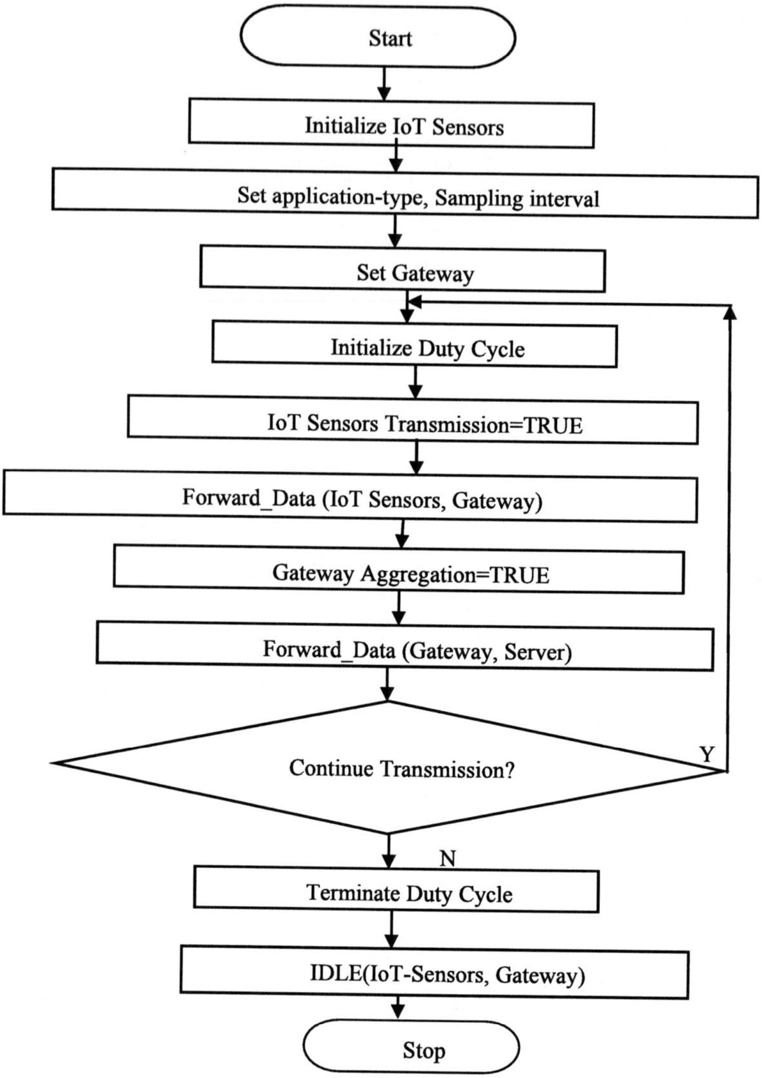

Flow Chart 1 Data aggregation for WSN-IoT using regulated duty cycle

sensors, it is 8 Kbps, 199Kbps for 200 IoT sensors, 396Kbps for 400 IoT sensors and 596Kbps for 600 IoT sensors.

In case of IoT-DA-r, with 100 IoT sensors, it is 174 Kbps, 357 Kbps for 200 IoT sensors, 718 Kbps for 400 IoT sensors and 1075 Kbps for 600 IoT sensors.

Figure 8 shows the Throughput of IoT-DA-n and IoT-DA-r scenarios. using non-periodic application w.r.t. IoT sensor density. Outcomes indicate that IoT-DA-r delivers higher Throughput as compared to IoT-DA-n. In case of IoT-DA-n, with

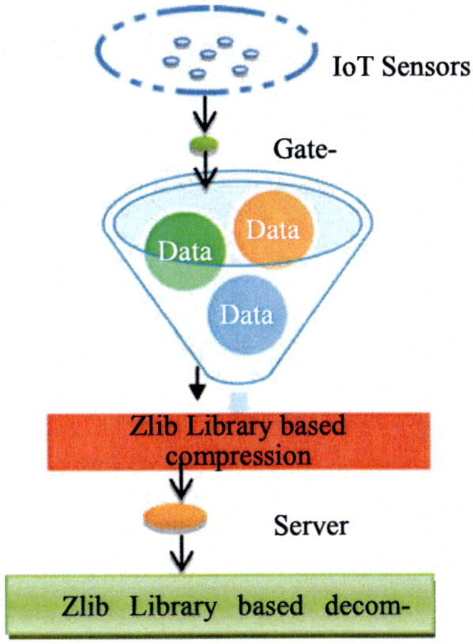

Fig. 5 Zlib library based data compression/decompression

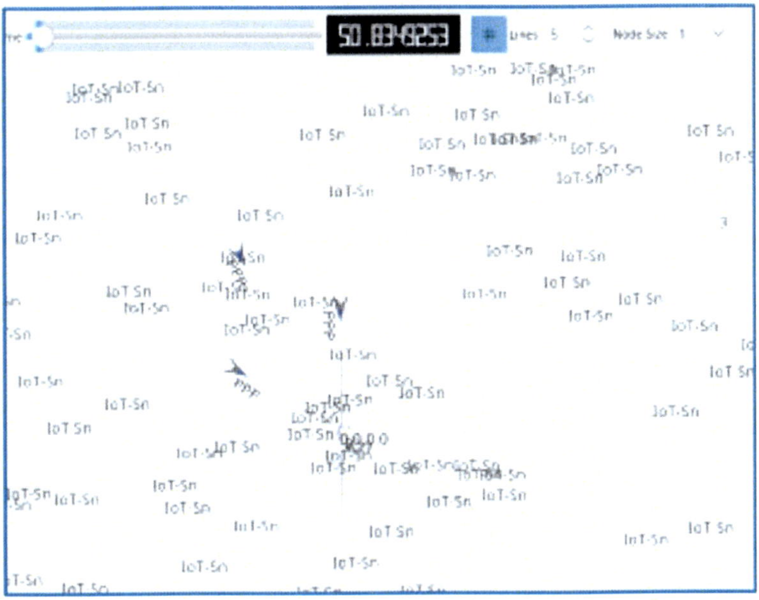

Fig. 6 Simulation scenario-Data aggregation- WSN-IoT

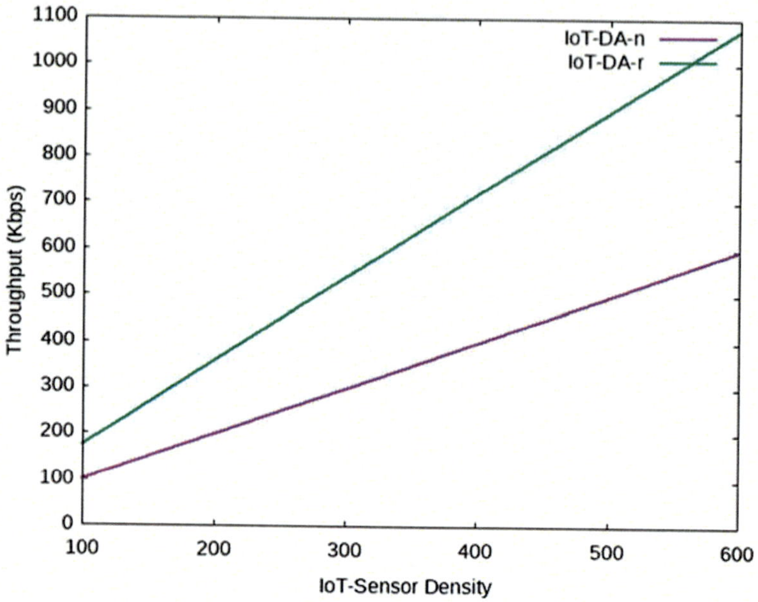

Fig. 7 Throughput-periodic application

100 IoT sensors, it is 208 Kbps, 421 Kbps for 200 IoT sensors, 799 Kbps for 400 IoT sensors and 1198 Kbps for 600 IoT sensors.

In case of IoT-DA-r, with 100 IoT sensors, it is 210 Kbps, 443 Kbps for 200 IoT sensors, 871 Kbps for 400 IoT sensors and 1202 Kbps for 600 IoT sensors.

Figure 9 shows the residual energy of IoT-DA-n and IoT-DA-r scenarios using periodic application w.r.t. IoT sensor density. Outcomes indicate that IoT-DA-r has higher residual energy level as compared to IoT-DA-n. In case of IoT-DA-n, with 100 IoT sensors, it is 166.261 J, 149.035 J for 200 IoT sensors, 115.678 J for 400 IoT sensors and 105.388 J for 600 IoT sensors.

In case of IoT-DA-r, with 100 IoT sensors, it is 288.138 J, 278.756 J for 200 IoT sensors, 270.052 J for 400 IoT sensors and 263.016 J for 600 IoT sensors.

Figure 10 shows the residual energy of IoT-DA-n and IoT-DA-r scenarios using periodic application w.r.t. IoT sensor density. Outcomes indicate that IoT-DA-r has higher residual energy level as compared to IoT-DA-n. In case of IoT-DA-n, with 100 IoT sensors, it is 284.329 J, 204.329 J for 200 IoT sensors, 201.683 J for 400 IoT sensors and 192.683 J for 600 IoT sensors.

In case of IoT-DA-r, with 100 IoT sensors, it is 290.941 J, 214.065 J for 200 IoT sensors, 205.329 J for 400 IoT sensors and 196.594 J for 600 IoT sensors.

Figure 11 shows the variations in SNR values at minimum level using non-Periodic Application w.r.t. IoT sensor density. It can be observed that in case of IoT-DA-n, with 100 IoT sensors, it is 3.88005 dB, 8.76028 dB for 200 IoT sensors, −7.0252 dB for 400 IoT sensors and −9.49441 dB for 600 IoT sensors.

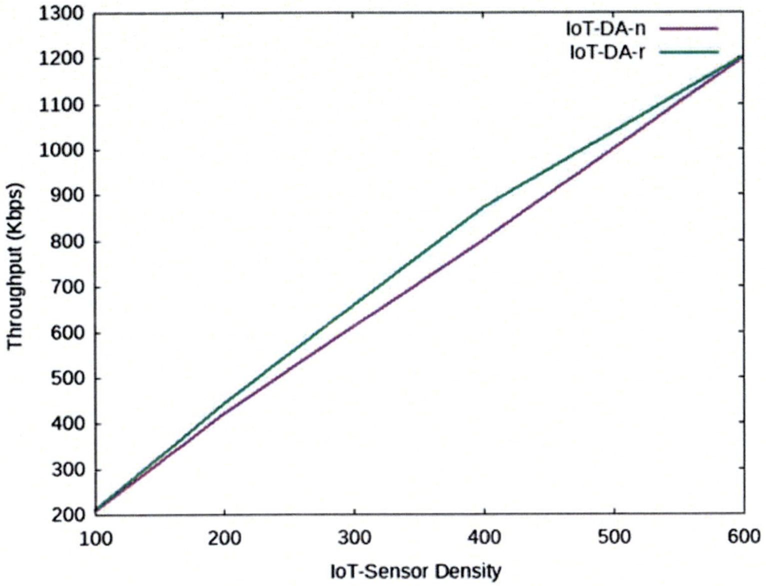

Fig. 8 Throughput-non-periodic application

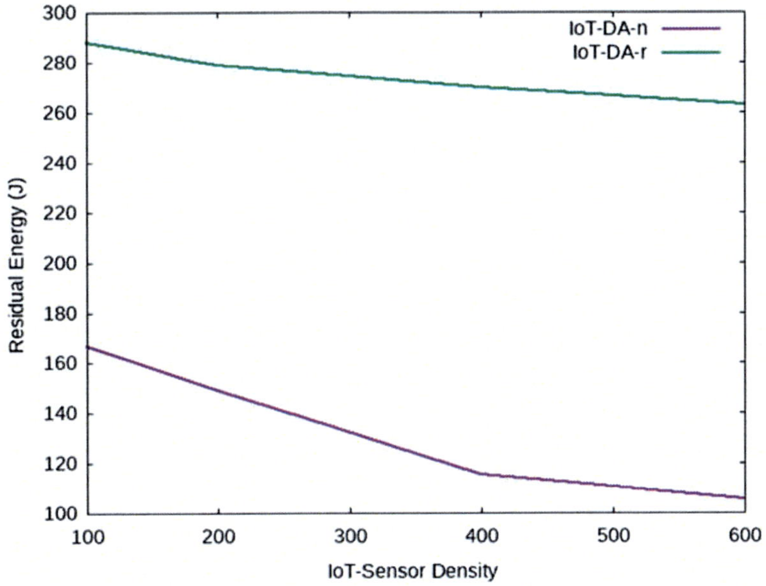

Fig. 9 Residual Energy-Periodic Application

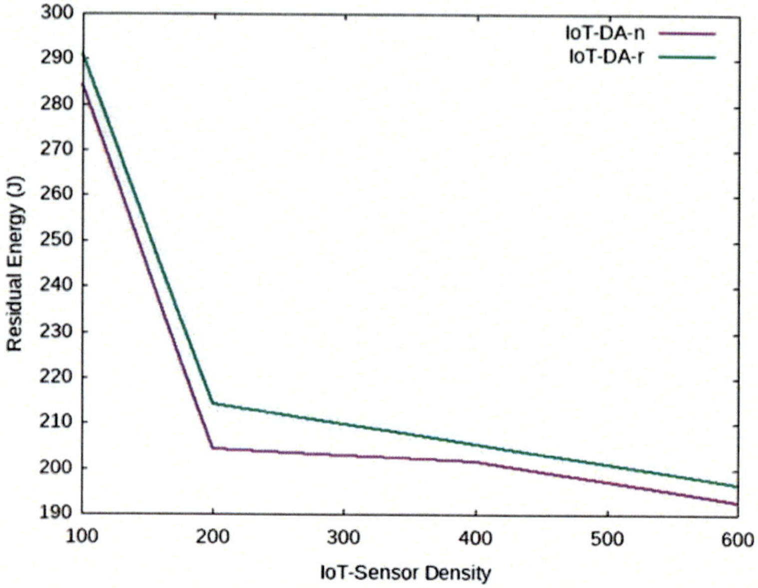

Fig. 10 Residual Energy-non-Periodic Application

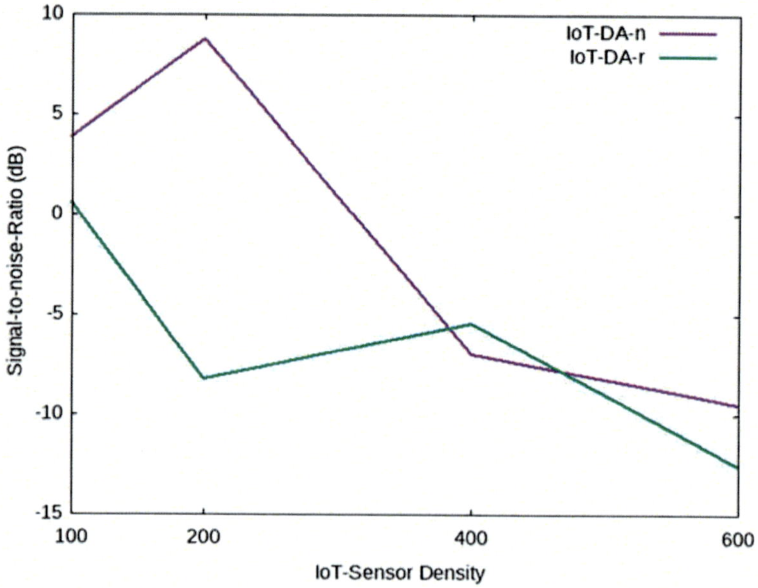

Fig. 11 Min-SNR- non-Periodic Application

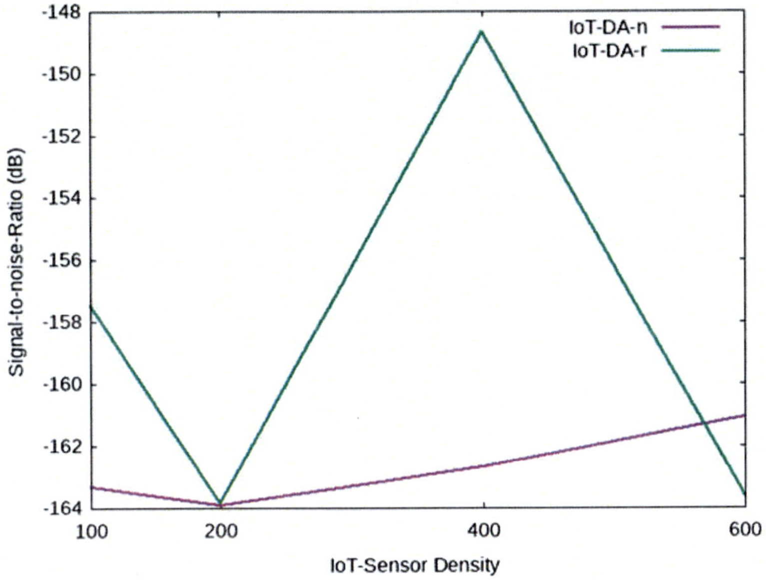

Fig. 12 Max-SNR-non-Periodic Application

In case of IoT-DA-r, with 100 IoT sensors, it is 0.593307 dB, −8.2685 dB for 200 IoT sensors, −5.4419 dB, for 400 IoT sensors and −12.5862 dB for 600 IoT sensors.

Figure 12 shows the variations in SNR values at maximum level using non-Periodic Application w.r.t. IoT sensor density. It can be observed that in case of IoT-DA-n, with 100 IoT sensors, it is −163.319 dB, −163.935 dB for 200 IoT sensors, −162.695 dB for 400 IoT sensors and −161.081 dB for 600 IoT sensors.

In case of IoT-DA-r, with 100 IoT sensors, it is −157.478 dB, −163.844 dB for 200 IoT sensors, −148.663 dB, for 400 IoT sensors and −163.642 dB for 600 IoT sensors.

Figure 13 shows the variations in SNR values at minimum level using Periodic Application w.r.t. IoT sensor density. It can be observed that in case of IoT-DA-n, with 100 IoT sensors, it is 3.88005 dB, 8.76028 dB for 200 IoT sensors, −7.0252 dB for 400 IoT sensors and −9.49441 dB for 600 IoT sensors.

Figure 14 shows the variations in SNR values at maximum level using Periodic Application.

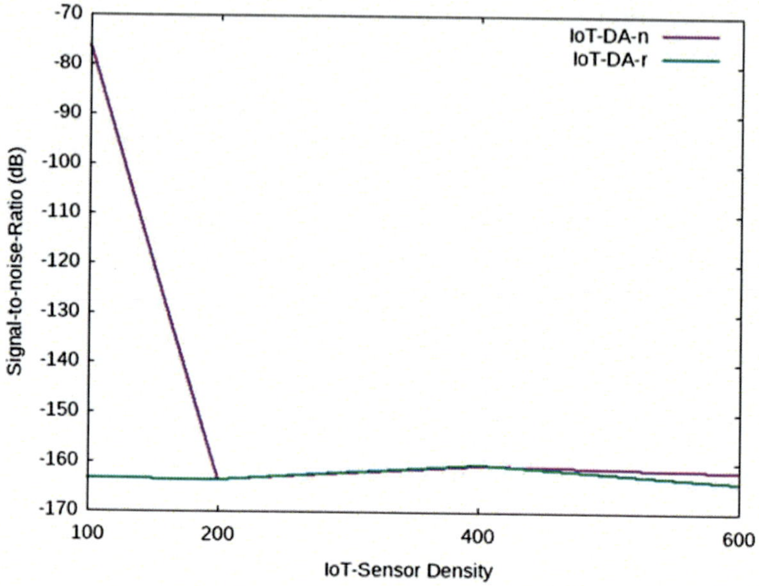

Fig. 13 Min-SNR-periodic application

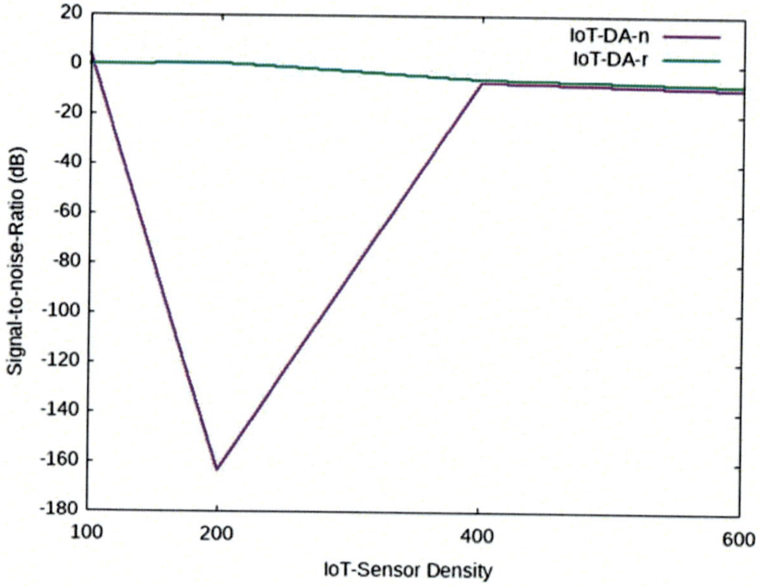

Fig. 14 Max-SNR-periodic application

4 Conclusion and Future Work

In this paper, an energy efficient data aggregation scheme for WSN-IoT networks has been introduced and its performance analysis was performed using different application types (periodic/non-periodic). It considers the application behavior and uses duty cycle for transmission as well as it also performs compression (using Zlib library) during aggregation.

Analysis illustrates that IoT-DA-r is significantly improved the network performance in terms of higher throughput and energy consumption as compared to IoT-DA-n. It can also be analyzed that periodic application has less throughput/residual energy in contrast of non- periodic application. However, in both application types, throughput varies w.r.t. IoT-sensor density thus results in higher energy consumption also. It is also be observed that SNR varies with different scenarios and proposed scheme performed under these constraints also.

As per discussion, it can be concluded that performance of WSN-IoT network varies with the sensor density as well as it is also affected by the application type. In future, it will be implemented for other network types (vehicular area network/ mobile ad hoc network).

Acknowledgements "This work was supported by the Slovak Research and Development Agency under the contract No. APVV-19-0581".

References

1. https://www.semtech.com
2. Dehkordi, S.A., Farajzadeh, K., Rezazadeh, J., Rezazadeh, J., Farahbakhsh, R., Sandrasegaran, K., Dehkordi, M.A.: A survey on data aggregation techniques in IoT sensor networks. Wirel. Netw. **26**, 1243–1263, (2020). Springer
3. Ibrahim, A.S., Youssef, K.Y., Eldeeb, A.H., Abouelatta, M., Kamel, H.: Adaptive aggregation based IoT traffic patterns for optimizing smart city network performance. Alexandria Eng. J. **61**(12), 9553–9568, (2022). Elsevier
4. Sirajudheen, M., Rekh, S.: An overview on data aggregation in IOT For wireless sensor network. Turkish J. Comput. Math. Educ. **12**(12), 3390–3395 (2021)
5. Nabil, Y., Sawy, H.El,, Al-Dharrab, S., Mostafa, H., Attia, H.: Data aggregation in regular large-scale IoT networks: granularity, reliability, and delay tradeoffs. IEEE Internet Things J. IEEE, 1–1(2022)
6. Jiao, X., Lou, W., Guo, S., Wang, N., Chen, C., Liu, K.: Hypergraph-based active minimum delay data aggregation scheduling in wireless-powered IoT. IEEE Internet Things J. **9**(11), 8786–8799 (2022). IEEE
7. Su, Y., Li, J., Li, Y., Su, Z.: Edge-enabled: a Scalable and decentralized data aggregation scheme for IoT. IEEE Trans. Indus. Inf. IEEE, 1–1 (2022)
8. Ibrahim, A.S., Youssef, K.Y., Eldeeb, A.H., Abouelatta, M., Kamel, H.: Adaptive aggregation based IoT traffic patterns for optimizing smart city network performance. Alexandria Eng. J. **61**(12), 9553–9568 (2022). Elsevier
9. Said, G., Ghani, A.: Light-weight secure aggregated data sharing in IoT-enabled wireless sensor networks. IEEE Access **10**, 3571–33585 (2022). IEEE

10. Saha, N., Misra, S., Bera, S.: Q-Flag: QoS-aware flow-rule aggregation in software-defined IoT networks. IEEE Int. Things J. **9**(7), 4899–4906 (2022). IEEE
11. VoV, V., LeD, T., Kim, M., Choo, H.: Data aggregation latency minimization in multichannel duty-cycled WSNs. International Conference on Information Networking (ICOIN), pp. 345–348. IEEE (2022)
12. Sharanappa, P.H., Kakkasageri, M.S.: Intelligent agents assisted information aggregation in internet of thing. IEEE Fourth International Conference on Advances in Electronics, Computers and Communications (ICAECC), pp. 1–7. IEEE (2022)
13. Li, K., Ni, W., Noor, A., Guizani, M.: Employing intelligent aerial data aggregators for the internet of things: challenges and solutions. IEEE Internet Things Mag. **5**(1), 136–141 (2022). IEEE
14. Yousefi, S., Karimipour, H., Derakhshan, F.: Data aggregation mechanisms on the internet of things: a systematic literature review. Internet Things **15**, 1–17 (2021). Elsevier
15. Ullah, A., Azeem, M., Ashraf, H., Alaboudi, A.A., Humayun, M., Jhanjhi, N.: Secure healthcare data aggregation and transmission in IoT—a survey. IEEE Access **9**, 16849–16865 (2021). IEEE
16. Zhang, M., Zhang, H., Yuan, D., Zhang, M.: Learning-based sparse data reconstruction for compressed data aggregation in IoT networks. IEEE Internet Things J. **8**(14), 11732–11742 (2021). IEEE
17. Ibrahim, M., Harb, H., Mansour, A., Nasser, A., Osswald, C.: All-in-one: Toward hybrid data collection and energy saving mechanism in sensing-based IoT applications. Peer-to-Peer Networking and Applications Vol. 14, pp. 1154–1173. Springer (2021)
18. Pappas, N., Dimitriou, I., Chen, Z.: On the benefits of network-level cooperation in IoT networks with aggregators. Performance Evaluation, Vol. 147, Elsevier, pp. 1–17 (2021)
19. Du, X., Zhou, Z., Zhang, Y.: Energy-efficient data aggregation through the collaboration of cloud and edge computing in internet of thing's networks. Procedia Comput. Sci. **174**, 269–275 (2020). Elsevier
20. Ullah, A., Said, G., Sher, M., Ning, H.: Fog-assisted secure healthcare data aggregation scheme in IoT-enabled WSN. Peer-to-Peer Networking and Applications, Vol. 13, pp. 163–174. Springer (2020)
21. Dehkordi, S.A., Farajzadeh, K., Rezazadeh, J., Farahbakhsh, R., Sandrasegaran, K., Dehkordi, M.A.: A survey on data aggregation techniques in IoT sensor networks. Wirel. Netw. **26**, 1243–1263 (2020). Springer
22. Alfarraj, O.: A machine learning-assisted data aggregation and offloading system for cloud–IoT communication. Peer-to-Peer Networking and Applications, Vol. 14, pp. 2554–2564. Springer (2021)
23. Wang, X., Garg, S., Lin, H., Kaddoum, G., Hu, J., Alhamid, M.F.: An intelligent UAV based data aggregation algorithm for 5G-enabled internet of things. Comput. Netw. **185**, 1–31 (2021). Elsevier
24. Gupta, A., Gulati, T., Bindal, A.K.: WSN based IoT applications: a review. In2022 10th International Conference on Emerging Trends in Engineering and Technology-Signal and Information Processing (ICETET-SIP-22) Apr 29, pp. 1–6. IEEE (2022)
25. Li, J., Siddula, M., Cheng, X., Cheng, W., Tian, Z., Li, Y.: Approximate data aggregation in sensor equipped IoT networks. Tsinghua Sci. Technol. **25**(1), 44–55 (2020). IEEE
26. Aziz, A., Osamy, W., Khedr, A.M., Ahmed, A., Sawy, A., Singh, K.: Grey Wolf based compressive sensing scheme for data gathering in IoT based heterogeneous WSNs. Wirel. Netw. **26**, 3395–3418 (2020). Springer
27. Alsharef, A., Sonia, K.K., Iwendi, C.: Time series data modeling using advanced machine learning and AutoML. Sustainability **14**(22), 15292. https://doi.org/10.3390/su142215292
28. Amarlingam, M., Durga Prasad, K.V.V., Rajalakshmi, P., Channappayya, S.S., Sastry, C.S.: A novel low-complexity compressed data aggregation method for energy-constrained IoT networks. IEEE Trans. Green Commun. Netw. **4**(3), 717–730 (2020). IEEE
29. Gupta, Z., Bindal, A.: Comprehensive survey on sustainable smart agriculture using IOT technologies. In: 2022 2nd International Conference on Advance Computing and Innovative Technologies in Engineering (ICACITE), Apr 28, pp. 2640–2645. IEEE (2022)

30. Sankar, S., Srinivasan, P., Luhach, A.K., Somula, R., Chilamkurti, N.: Energy-aware grid-based data aggregation scheme in routing protocol for agricultural internet of things. Sustain. Comput. Info. Syst. **28**, 1–8 (2020). Elsevier
31. Kumar, S., Champa, H.N.: IoT streaming data outlier detection and sensor data aggregation. Fourth International Conference on I-SMAC (IoT in Social, Mobile, Analytics and Cloud) (I-SMAC), IEEE, pp. 150–155 (2020)
32. Tanwar, S., Gupta, N., Iwendi, C., Kumar, K., Alenezi, M.: Next generation IoT and blockchain integration. J. Sens. 2022, Article ID 9077348, pp. 1–14 (2022). https://doi.org/10.1155/2022/9077348
33. Subbiah, S., Prema, R.: An optimized packet gathering scheme for sink transpose and data aggregation in WSN. 3rd Smart Cities Symposium, IEEE, pp. 408–414 (2020)
34. https://zlib.net/

Improving Data Centre Power Efficiency by Applying the VM Consolidation Approach

R. Karthikeyan, Ulligaddala Srinivasarao, Shanmugasundaram Hariharan, Prakash Kumar Sarangi, and Bhawani Sankar Panigrahi

Abstract Cloud computing enables businesses to scale up and down their resource utilization according to their requirements. A data centre uses as much electricity as a small city and houses thousands of servers. These servers need a tremendous amount of computational power to run, which creates a number of problems like high energy consumption and carbon emissions. Given that data centres use a lot of resources when they're operating, attention has recently been focused on how energy-efficient they are. One method to improve the usage of cloud properties is through resource consolidation. By combining virtualized servers into fewer physical servers, virtualization encourages the deployment of flexible resources and boosts energy efficiency because doing so would use less electricity. The Infrastructure as a Service (IaaS) concept, which deploys specialised virtual machines (VMs) on suitable servers in a data centre, is the subject of this study. By utilising a robust VM allocation process, it also improves the quality of services (QoS) and ensures the proper SLA. Additionally, it will examine the problems that will emerge during implementation and the employed performance metrics.

Keywords Cloud computing · Resource consolidation · Energy consumption · Virtual machines · Power efficiency

1 Introduction

According to service level agreements, one or more unified computing resources may be dynamically provisioned and shown as part of a parallel and distributed system called cloud computing, which is made up of networked and virtualized computers.

R. Karthikeyan (✉) · U. Srinivasarao · P. K. Sarangi · B. S. Panigrahi
Department of CSE (AI&ML), Vardhaman College of Engineering, Hyderabad, India
e-mail: karthkarthikhonda77@gmail.com

S. Hariharan
Department of CSE, Vardhaman College of Engineering, Hyderabad, India

C. Iwendi et al. (eds.), *Proceedings of ICACTCE'23—The International Conference on Advances in Communication Technology and Computer Engineering*, Lecture Notes in Networks and Systems 735, https://doi.org/10.1007/978-3-031-37164-6_7

Customers will have access to data and apps hosted in the cloud from anywhere on the earth.

In a cloud data centre, many servers are often placed, and the space is packed tightly to enhance efficiency. Instead of doing it yourself, customers have the option of hosting their programs or running their online service on centralised servers in data centres. One of the main benefits of cloud data centres is their ability to take use of economies of scale to spread out the cost of ownership and system maintenance across a large number of servers. The quantity of energy needed by data centres has significantly increased as a result of the recent rapid expansion in both the number and scale of cloud data centres, which is directly related to the number of hosted servers and their workload. Data centres' extensive use of electricity has a negative effect on the environment. Figure 1 shows the average amount of electricity used by a single server [1]. The amount of CPU power required depends significantly on the server processor type.

Per multi-core CPU, power consumption might range from 45 to 200 W. Recent Intel processors have power-saving features including Demand-Based Switching and Enhanced Speed Step technology. The energy efficiency of multi-core CPUs is substantially higher than that of earlier generations. Servers may run at 1.8 teraflops of peak speed while consuming less than 10,000 watts of electricity thanks to the most modern quad-core Intel® XeonTM processors. To achieve the same performance in 1998, Pentium® CPUs would have required around 800,000 watts of power. The workload on the server will also affect how much power the processors use.

Between 2005 and 2010, data centres' global energy consumption climbed by 56%. Between 1.1 and 1.5% of the energy utilised globally in 2010 was used by data centres. Data centres in the US made up between 1.7 and 2.2% of all usage, which is a much larger percentage. By 2011, the following trends and challenges will have emerged: Cloud data centers are predicted to consume more than 1,00,00,000 MW of electricity and emit more than 40,568,000 tons of CO_2. While virtualization technology increased data centre efficiency, the 2008 economic slump decreased mandate for data centre facilities [2–4]. The ability of virtual machine technology to separate, consolidate, and migrate workloads has recently made it a necessary component for data centres. When combined, these attributes enable a data centre to provide a wide range of users with a secure, adaptable service [5]. Therefore, it is believed that these virtualized infrastructures are essential to the development of the cloud computing paradigm.

The main method for increasing energy efficiency has been virtual machine migration, which involves moving a VM between real computers. Server consolidation through VM migrations, underutilized servers can be turned off. Normal migration and live migration are the two different sorts of migration. To transfer a virtual machine from one host to another, the first technique suspends the active server, replicates the memory contents, and then resumes the server at the new address. The second approach accomplishes the same goal without calling for a service interruption.

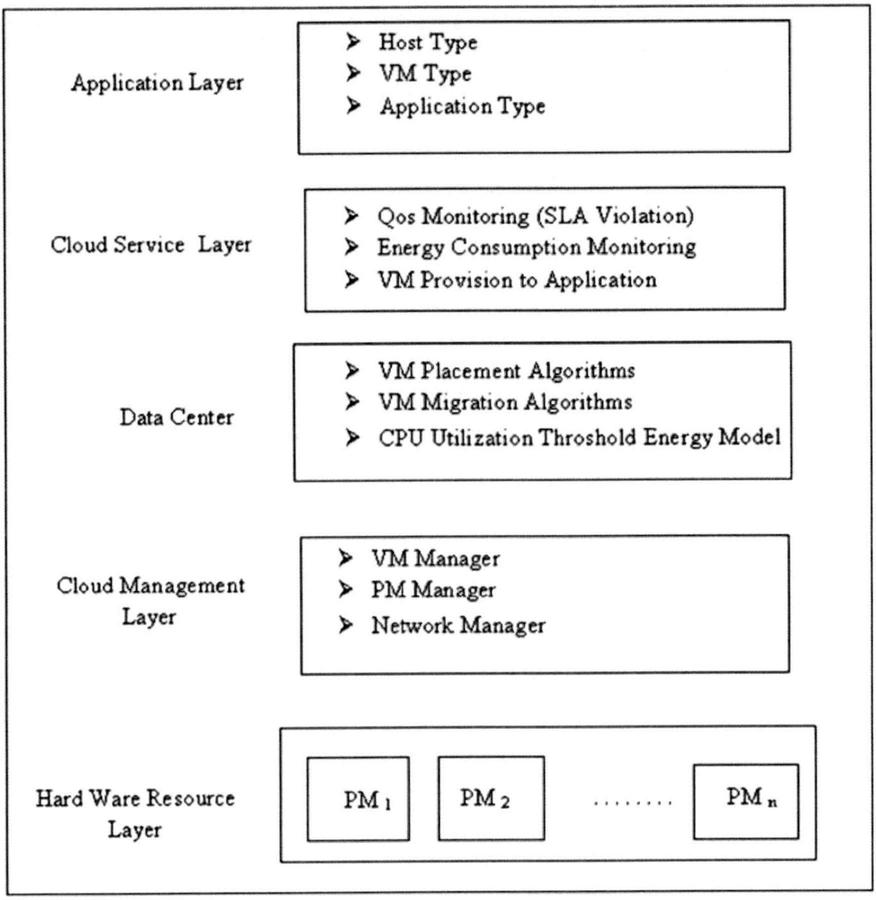

Fig. 1 Architectural frame work

In this study, we provide a novel strategy that includes:

- An architectural framework for cloud computing that is energy efficient.
- Energy-conscious virtual machine distribution.
- Energy-improved application management

2 Related Work

An energy-aware architecture is presented by Anton Beloglazov, Jemal Abawajy, and Rajkumar Buyya [1] for managing cloud power usage. This project's main objective is to use less energy in the data centre. In this article, the CPU, RAM, disc storage, and network interface were used to calculate the power factor. They talks about VM migration and the MBFD (Modified Bit Field Decreasing) algorithm for consistently allocating virtual machines to resources.

In the cloud, energy consumption is a huge issue. We can lower power consumption to maximize resource utilization because idle servers consume a significant amount of energy. Xiaomin, Laurence T. Yang, Huangke Chen, Ji Wang, Shu Yin, and Xiaocheng Liu are the authors of the rolling-horizon scheduling architecture [6]. Real-time controllers and virtual machine controllers play critical roles in this design. If the job given to the user is not finished by the deadline, the VM controller assigns the task to a new VM. The literature examines the VM allocation, migration, and deletion operations in this notion. Finally, they cut power usage by making effective use of the VM's resources. They cover real-time task scheduling and energy consumption in this work.

By executing virtual machine consolidation and migration activities, Hieu Trong Vu and Soonwook Hwan [7] present a strategy for minimizing cloud network congestion control. The VM placement method works as large capacity PMs are consolidated into VMs, and PMs are hosted on heavily-communicated VMs, each of which is located with each other, to provide a traffic and power-aware algorithm. These operations are carried out in order to maximize the server's capacity for use.

Awada Uchechukwu, Keqiu Li, and Yanming Shen disagree on mathematical formulations for determining cloud energy usage, according to [8], prompting them to propose a common model. In a broad model, the energy consumption of two different server types—idle servers and dynamic servers—is calculated. Based on his study, he attempts to demonstrate that 20% of the energy is saved.

The power consumption of the CPU, RAM disc, and network band width are all taken into account in Economou et al. [4] non-intrusive method for modelling the total system power consumption of two heterogeneous servers.

Karthikeyan and Balamurugan proposed the method for environmentally friendly cloud data centres, a tree hierarchical deep convolutional neural network optimised with sheep flock optimization algorithm based job load prediction is suggested [6]. Initially, the kernel correlation method is used to preprocess the historical data from the cloud data centre. In a dynamic cloud context, workload prediction is done using the proposed. The sheep flock optimization approach is used to optimise the T-CNN model's weight parameters. This approach has successfully decreased exorbitant power consumption at cloud data centres while properly forecasting the incoming demand.

It is important to develop a productive scheduling process in order to maintain cloud performance. Using the concepts of the directed acyclic graph, ant colony optimization, cross-entropy stochastic scheduling, and interlacing peak scheduling approach, Karthikeyan and Chitra suggested a method [9]. Higher execution times, overload difficulties, and increased power consumption were the main issues encountered with the enhancement of power efficiency. Author creates a fresh structure made up of crucial elements to overwhelm the issues.

3 Datacenter Power Model

The CPU, RAM, disc storage, and network interfaces all have a role in cloud resource power usage. The CPU consumes the most energy when compared to other resources. The goal of this project is to efficiently manage its power ingesting.

The Power model can be defined as

$$\mathrm{E(C) = I.E\max + (1 - I).E\max .C} \tag{1}$$

Emax is the amount of energy that a server will use at its highest capacity.

I represent the percentage of energy used by an idle server.

C represents the CPU usage.

Due to the variable nature of the workload, CPU utilisation may occasionally alter.

The total energy consumed by a physical node can be described as an integral of the power consumption function over a period of time, as shown below.

CPU usage is specified as a function of time and is represented by

$$\mathrm{C(t).\ E} = \int_{t0}^{t1} E(C(t))dt \tag{2}$$

3.1 VM Live Migration

Simply moving a virtual computer from one physical machine to another is a live migration of a virtual machine. This method allows you to move the VM to the destination node even after it has already been running on the source node, without interrupting any ongoing network connections. Executing a live migration has many benefits, one of which is the extremely minimal downtime, which is on the order of milliseconds.

We must consider the CPU status, RAM content, storage content, and network connectivity when performing a live VM transfer.

3.2 The Cost of VM Live Migration

You can migrate virtual machines between physical nodes using live VM migration with little to no downtime. Live migration, however, has a significant impact on the performance of apps running in a VM during a migration. To calculate the impact's value and create a model for it, Voorsluys et al. conducted an experiment.

They found that how many memory pages an application updates during execution, or the behaviour of the programme, affects performance degradation and downtime. However, it is believed that for online applications, the average performance degradation, including downtime, is about 10% of CPU consumption. This implies that any relocation could lead to a breach of the service level agreement (SLA). Therefore, it's crucial to reduce the amount of VM migrations. The duration of a live migration is determined by the amount of RAM used by the VM and the available network bandwidth. As a result, we will now report the performance degradation that VM j experienced in our work:

$$\text{Udj} = 0.1. \int_{t0}^{t0+Tmj} uj(t)dt, \tag{3}$$

$$\text{Tmj} = \frac{\text{Mj}}{\text{Bj}} \tag{4}$$

where

Udj is the total performance degradations caused by VM j,
t0 is the migration start time
Tmj is the migration completion time
uj(t) is the CPU usage of VM j,
Mj is the amount of memory VM j uses
Bj is the network bandwidth available

In a cloud setting, the process of assigning a virtual machine to an appropriate actual physical machine is known as VM placement. This approach involves categorising the hardware and resource needs of the virtual machines as well as the anticipated resource use and placement goal. This study aims to maximise the utilisation of available resources and improve power efficiency by moving low-used VMs and shutting down or hibernating idle servers.

4 Energy Efficient Cloud Computing Architecture

4.1 Application Layer

Applications, host machines, and virtual machine types can all be configured. The VM and host set configuration settings such as processing speed (in MIPS), memory, disc, and IO bandwidth after component definition.

4.2 Cloud Service

This layer monitors the overall cloud environment's energy usage and service quality. Additionally, it provides VM provisioning for the application layer.

4.3 Data Center

This layer consists of VM placement and migration algorithms. It will effectively allocate virtual machines to the appropriate PMs, transfer underutilized PM VMs to another PM, and shut down underutilized or idle servers.

4.4 Cloud Management Layer

Three significant modules are present in this layer. It's them,

- VM Manager: It contains details on the resource utilization and VM availability. The provisioning of new VMs and the redistribution of existing VMs among physical machines are responsible for adapting the placement, according to the VM placement algorithm and the VM migration algorithm.
- Node Manager: It will keep track of the historical data on resource utilization as well as the actual resource usage by VMs. Decisions about how to allocate resources will be improved.
- Network Manager: It gives the entire cloud data centre the network topology and bandwidth requirements. The network topology specifically outlines the connections between switches and hosts. The performance and cost of the cloud environments are both heavily influenced by the topology.

5 Energy Aware VM Allocation Algorithm

A crucial activity in cloud computing is VM placement, which is done to choose the best server or physical machine for the VM. Improving power efficiency, resource utilisation, and QoS in cloud infrastructure all depend on choosing the right host. The primary goals of VM placement are to reduce VM traffic, consume less energy, and improve security. Initial placement can change in dynamic VM placement due to alterations in the system load, etc. There are two stages to the VM placement process: the server placement of accepted VMs and the VM placement optimization during VM migration. Figure 2 displays the distribution of VMs among servers.

The suggested approach views the positioning of virtual machines as a bin packing problem with a variety of bin prices and sizes. The physical node is represented here

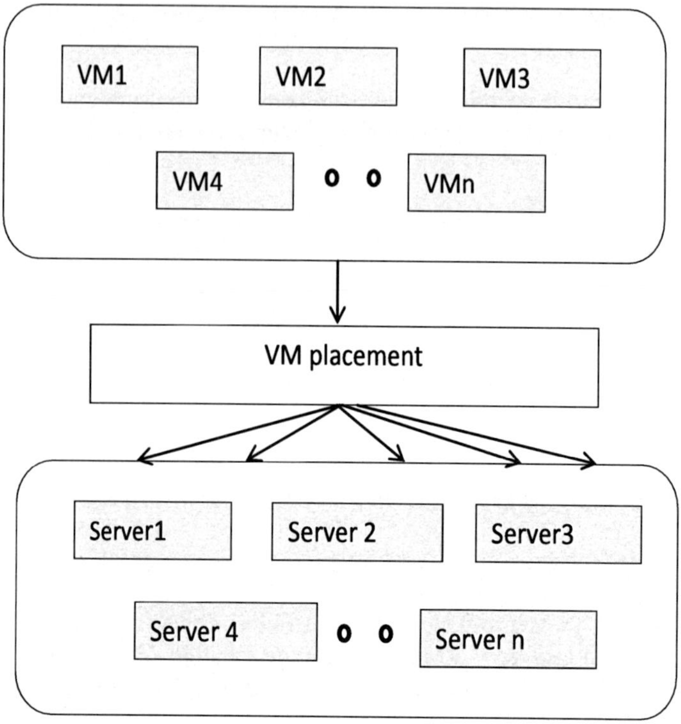

Fig. 2 VM migration

as a bin, while the assigned VMs are marked as items. The bin-packing issue is resolved using the Modified Best Fit Decreasing (BFD) method, which also reduces power usage. Based on CPU usage, the objects are sorted using modified BFD and placed in decreasing order. Additionally, it selects a bin (physical node) for allocating suitable objects (VMs). By giving the nodes with the lowest power usage priority, this permits utilising the heterogeneity of resources. Proposed modified BFD algorithm's pseudo code is as follows (Fig. 3):

The VM placement method effectively distributes VMs to the appropriate physical nodes while simultaneously minimizing the energy consumption of cloud environments (PM). The suggested method outlines the following methods for allocating VMs in an energy-efficient way: The VM and PM lists are first arranged by VM requirements in ascending order. The application then tries to install this VM on an appropriate actual machine for each request. The algorithms evaluate the resource capacity of the actual computers and contrast it with the needs of the virtual machines. The node's status is updated if the PM has enough resources and the VMi is successfully scheduled on that physical machine.

The algorithm chooses the VM (VMr) on the node with the lowest resource consumption if the schedule fails by first locating the node with the lowest resource usage. The algorithm then moves VMr to node PMn, which has the greatest utilisation

Algorithm 1 :VM Placement Algorithm

```
Input
    vm request { vm € VM}
    { pm1 ...... pmn } : Cluster
    Current VM List
Output
    Updated VM List
    Updated PM List
Ascending the VMlist order by utilization scale
Ascending the PMList order by available resource
for each vmi
    do for j = 1 to length [pm]
        do if pmj.available Resource = vmi
                then allocate the vmi to pmj
                    update pm list in ascending order
                    and return vmi is scheduled successfully.
        do else increment the j value and move to next pm
    if vmi is not scheduled
        then to get the value of pmn.available resource
            to find the vmr which has the MIN vm.r € pmn
            if pmn.avilable resource + vmr = vmi
                then if vmr is scheduled to another pm
                        then move vmr to pmx
                            allocate vmi to pmn
                            update pm list in ascending order
                            and return vmi is scheduled successfully.
            else vmi cannot be schedule now, wait for next turn.
```

Fig. 3 VM placement algorithm

right now and has space for VMr. The node status is being updated and PMn is now hosting VMi. If not, After re-adding the VMi to the list of VMs, the method waits for the succeeding replacement cycle to complete its requests before continuing.

6 Performance Metrics

6.1 Data Centre Overall Power Consumption

It is determined how much energy a data centre uses overall:

$$\Delta \text{ECtotal} = \Delta \text{ECsta} + \Delta \text{ECdyn} \tag{5}$$

where,

ΔECtotal denotes the total amount of energy consumed,
ΔECsta denotes energy used while in an idle state,
ΔECdyn denotes energy used when performing a processing activity.
The scheduling process is also increasing the energy consumption.

6.2 Violation of the SLA

In environments utilising cloud computing, quality of service (QoS) is essential. Since SLAs can be defined in terms of some of the qualities offered by the system, such as minimum throughput or maximum reaction time, they are widely used to codify QoS requirements. How these properties alter is determined by the application. In order to use it in our proposed inquiry, a generic measure must be defined. Carrying the SLA means always issuing the applications running inside a virtual machine with the performance they require, only constrained by the features of the virtual machine.

$$\text{SLA} = \frac{\sum_{j=1}^{m} \int_{t0}^{t1} U_{rj}(t) - U_{aj}(t)dt}{\sum_{j=1}^{m} \int_{t0}^{t1} U_{rj}(t)dt} \tag{6}$$

where,

The percentage of the difference between the MIPS that each VM requested is rep

resented by Ur j (t).

The MIPS that was actually assigned is Ua j (t).

The overall number of VMs is m.

This measured displays the fraction of the total CPU performance demand wasn't allocated when apps made requests for it.

By dividing the total number of instructions requested by all virtual machines by the actual number of MIPS allocated relative to the total number of MIPS requested over the lifetime of the virtual machine, we can calculate the general level of service violation that the system causes.

6.3 *How Long It Takes to Finish a VM Live Migration*

Performance metrics that must be taken into account during virtual machine live migration include the size of virtual machine memory, Workload parameters, data transfer rate, virtual machine migration technique, and precisely describing the memory access pattern of each running workload are all important factors.

7 Proposed Experimental Setup

To implement our suggested work, we plan to employ the Cloudsim simulator. A general and adaptable simulation framework called Cloudsim is used to model, simulate, and experiment with cloud computing infrastructure. It offers distinctive features including large-scale cloud infrastructure modelling and simulation, virtualized servers with customised policies, energy-aware compute resources, and user-defined virtual machine allocation criteria, claims our proposed study.

Our simulation system has 1500 nodes with different physical properties. Each node will include a single CPU core capable of performing 2000, 2500, or 3000 million instructions per second (MIPS), 32 GB of RAM, 10 GB/S of network connectivity, and 1 TB of storage. In accordance with this arrangement, a host uses 175 W when the CPU is not being used at all and 250 W when it is. One CPU core with a maximum of 1000, 2000, or 2500 MIPS, 2 GB of RAM, 100 MB/S of network speed, and 1 GB of storage are required for each virtual machine (VM) [10–14].

8 Conclusions and Future Work

Cloud computing's main objective is to help customers manage unexpected resource needs; as a result, data centres are now inherently dynamic. When a customer no longer requires the resources, immediate resource provisioning on demand and resource recovery take place in the IaaS cloud architecture. The heuristic method can produce a better, more believable solution faster than standard methods in a cloud data centre with hundreds of hosts. This study introduced and evaluated a dynamic VM consolidation technique for resource consolidation that is energy conscious. Our proposed work's main objectives are to effectively deploy virtual machines on the proper physical machines, consolidate host resources dynamically utilizing VM migration technologies, and switch inactive or underutilized hosts to power-saving mode. According to a lot of the literature, cloud resource usage and energy consumption are closely related. They demonstrate how job consolidation can increase resource utilization, and how doing so can result in energy cost savings. As a test bed, we plan to construct a small-scale genuine data centre with more than 50 processing nodes. In order to study energy consumption and conduct further research on workload and models where the dynamic behaviour of the data under consideration can change dramatically, a small-scale real data centre with more than 50 computing nodes will be set up in the future. This test bed will allow the resource management strategy to react quickly to changes in the workload and models [14–17].

References

1. Beloglazov, A., Abawajy, J., Buyya, R.: Energy-aware allocation heuristics for efficient management of data centres for Cloud computing. J. Future Gen. Comput. Syst. **28**, 755–768 (2012)
2. Vu, H.T., Hwan, S.: A traffic and power-aware algorithm for virtual machine placement in cloud data center. Int. J. Grid Distrib. Comput. **7** (2014)
3. Zhu, X., Yang, L.T., Chen, H., Wang, J., Yin, S., Liu, X.: Real-time tasks oriented energy-aware scheduling in virtualized clouds. IEEE Trans. Cloud Comput., 1–14 (2013)
4. Economou, D., Rivoire, S., Kozyrakis, C., Ranganathan, P.: Full-system power analysis and modeling for server environments. Workshop on Modeling, Benchmarking, and Simulation (MoBS) (2006)
5. Sundaravadivazhagan, B., Malathi, V., Kavitha, V.: A novel credit grounded job scheduling algorithm for the cloud computing environment. 2022 International Conference on Inventive Computation Technologies (ICICT), Nepal, pp. 912–919 (2022). https://doi.org/10.1109/ICICT54344.2022.9850650
6. Karthikeyan, R., Balamurugan, V., Cyriac, R., Sundaravadivazhagan, B.: COSCO2: AI-augmented evolutionary algorithm-based workload prediction framework for sustainable cloud data centers. Trans. Emerg. Tel. Tech. **34**(1), e4652 (2023). https://doi.org/10.1002/ett.4652
7. Bhadri, S., Bhat Ramakrishna, S.: Enhanced power monitoring for Dell PowerEdge servers. Dell Power Solut. (2011)
8. Corradi, A., Fanelli, M.: Corresponding, Luca Foschini "VM consolidation: A real case based on OpenStackCloud". Future Gen. Comput. Syst. (2012)
9. Karthikeyan, R., Chitra, P.: Novel heuristics energy efficiency approach for cloud data center. 2012 IEEE International Conference on Advanced Communication Control and Computing Technologies (ICACCCT), Ramanathapuram, India, pp. 202–207 (2012). https://doi.org/10.1109/ICACCCT.2012.6320771
10. Uchechukwu, A., Li, K., Shen, Y.: Energy consumption in cloud computing data centres. Int. J. Cloud Comput. Serv. Sci. **3**(3), 145–162 (2014)
11. Castane, G.G., Nunez, A., Llopis, P., Carretero, J.: E-mc2 A formal frame work for energy modeling in cloud computing. Simul. Model. Pract. Theory (2013)
12. Hsu, C.H., Poole, S.W.: Power signature analysis of the SPEC power_ssj2008 benchmark. Performance Analysis of Systems and Software (ISPASS) (2011)
13. Voorsluys, W., Broberg, J., Venugopal, S., Buyya, R.: Cost of virtual machine live migration in clouds: a performance evaluation. Proceedings of the1st International Conference on Cloud Computing, CloudCom 2009, Springer, Beijing, China (2009)
14. Calheiros, R.N., Ranjan, R., De Rose, C.A.F., Buyya, R.: CloudSim: a novel framework for modeling and simulation of cloud computing infrastructures and services. Tech. Rep., Grid Computing and Distributed Systems (GRIDS) Laboratory, Department of Computer Science
15. Shantz, J., Lutyya, H.: A Simulation Framework for Data Center Power Management Strategies. http://www.csd.uwo.ca/~jshantz4/projects/cs9842_project.pdf

16. Karthikeyan, R., Balamurugan, V.: Energy-aware and SLA-guaranteed optimal virtual machine swap and migrate system in cloud-Internet of Things. Concurr. Computat Pract Exper. **33**, e6171 (2021). https://doi.org/10.1002/cpe.6171
17. Sathishkumar, B.R., Sundaravadivazhagan, B., Martin, B., et al.: Revisiting computer networking protocols by wireless sniffing on brain signal/image portals. Neural Comput. Appl. **32**, 11097–11109 (2020). https://doi.org/10.1007/s00521-018-3919-x

Integration of a UHF Fractal Antenna into a 1U CubeSat for Low-Earth Orbit Mission

Raynell A. Inojosa, Celso B. Co, and Mengu Cho

Abstract While a number of novel patch antennas have been designed and integrated into a cube satellite (CubeSat), very few attempts have been made to exploit the self-similarity features of the so-called *fractals* in the antenna design to improve the satellite communications performance. This research work presents an initial attempt to utilize these *fractal* properties in the design of a patch antenna, adopting a geometry that allows it to be incorporated into a standard 10 cm × 10 cm × 10 cm (or 1U) CubeSat. Prior to fabrication, a preliminary design was analyzed and simulated using a commercial finite element method (FEM) solver for electromagnetic structures software, High Frequency Structure Simulator (HFSS)™. The proposed antenna was designed at a solution frequency, $f = 920$ MHz and resonates well from 800 MHz to 1 GHz. The antenna was fabricated on an FR4 substrate ($t = 1.6$ mm, $\varepsilon_r = 4.3$, and $tan\delta = 0.02$) of size 0.2λ × 0.2λ. Comparison of the simulated and measured results showed a good agreement in terms of reflection coefficient and voltage standing wave ratio (VSWR) of the proposed antenna at the frequency of interest. For comparison, a commercial off-the-shelf (COTS) antenna was also characterized and measured. The measured peak gain of the proposed *fractal* antenna is at 3.148 dBi as confirmed by the radiation pattern results. It also exhibits a reflection coefficient of −26.110 dB, a VSWR of 1.104, and a wide bandwidth (BW) across the ultra-high frequency (UHF) band. This type of antenna has the potential to provide a reliable link between a CubeSat and a ground sensor terminal (GST) as confirmed by the long-range communication tests. The proposed *fractal* antenna is expected to meet the satellite data rate requirement of a CubeSat-based Internet of Things (IoT) for low-earth orbit (LEO) mission.

R. A. Inojosa (✉)
Spacecraft Payload and Communications Systems Development Division, Philippine Space Agency, Quezon City, Philippines
e-mail: rainojosa@ieee.org; raynell.inojosa@philsa.gov.ph

R. A. Inojosa · C. B. Co
Graduate School of Engineering, Batangas State University, Batangas City, Philippines

R. A. Inojosa · M. Cho
Laboratory of Lean Satellite Enterprises and In-Orbit Experiments, Department of Electrical and Space Systems Engineering, Kyushu Institute of Technology, Kitakyushu, Japan

C. Iwendi et al. (eds.), *Proceedings of ICACTCE'23—The International Conference on Advances in Communication Technology and Computer Engineering*, Lecture Notes in Networks and Systems 735, https://doi.org/10.1007/978-3-031-37164-6_8

Keywords Cube satellite · *Fractal* geometry · Satellite antenna

1 Introduction

1.1 Cube Satellite (CubeSat)

There is a rapid advancement in satellite communications to employ a tiny cube-shaped small satellite, known as nanosatellite or cube satellite (CubeSat) as secondary payload in a launch vehicle and deployed in low-earth orbit (LEO) mission due to its low cost, relatively fast development, and smaller space requirement for launching. Various CubeSat missions are launched into space for different applications, such as space education, space tethers, for commercial sectors and for remote sensing like weather forecasting, natural disaster monitoring, maritime tracking, and multispectral earth imaging [1, 2].

Generally, the effectiveness of a satellite communications system depends on the link budget estimates, and one of the key components that is accounted to determine this is the antenna performance [3]. For internet of things (IoT) system based on CubeSat, the key feature that impacts the antennas is the link to the ground sensor terminal (GST). Therefore, the antenna should ensure a communication link between the CubeSat and the GST to support the required data rate. Also, antennas are utilized to control and monitor the satellite's health parameters. For small satellites, only a limited amount of space is allocated to the antennas. Hence, it is important to incorporate low-profile antennas to potentially reduce the volume of the satellite or allow for more space to be used for cameras, solar panels, and other payloads [4]. As a result, many antenna design engineers and researchers performed prior studies to implement various antenna architectures while considering the antenna size, CubeSat standardized structure, and the operating frequency.

1.2 Antennas for CubeSat

For small satellite applications, various types of antennas (e.g., deployable, inflatable, microstrip wire) have been utilized for both the uplink and downlink communications [5]. However, these types of antennas often come with a deployment complexity, which could become the point of failure. In some instances, deployable failures can result in a total or partial loss of the spacecraft and/or the mission. To address this, printed antennas are used due to their simple and easy integration into nanosatellites and their high design flexibility [6, 7]. Using the conventional printed circuit board (PCB) techniques, these antennas can be simply fabricated. Therefore, their low-profile characteristics make them excellent candidates for low-earth orbit CubeSat integration.

However, a challenging task is to design printed antennas on the CubeSat's limited surface area and allow them to resonate at relatively large wavelengths. As a result, fractal antennas have been recently investigated as the primary radiating structure for small satellites [8–10]. The geometry of fractal antennas is created by successively iterating a generator shape to a simple basis shape [11]. In the 1970s, the mathematical concepts of fractals were studied by Benoit B. Mandelbrot and were commonly defined by their self-similarity features where small regions of the geometry replicate the overall geometry on a miniaturized scale. This property of fractals is interpreted as a measure of the space-filling properties and complexity of its shape [12, 13].

1.3 Some Fractal Antennas for CubeSat Integration

So far, relatively few fractal antennas have been designed and optimized for CubeSat platforms and there are limited studies that integrate fractal geometries for the amateur radio band antennas like the one studied in [8]. In this work, a fractal slot antenna at second iteration was designed for CubeSat and employed for downlink telemetry application. This was optimized to operate at 458 MHz which exhibited a return loss of −16.53 dB, a bandwidth of 22.62 MHz, and an antenna gain of 2.24 dBi.

In [9], a dual-band fractal antenna based on *Minkowski* fractal geometry was investigated for LEO applications but operates in the L-band and S-band. It has a maximum gain of 7.36 dBi and a bandwidth of around 100–150 MHz. Another type of fractal antenna was explored in [10] that resonates at the S-band. The antenna has an omnidirectional radiation pattern with a maximum gain of 4.39 dBi at 2.3 GHz and a reflection coefficient of −39.56 dB.

Table 1 presents a brief survey of some CubeSat antennas adopting a fractal geometry. A summary of the key design parameters such as the gains and bandwidths are shown. The authors also confirmed whether the reported literature considers the cube satellite structure during the analysis and design.

In this work, a *Sierpinski* gasket fractal [14, 15] at fourth iteration is uniquely adopted as the main geometry of the patch antenna optimized at the 920 MHz band for CubeSat integration.

Table 1 Some fractal antennas for CubeSat

References	Antenna specifications			
	Type of fractal antenna	Gain (dBi)	BW (MHz)	Y/N*
[8]	Fractal slot antenna	2.24	22.62	N
[9]	Minkowski fractal	7.36	150	N
[10]	Koch snowflake fractal	4.39	190	N

design considers the cube satellite structure;* ***Y* *– Yes,* ***N*** *– No*

2 Methods and Procedures

2.1 Antenna Design and Simulation

The *Sierpinski* gasket fractal in Fig. 1 was simulated in HFSS™ to design the patch geometry which resulted from the continuous removal or etching of a similar triangle patch within the larger triangle. A microstrip line was added to feed the structure and a similar fractal patch was utilized as a ground plane as this provides better antenna performance based on the parametric analysis.

For the boundary, both the ground plane and the patch were assigned to finite conductivity as defined by the HFSS software. Then, an AutoCAD.*dxf* file of a 1U CubeSat structure was imported and oriented in a manner where the patch antenna radiates into free space as shown in Fig. 2. This is to simulate the possible effects of the CubeSat structure on the antenna performance.

The geometrical transformation of this geometry generated using HFSS™ is illustrated as follows.

The antenna was fabricated on an FR4 substrate (thickness, $t = 1.6$ mm, relative permittivity $\varepsilon r = 4.3$, and loss tangent $tan\delta = 0.02$) of size $0.2\lambda \times 0.2\lambda$. A python

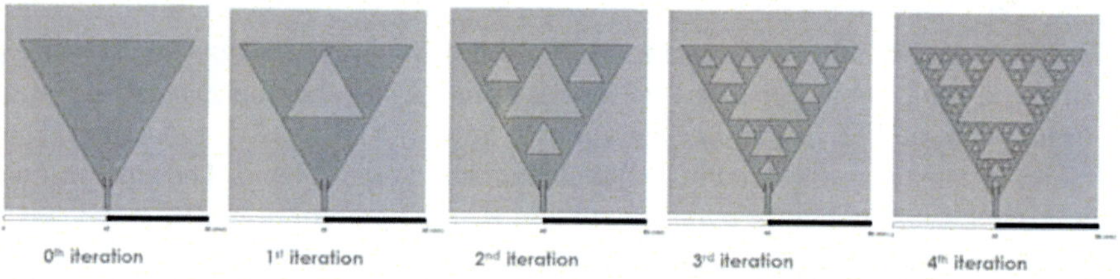

Fig. 1 Geometrical transformation of the fractal antenna up to fourth iteration

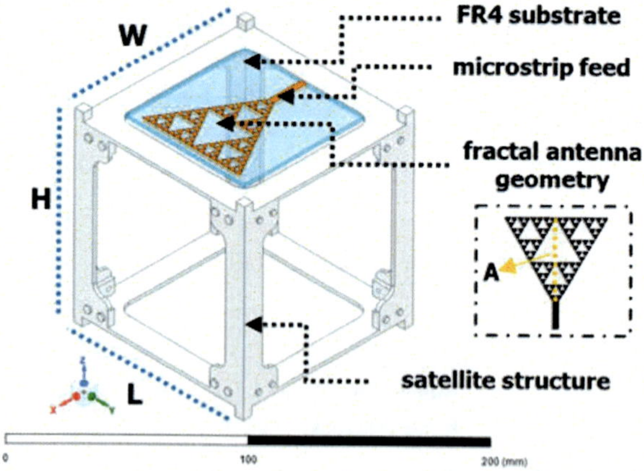

Fig. 2 Configuration of the simulated fractal antenna mounted on a 1U CubeSat structure (**L** = **W** = **H** = 10 cm) via HFSS™. Fractal patch antenna dimension is $0.15\lambda \times 0.15\lambda \times 0.006\lambda$ with the main triangle geometry altitude, **A** = 5 cm at fourth iteration

code was also generated and imported to HFSS™ to easily simulate the first four iterations of the fractal antenna when performing parametric analysis.

2.2 *Characterization of Antenna Performance*

To assess the performance of the designed fractal antenna, the simulated results of the reflection coefficient (S11), voltage standing wave ratio (VSWR), and radiation pattern were analyzed in comparison with a conventional UHF patch as a reference. The antenna performance was characterized and simulated based on finite element method.

The reflection coefficient S11 is a value expressed in dB that denotes the reflection of power towards the transmitter, and also determines if there is a mismatch between the Tx port and the antenna, which is expressed as,

$$\mathrm{S}_{11}(\mathrm{dB}) = 20\log_{10}|S_{11}| = 20\log_{10}\left|\frac{V_{reflected}}{V_{forward}}\right| \tag{1}$$

where *Vreflected* is the voltage that is reflected towards the transmitter and *Vforward* is the voltage that is incident towards the antenna. Commonly, the lower the value of S11, the lower the power reflection towards the transmitter, and at the same time a better impedance matching. In that case, a maximum amount of energy is radiated by the antenna into free space. Using the CubeSat's operating frequency to plot S11, the VSWR was also determined in terms of the reflection coefficient [16] by using the expression,

$$\mathrm{VSWR} = \frac{1 + |S_{11}|}{1 - |S_{11}|} \tag{2}$$

By means of simulations, the two-dimensional (2D) radiation pattern of the proposed fractal antenna was obtained to characterize the antenna response in terms of the far-field radiation. For the analysis, the co-polar plots of the radiation pattern were plotted, i.e., at 0° and 90° elevation planes.

2.3 *Field Tests and Measurements*

To emulate a more realistic environment, a radio anechoic chamber located at the Center for Nanosatellite Testing (CeNT) of Kyushu Institute of Technology (KyuTech) was used for measuring the radiation patterns of both the conventional and proposed fractal antenna. S11 and VSWR were also obtained to assess the actual antenna performance.

Figure 3 shows the configuration of the antenna far field measurement system in the anechoic chamber. The boresight gain of the antenna under test (AUT) was measured with the position controller changing the direction of the AUT in the azimuth plane. By using the external RF signal generator as the transmission source, this system was able to characterize the transmission performance of the fractal antenna mounted on a 1U nanosatellite at the specified frequency of interest.

A long-range test was also performed to verify the downlink communication between a ground sensor terminal (GST) and a satellite with the integrated fractal antenna. A flight model (FM) of a 1U CubeSat was setup at an elevated area in KyuTech while a GST was situated at the Takatoyama Park in Kitakyushu City, Japan. The distance between the transmitter and receiver was determined to be about 4 km. Figure 4 demonstrates the long-range test setup where a free space path loss (FSPL) at the operating frequency, $f = 920$ MHz, was measured to be approximately 102 dB.

To account for the additional free space path loss of the orbital altitude, which is 138 dBm in this long- range test, a 36 dB attenuator was attached between the GST module and the transmitting antenna, Tx. The GST sent a series of data packets to the CubeSat flight model. Upon successfully receiving the data without any corrupted packets, the ground station measured the received signal. The data rate or bit rate *Rb* was determined by using the formula below as presented in [17],

$$R_b(bps) = SF \times \frac{BW}{2^{SF}} \times CR \tag{3}$$

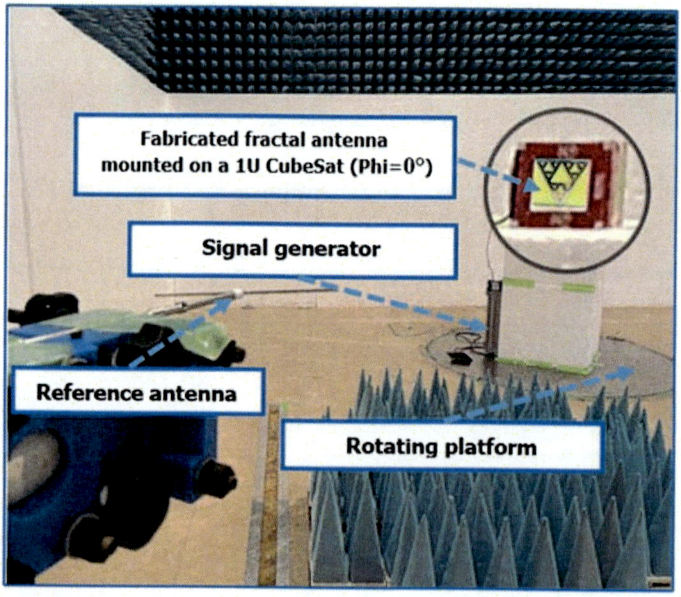

Fig. 3 Far-field measurement setup inside the anechoic chamber of KyuTech

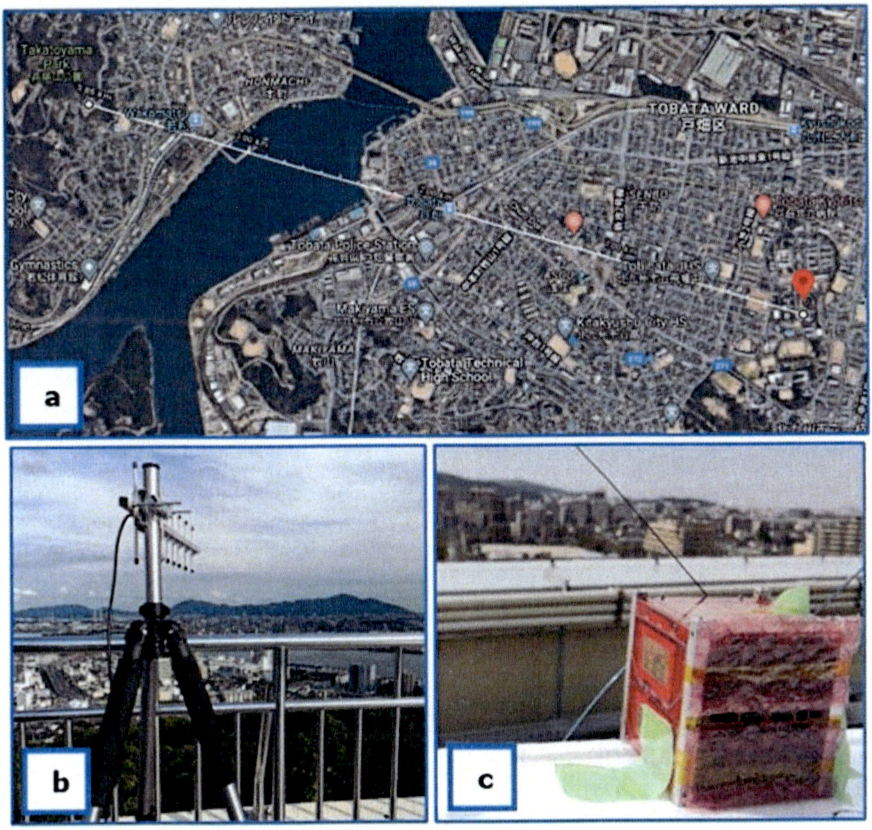

Fig. 4 Long-range test performed last October 2022 in Kitakyushu City, Japan: **a** distance between Tx and Rx; **b** ground station terminal, Tx; **c** CubeSat setup with the integrated fractal antenna, Rx

where *SF* is the spreading factor, *CR* is the coding rate, and *BW* is the bandwidth.

3 Results and Discussions

3.1 Frequency Characteristics

Figure 5 illustrates the frequency characteristics of the proposed fractal antenna at the simulated UHF band. It should be noted that the antenna might resonate well at higher frequencies above 1 GHz; however, for the context of this paper, only the nearby bands was shown for the purpose of comparing the data with the conventional patch antenna.

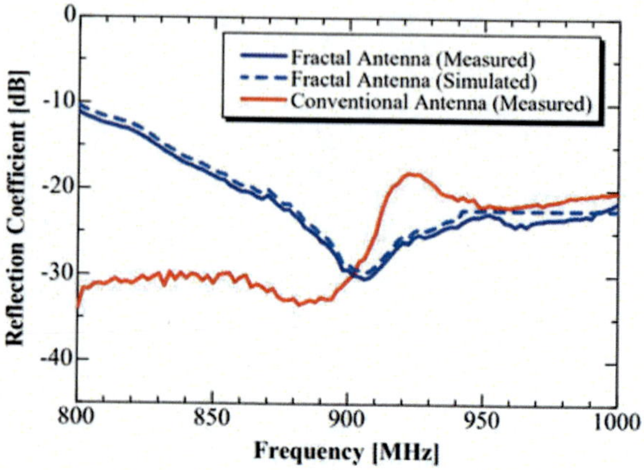

Fig. 5 Comparison of the reflection coefficient (S11) of the proposed fractal antenna against a conventional patch antenna within the simulated UHF band

Across the simulated frequency range, the fractal antenna resonates perfectly with S11 below −10 dB. From 900 MHz above, the proposed antenna exhibits better reflection performance compared with a conventional UHF antenna. Hence, when used as a radiating structure for higher bands, the proposed antenna is expected to be more efficient. At the satellite's operating frequency, $f = 920$ MHz, a reflection coefficient of −26.110 dB was determined.

The minimum measured S11 was −30.364 dB at 906 MHz while −21.425 dB was obtained at 1 GHz. The result provides an insight that the proposed fractal antenna is a good candidate for CubeSat missions that require multiband frequencies.

Figure 6 shows the comparison between the frequency characteristics of the measured and simulated VSWR of the UHF patch antennas. Similarly, the proposed fractal antenna offers better performance in terms of impedance matching at frequencies above 900 MHz in comparison with a conventional antenna. At 920 MHz, the fractal antenna's VSWR is at 1.104. Overall, the antenna's VSWR is within the acceptable value of less than 2.0.

3.2 *Radiation Pattern Profile with the CubeSat Structure*

The radiation pattern of the proposed fractal antenna was analyzed with a 1U CubeSat structure. The 2D radiation pattern is presented in Fig. 7. At Phi = 0°, the antenna achieved a peak gain of 4.009 dBi (simulated) and 3.148 dBi (measured) at the 920 MHz frequency band as presented.

When the antenna was rotated at Phi = 90°, the boresight gain was at 2.278 dBi (simulated) and at 2.973 dBi (measured). The simulated results show that the fractal antenna achieved a nearly omnidirectional radiation patterns in the azimuth plane. The difference between the simulated and measured data could be attributed to the fact that the simulation environment does not include the other modules of the actual CubeSat. Panel boards, electronic circuits, and other metal parts inside the CubeSat

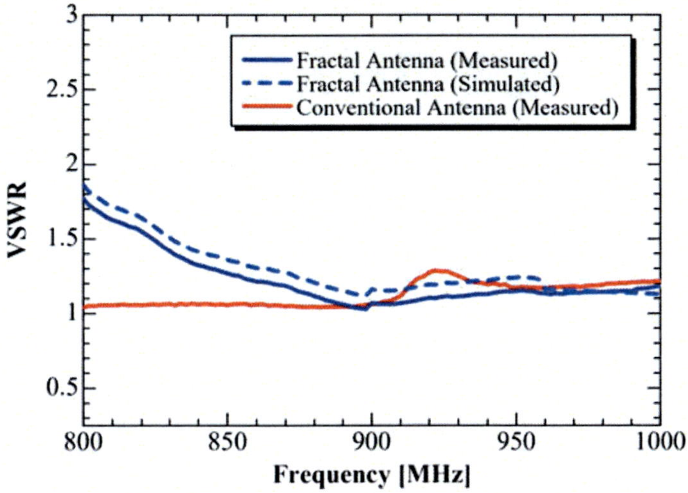

Fig. 6 Comparison of the VSWR of the proposed fractal antenna against a conventional patch antenna within the simulated UHF band

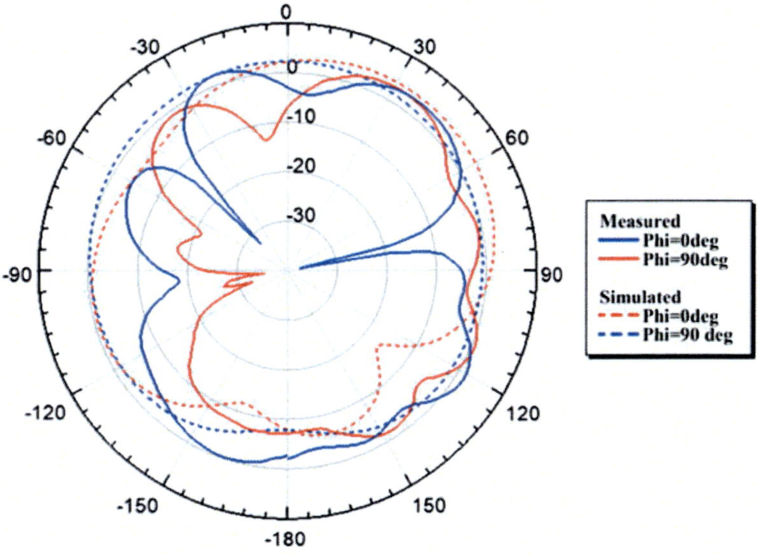

Fig. 7 2D radiation pattern of the proposed fractal antenna at 920 MHz

could degrade the electromagnetic performance of the antenna when it is rotated in the azimuth plane. In future designs, these structures should be considered as these components could attenuate the received signal and thereby affect the reliability of the measured results.

Table 2 Long-range test results

Receiver sensitivity (dBm)	Spreading factor	Coding rate	Bandwidth (kHz)	Data rate (bps)
−138	12	8	31.25	30.52
−136	10	8	31.25	101.73
−138	12	8	41.70	40.72

3.3 Long-Range Test Results

Table 2 also provides the long-range test configurations using a set of predetermined transmission parameters such as spreading factor (SF), coding rate (CR), and bandwidth (BW) that create a successful downlink between the GST and the CubeSat with the mounted fractal antenna. These parameters were selected since this research work adopts a satellite-based IoT system based on the heritage of the GST and the CubeSat platform presented in [18]. Additionally, a combination of lower SF and higher BW allows higher data rates with shorter transmission distance while higher SF and lower BW provide a longer range with reduced data rate.

A receiver sensitivity of −138 dBm was obtained from the field test for SF = 8 and CR = 12. For these configurations, the increase in bandwidth from 31.25 kHz to 41.70 kHz also increases the data rate as expected. With SF = 12 and CR = 8, the sensitivity is observed to decrease; however, it exhibited a data rate of 101.73 bps with a good satellite communication link.

Using the data above, a link budget analysis should be performed in future undertakings to properly design the communication system of a CubeSat. This should give an insight on how to determine power requirements, select appropriate equipment, and set up modulation parameters for signal transmission and reception. The analysis of the radio link should be done to account the gain of the proposed fractal antenna, the calculated data rate, the identified losses, and measured sensitivity among others, and to assess the link budget of the satellite communications system.

4 Conclusions and Future Works

This paper presents an initial attempt to design and analyze a *Sierpinski* gasket fractal antenna for low-earth orbit CubeSat integration, with a patch dimension of $0.15\lambda \times 0.15\lambda \times 0.006\lambda$. The proposed antenna resonates well from 800 MHz to 1 GHz with $f = 920$ MHz. The measured peak antenna gain is at 3.148 dBi as confirmed by the radiation pattern results. It also exhibits a reflection coefficient of −26.110 dB and a VSWR of 1.104 at the required operating frequency. Results confirmed that the proposed fractal antenna offers better performance against a conventional UHF antenna. This antenna also solves the challenge of designing non-deployable, low-profile, and compact antennas for the UHF band.

Other than modifying the patch size specifically for UHF band applications, the novelty of this research includes the consideration of the CubeSat frame on the antenna design. Additionally, the proposed structure consists of a ground plane with a similar geometry as the radiating patch, which has not been done yet in the context of CubeSat antennas. In future works, an investigation should be done for the optimized results for other patch geometries.

Acknowledgements This research is financially supported by the Philippine Space Agency Advanced Degrees for Accelerating Strategic Space R&D and Applications (PhilSA AD ASTRA) Research Grant. The authors would like to thank the Laboratory of Lean Satellite Enterprises and In-Orbit Experiments (LaSEINE) and the Center for Nanosatellite Testing (CeNT) of Kyushu Institute of Technology (KyuTech) for allowing the use of measurement equipment and facilities for testing the fabricated prototype. Special thanks to the members of LaSEINE for their assistance during laboratory measurements and field tests.

References

1. Alam, T. et al.: Lower ultra-high frequency non-deployable omnidirectional antenna for nanosatellite communication system. Nanomaterials **12**(18), 3143 (2022)
2. Woellert, K., Ehrenfreund, P. et al.: Cubesats: cost-effective science and technology platforms for emerging and developing nations. Adv. Space Res. **47**(4), 663–684 (2011)
3. Lokman, A. et al.: A review of antennas for picosatellite applications. Int. J. Antennas Propagation (2017)
4. Abulgasem, S., Tubbal, F., Raad, R., Theoharis, P. I., Lu, S., Iranmanesh, S.: Antenna designs for CubeSats: a review. IEEE Access **9**, 45289–45324 (2021)
5. Akan, V., Yazgan, E.: Antennas for space applications: a review. In: Advanced Radio Frequency Antennas for Modern Communication and Medical Systems, IntechOpen (2020)
6. Samsuzzaman, M., Islam, M.T., Kibria, S., Cho, M.: BIRDS-1 CubeSat constellation using compact UHF patch antenna. IEEE Access **6**, 54282–54294 (2018)
7. Podilchak, S.K., Murdoch, A.P., Antar, Y.M.M.: Compact, microstrip-based folded-shorted patches: PCB antennas for use on microsatellites. IEEE Antennas Propag. Mag. **59**(2), 88–95 (2017)
8. Simon, J. et al.: A second-iteration square koch fractal slot antenna for UHF downlink telemetry applications in cubesat small satellites. Inte. J. Antennas Propag. **2020**, Article ID 9672959 (2020)
9. Lahcene, H.A., Brahimi, A.: A new design of dual-band fractal antenna for LEO applications. In: Proceedings of the 8th International Conference on Systems and Network Communications (2013)
10. Palacios, O.F.G., Díaz Vargas, R.E., Heraud Perez, J.A., Correa Erazo, S.B.: S-band Koch snowflake fractal antenna for cubesats. In: 2016 IEEE ANDESCON, pp. 1–4 (2016)
11. Singh, K., Grewal, V., Saxena, R.: Fractal antennas: a novel miniaturization technique for wireless communications. Int. J. Recent Trends Eng. **2**(5) (2009)
12. Cohen, N.: Fractal antennas: Part 2. Commun. Quart. Summer, 53–66 (1996)
13. Mandelbrot, B.B.: The Fractal Geometry of Nature. W. H. Freeman and Company, New York, USA (1983)
14. Waqas, M., Ahmed, Z., Ihsan, M.B.: Multiband Sierpinski fractal antenna. In: IEEE 13th International Multitopic Conference, pp. 1–6 (2009)
15. Luintel, T., Wahid, P.F.: Modified Sierpinski fractal antenna. IEEE/ACES International Conference on Wireless Communications and Applied Computational Electromagnetics, pp. 578–581 (2005)

16. Balanis, C.A..: Modern Antenna Handbook. Wiley (2009)
17. Augustin, A., Yi, J., Clausen, T., Townsley, W.: A study of Lora: long range and low power networks for the internet of things. Sensors **16**(9), 1466 (2016)
18. Lepcha, P., Malmadayalage, T.D., Örger, N.C., Purio, M.A., Duran, F., Kishimoto, M., El Megharbel, H.A., Cho, M.: Assessing the capacity and coverage of satellite IoT for developing countries using a CubeSat. Appl. Sci. **12**, 8623 (2022)

A Novel & Innovative Blockchain-Empowered Federated Learning Approach for Secure Data Sharing in Smart City Applications

Tao Hai, Dan Wang, Tamizharasi Seetharaman, M. Amelesh, P. M. Sreejith, Vandana Sharma, Ebuka Ibeke, and Hong Liu

Abstract The very existence of smart cities forms the stepping stone in the evolution of many technological advancements in the future era. While smart cities have already grown in their way, the tremendous amount of data generated from them paves the way for new perspectives of development. This is because security and privacy remain to be the major constraint across smart city applications. Further, smart city applications such as smart homes, smart transportation, and smart healthcare are generating a huge amount of data every day and it is often complex to collect and manage all the data together at a single location. To address these constraints, this paper presents a novel and innovative blockchain-assisted federated learning approach for secure data sharing in IoT Smart Cities. Here, we implement a federated learning approach,

T. Hai
School of Computer and Information, Qiannan Normal University for Nationalities, Duyun 558000, Guizhou, China

Key Laboratory of Advanced Manufacturing Technology of the Ministry of Education, Guizhou University, Guizhou 550025, Guizhou, China

T. Hai · H. Liu
School of Information and Artificial Intelligence, Nanchang Institute of Science and Technology, Nanchang, China

D. Wang
School of Mathematics and Statistics, Qiannan Normal University for Nationalities, Duyun, Guizhou 558000, P. R. China

Key Laboratory of Complex Systems and Intelligent Optimization of Guizhou Province, Duyun, Guizhou 558000, P. R. China

T. Seetharaman (✉) · M. Amelesh · P. M. Sreejith
Department of AI & ML, Acharya Institute of Technology, Bangalore, India
e-mail: tamizharasi2539@acharya.ac.in

V. Sharma
Amity Institute of Information Technology, Amity University, Noida Campus, Noida, India

E. Ibeke
School of Creative and Cultural Business, Robert Gordon University, Aberdeen AB10 7AQ, UK

C. Iwendi et al. (eds.), *Proceedings of ICACTCE'23—The International Conference on Advances in Communication Technology and Computer Engineering*, Lecture Notes in Networks and Systems 735, https://doi.org/10.1007/978-3-031-37164-6_9

where the process of learning is made in a distributed fashion. The use of blockchain in turn adds more security and resilience to smart city applications. The security analysis proves that the proposed approach offers comparatively better performance and remains more resistant to various security threats and vulnerabilities.

Keywords Smart city · Federated learning · IOT · Secure data sharing · Blockchain

1 Introduction

In the last few decades, smart cities have grown significantly, and it has been expanded across different parts of the world with numerous advantages. The prime objective of smart cities is to provide optimized city functions and encourage economic growth while improving the quality of life through smart technologies and data analysis [1, 2]. It is often crucial for smart cities to manage various domains such as having access to clean energy, sufficient water access, smart buildings that meet the occupant's requirement, and the potential of travelling efficiently while maintaining the city greener. The growth of smart cities is closely associated with the fact that "how efficiently the technology is being used rather than how much of technological advancement is available." Building smart city infrastructure is often related to efficient data management. It is most vital to collect data from various smart applications, transfer it to information, extract insights, state actions, and develop strategies. It is the 'data' collected from various applications that make the smart city powerful. The effective integration of data from various smart city applications imposes a tangible benefit in improving people's quality of life and assisting in collectively building innovative solutions that assist in sustainable development [3, 4].

Since a billion 'things' are connected across the smart city infrastructure. The widespread adoption of IoT technologies across smart cities increases the risk of security vulnerabilities exploited by malicious actors and cybercriminals. Any compromise in the security of the smart city applications will not only impact the productivity and efficiency of the city infrastructure but also greatly affect the residents and authorities [5]. Smart city applications are subject to several security vulnerabilities, some of the potential threats include privacy, data & identity theft, permanent denial of service attacks, device hijacking, Man in the middle attack, and application-level distributed denial of service attacks. These issues can be resolved through the effective implementation of security countermeasures such as authentication, access control, encryption, security monitoring and analysis, and security life cycle management. Security concerns are constantly raising as smart city applications continuously seek to modernize their services by connecting through the internet, easily creating an attack surface for cyber attackers. Hence, there is a crucial need for connected smart city devices to be protected by comprehensive IoT security solutions [6] (Fig. 1).

Further, the ultimate goal of smart cities is to apply cutting-edge technologies and intelligence to a wide range of services to enable them to operate independently with

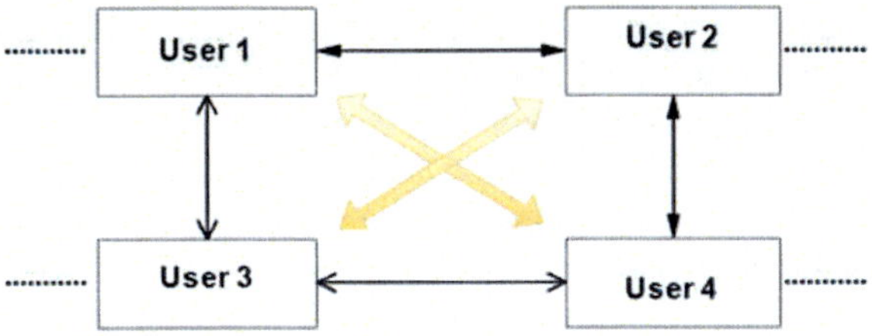

Fig. 1 Proposed model of decentralized architecture

real-time response and operations [7, 8]. This can attract people towards smart cities and at the same time attaining this objective requires a complex system of architecture that has to be constantly monitored from the technology and security standpoint. One of the major concerns associated with the existing smart city application is that most of them are dependent on a central server and third-party cloud system to store and access the data. In general, all smart city applications aim to provide smart, safe, and efficient services to their end users through the exploitation of modern technologies. It is predicted that the smart city application may approximately produce more than 4,000 GB of data soon. Effective utilization and mining of the data will greatly assist in the automation of various services such as smart healthcare, smart transportation, smart homes, and many more. Despite technological advancements, the present-day smart city infrastructures are still traditional and require greater improvement in terms of security and privacy. Many of the security problems remain open and extensive efforts are required to address them efficiently. For example, let us consider a smart transportation application, real-time traffic status assessment is vital to prevent road accidents, manage traffic, and road closure, which saves people time. Similarly, the process of collecting and mining real-time travel data from a wide range of smart devices will significantly improve the quality of services relating to smart city applications. At the same time, such a process will result in numerous security concerns [1, 9].

1. Most of the existing smart city applications are centralized in nature and they are subject to 'single point of failure' and congestion.
2. Automation of smart city applications requires users' real-time data, which may sometime result in the compromise of user data privacy.
3. Smart city users do not have any control over how their data is being used. If compromised, the smart city applications will not only behave maliciously but also affect user privacy and result in serious security concerns.

To overcome these issues efficiently, this paper presents a promising solution with blockchain and federated learning technologies. The proposed approach makes use of blockchain and federated learning to devise a real-time smart city information collection system called FedChain that is both secure against the vulnerable threat on the system and protects the user's private information. The major contribution of the proposed approach is twofold. First, we define a blockchain and federated learning-based decentralised architecture for secure data sharing in smart city applications. Next, we define a security scheme called FedChain to increase security and resilience in smart city infrastructure. Moving towards the growth plan of developing and developed countries, it is envisaged that smart cities will encounter numerous advancements. Which in turn creates the need for security and privacy requirements. In this regard, the proposed FedChain model is practical and efficient enough to share the data across the smart city environment in a secure manner. It enables safer training of the user data and sharing it securely among users through blockchain. Since, both the blockchain and federated learning follow the decentralized architecture, its application in real-time is highly feasible and efficient.

The rest of the paper is organized as follows: Sect. 2 provides a brief literature review of the proposed system. Section 3 defines the proposed architecture in detail. Section 4 depicts the security analysis of the proposed scheme and Sect. 5 concludes the paper.

2 Related Works

Securing smart city applications is an emerging area of research. In this section, we precisely explore some of the related technologies and frameworks that are closely associated with the proposed system.

Haque et al. [10] provide a brief conceptualization of smart city applications. The prime objective of this paper is to provide seamless access to smart city services with intelligent decision-making and optimized resource management functionalities. Various security and privacy threats that emerge from smart city applications are briefly illustrated regarding future work.

In [11] the authors briefly discuss the problem of secure computation outsourcing of IoT data across cloud computing platforms. The author emphasises the need for a global common approach to secure IoT data. Further, the impact of cloud computing and IoT technologies on the deployment of smart applications are also discussed in detail. The author focuses on both the technologies of cloud and IoT. They also explore the list of security issues that occurs with the integration of these technologies with appropriate countermeasures.

The authors Yang, Qiang, et al. in [14] have briefly explored the concept and application of federated learning. They have introduced a secure federated learning approach safely share data across IoT applications without compromising user privacy.

In [12] the author proposes a blockchain-enabled federated learning approach for securing critical IoT infrastructures. The prime focus of this work is to establish trustworthiness among the users. In general, network trustworthy solutions are highly challenging, especially for critical infrastructures as IoT devices can be easily compromised for security vulnerabilities. This approach integrates blockchain and federated learning approaches to establish the factor of network trustworthiness and security. The improved accuracy and detection rate of security vulnerabilities.

The authors Cui, Lei, et al. in [13] present an efficient approach for anomaly detection in IoT infrastructures. The objective here is to focus on robustness, security, and efficiency-related challenges. To address these issues the authors have proposed a blockchain-powered asynchronous federated learning framework. The results are evaluated with real-time datasets, and it attains comparatively better results.

In [15], the authors have attempted to effectively utilize blockchain technology for securing smart city applications. This work has effectively addressed some of the imperative research challenges in the IoT environment. However, this approach was unable to address all the security concerns associated with IoT systems as they vary dynamically from one context to another.

In [16], the authors have presented a deep block scheme for securing smart city applications. They have attempted to integrate deep learning and blockchain technologies for securing smart city applications. The key objective of this approach is to preserve security and privacy factors.

Moving forward, a quantum-inspired cybersecurity-based blockchain protocol is given by Abd El-Latif, Ahmed A., et al. [17]. They have proposed an authentication protocol based on Quantum Inspired Quantum Walk (QIQW). This approach can be extensively used for securing IoT devices. Rather than using the conventional hash functions to manage the blockchain paradigm, this approach makes use of quantum hash functions.

In [18] a blockchain-based efficient authentication and authorization protocol is presented for smart city applications. This work offers a novel solution for the decentralized management of user identity and authorization policies by leveraging blockchain technology. The experimental results indicate that this approach provides comparatively better results than the existing algorithms.

In [19] blockchain-based hybrid network architecture is proposed for smart city systems. This approach receives the benefits of both the centralized and decentralized architecture. The proof-of-work deployed in this scheme ensures better security and privacy measures.

Summary of Literature: Although much of the existing literature [20–24] focuses on the analysis of various cyberattacks and tries to implement blockchain and federated learning-based solutions, many of them face the challenge of higher computational complexity and privacy concerns (leakage of sensitive user information). Further, with the decentralized system models, most of the existing approaches were using linear models. Secure federated learning for smart city applications with heterogeneous models remains to be an open issue that requires more in-depth research. An effective strategy to address these constraints will be discussed throughout the remaining section.

3 Blockchain-Empowered Secure Federated Learning

This section provides a detailed description of the proposed system. It provides the proposed system architecture, system entities and the working of the proposed approach in a detailed manner.

3.1 System Architecture

To attain the design goals and to effectively address the security challenges defined in the threat model, we design an efficient architecture, which is clearly illustrated in Fig. 2. The proposed architecture is mainly composed of three major components. First is the data collection module which is closely related to the users involved with the system. Next, is the feature extraction module that extracts meaningful features from the data collected from IoT devices. Next is the Fedchain-based privacy preserving aggregation module that optimize the interaction between the users of s mart city applications and the cloud [25]. This module also ensures the data integrity and confidentiality measures between the users/IoT devices and the cloud systems.

Given the proposed system architecture, each entity could be an IoT device or a personal device such as a desktop or smartphone. Initially, every user will have a local model and that model is trained on a private dataset. Since the proposed approach implements a heterogenous model, information such as the type of model, parameters associated with the model, learned classes, and model information is stored across the blockchain. The user can search for the calibration of the data sample and can test the model on the blockchain only during their allocated time slot [26, 27]. If a valid sample is obtained, the user can broadcast it over the blockchain and they can also update the models on the chain. This procedure is also called Fed-off-chain data mining. The users or miners associated with the system acquire the calibration sample which is broadcasted by the learning users and validated based on the design protocol. If a sample is validated by the user who possesses the authority to write the next block, then they can update the model on the blockchain and this is called the calibration process. Depending upon various pre-defined model update protocols, the user can construct a legal block header, update model information, and create a new block. All these processes including calibration, block construction, and sample validation are called "Fed-on-chain mining." Since the updated information of the model is stored across blocks and can be viewed by every participant across the system, all the users including both fed-on-chain and fed-off-chain can get the privilege to build a new block.

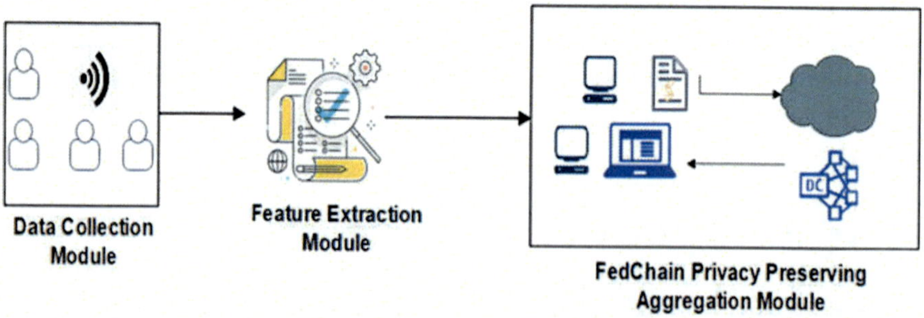

Fig. 2 System architecture of the proposed scheme

There are mainly three different kinds of users associated with the system. Original users (OU) start the blockchain and they collaboratively work with one another to acquire better local models without sharing the identity of the local data. Regular Users (RU) make use of the system after initialization. Each RU possesses a local model and that model is trained based on their local dataset. This enables the system to learn more robust local models and assist in the efficient classification of data even from unknown classes. Miner builds the blockchain based on consensus protocol. They are usually IoT devices with an adequate computing device or an edge device.

3.2 System Definition

Let us consider a decentralized learning system with N users. Such that every user a $\in [1, N]$ posses a local data $F_i = (u_i, v_i) = \{((u_a^1, v_a^1), \ldots, (u_a^{na}, v_a^{na})\}$ with n_a samples of data including their labels. For any user $a \in [1, N]$, $F_i \subseteq F$. Thus F $= \{u,v\}$ remains to be the global dataset and $v = \{i, \ldots, e\}$ represent the set of total e classes. In consideration of the learning system, every user has to possess a local model d_a that is already trained F_a. Just because the classifier is trained with a partial set of F, there is a probability it may misclassify unknown data samples into the wrong class $V\prime \in V_a$ and also from its unknown space $V \not\subset V_a$. For every user a, its classifier $d_a : u_a \rightarrow v_a$ can be attained through training of a local model w_a with model parameter θ_i, which significantly varies from other users. For a particular data sample a, the output d_a is defined as

$$d_a(u) = arg_{v \in v_a}^{max} j_a(\theta_a, u, v)$$

Here, $j_a(\theta_a, u, v)$ can be represented as a score function associated with the local model d_a. This, in turn, return the predicted possibility of u. It is also well known that the traditional federated learning system is dependent on a central server to initiate the process of learning and calibrate local models. There is no doubt that it turned out to be a single point of failure. Thus, to solve this issue proposed approach makes use of a decentralized federated learning system that does not have any central server.

For a given set of models $D = \{d_1, \ldots, d_N\}$ the value of max-model predictor is obtained as

$$d_D(u) = arg_{v \in V, a \in [1,N]}^{max} j_a(u, v)$$

The value of multi-party multi-class margin for a data sample is computed as,

$$\rho\big(u, v, v^-\big) = j_i(u, v) - j_m\big(u, v^-\big)$$

where v^- is an incorrect class label for u, and $i = arg_a max_{j_a}(u, v)$, $v \in v_a$, m, $v^- = arg_{a,v\prime} max_{j_i(u,v\prime)}$, $v\prime \in v_a | \{v\}$.

Such that each user knows about all the other users in the learning model. At every iteration, each user will make use of their local data to acquire a valid calibration sample (u, v, v^{-}). Such that $\rho \leq 0$ and transfer the sample to both the parties i and m based on the equation defined above. Further, with the help of various model update methods, user I can amplify the significance of (u,v) by raising $j_m\left(u, v^{-}\right)$. Once the iterations of model calibration are completed, all the users can update their local model and at least they make use of the max-model predictor to make a classification.

3.3 Threat Model

The most challenging cyberattack associated with a decentralized system is a byzantine attack. In this regard, the attacker follows the system protocols but introduces malicious information to system users. This will significantly degrade the system's performance. Type I and Type II byzantine attacks are the most common among the existing models.

3.4 Secure Blockchain-Based Federated Learning (Fedchain)

The proposed scheme consists of three important steps namely system initialization, Fed-off-chain data mining and Fed-on-chain mining. During the phase of system initialization, OU enrols their identities and model information to the blockchain. Next, the RU registers their credentials to the blockchain. Then it is followed by Fed-of-chain and Fed-on-chain mining. The Fed-off-chain is used to appropriate data samples for model calibration, while Fed-on-chain mining is used to calibrate the exact local model depending on the samples and records found on the new block.

System Initialization

Let us consider there are a total of W(W < N) OU, with each having a local model, such that $a \in [1, w]$, $V_i \complement V$. Here, jV_a denotes the trained model over the data from the class V_a. Such that, every O ∪ broadcast their model information with the corresponding ID. The model information of the first block is computed as,

$$\left[\left\{j_{v_i}^{Q^P}, j_{w_a}^{Q^P}, v_1, AF_Q P_1\right\} \cdots \left\{j_{w_b}^{Q^P}, V_w, AF_Q P_b\right\}\right]$$

Once the above condition is satisfied, OU will prove the initialization model $j_{w_b}^{Q^P}$ for R ∪ h on the blockchain. As a result, the message $<j_{v_l}^{TP}, IF_{TP_h}, SignOU_a>$ is broadcasted. This is also included as a new block in the chain. Otherwise, the model is found to be unfit for classification, and this process cannot be registered into the system.

Fed-off-chain Mining

In this phase, every user associated with the system is stimulated to make use of the local model on the blockchain using their dataset, with the intent to find a data sample that helps to calibrate a few of the models. The proposed approach makes use of more secure methods such that it is resistant to byzantine attacks.

$$\rho\left(u, v, v^{-}\right) = j_i * (u, v) - i_m * \left(u, v^{-}\right), v^{-} \epsilon \mathrm{V} \backslash \{\mathrm{v}\}$$

$$j, (u, v) = \frac{1}{S} \sum_{S_{d \to j}} j_d(u, v)$$

$$j_{m^*}\left(u, v^{-}\right) = \frac{1}{S} \sum_{S_{d \to m}} j_d\left(u, v^{-}\right)$$

$$i = arg_{h \in m(v\prime)}^{min} \sum_{S_{d \to h}} |j_d(u, v) - j_h(u, v)|$$

$$V^{-} = arg_{v\prime \in v | \{v\}}^{max} j_a(u, v\prime)$$

$$a = arg_{h \in m(v\prime)}^{min} \sum_{S_{d \to h}} |j_d(u, v\prime) - j_h(u, v\prime)|$$

Here, $d \xrightarrow{s} h$ represent S closest value to $j_h(., .)$, m(v) and m(v^{-}) denotes the set of users who possess the data that belong to class V and also in class v^{-} correspondingly. A($0 < \alpha \leq 1$) represents the control parameter.

Fed-On-Chain-Mining

For every broadcasted sample (u,v,v^{-}) desires to be validated by the miner before it is utilised for model calibration. For a calibration sample (u,v,v^{-}), the miner check is validated as,

$$\rho(u, v, v^{-}) \leq 0$$

$$j_{i^*}(u.v) \geq \delta$$

$$MAX_{d \xrightarrow{s} i} |j_d(u, v) - j_{i^*}(u, v)| \sigma$$

where δ and σ represent the predefined thresholds. And the above equation confirms that (u,v,v^{-}) is eligible for calibration. Once, all the conditions mentioned above are satisfied, the miner can adapt the sample for model calibration.

4 Security Analysis

The security analysis of the proposed Fedchain model is examined in this section. In general, there exist two types of byzantine attacks, when it comes to the implementation of decentralized multiparty federated learning. In the case of the type1 byzantine attacks, RU's are validated by taking advantage of OUs. The control parameter M specifies the requirement on the model of R U_n demanding the users to join the system. To be more specific, R $U_{n/s}$ model, i.e., Y_a^{RU}, needs to act based on the appropriate OU's local models that include Y_n across their learning space. Further, with the proposed model Fedchain assisted system, the process of model update is performed publicly and is later validated for every model update process, irrespective of the fact that the newly registered RU's models can be slightly malicious at the commencement.

TypeII byzantine attacks are the additional major security threat across the proposed system. In such cases, the bogus calibration sample will seriously interrupt the process of updating a model. We prove that the proposed fed chain algorithm is resistant against type II byzantine attacks and has bounded performance through the following proof:

Let us consider the set of constant models (p) $= \{q_1, \ldots, q_{|S(P)|}\}$. The objective here is to classify the data across class P. Let |S(P)|=$N^{S(P)}$+$M^{S(P)}$ denotes the number of honest models and malicious models. Let us assume $M^{S(P)}$< f < $N^{S(P)}$. Also, for a given data sample (a,b), let $q_{min}(a, b)$ and $q_{max}(a, b)$ represent the minimum and maximum output of the honest models, respectively. Then the value of $q_x * (a, b)$ in the worst case is computed by

$$q_{\min}(a, b) - \frac{m^{s(b)}}{f}(q_{\max}(a, b) - q_{\min}(a, b))$$

$$< q_x * (a, b) < \frac{q_{\max}(a, b) + M^{s(p)} / f}{q_{\max}(a, b) - q_{\min}(a, b)}$$

$$< q_{max}(a, b) + \frac{M^{S(P)}}{f} / q_{max}(a, b)$$

In the same way, let us assume the set of models S(b^-) that can categorize the data in class b^-. For a data sample (a,b^-), q_y*(a, b), the worst-case scenario is bounded by,

$$q_{min}\left(p, q^-\right) - \frac{M^{(b^{-1})}}{f}$$

$$\left(q_{\max}\middle|a, b^-\right) - q_{\min}\left(a, b^-\right))$$

To analyze the bounds in the scenario, such that we assume the byzantine models are included in the aggregation. Such that,

$$q_{x^*} = \frac{1}{f}\left(\sum_{f-m}^{m} q_i^B + \sum q_n^o\right)$$

Further, the ground truth is defined as $\frac{1}{N}\sum_{i=1}^{N} q_i + [q_{min}, q_{max}]$, consequently the additional term in q_{x^*} is defined as $\sum^{f-m} q_h^o \in [f - m]q_{min}, (f - m)q_{max}]$.

Here, the major objective is to direct the aggregated value q_{x^*} as possible as smaller or larger. Henceforth, we will deliberate the attack in two worst cases.

Case 1: In this case, the threat models intend to steer q_{x^*} smaller, where $q_m < q_{min} \leq q_x$.

$$\left|q_1^\beta - h_x\right| + \cdots . + |q_m^\beta - q_x + \ldots |q_{f-m}^0 - q_x| < |q_{m+1}^0 - q_a| + \cdots$$
$$+ |q_f^0 - q_x| + |q_1^0 - q_x| + \ldots + |q_{f-m}^0 - q_x|$$

Due to $q_m^\beta < q_x < q_{f-m+1}^0$, the inequality associated here is rewritten as

$$M_{q_x} - \sum_{t=1}^{m} q_j^\beta < \sum_{t=f-m+1}^{f} q_1^b - m^{q_a}$$

$$\sum_{t=1}^{f} q_1^\beta > 2mq_x - \sum_{l=f-m+1}^{t=f} q_1^0 > 2fmq_{min} - mq_{max}$$

$$\sum_{l=1}^{m} q_1^\beta + \sum_{l=1}^{f-m} q_1^0 > 2mq_{min} - mq_{max} + (f - m)q_{min}$$

$$= (\mathrm{f} + \mathrm{m})q_{min} - f_m q_{max}$$

$$= q_x^* > q_{min} - \frac{m}{f}(q_{max} - q_{min})$$

Case II: All the threat models intend to steer q_x^*, greater, where $q_1^\beta > q_{max} \geq q_a$. Likewise, we obtain

$$\left|\left|q_1^\beta - q_a\right| + \cdots + \left|q_m^\beta - q_a\right| + \left|q_{n-f}^0 + m + 1 - q_x\right| + \cdots + \left|q_{n-f+m}^0 - q_{a1} + \right|q_{n-f}^0$$
$$+ m + 1 - q_x| + \cdots + \left|q_n^o - q_x\right|$$

Due to the reason, $q_1^\beta > q_x > q_{n^{n-f}}^0 + m$, we have

$$\sum_{l=1}^{n-f+m} q_l^\beta < 2m^h x - \sum_{l=N-f+1}^{l=N-f+m} q_1^0 < 2mq_{mx} - m_{min}^q + (f - m)q_{max} - mq_{min}$$

$$q_x^* < q_{max} + \frac{m}{f}(q_{max} - q_{min})$$

5 Results and Discussions

The simulation of the proposed model is made with the help of the MNIST dataset. The dataset contains 60,000 training samples and 10,000 testing samples. Every image has a fixed pixel size of 28 × 28. To perform the simulation with multiple parties, we have separated the dataset with various data distributions and four cases are developed for simulation. In the proposed model of a decentralized system, every party posse a LeNET-5 that is the same as the centralized model. The major difference is that in the proposed method we adopt both fully connected neural networks and convolutional neural networks. The setup of the calibration model is based on the user updates on the corresponding local models. A comparison is made with the baseline models in terms of prediction accuracy and attack resilience measures.

The results of the comparison of the prediction accuracy measure are clearly illustrated in Fig. 3. The performance of the proposed decentralized model is observed to be comparatively better than the conventional baseline methods. The competitive security measure value of every security scheme is analyzed in detail, and it's found to be relatively greater than the existing schemes. In the case of observation between every 5 to 10 parties, the prediction speed measure is also found to be better than traditional methods. From this, we can conclude that the proposed decentralized scheme can act as a genuine alternative to the baseline methods. Further, the security resilience measure of the proposed scheme is also observed to be relatively greater than conventional methods.

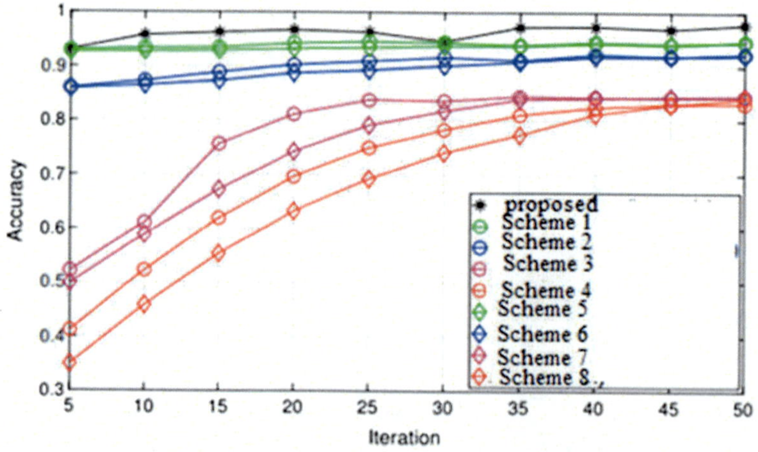

Fig. 3 Comparison of prediction accuracy with the proposed scheme

6 Conclusions

In this article, we proposed a scheme called the secure and privacy-preserving FedChain model for securing smart city applications. This approach attains the data from IoT devices associated with the smart city applications and applies blockchain and federated learning techniques for securing it efficiently. The aggregation model is designed in a decentralized manner that prevents emerging security threats across the smart city environment. The security feasibility of the proposed approach is evaluated in detail with the security analysis model and this approach is found to be more secure against major security threats.

In future, the scope of the proposed approach could be extended towards various applications including autonomous vehicles, wearable devices, etc. It finds an appropriate place in applications where the secure transmission of data sharing has become inevitable. It further, helps efficiently to learn data across organizations, especially across heterogeneous networks.

Further, as discussed earlier the proposed model of FedChain enhances the security and privacy features of the smart city applications, and it paves the way for several new questions that have to be addressed efficiently. One major limitation is that learning over organizations and mobile phones is quite difficult and often a complicated process. This is because the training data is distributed across the edge. The other limitations include statistical heterogeneity, expensive communication, and system heterogeneity.

References

1. Moreno, Victoria, M., et al.: Applicability of big data techniques to smart cities deployments. IEEE Trans. Indus. Inf. **13**(2), 800–809 (2016)
2. Odendaal, N.: Everyday urbanisms and the importance of place: exploring the elements of the emancipatory smart city. Urban Studies **58**(3), 639–654 (2021)
3. Zhang, K. et al.: Security and privacy in smart city applications: challenges and solutions. IEEE Commun. Mag. **55**(1), 122–129 (2017)
4. Lacinák, M., Ristvej, J.: Smart city, safety and security. Procedia Eng. **192**, 522–527 (2017)
5. Braun, T., et al.: Security and privacy challenges in smart cities. Sustain. Cities Soc. **39**, 499–507 (2018)
6. Habibzadeh, H., et al.: Sensing, communication and security planes: a new challenge for a smart city system design. Comput. Netw. **144**, 163–200 (2018)
7. Cui, L., et al.: Security and privacy in smart cities: challenges and opportunities. IEEE Access **6**, 46134–46145 (2018)
8. Li, Y., et al.: Privacy protection for preventing data over-collection in smart city. IEEE Trans. Comput. **65**(5), 1339–1350 (2015)
9. Qi, L., et al.: Privacy-aware data fusion and prediction with spatial-temporal context for smart city industrial environment. IEEE Trans. Indus. Inf. **17**(6), 4159–4167 (2020)
10. Haque, Bahalul, A.K.M., Bhushan, B., Dhiman, G.: Conceptualizing smart city applications: requirements, architecture, security issues, and emerging trends. Expert Syst. **39**(5), e12753 (2022)
11. Singh, D., et al.: Security issues in IoT and their countermeasures in smart city applications. Advanced Computing and Intelligent Engineering. Springer, Singapore, pp. 301–313 (2020)

12. Otoum, S., Al Ridhawi, I., Mouftah. H.: Securing critical IoT infrastructures with blockchain-supported federated learning. IEEE Internet Things J. **9**(4), 2592–2601 (2021)
13. Cui, L., et al.: Security and privacy-enhanced federated learning for anomaly detection in IoT infrastructures. IEEE Trans. Indus. Inf. **18**(5), 3492–3500 (2021)
14. Yang, Q., et al.: Federated machine learning: Concept and applications. ACM Trans. Intell. Syst. Technol. (TIST) **10**(2), 1–19 (2019)
15. Hakak, S., et al.: Securing smart cities through blockchain technology: architecture, requirements, and challenges. IEEE Netw. **34**(1), 8–14 (2020)
16. Singh, Kumar, S., et al.: DeepBlockScheme: A deep learning-based blockchain driven scheme for a secure smart city. Hum. Centric Comput. Inf. Sci. **11**, 12 (2021)
17. El-Latif, A., Ahmed A., et al.: Quantum-inspired blockchain-based cybersecurity: securing smart edge utilities in IoT-based smart cities. Inf. Process. Manag. **58**(4), 102549 (2021)
18. Esposito, C., Ficco, M., Gupta, B.B.: Blockchain-based authentication and authorization for smart city applications. Inf. Process. Manag. **58**(2), 102468 (2021)
19. Sharma, Kumar, P., Park, J.H.: Blockchain-based hybrid network architecture for the smart city. Future Gen. Comput. Syst. **86**, 650–655 ((2018))
20. Wang, Z., Hu, Q.: Blockchain-based federated learning: a comprehensive survey. arXiv preprint arXiv:2110.02182 (2021)
21. Demertzis, K.: Blockchained federated learning for threat defense. arXiv preprint arXiv:2102.12746 (2021)
22. Li, D., Luo, Z., Bo., Cao: Blockchain-based federated learning methodologies in smart environments. Clust. Comput. **25**(4), 2585–2599 (2022)
23. Yu, F., et al.: Blockchain-empowered secure federated learning system: architecture and applications. Comput. Commun. **196**, 55–65 (2022)
24. Issa, W., et al.: Blockchain-based federated learning for securing internet of things: a comprehensive survey. ACM Comput. Surv. (CSUR) (2022)
25. Ngabo, D., Dong, W., Ibeke, E., Iwendi, C., Masabo, E.: Tackling pandemics in smart cities using machine learning architecture. Math. Biosci. Eng. **18**(6) (2021)
26. Mantey, E.A., Zhou, C., Mani, V., Arthur, J.K., Ibeke, E.: Maintaining privacy for a recommender system diagnosis using blockchain and deep learning. Human-centric computing and information sciences (2022)
27. Latif, S.A., Wen, F.B.X., Iwendi, C., Li-li, F.W., Mohsin, S.M., Han, Z., Band, S.S.: AI-empowered, blockchain and SDN integrated security architecture for IoT network of cyber-physical systems. Comput. Commun. **181**, 274–283 (2022)

A Qualitative Study of the Effects of Socio—organizational Factors on the Information Security Culture of Employees in a Financial Institution

Kolawole Emmanuel and Thaier Hamid

Abstract Information communication technology (ICT) is constantly evolving, it is currently used in virtually every aspect of human life, such as financial transactions. Due to its ongoing development, the internet serves as the basis for all service platforms. Financial institutions are responding to the development of ICT by following suit in an effort to improve services and maintain existing technology. Similarly, worldwide cybercrime has increased, resulting in financial losses. As a result, information security has become a pressing concern for any institution tasked with providing adequate services. Developing complex security methods that address people, processes, and technology is required to strike a balance between the technical and behavioural components of information security. This qualitative study will examine how socio-organizational factors can enhance employees' attitudes toward information security culture at any financial institution. To fulfil the above aim, a qualitative methodology is utilised. Interviews (prior and main) and document analysis (Review & Analysis of Internal Materials in the banking sector of Nigeria) will be used to collect data. The *Quota Sampling* method is used in recruiting participants for interviews. With the assistance of academic experts, fellow researchers, and information security professional experts, the interview questions will be pretested. *The Thematic analysis* will be utilised to deduce occurrences, patterns, and use of concepts. Based on the analysis of data gathered from 38 top management of a financial institution, the study concludes that improving information security culture in any institution requires strong leadership, accountability, education/awareness/training/experience, commitment & trust, monitoring & internal control (audits), process integration, technology, permissible/institutional pressure, societal pressure, personal beliefs, values, and attitudes. The study expands on'*behavioural information security*', a crucial area of study. It bridges the gap in research most especially in Africa as it expands knowledge of information security by critically exploring human factors in the form of employees in the financial sector of a cross-cultural country Nigeria. From a theoretical perspective, the study presents socio-organizational factors that

K. Emmanuel · T. Hamid (✉)
University of Bolton, Greater Manchester, Bolton, UK
e-mail: t.hamid@bolton.ac.uk

C. Iwendi et al. (eds.), *Proceedings of ICACTCE'23—The International Conference on Advances in Communication Technology and Computer Engineering*, Lecture Notes in Networks and Systems 735, https://doi.org/10.1007/978-3-031-37164-6_10

can enhance employees' attitudes toward information security culture. From a practical perspective, it is a study which provides guideline for better articulation, formation, and implementation of an organisation's Information security laws, policies, and procedures as well as to create a framework which can aid top management in recruiting. The usefulness of this thesis is not limited to the financial institution of Nigeria/Africa.

Keywords Information security · Information security culture · Socio-organisational · Employees · Top management · Human factor

1 Introduction

The speed of ICT development, sometimes known as the "*information age,*" which is characterised by the rapid development of the internet and the use of information systems, has increased the complexity and susceptibility of people and organisations employing ICT. This 'information age' has rebuilt the way individuals communicate. For example, the internet has assumed control over public relations, promoting or marketing, advertising and so on. The internet limited the impact of worldwide distance, empowered overall trade through superior correspondence platforms. So, organisations these days have enthusiastically embraced ICT inventions so that they thrive in the really serious and competitive market in which great strategies and successful functional control needs accessible, essential, and quality information. As ICT continued to progress, so did critical information security issues like confidentiality, integrity, and availability [1]. Technical information security solutions are crucial but not adequate, people who make use of ICT innovations must be taken into consideration [2, 3].

Since individuals are the weakest link in the security chain, it is essential to comprehend their potential contribution to security [4, 5]. The efficiency with which individuals carry out these responsibilities as well as their individual behaviour form the foundation of the rounded approach to information security management. Hence the reason for an increasing focus on developing and improving information security culture for a better information security. An institution's information security culture can be best described as the sum of all behaviours that contribute to the protection of all types of information [6]. Instilling a culture of information security is not easy, and it can take time [6].

This qualitative study will examine how socio-organizational factors can enhance employees' attitudes toward information security culture at any financial institution. Based on the analysis of data collected from employees at managerial level of a financial institution, the study indicates that leadership, accountability, education/awareness/training/experience, and commitment & trust, monitoring & internal control(audits), process integration, technology, permissible/institutional pressure, societal pressure, personal beliefs, values, and attitudes have a significant role in improving information security culture in any institution.

The study is organised into five sections as detailed below:

1. *Introduction: it presents an overview into reason for this research.*
2. *Review of related works: It examines Nigeria's banking and information security; related research works and documents in order to identify gaps and expand understanding.*
3. *Methodology: it describes the full research methodology. Implementation of methodologies is discussed to cover population sample size and technique, data collection process and analysis process. This chapter is very important to achieving the research aim.*
4. *Data analysis, Results and Findings: it focuses on analysis of data qualitative collection approach—interview and document analysis. Procedures utilised to analyse all data are illustrated. Thematic analysis is utilised to present results and findings.*
5. *Conclusion: it summarises research results, contribution of this research and suggestions for future endeavours.*

2 Review of Related Works

How employees view an institution is defined by organisational culture [6]. Institutions have separate organisational culture. An organisational culture grows and changes with time, it influences an employees' behaviour and explains an institution's view of conduct or activities [6]. Organisational culture defines collectively the attributes which reflects behaviour of employees in an establishment and other manifestations required [6]. The best way to describe a company's information security culture is as the whole set of behaviour that help protect all types of information; It is completely ingrained in the organizational culture of a company [6]. Information security culture is best defined as the entirety of forms of behaviour in a business which adds to the protection of all types of information; it is totally embedded in a business' culture (organisational culture). Organisational culture defines collectively the attributes which reflects behaviour of employees in an establishment and other manifestations required [6].

It is impossible to overstate the significance of the human factor in information security. According to studies, behavioural information security research has recently been recognized as important and has attracted attention [7]. Behavioural information security research is a subfield of the larger field of information security research that focuses on the human factor [7]. The attitudes and actions of those who are in charge of safeguarding information, its system, and its resources [8], which include information system equipment, networking infrastructure, and the entire establishment's assets [9, 10] are available.

Generally, in the world, financial institutions play vital roles in the advancement of an economy. The Level of advancement of the financial institutions determines the efficiency and effectiveness in performing crucial tasks, especially the intermediation between the profit and deficit unite of an economy [11]. Financial institutions are undergoing a global shift thanks to information and communication technology (ICT). Nowadays, financial institutions provide ICT-based services. These services are all connected to the internet, networks, and electronic platforms, allowing customers of all locations, times, and distances to access information quickly, effectively, and reliably. In 2014, Nigeria Interbank Settlement System (NIBSS) revealed that there's a massive increase in internet banking [14]. Despite the positives of ICT in the financial institutions, it has also produced a negative effect globally known as *Cybercrime.* Cybercrime has made it easier to steal [12]. Operation in financial institutions is highly based on trust from all stakeholders hence reason why security will always be of relevant concern in the sector and a major need to understand all aspect of information security [13].

The confidentiality, integrity, availability, and safety of information are all dependent on information security [14]. Information security ensures that information systems, their supporting processes, and infrastructures are adequately protected from any threat or vulnerability. It covers both physical and logical access control protection and authorization. The Nigerian financial system faced some challenges towards the final stages of transitioning into a cashless society which was due to factors such as exclusion of substantial number of adults from a formal financial system, poor infrastructure for support, accurate data, and information security [14]. The financial system in Nigeria varies in terms of its level of security and is susceptible to numerous risks due to their vulnerability to attacks and limitations [14]. The NIBBS 2015 e-fraud reports revealed that there's an increase in e-payments fraud as Nigeria changed its modes of payment [14]. Platforms like internet banking, ATMs and others accelerated fraudulent activities and are leading channels for e-frauds in 2014 [14]. Also, e-fraud has become sophisticated and more concise due to over-the-counter frauds and cheques [14]. There are various research on behavioural information security relating to this area of study. For the purpose of this paper, important works are presented in the Table 1.

The Table 1 summarises related literature in this study area.

3 Methodology

Research methodology is a controlled and organized approach to problem-solving [46]. It is a structured approach in which research issues are investigated from the theoretical foundation to data collection, analysis, and outcomes [47].

This research will make use of an Interview and document analysis (Review and analysis of internal materials) which will enable the development of a comprehensive understanding and enhancing credibility of the study. One type of qualitative research is a case study. This study will be able to apply the findings to other cases because it used a case study. In this instance, Nigeria's financial institution using one of her oldest banks served the purpose.

Table 1 Summary of other related thesis/paper

Author/researcher	Findings/contributions
Ryan, J. E. (2006)	This demonstrated that higher user-level ISA measures can positively influence user information security practices at work and at home [15]
McFadzean, E., Ezingeard, J. N., & Birchall, D. (2006)	The study examined the information security roles and responsibilities of organizations' boards [16]
Von Solms, R., & von Solms, S. B (2006)	ISM self-evaluation of top management roles and responsibilities was found in the study. Also demonstrated that a top-to-bottom governance structure for managers (from the board to lower management) has been proposed [17]
Mancha, R., & Dietrich, G. (2007)	A model that investigates the connection between personality traits and environmental factors that influence peer influence as a factor in IS users' attitudes and actions toward ISS [18]
Dinev, T., & Hu, Q. (2007)	An individual's awareness of risks, threats, and harmful technologies or activities strongly influences their use of preventive IS security technologies, according to the findings [19]
D'Arcy, J., & Hovav, A. (2007)	Awareness of countermeasures and preventive security software are effective against information security misuse behaviour, according to the general deterrence theory (GDT). Additionally, it appears that IS users are only aware of severe forms of IS misuse [20]
Hazari, S., Hargrave, W., & Clenney, B. (2008)	This study used Ajzen's Theory of Planned Behaviour to investigate factors that influence information security awareness among WRHC (work-related home computing) users. In order to determine the factors that influence behaviour intention, the study also looked into demographic characteristics, attitudes, subjective norm, and perceived behavioural control [3]
Choi, N., Kim, D., Goo, J., & Whitmore, A. (2008)	According to the findings of this study, higher levels of managers' ISA result in an organization's performance toward information security that is both more efficient and effective [21]
D'Arcy, J., & Hovav, A. (2009)	High levels of computer self-efficacy and virtual status significantly increase the deterrent effectiveness of SETA programs and computer monitoring, which results in fewer intentions to misuse IS [22]
Dinev, T., Goo, J., Hu, Q., & Nam, K. (2009)	Dinev and Hu (2007) used the same model to study the cultural differences between the United States of America and South Korea. The study found that South Korean users' attitudes and intentions to use anti-spyware were less affected by technology awareness than US users' attitudes and intentions [23]
Galvez, S. M., & Guzman, I. R. (2009)	To explain the individual and environmental factors that influence corporate information security behaviour, a model that combines Social Cognitive Theory and Control Theory was developed [24]

(continued)

Table 1 (continued)

Author/researcher	Findings/contributions
Bulgurcu, B., Cavusoglu, H., & Benbasat, I. (2009, August)	Based on insights from Fishbein and Ajzen's theory of planned behaviour from 1975 and 1991, this study found that ISA and perceived fairness of the ISP requirements positively affect attitude, as well as attitude positively affects intention to comply [9, 26]
D'Arcy, J., Hovav, A., & Galletta, D. (2009)	An empirical test of a GDT-based model suggested that a user's awareness of security countermeasures has a direct impact on employees' perceptions of the certainty and severity of organizational sanctions for IS misuse, which in turn reduces intention to use IS [25]
Werlinger, R., Hawkey, K., Botta, D., & Beznosov, K. (2009)	The study revealed that organizational ISM is influenced by 18 factors. It revealed that businesses struggle with both security and business development and place security at a lower priority [27]
Bulgurcu, B., Cavusoglu, H., & Benbasat, I. (2009)	This study found that employees' intentions to comply with ISPs are positively influenced by TPB constructs like self-efficacy, attitude toward ISP compliance, and normative beliefs. Employee attitudes were also shown to be significantly influenced by outcome beliefs regarding an overall assessment of the consequences of compliance and non-compliance. Through attitude and outcome beliefs, it was also discovered that ISA increases intentions both directly and indirectly [9, 26]
Dinev, T., Goo, J., Hu, Q., & Nam, K. (2009)	The multifaceted differences in user attitudes toward protective information technologies between South Korea and the United States were the subject of this study. The study found that the quality of the connections in the behavioural model of protective information technologies was moderated by cultural factors [23]
Liang, H., & Xue, Y. (2010)	The purpose of this paper was to acquire sufficient knowledge regarding the IT threat avoidance behaviours of personal computer users, investigate how PC users deal with IT threats, explain the development of threat perception in individuals, evaluate safeguards for individuals, and also examine how individuals engage in avoidance behaviour [37]
Kruger, H. A., revin, L., & Steyn, T. (2010)	Empirical research was conducted to determine the connection between information security behaviour and higher levels of ISA [28]
Spears, J. L., & Barki, H. (2010)	This demonstrated that IS users' participation in the ISS risk management process and high level of awareness of security risks and controls contribute to the improvement of control development and performance [10]
Bhattacharya, D. (2011)	Transactional leadership was found to be the most efficient method for resolving information security issues in this study. When it comes to managing information security, leadership is crucial. When a company has more transactional leadership, there are more concerns about information security [29]

(continued)

Table 1 (continued)

Author/researcher	Findings/contributions
Claar, C. L. (2011)	It explored the factors affecting the adoption of computer security. It emphasises that the factors affecting usage and acceptance of computer security in homes is not effectively reflected by present models used in research regarding technology acceptance as at that time. Also, the health belief model's constructs were looked at to see if they helped improve methods for finding behavioural antecedents that affect how people accept and use protective technologies [38]
Monfelt, Y. (2011)	It was discovered that organizations' lack of communication regarding information security results in ineffective adoption and management issues [30]
Takemura, T. (2011, October)	In order to discuss the connection between information security awareness and behaviour, data from a web-based survey on information security measures in Japan were gathered and analysed. It was found that employees' ISS behaviour is significantly influenced by their level of ISA [31]
Yayla, A. (2011)	The control over insider threat was examined using a model. Unintentional insider threats to information security can be reduced by increasing user awareness [32]
Howe, A. E., Ray, I., Roberts, M., Urbanska, M., & Byrne, Z. (2012, May)	The four sections of the study are a review of previous research on the factors that influence security decisions made by home computer users: comprehension of threats, attitudes toward security interventions, efforts to avoid security breaches, and perceptions of risky behaviour All of the studies that were looked at revealed a variety of reasons why the security measures that are currently in place might not be able to meet the needs of home computer users [4]
Al-Omari, A., ElGayar, O. F., Deokar, A., & Walters, J. (2012)	A security acceptance model (SAM) based on Davis's technology acceptance model (TAM) is used in this study to investigate users' behavioural intentions to comply with ISPs. According to the findings of the study, ISA has an indirect effect on behavioural intentions to comply [33]
Lee, M., & Lee, J. (2012)	It was discovered that failures in information security can result in significant customer loss and retreatment [34]
Pahnila, S., Karjalainen, M., & Siponen, M. T. (2013)	The purpose of this study was to see if various factors could explain or predict the behaviour of employees who knew about ISP (information security policies) from those who didn't [39]
Guo, K. H. (2013)	It re-evaluated and orchestrated the various conceptualizations of security-related behaviour and provide some possible explanations for the findings of the current IS security study that appear to go beyond reconciliation. A conceptual framework for defining behaviour that is related to security: Based on research in psychology and organizational behaviour, the terms security damaging behaviour (SDB), security risk-taking behaviour (SRB), security assurance behaviour (SAB), and security compliant behaviour (SCB) were proposed [40]

(continued)

Table 1 (continued)

Author/researcher	Findings/contributions
Hassan, N. H., Ismail, Z., & Maarop, N. (2015)	The study looked into several past work. One of which was factors influencing information security culture [35]
Shropshire, J., Warkentin, M., & Sharma, S. (2015)	This study found a link between self-reported behavioural intent and actual use of security programs, as well as a role for personality in moderating the former relationship, with the goal of establishing the role of personality factors in determining whether or not a person will carry out and be committed to engaging in protective behaviours [44]
Safa, N. S., Sookhak, M., Von Solms, R., Furnell, S., Ghani, N. A., & Herawan, T. (2015)	This study aims to change users' behaviour in the security domain to one of conscious care and to emphasize the human aspects of information security to reduce breaches. In addition, the study investigates whether users' attitudes shift as a result of Infosec awareness; Subjective norms are influenced by the information security policy (ISP) of the organization; Users' information security conscious care behaviour is influenced by information security experience, involvement, and, finally, whether assessment of information security threat and information security self-efficacy influence information security conscious care behaviour [42]
Connolly, L., Lang, M., & Tygar, J. D. (2015, May)	By looking at how the combination of security measures and cultural factors affect employee security behaviour in organizations, this paper aimed to contribute to the field of behavioural information security research. In addition, it investigated how employee actions regarding information security are influenced by organizational and national culture [41]
Chatterjee, S., Sarker, S., & Valacich, J. S. (2015)	The goal of this study was to find and describe the observed variations in unethical IT use, as well as the factors that link them together and the conditions and contexts in which they occur [43]
Da Veiga, A. (2016)	Employees who had read the information security policy and those who had not are compared in the study's information security culture [36]
Al-Kalbani, A. (2017)	It aimed to develop a framework for information security compliance, identify the essential factors for effective information security compliance in public organizations in Oman, and offer the government recommendations for effective information security compliance in public organizations for the development of e-government [45]

Document analysis—'Documents are standardized [47], it will serve as a secondary source of data to support interviews—data will give research background information of what exists in the financial institution of Nigeria. The research hopes that this method will lead to questions needed to be asked or to situations that should be examined [48]. Before document analysis can be carried out, it must be well planned [49]. For this research the quality of documents won't be questioned as they will be from reliable source typically international standard as financial institutions in Nigeria are well regulated and adhere to international laws and regulations.

The interview, is one of the most crucial methods for data collection in qualitative research [50]. It will be in-depth, semi structured, and based on findings from internal materials, reviews, and a comprehensive literature review. The core concepts of ISO/IEC 27,001: 2013 standard and keywords deduced from internal documents will also be used to structure the interview questions. Interviews are appropriate for this kind of research due to several reasons [51]. It is a method of data collection that is suitable for studying behaviour, particularly when participants are given the opportunity to describe their experience [51]. Hence, Interviews in this research will be held in an informal way to limit participant from withholding answers, promote honesty and freedom in a relaxed way which will further show that interviews can allow deep probing, open various dimensions of a problem [52]—Interviews can be face to face or through video calls depending on the participants availability and open question or discussion technique used in order to get concrete feedbacks that could lead to other findings or areas to explore. Answers will be noted down, and audio backups will be made based on participant's consent. In order to avoid the limitations identified in some studies [53], such as the issue of "minimizing social dissonance," i.e., building trust between the researcher and the participant is essential, this research will limit interviews to "Face to Face" or "Video call" only. A 'Face to Face' or 'Video call' interview allows development of trust—eye contacts and gestures are observed.

There will be 2 stages of the interview, *The Prior* and *the main.*

The prior stage will consist of introductory questions typically focusing on certain demographic information like age, occupational role, education level, religion, e.tc. this will enable participants to be known properly—it will identify the most qualified and experienced ones ideally suitable for the interview, increase chances of identifying certain areas that needs to be questioned or investigated in the main stage of the interview.

The main stage will focus on the purpose of the interview, which will contribute to achieving this research's aim and objectives.

With the assistance of academics, fellow researchers, and information security industry professionals, the interview questions will be pretested.

The *Quota Sampling* method is used in recruiting participants for this interview [54]. The purpose of the quota sampling method is to provide flexibility in determining the number of participants to interview and an opportunity to concentrate on selected participants based on particular characteristics, such as experience and location in this case [54]. A simple criterion was used in the quota sampling procedure.

As there are 36 headquarter branches of *the selected financial institution in Nigeria with its governing body*, it is ideal to interview Information security manager or IT (information technology) manager or the professional who leads the IT team or department of each branch. Demographics information will play a vital role in identifying professionals with most experience, qualification, years with the bank, e.tc. The researcher will interview 38 professionals. Before the interview can be conducted, contact will be established via telephone calls and official letter before emails are sent to invite and enlighten participants. Due to sensitivity of this research, interviewees will be offered a chance to meet outside work if not comfortable with office environment, anonymity will be guaranteed.

Interview Design—An extensive literature review, core concepts of ISO/IEC 27001:2013 standard and document analysis plays major role in the interview design and format will be made available to target: professionals in managerial roles, technical roles and those with background knowledge or experience in both roles.

There are various ways that qualitative data can be analysed [55]. Such as Grounded theory, Descriptive analysis, Thematic analysis, Narrative analysis and many more. The interview data will be transcribed and analysed using the thematic analysis approach to achieve the research's objectives [56]. In this section, the use of concepts and occurrences will be identified through the use of *thematic analysis.* It will make it possible to have a broader perspective on the current state of information security in Nigeria's financial institution. It will provide information that is both meaningful and pertinent [56]. Emerging themes will be organized into categories for comparison and integration with interview data [48, 49].

The sample demographics is reported in the Table 2.

4 Data Analysis Results and Findings

The findings of this study regarding the socio-organizational factors identified in the interview/analysis are summarized in the Table 3. The socio-organizational factors are leadership, accountability, education/awareness/training/experience, and commitment & trust, monitoring & internal control (audits), process integration, technology, permissible/institutional pressure, societal pressure, personal beliefs, values, and attitudes. Each factor consists of relevant sub-factors that are recognised in this study.

The transcripts of the 38 interviews completely support the idea that employees' overall information security behaviour can only be developed or improved by having a strong information security culture. The better and more effective an institution's information security will be, the more attention that is paid to the information security culture of its employees.

Leadership: This demonstrates how the Nigerian financial institution's top management has supported the information security culture. The interview's findings showed that a strong information security culture relies heavily on managerial commitment and perspectives. Having regular managerial support in taking decisions, having them involved in every aspect of security, having them listen to your ideas, give suggestions, give recommendations contact feedbacks and re-iteration of goals and policies are means of motivating employees towards believing and valuing information security culture within any institution.

Table 2 Demographic characteristics of Participants. (Source—Interview data)

Q/N		Frequency
1. Sex	Female	11
	Male	27
	Total	38
2. Status of Marriage	Widowed	1
	Separated	1
	Divorced	1
	Single	13
	Married	22
	Total	38
3. Ages	Below 25 years	0
	26 to 30 years	0
	31 to 35 years	6
	36 to 40 years	12
	41 to 45 years	10
	46 to 50 years	7
	51 to 55 years	2
	Above 56 years	1
	Total	38
4. Length of service with organisation	Less than 5 years	2
	6 to 10 years	2
	11 to 15 years	14
	16 to 20 years	17
	21 to 25 years	2
	26 years and above	1
	Total	38
5. Highest education qualification	Diploma	0
	First Degree (Bachelors)	5
	Second Degree (Masters)	24
	Doctorate Degree (PhD)	9
	Total	38
6. Professional Qualification relevant to Job role	Yes No	33 5
	Total	38
7. Degree in IT or Infosec	Yes No	21 17
	Total	38

Table 3 In any financial institution, these socio-organizational factors were found to play a crucial role in enhancing the information security culture of employees

Factors	Sub-Factors	Statements/Evidence
Leadership	Regular support, Contribution & Active communication	*"…The employees in the banks of Nigeria naturally look up to its leadership hence it is essential for its leadership to always lead by example and provide necessary support…"* *" …we allow employees to raise issues freely, we want to get ideas from everyone and remind ourselves of duties and information security…"*
Accountability	Delegation of responsibility, Legal & regulatory enforcement, and Penalty & Reward	*"…Regular monitoring is one of the enforcement techniques in place which the leadership use to ensure employees would have no choice but to carry out their delegated responsibilities towards information security…"*
		" …assigning task makes it easier to detect lapses in conduct or duty…"
Education/Awareness/ Training/Experience	Efficacy of education, experience, and training/awareness program; Distinctness of training/awareness program and Benefits of education, experience, and training/awareness program	*"…Our training/awareness programs are simplified, made available to all and updated regularly to meet recent information security trends…"* *"…Employees with years in the industry or role show great knowledge of basic information security…"*
Commitment & Trust	Career progression & opportunities; Career or personal needs fulfilment	*"…We help our employees by giving them training, listening to what they have to say, and rewarding them. We know this helps them be more committed to the bank, gain trust in it, and have a positive attitude about information security…"* *"…We show trust in employees who are dedicated and have no major history of information security errors in order to boost their morale and push others…"* *"…Personally, I notice that employees who believe in moving higher in their career or moving to another position within the business take information security seriously and show desire to know more about it…"*

(continued)

Table 3 (continued)

Factors	Sub-Factors	Statements/Evidence
Monitoring & Internal control (audits)	Perceived benefits, Documentation/ Reports and Suitability of audit & monitoring activities and Awareness of the outcome of monitoring and audit activities	*"…The level of information security in the banks can be improved with good audit, monitoring and incident reports or documentation…"* *"…Audit, monitoring, and incident management activities has improved the security level…in fact it puts everyone on the right part…"* *"…Due to the constant monitoring and auditing, bank employees are forced to adhere to information security policies…"*
	Perceived benefits, Documentation/ Reports and Suitability of audit & monitoring activities and Awareness of the outcome of monitoring and audit activities	
Process Integration	The creation and implementation of information security requirements, perceived advancements, and compliance	*"…At times information security makes it hard to perform certain task essential to the business… It is at times challenging to create/ find that balance between information security and business-related processes…"* *"…once information security controls are properly merged with business processes, organisations can reduce mitigation and do more of avoidance which will improve information security compliance, culture and behaviour through business-related process…"*
Technology	Technology proficiency, Technology affinity, Technology reliability,	*"…if employees are aware of various tools and services that are effective and easy to use or setup, there is a higher rate of acceptance—in terms of general usage and compliance…"* *"…we ensure the affinity of any information security technology to be implemented with current work practices, organisation values and with desired work…"* *"…the financial sector of any country is sensitive. Information security technologies that are robust, easy to use and maintain should always be available…"*

(continued)

Table 3 (continued)

Factors	Sub-Factors	Statements/Evidence
Permissible/ Institutional pressure	Availability of laws and regulations, Control of law, and Severity of infringements	*"...We take the country's laws, policies, standards and cultural reactions into consideration before making any decision..."* *" ...The news regarding what companies and individuals face due to breaches, forces, and pressures us as a company or individually to take information security seriously..."*
Societal pressure	Societal Responsibilities, Establishment of Trust, and comprehension of information security issues	*"...Nigeria is working towards having an all-inclusive financial sector and as such is always doing a lot to ensure banks provide secured services so that prospective and current stakeholders have confidence and trust..." "...In this country, the knowledge of security challenges makes individuals careful and tend to adhere to advice from the banks or experts..."* *"...it is easier dealing with security cautious people..."*
Personal beliefs	Ethnicity (Religion, Knowledge, culture)	*"...The bank's information security culture, policies, guides and other relevant strategy the business puts forward is all about moral intuitions about what's right or wrong which religious doctrines further strengthens..."* *"...Religious teachings cover morals like obedience, respect, loyalty, honesty, and many more which must be showed in all aspect of life and as such influence decisions or behaviours at work..." "...Religion provides a common value system for a society..."*
Values	Loyalty, Career progression, Family, Health, Wealth, public opinion	*"...we do character evaluation during recruitment; this is just to ensure we have the best personnel..."* *"...Employees that aspire to progress faster within the business are usually hardworking, passionate and follow instructions, policies, standards... I would love to say they are compliant..."* *"...Job titles, salary, health benefits, family care plan & pension motivates us a lot, we all love good things, we want honour, rewards, recognition to show for our hard work..."* *"...Family is important...once we can cater for them, we are mentally ok and motivated to work..."*

(continued)

Table 3 (continued)

Factors	Sub-Factors	Statements/Evidence
Attitudes	Mutual respect, Professionalism, Psychological stressors	*"...Mood can affect work, when a customer or employee is sad or happy it is known by others, and it reflects in their attitude to work or work output...even if it is unknown to the person in that state..."* *"...we encourage polite communication over business issues/ideas and disregard judgemental tendencies..."* *"...We have counselling units within the business to assist with mental health issues, relationship issues and general career advice..."* *"...Health is wealth...we ensure employees are fit to work..."*

Accountability: This is another factor which highlights employees' responsibilities towards information security and other duties. Accountability generates a work environment which indoctrinates that an employee is responsible for adherence towards every aspect of information security. Findings revealed that proper delegation of responsibilities enables all aspect of information security to be taken seriously and viewed as very important. Hence a good institution should properly delegate responsibilities and define positions which must all be included in its information security policy and standard documentations. In financial institutions, it is required to inform employees that failing to carry out their assigned duties will result in a fine or other appropriate punishment; It is discovered in this study that improving employees' information security culture is sparked by the availability of legal and regulatory enforcement for accountability. A lack of incidence response or delay in reporting has been revealed in this research to deteriorate the accountability in organisations. Delays can lead to risks. Penalties and rewards are relevant in terms of encouraging employees.

Education/Training/Awareness & Experience: Fostering a culture of information security necessitates this. By providing appropriate training and awareness, which will further enhance employees' knowledge and understanding of information security, it is essential to strike a balance between education and experience. Employee compliance will be achieved, risk will be reduced or eliminated, and many other benefits will result from this improvement in information security culture. The efficacy of education, experience, and training/awareness program in the financial institution of Nigeria ensures that employees are qualified and experienced where required to take a role. It also ensures that employees are well informed key elements like rules, laws, policies, and procedures. The distinctness of a training/awareness program is essential within any institution to ensure all employees have access to the right and recent information security training. Furthermore, this study revealed that employees knowing or noticing the benefits of education, experience and training/awareness programs further strengthens their desire to acquire knowledge and understanding of information security concepts hereby improving their information security culture.

Commitment and Trust: This reflects the level of dedication of an employee to an institution which can influence leadership's trust in them. Likewise, leadership dedicating time and resources to employee's general development especially towards information security will influence employees trust in the leadership/organisation. This concept is essential if any institution is to have an effective or improved information security culture that encourages ideal information security behaviour. This study reveals that employees who have a feeling of belonging due to provision of related training to develop their skills, chances of gaining higher paying roles or the ability to get to the highest level in an institution with job security as well usually follow or take information security as key. The satisfaction of personal or professional needs can increase an employee's level of commitment and trust, thereby decreasing the likelihood of an information security breach and improving an institution's information security culture.

Monitoring & Internal control (Audit): Monitoring helps to achieve internal control requirements [57, 58]. Internal control guarantees the necessary and rightful implementation of control policies, procedures, technologies and even deployment of experts to mitigate risks [57, 58]. Internal control also ensures compliance within an organisation. According to the transcript of the interview, regular monitoring and auditing is one method for ensuring that employees adhere to information security policies and standards and act accordingly. Effective audit and monitoring activities encourages the acceptance of policies, regulations, and laws. It also ensures good information security culture, compliance, and ethical behaviour within any type of institution.

Process Integration: This factor influences an institution's information security culture and behaviour in relation to information security strategies. It entails the proper merging of information security procedures and processes in order to achieve efficient strategies that will support an institution's business-related objectives [61]. This will further develop performance, speed up the realization of information security techniques (process/procedures) thus boosting employees' information security culture and conduct. This research shows that it is fundamental for an institution to support a sound harmony between information security process/procedures and its functional or business-related processes. Interviews confirmed that the information security compliance-based approach should be improved in order to avoid difficulties in the creation and implementation of information security strategies (process/procedure). It also demonstrates that information security culture and behaviour can be influenced by employee perceptions of adoption of a process or procedure. The top management suggested the automation of information security processes/procedures within business-related processes which will likewise diminish rate at which humans mediate for information security. This will in turn reduce complexity and costs within the institution as regards to security. Information security processes especially that which is well integrated with business-related process is confirmed by this research to be a vital factor for creation of an improved information security culture.

Technology: This is a remarkable factor; It describes services and devices used to guarantee that confidentiality, integrity, and availability is kept up with or not disregarded. Technology tools and services upholds the authorization of approved polices in the financial institution of Nigeria and furthermore the processes that enhances information security [59]. *Technology proficiency* reveals the capacity of information security technologies to fulfil information security requirements in accordance with business-related processes, policies, and standards [60]. Top management revealed that having top-notch information security tools and services which fulfil technical security requirements works on employees' and stakeholders' certainty and trust which thus helps information security consistency, compliance, culture, and conduct. This study reveals that having adequate technical support is crucial to maintaining information security technologies which will also impact information security culture. *Technology affinity* is the capacity of information security technologies to enforce information security requirements across various technologies in any financial institution [60]. The perceived impact of technology affinity on information security requirements, technical responsibilities, and the affordability of implementing information security technologies has an impact on the culture and behaviour of an institution's employees regarding information security. *Technology Reliability* refers to a technology's capacity to continuously perform in accordance with its specification, which is based on established information security requirements [60]. A financial institution's information security is improved by the reliability of information security technologies, according to top management. It improves information security culture as a whole and increases trust, confidence, and compliance. Top management emphasise that an improved or good information security culture is achieved when employees are aware that technologies are working perfectly.

Permissible/Institutional Pressures: This is a situation in which current laws and regulations have a legal hold on any institution of a country in other to meet up with her information security requirements [62]. This study shows that permissible/institutional pressure can inspire top management to make drastic or extensive change across all levels to ensure the financial institution meet their legal obligations. This could involve formalising employees' ideas deemed fit by experts to enhance information security culture. This research obviously shows the significance of laws and regulations. All top management through interview confirmed that the institution's regulation guide, regulatory bodies, and current laws of the country already stipulates punishment for infringements which is carried on into the institution's system alongside their own in other to advance great information security culture. The analysis revealed that employees' awareness with the severity of infringement gives them a positive mindset, it increases their intention to act accordingly when it comes to information security.

Societal Pressure: This deals with an individual's desired needs [63]. Any investor or stakeholder interested in joining a financial institution should prioritize the confidentiality, integrity, availability, and quality of services provided, according to this study. It is also shown that a lack of faith in a financial institution like a bank can result in fewer investments, slower sector growth, and even bank failure, which can have an impact on a nation's economy. The cognizance of information security issues by all permits institutions to present best information security practises effortlessly and simultaneously have people comply to it.

Personal Beliefs: This refers to an idea an individual holds or considers to be right, true, or ideal [64]. According to the findings of this study, religion is found to be primarily a sense of moral obligation that, in turn, creates a sense of social constraint based on the observance and upkeep of a set of rules based on a set of shared beliefs. Also, it is revealed that a person's character is positively influenced by their knowledge and acceptance of religious doctrines, culture, or background. Religious doctrines are projected by spiritual leaders and encompasses acceptable practices to ensure high moral standard; it covers various scenarios and how to handle it or what was done that can be applied if it should become a reality. In this study, findings show that employees with strong knowledge or religious belief tend to follow doctrines such as "obeying higher authorities", "abiding by rules/regulations", "doing good at all times" and many more which can impact them in their work environment—they are well disciplined, abide by rules/regulation and will not jeopardise the security of the institution intentionally. This will improve information security culture.

Values: This deals with an individuals' stable and long-lasting beliefs about what is or should be important [65]. This can determine how they act or their decisions. Beliefs usually develop into values when the commitment to it grows. This study shows that for a person to act morally, consistently make clear and rational decisions such person must be able to articulate their values. The finding from interview of top management shows that character values such as loyalty, commitment are core characteristics expected to be discovered in an individual during recruitment.

The interview data were also analysed thematically, and it turned out that certain work-related values like career progression can help a person to discover what they want in a job while also increasing job satisfaction. The top management also revealed that employees bring their personal goals or ambitions into the work environment. These goals or ambitions are like guiding principles of their life, they can be referred to as personal values. Personal values assist individuals in defining their life, it determines their state of mind, happiness and how they value life generally.

Attitudes: It is the mental disposition a person has towards others, views, or anything before making decisions that results into a behaviour [66]. Generally, a person forms his/her attitude from underlying values and beliefs. This study revealed that attitudes can be explicit and implicit. There are attitudes that people are conscious of and clearly influence behaviours (Explicit) while there are some which people are not conscious of and still clearly influence behaviours (Implicit). The interview of top managements also revealed that attitudes can be influenced or improved through learnings, guides, code of conducts, rules, and regulations. It is revealed that borrowing concepts from other industries to teach, correct or guide employees are common. This study shows that an employee's wellbeing can greatly affect attitude, work output and general behaviour within and outside the work environment. The lack of attention or evaluating an employee's wellbeing can affect the business badly. Top management revealed that the financial sector is forced to always access their employee's wellbeing especially mental health. They also ensure the work environment is safe and conducive and have a policy which encourages everyone to open up on their challenges as there are various help options available in the business.

5 Conclusion

Proper management of information security in any institution is very essential to its success. It boosts stakeholders' confidence and trust in such institution.

When it comes to protecting information security, the qualitative study enables a deeper comprehension of the socio-organisational factors that influence information security culture. This study contributes to existing literature, it will enhance the information security culture in Nigeria's financial institution, thereby increasing output and customer satisfaction. It also demonstrated that information security is more than just a technical issue and shed light on the current state of behavioural information security research. By using a sensitive institution as a case study, the study provides and investigates several socio-organisational factors that can improve or used in developing employees' information security culture. Creating or improving information security culture is not an easy task and can take time due to its complexity, this study has been able to simplify factors that can assist with this.

The fact that the top management of the institution interviewed understood this study's findings lends credence to its validity. Also, most of the socio-organisational factors in this study resonated with findings and theories from previous studies. This study's findings have been validated in a multicultural country; *Nigeria* and only collected data from top management of one of her first-generation banks with several branches across the country and some West-African countries like Ghana, Benin republic.

Future works or studies should consider the following:

- Another population sample—having this type of study in another multicultural country.
- Focus on a wider population size—employees.
- using a quantitative methodology or using a mixed-methodology approach.
- consider another sensitive institution like the health care.
- Have a comparative analysis between two institutions.
- Identify the major challenges of creating or improving information security culture in start-up enterprise.

The limitations of this study are data collection was mainly from interviews, didn't look deeply into national culture and also survey respondents were limited to 38 top management who are information security conscious and likely have higher understanding than other staff of the institution.

This study concludes that improving or creating information security culture in any institution requires strong leadership, accountability, education/awareness/training/experience, commitment & trust, monitoring & internal control (audits), process integration, technology, permissible/institutional pressure, societal pressure, personal beliefs, values, and attitudes.

References

1. Ng, B.Y., Kankanhalli, A., Xu, Y.C.: Studying users' computer security behaviour: a health belief perspective. Decis. Support Syst. **46**(4), 815–825 (2009)
2. Stanton, N.A., Hedge, A., Brookhuis, K., Salas, E., Hendrick, H.W.: Situation awareness measurement and the situation awareness global assessment technique. In: Handbook of Human Factors and Ergonomics Methods, pp. 445–453. CRC Press (2004)
3. Hazari, S., Hargrave, W., Clenney, B.: An empirical investigation of factors influencing information security behaviour. J. Inf. Privacy Secur. **4**(4), 3–20 (2008)
4. Howe, A.E., Ray, I., Roberts, M., Urbanska, M., Byrne, Z.: The psychology of security for the home computer user. In: 2012 IEEE Symposium on Security and Privacy, pp. 209–223. IEEE (2012)
5. Huang, D.L., Rau, P.L.P., Salvendy, G.: A survey of factors influencing people's perception of information security. In Human-Computer Interaction. HCI Applications and Services: 12th International Conference, HCI International 2007, Beijing, China, July 22-27, 2007, Proceedings, Part IV 12, pp. 906–915. Springer, Berlin Heidelberg (2007)
6. Connolly, L., Lang, M.: Investigation of cultural aspects within information systems security research. In: 2012 International Conference for Internet Technology and Secured Transactions, pp. 105–111. IEEE (2012)
7. Zafar, H., Clark, J.G.: Current state of information security research in IS. Commun. Assoc. Inf. Syst. **24**(1), 34 (2009)
8. Posey, C., Roberts, T.L., Lowry, P.B., Bennett, R.J., Courtney, J.F.: Insiders' protection of organizational information assets: development of a systematics-based taxonomy and theory of diversity for protection-motivated behaviors. Mis Quart., 1189–1210 (2013)
9. Bulgurcu, B., Cavusoglu, H., Benbasat, I. : Roles of information security awareness and perceived fairness in information security policy compliance. AMCIS 2009 Proceedings, 419 (2009)
10. Spears, J.L., Barki, H.: User participation in information systems security risk management. MIS Quart., 503–522 (2010)
11. Abimiku, A.C., Dabwor, T.D., MJ, M.: Financial Deepening and Economic Development in Nigeria: 1990–2014
12. Danquah, P., Longe, O.B., Lartey, J.D., Tobbin, P.E.: Towards a theory for explaining socially engineered cyber deception and theft. In: Modern Theories and Practices for Cyber Ethics and Security Compliance, pp. 44–58. IGI Global (2020)
13. Tunmibi, S., Falayi, E.: IT security and e-banking in Nigeria. Greener J. Int. Inf. Commun. Syst. **1**(3), 061–065 (2013)
14. Modibbo, A., Aliyu, Y.: Cashless society, financial inclusion and information security in Nigeria: the case for adoption of multifactor biometric authentication. Int. J. Innov. Sci. Res. Technol. **4**(11), 872–880 (2019)

15. Ryan, J.E.: A comparison of information security trends between formal and informal environments. Auburn University (2006)
16. McFadzean, E., Ezingeard, J.N., Birchall, D.: Anchoring information security governance research: sociological groundings and future directions. J. Inf. Syst. Secur. **2**(3), 3–48 (2006)
17. Von Solms, R., von Solms, S.B.: Information security governance: due care. Comput. Secur. **25**(7), 494–497 (2006)
18. Mancha, R., Dietrich, G.: Development of a framework for analyzing individual and environmental factors preceding attitude toward information security (2007)
19. Dinev, T., Hu, Q.: The centrality of awareness in the formation of user behavioral intention toward protective information technologies. J. Assoc. Inf. Syst. **8**(7), 23 (2007)
20. D'Arcy, J., Hovav, A.: Deterring internal information systems misuse. Commun. ACM **50**(10), 113–117 (2007)
21. Choi, N., Kim, D., Goo, J., Whitmore, A.: Knowing is doing: an empirical validation of the relationship between managerial information security awareness and action. Inf. Manag. Comput. Secur. **16**(5), 484–501 (2008)
22. D'Arcy, J., Hovav, A.: Does one size fit all? Examining the differential effects of IS security countermeasures. J. Bus. Ethics **89**, 59–71 (2009)
23. Dinev, T., Goo, J., Hu, Q., Nam, K.: User behaviour towards protective information technologies: the role of national cultural differences. Inf. Syst. J. **19**(4), 391–412 (2009)
24. Galvez, S.M., Guzman, I.R.: Identifying factors that influence corporate information security behaviour. AMCIS 2009 Proceedings, 765 (2009)
25. D'Arcy, J., Hovav, A., Galletta, D.: User awareness of security countermeasures and its impact on information systems misuse: a deterrence approach. Inf. Syst. Res. **20**(1), 79–98 (2009)
26. Bulgurcu, B., Cavusoglu, H., Benbasat, I.: Effects of individual and organization based beliefs and the moderating role of work experience on insiders' good security behaviors. In: 2009 International Conference on Computational Science and Engineering, Vol. 3, pp. 476–481. IEEE (2009)
27. Werlinger, R., Hawkey, K., Botta, D., Beznosov, K.: Security practitioners in context: Their activities and interactions with other stakeholders within organizations. Int. J. Hum. Comput. Stud. **67**(7), 584–606 (2009)
28. Kruger, H.A., Drevin, L., Steyn, T.: The use of an information security vocabulary test to assess information security Awareness-An exploratory study. In: Proceedings of the South African Information Security Multi-Conference, Port Elizabeth, South Africa 17–18 May. Plymouth: Centre for Security, Communications & Network Research, University of Plymouth (2010)
29. Bhattacharya, D.: Leadership styles and information security in small businesses. Inf. Manag. Comput. Secur. **19**(5), 300–312 (2011)
30. Monfelt, Y.: Coherence in control of information security: COINS. Manag. Prod. Eng. Rev., 2 (2011)
31. Takemura, T.: Empirical analysis of behaviour on information security. In: 2011 International Conference on Internet of Things and 4th International Conference on Cyber, Physical and Social Computing, pp. 358363. IEEE (2011)
32. Yayla, A.: Controlling insider threats with information security policies (2011)
33. Al-Omari, A., El-Gayar, O.F., Deokar, A., Walters, J.: Information Security Policy Compliance: An Ethical Perspective (2012)
34. Lee, M., Lee, J.: The impact of information security failure on customer behaviors: a study on a large-scale hacking incident on the internet. Inf. Syst. Front. **14**, 375–393 (2012)
35. Shropshire, N.H., Ismail, Z., Maarop, N.: Information security culture: a systematic literature review (2015)
36. Da Veiga, A.: Comparing the information security culture of employees who had read the information security policy and those who had not: Illustrated through an empirical study. Inf. Comput Secur. **24**(2), 139–151 (2016)
37. Liang, H., Xue, Y.L.: Understanding security behaviors in personal computer usage: a threat avoidance perspective. J. Assoc. Inf. Syst. **11**(7), 1 (2010)

38. Claar, C.L.: The adoption of computer security: An analysis of home personal computer user behaviour using the health belief model. Utah State University (2011)
39. Pahnila, S., Karjalainen, M., Siponen, M.: Information security behaviour: towards multi-stage models (2013)
40. Guo, K.H.: Security-related behaviour in using information systems in the workplace: a review and synthesis. Comput. Secur. **32**, 242–251 (2013)
41. Connolly, L., Lang, M., Tygar, J.D. Investigation of employee security behaviour: a grounded theory approach. In: ICT Systems Security and Privacy Protection: 30th IFIP TC 11 International Conference, SEC 2015, Hamburg (2015)
42. Safa, N.S., Sookhak, M., Von Solms, R., Furnell, S., Ghani, N.A., Herawan, T.: Information security conscious care behaviour formation in organizations. Comput. Secur. **53**, 65–78 (2015)
43. Chatterjee, S., Sarker, S., Valacich, J.S.: The behavioral roots of information systems security: exploring key factors related to unethical IT use. J. Manag. Inf. Syst. **31**(4), 49–87 (2015)
44. Shropshire, J., Warkentin, M., Sharma, S.: Personality, attitudes, and intentions: Predicting initial adoption of information security behaviour. Comput. Secur. **49**, 177–191 (2015)
45. Al-Kalbani, A.: A compliance-based framework for information security in e-government in Oman (Doctoral dissertation, RMIT University) (2017)
46. Almalki, S.: Integrating quantitative and qualitative data in mixed methods research-challenges and benefits. J. Educ. Learn. **5**(3), 288–296 (2016)
47. Kothari, C.R.: Research methodology: methods and techniques. New Age Int. (2004)
48. Bowen, G.A.: Document analysis as a qualitative research method. Qual. Res. J. **9**(2), 27–40 (2009)
49. O'Leary, D.E.: Knowledge management: an empirical analysis of reuse and productivity. J. Decis. Syst. **23**(3), 249–265 (2014)
50. Creswell, J.W.: Mixed-method research: Introduction and application. In: Handbook of Educational Policy, pp. 455–472. Academic Press (1999)
51. Maykut, P., Morehouse, R.: Beginning Qualitative Research: A Philosophical and Practical Guide. Routledge (2002)
52. Jones, C.: Qualitative interviewing. In: Handbook for Research Students in the Social Sciences, pp. 203–214. Routledge (2020)
53. Myers, M.D., Newman, M.: The qualitative interview in IS research: examining the craft. Inf. Organ. **17**(1), 2–26 (2007)
54. Yang, K., Banamah, A.: Quota sampling as an alternative to probability sampling? An experimental study. Sociol. Res. Online **19**(1), 56–66 (2014)
55. Fetters, M.D., Curry, L.A., Creswell, J.W.: Achieving integration in mixed methods designs—principles and practices. Health services research, 48(6pt2), 2134–2156 (2013)
56. Howitt, D., Cramer, D.: Introduction to qualitative methods in psychology (2010)
57. Alawonde, K.O.: Tailored Information Security Strategies for Financial Services Companies in Nigeria (Doctoral dissertation, Walden University) (2020)
58. Bustamante, F., Fuertes, W., Diaz, P., Toulqueridis, T.: Methodology for management of information security in in-dustrial control systems: a proof of concept aligned with enterprise objectives. Adv. Sci. Technol. Eng. Syst. J. **2**, 88–99 (2017)
59. Amin, R., Shah, N., Shah, B., Alfandi, O.: Auto-configuration of ACL policy in case of topology change in hybrid SDN. IEEE Access **4**, 9437–9450 (2016)
60. Ahmad, A., Hadgkiss, J., Ruighaver, A.B.: Incident response teams–challenges in supporting the organisational security function. Comput. Secur. **31**(5), 643–652 (2012)
61. Flowerday, S.V., Tuyikeze, T.: Information security policy development and implementation: what, how and who. Comput. Secur. **61**, 169–183 (2016)
62. Armstrong, D., Farrell, T., Lambert, H.: International Law and International Relations. Cambridge University Press (2012)
63. Linton, R.: Cultural Background Personality ILS 84. Routledge (2013)
64. Kohlberg, L.: Stages of moral development. Moral Educ. **1**(51), 2392 (1971)

65. Georgel, J.M., Jones, G.R.: Experiencing work: Values, attitudes, and moods. Human Relations **50**(4), 393–416 (1997)
66. Allport, G.W.: 1. Attitudes. Terminology, 219 (1933)

Embedded Systems and Smart Applications

An Overview of Concentrated Photovoltaic Technology

Sarah El Himer, Mariya Ouaissa, Mariyam Ouaissa, and Zakaria Boulouard

Abstract Concentration photovoltaic is based on the use of optical elements to focus incident solar radiation on a small area of the size of the photovoltaic cell, most of the time, multijunction. These elements determine the concentration ratio, the acceptance angle, the uniformity of the solar irradiance and, finally, the efficiency of the module. In this chapter, we will discuss an introduction to CPV technology and its composition.

Keywords Concentrated photovoltaic · Solar cell · Multijunction · Optic

1 Introduction

Demographic development accompanied by the development of industry, transport and communication systems is among the factors that justify the strong growth in the consumption of electrical energy worldwide. The traditional sources of energy (natural gas, uranium, oil, coal, etc.) have been the most used up to now for the production of electrical energy. However, it should be noted that these sources of non-renewable energy are doomed to exhaustion and their consumption is not carbon neutral and has a harmful effect on the environment (pollution, global warming, acid rain, …). All these factors have prompted many countries to redouble their efforts to increase the share of renewable energies in their energy mix. Morocco is among the first countries to have adopted strategic plans accompanied by a high-level political will aimed at developing its electricity production capacities from renewable energies

S. El Himer
Sidi Mohamed Ben Abdellah University Fez, Fes, Morocco

M. Ouaissa (✉) · M. Ouaissa
Moulay Ismail University, Meknes, Morocco
e-mail: mariya.ouaissa@edu.umi.ac.ma

Z. Boulouard
LIM, Hassan II University of Casablanca, Casablanca, Morocco

C. Iwendi et al. (eds.), *Proceedings of ICACTCE'23—The International Conference on Advances in Communication Technology and Computer Engineering*, Lecture Notes in Networks and Systems 735, https://doi.org/10.1007/978-3-031-37164-6_11

by setting itself the objective of increasing their share to 42% in 2020 (14% solar, 14% wind, 14% hydraulic) and 52% in 2030 in the Moroccan energy mix [1]. It is even very likely to exceed these objectives. The percentage of solar energy in this energy mix can be significantly increased by adopting the emerging technology of concentrated solar photovoltaic (CPV), in fact, this technology has the best industrial integration rate in Morocco: 40% in the short term. and more than 80% in the medium term [according to a study published in the book: Photovoltaic: Technologies, Markets and Industrial Integration [2].

The main idea of CPV is the use of optical elements to concentrate solar radiation on special photovoltaic cells (multi-junctions) and thus greatly increase their efficiency. This also makes it possible to considerably reduce the surface of the cell and consequently its price. A CPV system generally consists of three main elements: concentration optics which is the heart of the system, solar cells, generally multi-junctions, and a tracker (solar tracker) to position the structure (optical devices) facing the sun. CPV systems are classified according to their concentration factors [3]:

- Low concentration (LCPV): $(1-10\times)$ (The symbol "nn ×" is read nn times the sun).
- Medium concentration (MCPV): $(10-100\times)$
- High concentration (HCPV): $(100-2000\times)$) (Table 1).

Table 1 Summarizes some strengths and weaknesses of the CPV

Strengths of the CPV	Weaknesses of the CPV
High yields	CPV can only use direct solar radiation (DNI)
No need for high temperature	Tracking with precision and reliability sufficient is required
Modular-kW to GW scale	Need for frequent cleaning to reduce performance losses due to clogging (dust, dirt, etc.)
Increased and stable energy production throughout the day thanks to tracking	Limited market-can only be used in regions with high DNI, cannot be easily installed on roofs
Opportunities for cost-effective local manufacturing of certain system components	New generation technologies, without production history (therefore risky)
Very low energy recovery time [4, 5]	Additional optical losses
Less sensitive to semiconductor price changes	Lack of technological standardization

2 Historical Landmarks

The strong point of concentrated photovoltaics is the increase in the efficiency of solar cells. In fact, Shockley and Queisser defined, in their article published in 1960 and entitled "Detailed Balance Limit of Efficiency of p–n Junction Solar Cells" [6], a maximum conversion efficiency of about 30% for single-junction solar cells under an illumination of 1000 W/m^2. It is also in this article that they lay the first foundations of the CPV concept since they mention that the efficiency should increase with the incident light intensity on the cell.

The development of CPV technology began in 1976 with the development of a panel delivering a power of 1 kWp in the Sandia national laboratory in New Mexico [7] (Fig. 1). This initiative was immediately followed by the Polytechnic University of Madrid (UPM) which developed a similar panel in 1980 (Fig. 2). All of these panels used Fresnel lenses, silicon-based solar cells with 40× concentration, and a two-axis solar tracker.

The first significant industrialization took place in 1990, in Texas by ENTECH which installed a few hundred kW, using curved Fresnel lenses, under a concentration of 20×. An opportunity for this technology opened up in 1995 with the EUCLIDES prototype developed by IES/UPM and BP Solar which was installed in Madrid. This project has enabled the installation of a 480kWp solar farm in Tenerife (Fig. 3).

In 2001, this progression will become more rapid with the appearance of the single-junction GaAs cell which operates at a concentration of 1000× with a yield of 26%.

In 2005, appears the cell with triple junctions reaching 35.2% yield under a concentration of 77× [7].

High concentrations are, however, only possible if the cells are very small, around 1 mm^2, so that current extraction becomes easy. This small size is not possible in silicon because of its long diffusion length, which makes the cells very sensitive and

Fig. 1 Sandia-II panel, 1977

Fig. 2 Ramon arces array (Madrid)

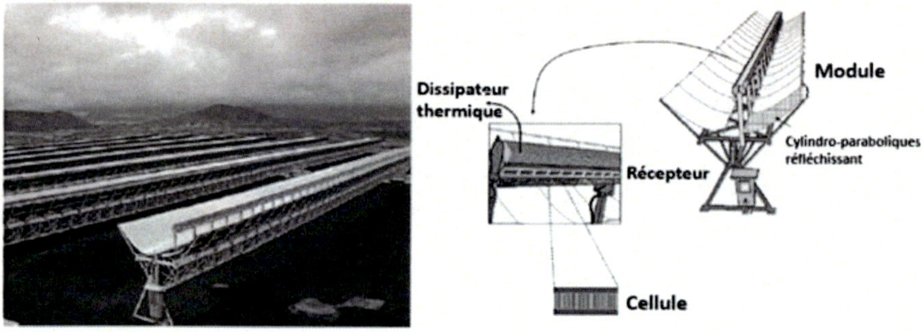

Fig. 3 The EUCLIDES demonstration plant in Tenerife (1998) subsidized by the EU to industrialize the Madrid prototype (1995)

the non-radiative recombination more important on the edges of the cell because of the dangling bonds which are there. find [8].

The first project that exceeded 1 MW was developed in Spain in 2006 (Fig. 4), and since then a significant number of large projects have emerged. By way of example, we can mention.

The "Golmud 1" project, with a capacity of 58 MW, which was installed in China in 2012 and its capacity was increased in 2013 to 79.8 MW, and the "Touwsrivier" project installed in South Africa (2014), with a power that reaches 44.2 MW. Every day, the market diversifies and different projects are launched in several countries (China, Italy, United States, ...).

In 2016, Sumitomo installed a 1MWp CPV plant in Morocco.

Cost is seen as a main challenge to manufacturers. From the literature as well as the industry, CPV system prices including installation for CPV power plants with a capacity of 10 MW have been identified between 800 and 1400e/kW. According to the literature [9], the Levelized Cost of Electricity (LCOE) values range from

Fig. 4 Examples of large CPV power plants since 2006. From top to bottom: (a) 30MW plant in Alamosa, Colorado, USA (Amonix); (b) 44MW in Touwsrivier, South Africa (Soitec); (c) 140MW at Golmud, China (c Suncore)

6.7e/kWh, in the case of the United Arab Emirates, to 62e/kWh in Vietnam. The areas with the lowest LCOE are:

North America, Chile, Australia, Northern and Southern Africa and Southern Europe.

Morocco is among the countries whose LCOE is relatively low between 10 to 15 e/kWh, this is shown in the global map of the LCOE (Fig. 5).

3 Composition of a CPV System

A CPV system is based on the combination of three components, the first two serving as the basis for what is called a CPV module:

- The photovoltaic receiver: it consists of solar cells, generally multi-junction made from III-V materials mounted on a mechanical support serving as an electrical and thermal interface. The cell being placed at the focal point of the concentration optics.

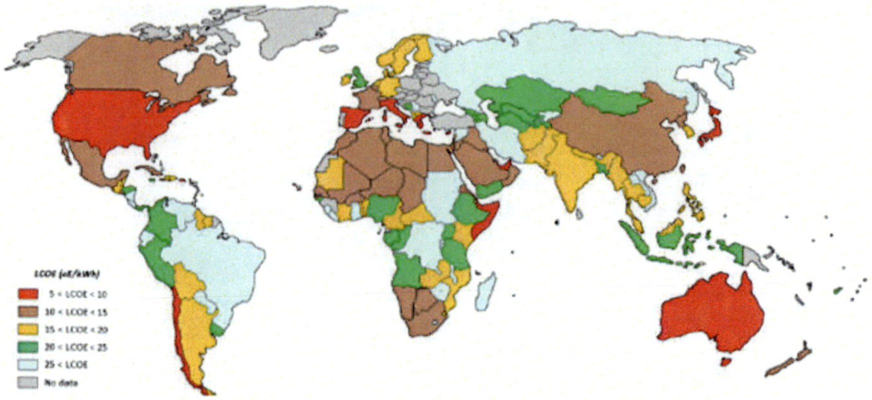

Fig. 5 Global leveled cost of electricity (LCOE) map [9]

- Concentration optics: Refractive or reflective elements, which can contain several "concentration stages" with first a primary optical element (POE) and depending on the design criteria, a secondary optical element (SOE).
- A solar tracker to position the optical element facing the sun throughout the day.

But before starting the design and operation of a module, we will define some concepts related to the use of the solar resource in concentrated photovoltaic applications.

3.1 Solar Radiation and Its Composition

a. **Solar Radiation**

Solar radiation consists of photons whose wavelength ranges from ultraviolet (0.2 μm) to far infrared (2.5 μm) as shown in Fig. 6.

The energy associated with this solar radiation consists of:

- 9% in the ultraviolet band (<0.4 μm),
- 47% in the visible band (0.4 to 0.8 μm),
- 44% in the infrared band (>0.8 μm).

b. **Composition of Solar Radiation**

Solar radiation on the ground consists of a diffuse flux and a direct flux.

The direct component corresponds to the rays coming from the angle under which the solar disk is seen, i.e. ± 0.27°. The diffuse flux comes from reflections and scattering by the Earth's atmosphere and by the ground, while the "global" radiation represents the sum of the direct and diffuse components. CPV modules convert

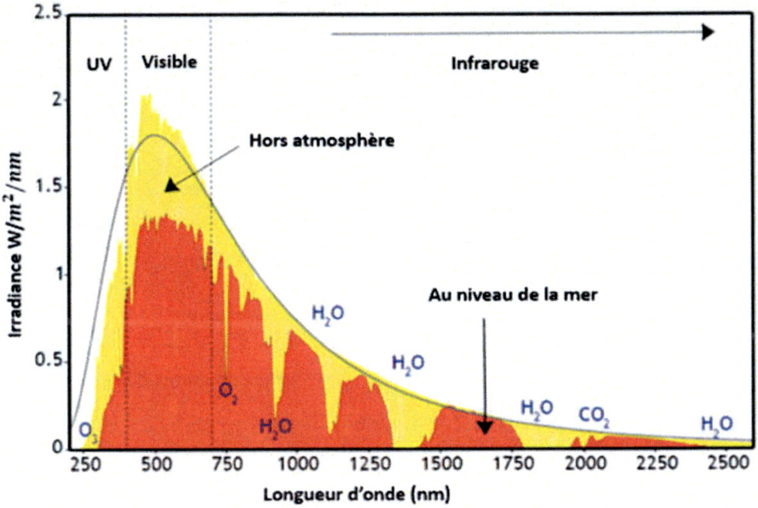

Fig. 6 Solar spectrum [10]

direct solar radiation with an angle of incidence below a certain value (less than 1) for (HCPV) modules.

3.2 Multi-junction Cell

a. **Operating Principle of a Photovoltaic Cell**

A photovoltaic cell is based on the physical phenomenon called the photovoltaic effect which consists in creating an electromotive force when the surface of this cell is exposed to light. The voltage generated is between 0.3 and 0.7 V [11] and depends mainly on the material used, it also depends on the temperature and the aging of the cell. The transformation of solar energy into electrical energy is based on the following three mechanisms: the absorption of photons (the energy is greater than the gap) by the material that constitutes the device; the conversion of photon energy into electrical energy, which corresponds to the creation of electron/hole pairs in the semiconductor material and the collection of the particles generated in the device.

In order to collect the generated particles, an electric field to separate the created electron/hole pairs is needed. For this, a p–n junction is most often used. The operation of photovoltaic cells is illustrated in Fig. 7 [12]. The incident photons create carriers in the n, p and space charge regions. Carrier photons behave differently depending on the region.

- In the space charge zone, the electron/hole pairs created by the incident photons are dissociated by the electric field, the minority carriers are sent by the electric

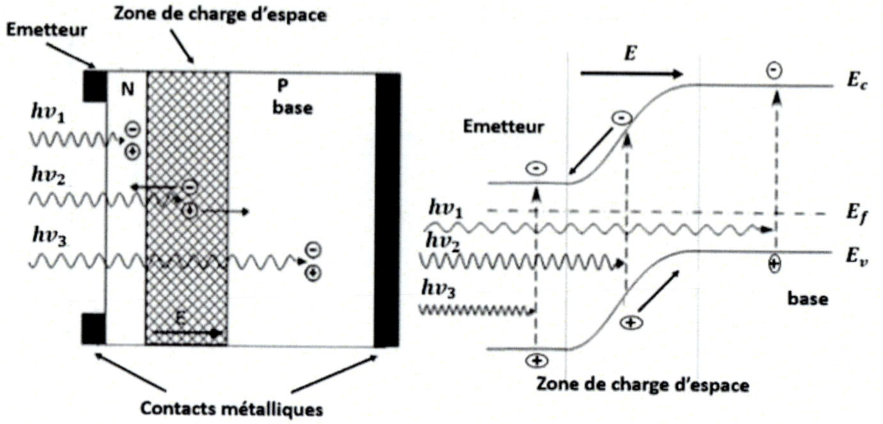

Fig. 7 Structure (left image) and band diagram (right image) of a photovoltaic cell [12]

field in the p zone (for holes) or in the n zone (for the electrons) where they will be in the majority. We will have a current broadcast photo;

- In the n or p region, the electrons will go to the n region, the holes to the p region. This then creates a potential difference between the two layers. This potential difference is measurable between the connections of the positive and negative terminals of the cell.

Through an external load, the carriers can be recovered. The maximum cell voltage is about 0.6 V [11] for zero current in the case of silicon. This voltage is called open circuit voltage (Voc). The maximum current occurs when the cell terminals are short-circuited, it is called short-circuit current (Isc) and strongly depends on the level of illumination.

A PV cell can be modeled from the equation defining the static behavior of the PN junction of a conventional diode. Figure 8 illustrates the equivalent circuit diagram of a real PV cell. In this figure, Ip is the current through the shunt resistor, Id is the current through the diode, Iph is the photocurrent, and Isc is the short-circuit current of the cell.

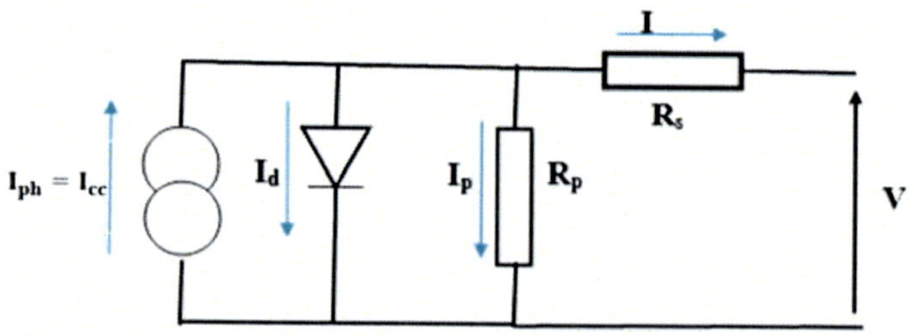

Fig. 8 Equivalent circuit of a photovoltaic cell

3.3 Current–Voltage and Power-Voltage Characteristics

If a thin PN junction is exposed to a light flux, the electron–hole pairs are created by the photons whose energy is greater than the bandgap width of the material. Consequently, the increase in the reverse saturation current is proportional, for a given spectrum, to the number of photons, therefore to the power of the luminous flux. Physically, this effect is reflected in two ways according to the operating quadrant of the device in Fig. 9.

From this figure, it can be concluded that:

- In quadrant III, the system operates as a receiver whose resistance varies greatly with the illumination at a given voltage, it behaves as a photoresistor;
- In quadrant IV, the system operates as a generator whose short-circuit current is directly proportional to the power of the illumination and whose no-load voltage V_{oc} is that of the forward-biased diode (0.5 to 0.8 V for Si). It behaves as a photovoltaic converter.

A photovoltaic cell with a P–N junction is thus obtained, an elementary direct current generator which directly converts the light energy into electrical energy delivered under voltage less than or equal to the direct bias voltage of the junction, therefore a very low voltage. The other power losses mentioned are generally taken into account by adding to the model of the ideal cell two dissipative elements

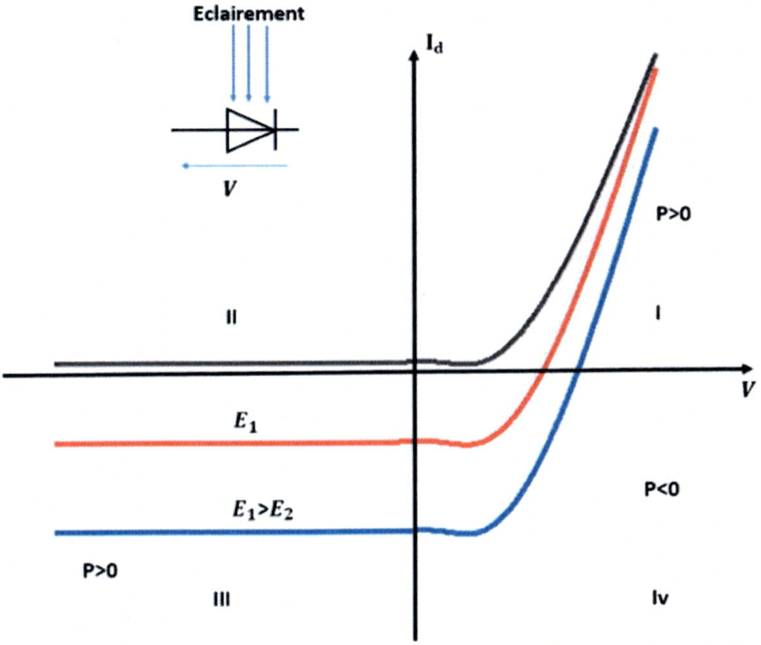

Fig. 9 Receiver convention characteristics of a dark and illuminated PN junction (E)

as indicated in Fig. 8: a series resistance Rs representing the various resistances of contacts and connections, and a shunt resistance Rp in parallel to the current generator which characterizes the various leakage currents.

According to the electrical model (Fig. 8), the following expressions can be deduced:

$$I = I_{ph} - I_d - I_p \tag{1}$$

With

$$I_d = I_s\left(\frac{V + R_s I}{nV_T} - 1\right) \tag{2}$$

And

$$I_p = \frac{V_d}{R_p} I_s = \frac{V + R_s I}{R_p} \tag{3}$$

Then

$$I = I_{ph} - I_s\left(\frac{V + R_s I}{nV_T} - 1\right) - \frac{V + R_s I}{R_p} \tag{4}$$

With

$$V_T = \frac{KT}{q} \tag{5}$$

Is is the reverse saturation current of the diode.

These expressions show that the electrical characteristics of a photovoltaic cell depend on the material, the temperature and the illumination.

The photo-created current Iph determines the short-circuit current Isc of the cell.

In Fig. 10 are represented two characteristics corresponding to those of the variations of the current and of the power as a function of the voltage. The associated power curve shows that the power goes through a maximum. This maximum corresponds to the ideal operating point of the cell.

3.4 *Multi-junction Cells*

Sunlight is composed of a spectrum of continuous waves very rich in photons whose energy varies between 0.49 and 6.2 eV [13]. In monojunction solar cells, the incident photons having an energy (h) lower than the gap (Eg) of the semiconductor are not absorbed and those whose energy is higher by restoring a part in pure loss in the form of heat.

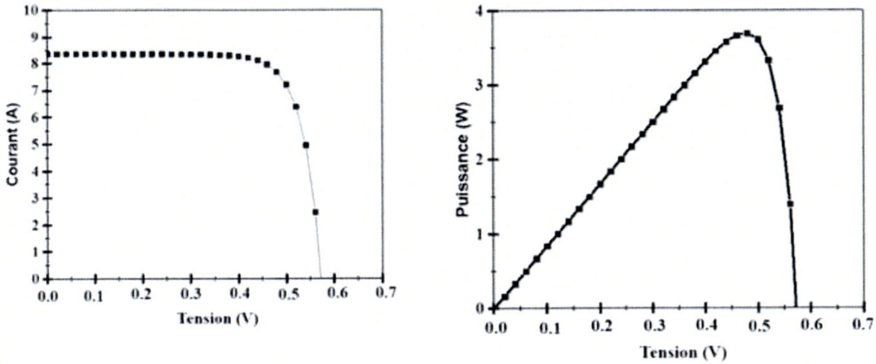

(a) I-V characteristic (b) P-V characteristic

Fig. 10 Electrical characteristic of a solar cell

Masden and Backus studied the GaAs monojunction solar cell [14] and showed that the sources of major losses are due to either an excess or a weakness of the photon energy.

In order to minimize these types of losses and to have a wide absorption of the solar spectrum, it suffices to choose a material with a very low gap [1 eV – 1.5 eV] [7]. This will increase the photo current and decrease the voltage and the duty cycle, hence a decrease in power. Indeed, a high-gap material can absorb less light, resulting in low short-circuit current and high open-circuit voltage [15]. For this, the use of multi-junction cells is more promising, because this approach is based on the stacking of several solar cells in decreasing order of their band gap, and each cell will convert the part of the spectrum which corresponds to its own gap. By this method, we will avoid losses by thermalization which come from high energy photons, or by transmission from low energy photons.

To increase the efficiency of the cell, three configurations exist: monolithic solar cells, mechanically stacked solar cells and the multi-spectral system.

- Monolithic solar cells:

This approach is based on the assembly of solar cells in series and on the same optical path with a decreasing order of the different gaps (Fig. 11). Each cell acts as a filter for the next cell.

- Mechanically stacked solar cells:

For this category, the cells are also placed on the same optical path but these cells are independent and separated by an adhesive material to guarantee good electrical insulation (Fig. 12). Each cell will be electrically isolated from the others, but these cells increase shadow losses and this reduces their yields.

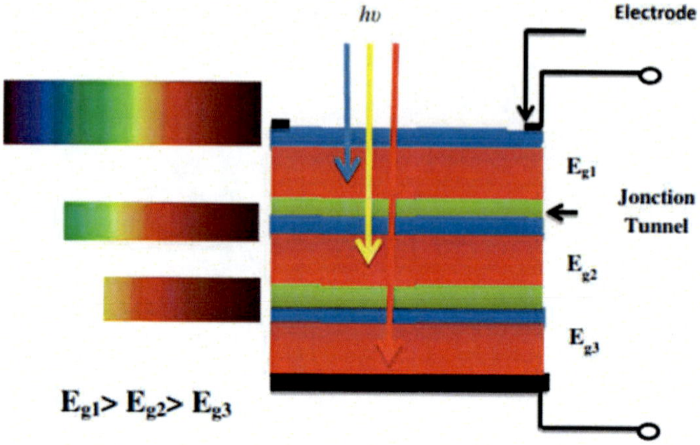

Fig. 11 Block diagram of a monolithic solar cell

Fig. 12 Diagram of the principle of mechanical stacking of solar cells

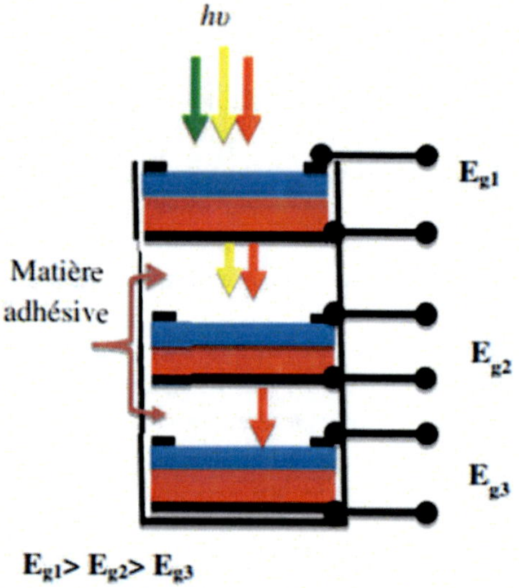

- Multi-spectral system:

This design is considered the simplest. This approach uses dichroic filters to send each cell the part of the spectrum corresponding to its band gap as shown in Fig. 13. The filter has the role of having a better adaptation of the solar spectrum, and the connection of these cells is independent. The alignment, filter efficiency, and wiring independence of each cell make this method of stacking undesirable.

NREL (National Renewable Energy Laboratory) has listed the evolution of the yields of the different generations of solar cells in a graph shown in the Fig. 14. It is found that the use of solar concentrators helps to increase the efficiency of the cells,

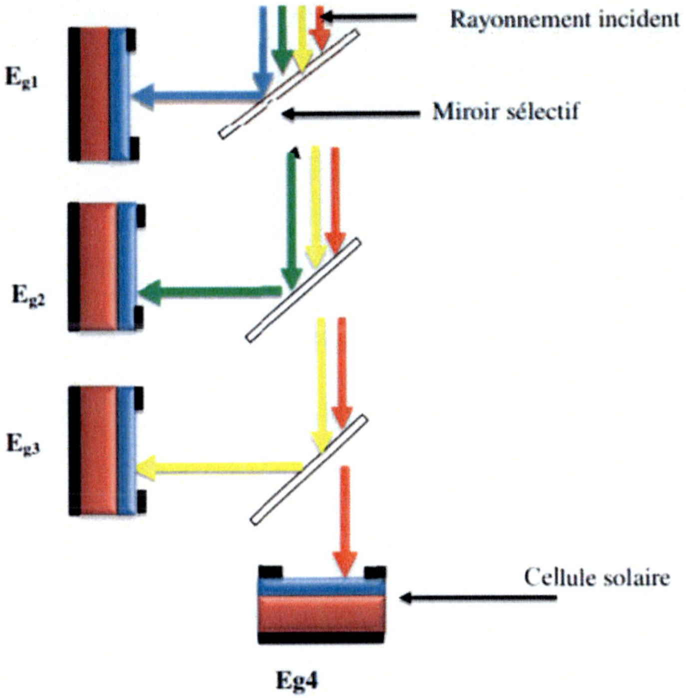

Fig. 13 Diagram of the principle of association for a multispectral system

indeed, the efficiency of the four-junction cells has been increased from 39.2% to 46%, and the same for the three-junction cells, it is increased from 37.9% to 44%.

3.5 Solar Trackers

A tracker or solar follower is a system using the heliostat principle. It is a structure that facilitates the orientation of the solar collectors constantly towards the sun. This increases productivity substantially. Rigorous monitoring of the sun must be done on two axes: in azimuth (from East to West, as the day progresses) and in inclination of the sensors (according to the progress of the day). However, there are single-axis trackers, typically in azimuth; the angle of inclination being fixed according to the local optimum.

There are mainly two main families of solar trackers: passive and active which include single-axis and dual-axis trackers. In our laboratory, Benmohammadi [17] and Kaddouri [18] have already worked on solar trackers in their graduation projects. Their objectives were the development of motor control of a two-axis solar tracker

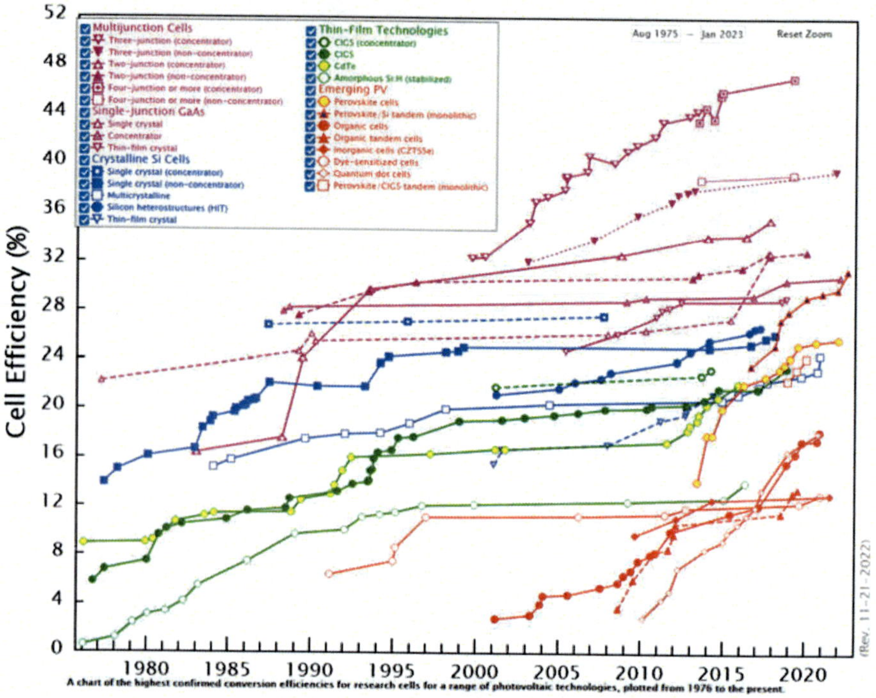

Fig. 14 Yields achieved by different solar cell technologies [16]

based respectively on LDR sensors and on the astronomical equations of the motion of the sun.

4 Conclusion

In this chapter, we have covered an introduction to CPV technology. The principle of concentrated photovoltaic is based on the use of inexpensive concentration optical elements concentrating a large surface of incident light on high-efficiency cells. The advantage is twofold: a reduction in the cell surfaces required and a higher conversion yield, all the more so at higher concentration.

References

1. Benkhadra, A. : Mobilisation pour une croissance verte au maroc. In Annales des Mines-Responsabilité et environnement, vol. 1, pp. 25–31. ESKA (2011)

2. Feynman, R.: Badr Ikken, El Ghali Bennouna, Abdellatif Ghennioui, Moussa El Khal. Iresen (2013)
3. Khamooshi, M., Salati, H., Egelioglu, F., Hooshyar Faghiri, A., Tarabishi, J., Babadi, S.: A review of solar photovoltaic concentrators. Int. J. Photoenergy (2014)
4. Peharz, G., Dimroth, F.: Energy payback time of the high-concentration pv system flatcon R. Prog. Photovoltaics Res. Appl. **13**(7), 627–634 (2005)
5. Fthenakis, V.: Life Cycle Analysis of CPV Systems. Handbook of Concentrator Photovoltaic Technology, pp. 685–704 (2016)
6. Shockley, W., Queisser, H.J.: Detailed balance limit of efficiency of p-n junction solar cells. J. Appl. Phys. **32**(3), 510–519 (1961)
7. Fahrenbruch, A., Bube, R.: Fundamentals of Solar Cells: Photovoltaic Solar Energy Conversion. Elsevier (2012)
8. Díaz, V., Algora, C.: The influence of perimeter recombination in the design of very high concentrator gaas solar cells. In: Proceedings of the16th European Photovoltaic Solar Energy Conference, Glasgow, UK, pp. 1034–1037 (2000)
9. Talavera, D.L., Ferrer-Rodríguez, J.P., Pérez-Higueras, P., Terrados, J., Fernández, E.F.: A worldwide assessment of levelised cost of electricity of hcpv systems. Energy Convers. Manage. **127**, 679–692 (2016)
10. Nyamsi, W.W.: Vers une méthode automatique d'estimation de la distribution spectrale du rayonnement solaire. Cas du ciel clair. : Applications à la lumière du jour, photosynthèse et ultraviolet. PhD thesis, Ecole Nationale Supérieure des Mines de Paris (2015)
11. Pastor, A.C.: Conception et réalisation de modules photovoltaïques électroniques. PhD thesis, INSA de Toulouse (2006)
12. Damon-Lacoste, J. : Vers une ingénierie de bandes des cellules solaires à hétérojonctions a-Si : H/c-Si. Rôle prépondérant de l'hydrogène. PhD thesis, Ecole Polytechnique X (2007)
13. Olson, J.M., Friedman, D.J., Kurtz, S.: High-efficiency III-V Multijunction Solar Cells. Handbook of Photovoltaic Science and Engineering, pp. 359–411 (2003)
14. Masden, G.W., Backus, C.E.: Increased photovoltaic conversion efficiency through use of spectrum splitting and multiple cells. In: 13th Photovoltaic Specialists Conference, pp. 853–858 (1978)
15. Nelson, J.: The Physics of Solar Cells. World Scientific Publishing Company (2003)
16. Fonash, S.: Introduction to Light Trapping in Solar Cell and Photo-Detector Devices. Elsevier (2014)
17. Benmohammadi, Z.: Etude, conception et réalisation d'une commande d'un tracker solaire (suiveur solaire). PhD thesis, Université Sidi Mohammed Ben Abdellah, Faculté Des Sciences et Techniques Fès (2014)
18. Kaddouri, A.: Conception et réalisation d'un suiveur solaire à deux axes à base des équations astronomiques. PhD thesis, Université Sidi Mohammed Ben Abdellah, Faculté Des Sciences et Techniques Fès (2015)

Electronic Prescription Service for Improved Healthcare Delivery

Tao Hai, Shaoyi Li, Afolake O. Adedayo, Shirin Abolfath Zadeh, Jiuping Cai, and Celestine Iwendi

Abstract The aim of this research paper is to explore the potential of machine learning techniques in predicting the utilization of the Electronic Prescription Service (EPS) and Electronic Repeat Dispensing (eRD) items to categorize General Practitioner (GP) practices based on their usage patterns. The study utilized raw data related to dispensaries, EPS, and eRD acquired from the National Health Service online medical database. To achieve this objective, exploratory data analysis was conducted on the dataset, which was then split into a training set and a testing set. Various machine learning algorithms, including linear regression, decision tree regression, and random forest regression, were applied to the training set to develop a predictive model. The models were evaluated using measurements such as the "Score", "Mean Squared Error (MSE)", "Mean Absolute Error (MAE)", "Sqrt Mean Absolute Error (MAE)" and "Coefficient of determination (R^2)". The study found that the machine learning models developed were effective in predicting EPS utilisation and could categorize GP practices based on their usage patterns. This categorization could help identify high-utilization practices, leading to more efficient resource allocation and ultimately improved healthcare delivery. The results also indicate the potential for machine learning techniques to predict the utilization of other healthcare services and could pave the way for more personalized and targeted healthcare services in the future.

Keywords Electronic Prescription Service (EPS) · Electronic Repeat Dispensing (eRD) · National Health Service (NHS) · Machine learning · Regression models · ePrescribing systems · Exploratory data analysis · Web applications

---

T. Hai
School of Computer and Information, Qiannan Normal University for Nationalities, Duyun, Guizhou 558000, China

A. O. Adedayo (✉) · S. A. Zadeh · C. Iwendi
School of Creative Technologies, University of Bolton, Bolton BL3 5AB, UK
e-mail: afolake.adedayo@ieee.org

T. Hai · S. Li · J. Cai
School of Information and Artificial Intelligence, Nanchang Institute of Science and Technology, Nanchang, China

C. Iwendi et al. (eds.), *Proceedings of ICACTCE'23—The International Conference on Advances in Communication Technology and Computer Engineering*, Lecture Notes in Networks and Systems 735, https://doi.org/10.1007/978-3-031-37164-6_12

1 Introduction

Web applications are an essential part of our daily lives, as we rely on them to meet our needs and find information. With the right use and innovative ideas, these applications can also help reduce human workload. Advancements in technology have provided new opportunities for improving healthcare infrastructure and data management, such as electronic health records, telemedicine, and precision medicine [1]. The Digital and Interoperable Medicines Programme in the United Kingdom, for example, supports the development and adoption of IT systems that enable the seamless sharing of information across health and care services. This includes the Clinical Data Sharing program and the Electronic Prescription Service (EPS), which has led to more efficient use of resources and improved delivery of healthcare.

The Electronic Prescription Service (EPS) enables healthcare providers to electronically send prescriptions to a selected dispensing location (such as pharmacies) for the patient. This results in a more efficient and convenient process for patients and staff and reduces the errors that can occur during manual transcription. The benefits of EPS include faster processing of prescriptions for health care providers, reduced paper use for dispensing locations, improved stock control, and easier access to repeat prescriptions for patients without visiting their GP, ultimately improving the efficiency of prescription delivery [2]. Electronic Repeat Dispensing (eRD) allows for routinely prescribed medicine to be prescribed in batches and stored securely on the Spine, which delivers the prescriptions to the patient's nominated pharmacy or contractor at regular intervals set by the prescriber. According to the National Health Service (NHS) Digital Service, most prescriptions issued in primary care are repeat prescriptions, approximately 410 million repeat prescriptions are produced annually, averaging over 375 per GP each week. These account for about 330 million repeat prescriptions, or 80% of the NHS's medicine costs. The use of eRD, as part of the EPS system, can replace these repeat prescriptions, potentially saving 2.7 million hours of GP and practice time worth £90 million to the NHS.

The objective of this research is to utilise machine learning methods to develop a model that categorizes GP practices based on their usage patterns and predicts the utilisation of EPS and eRD items, with the goal of maximising the benefits of the Electronic Prescription Service.

2 Literature Review

While international and regional communities have recognised the importance of ePrescribing systems, the European Union (EU) has been at the forefront of its implementation, recognising its potential to improve healthcare delivery [3]. The implementation of EPS systems varies in different communities, but the potential to improve patient safety and healthcare efficiency is a common theme.

The growth of e-prescribing in the United States (US) has been facilitated by several factors, including the enactment of the Medicare Prescription Drug, Improvement, and Modernization Act (MMA) of 2003 and the Medicare Improvements for Patients and Providers Act (MIPPA) of 2008. A review of its benefits and barriers in the ambulatory care setting in the US highlights the advantages of e-prescribing in improving the safety, quality, and cost of the prescribing process in ambulatory care settings. The research suggests that e-prescribing has decreased medication errors to only one-seventh of their previous rate, resulting in potential cost savings of $140 billion to $240 billion over a decade for practices that adopt e-prescribing [4].

The increasing number of adverse medical and drug events and financial losses in the healthcare industry have put pressure on the industry to improve the quality-of-care delivery. Chu [5] focuses on the development and testing of an electronic prescription application with the aim of improving the quality of medication prescription and reducing medication errors. The research highlights the key challenges that need to be overcome for its wider adoption, including the legal status of electronic signatures, mobile connectivity, cost of technology, security, and standard drug terminology and dictionary.

Bertalan and Heja [6] focus on the implementation of the Hungarian Electronic Health system (EEsZT), with its central component being the e-prescription system. The e-prescription system allows doctors to upload their prescriptions to a central IT cloud that can be viewed by pharmacists throughout the country and provides access to patient medication history and basic medical information to improve control of adverse drug reactions (ADR) and medication compliance. The aim of this system is to address the challenges in the Hungarian healthcare industry to improve the quality-of-care delivery, including reducing medication errors and increasing medication safety, while providing benefits to all participants including medical doctors, patients, the health insurance fund and pharmacies. The implementation was preceded by a long planning process that involved consultations with partner authorities, software developers, and health professional organizations, and will occur alongside the continued use of paper-based prescriptions.

Healthcare fraud is estimated to cause billions of euros in damage each year, prescription fraud is a major issue that needs to be addressed. Tagaris et al. [7] present an electronic system aimed at preventing healthcare fraud by detecting "suspicious" prescriptions. The system was created using the Microsoft Access 2002 relational database system and tested using data from the Greek Social Security Body in the healthcare industry, with all personal information removed. The system uses rules based on medical and pharmaceutical sciences, including ATC coding, to verify the medical validity of prescriptions and prevent fraudulent activity.

Similar to other researchers, Ionescu [8] proposes an electronic prescription system that would improve the management of medical prescriptions by publicly registering them using blockchain technology. The aim of this system is to facilitate the validation process of electronic prescriptions by any pharmacist or healthcare provider. The proposed implementation is suitable for issuing and keeping track of these prescriptions, making it easier for all parties involved to access and validate the information.

Another set of researchers, Puustjarvi and Puustjarvi [9] use XML-based technologies to improve the interoperability of electronic prescription systems in the healthcare sector. Their goal is to simplify the prescription process, reduce medication errors, speed up the ordering process, improve data for research, and make financial savings. The researchers propose using Business Process Modelling Notation (BPMN) and Business Process Execution Language for Web Services (BMEL4WS) to automate the coordination of interoperability within electronic prescription systems. By illustrating how BPMN and BMEL4WS can be used to model and implement interoperability, they aim to show that these technologies are suitable for the healthcare sector as they are easily understood by employees and business analysts.

Prescribing errors are a common issue in healthcare, affecting around 9% of all medication orders. Although not all of these errors put patients' lives at risk, a harmless error could lead to undesirable side effects and affect patients' confidence in the medical treatment. To address this problem, Santos et al. [10] propose an unsupervised method called density-distance-centrality (DDC) to automatically detect outlier prescriptions in electronic health records that could lead to harmful medication dosages and frequencies for patients. This method uses graph models and previous prescriptions to learn the threshold between normal and abnormal medication doses for each medication, making it context-aware and reducing computational costs. The research is important as there is a lack of previous studies on the automatic detection of wrong medication dosages and frequencies in electronic prescriptions.

To optimise patient safety, reduce hospital costs, and optimize medical resources related to medication prescriptions, Fernandes and Belo [11] conducted a study in the northern region of Portugal. They used data mining techniques, specifically association rules, to analyze electronic prescription data of medication, doctors, and laboratories to discover medication prescription patterns.

To overcome the limitations of previous efforts to migrate to electronic dispensation of prescriptions, Ghinea et al. [12] propose a ubiquitous computing technology solution. Their research focuses on addressing the widespread use of paper-based medical prescriptions and presents a Jini-based solution for electronic prescriptions. The solution includes a wireless prototype that allows for wireless transmission of electronic prescriptions to nearby pharmacies and provides patients with information about queue lengths and estimated waiting times. Clinical and user evaluations have shown high levels of agreement regarding the prototype's effectiveness, ease of use, and usefulness.

The application of machine learning techniques in healthcare has shown promising results in improving healthcare delivery. One such area is COVID-19 telemonitoring for smart hospitals, as discussed by Shabbir et al. [13] who illustrate the use of exploratory data analysis, classification, and comparative analysis to predict the severity of COVID-19 cases. Machine learning algorithms were trained on patient data, including demographics, comorbidities, and symptoms, to predict the severity of cases. The researchers suggest that these predictive models could help identify high-risk patients and allocate resources more effectively, ultimately improving healthcare delivery.

In their 2021 study, Hossain et al. [14] aimed to identify the challenges and opportunities of effectively utilizing electronic health data for disease prediction. They argue that this approach can lead to improved patient care, cost savings, and better health outcomes. To achieve this goal, the study employs a systematic review approach to identify relevant studies and discusses issues related to data quality, privacy, and security. Additionally, the study analyzes various techniques, including machine learning algorithms and statistical models, used for disease prediction. The authors stress the importance of collaboration among stakeholders to ensure successful implementation.

Bowman and Acharya [15] utilised the NIST risk model to collect and analyse data on the top challenges in the pharmacy domain, including loss and theft of portable devices, email phishing attacks, and lack of essential software functionalities. Their study provides insight into the current state of e-prescription technology and informs the development of solutions to address the identified risks and concerns, such as human error, software integration issues, and negative patient perceptions, while also maximizing its benefits like increased patient safety and financial savings.

In 2022, Rubio et al. [16] focused on improving the safety of obstetric care by reducing medication errors in the Obstetrics and Gynaecology Service at a university hospital in Colombia. This was achieved by utilising machine learning models to predict prescriptions for 7 types of drugs for 4 of the most frequent diagnoses, using electronic medical records with approximately 10,000 hospital admissions as the database. Before the models were trained and evaluated using the evaluation metrics Sensitivity and Specificity, the textual data from the records underwent natural language processing. The objective was to decrease the frequency of medication errors and enhance patient outcomes, as errors in prescriptions and treatments during pregnancy can harm both the patient and the foetus. The findings suggest that artificial intelligence techniques, particularly machine learning, have the potential to reduce medication errors in the healthcare field.

Privacy concerns can however be a significant barrier to the adoption of machine learning techniques in healthcare systems. To address this, Padinjappurathu Gopalan et al. [17] propose a privacy-preserving scheme for disease prediction that allows hospitals to train machine learning models on their local data without sharing it with external parties. The feasibility of their approach was demonstrated using a dataset of patients with diabetes and cardiovascular disease, suggesting that privacy-preserving machine-learning techniques could improve disease prediction without compromising patient privacy.

Alhorishi et al. [18] explore the use of machine learning algorithms to predict the early preparation of pharmacy prescriptions at PSMMC hospital in Riyadh, Saudi Arabia. They investigate the predictive ability of four machine learning algorithms to predict the early preparation of pharmacy prescriptions and found that XGBoost had the best predictive performance. This suggests that machine learning algorithms can be used to accurately predict the early preparation of pharmacy prescriptions, potentially improving patient outcomes and reducing healthcare costs.

This research paper aims to use three different machine learning techniques to develop predictive models for EPS (Electronic Prescription Service) and determine which technique performs best in terms of prediction accuracy.

3 Materials and Methods

For this research, a selection of raw data related to dispensaries (such as pharmacies and chemists), EPS (Electronic Prescription Service), and eRD (Electronic Repeat Dispensing) was acquired from the National Health Service (NHS) online medical database, specifically 1,498 records across the UK. The data was split into training (80%) and testing (20%) sets and the same records were used for training and testing in all the machine learning techniques applied.

Figure 1 shows the process by which the dataset will be put through to arrive at an appropriate model for predicting EPS Utilisation.

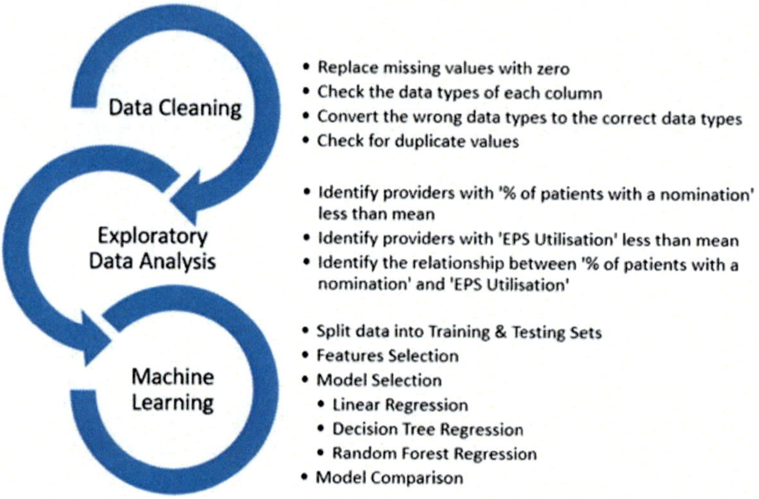

Fig. 1 Data manipulation process

3.1 Data Description

The dataset contains 1,498 records related to dispensaries, EPS, and eRD acquired from the National Health Service (NHS) online medical database in the UK. Each record includes the following variables:

- Region: The region of the UK where the dispensary is located.
- Place Code: The code assigned to the place where the dispensary is located.
- Place: The name of the place where the dispensary is located.
- ODS Code: "Organization Data Service", a unique identifier assigned to the dispensary.
- GP Practice (ODS Code): The ODS code of the GP practice associated with the dispensary.
- Registered Patients: The total number of patients registered with the dispensary.
- % of patients with a nomination: The percentage of registered patients who have nominated the dispensary as their preferred dispensing site.
- EPS Items: The number of prescription items that have been processed using the Electronic Prescription Service (EPS) at the dispensary.
- eRD Items: The number of prescription items that have been processed using the Electronic Repeat Dispensing (eRD) service at the dispensary.
- All Items: The total number of prescription items dispensed at the dispensary.
- EPS Utilisation: The percentage of prescription items processed using EPS out of all items dispensed.
- eRD Utilisation (EPS Items): The percentage of prescription items processed using eRD out of all items processed using EPS.
- eRD Utilisation (All Items): The percentage of prescription items processed using eRD out of all items dispensed.

Figure 2 is a view of the first few rows of the dataset used for this study.

3.2 Exploratory Data Analysis

Following the data cleaning process, the dataset underwent exploratory data analysis (EDA). The preliminary analysis involved:

1. Calculating the mean percentage of patients with a nomination.
2. Identifying healthcare providers with a percentage of patients with a nomination lower than the mean.
3. Determining the average EPS Utilisation.
4. Identifying healthcare providers with an EPS Utilisation lower than the mean.
5. Investigating the relationship between the percentage of patients with a nomination and EPS Utilisation.

	Region	Place Code	Place	ODS Code	GP Practice (ODS Code)	Registered Patients	% of patients with a nomination	EPS Items	eRD Items	All Items	EPS Utilisation	eRD Utilisation (EPS Items)	eRD Utilisation (All Items)
0	North East and Yorkshire	16C	Tees Valley	A81001	The Densham Surgery (A81001)	4019.0	81.7%	8120	607	8172	99.4%	7.5%	7.4%
1	North East and Yorkshire	16C	Tees Valley	A81002	Queens Park Medical Centre (A81002)	18288.0	86.3%	37318	10937	37357	99.9%	29.3%	29.3%
2	North East and Yorkshire	16C	Tees Valley	A81004	Acklam Medical Centre (A81004)	11209.0	81.4%	18558	7163	18561	100.0%	38.6%	38.6%
3	North East and Yorkshire	16C	Tees Valley	A81005	Springwood Surgery (A81005)	8032.0	83.5%	16682	7561	16727	99.7%	45.3%	45.2%
4	North East and Yorkshire	16C	Tees Valley	A81006	Tennant Street Medical Practice (A81006)	14407.0	81.9%	30425	3378	30442	99.9%	11.1%	11.1%

Fig. 2 Overview of the dataset

The findings can be summarized as follows:

1. The analysis showed that the average (mean) percentage of patients with a nomination is 0.730. Out of the total number of providers, 2,224 had a percentage of patients with a nomination less than the mean. While the mean value of 73% is relatively high, there is still room for improvement in increasing the number of patients with a nomination for those providers below the mean.
2. Based on the exploratory data analysis, the average (mean) EPS utilization is 0.917. Further analysis showed that 1,018 providers had an EPS utilization rate less than the mean value. This suggests that there is potential for these providers to increase their utilization of the Electronic Prescription Service, which could ultimately lead to improved patient care and outcomes.
3. A visual examination of the relationship between '% of patients with a nomination' and 'EPS Utilisation' using a scatter plot showed a strong positive correlation between the two variables. However, some outliers were detected in the data. See Fig. 3.
   To address outliers, two functions were defined to identify them using the Z-score and IQR methods, and applied to each column in the dataset. Outliers were then clipped at the 1st and 99th percentiles using another function, and the resulting data was stored in a new dataframe. A loop was used to clip the outliers in each column and an empty list was created to store the results. These measures were taken to ensure that any extreme values in the dataset did not unduly skew the analysis or results.
4. The strong positive relationship between '% of patients with a nomination' and 'EPS Utilisation' is further supported by a correlation coefficient of 0.80258. Figure 4 shows the correlation between the variables using a Heat Map.

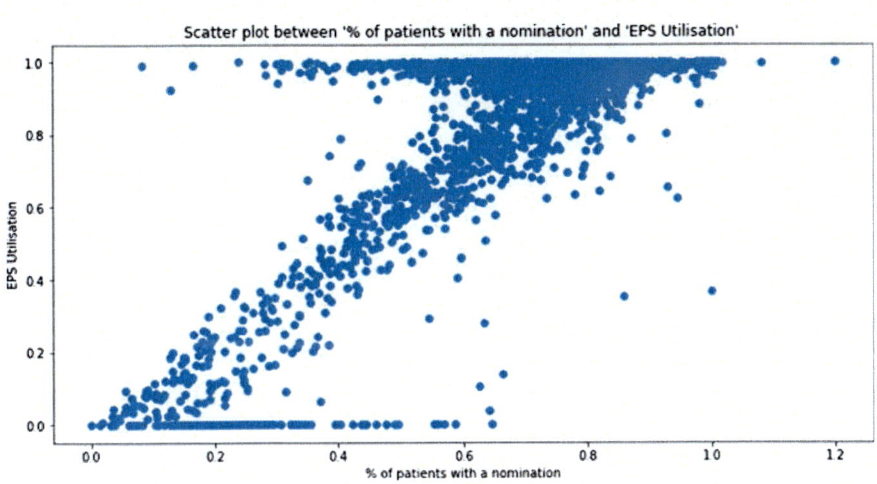

Fig. 3 Relationship between '% of patients with a nomination' and 'EPS Utilisation'

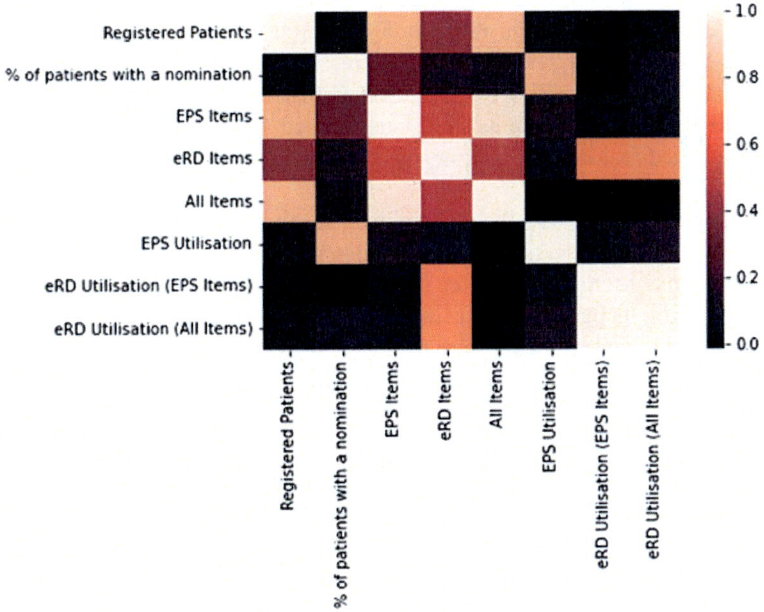

Fig. 4 Heat map showing the correlation between the selected variables

3.3 *Label Encoding*

The LabelEncoder function was used to encode categorical data in five columns ('Region', 'Place Code', 'Place', 'ODS Code', and 'GP Practice (ODS Code)'). The fit_transform() method was applied to each column separately using the same label_encoder object to transform categorical data into numerical form. After encoding, each unique category in a column is assigned a unique integer value. This process makes it easier for machine learning algorithms to process and analyze the data.

3.4 Data Preparation and Machine Learning

In order to develop a machine learning model to predict EPS utilization, the dataset was split into two sets - a training set (80%) and a testing set (20%). It is important to note that the same records were used for both training and testing in all the machine learning techniques applied.

Features selection was performed (using machine learning) to identify the most relevant features that contribute to the prediction or classification of the target variable. Two methods of feature selection are used - p-value and regularization. The SelectKBest class is used for p-value feature selection with f_regression scoring function and k value of 9. The SelectFromModel class is used for regularisation feature selection with a Ridge regression estimator with alpha value of 1.0 and maximum of 9 features. The selected features were '% of patients with a nomination','eRD Utilisation (EPS Items)','eRD Utilisation (All Items)','ODS Code', 'Registered Patients','EPS Items','eRD Items','All Items','EPS Utilisation'. The resulting selected features are concatenated and visualised using a Seaborn bar plot as seen in Fig. 5.

After splitting the data, the appropriate machine-learning algorithms were selected. In this study, linear regression, decision tree regression, and random forest regression were chosen. The next step involved training the model using the training set and evaluating its performance using the test set.

The following measurements were then used to see how accurate the predictions were. The "Score" indicates how much of the difference between the predicted and

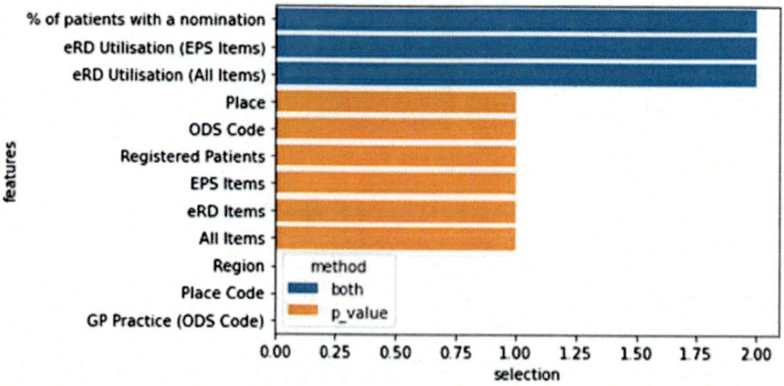

Fig. 5 Selected features

actual value can be explained by the model. The "Mean Squared Error (MSE)" and "Mean Absolute Error (MAE)" measurements indicate how close the predictions are to the actual values. Lower values of MSE and MAE mean that the program is making more accurate predictions. The "Sqrt Mean Absolute Error (MAE)", which is basically the square root of the MAE, is also used. Lastly, the "Coefficient of determination (R^2)" indicates how well the model can explain the variation in the predicted value based on the input data.

4 Results

The "Score", "Mean Squared Error (MSE)", "Mean Absolute Error (MAE)", "Sqrt Mean Absolute Error (MAE)" and "Coefficient of determination (R^2)" was calculated for each of the regression models. The respective values are shown in Fig. 6.

The Random Forest Regression model outperformed both the Linear Regression and the Decision Tree Regression models, as it had the lowest RMSE value (0.05), and the highest accuracy score (0.89). This suggests that the Random Forest model

	Model	Score	r2_score	MAE	MSE	RMSE
0	LinearRegression	0.846	0.85	0.04	0.0	0.06
1	DecisionTreeRegressor	0.849	0.85	0.03	0.0	0.06
2	RandomForestRegressor	0.888	0.89	0.03	0.0	0.05

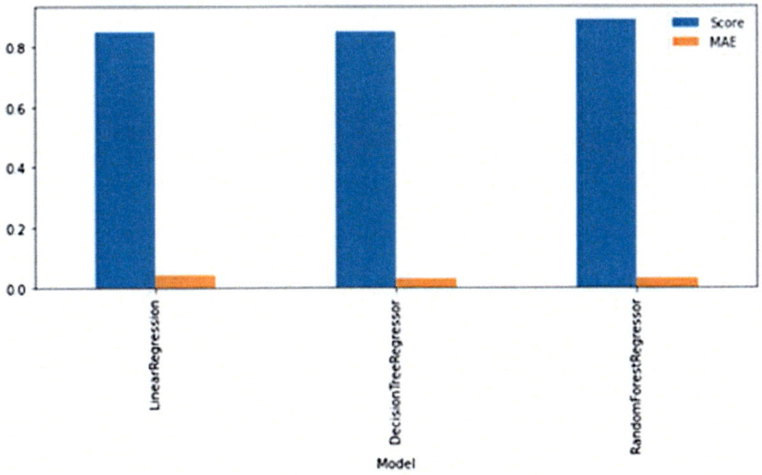

Fig. 6 Comparing the score and root mean squared error (RMSE) of the 3 models

was able to predict the target variable (EPS Utilisation) more accurately using the input features.

5 Conclusion

To summarize, this study analyzed data on dispensaries, Electronic Prescription Service (EPS), and Electronic Repeat Dispensing (eRD) from the UK's National Health Service (NHS) online medical database. The results of exploratory data analysis showed that healthcare providers could increase their use of EPS, potentially leading to better patient care and outcomes. The study also developed machine learning models to predict EPS utilization and found that the Random Forest Regression model was the most accurate, with an RMSE value of 0.05 and an accuracy Score of 0.89. Overall, the study provides insights into how healthcare providers can improve their utilization of electronic prescription services to enhance patient safety and healthcare efficiency. To improve upon this study, future research should aim to investigate the correlation between the "percentage of patients with a nomination" and "EPS Utilisation" in greater detail. Additionally, similarities between healthcare providers with a nomination percentage or EPS utilization rate lower than the mean should be identified, possibly by examining factors such as location. Obtaining additional data on repeat dispensing and comparing it to eRD utilization can help determine the proportions and ways to enhance eRD if it is found to be deficient.

Acknowledgements This work was supported by Key Laboratory of Advanced Manufacturing Technology of the Ministry of Education of Guizhou University (GZUAMT2022KF[07]), the National Natural Science Foundation of China (No.61862051), the Science and Technology Foundation of Guizhou Province (No.[2019]1299, No.ZK[2022]449), the Top-notch Talent Program of Guizhou province (No.KY[2018]080), the Natural Science Foundation of Education of Guizhou province(No.[2019]203) and the Funds of Qiannan Normal University for Nationalities (No. qnsy2019rc09). The Educational Department of Guizhou under Grant NO. KY[2019]067.

References

1. Burney, S.M.A., Mahmood, N., Abbas, Z.: Information and communication technology in healthcare management systems: prospects for developing countries. Int. J. Comput. Appl. **4**, 27–32 (2010). https://doi.org/10.5120/801-1138
2. Matheson-Monnet, C.: An evaluation of electronic repeat dispensing services in Wessex: exploring perceptions and experiences of those working in general practice, community pharmacy and people who receive medication by Electronic Repeat Dispensing Final Report (2020)
3. Leqi, S.: Master Thesis in European Union Law Patients' right and data protection-use of electronic health records in cross-border healthcare at the European Union. https://jmchaireulawglobalcontext.fll.um.edu.mo/wp-content/uploads/2022/11/research-thesis-2020-6-sha-leqi.pdf (2020)

4. Porterfield, A., Engelbert, K., Coustasse, A.: Electronic prescribing: Improving the efficiency and accuracy of prescribing in the ambulatory care setting. Perspect. Health Inf. Manag. **11**(Spring) (2014) 1g. PMID: 24808808; PMCID: PMC3995494
5. Chu, S.: EPrescription: road map from wired to wireless point-of-care order entry. IEEE Xplore (2004). https://doi.org/10.1109/HEALTH.2004.1324460
6. Bertalan, L., Heja, G.: The Hungarian Electronic Prescription—a long way from planning to introduction. IEEE Xplore (2017). https://doi.org/10.1109/NC.2017.8263271
7. Tagaris, A., Mnimatidis, P., Koutsouris, D.: Implementation of a prescription fraud detection software using RDBMS tools and ATC coding. IEEE Xplore (2009). https://doi.org/10.1109/ITAB.2009.5394458
8. Ionescu, S.V.: E-Prescription using blockchain technology. IEEE Xplore (2022). https://doi.org/10.1109/SmartBlock4Health56071.2022.10034520
9. Puustjarvi, J., Puustjarvi, L.: Automating the coordination of electronic prescription processes. IEEE Xplore (2006). https://doi.org/10.1109/HEALTH.2006.246436
10. Dos Santos, H.D.P., Ulbrich, A.H.D.P.S., Woloszyn, V., Vieira, R.: DDC-outlier: preventing medication errors using unsupervised learning. IEEE J. Biomed. Health Inform. **23**, 874–881 (2019). https://doi.org/10.1109/JBHI.2018.2828028
11. Fernandes, J., Belo, O.: Discovering patterns on medication prescriptions. IEEE Xplore (2010). https://ieeexplore.ieee.org/document/5556647
12. Ghinea, G., Asgari, S., Moradi, A., Serif, T.: A jini-based solution for electronic prescriptions. IEEE Trans. Inf. Technol. Biomed. **10**, 794–802 (2006). https://doi.org/10.1109/TITB.2006.879586
13. Shabbir, A., Shabbir, M., Javed, A.R., Rizwan, M., Iwendi, C., Chakraborty, C.: Exploratory data analysis, classification, comparative analysis, case severity detection, and internet of things in COVID-19 telemonitoring for smart hospitals. J. Exp. Theor. Artif. Intell. 1–28. (2022). https://doi.org/10.1080/0952813X.2021.1960634
14. Hossain, M. E., Khan, A., Moni, M.A., Uddin, S.: Use of electronic health data for disease prediction: a comprehensive literature review. IEEE/ACM Trans. Comput. Biol. Bioinf. 18 (2), 745–58 (2021) https://doi.org/10.1109/tcbb.2019.2937862
15. Bowman, M., Acharya, S.: Risk assessment of pharmacies & electronic prescriptions. In: Proceedings of the 2019 IEEE/ACM International Conference on Advances in Social Networks Analysis and Mining (2019). https://doi.org/10.1145/3341161.3343697
16. Rubio, A.M., Alzate Grisales, J.A., Marín Barrera, A.Y., Ruiz Delgado, A., Aguirre Ospina, O.D., Zuluaga, N.A., Restrepo Franco, A.M.: Using machine learning techniques to predict a list of prescription medications in the obstetrics and gynecology service. IEEE Xplore (2022). https://doi.org/10.1109/ColCACI56938.2022.9905350
17. Padinjappurathu Gopalan, S., Chowdhary, C.L., Iwendi, C., Farid, M.A., Ramasamy, L.K.: An efficient and privacy-preserving scheme for disease prediction in modern healthcare systems. Sensors **22**(15), 5574 (2022). https://doi.org/10.3390/s22155574
18. Alhorishi, N., Almeziny, M., Alshammari, R.: Using machine learning to predict early preparation of pharmacy prescriptions at PSMMC—a comparison of four machine learning algorithms. Acta Informatica Medica **29**(1), 21 (2021)

A Review on Blockchain Smart Contract Applications

Sarvesh Tanwar, Neelam Gupta, Jai Singh Tomar, Karan Kumar, Silvia Treľová, and Iryna Ivanochko

Abstract Due to the high level of customization they provide to transactions, smart contracts have become one of the many innovations since the advent of the blockchain. This has led to the development of numerous smart contract applications in industries as diverse as financial services, voting, and the life sciences. However, because they are still in their infancy, smart contracts continue to present several difficulties for all parties involved in their interaction, encompassing smart contract consumers, programmers, and enterprises. This paper aims to add to the source of information regarding blockchain technology's use of smart contracts. This document's main goal is to act as a reference and manual for ongoing research initiatives. The study looks into significant uses of smart contracts that have previously derived from them. From the standpoint of these applications, we also highlight the great possibilities of blockchain-based smart contracts.

Keywords Blockchain · Smart contracts · Cryptocurrency · Applications · Traditional contracts

S. Tanwar · N. Gupta · J. S. Tomar
Amity Institute of Information Technology, Amity University, Uttar Pradesh, Noida, India

K. Kumar (✉)
Electronics and Communication Engineering Department, Maharishi Markandeshwar Engineering College, Maharishi Markandeshwar (Deemed to be University), Mullana, Ambala 133207, India
e-mail: karan.170987@gmail.com

S. Treľová · I. Ivanochko
Faculty of Management, Comenius University in Bratislava, 25, Odbojárov 10, 82005 Bratislava, Slovakia

I. Ivanochko
Department of Management and International Business, Institute of Economics and Management, National University "Lviv Polytechnics", Lviv, Ukraine

C. Iwendi et al. (eds.), *Proceedings of ICACTCE'23—The International Conference on Advances in Communication Technology and Computer Engineering*, Lecture Notes in Networks and Systems 735, https://doi.org/10.1007/978-3-031-37164-6_13

1 Introduction

In our society, trust has traditionally been built through middlemen. We rely on these third-party organizations because we believe they will be able to preserve and safeguard our products although sending the appropriate quantity at the appropriate time. By shifting trust to decentralized systems, blockchain eliminates the need for middlemen. Central banks or law firms are most impacted by blockchain innovation, as it changes their marketing strategy from charging per hour to charging per item. It is now required to charge by the thing rather than by the hour because it shifts their business model from charging by the session to charging by the item. The concept of smart contracts is a possible approach with several uses that advances blockchain technology. Due to the near-instantaneous nature of blockchain transactions, individuals are charged per transaction. Traditional transactions [1], however, take longer, cost more money, and pose a security risk because they have only one point of failure. a glimpse of a text corpus that was built up utilizing the writers' most used terms throughout the study, as shown in Fig. 1.

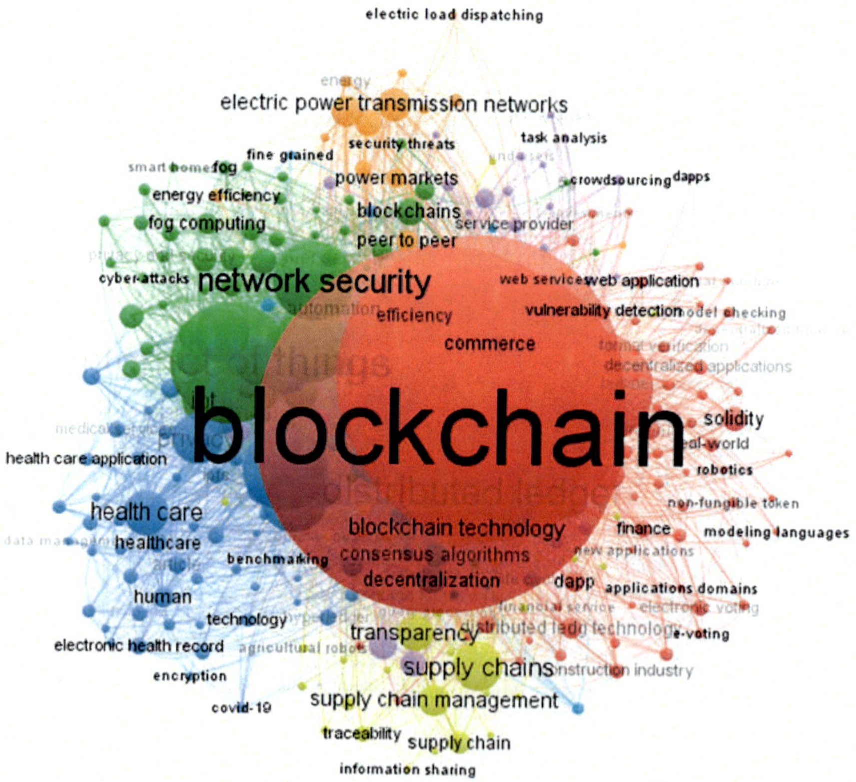

Fig. 1 Word cloud created throughout the investigation using the authors' most popular keywords

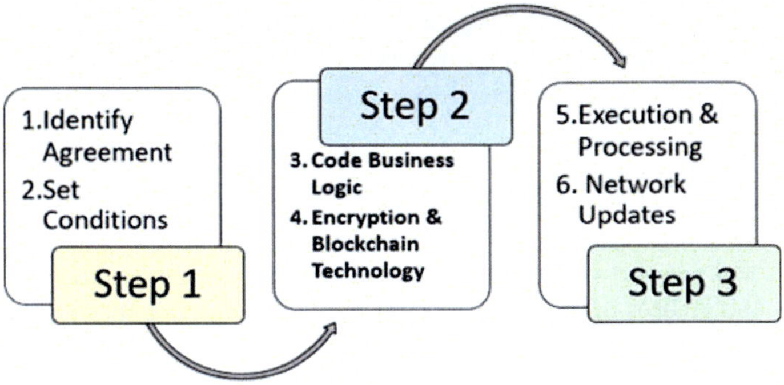

Fig. 2 Process of smart contract in blockchain

An arrangement among unreliable parties can be facilitated, carried out, and enforced via a smart contract, which is blockchain-based code. Figure 2 depicts the smart contract's workflow on the blockchain. One way to conceptualize it is as a mechanism for distributing cryptographic certification to either all or certain of the necessary parties. Due to the absence of a trustworthy third party, smart contracts function differently than traditional contracts and have reduced transaction costs. The most popular blockchain platform for developing smart contracts is Ethereum, which enables more complex and customized contracts. These include smart properties, e-commerce, and music rights administration. Ethereum's language is Turing-complete, allowing for the creation of more complex and individualized contracts.

This essay tries to offer a thorough evaluation of current research on blockchain-based smart contracts. We want to look into problems with smart contracts on the blockchain. The approach chosen to deal with this is a methodical mapping strategy. Through the mapping study, we can learn about the issues and pinpoint potential fixes for those issues [2]. It can therefore offer crucial advice to all types of stakeholders who seem to be active in the problem under study. The platforms, programming languages, and blockchain technology all contribute to the issues.

This paper is organized as follows: Sect. 2 provides background of features of blockchain based smart contract and platforms for smart contracts. The literature review is presented in Sect. 3. The comparison of traditional contracts and smart contracts is displayed in Sect. 4. Section 5 addresses smart contract applications. This paper is concluded in Sect. 6.

2 Background

Smart contracts were first introduced by Nick Szabo and are now used across many industries. Ethereum is one of the best-known platforms. ERC20 tokens and other financial applications were the only ones for which smart contracts were first designed. Smart contract infrastructures had already changed throughout time as a result of changing industrial requirements.

2.1 Feature of Blockchain-Based Smart Contracts

Smart contracts use software protocols and intellectual machinery to implement the conditions and clauses of a particular agreement. They are identity and conscience programs. Smart contracts [3] are decentralized applications that augment the functionality of the blockchain connectivity. The features of blockchain based smart contracts in blockchain is shown in Fig. 3.

Smart contracts are immutable and cryptographically verified, allowing them to be used in a variety of situations. They operate in peer-to-peer mode, free from the meddling of a central third party, and allow automated transaction execution when certain conditions are met. These features are key features of blockchain-based smart contracts. They deliver services without needing a centralized server. The blockchain technology that powers smart contracts imparts some traits on them. Because of these features, smart contracts can be applied in a number of situations.

2.1.1 Separation of Autonomous Execution and Trusted Third Party

Blockchain-based smart contracts provide decentralisation, eliminating the need for middlemen and reducing management fees and the legitimacy of substances. The need for dependable middlemen like dealers, intermediaries, or network operators can be eliminated when a system is connected with blockchain-based smart contracts. The management fees and legitimacy of substances in the form will decrease by

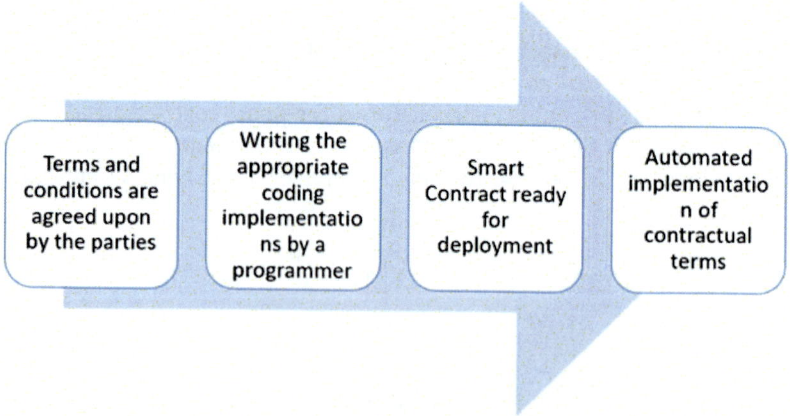

Fig. 3 Features of smart contracts in blockchain

getting rid of a reliable third party. One model of a permissioned blockchain that modifies the role of dependable third parties is Bitcoin [4], such as central banks. The centralised third parties act as the final regulatory authority and incur high transaction costs. Bitcoin users are required to abide by the guidelines established by the centralised authority, and the execution is automated so that a centralised third party is not required.Because of their community functioning and lack of dependence on a central third party, smart contracts are a great alternative for scenarios that call for substitutes for third-party trust. Because the administration is completely impartial rather than relying on a single third party, service availability is ensured. Without prejudice or human error, autonomous execution in accordance with the circumstances ensures operation accuracy [5].

On the other hand, smart contracts increase democracy by allowing participants to choose their own agreement process. The rules and guidelines for creating smart contracts are decided upon by the parties, and the predicted circumstances and flow of actions will take place after the blockchain has reached a specified, predefined state. The particular state will be generated in the smart contract once all blockchain network users have consented. The rules and guidelines for creating smart contracts [6] are decided upon by the parties, who then put them into effect. Smart contracts are a promising solution for applications that require substitutes for third-party trust. This state could be any circumstance, like a predetermined wallet balance or a set amount of time, for instance. After that, the execution is automated so that no outside help is needed. Service availability is guaranteed because the operation is community-based [7] and does not depend on a centrally managed third party. Autonomous execution in accordance with the circumstances ensures operation accuracy.

2.1.2 Forge Resistance and Immutability

In distributed ledgers, cryptographic algorithms are used to confirm the accuracy of transaction records.Before being recorded in the ledger, particular trades have also been reviewed and approved. The ever-expanding ledger's immutable transactions are approved transactions. People are unable to bring about the change. Code for smart contracts is unchangeable on the blockchain. Various strategies [8] can be used to deploy the code on each node. As an illustration, consider the executable that the container contains. Tampered smart contracts cannot be performed because the smart contract code is hacker-proof. Smart contracts are hacker-proof and can be modified with the support of blockchain network nodes. The executed code incorporates the logic of each bitcoin blockchain member, allowing all parties to trust the smart contract.

2.1.3 Transparency

It is one of the key characteristics that smart contracts get from blockchain. The smart contract's transparency is twofold. To begin with, the code written in smart contracts is visible to both concerned parties [9] and the public. Second, the blocks' transactions are likewise visible to the public.

As a result, the blockchain network's intervening players can have faith in its reasoning and transactions. According to one particular situation, if the smart contract philosophy is created by a ruling authority that is a member of the Ethereum blockchain [10], the work done on the rationale can be viewed as trustworthy and unbiased because the source is available to the public.

2.2 *Smart Contract Platforms*

Smart contracts can be created and used on a number of blockchain platforms, such as NXT, Cryptocurrency, and Hyperledger Fabric. Using high-level programming languages, which offer capabilities like contract programming abilities, code execution, and authentication methods, smart contracts can be created and put into use. The development of smart contracts utilizing high-level programming languages is possible on a number of platforms.

- Although it has a very small amount of CPU capacity, Bitcoin is a public blockchain platform for processing bitcoin transactions. Stacks serve as the foundation of the scripting language for Bitcoin [11]. Stacks serve as the foundation of the scripting language, but the Bitcoin programming language has few capabilities for creating logic-based smart contracts. Mining operations and compensation packages need to be adjusted to support smart contracts.
- Open-source blockchain platform NXT only has a methodology for proof-of-stake confirmation.

Its dependency on pre-made templates prevents it from being Turing-complete and prevents the deployment of customized smart contracts. There is a list of active smart contracts included in it.

- The Ethereum [12] virtual processor, a Turing-complete virtual machine, is the first blockchain platform for building smart contracts (EVP). Every node in the Ethereum network runs an implementation of the EVM and executes the same commands, acting as the execution environment for smart contracts. Smart contracts are created using the high-level programming language Solidity and distributed on the blockchain for execution after being converted to EVM bytecode [13]. The most well-known smart contract platform is Ethereum, which enables programmers to create decentralized apps (DApps) in a variety of industries.
- Hyperledger Fabric: Smart contracts are supported by the open-source, corporation-decentralized blockchain proprietary technology known as Hyperledger Fabric. The network of this permissioned blockchain is made up of peers [14] that are owned and contributed by the set of enterprises that can only join through such an affiliation service provider. Using plug-and-play components, it provides adaptation and versatility for a number of business applications. The

modular architecture of Hyperledger Fabric allows a range of enterprise use cases using plug-and-play components. Compared to publicly accessible blockchains like Bitcoin and Ethereum, where anybody can join the network, Hyperledger Fabric is a permissioned blockchain.

There are numerous differences between smart contracts on Ethereum and Hyperledger Fabric. Hyperledger Fabric supports smart contracts written in a variety of languages, unlike Ethereum, which only supports Solidity. Any miner who receives a transaction containing contract code in Ethereum [15] can execute it in their own virtual machine, and transactions are transmitted throughout the peer-to-peer network. A transaction made by an application using Hyperledger Fabric has only been examined and acknowledged by particular peers (endorsing peers). Following receipt of the application's transaction proposal, each of these endorsing peers independently runs the transaction's chain code [16]. For security and isolation, chain code operates in a container environment (like Docker).

3 Literature Review

Nick Szabo coined the phrase "smart contracts" in 1994 to describe mechanized transaction protocols that execute contractual terms. According to some authors, smart contracts are software that automatically applies parameters that two or more parties can agree to when agreeing to something in an unreliable environment, using the reliable computation capabilities of a blockchain network. Another definition of smart contracts is computer programmes that can be operated even without the engagement [17] of a dependable authority in a network of generally trustworthy nodes, which includes the blockchain. They are useful in sectors like financial services that involve money transfers based on norms that are mutually agreed upon. These smart contracts are helpful in a number of situations due to their resilience to manipulation.

Digital contracts, known as smart contracts," enable the creation of temper-proof clauses based on a decentralized agreement and through autonomous system, are conscience. However, they lack the legal standing of conventional contracts and do not make use of artificial intelligence resources. Smart contracts are simple programs that try to automate actions based on if–then conditions. They are recorded and implemented in various devices that are connected in a peer-to-peer network and are capable of fulfilling contractual obligations. Software called "smart contracts" [18] implements a logical flow of events in line with predefined terms and regulations. They are made up of three parts: the mathematical code that represents the contractual logic, the collection of messages that can be used to receive and represent events, and the set of procedures that will trigger the reactions foreseen by the contract logic [19].

One advantage of smart contracts over conventional contracts is the reduction of risk because smart contracts cannot be amended arbitrarily once they have been issued because blockchains are immutable. Additionally, all transactions conducted and maintained on distributed blockchain systems are auditable and traceable. Hence, malicious activities [20–23], for instance, financial cybercrime, can be quitedecreased. Administrative and service costs are reduced because smart contracts stored on these networks can be automatically activated in a decentralized manner and because blockchains guarantee the integrity of the entire system through distributed consensus mechanisms that don't go through a central mediator or intermediary. As a result, costs associated with the operation and upkeep could be greatly decreased. The efficiency of company processes can be increased by reducing administration and maintenance costs by eliminating intermediaries. For example, cash payments between users of the peer-to-peer network can happen instantly when the predefined conditions in smart contracts on a blockchain are satisfied, reducing transaction return time.

Although there is still debate in the literature as to whether financial intermediaries can be given the same legal standing as traditional contracts, some authors think that they can serve as a sort of tutelage without the need for judicial intervention. When particular requirements are met, the provisions will be implemented. Smart contracts are instruments [24] in the judicial framework used to create, alter, or terminate a contractual entity between multiple participants in the memorandum of understanding, and reference implementation interactions have legal consequences that are analogous to those acquired via traditional contracts. The parties' statement of will must be free of vices and be a requirement for the legal enforceability of smart contracts. Additionally, smart contracts have a fiduciary aspect, which ensures that the conditions of the arrangement will be upheld. They have legal effects comparable to conventional contracts but must be free of vices and have a fiduciary aspect. They are also fully digital, allowing them to be used to carry out non-digital assets like real estate, intellectual property rights, and so forth. The fact that this more recent version of the contract is properly integrated [25] in software and smart contracts have a dual legal nature allows them to be regarded as both intellectual property rights and legally binding documents. They increase the likelihood of contractual clauses being carried out, are development tools, and are self-sufficient, meaning they do not need to be registered with intermediaries.

Our data shows that the publishing of blockchain-based smart contract research at conferences and journals has been increasing since 2019. This document compiles 325 publications together with the present condition of the research in the topic. It is arranged according to the most significant publication-related sources, platforms, techniques, and strategies. The search approach should be thorough, impartial, simple, and repeatable (Fig. 4 and Table 1).

4 Traditional Contracts Versus Smart Contracts

Consider the definition of a contract for a moment before we begin the comparison. A contract is a signed agreement between two or more parties that outlines the conditions under which one commodity of value is exchanged for another [26].

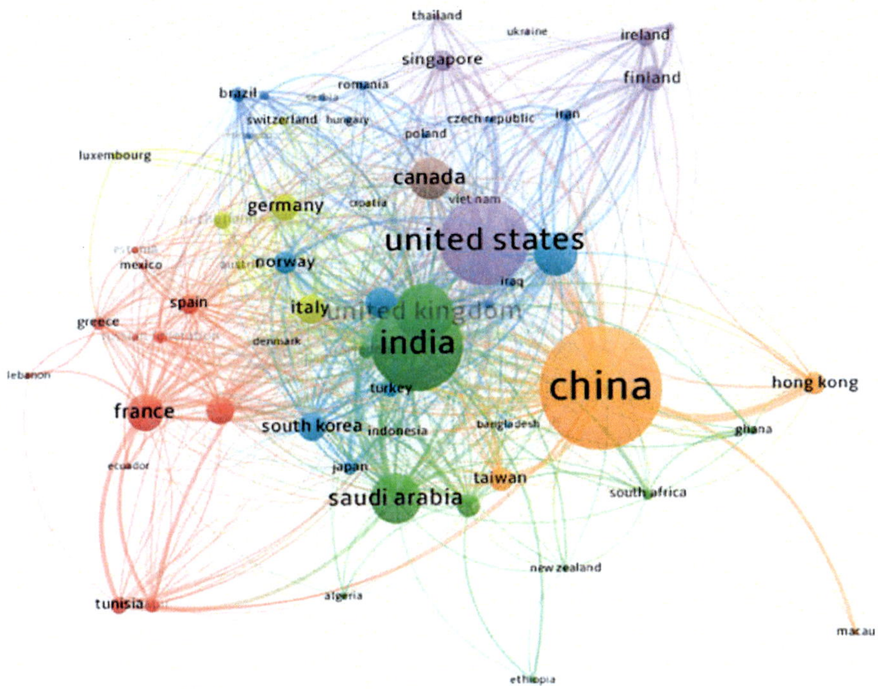

Fig. 4 Total number of documents of an author by country

Table 1 Total number of documents by year

Year	Number of documents
2023	68
2022	973
2021	884
2020	737
2019	573
2018	262
2017	56
2016	13

Theoretically, it may be enforced by law, but this depends on the form of the document (oral or written) and its status. For example, Unlike an employment contract, which must be legally binding, an agreement involving acquaintances is not required to be so. Moreover, a contract must be made legally in order to be valid [27]; otherwise, it cannot be enforced. The advantage of smart contracts is that they do not need the approval of a foreign entity to verify their legality. This permits the parties to enter into a contract that would save both time and funds for that particular transaction.

Furthermore, smart contracts also offer a considerable advantage in terms of validity. Conventional contracts can be falsified or altered if they are not properly preserved or evaluated by a qualified expert., as shown in Fig. 5. The final agreement can be significantly impacted by a few word changes. The immutability of smart contracts [28] contrasts with this because they are stored on a blockchain. All agreements are consequently distributed among the network's numerous nodes and timestamped. As a result, it is impossible to modify the words once they have been uploaded to the system. The comparison between traditional contract and smart contract as shown in Table 2. Traditional contracts also have the drawback of allowing one party to sign them at both ends without informing the other. On the other hand, smart contracts [29] cannot be implemented since they necessitate the use of a biometric key authentication that is known only to the person in control of it.

Fig. 5 Traditional contract versus smart contract

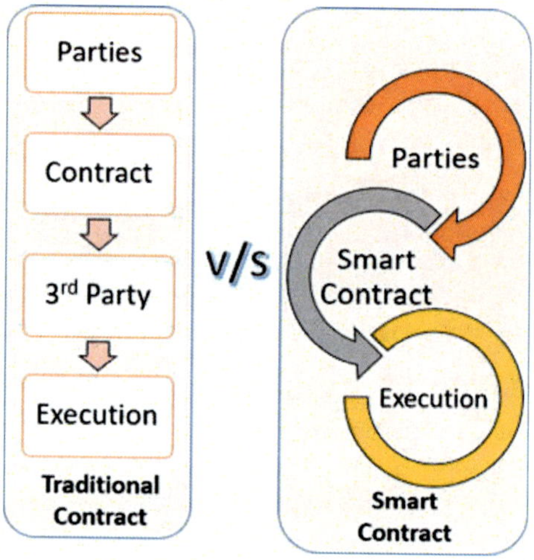

Table 2 Comparison between smart contract and traditional contract

Smart contract	Traditional contract
Immutable	Mutable by anyone
Digital signature	Wet signature
Automatic	Manual
Expensive	Cost effective
Fast and efficient	Time consuming
Decentralised	Centralised
No middlemen	Needs human intervention
Layers not necessary	Layers necessary

5 Smart Contract Applications

A variety of applications are possible for smart contracts. Here are a few of these applications:

- **Internet of things and intelligent real estate**: The Internet connects billions of nodes, and smart contracts built on blockchain technology can be used to share or use digital assets without a middleman.This application is being looked into by numerous companies [29]. Slock is a German company that uses Ethereum smart contracts to rent, buy, and sell anything without resorting to a third party.
- **Music right enforcement**: Blockchain-based smart contracts can be used to record ownership and control of a piece of music and to pay producers when their works are used for profit [30]. It also makes sure that the money is distributed equally among the proprietors of the song. A startup called Ujo is investigating the potential applications of these smart contracts in the music industry [31].
- **E-commerce**: Smart contracts may be implemented to facilitate trade between untrustworthy parties without the involvement of a trustworthy third party, lowering trading costs. Trading costs [32] would go down as a result. Only in the event that the buyer is satisfied with the product or commodity they acquired can smart contracts release the vendor of their money.

Additional applications [33] include digital rights management, supply chain management, distributed file storage, voting systems, down payment, and vehicle insurance, as shown in Fig. 6.

Fig. 6 Applications of smart contract

6 Conclusion and Future Work

In their most basic form, smart contracts are computer programmes that execute conditional operations within computational algorithms to carry out the provisions of a contract. As can be seen, the term "smart contract" was first used in 2008, but it wasn't until Blockchains, like the Ethereum platform, became the most popular technology for this kind of electronic agreement. Other solutions still permit its use. Other solutions, like the well-known Bitcoin [34, 35], still permit its use, but with some restrictions. Smart contracts are used to transfer value, register assets, manage blockchain addresses, and create utility functions. They can also be used to establish, alter, or end legal relationships [36–38]. Some of the sections of smart contracts listed in the literature include economic, user-written, asset allocation, and library smart contracts, which are used for value transfer operations, the registration of assets and immaterial rights, the management of operations involving blockchain addresses, and the creation of utility functions reusable by other smart contracts. Digital contracts are similar to conventional contracts, but they are entirely digital, implemented properly in development tools, and dual in nature, with the source code serving as both a trademark and the contract serving as a binding legal document. Additionally, they increase the level of execution confidence, use software to protect the result, are self-executing and self-sufficient, and do not require prior registration.

References

1. Wood, G.: Ethereum: a secure decentralised generalised transaction ledger. Ethereum Project Yellow Paper **151**, 1–32 (2014)
2. Zheng, Z., Xie, S., Dai, H.-N., Chen, W., Chen, X., Weng, J., Imran, M.: An overview on smart contracts: Challenges, advances and platforms. Futur. Gener. Comput. Syst. **105**, 475–491 (2020)
3. Bashir, I.: Mastering blockchain: distributed ledger technology, decentralization, and smart contracts explained (2018)
4. Norta, A.: Designing a smart-contract application layer for transacting decentralized autonomous organizations. In: International Conference on Advances in Computing and Data Sciences, Springer (2016)
5. Mancini-Griffoli, T., Peria, M.S.M., Agur, I., Ari, A., Kiff, J., Popescu, A., Rochon, C.: Casting light on central bank digital currency (2018)
6. Shapiro, E.: Point: foundations of e-democracy. Commun. ACM **61**(8), 18–28 (2018)
7. Gupta, S.S.: Blockchain. John Wiley & Sons, Inc (2017)
8. Rizal Batubara, F., Ubacht, J., Janssen, M.: Unraveling transparency and accountability in blockchain. In: Proceedings of the 20th Annual International Conference on Digital Government Research (2019)
9. Nugent, T., Upton, D., Cimpoesu, M.: Improving data transparency in clinical trials using blockchain smart contracts. F1000Research **5** (2016)
10. Yiannas, F.: A new era of food transparency powered by blockchain. Innovations Technol. Governance Globalization **12**(1–2), 46–56 (2018)
11. Farnaghi, M., Mansourian, A.: Blockchain, an enabling technology for transparent and accountable decentralized public participatory GIS. Cities **105** (2020)
12. Nakamoto, S.: Bitcoin: a peer-to-peer electronic cash system (2008)

13. Lewis, A.: A gentle introduction to smart contracts (2016)
14. Nxt community, Nxt whitepaper (2016)
15. Androulaki, E., Barger, A., Bortnikov, V., Cachin, C., Christidis, K., De Caro, A., Enyeart, D., Ferris, C., Laventman, G., Manevich, Y., et al.: Hyperledger fabric: A distributed operating systemfor permissioned blockchains. In: Proceedings of the Thirteenth EuroSys Conference (2018)
16. Szabo, N.: Smart contracts (1996)
17. Pinna, A., et al.: A massive analysis of ethereum smart contracts empirical study and code metrics (2019)
18. Bartoletti, M., Pompianu, L.: An empirical analysis of smart contracts: platforms, applications, and design patterns (2017)
19. Cong, L.W., He, Z.: Blockchain disruption and smart contracts (2019)
20. Jaiswal, M.: Cryptocurrency an era of digital currency. Int. J. Creative Res. Thoughts **8**(1), 60–70 (2020)
21. Sharma, R., Mogha, M., Tanwar, S., Rana, A.: Emerging part of industry 4.0: The digital and physical technology. In: 2020 9th International Conference System Modeling and Advancement in Research Trends (SMART), pp. 149–154. IEEE (2020)
22. Gundaboina, L., Badotra, S., Tanwar, S.: Reducing resource and energy consumption in cryptocurrency mining by using both proof-of-stake algorithm and renewable energy. In: 2022 International Mobile and Embedded Technology Conference (MECON), pp. 605–610. IEEE (2022)
23. Tanwar, S., Badotra, S., Gupta, M., Rana, A.: Efficient and secure multiple digital signatures to prevent forgery based on ECC. Int. J. Appl. Sci. Eng. **18**(5), 1–7 (2021)
24. Zheng, Z., et al.: An overview on smart contracts: Challenges, advances and platforms (2019)
25. Savelyev, A.: Contract law 2.0: 'Smart' contracts as the beginning of the end of classic contract law (2017)
26. Oliva, G.A., Hassan, A.E., Jiang, Z.: An exploratory study of smart contracts in the Ethereum blockchain platform (2020)
27. Christidis, K., Devetsikiotis, M.: Blockchains and smart contracts for the internet of things. IEEE Access **4**, 2292–2303 (2016)
28. Egbertsen, W., Hardeman, G., van den Hoven, M., van der Kolk, G., van Rijsewijk, A.: Replacing paper contracts with ethereum smart contracts (2016)
29. Banasik, W., Dziembowski, S., Malinowski, D.: Efficient zero-knowledge contingent payments in cryptocurrencies without scripts. In: European Symposium on Research in Computer Security (2016)
30. Bank, E.C.: Virtual currency schemes—a further analysis (2015)
31. Zheng, Z., Xie, S., Dai, H., Chen, X., Wang, H.: An overview of blockchain technology: architecture, consensus, and future trends. In: IEEE International Congress on Big Data (BigData congress) (2017)
32. Macrinici, D., Cartofeanu, C., Gao, S.: Smart contract applications within blockchain technology: a systematic mapping study. Telematics Inform. **35**(8), 2337–2354 (2018)
33. Wang, S., Yuan, Y., Wang, X., Li, J., Qin, R., Wang, F.Y.: An overview of smart contract: architecture, applications, and future trends. In: 2018 IEEE Intelligent Vehicles Symposium (IV), pp. 108–113. IEEE (2018)
34. Hewa, T., Ylianttila, M., Liyanage, M.: Survey on blockchain based smart contracts: applications, opportunities, and challenges. J. Netw. Comput. Appl. **177**, 102857 (2021)
35. Alharby, M., Aldweesh, A., Van Moorsel, A.: Blockchain-based smart contracts: A systematic mapping study of academic research. In: 2018 International Conference on Cloud Computing, Big Data and Blockchain (ICCBB), pp. 1–6. IEEE (2018)
36. Tanwar, S., Gupta, N., Iwendi, C., Kumar, K., Alenezi, M.: Next generation IoT and blockchain integration. J. Sens. **2022**, 9077348 14 pages, (2022). https://doi.org/10.1155/2022/9077348
37. Reiff, N.: What is ERC-20 and What does it mean for Ethereum? (2020)
38. Macrinici, D., Cartofeanu, C., Gao, S.: Smart contract applications within blockchain technology: a systematic mapping study (2018)

Carbonic Acid Gas Emission Rating by Vehicles Using Datascience Techniques

C. Prathiksha, Christy Jackson Joshua, Md. Abdul Quadir, Senthilkumar Mohan, and Celestine Iwendi

Abstract One factor contributing to the warming of the upper orbit is the rollout of man-made pollutants into the eco system (biogas, Dioxide, laughing gas, and so on). Approximately 14% of total worldwide carbon dioxide emissions are attributed to the road transport. Wheels dust are dangerous to us and contain global warm gases that leads to changes in climate. Products of gas and diesel fuels that include NO_2, CO, CH, C_6H_6, CH_2O. Wheels also emit CO_2, common human-caused global warm gas. It has been set emission targets to dramatically reduce highway's contribution to Dioxide. These are inferred from the global weather conference's goal of keeping the peak warming of the planet to a maximum of 2 degrees Celsius until 2100. In order to accomplish, in this study, a machine learning hybrid algorithm was developed in the combination of many classifications' algorithm to find the vehicle CO_2 emission with high accuracy rate. The results show that hybrid models can produce more accuracy with a lower error rate when developing an application for emission rating. Accurate carbon emission prediction models can aid in the development of emission-reduction policies.

Keywords Carbon dioxide · Emission value · Hybrid models · Classifications

---

C. Prathiksha · C. J. Joshua (✉) · Md. Abdul Quadir
School of Computer Science and Engineering, Vellore Institute of Technology, Vellore, India
e-mail: christyjackson.j@vit.ac.in

S. Mohan
School of Information Technology and Engineering, Vellore Institute of Technology, Vellore 632014, India

C. Iwendi
School of Creative Technologies, University of Bolton, Bolton BL3 5AB, UK

C. Iwendi et al. (eds.), *Proceedings of ICACTCE'23—The International Conference on Advances in Communication Technology and Computer Engineering*, Lecture Notes in Networks and Systems 735, https://doi.org/10.1007/978-3-031-37164-6_14

1 Introduction

1.1 An Overview

Our personal automobiles are a major contributor to global warming. Automobiles account for nearly one-fifth of total emissions, emitting approximately 2 pounds of carbon dioxide and other greenhouse gases per gallon of gasoline. The extraction, production, and delivery of fuel contribute about 5 pounds, but the majority of heat-trapping emissions (over 19 pounds per gallon) come directly from the car's tailpipe. Today's average gas mileage is thought to be around twenty two miles per gallon, or about eleven five hundred miles/year. One gal of gas emits around 8,887 grammes of CO_2. The automotive industry made a voluntary commitment to reduce emissions from new vehicles by 25% by 2008 more than 20 years ago (1998). At the time, CO_2 emissions from a new car were around 203 g/km. It currently stands at around 170 g/km and is not expected to rise above 10 g/km by 2020. A typical passenger vehicle emits approximately.6 tons of CO_2 per year. This figure is subject to change based on fuel consumption, fuel consumption, and annual vehicle mileage. The lower the prediction accuracy, the more controlled and uncontrolled factors affecting CO_2 properties there are. Nonetheless, several experimental designs have been proposed that take into account controllable effect variables and their interaction terms. A machine learning model application that uses the car's attributes to indicate if the car has exceeded his CO_2 threshold was developed to predict a car's gas emissions. Carbon dioxide and other gas-es trap heat in the atmosphere, raising global temperatures. Although there are natural sources of carbon dioxide, such as volcanic eruptions, human activity has increased it from 280 ppm before industrial development to 370 ppm today. Preparing to tests various original model takes a worthy time and effort.

Moreover, method is likely to manmade error, small minor errors can significantly raises latency. Cars smoke has fast-reaching consequences, reflecting earth quality. Carbon dioxide, another exhaust gas, because it interferes with the ability of blood to carry oxygen, it is especially dangerous for infants and heart patients. Sulfur dioxide, benzene and formaldehyde is another harmful automotive pollutant to human health, Car noise is also harmful, impairing hearing and causing mental illness. ML techniques have grown in popularity in recent years because they enable researchers to improve the accuracy of their predictions of fuel burn and CO_2 emission thresholds. ML methods were used to improve the prediction accuracy of fuel emissions and data obtained from literature sources. A classification model is typically used to see future CO_2 from automobiles. The ML model predicts CO_2 emissions based on input data such as vehicle details. In addition, if his predicted CO_2 emission level exceeds the threshold, RTA personnel will request that the registered car owner repair it. By combining multiple classifiers, ensemble method improves the learning results. When comparing to each model, it can produce robust performance. Main challenge to obtain a base model that introduces various types of errors rather than a highly accurate base model. When using an ensemble for classification, for example, high accuracy can be obtained even if the base classifier has low accuracy if different base models misclassify different training samples. Method Ensembling is classified into 2 types: boost and bag. These two combine estimates from various models, reducing the variance of a single.

To see results, a more stable model can be developed. Bagging is an ensemble meta-machine learning algorithm designed to improve the stability and accuracy of ma-chine learning algorithms used in statistical classification and regression. It aids in reducing variance and avoiding overfitting. Typically used in DT methods. Bagging is a substitute of the model average method. Assume a set T of t-tuples is chosen for each loops I of the set DI of di-tuples over row scans using a permutation method from D (note that elements from different d-tuples may be repeated) (i.e. Bootstrap). The classifier model Mi is then trained for each training set D and It. I. For each classifier Mi, the class predictions are returned. M*, the bager always does vote-counting and allot each class with the highest number of votes. Boost is a model which has ensembe techniques that are attempted to construct a powerful learning algorithm from just a huge array of weak concepts. It was accomplished by constructing a model from a set of weaker configurations. First, create a model with training sample. To attempt to fix the mistakes in the first method, a supplementary model is created. Process is repeated until the entire training dataset is correctly predicted or the maximum number of models is reached.

2 Literature Review

A novel GDP prediction technique based on transfer learning using CO_2 to analyse relationship between gas emission and gross domestic product per capita GDP of a nation using CO_2 emission, soft computing technique is used to predict a compact solution for multivariable parameter and compare the accuracy of the algorithm. They proposed new technique called transfer learning is termed as domain adapted transfer learning, it's costly and it increase competition between food and fuel. The result is extreme learning machine is more effective compare to other algorithm... [1]. They used four levels of analytics methodology to study and com-pare fuel consumption and CO_2 emissions from different models, the use of description of statistics analysis, hypothesis testing analysis, advanced analytics, and supervised learning are all examples of quantitative tests. They used polynomial regression, multiple linear regression, multivariate polynomial regression, convolutional neural network, linear regression, univariate polynomial regression, it takes more time to run, demonstrates that univariate polynomial regression is the best model for predicting from a single vehicles which featured input is up to 90% accuracy [2]. Prediction carbon dioxide emission using ML, this paper tries to give info about CO_2 level gas emission using ml, they used support vector machine and random forest given dataset is not correct,it leads to model failure, multiple model decreases efficiency,less quantity of data achieves more accuracy, both model gives almost better accuracy to the predicted topic.

In [3] CO_2 emission information in yangte-River economics, it analyses the yangte River economics bend as well as its carbo highs and emission. They used support vector machine to train and realize the experimental prediction of CO_2, its bit complicated to solve but achieve high accuracy, they implemented COANN (cuckoo optimization algorithm neural network) prediction model, the proposed method SVM A classifier for solving complicated non-linear method can predict with reasonable accuracy [4] Carbon dioxide emission prediction using SVM, they used cross validation technique to train and test sets to find parameterized SVM model were used to trial–error method by epsilon and also contributes for an management in making useful decision for B to B operation were looking for expenditure in CO_2 as always SVM is bit complex to use, it also monitored electric energy and coals which affect the amount of carbo emission, it achieves high accuracy with low error rate in trial period [5]. System to predict CO_2 emission levels in cars to mitigate air pollution regression model used for the prediction of emission of CO_2 cars, it also demonstrates the concrete compressive strength depends on mixing ratio, it needs to observe 24/7 vehicle activities to give greater result, they used random forest regression using ml illegal issuing of pollution under certificates from the pollution check vehicles can be eradicated. RTA will seize the high emitting cars.

Zeng et al. [6] it is an Internet-of-things proposal, thus employing a cloud provider and an internet Wireless sensor monitoring centre console to from anywhere assemble, showroom, and interpret the certainly significant, and the correlation coefficients of dioxide based on these criteria were assessed with statistical toolkits, line graphs, and frequency distributions. they used PEMS (portable emission measurement system) on board emission testing equipment, it is costly bit complex to fetch data from PUV broadcast web server, this study recommends further study of quantifying the CO_2 emission based on vehicle idling time [7]. Fuel consumption and CO_2 emission from passenger cars in Europe laboratory versus real world emission, The term "advanced technology" refers to the use of advanced technology in the development of new products., they used WLTP approach (International Standardised Light Truck Testing Method) is a processes used to appraise the combustion, It reiterated that an only one pan-European Greenhouse gas target and chasm contextually may not be the most successful means of lowering Emissions of Co_2 from transportation. Emitter, which requires further investigation, as well as a shortage of consistent information dissemination and datasets, their contribution on Co_2 emission is minimal.

Fan et al. [8] based on sample data, they used three approaches finding high duty vehicle Fuel combustion from fleet vehicles. (a) sample immediately from vehicle records (b) sample from vehicle component data to identify use key fleet composition statistics that are available employing a sampling methods encourages the arithmetic of the vehicle dioxide (CO_2) emissions, avoiding the need to initiate substantial experiments with a perfect collection of license information, which could be time intensive marginal specific distribution approach show better performance among sampling methodology [9]. ML based clock series model for CO_2 prediction in India-,they uses univariate time series data to predict the CO_2 emission used 3 statistical model–autoregressive integrated moving average model, seasonal Arima with exogeneous model, holt winter model, linear regression and random forest, it apply clock series foreseeing in the economy and finance fields, they developed a method also combines of Ari-ma and grey metabolic non-linear model, they reduce residual error and decrease forecasting error the performance analyses shows SARIMAX and holt are most accurate model compared to others [10]. Prediction of CO_2 using ANN, they implemented back propagation artificial neural network model to predict the expenditure of CO_2 emission, objective is following fuel used to the boil machine, they used trial/error approach technique to find parameter of an method, this model includes testing, cutting, mechanics, signal processing, image processing, the main ambition is to have the low root error value, as a result that low error value, high amount of prediction [11]. Prediction of vehicle CO_2 emission and its application to eco-routing navigation an eco-routing approach combining the weighting method and k-shortest path algorithm is developed to find the optimal path along the pareto frontier, it is carrying out a sensitive analysis in network wide scale, it is difficult to develop this due to constrained terms so, they used path ranking technique regarded as solution for satisfying the con-strains it solved the routing problem and applicable to use to all eco routing problems, In their analyses, they conduct cost-effective carbon emission reduction research. Electric vehicles should be used to reduce CO_2 emissions because electric vehicles play a positive role in reducing carbon emissions, and the unit cost of reducing carbon emissions is decreasing over time [12]. In [13–39] the authors proposes different models based on QoS parameters.

3 Materials and Methods

3.1 System Study

3.2 Aim

Co_2 emission is one of the major factors in our global warming. There are lot of vehicles that are actively present in the world. It is difficult to find the vehicles emission. So, this project can easily find out the CO_2 emission.

3.3 Objectives

Objective is to build a ML model for CO_2 Prediction, to strongly replace the updated supervised ML classifying models by getting results in the form of high accuracy to compare all supervised algorithms.

3.4 *Scope of the Project*

'Combining vehicle pollution with computer-based prediction could reduce errors and improve prediction outcomes, according to the project's scope. This suggestion is encouraging because data modelling and analysis tools, such as data mining, have the potential to create a knowledge-rich environment that can significantly improve vehicle pollution prediction accuracy.

Packages

SKLEARN:

- Sklearn is a ML packages in Python that are having many Machine algorithms. On top of it that the modules are used here are doing as train to test and also split data's, Decision-node Classifier, Logistic Regression, and result chart.

NUMPY:

- This Python has some model that gives us a quick maths formula for all kind of problematics. Also it reads and manipulate data from numpy package.

PANDA:

- These packages used to write or read files. Frame lines often used to manipulate files.

MATPLOT-LIB:

- Set visuals is a useful tool for finding patterns dataset given. It is easy to fuzz function data.

4 Proposed Analysis Approach and Results

1. *Data Pre-processing*:

ML validates the techniques which they are utilized to achieve the ML model's mistake amount that is closest to the information's true error rate. These values critical examination are not required if the volume of information is sufficiently big to accurately represent the inhabitants. Even in today's world, it is essential for working with data that is not always provided by the dataset. Missing value, dupe value, and types of da-ta description, regardless of whether the variable is which type of variable. While knob twiddling model hyper-parameters, a survey dataset is used to get an impartial models suitable to prepare synthetic data. Even though expertise just on confirmation data is imbued into the simulation model, this becomes more slanted. A validation set is used to evaluate a given model on a routine basis. This data is used by ML engineers to fine-tune hyper-parameters models. Collection of data, analyzation, and handling with metadata, efficiency, and framework can all add up to a long list of tasks. During the data detection phase, recognising your data and its characteristics will help you choose which algorithm to use to construct your model. Python's Pandas repository is used to undertake a variety of data daily cleaning; specially, it tends to focus on one of the most shared data cleaning job at hand, seriously lacking values, and is able to remove data so quickly. It would rather experiment and model rather than spend time cleaning data. Some of this are unintentional errors which has other times, data may be missing as a result of a more difficult problem. The various types of missing data is critical from a standpoint. This category of incomplete information will sway how numerical data are filled in and detected, perform some basics, and a detailed statistical approach to dealing with missing data.

2. *Data Exploration*:

Data visualisation is a critical skill in applied statistic and ML. Statistics, is concerned with quant data descriptions and estimations. Data visualize can help you gain a quants understanding of your data. When a keep exploring learns about a data-set, it can help identify registry errors. Statistics can depict to express and practise keys in plot points and maps that are more intense and relevant holders to quantify with small contexts (Figs. 1 and 2).

There are two types of field called visual and exploring data in their own field. Data's you don't understand unless or until they made visually, such as in charts, bars and plots. The ability to quickly see and under-stand the significance of information specimen as well as other facts and figures and tried to apply ML cannot be overstated. It will actually taught you about the various types of plotlines you have to know when goals of this study in a computer language, as well as how to use people for our own comprehension (Figs. 3 and 4).

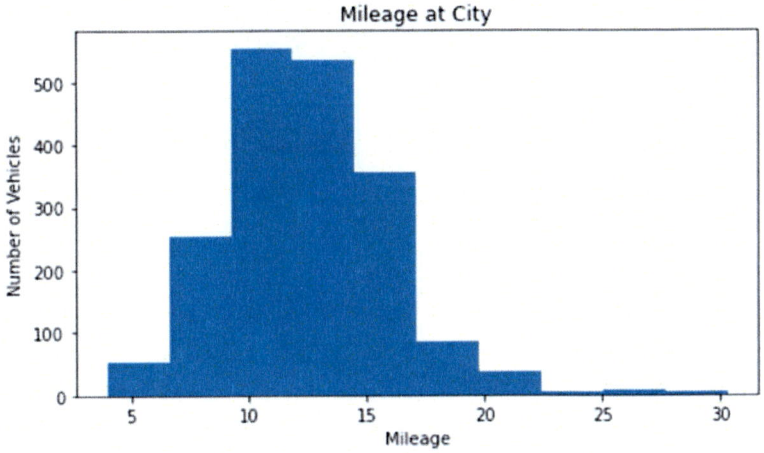

Fig. 1 Mileage at city

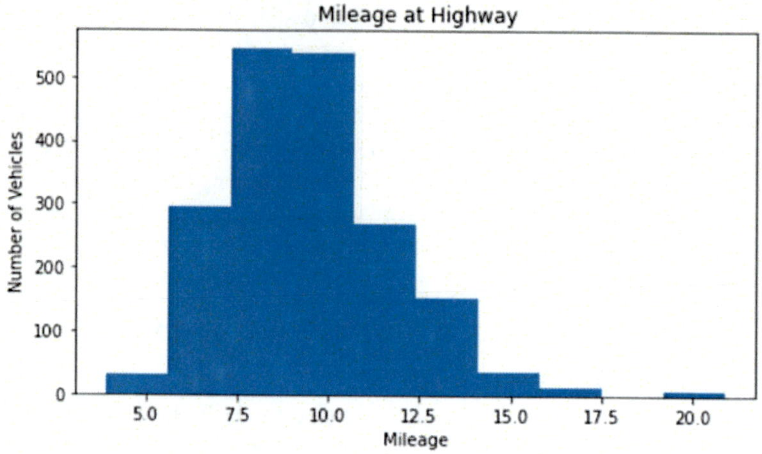

Fig. 2 Mileage at highway

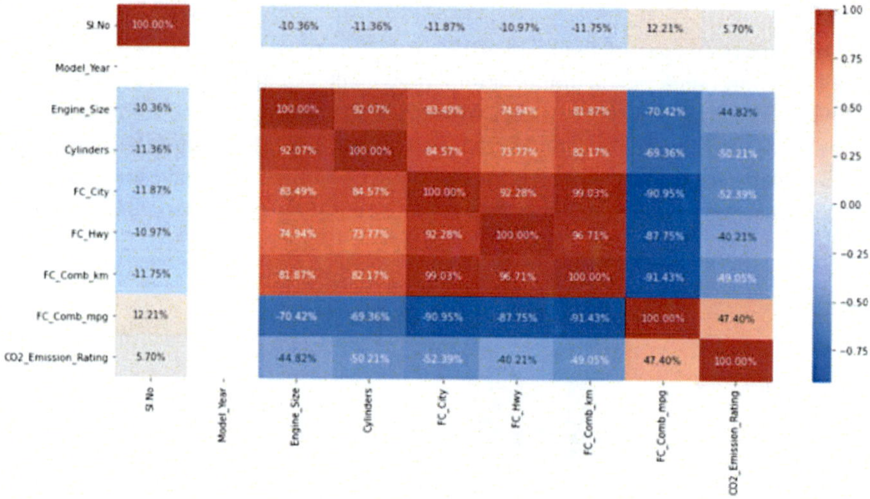

Fig. 3 Data vizualization of emission

On a mono data frame, ML and DL algorithms are used.

- Untrue-positive (FP): A creditor is someone who is anticipated to make payments. When the number of classes is present but the number of classes is not. For instance, suppose the actual data suggests that this rider failed to make it but the number of classes indicates that this commuter will.
- Falsified-negative (FN): Individuals who predict the fall back player. Once the pure class is positive but the constant term is negative. For instance, suppose the

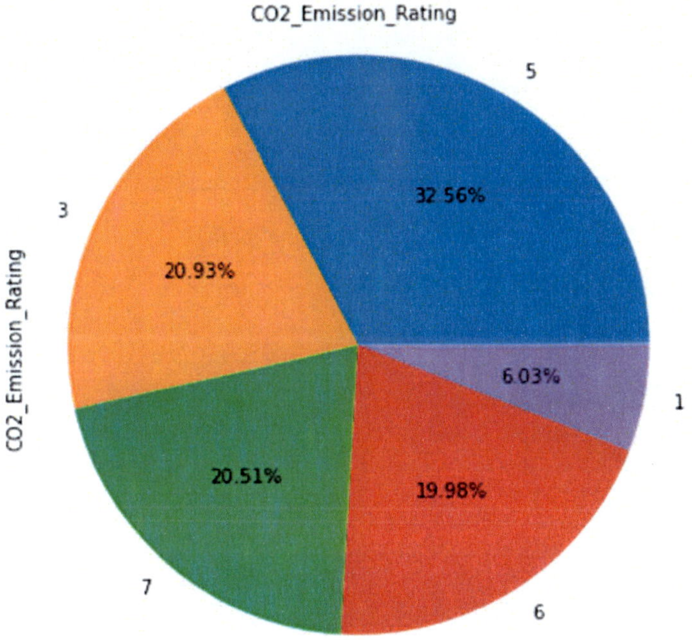

Fig. 4 CO_2 emission chart

current class figure suggests that this traveller stayed alive and the number of classes implies that the rider dies.

- Real-positive (TP): A fraudster is a person fails to pay their bills on time. These have been the correct predictions positive values, denoting that the actual class worth is also true. For instance, suppose the current class value means that this rider survived and the correctly classified value indicates that there is something happened.
- Real-negative (TN) are folks who accurately predicted low traits, denoting that the current class worth is zero, while the correctly classified worth is also absolutely zero. For illustration, if the class value says this rider died and the number of classes states the same thing.

4.1 Data Acquisition

The dataset was collected through online resource to establish the project. It consists of model, vehicle class, engine size, fuel consumption by city, highway and comb of (1/100 km) which helps to predict the accuracy.

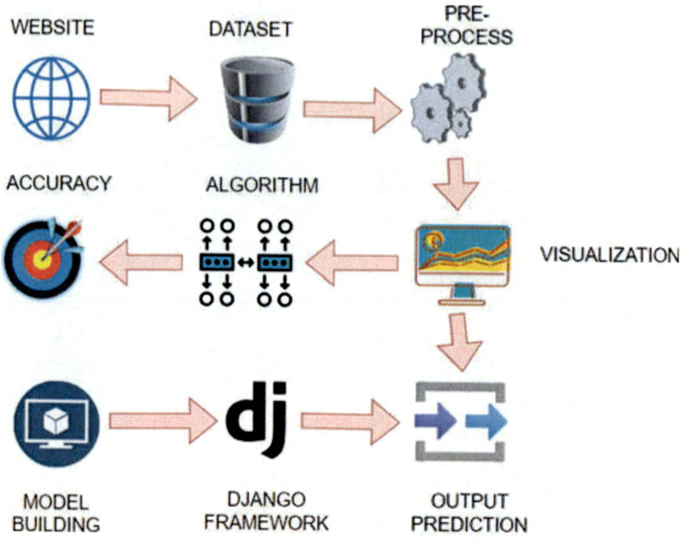

Fig. 5 Architecture system

4.2 *Architecture of the Application*

ML is the study of programming algorithms that self-improve over time. It is considered a substitute of AI. ML algorithms built a maths model from sampled dataset, known as "trained data," in order to foresee or make decisions without being explicitly programmed. Machine learning algorithms are used in a variety of apps, including mail filtering and computerizes vision, where long follows algorithms would be complicated or impossible to develop. ML is closely related to computerized statistics, which focuses on making predictions using computers. For building any application, the machine learning model follows the prototype below. After preparing the data, dealing with missing values, and developing the model with categorical values (Figs. 5 and 6).

4.3 *Problem Statement*

The main problem in the existing system is they used build in algorithms of automation based learning and DL methods of mod to analyse to train the data for accuracy which is average/moderate percentage shows in result. The proposed architecture is developed by hybrid model which is a combination of two or more algorithm to give accurate result by using boosting and bagging strategies which make it more efficient to use and to learn for further studies.

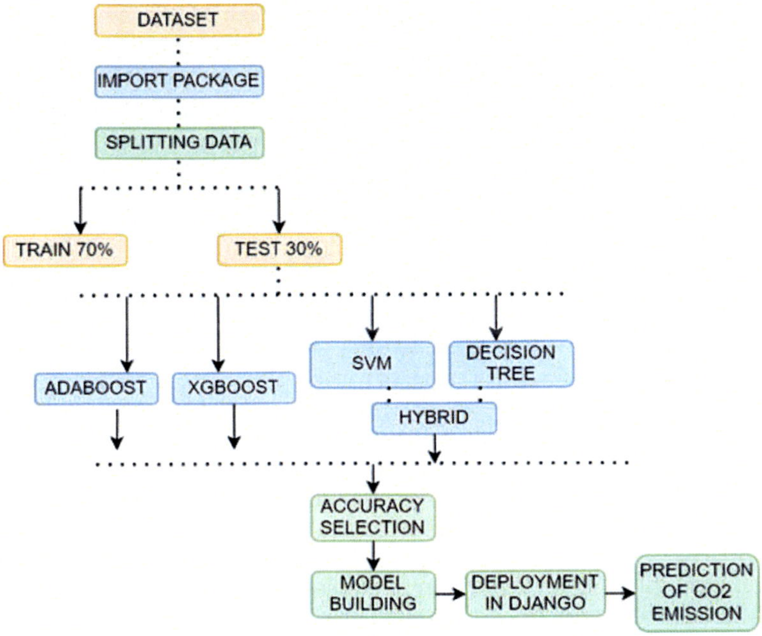

Fig. 6 Flow architecture of the system

4.4 Evaluation Metrics/Performance

The logistic regression algorithm's linear equation and independent predictors are also used to forecast a value. The expected value can range from −infinity to infinity. The algorithm's output requires variable data classification. The logistic regression model, which has the highest accuracy, predicts the end result better.

A CONFUSION MATRIX is a table design that aids in visualising the potential scenarios of a categorization original problem forecasting and results. It produces a table that contains all of the anticipated and actual outcomes of a classifier (Fig. 7).

REPORT OF CLASSIFIERS a categorisation study is a supervised ml measuring performance criterion. It is used to display your implementing this model's exactness, recall, F1-measure, and support.

PRECISION-RECALL: Percentage of positive predictions that are correct (i.e.,) Precision is defined as the proportion of correctly predicted positive observations to all predicted positive observations. The proportion of accurately predicted optimistic parameter estimates. (The prototype accurately predicted the fraction of actual defaulters.) Sensitivity (Recall)-Yes, the proportion of accurately guessed be use to all findings in the full class is referred to as sensitivity.

F1 SCORE: The harmonic mean of precision and recall is used to calculate the F1 score. To refresh your memory, the harmonic mean is an alternative metric to the

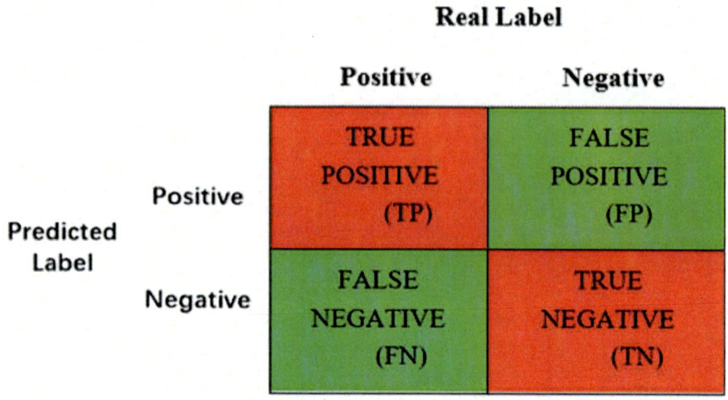

Fig. 7 Confusion matrix

more commonly used arithmetic mean. It is frequently useful when calculating an average rate.

ACCURACY_SCORE—The quantity of positive predictions pieces of data among all the sample points is did refer to as exactness.

HAM LOSS—Proportion of incorrect labels to total labels. Hamming loss is calculated in multi-class classification as the hamming distance between y true and y pred. Ham loss penalises only the seperated labels in various label classifications.

DATAFRAME—Data-Frame is structure of data's that are, like a spread-sheet, organises data into a two dimensions table of rows to columns. It is also used in current data analysis because it is a versatile and user-friendly method of storing also working with data.

3. *Implementation of Machine Learning Models*:

ML, a substitute of AI, which is a branch of computer science that focuses on analyses and interpret same pattern and structure in various data's in order to empower teaching, rationalisation, and judgement without the assistance of human created engagement. Users can use computer vision to feed immense quantities of information to an up with a valid, and the system will make appropriate data-driven choices and suggestions ac-cording to the input data. If any corrections are discovered, the algorithm will be able to use this information to make better decisions in the future. Models are fed parameter of data for which there's a proposed solution. The algorithm is after compiled, and adjust until the algorithm's output (learning) agrees with the pro-posed answer. At this point, raising amounts of data are fed into the system to help it learn and process higher-level computational decisions. Data are driven decisions are increasingly making the difference between Aspects of working while falling farther and farther behind. Computer learning has the capacity to unveil the worth of business and buyer info, allowing businesses to make choices that continue to put them in front of their rivals (Table 1).

Table 1 Comparative analyses in terms of accuracy among the proposed and existing models

Reference	Year	Method	Accuracy/result
Swapnil Wani and Akash Yadav	2022	Analyze between algorithms	86%
Huafang Huang	2021	DL-LSTM algorithm	No information of accuracy
Nur Rachman	2019	SVM method	73%
Proposed model (own algorithm)	2023	Hybrid classifiers of ml technique	98%

4.5 Adaboost Algorithm

In 1996, Yoav Freund and Robert Schapire proposed the Ada-boost or Adaptive Boosting ensemble boosting classifier. It combines several classifiers to increase classifier accuracy. AdaBoost is an algorithm of ensemble iteration. It combines a number of insignificant classifying model to produce a robust classifier with an increase degree of result. The basic idea behind Ada boost is to trained data of set and set the classifier into each iteration to ensure accurate predict uncommon observations. Basis classifier can be any ML method that accepts weigh from the trained set. AdaBoost is a specific way to train an enhanced prediction model. A greatly in-creased classification model is a type of classification.

$$F_t(x) = \sum_{t=1}^{T} ft(x) \tag{1}$$

$$weight = \frac{1}{total\ number\ of\ rows} \tag{2}$$

$$performance\ stump = \frac{1}{2}\log e\left(\frac{1 - total\ error}{total\ error}\right) \tag{3}$$

Thus every ft is a sluggish learner that takes an entity x as its input and provides a scale that measures the object's category. For all specimen in the strength test, each weak classifier produces a yield hypothesis h, which ultimately resulted in a forecasting c(Xi). A new instance is called and assigned at each iteration t. co-efficient t in order to reduce the total training loss Et of the eventually results t-stage boosted classifiers. Here $Ft - 1\ (x)$ is the boosting classifiers that been built to old stages of trains, e(f) is fault that functions and $ft\ (x) = \alpha th(x)$ is the insufficient learner under consideration for inclusion in the final classifier. Weighs: A weight wi, t is assigned to all sample in the trained set at separate loop of the trained progress, equal of current error E(Ft1 (xi)) on that sample. These can be used to train the slow learner.

Algorithm 1: AdaBoost Pseudo-code

Input: (X, y), H (i,e., Adaboost)
Output: Ensemble classifier H(x)
Step 1: H=0
Step 2: $\vec{w} = \{\frac{1}{n}, \ldots, \frac{1}{n}\}$
Step 3: **for** t=0: T-1 **do**
Step 4: find ht by training classifier on weighted data
Step 5: calculate error: $\sum_{i:h}(xi) \neq y(xi)^{wi}$
Step 6: if $\epsilon < \frac{1}{2}$ *then*
Step 7: calculate iteration classifier weight: $\alpha = \frac{1}{2} ln \frac{1-\epsilon}{\varepsilon}$
Step 8: add classifier to ensemble: $H^{t+1} = H^t + \alpha^t h^t$
Step 9: update point weight: $w^{t+1} = w^t \frac{e^{-\alpha t h t(x) \overrightarrow{y(x)}}}{2\sqrt{\epsilon(1-\epsilon)}}$
Step 10: else
Step 11: return H;
Step 12: end
end

4.6 Decision Tree Algorithm

Decision Trees one of the supervised methods which used to solve classifier and regressed problems, it can also solves classificaters too. It is a branch like classificaters, which has internal nods presenting data-set features, also presenting law, and leaf nods presenting classification results. A tree has network links: the Judgment Node and the Child Nodes. Decision has many branches used to make decision, other hand Leaf nodes gives only result of those decisions and have no extra branches. The features of the given dataset are used to make decisions or run tests. Graphical visuals of all possible answers to a problem or decision under conditions. A decision-tree is so named because, it is a tree, and it starts with node of roots and the branch forms a tree structure. To construct a tree, the CART method, which has Classification and Regression Tree algorithm, is used. A decision tree questions and then divides the tree in many sub tress based on answer (Yes/No).To find the leaf node, we must use entropy. The pure leaf node has an entropy value of 0 and the impure leaf node has an entropy value of 1.

$$Gini\ Impurity = 1 - \sum_{i=0}^{n}(p2) \tag{4}$$

Formulae for the pi function. pi function is maximised were we chose the featured that has homogeneous split with approximately the same number of sample of each split (s, t) = (2 PL PR) Q(s|t) This is the formula for the informs gain function.

According to this, information has its function of the entropy of a decision branch negatives the entropies of a candidature splited at node t of a decision makers. To find the root node,

$$Information\ Gain(s, f1) = H(s) \sum v \leftarrow val \frac{|sv|}{|s|} H(SV) \tag{5}$$

where $|SV|$, is the total categories yes and no $|S|$, is the total of feature yes.

Algorithm 2: Decision Tree Pseudo-code

Decision Tree (Sample s, Feature F)

Step 1: $\boldsymbol{If\ Stopping\_Condition(S, F) = True\ then}$

a. $\boldsymbol{leaf = create\ Node()}$

b. $\boldsymbol{leaf.label = classify(s)}$

c. $\boldsymbol{return\ leaf}$

Step 2: **root** = create Node()

Step 3: $\boldsymbol{root.test\_condition = findBestSplit(S.F)}$

Step 4: $\boldsymbol{V = \{v|\ v\ a\ possible\ outcome\ of\ root.test\_condition\}}$

Step 5: **for each** value $v \epsilon V$

a. $\boldsymbol{S_v := \{s\ |root.test_{condition(s)} = v\ and\ s\ \epsilon\ S\};}$

b. $\boldsymbol{child = Tree\ Growth(S_v, F);}$

c. Adding a child as a decent roots and also label 1 the edges (roots -> child) as p

Step 6: **return the root**

4.7 *Classifiers of Support Vector*

SVM models are a strong instrument to recognize forecasting models or predictor variables, they also facilitate scattered information well, but also because individuals has classifying gatherings or to generate predict guidelines for data that piecewise linear functions could really easily classifies. SVM's main goal is to find the suitable hyperplane to classify correctly data points of various classes. The SVM, is a machine learning algorithm that is used to solve issues with regression and classification. It is, however, used mostly in machine learning for classification difficulties. The goal of the SVM algorithm is to find the best quote or distance measure that can divide n-dimensional area into classes, enabling us to categories new data points rapidly in the future. SVM chooses the extreme vectors and points that will help create the hyper—plane. The SVM technique relies on support vectors, which are employed to portray these severe situations.

$$y = mx + n$$

where m is slope, c is intercept

$$y = w1x + w2x + \cdots \ldots + b \tag{6}$$

where w1, w2 are weights

$$y = wTx + b(Hyperplane) \tag{7}$$

$$wTx + b = -1(Negative\ hyperplane) \tag{8}$$

$$wTx + b = 1(Positive\ hyperplane) \tag{9}$$

Symmetrically, the distance that separates these 2 hyper-plane is 2 w, so our goal is to minimize w to maximize the distance here between planes. The distance is computed using the equation. To avoid measured values falling into to the edge, we add the following limitation:

.for all each i either

$$wTxi - B \geq 1, if\ yi = 1 \tag{10}$$

This can be rewritten as,

$$yi(wTxi - b) \geq 1,\ for\ all\ 1 \leq i \leq n \tag{11}$$

People can use this to resolve the issue of optimization.

$$Minimize\ \|w\|\ SUB\ to\ Yi(WTxi - B) \geq 1 for\ I = 1, \ldots, n \tag{12}$$

The W and B that solve all problems that determine our classifier, $x \to sg\ (wTx$ – B) so Sgn (.) sign function of it.

$$max(0, 1 - wTxi - b) \tag{13}$$

yi is the I-th aim (i.e., all cases, 1 or -1), and is $wTxi - b$ I-th output.

If the limitation in (1) seems to be satisfactory, that is, if xi occupies the right side of the edge, this role is zero. The method's values is inversely proportionately distancing from the edge for information on the losing end of the edge.

The aim of the optimization then is to minimization of formulas:

$$\lambda\ \|w\|\ 2 + \left[1n \sum max(0, 1 - yi(wTxi - b))nI\ = 1\right] \tag{14}$$

in which the variable > 0 specifies the good compromise among continuing to increase the profitability size and making sure that the xi are on the right side of the edge.

Algorithm 3: SVM Pseudo-code

Step 1: **START**
Step 2: **INPUT:**
Dataset: Training Set p*
Parameter: C
Step 3: Process $\boldsymbol{P} \leftarrow \boldsymbol{P}^{*}$;
Step 4: while $\boldsymbol{p} \geq \boldsymbol{2}$ ***do***
SVM – SVM with the optimized tuning parameters for the p variables and Observations in data;
Step 5: $\boldsymbol{wp} \leftarrow \boldsymbol{calculate\ weight\ vector\ of\ the\ svm}(\boldsymbol{w_{p1,\dots,}w_{pp}})$
Step 6: $\boldsymbol{rankcriteria} \leftarrow (\boldsymbol{w_{p1}^{2}}, \dots \dots, \boldsymbol{w_{pp}})$;
Step 7: minRankcriteria – variable with lowest value in Rankcriteria vector;
Step 8: remove minRankcriteria from data;
Step 9: $\boldsymbol{p} \leftarrow \boldsymbol{p} - \boldsymbol{1}$
Step 10: **end**
Step 11: $\boldsymbol{Rank1} \leftarrow \boldsymbol{variable\ in\ data} \notin (\boldsymbol{Rank_{2,\dots,}Rank_{p^{*}}})$;
Step12: return $(\boldsymbol{Rank_{1,\dots,}Rank_{p}})$;
OUTPUT
END

4.8 XGboost Algorithm

The term XG Boost can relate to a tree based gradient enhancement of the quality that solves many data science problems quickly and accurately, as well as an accessible structure that enforces that automated system. To differentiate between both of the XG Boost understandings, the method will be referred to as XG Enhance the Automated system, and the guideline will be referred to as XG Boost the Foundation. The XG Boost the Algorithm relies on decision tree branches, which are models that construct a graph by using variety "if" assertions to analyse the input (vertices in the graph). If the "if" requirement is met, the next "if" condition and ultimate prediction are influenced. XG Boost the Automated system linear relation an increasing number of "if" conditions to the decision tree to create a more powerful model. XG Enhancement algorithm differentiates itself from different gradient boosting methods by using a 2nd estimation of the scoring function. Gradient Boosting can quantify the best "if" situation and its performance impact to use this estimation. Introducing the XG Boost. The Method can then store these in recollection to use in the subsequent decision tree, avoiding updating. XG Enhance the Algorithm works well on its own, yet it additionally works ok in tandem with other machine learning instruments in your toolbox. Take into account feature extraction, in which the supervised learning engineer pre-processes the baselines to generate new input features before allowing the model to run. The XG Boost the Method makes use of engineered features to create a user-friendly and strong algorithm.

Algorithm 4: XGBOOST Pseudo-code

Step 1: Data and Hyper parameters
Step 2: Initialize $\boldsymbol{f_0(x)}$;
Step 3: **for** k=1, 2, …, M **do**
Step 4: $\boldsymbol{Calculate\ g_k = \frac{\partial L(y,f)}{\partial f}}$;
Step 5: $\boldsymbol{Calculate\ h_k = \frac{\partial^2 L(y,f)}{\partial f^2}}$;
Step 6: Determine the structure by choosing splits with maximized gains
Step 7: $A = \frac{1}{2}\left[\frac{G^2 L}{H_L} + \frac{G^2 R}{H_R} + \frac{G^2}{H}\right]$;
Step 8: Determine the leaf weights w* = -G/H;
Step 9: Determine the base learner $\boldsymbol{\hat{b}(x) = \sum_{j=1}^{T} wI}$;
Step 10: Add trees $\boldsymbol{f_k(x) = f_{k-1}(x) + \hat{b}(x)}$;
Step 11: **end**
OUTPUT: $\boldsymbol{f(x) = \sum_{k=0}^{M} fk(x)}$

4.9 Hybrid Algorithm

The proposed classifier is a combination of Decision Tree and Support Vector Machine implemented using python packages like pandas, NumPy. It was built with numerical datatypes. Hybrid algorithms is powerful on its own for higher accuracies. It also gives a good model that is easy to understand and performs well. Because of significant variability or partiality, a solitary model, as well known as the base or insufficient learner, may still not operate well on its own. However, when weak learner combines with strong learner it reduces high bias or high variance leads to yielding better model performance. It trains data on an ensemble model to predict the output. It employs multiple classifiers to make accurate prediction. Estimator is a equation to compare one another for best quality or observation. It has enough performance metrics to show less error rate and correlation of the data. Together, several straightforward algorithms support and improve one another. They can address things which they were n't yet meant to address alone. There really are numerous methodologies inside of HML that communicate with the information in different ways. A hybrid method is one that incorporates 2 or many distinct methodologies that finishes the very same error, either by selecting one based on certain feature of the files or by swapping among them throughout the automated system. In overall, this is completed to incorporate desirable characteristics of single overview of the algorithm outperforms the individually parts. The term "hybrid heuristic" doesn't really make reference to together with various algorithms to finish a unique set of problems—many methods can be thought of as mixtures of same pieces-but by only togethering models that ends the same error but vary in different aspects, most noticeably achievement.

Algorithm 5: HYBRID Pseudo-code

Step 1: **START**
Step 2: vot_clf = Voting Classifier (estimators= [('Decision Tree Classifier',clf), ('SVC', sv)], voting='hard')
Step 3: Weights= [0,3,5,6,7]
Step 4: vot_clf.fit (X_train, y_train)
pred_vtng = vot_clf. predict(X_test)
Step 5: accuracy = accuracy_score (y_test, pred_vtng)
Step 6: loss = hamming_loss (y_test, pred_vtng)
Step 7: cr = classification_report (y_test, pred_vtng)
Step 8: label = Label Encoder ()
for i in col:
df[i] = label.fit_transform(df[i]).as type(int)
Step 9: plot_confusion_matrix (vot_clf, X_test, y_test, ax=ax)
Step 10: plt. plot (DF["predicted"] [:100], marker='x', linestyle='dashed', color='red')
plt. plot (DF["y_test"] [:100], marker='o', linestyle='dashed', color='green')
Step 11: clf = DecisionTreeClassifier(random_state=0)
sv = SVC (max_iter=0, random_state=0)
OUTPUT

4.10 2-Dimentional Graph Loss Figures

See Figs. 8, 9, 10, 11 and 12.

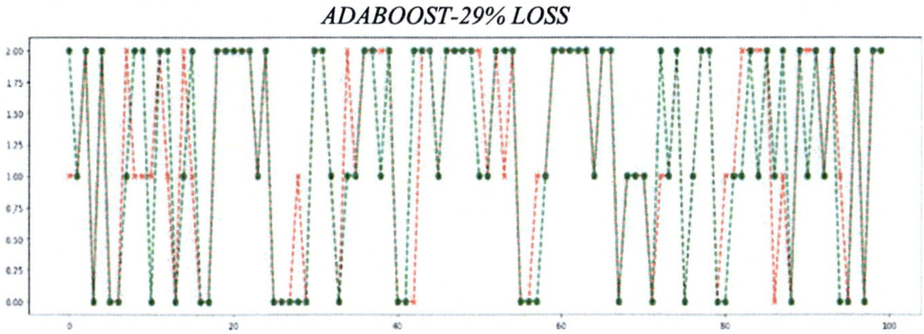

Fig. 8 AdaBoost loss

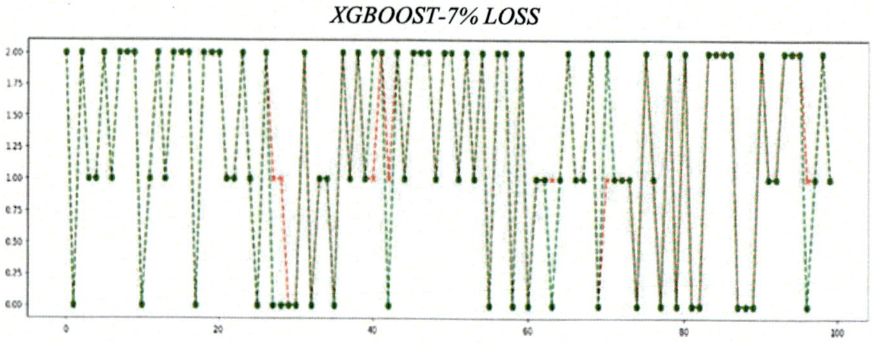

Fig. 9 XGBoost loss

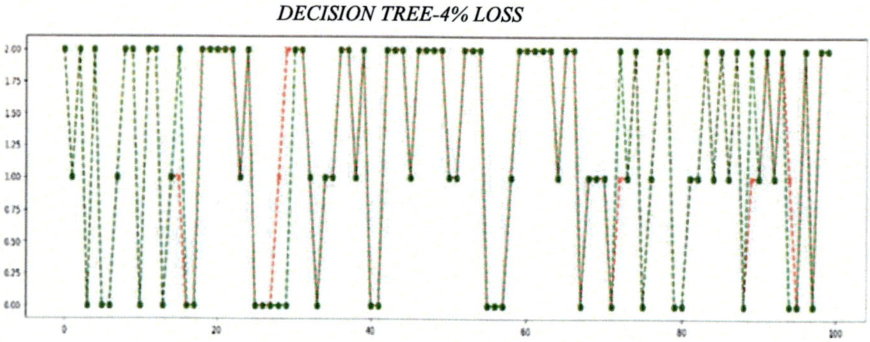

Fig. 10 Decision tree loss

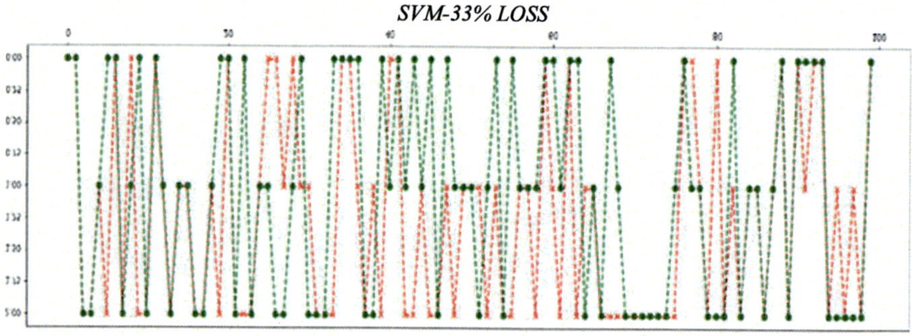

Fig. 11 SVM loss

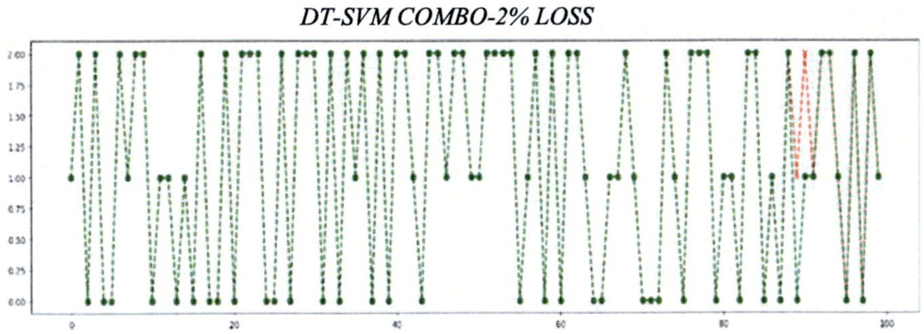

Fig. 12 Dt-Svm loss

4.11 Error Evaluation

A crucial phase of creating a machine learning algorithm is determining the performance of the models. In linear model, the statistics Overall Complete and utter Error, Root Average Square Mistake, Rooted Root-mean-square Erroneous, and defined as the propensity they're utilized to evaluate the effectiveness of the algorithm. The total of the discrepancies among both the dataset's real and envisioned value systems is the mean percentage error. It computes the mean of the dataset's remnants.

$$MAC = \frac{1}{2}\sum_{I=1}^{N}|xi = x| \tag{15}$$

In which xi is the estimated output.

X—imply x benefit [MSE].

The average total squared distinction between both the present and projected values of the large data is described as the mean-squared error (mse). It computes the variability of the error terms.

$$MSE = \frac{1}{2}\sum_{I=1}^{N}(xi - x) \tag{16}$$

The quadratic formula of Root Average Square (rms Error equals Minimum Glitch in the geometric mean. It computes the residual random deviation. The determination coefficient, also known as R-squared is indeed the percentage of a variation in the dependent variable that can be explained by a regression model using linear regression. It is a magnitude rating, which means that regardless of how small or large this same morals are, R-squared is never greater than thing.

Correlation:

The population correlation coefficient $\rho x, y$ between two random variables X and Y with expected values μx and μy and standard deviations σx and σy is defined as:

$$\rho X, Y = corr(X, Y)\frac{cov(X, Y)}{\sigma x \sigma y} = \frac{E\left[(X - \mu x)(Y - \mu y)\right]}{\sigma x \sigma y}, \text{if } \sigma x \sigma y > 0 \quad (17)$$

where E is the expected value operator, *cov* means covariance, and *corr* is a widely used alternative notation for the correlation coefficient.

4.12 Django Framework

Django is an elevated Django framework that enables the quick creation of safe and sustainable internet sites. Django, which has been generated by application experts, handles much of the tedium of web design, so you can focus on composing your plugin rather than trying to reinvent the motor. The Django web framework is a free, open-source framework that can help to accelerate the development of a web application written in python. Django, titled "after the popular jazz guitar player Django Reinhardt, is an unrestricted, accessible structure which was initially officially disclosed in 2005. Django is a Python-based programming framework. It's a Python coding tool that adds functionality and speeds up the process. Django tends to favour the reuse of parts, also renowned as DRY (Don't Repeat Yourself), and involves willing features such as an user account, authentication process, and CRUD functions like create then read, updates or deleted. Django follows the MVC design pattern. That is, all entities such as db (database), back-end and front-end code are distinct entities. Django allows us to separate code from the static media that makes up your website, such as images, files, CSS, and JavaScript.

SYNTAX: $ python-m pip install Django.

5 Discussion on Findings

The atmospheric concentration of carbon dioxide contributes to global warming, resulting in global warming. The actions of humans have raised the dioxide content of the environment by 50% in far less than 200 years. Propane contains atoms of hydrogen and carbon. When oxide (C) from the energy reacts with oxygen (O_2) first from wind all through ignition, carbon dioxide is generated (CO_2). In this work, a hybrid comparative algorithm of Decision Tree and Support Vector Machine is used to produce high accuracy for accurate prediction, and to make it unique, it has performance metrics to show the error rate, as well as mean square and root mean square methods. Our primary objective is to track Dioxide (CO_2) emissions from power

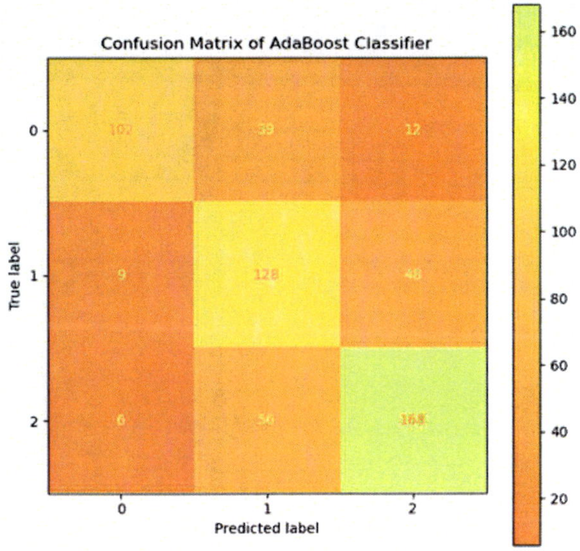

Fig. 13 Plot Adaboost

usage by attempting to make the most precise forecast with the least number of errors. Bridge was used in this experiment, as previously stated, to split the information into a training dataset or equal parts: 70% and 30% for the test dataset. Ensemble methods improves machine learning outcomes by combining multiple models. This approach outperforms a solitary model in terms of predictive performance. The basic idea is to train a group of classifiers (experts) and then let them cast a ballot on the results. Advantage: enhanced forecasting ability (Figs. 13, 14, 15, 16, 17 and Table 2).

6 Conclusion and Future Scope

The system is adaptable, secure, portable, and efficient, with service availability 24 h a day, seven days a week. The user RTA can predict the Co_2 emission level of any particular car by using the proposed model. As a result, if the Co_2 emission level exceeds the threshold value, the RTA personnel will notify the registered car owner to service the vehicle. As a result, the level of air pollution can be reduced. Furthermore, the illegal issuance of pollution certificates from pollution check vehicles can be eliminated. More features could be added in the future. The proposed work can be expanded further by developing a model capable of predicting the Co_2 emissions of all other vehicles, including automobiles. Real-time data from vehicles equipped with sensors can also be used to improve model development. The registered car details

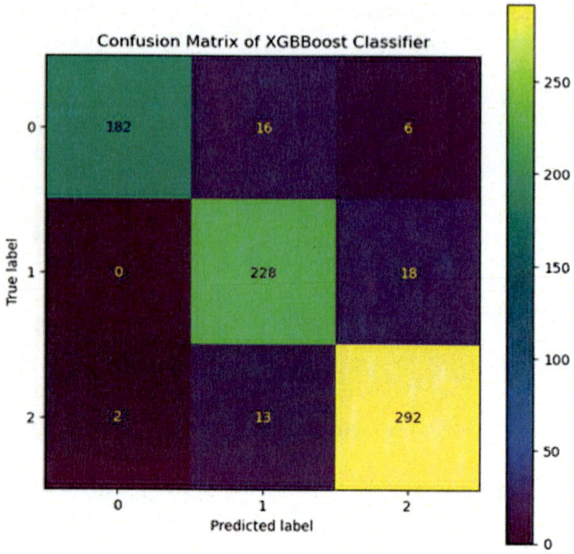

Fig. 14 Plot XGboost

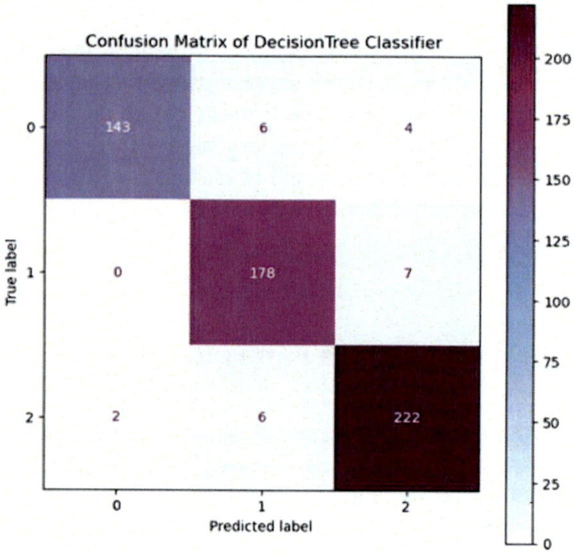

Fig. 15 Plot DT

can be obtained from them with the permission of the government for more efficient results. The work can be expanded further by deploying it in a cloud environment to store large amounts of data and to provide cloud services to it.

Fig. 16 Plot SVM

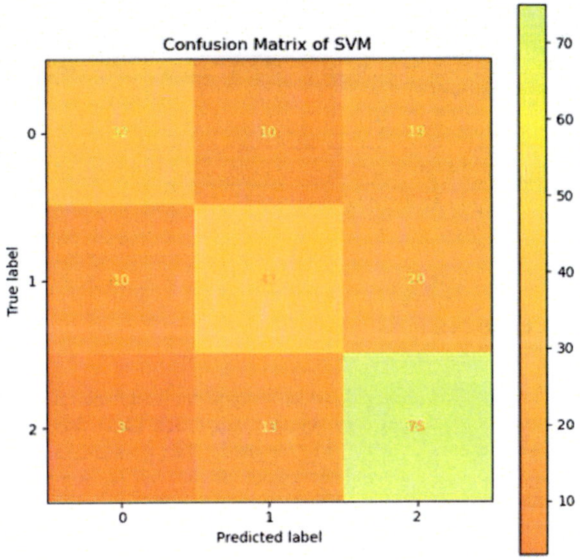

Fig. 17 Plot hybrid

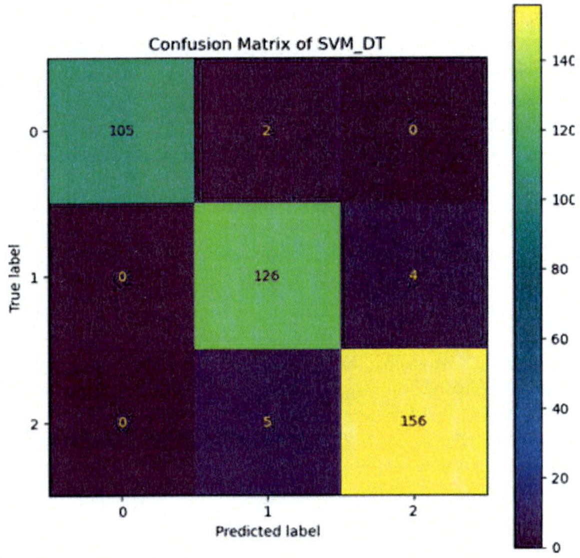

Table 2 Classification metrics of different models

Sn	Model	Precision	Recall	F1 score	Accuracy (%)	Wgd avg
1	Adaboost	0.87	0.67	0.76	71	568
2	XGboost	0.99	0.89	0.94	93	757
3	Decision tree	0.99	0.93	0.96	96	568
4	Support vector machine	0.71	0.52	0.60	67	225
5	DT-SVM	1.00	0.98	0.99	98	398

References

1. Hien, N.L.H., Kor, A.L.: Analysis and prediction model of fuel consumption and carbon dioxide emissions of light-duty vehicles. Appl. Sci. **12**(2), 803 (2022)
2. Magazzino, C., Mele, M., Schneider, N.: A machine learning approach on the relationship among solar and wind energy production, coal consumption, GDP, and Co_2 emissions. Renew. Energy **167**, 99–115 (2021)
3. Huang, H., Wu, X., Cheng, X.: The prediction of carbon emission information in Yangtze River economic zone by deep learning. Land **10**(12), 1380 (2021)
4. Saleh, C., Dzakiyullah, N.R., Nugroho, J.B.: Carbon dioxide emission prediction using support vector machine. In: IOP Conference Series: Materials Science and Engineering, Vol. 114, No. 1, p. 012148. IOP Publishing (2016)
5. Fontaras, G., Zacharof, N.G., Ciuffo, B. (2017). Fuel consumption and CO_2 emissions from passenger cars in Europe–Laboratory versus real-world emissions. Progress Energy Combust. Sci. **60**, 97–131
6. Zeng, W., Miwa, T., Morikawa, T.: Prediction of vehicle CO_2 emission and its application to eco-routing navigation. Transp. Res. Part C Emerging Technol. **68**, 194–214 (2016)
7. Zang, L., Xiao, B., Ma, J., Liu, Y., Tian, P., Zhang, L.: Research on the cost effectiveness of carbon emission reduction in the full life cycle of electric vehicles based on grey prediction. In: MATEC Web of Conferences, Vol. 355, p. 02031. EDP Sciences (2022)
8. Fan, Y., Liu, L.C., Wu, G., Wei, Y.M.: Analyzing impact factors of CO_2 emissions using the STIRPAT model. Environ. Impact Assess. Rev. **26**(4), 377–395 (2006)
9. Kumari, S., Singh, S.K. : Machine learning-based time series models for effective CO_2 emission prediction in India. Environ. Sci. Pollut. Res. 1–16 (2022)
10. Ishii, J., Osuga, M., Okada, T., Miyazaki, H., Koseki, M., Tanikoshi, K.: Reduction of CO_2 emissions for automotive systems. Hitachi Rev. **57**(5), 185 (2008)
11. Dong, Y., Xu, J.: Estimation of vehicle carbon emissions in China accounting for vertical curve effects. Math. Probl. Eng. **2020**, 1–20 (2020)
12. Burns, T.D., Pai, K.N., Subraveti, S.G., Collins, S.P., Krykunov, M., Rajendran, A., Woo, T.K.: Prediction of MOF performance in vacuum swing adsorption systems for postcombustion CO_2 capture based on integrated molecular simulations, process optimizations, and machine learning models. Environ. Sci. Technol. **54**(7), 4536–4544 (2020)
13. Priyadarshini, R., Md, Q.A., Rajendran, N., Neelanarayanan, V., Sabireen, H.: An enhanced encryption-based security framework in the cps cloud. J. Cloud Comput. **11**(1), 64 (2022)
14. Md, A.Q., Varadarajan, V., Mandal, K.: Correction to: Efficient algorithm for identification and cache based discovery of cloud services. Mobile Netw. Appl. **24**(4), 1198–1198 (2019)
15. Panja, A., Jackson, C., Md, A.Q.: An approach to skin cancer detection using Keras and Tensorflow. J. Phys. Conf. Series **1911**(1), 012032 (2021)
16. Md, A.Q., Vijayakumar, V.: Combined preference ranking algorithm for comparing and initial ranking of cloud services. Recent Adv. Electr. Electr. Eng. (Formerly Recent Patents Electr. Electr. Eng.) **13**(2), 260–275 (2020)

17. Prassanna, J., Md, A.Q., Christy Jackson, J., Prabakaran, R.: SecrecyProtector: a novel data analytics based credit score management system. Int. J. Sci. Technol. Res. **9**(03) (2020)
18. Matthew, K.M., Quadir, A.: Computer network and information security. Comput. Netw. Inf. Secur. **2**, 36–42 (2017)
19. Md, A.Q., Vijayakumar, V.: Dynamic ranking of cloud services for web-based cloud communities: efficient algorithm for rating-based discovery and multi-level ranking of cloud services. Int. J. Web Based Communities **15**(3), 248–270 (2019)
20. Mathew, S.A., Md, A.Q.: Evaluation of blockchain in capital market use-cases. Int. J. Web Portals **10**(1), 54–76 (2018)
21. Md, A.Q., Kapoor, S., AV, C. J., Sivaraman: Novel optimization approach for stock price forecasting using multi-layered sequential LSTM. Appl. Soft Comput. 134, 109830 (2023)
22. Basit, Z., Tabassum, M., Sharma, T., Furqan, M., Quadir Md, A.: Performance analysis of OSPF and EIGRP convergence through IPsec tunnel using Multi-homing BGP connection. Mater. Today Proc. (2022)
23. Md, A.Q., Jaiswal, D., Daftari, J., Haneef, S., Iwendi, C., Jain, S.K.: Efficient dynamic phishing safeguard system using neural boost phishing protection. Electronics **11**(19), 3133 (2022)
24. Matthew, K.M., Muhammed, A.Q., Varadarajan, V.: An improved key management scheme in cloud storage. Int. J. Adv. Intell. Paradigms **14**(3/4), 197 (2019)
25. Matthew, K.M., Md, A.Q.: Analysis framework for logs in communication devices. Int. J. Web Portals **10**(1), 15–26 (2018)
26. Srinivasan, A., Quadir, M.A., Vijayakumar, V.: Hybrid cloud for educational sector. Proc. Comput. Sci. **50**, 37–41 (2015)
27. Quadir, M.A.: An efficient algorithm to detect DDoS amplification attacks. J. Intell. Fuzzy Syst. **39**(6), 8565–8572 (2020)
28. Md, A.Q., Narayanan, S.S.S., Sabireen, H., Sivaraman, A.K., Tee, K.F.: A novel approach to detect fraud in Ethereum transactions using stacking. Expert Syst. e13255
29. Gupta, B., Madan, G., Abdul Quadir Md.: A smart agriculture framework for IoT based plant decay detection using smart croft algorithm. Mater. Today Proc. (2022)
30. Sathish, S., Ashwin, S., Abdul Quadir, Md., Pavithra, L.K.: Analysis of convolutional neural networks on indian food detection and estimation of calories. Mater. Today Proc. (2022)
31. Abdul Quadir, M., Prassanna, J., Christy Jackson, J., Sabireen, H., Gupta, G.: Efficient algorithm for CSP selection based on three-level architecture. In: Artificial Intelligence and Technologies, pp. 515–531 (2021)
32. Md, A.Q., Agrawal, D., Mehta, M., Sivaraman, A.K., Tee, K.F.: Time optimization of unmanned aerial vehicles using an augmented path. Future Internet **13**(12), 308 (2021)
33. Rajakumaran, G., Venkataraman, N., Quadir, A.: Early detection of LDoS attack using SNMP MIBs. In: ITM Web of Conferences, vol. 37, p. 01025 (2021)
34. Md, A.Q., Jha, K., Haneef, S., Sivaraman, A.K., Tee, K.F.: A review on data-driven quality prediction in the production process with machine learning for Industry 4.0. Processes **10**(10), 1966 (2022)
35. Joshua, C.J., Jayachandran, P., Md, A.Q., Sivaraman, A.K., Tee, K.F.: Clustering, routing, scheduling, and challenges in bio-inspired parameter tuning of vehicular Ad Hoc networks for environmental sustainability. Sustainability **15**(6), 4767 (2023)
36. Hai, T., Zhou, J., Padmavathy, T.V., Md, A.Q., Jawawi, D.N., Aksoy, M.: Design and validation of lifetime extension low latency MAC protocol (LELLMAC) for wireless sensor networks using a hybrid algorithm. Sustainability **14**(23), 15547 (2022)
37. Sivakumar, J., Md, A.Q., Vigneswaran, T., Punitha, K., Sivaraman, A.K.: Reliability of smart-wearables using PSO-GA optimized algorithm in terms of data analysis. In: 2022 Third International Conference on Intelligent Computing Instrumentation and Control Technologies (ICICICT). pp. 1614–1620. IEEE (2022)
38. Jojo, A., Sunil, G.K., Md, A.Q., Vigneswaran, T., Punitha, K., Sivaraman, A.K.: Intruder detection system using IoT with adaptive face monitoring and motion sensing algorithm. In: 2022 Third International Conference on Intelligent Computing Instrumentation and Control Technologies (ICICICT), pp. 1628–1634. IEEE (2022)

39. Setia, P., Singh, Y., Md, A.Q., Vigneswaran, T., Punitha, K., Sivaraman, A.K.: An efficient dynamic K-papa architecture for communicating with the things using collision-free algorithm. In: 2022 Third International Conference on Intelligent Computing Instrumentation and Control Technologies (ICICICT), pp. 1640–1646. IEEE (2022)

Robotics Cyber Security Issues

Hadi Dastan Elikhchi, Thaier Hamid, and Maria Akpoduado

Abstract Robotics is becoming more ubiquitous, and this technology has existed for almost a century and also keeps continuously evolving with better capabilities and advancements in an environment which includes both hardware and software. Nowadays, Robotics come in different shapes with different capabilities. These days, a robot can be small, and simple and still have increased performance and productivity. Additionally, this technology has become an essential component of daily life, and robotics automation usage cases have expanded from household utilities to medical, or space projects, defense, and education. Robotic functionalities are similar to computer systems running by a program that is defined to repeat tasks or operations. Moreover, they experience security problems that are comparable to those that computer systems have had for years. Hence, when robots are compromised, it immediately affects two different areas.: I. Physical elements may raise concern in three key aspects; integrity, confidentiality, and availability of the robot system's operations and functions. It can also lead to attacks or hacks. I. Virtual security vulnerabilities can be found in provider platforms, misconfigurations, built-in weaknesses in robot systems, and risks in data communications channels. For instance, the WIFI connection may be designed with a poor encryption system. Therefore, both physical and virtual problems in robotics systems are part of cybersecurity in Robotics which is rapidly growing. In this case, the paper focuses on cybersecurity in Robotics and identifies security gaps, vulnerabilities, and risks.

Keywords Cyber-attacks · Cybersecurity in robotics · Intrusion detection and prevention system · Privacy · Safety · Accuracy · Robot threats · Data vulnerabilities

H. D. Elikhchi · T. Hamid (✉) · M. Akpoduado
Department of Creative Technologies, University of Bolton, Bolton BL3 5AB, UK
e-mail: t.hamid@bolton.ac.uk

C. Iwendi et al. (eds.), *Proceedings of ICACTCE'23—The International Conference on Advances in Communication Technology and Computer Engineering*, Lecture Notes in Networks and Systems 735, https://doi.org/10.1007/978-3-031-37164-6_15

1 Introduction

Nowadays, the field of robotics is emerging with different technologies such as human–machine collaboration, the Industrial Internet of Things (IIoT), Artificial Intelligence (AI) and Machine Learning (ML) to provide better services and at the same time increase performance and productivity [8]. In addition, most robots have become intelligent with different capabilities, and have become more beneficial offering important resources for digitization in the manufacturing industry.

According to the International Federation of Robotics (IFR), the current market value for robots increased by 30% to US$ 9.5 billion in 2019 and at the same time, sales increased. Moreover, most of the robot's systems are used in different environments (non-manufacturing) such as professional robots, transportation sectors, hospitals, and homes. In this case, some critical requirements must be implemented like overlooking cybersecurity factors during the production phases and during the design. However, some robotic applications—such as those used in autonomous cars, entertainment robots, educational robots, and medical robots—are susceptible to cybersecurity vulnerabilities.

Therefore, it is important to understand, identify and apply mitigations in the robotics systems to reduce security gaps, threats, and potential risks. According to Kragic [9], the most critical challenges in robotics fields are related to cyber security risks which are looming over the dawn of yet immature industry and require adapting with capable safety measurements and adding additional security layers.

This paper aims to research, collect relevant information, review and analysis robotics security issues, and also implement and apply penetration testing to identify security gaps and vulnerabilities leading to a conclusion with suggestions to overcome the problems.

2 Robotic Systems

According to Pratt [14] nowadays, the rapidly increasing power of advanced technology like Robots had a huge impact on society, operations, and manufacturing as well as significantly increasing their level of functionalities. At the same time, Robotics used to be considered "the art of system integration" as most Robots offer a wide range of different capabilities, and functionalities with usage options of the machine to perform tasks by automation. In general, a Robot can provide simple functions via designed tools and technology use such as:

i. **User interface**: This is the environment that allows the user to set or command the system to function and enables options to take control of operable functions through a visible framework.
ii. **Sensor**: It allows the robot to detect, detect, send, and receive signals as well as measure its environmental conditions. The robot sensor converts the physical and monitored signals in this case. For instance, converting temperature, motion,

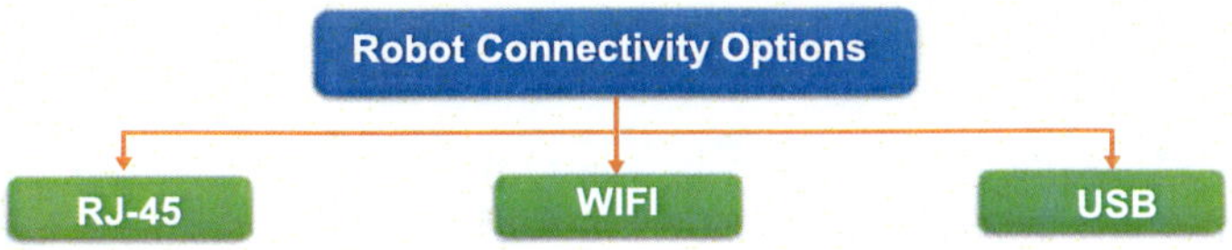

Fig. 1 Example of robot port connectivity options

and light intensity into electrical signals. These signals are sent to the receiver using the available wireless or wired technologies.

iii. **Connectivity**: According to Constantinescu [4], robot port connectivity allows it to take control of the entire system, and access system files through different options (see Fig. 1) like Wireless Fidelity (WIFI), Universal Serial Bus (USB) ports or even Registered Jack (RJ-45), etc.

Most robots are provided with the ability to act, process and sense the environment around them. This field of robotics technology benefited from a variety of disciplines via continued advancements in manufacturing techniques, sensor fabrication, material science, computer science, and mechanical engineering.

In addition, some Robots are only designed for specific environments and/or different tasks. For example, repairing, assembling, and mechanical cutting, may not be suitable for medical use or may not be readily adaptable for other applications as well. Furthermore, when it comes to Robot physical/wireless connectivity (data communications), it is important to classify and consider particular aspects of robotic systems with CIA triad[1] [1] data security such as:

i. **Availability**: This section ensures authorised users are able to access resources in a reliable manner. Data may not be accessible due to a hack, attack, or malicious action like a distributed denial of service attack or a natural disaster like a fire. In addition, any failures of hardware or software can comprise availability as well.
ii. **Integrity**: It is the other key element which may undermine the security of robotic machines with a violation of their integrity, which implies the possible deletion or modification of stored data. This is a big risk to robots that constantly expose, it can be manageable via Intrusion Detection and prevention systems.
iii. **Confidentiality**: It is offering to keep control of the access data and ensures that only those who are authorised have access to specific assets and are also, able to avoid unauthorised disclosure. In this case, the robot's data can be violated in many ways such as man-in-the-middle attacks, etc.

[1] CIA: This model of security guidance is designed to avoid any confusion and is able to aid originations to ensure confidentially, integrity and availabilities (data security) [1].

3 Robotic Systems and Cybersecurity Issues

The history of robotics security can be traced back to the early days of robotics development when researchers and engineers first began to explore the possibilities of using robots in various industries and applications. Over time, as robots became more widespread and sophisticated, concerns about their security and potential malicious use also grew.

In the early 2000s, security researchers began to focus more closely on the security of robots and their associated systems. This included exploring potential vulnerabilities in robotic systems and developing methods for mitigating these risks. For example, researchers developed methods for protecting robots from hacking and other types of cyber-attacks, as well as developing new protocols for ensuring the security of robotic communication networks.

As the use of robots in various industries and applications has continued to grow, so too has the need for effective and reliable security measures. Today, robotics security is a critical concern for organizations that use robots in their operations, as well as for government agencies and military organizations that rely on robots for various purposes.

In recent years, the focus on robotics security has only increased, with researchers and organizations investing more resources into developing new security measures and technologies. This has resulted in a rapidly evolving field, with new breakthroughs and innovations being made all the time. As a result, the future of robotics security is likely to be one of continued growth and development, as researchers and organizations work to ensure the safe and secure use of robots in a wide range of industries and applications.

According to Bilberg [3], a robotic system is required to have additional protection layers such as:

i. **Integrity**: Robotic controllers should continually seek to minimise the impact of potential risks and any incidents that can involve physical parts such as avoiding collisions.
ii. **Safety**: The robot system should be able to offer information available to users and operators. This section of requirement enables human operators to make decisions based on safety or even perform emergency procedures when required.
iii. **Accuracy**: Any is robot required to send actuators commands and functions.
iv. **Security and Privacy**: Any robot is required to provide necessary security and privacy by following The data protection act, ISO standards, provided policy and guidance.

Therefore, it is necessary to highlight and consider all the above key elements that are required for robotic operations and functions otherwise, any violation of the CIA Triad model can lead to exposing robots to cybersecurity threats, attacks, and hacks. This is including the potential of compromising the security and privacy of the operator or the safety of the environment. These additional protection layers are strongly connected to integrity, availability, confidentiality and operator safety i.e., the possibility of the robot machine harming humans or even injury. Clearly, these key concepts are strongly interconnected with cyber security.

3.1 Robots Vulnerabilities

There are several tools and techniques that can be used to discover vulnerabilities in robots and their associated systems. Some of the most commonly used tools and techniques include.

i. **Penetration Testing**: This involves attempting to penetrate a robot's systems to find vulnerabilities and assess its security. This is typically done by simulating attacks that could be launched by a real attacker and evaluating the robot's response.
ii. **Vulnerability Scanning**: This involves using specialized software tools to scan a robot's systems for known vulnerabilities and potential security threats.
iii. **Code Analysis**: This involves reviewing the source code of a robot's software to identify any potential vulnerabilities or security risks.
iv. **Physical Security Testing**: This involves attempting to physically access or manipulate a robot to determine its physical security and the robustness of its control systems.
v. **Threat Modelling**: This involves using a systematic approach to identify and evaluate potential security threats to a robot and its associated systems, and to determine the best methods for mitigating those risks.

The choice of tool or technique will depend on the specific robot and its associated systems, as well as the specific security goals and requirements of the organization. In some cases, multiple tools or techniques may be used in combination to provide a more comprehensive view of the robot's security posture.

3.2 Cyber Threats

Cyber security and potential physical attacks on robots result in massive risks to properties and life [6]. In general, mechanical components of robots such as wheels, gears, motors, and grippers which enables a robot machine to operate, lift and grab items, can pose serious cyber threats when controlled by any malicious attacker to the robot's system. According to Bhavsar [2], the robot data communications via ports, and connectivity network links, allow attackers to take advantage and exploit vulnerabilities in the systems. This is including robotic platforms vulnerabilities, misconfigurations, use of insecure protocols, denial of service (DoS) attacks, malware, etc. For example, recent attacks associated with physical attacks and cyber security issues are listed below in Table 1.

Table 1 Examples of physical security attacks and cyber security issues in robots

Cyber security and physical attacks by robots			
Date	Locations	Investigations	Event
2018 (Jun)	Japan, Tokyo	Robotic systems misconfigured [11]	Robots by using AI and Detected frogs for trucks
2018 (March)	UK, Cambridge	Developers made 3D images to fool robots which attacked by robots [12]	Robot by using AI and attacked to 3D Turtle
2017 (December)	US, shopping mall	The robot platform was misconfigured, and attackers exploited vulnerabilities [15]	Floor cleaning robot (Toddler run over)
2016 (April)	US, military	Malware in robot system [7]	Soldiers shot dead by robots

4 Findings (Weakness and Security Gaps)

According to Mahler [10], Robotics issues are not only limited to one area, but also to many factors that can impact or even lead to exploiting security gaps, and vulnerability to attack robotics platforms, applications, and entire system functionalities. Therefore, this paper aims to identify security gaps, vulnerabilities and potential risks. This includes the detection of weaknesses, classifying them into different categories and overcoming problems.

These findings are based on the gathered data from research and investigations of the robotics project via Eva arm robot[2] with six axes (see appendix 1) at the University of Bolton. Moreover, the following are the top fifteen weaknesses and security issues that were detected during the research (selected resources) and investigations (by using selected network vulnerability testing tools. See appendix 2) such as:

1. **Lack of Collaboration**: When it comes to configurations which are required to have a cloud-based account (Eva arm robot platform), it is clear there is a lack of collaboration between humans and machines with limited functions. It can impact performance, and activities or even lead to misconfigurations.
2. **Lack of Patches**: This is an increasing chance of being hacked or advanced attacks. For example, rootkit, and remote access to sensitive data.
3. **Lack of Advanced Intrusion Detection System**: It is a major security issue, especially when it comes to IDS which mostly depends on the signature, behaviour and anomaly, rather than advanced technology like AI-based IDS.
4. **Lack of AI or ML Based Designs**: This can affect the functionality and operations of the robotic system especially when it comes to assigning a task. In this case, robot system performance and accuracy are affected and limited.

[2] Eva Arm Robot: it is a kind of robot that offers automation in different areas such as manufacturing, sorting default processes, and lightweight picking, and can involve inspections, and product testing processes. This robot functions through six different axes and at the same time supports Python programming language as well [16].

5. **Lack of Advanced Security Features (By Design)**: This kind of weakness can lead to attacking the system and accessing the robotic system's architecture. Also, it will enable attackers to scan and exploit security gaps. For example, attackers can modify data or inject malicious data.
6. **Lack of Advanced Safety Feature (By Design)**: It is very risky and has been proven in the many different real cases of incidents threatening users' lives, financial losses, and a remarkable number of casualties.
7. **Lack of Self-Assessing Functions**: This is leaving the robot machine prone to the possibility of flooding attacks. Therefore, the inability to self-assess, react and retrieve on time reduces further functions and causes degradation in its performance. According to George [5], Robotic self-assessing is necessary to make sure that the robot machine can detect disruptions, can sense faults and at the same time able to reconfigure from backup resources.
8. **Lack of Hardware and Software Resistant**: This kind of weakness can also lead to the loss of the robot system's operational and functional capabilities.
9. **Lack of Multi-Factor Authentication**: It can increase security controls by offering an extra layer of protection to robot systems' accessibilities, by verifying users' identity to log in.
10. **Lack of Integrity in the Robot System**: It is due to the use of very weak message authentication protocols and encryption systems which can be easily compromised and also manipulate stored, in transit data by attackers to reach robotic sensitive data too.
11. **Lack of Privacy in the Robot System**: This kind of weakness can result in the exposure of the entire network in an organisation (e.g., the university of Bolton) and at the same time can affect the organisation's reputation.
12. **Lack Of Confidentiality in the Robot System**: This is a result of the deployment of vulnerable weak algorithms that expose the robotic system and allow the interception of private data.
13. **Lack of Advanced Configurations**: When it comes to robot applications, including insecure features and settings (weak default configurations), this type of weakness allows attackers to easily take control of robot functionalities or even disable users' accessibility into the system. According to Moran [13], hackers or attackers exploit these weak features that operate in the hardware or software of robot systems. For instance, hackers can use default passwords to access systems before users can manage to change them.
14. **Robots' Platform Vulnerabilities**: This includes the lack of constant application updates, security patches, firmware patches and platform vulnerabilities as well. For example, during research and investigations, the robot platform did not function properly. (See appendix 3).
15. **Security Gaps in The Eva Arm Robot Connectivity Options (WIFI)**: The security gap was identified after detecting WIFI access points and applying a four-way handshake process which was an encrypted key that was shared between the client and access point. In this case, during the key transmission, we were allowed to sniff out and obtain the password of Eva arm robot WIFI connection. It is clear that the lack of a strong encryption system and the public

accessibility of WIFI makes robotic systems more vulnerable to attacks or hacks through different connectivity options as well.

5 Conclusion

This paper reviewed, researched, and investigated the current status of cybersecurity and vulnerabilities in robotics systems. It is very clear, there are ongoing challenges, security gaps, and vulnerabilities which put users in critical conditions. Moreover, the majority of robotic systems suffer from severe security weaknesses which make it easier to exploit or deploy dangerous attacks. Also, some of those weaknesses are very dangerous to users as robotic systems rely on them to establish data communications. For example, identified vulnerabilities in wireless communication channels. Therefore, the gathered findings show that robotic systems require additional protection and detection layers which is necessary to secure operational and functional robotics automation systems.

Suggestions to Improve Robotics Security

There are several strategies that organizations and individuals can use to improve the security of robots and their associated systems. Some of the most effective approaches include:

1. **Regular Software Updates and Patches**: Keeping software up to date with the latest security patches and bug fixes can help protect against known vulnerabilities and potential security threats.
2. **Encryption and Authentication**: Encrypting sensitive data and using secure authentication methods (such as passwords or biometric authentication) can help protect against unauthorized access to a robot's systems.
3. **Physical Security Measures**: Physical security measures, such as enclosing robots in secure areas or using security cameras, can help prevent unauthorized access and tampering with the robot.
4. **Network Security**: Securing the network that the robot operates on can help prevent unauthorized access and protect against potential attacks.
5. **User Education and Training**: Training users on the importance of cybersecurity and how to use robots securely can help prevent security incidents and reduce the risk of attack.
6. **Incident Response Planning**: Having a plan in place for responding to security incidents can help organizations quickly and effectively address security threats and minimize the impact of a breach.
7. **Regular Security Audits**: Regular security audits can help organizations identify potential security risks and weaknesses and take steps to address them.

It's important to note that the specific security measures used will depend on the specific robot and its associated systems, as well as the organization's specific security goals and requirements. In some cases, multiple strategies may be used in combination to provide a more comprehensive view of the robot's security posture.

References

1. Bhardwaj, A.: Cyber security attacks on robotic platforms, Cyber security attacks. https://doi.org/10.1016/S1353-4858%2819%2930122-9 (2021). Last accessed 25 Jan 2023
2. Bhavsar, C.: Robotic arm control based on internet of things, Robotic Arm Control. https://ieeexplore.ieee.org/abstract/document/8817333/ (2019). Last accessed 26 Jan 2023
3. Bilberg, A.: Developing a reference model for human–robot interaction, SpringerLink. Springer Paris. https://doi.org/10.1007/s12008-019-00591-6 (2019). Last accessed 25 Jan 2023
4. Constantinescu, D.: Multirobot system-IEEE Xplore, Multirobot System with Connectivity. https://ieeexplore.ieee.org/abstract/document/9463459 (2022). Last accessed 22 Jan 2023
5. George, T.: A review on self-healing polymers for Soft Robotics. Mater. Today, Elsevier. https://www.sciencedirect.com/science/article/abs/pii/S1369702121000080 (2021). Last accessed 25 Jan 2023
6. Ghelani, D.: Cyber security, cyber threats, Authorea. https://doi.org/10.22541/au.166385207.73483369 (2022). Last accessed 25 Jan 2023
7. Goyal, N.: Robotics, AI, and the IOT in defence systems, Robotics, AI. https://doi.org/10.1002/9781119711230.ch7 (2021). Last accessed 25 Jan 2023
8. Howard, D.: A review of jamming actuation in soft robotics. MDPI, Multidisciplinary Digital Publishing Institute. https://www.mdpi.com/2076-0825/9/4/104 (2020). Last accessed 25 Jan 2023
9. Kragic, D.: Challenges in robot, trends and challenges in robot. https://doi.org/10.1126/science.aat8414 (2019). Last accessed 25 Jan 2023
10. Mahler, T.: Cybersecurity, safety and robots: Strengthening the link between cybersecurity and safety in the context of Care Robots. Comput. Law Secur. Rev. Elsevier Advanced Technology. https://www.sciencedirect.com/science/article/pii/S0267364921000017 (2021). Last accessed 25 Jan 2023
11. Mao, Z.M.: Automated Runtime Mitigation for Misconfiguration Vulnerabilities, ACM Other conferences. https://doi.org/10.1145/3545948.3545954 (2022). Last accessed 25 Jan 2023
12. Maple, C.: Identifying attack, identifying attack surfaces. https://ieeexplore.ieee.org/abstract/document/9172785/ (2020). Last accessed 25 Jan 2023
13. Moran, J.: Preliminary examination of emergent threat and risk landscapes in intelligent harvesting robots. MDPI, Multidisciplinary Digital Publishing Institute. https://www.mdpi.com/2076-3417/12/24/12931 (2022). Last accessed 25 Jan 2023
14. Pratt, G.: Robotics. J. Econ. Perspect. https://www.aeaweb.org/articles?id=10.1257%2Fjep.29.3.51 (2021). Last accessed 20 Jan 2023
15. Serafino, A.: Cybersecurity issues in robotics, Issues in Robotics. https://isyou.info/jowua/papers/jowua-v12n3-1.pdf (2021). Last accessed 25 Jan 2023
16. Villa, L.: Incorporating Affective Proactive Behaviour to a Social Companion Robot for Community Dwelling Older Adults. Springer Nature Switzerland. https://doi.org/10.1007/978-3-031-19682-9_72 (2020). Last accessed 25 Jan 2023

Computational Intelligence

A Comparative Study of GANs (Text to Image GANs)

B. Thamotharan, A. L. Sriram, and B. Sundaravadivazhagan

Abstract Synthesizing high-quality images from text descriptions is a challenging problem in computer vision and has many practical applications like creating cartoon characters from the description, creating the room interiors, and so on. Deep Convolutional Generative Adversarial Networks (GANs) are very powerful and have simple architecture. It generates images from the text description and random noise. Combining the random noise and text embedding creates a multi-modal problem. The generator generates the image from the random noise and the discriminator identifies that the generator image is fake and rejects it. This model of generator and discriminator is like the Min–Max problem. The major problem with this DCGAN is that it takes more time and computation power. In our architecture, we proposed that, instead of initializing the random noise vector, we take the basic image features from the pre-trained VGG-19 layers. It makes the whole computation simple and reduces the time complexity. In the final result, we are going to compare the computation and time complexity between Modified DCGAN (i.e. our architecture) and Stackgan.

Keywords Generative adversarial network (GAN) · Deep convolutional generative adversarial network (DCGAN) · Visual geometry group (VGG)

1 Introduction

Generative adversarial networks are the most popular and exciting innovation in machine learning. Generative Adversarial Networks (GAN) are generative models which generate their instances of the data from the given training dataset. The generative model involves an unsupervised learning task, which automatically learns and discovers the patterns in the given input data and outputs new data which looks like

B. Thamotharan (✉) · A. L. Sriram
SASTRA Deemed University, Thanjavur, India
e-mail: thamotharan@ict.sastra.edu

B. Sundaravadivazhagan
Department of IT, University of Technology and Applied Sciences-Almussanah, Muladdah, Oman

C. Iwendi et al. (eds.), *Proceedings of ICACTCE'23—The International Conference on Advances in Communication Technology and Computer Engineering*, Lecture Notes in Networks and Systems 735, https://doi.org/10.1007/978-3-031-37164-6_16

the data from the original dataset. GANs are exciting and have the ability to generate more realistic examples across a range of problem domains, notably in text-to-image generation, and text-to-text generation.

The Generative Adversarial Network has two main components namely the generator and discriminator. The generator and discriminator play an important role in the GANs. These two convolutional neural networks compete with each other like two players in the min–max game. The generator's main role is to produce new instances of the data by learning the patterns of the given input data. The discriminator's role is to distinguish between the data that has been produced by the generator and the real data set. The Architectures of GANs are proposed and one among them is the DCGAN which is the Deep Convolutional Generative Adversarial Network (1). DCGAN is a very powerful and simple architecture that is used for the text-to-image generation of any dataset. Given the text description, the model can able to photo-realistic images which humans cannot tell are fake [1]. The text embedding is given into the generator as the input with the sampled noise vector and the discriminator can able to tell which is real and fake. Here, we are going to use the modified DCGAN architecture and the STACKGAN architecture for comparing the complexities of generating the image from the input. Deep learning methods like Long Term Short Memory (LSTM), and Visual Geometry Group (VGG-19) are used.

In the DCGAN architecture, text embedding and the noise vector are concatenated and fed into the generator network. The text description is encoded using the text encoder and sampled with the noise vector. The noise vector and the embedding are concatenated and form an input to the generator. The generator network generates the images on its own from the pre-trained features from the VGG architecture, which has a series of four fractionally strided convolutional layers [2]. Now the generator produces an image of pixel 64 * 64. The output from the generator is fed as input into the discriminator. In the discriminator, several layers of strided convolutions along with batch normalization is carried out. The discriminator always tries to classify the image as real or fake, see the Fig. 1.

When the generator can able to produce a more realistic image then the discriminator will be fooled and lose the game. In this work, the zero-sum game is made between the generator and the discriminator until the generator wins.

2 Methodology

2.1 Pre-processing

The images and the text descriptions in the dataset should be pre-processed first before giving them as input to the generator. Datasets may contain incomplete data, noisy data (outliers), and inconsistent data. The unwanted noise such as the variation of brightness or color information is removed by the pre-processing. Since we have categorical data, it is not possible for the machine learning algorithms to directly

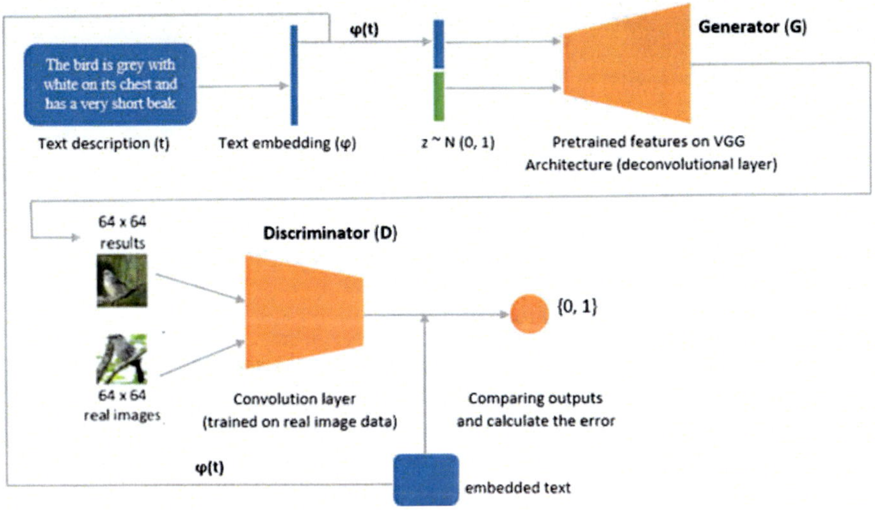

Fig. 1 The architecture of the modified DCGAN

work with them. We use "one hot encoding" for the text. The text description must be mapped to certain index values. The mapping of the index value is taken care of by the library that we use. Then the integer value is converted to a binary vector containing 0s and 1's where the index for the integer is marked as 1 and all others are marked as zeros. Next, the image dataset should be converted to the numpy array so that it will be easy for us to do scientific computation. These array objects can be used for matrix operations like multiplying or adding matrices. These matrix operations aid in classifying the images, aligning the images, etc. The images will be converted to the list of arrays having values (row, column, channel). Now, the pre-processed image dataset and the text description can be fed as the input to the neural network.

Deep Symmetric Structured Joint Embedding

The text embedding is to be created from the given text description for the generator's input. Here, the unique embedding is formed using deep convolutional and recurrent text encoders. These two networks learn correspondence function with the images (3). The unique text embedding has the visual attributes of the images. The vector for the generator input is constructed using the following

$$\frac{1}{N}\sum_{n=1}^{N}\Delta(y_n, f_v(v_n)) + \Delta(y_n, f_t(t_n))$$

where $\{(v_n, t_n, y_n) : n = 1, \ldots, N\}$ is the training data set, Δ is the 0–1 loss, v_n are the images, t_n are the corresponding text descriptions, and y_n are the class labels. Classifiers f_v and f_t are parametrized as follows:

$$f_v(v) = \arg\max_{y \in Y} E_t \sim T_{(y)}[\varnothing(v)^T \varphi(t))]$$

$$f_t(t) = \arg\max_{y \in Y} E_v \sim V_{(y)}[\varnothing(v)^T \varphi(t))]$$

where

φ Text encoder (LSTM),
ϕ Image encoder (deep convolutional neural network),
T(y) is the set of text descriptions of classy,
V(y) is the set of image descriptions of classy.

This says that a "higher compatibility score must be possessed by the text encoding with the images of its corresponding class than any other class". The dimension of the image is compressed to the 1024 × 1 vector by the classifier [3]. The objective function aims to minimize the distance between the image created by the image encoder and the text produced by the text encoder.

LSTM Architecture

Long Short-Term Memory (LSTM) Network is a particular kind of Recurrent Neural Network (RNN) that can learn long-term dependencies. The output of the Embedding layer is the input to the LSTM layer, see the Fig. 2.

This is the diagrammatic representation of the LSTM architecture. While performing backpropagation, the weights of the parameters are updated i.e., the values will get decreased which may also lead to getting some negative values this is referred to as the vanishing gradient problem [4]. LSTM is focused on sorting out this kind of vanishing gradient problem.

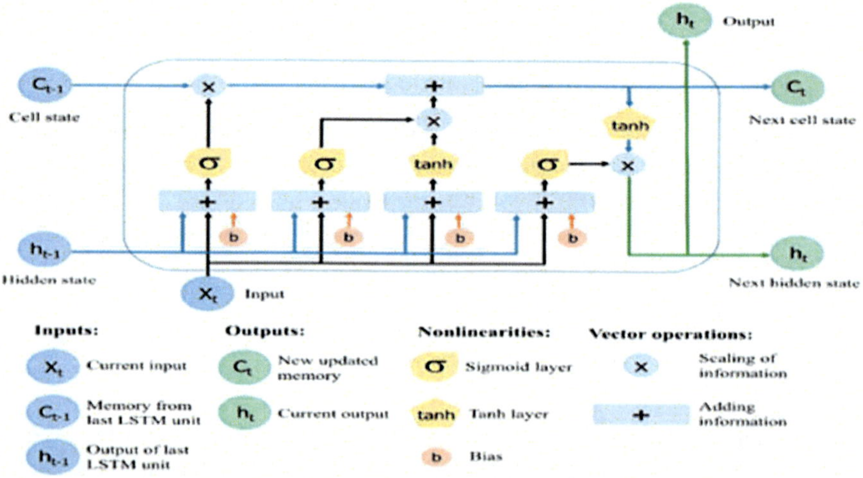

Fig. 2 LSTM architecture

This overall architecture can be divided into four components.

- Memory cell
- Forget gate
- Input gate
- Output gate

Memory Cell

It is fundamentally used for recalling and forgetting based on the conditions of the input. The text description which is given as input contains many kinds of context i.e., the context of the description may change so in that case the memory cell is used for recalling the previous context and it is ready to add some more context. So, mathematically in this memory cell two operations are involved one is pointwise multiplication and the other is addition operation, see the Fig. 3.

Forget Gate

See Fig. 4.

In this formula, wf represents the combined weights of the previous output and the present input. After performing concatenation of both the weights and addition

Fig. 3 Representation of memory cell

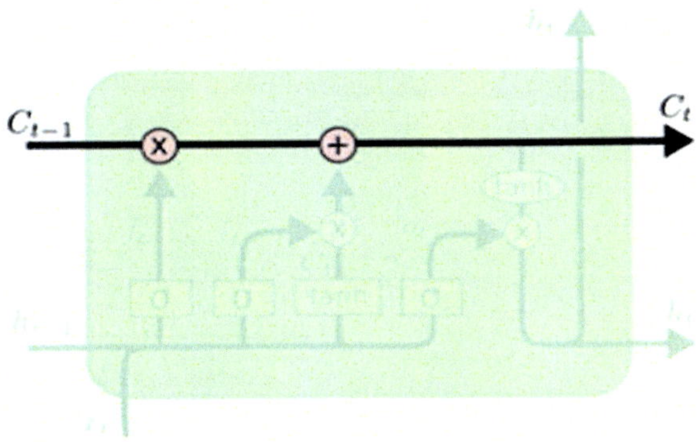

Fig. 4 Forget gate

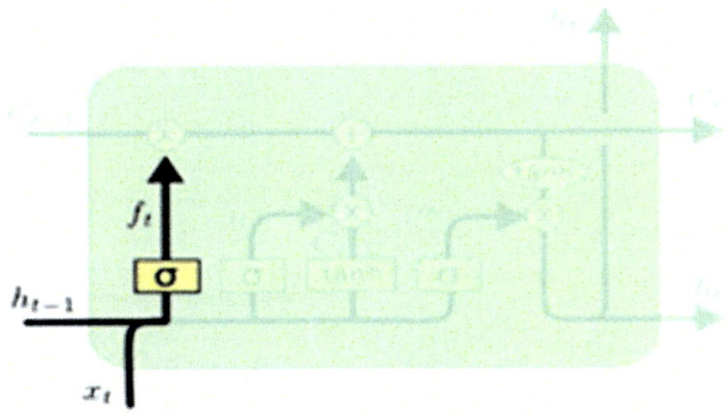

of the bias, if the previous output is not similar to the present input i.e., if there is a change in the context of the text then after applying the sigmoid activation function then the output will be 0 or 1 which means that it will be nearer to 1 or it will be nearer to 0. If the value is 1 then it means that the context is not changed but when it is a combination of 1's and 0s this means that it has forgotten some of the information [5].

Input Gate

See Fig. 5.

In this input gate, information is added after performing the point-wise operation. As already explained in the previous step the value of the sigmoid function will be 0 to 1. Here one more thing tanh function is added, the reason for using the tanh function is that it helps to keep the values between −1 to +1. Finally, after performing a point-wise multiplication operation the value of it is added to the memory cell.

Output Gate

See Fig. 6.

The same step as explained in the forget gate is carried out here and one more step is added here. After performing tanh operation for the output of the memory cell, it

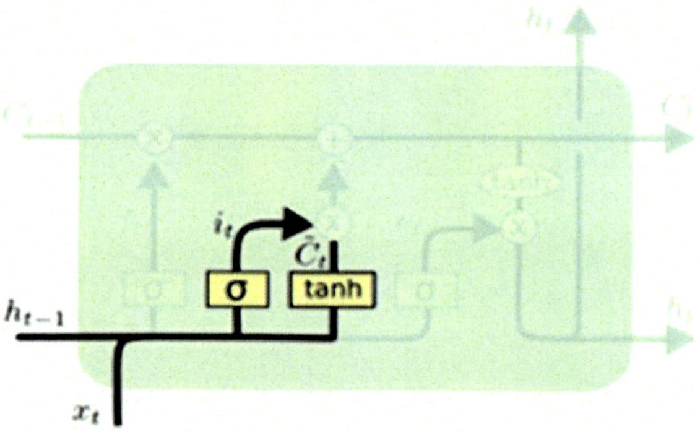

Fig. 5 Input gate

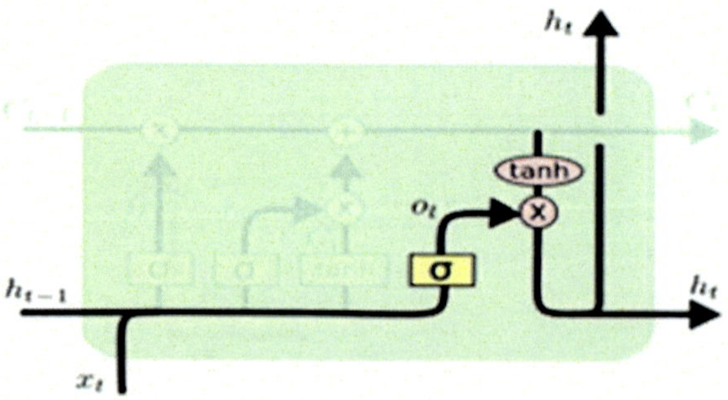

Fig. 6 output gate representation

gets multiplied with the sigmoid function [6]. An important point to note here is that only meaningful full context is retrieved.

Visual Geometry Group (VGG)

VGG 19 is a convolutional network used for classification and detection. Among the four layers, only two layers are used here. They are

- Convolutional layer
- Softmax layer

Convolutional Layer

The 224 × 224 image is the input to the convolutional layer. The convolutional layer has filters and stride. The image features are learned using the small input data and the relation between the pixels is preserved here. The image has been converted to the image matrix [7]. The input for this layer is the matrix and the filter. The stride is defined as the pixel shifting mechanism which defines the number of pixels that should be shifted by the filter while sliding through the array. The convolutional layers have a filter size of 3 × 3 and a stride of 2.

Softmax Layer

For Calculating, the possibility or confidence score, the Softmax function is used. Mathematically how the softmax function works in that takes the exponent of the values which are to be calculated and then normalizes those values i.e., the exponent by the sum. The output of the softmax function will be between 0 and 1. The sum of the output will be 1. If the values which we are calculating are not equal or nearer, then the probability will not be the same [8]. For negative values too, the output will be between 0 and 1.

Generative Adversarial Network (GAN)

The GAN consists of two main components namely the generator (G) and discriminator (D). The generator and discriminator are two neural networks that compete with each other in the training phase. The GANs are designed in a way such that the generator always wins and fools the discriminator by generating new instances of data that look alike the real data [9]. The min–max game is made between the generator and discriminator. The V(G, D) formula is given as follows

$$\min_G \max_D V(D, G)$$

$$V(D, G) = E_{x \sim pdata(x)}[\log D(x)] + E_{Z \sim p_z(z)}[\log(1 - D(G(z))]$$

G	Generator,
D	Discriminator,
Pdata(x)	Distribution of real data,
P(z)	Distribution of generator,

x	Sample from Pdata(x),
z	Sample from P(z),
D(x)	Discriminator network,
G(z)	Generator network.

Generator

The generator in the DCGAN architecture is a special kind of neural network which always tries to fool the discriminator by generating its images from the dataset. It learns on its own to generate the images from the given dataset. The input of the generator is text embedding, which is a combination of noise vector and text embedding φ(t). The noise vector (z) is obtained by sampling using the normal distribution. Then the text description is encoded using the text encoder and forms text embedding. The text embedding is compressed to a small dimension using the fully-connected layer followed by leaky-ReLU. Then the text embedding is concatenated with the sampled noise (z) and forms the input to the generator (Fig. 7).

The generator network has a series of four fractionally-strided convolutions and it has no max-pooling and a fully-connected layer. The transpose convolution takes place in the generator where it is used to upscale the input image to higher resolutions [10]. The first layer of the generator expands the random noise vector of dimensions 100×1 to $1024 \times 4 \times 4$. Then the transpose convolution takes place with the filter size of 5×5 and the stride of 2 without any padding. The final layer of the generator produces the output image of pixel $64 \times 64 \times 3$.

Discriminator

The job of the discriminator is to find whether the image is a real image or generated image. The discriminator has two expectations, one is to assign a minimum value for the generated image and the other is to assign a maximum value for the real

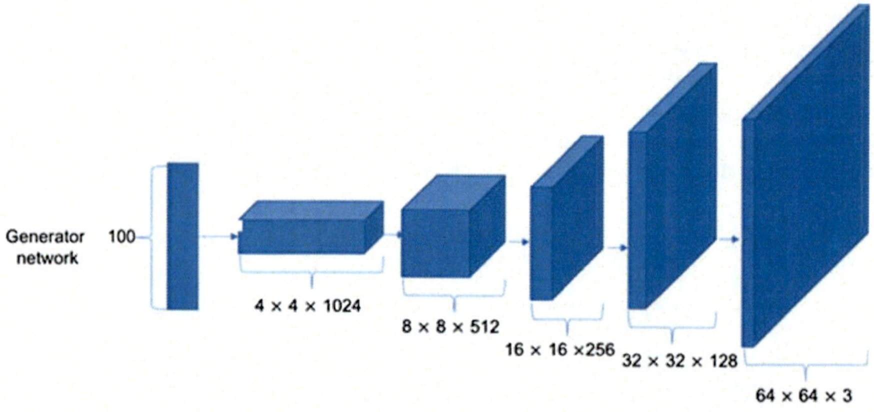

Fig. 7 Generator architecture

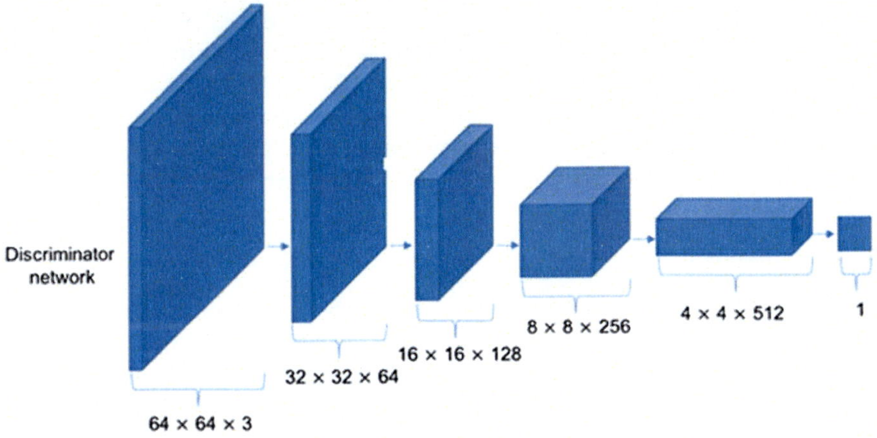

Fig. 8 Architecture of discriminator

image [11]. Mathematically this can be represented as Maximize log D(x) + log (1 – D(G(z))).

Where x is the real data and z is the generated data. Maximize of log D(x) represents that it tries to assign maximum value forthe real image and maximize of log (1 – D(G(z))) represents it tries to assign minimum value for the generator, see the Fig. 8.

This is the overall architecture of the discriminator. In this discriminator, Visual Geometry Group (VGG) architecture is used [11]. Strided-Convolution, the one which is used for compression is helpful for the discriminator. To advance the speed and performance batch normalization (Batchnorm) is used. 64 * 64 output of the generator and the real image is fed as an input to the discriminator. After applying convolution, the final output of the discriminator is the single value which is between 0 and 1. At last, the Back propagation algorithm is carried out. The discriminator is already pre-processed with the real image so that mathematical it assumes some weight for the real image. The output which is obtained now is compared with the weight of the real image, if it matches the weight then the image is a real image.

If the weight doesn't match the weight of the real image then a back propagation algorithm is carried out which means that mathematically the weight of the previous steps is updated and then again the same process is carried out until it matches the weight of the real image. As already mentioned the output value is between 0 and 1 which means that if the value is above 50% (i.e., above 0.5) it is assumed a real image, if it is below 50% (i.e., below 0.5) then it is assumed as a fake image that is the image generated by the generator.

Activation Function

Activation functions introduce non-linearity to the model which allows it to learn complex functional mappings between the inputs and the response variable. There are quite different activation functions like ReLU, sigmoid, tanh, Leaky ReLU, etc.

In the discriminator, the Leaky ReLu function is used. Even for negative values Leaky ReLU allows us to pass the small gradient values [12]. Instead of passing a zero value, we pass the negative small which is small for the backpropagation. In the generator, ReLU is used except in the last layer and tanh activation functions are used in the final layer. ReLU is a piecewise linear function that outputs the input directly if it is positive, otherwise zero is the output for the negative values. The formula for the ReLu = max(0, x).

Batch Normalization

Batch normalization is a technique where it is used to increase the stability and performance of the neural network. Each layer in the neural network can learn independently while we use this normalization. For each mini-batch, the input is standardized by it. It greatly reduces the number of epochs used for training the neural network.

Drop Out

Dropout is a regularization technique that prevents and reduced overfitting in the artificial neural network. Dropout means ignoring certain neurons at the training phase. At the time of the forward and backward pass, instead of using all the neurons some neurons are chosen randomly and used.

ADAM

ADAM is an adaptive learning algorithm used in deep learning. This optimization algorithm is a combination of Stochastic Gradient Descent and RMS prop. This is mainly used to train the neural network. The learning rate of the model is scaled by the squared gradient using the RMS prop and the average of the gradient is used in the SGD instead of the gradient.

3 Results and Discussion

In Fig. 9. the plot shows the change in accuracy of the real and fake data with training steps. Here the training steps are 450 and the data used here is handwritten digits, because the training on bird and flower images with high resolution needs more computation power (GPU) and time. So, we did this in a small pixel image of 28 * 28 on a grey scale [13]. The hyper-parameters like epochs, batch size, and learning rate are changed and analyze the accuracy rate.

In Fig. 10. plot, the number of training steps increased and the optimizer used is 'ADAM', The accuracy in the second plot is more consistent and reached the accuracy faster than the first plot. In both models, they use the VGG network in the discriminator. But in the second case, the deconvolutional layers use the pre-trained basic featured layer, where the noise image goes. By comparing the normal GAN with the modified DCGAN. The accuracy and consistency of performance are improved in the modified one. But in comparison with the complicated Stack GAN

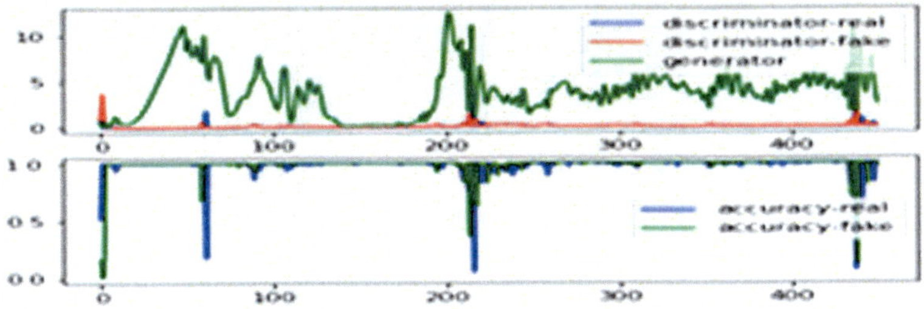

Fig. 9 Implementation of GAN without modification

architecture, the modified model will produce a less-resolution image only, refer Fig. 11 and Fig. 12 [14].

As we can see the image generated from the modified is more accurate and looks like the real image.

The flower image shown has a yellow anther red pistil and bright red petals.

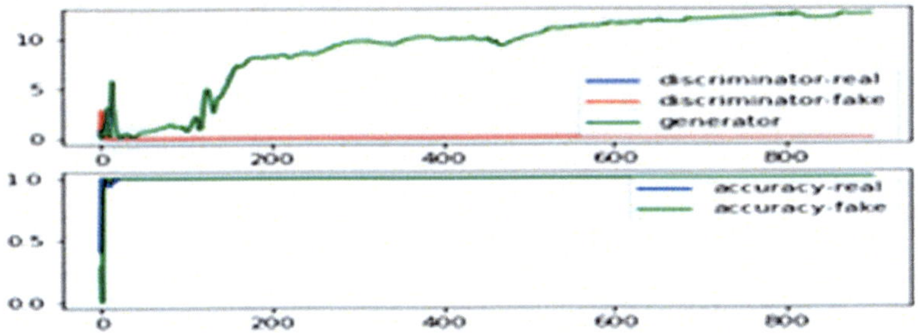

Fig. 10 Implementation of modified GAN

Fig. 11 Real image from the dataset

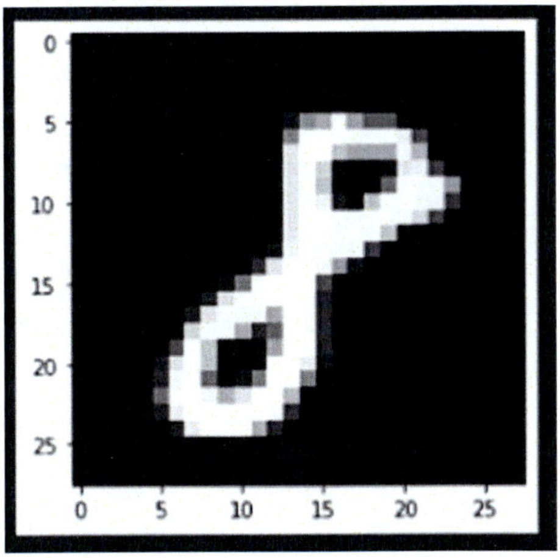

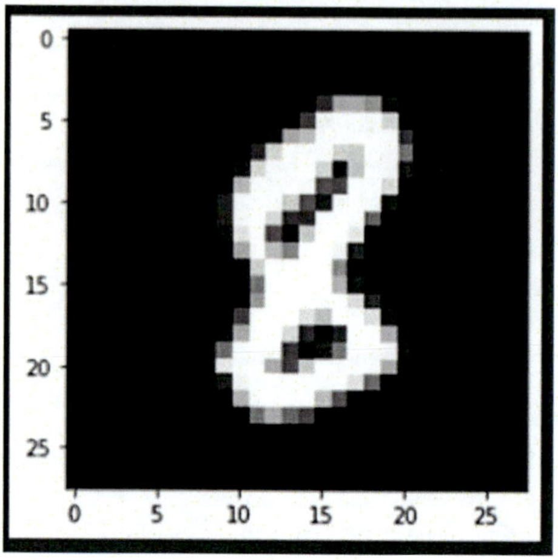

Fig. 12 Generated image from modified GAN

This flower has petals that are yellow, white, and purple and has dark lines.

The petals on this flower are white with a yellow center.

4 Conclusion and Future Work

We have used a simple model yet an effective one for synthesizing the images from the visual description given in detail. Plausible image creation from the detailed text description is demonstrated using the generator and discriminator network.

New Variations of GANs are developing nowadays, it is an energetic area in research. There are many domains where GANs can be applied. In the medical domain, GANs can be used for producing new chemical compounds. In computer vision, we can even blend two images and form a new image. A new video sequence can be created by recognizing the background of the image and time sequences are newly created for the foreground action. Shortly we may even see the expressions of the portraits which has become possible using Deepfake GANs.

References

1. Akata, Z., Reed, S., Walter, D., Lee, H., Schiele, B.: Evaluation of output embeddings for fine-grained image classification. In: CVPR (2015)
2. Radford, A., Metz, L., Chintala, S.: Unsupervised representation learning with deep convolutional generative adversarial networks. In: Proceedings of the ICLR (2016)
3. Bengio, Y., Mesnil, G., Dauphin, Y., Rifai, S.: Better mixing via deep representations. In: ICML (2013)
4. Donahue, J., Hendricks, L.A., Guadarrama, S., Rohrbach, M., Venugopalan, S., Saenko, K., Darrell, T.: Longterm recurrent convolutional networks for visual recognition and description. In: CVPR (2015)
5. Dosovitskiy, A., Tobias Springenberg, J., Brox, T.: Learning to generate chairs with convolutional neural networks. In: CVPR (2015)
6. Kingma, D.P., Ba, J.: Adam: a method for stochastic optimization. In: Proceedings of the ICLR (2015)
7. Gauthier, J.: Conditional generative adversarial nets for convolutional face generation. Technical report (2015)
8. Hochreiter, S., Schmidhuber, J.: Long short-term memory. Neural Comput. **9**(8), 1735–1780 (1997)
9. Arjovsky, M., Bottou, L.: Towards principled methods for training generative adversarial networks. In: Proceedings of the ICLR (2017)
10. Radford, A., Metz, L., Chintala, S.: Unsupervised representation learning with deep convolutional generative adversarial networks (2016)
11. Reed, S., Akata, Z., Lee, H., Schiele, B.: Learning deep representations for fine-grained visual descriptions. In: CVPR (2016)
12. Reed, S., Akata, Z., Yan, X., Logeswaran, L.: B text-to-image synthesis. In: Proceedings of the 32nd International Conference on International Conference on Machine Learning, pp. 1060–1069 (2016)
13. Wah, C., Branson, S., Welinder, P., Perona, P., Serge Belongie, S.: The caltech-UCSD birds-200-2011 dataset
14. Mansimov, E., Parisotto, E., Ba, J.L., Salakhutdinov, R.: Generating images from captions with attention. In: ICLR (2016)

QCNN—A Conceptual Framework for Duplicate Removal in Big Data Using Quality Assured Crowd Sourcing Coupled with Neural Networks

K. Shantha Kumari, Mohamed Ashik, K. Gowrishankar, and P. Kanmani

Abstract Success of an Enterprise depends on the quality of the data available in its backup. The data need to be consistent, error-free and maintain its integrity irrespective of any threats in any form. Quality data is aka clean data helps in the decision-making process of the Enterprise. However due to multiple entry points available in populating a database, there exists the problem of duplication data, which is one form of dirty data. This duplication data has to be removed without compromising the privacy and integrity of the data. The automated computing algorithms have been limited in its capability for the tasks such as labelling, filtering, translating or transcribing data etc. Hence, crowd sourcing famously referred as human computing are used as supplement to identify and eliminate the duplications available in data. However, it is important to understand that the human crowd is prone in error making and it is essential to choose a qualified crowd that does less error. Because the data labelled by the human crowd is fed to the Neural networks further to identify and eliminate the duplication. This paper provides a conceptual framework that aims to identify the duplicate data through qualified crowds and remove the identified duplication using neural network techniques. The framework explains how to qualify the best crowd to do work, so that clean data is assured with the help of ANNs.

Keywords Crowd sourcing · Quality · Data duplication · Human computing · Neural networks

K. Shantha Kumari (✉) · P. Kanmani
Data Science and Business Systems, SRM Institute of Science and Technology, Chengalpattu, India
e-mail: shanthak@srmist.edu.in

M. Ashik
College of Computing and Information, University of Technology and Applied Sciences, Salalah, Oman

K. Gowrishankar
Electrical and Electronics Engineering, AMET University, Chennai, Tamil Nadu, India

C. Iwendi et al. (eds.), *Proceedings of ICACTCE'23—The International Conference on Advances in Communication Technology and Computer Engineering*, Lecture Notes in Networks and Systems 735, https://doi.org/10.1007/978-3-031-37164-6_17

1 Introduction

Enterprises are bound to take decisions every day for their sustainment in business activities. The decisions require evidence, which are provided by the enormous data that finds its way to the enterprises. For example, if the enterprise wishes to solicit customer's preferences for new products, it can collect through email, phone, live chats, SMS, forms, interviews etc. Same customer can interact with the enterprise feedback collecting system via several entry points such as email, WhatsApp, twitter etc. The data now many be duplicate due to several entries by the same customer. The data collected through such feedback mechanism might be helpful for an enterprise in terms of being voluminous, yet not unique [1–3]. Unique entries are meaningful.

Hence it is important to identify the duplicates in data and remove them. Computers and computing algorithms have been pathbreaking for decades in solving many human impossible tasks. However, the machines are limited when it comes to the task for imaging, translating, transliterating, labelling, filtering, identifying duplicates etc., [4]. There is a need for a supplementary technique which would help the machine to do its work correctly. That kind of technique is human computing [5, 6]. Crowd sourcing is a tool that a human computing uses to distribute the tasks among the crowd to get the work done [7, 8]. Crowd sourcing is the name coined by Jeff Howe and Mark Robinson. It is defined as concept that solves issues which are difficult for a machine to complete by itself. Crowd sourcing integrates the capabilities of both humans and machines.

Crowdsourcing has found its success in many areas such as image processing, sentiment analysis, opinion mining viz [9–11]. Commercial tools for implementing crowdsourcing has found its success and this shows the capabilities of crowdsourccing [12]. Crowdsourcing had proven its capability in detecting duplication in data thereby improvising the process of duplication removal [13, 14]. Several commercial tools for crowdsourcing such as Crowdclustering [15], CrowdER [16] are availabe from the start of 2010 itself. However, the efficiency of a crowdsourcing application or task completion depends on the quality of the crowd that performs the task. If the error generated by the crowd while doing the task is minimum, then the output has its own accuracy. The reverse will worsen the task expectations thereby affecting the purpose for which the crowdsourcing is considered as the supplementary technique. Hence it is important to monitor the quality of the crowd that performs the given task.

Data duplication identification and its removal are considered to be primary tasks in the decision making process of an Enterprise policy management. This paper discusses about a conceptual framework—QCNN that harness the power of both crowd sourcing and machine learning algorithms in effectively removing the duplication in data. The contribution of this paper includes a quality model definition that finds the well-performing workers, an duplicate identification with annotation algorithm based on consenus among the good works and duplicate removal methodolgy using ANN. The paper proceeds by describing the conceptual model that can be used for Data duplication detection and removal.

2 Background

The conceptual model proposed in this paper is based on the case study given as follows—an education institution has so many stakeholders whose feedback data plays vital role in policy decision. The education institution is in process of implementing new policies and products to tackle the competitions in education forum. The new policies are shaped based on the stakeholder feedback. These new policies and product are developed for the benefit of students during this online new normal way of learning process. The whole process is briefed as follows:

- The educational institution opens its feedback channels to its stakeholders.
- The feedback channels include social media, email, texts, telephone conversations, blogs, forms viz.
- A stakeholder can participate in any feedback channel for any number of times.
- A database at the backend is populated by the feedback data.
- The management of the educational institutions takes the feedback data to decide upon new policies and products.
- The feedback data is filled with duplicate data since, same stakeholders can give feedback for several times.
- The educational institution now wants to have concrete data so that the new policies and products are decided seamlessly.
- Hence, there is a need for data duplication detection and removal framework.

The following subsection provides the basics of the conceptual framework—QCNN that can be employed to the above case study.

2.1 Characterizing the Strength of the Crowd

The conceptual model to be proposed in this paper has its success owing to the performance of the crowd, that is assigned to identify the duplicate data [12]. The main obstacle of the crowdsourcing projects is the poor performance of the crowd. This indirectly affects the quality of the data stored in the database [5]. Many techniques based on management of human resources are proposed to identify the quality workers [4]. Along with those techniques, there exists a statistical methodology of quality assurance based on manufacturing process lines is also being used in recent times [5].

In our conceptual Model, a Quality evaluation algorithm is proposed based on the quality model given in [12] *is used. The quality model used in this conceptual model is three-fold—it comprises of definition of quality parameters, assessment of the quality parameters and assurance of quality.*

2.2 Data Duplication Identification

Duplicates in data is the kind of dirty data. This data will affect any decision-making process that happens in an enterprise. Before removing the duplication, it is important to identify which pairs of words or entries are actually duplicate. For this purpose, many attributes of the data are considered. They include date of creation, date of entry in the database, tokens used in the data, textual presentation in the data, context of the data stored etc. [5]. However, the machine computing algorithms are found to be limited to identify the duplicates in the data. In literature, crowd sourcing is found to be much more efficient supplementary method in identifying and labelling the duplicated in the data stored in the database [9].

In our conceptual Model, the duplication in pairs is identified by adapting in-house crowdsourcing methodology as inspired from [5]. *The records in which duplication need to be checked are divided into batches and it is defined as Human intelligent task [HIT]. These batches are given to the crowd to find the similarity. The human crowd will categorize the records based on similarity as Duplicate Record [DR], No Duplicate Record [NDR], Likable Duplicate Record [LDR]. It is obvious that human errors do occur in categorization. In order to mitigate the errors, the batches are distributed to multiple set of crowd and majority voting is considered for final labelling. The crowds are rewarded if they could categorize correctly.*

2.3 Data Cleansing—Data Duplication Removal

Data cleansing is the approach used to clean the dirty data. This is done by various traditional methods as given in [17]. The data cleansing step is also same as the data reduction in terms of voluminous in big data. Data cleansing includes identifying, detecting, removing the errors, and thereby improving the quality of the data. Considering the voluminous data, it is important to have faster data cleansing methodologies. Usually, enterprises employ professional knowledge experts to achieve the faster data cleansing. This sometimes ends up costly by allowing a third party to access the propriety data of the company which could break the security and integrity of the data.

The qualified crowd comes as the replacement of professional third party. This crowd is picked from the in-house. The data is analyzed by the crowd and annotated as duplicate, not duplicate and may be duplicate based on various data attributes. The annotated data can be manually cleaned for a volume that is manageable. However, as an enterprise data, the volume is very Big. Hence the human computing is not possible as a wholesome. The combined effect of machine computing is needed now [5, 18]. Many supervised and unsupervised computing algorithms are available to clean the duplicate data. Artificial neural networks are one such machine computing paradigm that perfectly fits in data cleansing.

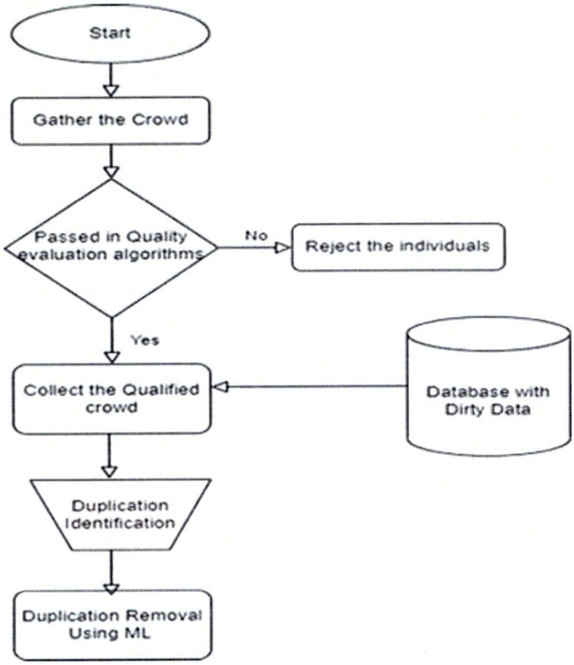

Fig. 1 Process of duplication removal with human computing and machine learning

This conceptual model uses the Deep neural networks with whale optimization [18] *as machine computing process in eliminating the data duplication removal with much success rate.*

Thus, the briefed process is presented as following (Fig. 1).

3 Description—The Conceptual Model

QCNN is conceptual proposal designed to have an effective duplicate removal scheme in Big Data. The advantages of QCNN are truly empowered with merits of crowd sourcing and machine learning algorithms. These two different computing paradigms are combined in QCNN to achieve higher accuracy. This section details various phases in QCNN-three-tiered conceptual model. First tier of QCNN details about the Qualifying the crowd, the second tier about duplication identification through the identified crowd. The Third tier of QCNN involves the Machine learning algorithms specifically neural networks as given in [18].

3.1 Tier-1: Qualifying the Crowd

Crowd sourcing is the outsourcing a task to the crowd people through open calls. There are two types of peoples available in Crowdsourcing-One is the Requestor, the person who wants the tasks to be done, another one is the Worker, the person who performs the tasks. The Requestor places the task in any crowdsourcing platform. The output is obtained by solving the tasks. Requestors check the output. If the output is same as the Requestors expected level, the workers are given with incentives or rewards such ass money, badges, coupons etc. A pre-informed negotiations must be made between the Requestor and the Workers regarding the acceptance or rejection of the output based on the quality levels.

According to the authors in [19] the quality of a crowdsourced task depends on the quality of the workers involved in it. Existing methodologies available in determining the quality of the crowd and ultimately the quality of the tasks completed by the mixture of crowd with respect to their capabilities and nature are limited for many reasons as stated in [12] and the following representation is adapted from same literature highlights the crowdsourcing participants with respect to the quality parameters (Fig. 2).

Another illustration adapted from [12] given in Fig. 3 depicts the various components that are required to develop a quality model that measures the quality of the crowd.

- Quality Model—it captures the various parameters that define quality of the product, data or task executed by the crowd. The parameters include accuracy of the worker, speed in execution etc. QCNN uses the following quality model for its quality checking (Fig. 4).

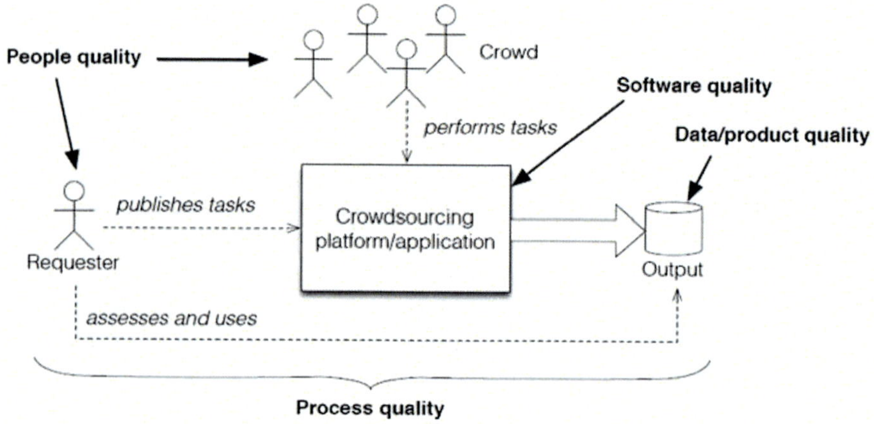

Fig. 2 Crowdsourcing basic model

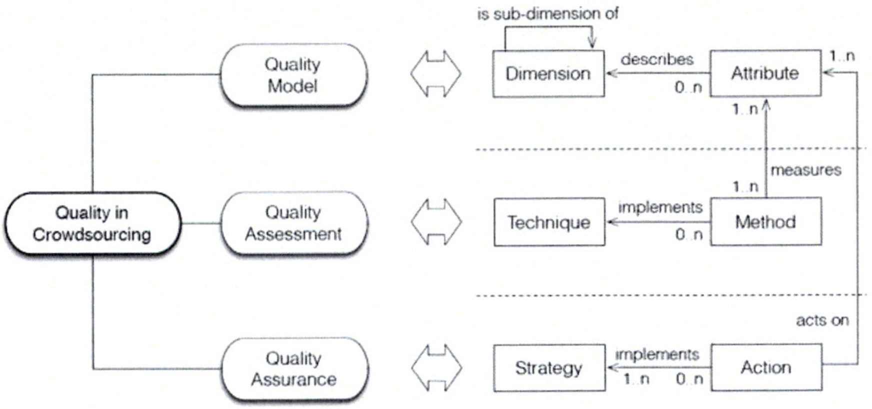

Fig. 3 Components of measuring quality in crowd

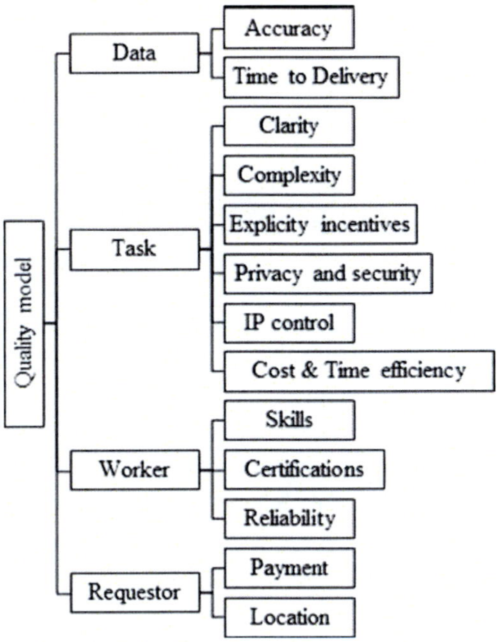

Fig. 4 Quality model-attributes

- Quality assessment—it defines the various methodologies used to assess or evaluate the parameters of the quality. Examples of assessment includes surveys, questionnaire, voting etc. The Quality assessment proposed in QCNN is as follows (Fig. 5).

  The qualification test as given in the above assurance model is done as based on the quality evaluation algorithm given in [4, 5]. The Quality evaluation algorithm is as follows:

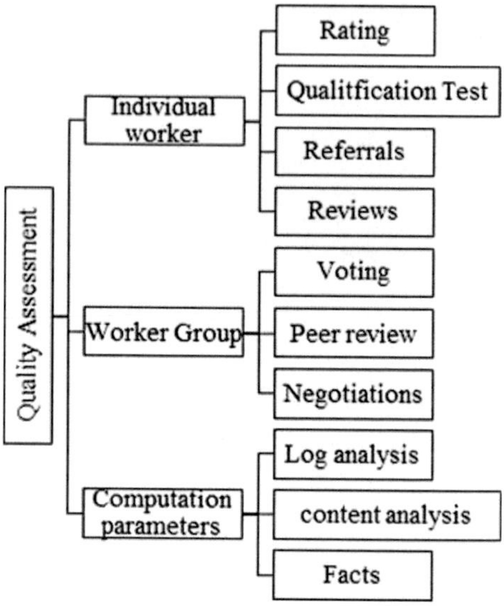

Fig. 5 Quality assessment attributes

Input: Tasks whose outputs are known and standard [Gold tasks]

Process:

Given 'n' crowd and 'm' Gold tasks:
For 'i'th worker in crowd, assign 'n' the gold task
Compare the output obtained with actual output
Compute error
Evict the worker if the error is above threshold
Check the Quality model for each task done

Output: Qualified crowd

- Quality assurance—it consists of the various strategies and methods used to boost the quality of the worker. Trainings, feedback analysis, rewarding etc. are some of the quality assurance tasks. The Quality assurance proposed in QCNN is as follows (Fig. 6).

By following the above proposed quality framework, it is possible to have a well-qualified crowd at work.

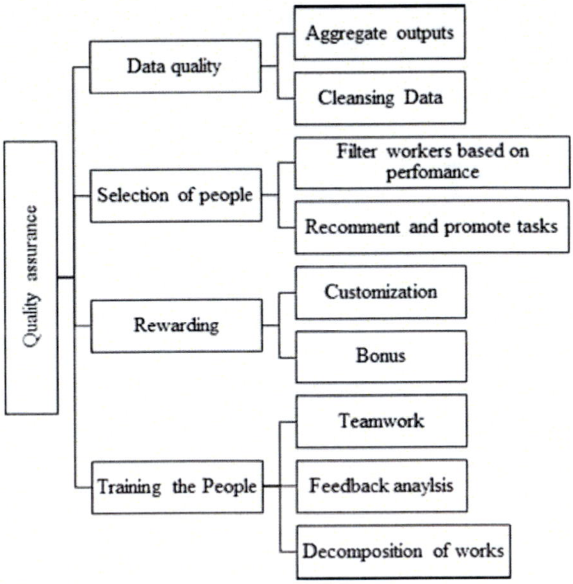

Fig. 6 Quality assurance attributes

3.2 *Tier-2: Data Duplication Identification by Qualified Crowd*

This marks the second tier of QCNN. In this tier, Data Duplication identification is done by the humans who have qualified from the first tier. QCNN uses the methods identified in [5, 20].

- The pairs of data to be identified as duplicates are chosen based on the date, time, any other common information such as phone number, name etc. The similarity index is calculated as given in [5].
- If the similarity index is greater than the threshold, the pair is annotated as Duplicate by the crowd.
- Then the same pairs of data are given to another crowd and the process repeats. Finally majority decision yields and the awards are given to correct crowd.

3.3 *Tier-3: Duplication Removal*

Data duplication is the Data cleansing methodology that is used to make the data dirty-free [17] lists various methodologies used to clean the dirt data. MapReduce is one such very famous crowd-based algorithm used to deduplicate the data [20]. In QCNN, modern and innovative yet so strong Mimic Deep neural networks with

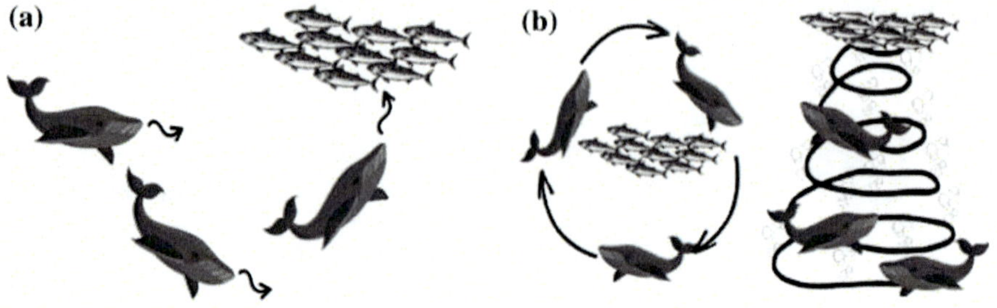

Fig. 7 **a** Searching the prey, **b** encircling and attacking prey

Chaotic whale optimization (CWO) [18] is proposed to clean the annotated data received form tier 2.

- Chaotic Whale optimization [21]—The social behaviour of the Whales is the inspiration of this optimization technique. This is an improvised version of Whale optimization that is based on the hunting strategy followed by humpback whales. The three phases of hunting such as searching the prey shown in Fig. 7a, encircling the prey, and then attacking the prey is used in optimization shown in Fig. 7b. However, the whale optimization suffers with slow convergence rate. Hence Chaos is introduced to improve the convergence rate as well as avoiding the local optima decisions.
- Mimic Deep Neural networks- This Machine learning model N number of has Input, Hidden and Output layers and use relu as activation function for triggering the neurons. Weights are appropriately assigned to every layer. Feedbacks through back-propagation ensures proper weight adjustment. MDNN architecture is inspired from [5] and shown in Fig. 8:
  - The crowd-sourcing based annotations schemes produce a training set to the Mimic Deep neural networks, MDNN. These networks progress with regular time interval in training and there by recognizing any given pairs as duplicate or not. This acts as complementary technique to the crowd-sourced annotation technique.
  - The Chaotic Whale Optimization (CWO) algorithm is used for trust computation of data using multiple decision metrics. The computed trust values and the nature of the data are sequentially applied to the training process by the Mimic Deep Neural Network (MDNN).
  - Based on the classification system finds the duplicated files and deletes it; otherwise, the user needs to investigate again to perform the duplicate identification.

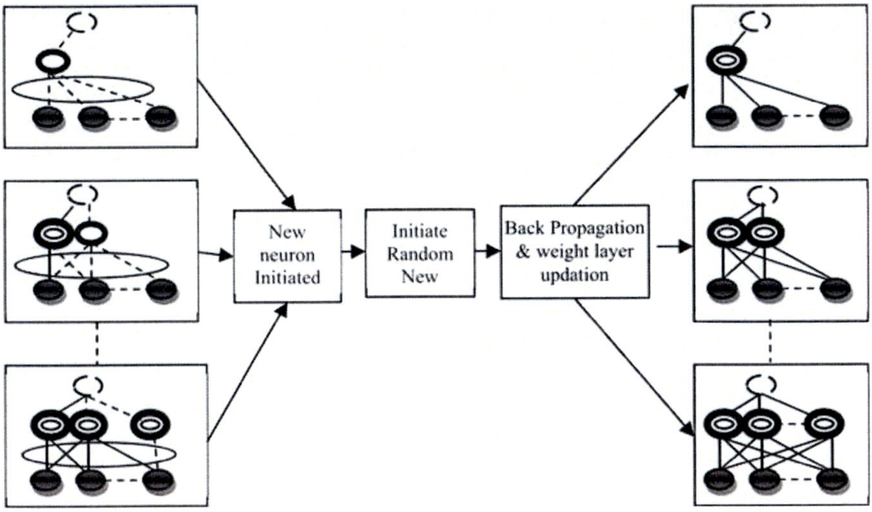

Fig. 8 MDNN overall architecture [18]

3.4 QCNN—Complete System

The following section delivers the conceptual model used in QCNN to deduplicate the data in Big Data.

- Selection of Quality crowds using Quality evaluation algorithm.
- Categorization of Batches as DR, NDR, LDR by the Quality crowd. The Categorized batch of records act as training set to the MDNN.
- MDNN performance is optimized by using CWO, thereby training errors, generalization errors are minimized, and accuracy is improvised, and the duplicates are finally removed.

The entire process is illustrated as follows (Fig. 9).

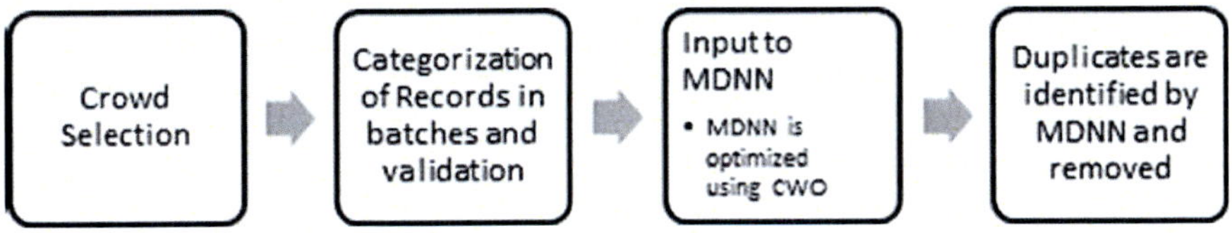

Fig. 9 QCNN—flow

3.5 Simulation Setup

The case study discussed in Sect. 2 is simulated with QCNN. Approximately 4000 records via social media, email, feedback forms etc., are collected from the stakeholders of the educational institution. In order to categorize the records, set of in-house crowds [75 employees] were chosen. They were initially trained and explained about the process. Then they are assessed based on the quality evaluation model. The selected crowd [47 employees] is now practiced for categorization of records into DR, NDR and LDR. Appropriate rewards are given to correct categorization. Thus, the entire record is now categorized. It is fed to MDNN enhanced with CWO. DNN with CWO is implemented in python using Juptyer notebook under Windows 10 operating systems in an Intel i7 processor.

4 Results

This section explains the results obtained from the simulation. The simulation is a minimal version of big data. Hence the results can be utilized for validation while employing much more voluminous data.

4.1 Selection of Quality Crowd

The selected crowd is evaluated based on Quality model, Quality assessment and Quality assurance. The following graph narrates the filtering process (Fig. 10).

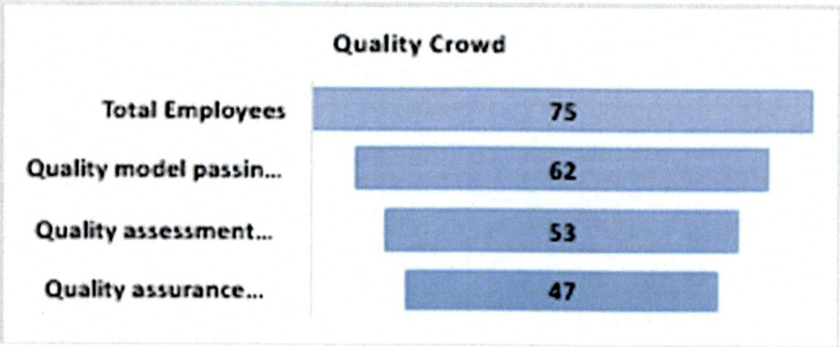

Fig. 10 Quality crowd selection

Table 1 Majority selection and categorization

	Record 1	Majority	Categorization
Employee 1	NDR		DR
Employee 2	DR	DR	
Employee 3	DR	DR	
Employee 4	NDR		
Employee 5	LDR		
Employee 6	LDR		
Employee 7	DR	DR	
Employee 8	DR	DR	
Employee 9	DR	DR	
Employee 10	NDR		

Table 2 Summary of categorization

Number of employees	47	
Total record	4196	
Number of records in a Batch	90	
Duplicate records	1725	Training set to MDNN
Non-duplicate records	1877	
Likeable duplicate records	594	

4.2 Categorization of Batches

The selected crowd is now given batch of records. Based on the similarities, the crowd must categorize the records as DR, NDR, LDR. The batches are interchanged between various crowds and majority decision is used for final categorization. Each employee must evaluate a batch of 90 records. For e.g., a record from a batch is given to Employee 1 to Employee 10. The results are obtained as follows (Table 1).

Every record is similarly labeled. The summary of the categorization is given as follows (Table 2).

4.3 Duplication Removal Using MDNN with CWO

The training set is fed to deep neural network. In this case, Convolutional neural networks (CNN) in its simplest form are considered. CNN process the data in the form of grids. The input, hidden and output layers are suitably considered. The Chaotic values for Whale optimization are used to minimize the training errors, thereby increasing the accuracy of the MDNN. The Chebyshev chaotic map $u_{i+1} =$

Table 3 Performance of MDNN

	Time taken(sec)
Without CWO	15.6
With CWO	10.8

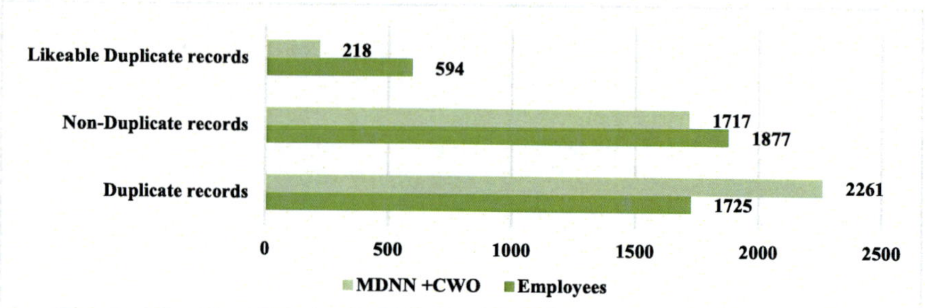

Fig. 11 Performance comparison

$\cos(icos^{-1} - 1(u_i))$ is used for the above said purpose. Here in the equation 'u' is the initial state that progresses to minimize the training errors in MDNN. Following table shows the summary of duplicate detection by MDNN with and without CWO (Table 3).

Finally, the following graph shows the performance comparison between crowd and the machine. The MDNN finds more LDR to be DR and certain NDR is also categorized to DR. Thus, the accuracy for duplication identification and subsequently removal is improvised when compared with crowdsourcing (Fig. 11).

5 Conclusion

The Data Deduplication in Big data is a challenging task. Many solutions have been provided to address the deduplication processes. However, those solutions stagger with poor accuracy in the process. The accuracy can be improvised by the conceptual framework given—QCNN. This framework used both crowd sourcing a format of Human computing and the Machine learning algorithms to identify the duplications correctly and remove it. The success of QCNN depends of the quality of the crowd, who can produce higher accuracy in deduplication. The future of this study is the real time implementation of QCNN thereby comparing the conceptual and practical QCNN model with the advantages and barriers in implementation.

References

1. Karthikeyan, R., Balamurugan, V., Cyriac, R., Sundaravadivazhagan, B.: COSCO2: AI-augmented evolutionary algorithm based workload prediction framework for sustainable cloud data centers. Trans. Emerging Tel. Tech. **34**(1), e4652 (2023). https://doi.org/10.1002/ett.4652
2. Mantey, E.A., Zhou, C., Srividhya, S.R., Jain, S.K., Sundaravadivazhagan, B.: Integrated blockchain-deep learning approach for analyzing the electronic health records recommender system. Front. Public Health **10**, 905265 (2022). https://doi.org/10.3389/fpubh.2022.905265
3. Saleem Raja, A., Sundarvadivazhagan, B., Vijayarangan, R., Veeramani, S.: Malicious webpage classification based on web content features using machine learning and deep learning. In: 2022 International Conference on Green Energy, Computing and Sustainable Technology (GECOST), Miri Sarawak, Malaysia, 2022, pp. 314–319. https://doi.org/10.1109/GECOST55694.2022.10010386
4. Sorokin, A., Laughlin, G., Hester, V., Le, J., Biewald, L.: Programmatic gold: targeted and scalable quality assurance in crowdsourcing. In: AAAIWS'11-11: Proceedings of the 11th AAAI Conference on Human Computation (2011)
5. Saberi, M., Hussain, O.K., Chang, E.: Quality management of workers in an in-house crowdsourcing-based framework for deduplication of organizations' databases. IEEE Access **7**, 90715–90730 (2019)
6. Pantic, M., Pentland, A., Nijholt, A., Huang, T.S.: Human computing and machine understanding of human behavior: a survey. In: Artificial Intelligence for Human Computing. Lecture Notes in Computer Science, vol. 4451, pp. 47–71 (2007)
7. Lease, M., Alonso, O.: Crowdsourcing and human computation: introduction. In: Encyclopedia of Social Network Analysis and Mining. Springer, New York, NY (2018)
8. He, H.R., Liu, Y., Gao, J., Jing, D.: Investigating business sustainability of crowdsourcing platforms. IEEE Access **10**, 74291–74303 (2022)
9. Neto, F.R.A., Santos, C.A.: Understanding crowdsourcing projects: a systematic review of tendencies, workflow, and quality management. Inf. Process. Manag. **54**, 490–506 (2018)
10. Jing, Z., Guo, W., Liang, R., Wang, L., Fu, Z., Sun, J.: How to find the key participants in crowdsourcing design? Identifying lead users in the online context using user-contributed content and online behavior analysis. Sustainability **14**, 4 (2022)
11. Park, I., Vel, V., Liu, J.: The role of enterprise crowdsourcing systems on knowledge application. J. Comput. Inf. Syst. **62**(3), 587–597 (2022)
12. Daniel, F., Kucherbaev, P., Cappiello, C., Benatallah, B., Allahbakhsh, M.: Quality control in crowdsourcing: a survey of quality attributes, assessment techniques and assurance actions. ACM Comput. Surv. **51**, 1 (2018)
13. Yang, Y., Singh, P., Yao, J., Yeung, C.M.A., Zareian, A., Wang, X., Cai, Z., Salvadores, M., Gibbins, N., Hall, W., Shadbolt, N.: Distributed human computation framework for linked data co-reference resolution. In: The Semantic Web: Research and Applications, pp. 32–46 (2011)
14. Abboura, A., Sahrl, S., Ouziri, M., Benbernou, S.: CrowdMD: crowdsourcing-based approach for deduplication. In: 2015 IEEE International Conference on Big Data (Big Data), Santa Clara, CA, USA (2015)
15. Gomes, R., Welinder, P., Krause, A., Perona, P.: Crowdclustering. In: 24th International Conference on Neural Information Processing Systems (NIPS'11), Red Hook, NY, USA (2011)
16. Wang, J., Kraska, T., Franklin, M.J., Feng, J.: Crowder: crowdsourcing entity resolution. In: The 38th International Conference on Very Large Data Bases, Istanbul, Turkey (2012)
17. Ridzuan, F., Zainon, W.M.N.W.: A review on data cleansing methods for big data. In: The Fifth Information Systems International Conference 2019 (2019)

18. Praveena, A., Bharathi, B.: An approach to remove duplication records in healthcare dataset based on mimic deep neural network (MDNN). Concurrent Eng. Res. Appl. **29**(1), 58–67 (2021)
19. Kritikos, K., Pernici, B., Plebani, P., Cappiello, C., Comuzzi, M., Benrernou, S., Brandic, I., Kertés, A., Parkin, M., Carro, M.: A survey on service quality description. ACM Comput. Surv. (CSUR) **46**(1) (2013)
20. Priya, B.N., Devi, D.G.: Crowd sourcing-based deduplication in big data environment. Int. J. Recent Technol. Eng. **8**(4) (2019)
21. Gaganpreet Kaur, S.A.: Chaotic whale optimization algorithm. J. Comput. Des. Eng. **5**(3), 275–284 (2018)

Design and Implementation of Camera-Based Security System (CBSS) for Detecting Missing Persons

Mohamed Jasim Alhammadi, Hind Saeed AlMehrezi, Ahmed Mohamed AlAmeri, Saeed Khamis Alhammadi, Joseph Henry Anajemba, and Celestine Iwendi

Abstract An efficient AI-based Camera security system is important for locating missing persons. In simple words, if any suspicious face is detected by the system, it sends an alert to the authorized people. Thus, this eliminates the risk of a certain theft from the organization. The focus in our project is implementing and designing a camera-based security system (CBSS) for locating missing persons. The CBSS system is designed using a high resolution camera for detection and several facial recognition features were implemented in the model. In the end, the experiments of our program showed that the system had a more accurate detection notwithstanding the proximity of detection and in almost all the angular positions of detection.

Keywords Security system · Camera-based system · Face-recognition · Face-detection · Databases

All authors are equally contributed to this work.

M. J. Alhammadi · H. S. AlMehrezi · A. M. AlAmeri · S. K. Alhammadi · J. H. Anajemba (✉)
Department of Information Security Engineering Technology, Abu Dhabi Polytechnic, Abu Dhabi 11101, Abu Dhabi, United Arab Emirates
e-mail: joseph.anajemba@adpoly.ac.ae

M. J. Alhammadi
e-mail: A00054745@adpoly.ac.ae

H. S. AlMehrezi
e-mail: A00053989@adpoly.ac.ae

A. M. AlAmeri
e-mail: A00051922@adpoly.ac.ae

S. K. Alhammadi
e-mail: A00053930@adpoly.ac.ae

C. Iwendi
School of Creative Studies, University of Bolton, Bolton, UK
e-mail: celestine.iwendi@ieee.org

C. Iwendi et al. (eds.), *Proceedings of ICACTCE'23 – The International Conference on Advances in Communication Technology and Computer Engineering*, Lecture Notes in Networks and Systems 735, https://doi.org/10.1007/978-3-031-37164-6_18

1 Introduction

Face recognition is an important research problem spanning numerous fields and disciplines. This is because face recognition, in additional to having numerous practical applications such as bankcard identification, access control, Mug shots searching, security monitoring, and surveillance system, is a fundamental human behavior that is essential for effective communications and interactions among people [1]. In recent time, there have been different experimentation's conducted by several researchers in the field of security cameras Surveillance systems, the tracking of the camera and also the Change detection areas and how to operate them and the safety of the place that must be protected. With the advancement of technology and security tools, it is important that the authors mention that surveillance cameras have become an integral part of safety and security [2]. The more intelligent the surveillance cameras became, the more security Provided, and the researchers praised that the security of these cameras can be increased in several ways, and from these ways, we will mention to you after we mention the problems that hinder the safety of tracking surveillance cameras and the problems of the research, the camera coordination and detection is been foreground object detection becomes difficult when the background includes contrast due to light and shadow and It consumes more memory, which causes the problem of more use of the CPU, which needs a lot of money to compensate for the shortage of other cameras. The researchers praised that the surveillance cameras may drain a lot of time and effort to address the subject the objectives and The goal of this project are to design artificial intelligence through surveillance cameras and this is to increase a large number of surveillance cameras to cover a large number of space the results of the and The result of this research is that these limitations can be overcome by modelling the value of each pixel as a Gaussian mixture [3]. These limitations can be overcome. The value of each pixel is modelled as a mixture of Gaussians. According to the tenacity and variance of each combination of Gaussians, a pixel is determined by whether it belongs to the foreground or background. Once an object is detected, tracing is required to estimate its path in the image plane. In other words, the tracker assigns consistent labels to the tracked objects across different frames of the video. Laplacian and Gaussian filters; It is often implemented using a simple kernel [4]. In our project we are going to demonstrate of using Face-Recognition Technology to implement a system that can locate the people based on their face fingerprint. Some of the benefits of Face-Recognition and camera-based security system is helps find missing people, helps to minimize the time consumed on finding missing people and ease of use.

1.1 Face Recognition

To automatically and without human involvement identify objects in an image is a fundamental challenge in computer vision. Face detection is one such issue where we must identify human faces in an image. Although there may be very minute variations among human faces, it is safe to state that all human faces share a number of characteristics. Although there are several face detection algorithms, the Viola-Jones Algorithm is one of the first and is still in use today. We will utilize it in the following sections of the study. After reading this article, you can access the Viola-Jones Algorithm via the link at the bottom [5, 6].

1.2 Aim and Objective

The aim of our research is to design and implement a software that are based on python programming language. And is going to help us on finding missing people based on their face fingerprint [2]. Moreover, to help the organization to find the missed people based on their face fingerprint. Likewise, Objectives of our research is the following:

- Design a standalone software that will help finding a missing people.
- Implementation phase where the software will take the picture by a security camera.
- Those pictures will be matched by the other users in the database using the face fingerprint.
- Moreover, if the target face founded there will be an alarm will be sending the authorized admin.
- There will be location sent by the alarm device automatically.

1.3 Research Contributions

CBSS will serve a major contribution to our society by protecting the environment with recording video all the time and detecting faces for evidence if anything happens. Our project will contribute to the world in many different areas. Firstly, it's going to protect the organizations against theft and unauthorized users. As well as reducing the amount of time, cost, effort of the police officers. It will also provide better security measures in different aspects of work fields [5]. As well as ensuring the safety to people which will make them more comfortable and safety in their own environment.

2 Literature Review

The research of [7] presented a face recognition methods and applications such as Eigenfaces, Neural Networks, and Graph Matching. All types of these techniques are presented and discussed in detail as well it have been developed in their system.

The authors stated they have took out a study. Then used to perform and gather an information on facial recognition systems that are developed with the machine learning algorithms. On the other hand, the objective of the [8] was to develop a model and a term for understanding and discussing how the systems are going to identify and recognize systems. As well it illustrated about the relationship between the recognition and the other aspects of face processing. The proposed study of [9] Multi-Camera Based Surveillance System is to develop a system that can create and provide a system that are able to locate and track object while is it moving. There were different Results in the study which is they provided system that detect and track the objects in the camera. As well they provided an automatically warning in case of suspicious activity. Moreover, we have found some of the pitfalls in the system which is there were no Security and integrity and encryption methods [10]. Furthermore, the research of [9] illustrated about The Facial Recognition Technology. Their system was implemented to provide a detailed instructions about face recognition technology and the way of working the technology. Their results it is about user friendlies about using their technology using the Graphical User Interface (GUI) so that the use of the technology becomes easy. There were disadvantages that are previewed in the system which is there were no specific framework that can be followed in the paper. As well there were some factors that are used to affect the performance of their recognizer that they used. As well lastly, they didn't use specific algorithm to implement them. Moreover, the study of [11] show the way of integrating the Databases with the Facial recognition technology and their benefits. Their result was about providing interface that is efficient and has an access with more than one database. Their pitfalls was the functions that are not implemented to provide the interface to connect to the database. The experiments presented in [11] was about how the Reidentification people on the wide area camera networks where is used to prepare a study about creating Facial Recognition System with Uses a wide area camera so to make sure that their system is efficient. Their results provided a high resolution of taken images. Their pitfalls were about the system requirements was accepting all the images with no specification which can leads to false positives results [4]. T. Deshpande and S. Ravishankar [12], their proposed methodology depends on to detect and identify the human face in the given image, using Viola-Jones object detection framework and Linear Discriminative Analysis and Artificial Neural Network to detect and identify the human face in a clustered image. Several recent studies [13], carried out the new study about the parts of the human face information analyzer, by dealing with the face as a promising model they depending on locating the facial features in images for detecting faces. They build from complex background images a fully automated human facial measurement system. Their work detects facial features such as nose, eye, and mouth is an important step for analyzing human face tasks. Their study presents a simple model approach depends on Viola-Jones object detection framework algorithm concerted with symmetric information and geometry of the face parts of the image. Several recent studies [13], carried out a new proposes which improve the approach for face detection by using Open CV, Viola-Jones algorithm based on coding eyes which removes the falsely detected faces. The Haar training module in Open CV is an implementation of the Viola-Jones framework, the input represents a training group of positive and negative images, the algorithm work to generate strong features in the format of an XML file for detecting the wanted face and eyes in a given image, to speed up Haar-like features calculation for each image the integral image is used to collect the weak classifiers and produce a strong classifier dataset and the Adaboost algorithm is implemented. By using a classifier cascade process, the speed and accuracy of the face detection system are increased. Singh and Kaur [14], they carried out that human face detection is the process of detecting region of the face from a picture of one or multiple persons together [15]. That depends on the viola-Jones algorithm to detect and process the face. The correlation model is the use of recognition face in the proposed model. The face recognition process can detect the person among the database of faces without knowing any other details about the person-specific.

3 Research Design

In this section, the overall research design and all the applicable tools for the research are discussed.

3.1 System Implementation

3.1.1 Proposed System

According to Proposed system (Fig. 1) which illustrates about the way that the system architecture and how is going to work. Firstly, we have created the portal which is a website that contains all the user's information and results. The main function of the portal is to connect the users to the system so that it can be used by all users with no difficulties. Moreover, we have chosen a website for our portal it has major advantages such as it is accessible to anyone, anywhere, and anytime. As well it can be accessed during out of business hours. Secondly, Two Factor Authentication is implementing while a user logging in into the system so that we make sure that he is using his real account. This type of authentication is used to verify the identity of the user by using one of the factors such as something that we know such as a passcode, and something that we have such as a key, and something that we are such as a fingerprint. While the user has access to the website, he could create a new search that he is going to look for. Then he should provide an information about the missed user so that we able to find the face that we are looking for. Moreover, the Website is going to check for the image resolution. If it was success the user will be able to submit the form and they we are going to the python program phase. Lastly, if the face is found and match there will be an alarm will be sent to the user using

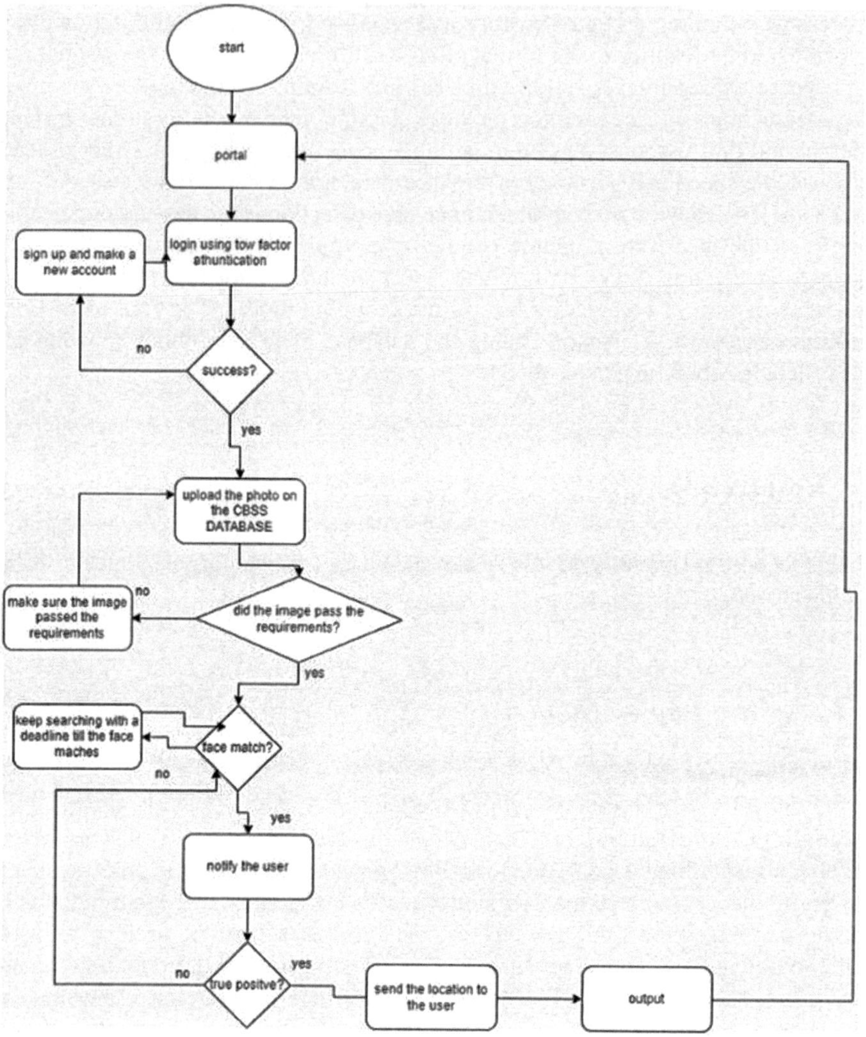

Fig. 1 Proposed system architecture

the portal (website) then the user will verify if it is true the face that are found or no. if it was found our system is going to send a notification to the user using an SMS message or Email that he chooses since creating his account.

4 Results and Implementation

A website is a set of web pages and related material that is published on at least one instance and given a shared web address. Websites like Google, Facebook, Amazon, or Wikipedia are examples of well-known sites.

A web application main page, also known as the home page, is the first page that a user interacts with when accessing a website. It serves as the gateway to the rest of the application, and its design and functionality should be carefully thought out in order to provide a positive user experience (Figs. 2 and 3).

The "Create a New User" page is designed to be simple and straightforward, with clear instructions and intuitive navigation. It includes fields for the user to enter their basic information, such as their name, email address, and password. Depending on the web application, additional fields may also be required, such as a billing

Fig. 2 CBSS GUI

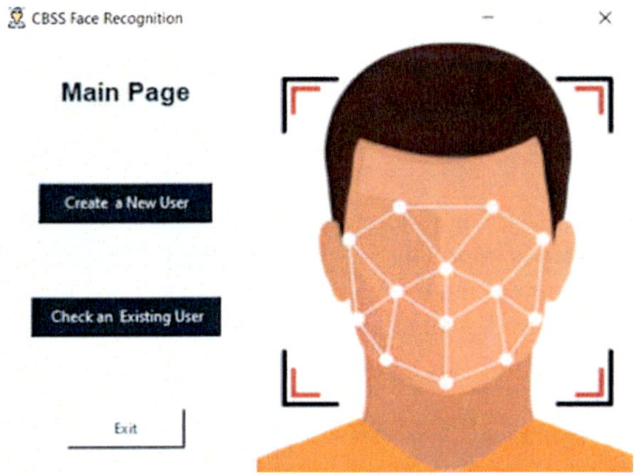

Fig. 3 CBSS user detection page

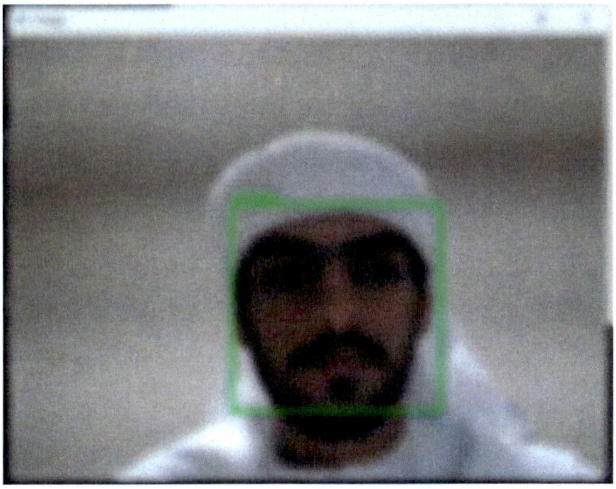

address, phone number, or answers to security questions. After creating a user profile, users can go ahead to upload the photo of the missing person based on the standard acceptable picture format.

5 Programming Related Code

5.1 *Face Detection*

The first step of the CBSS is to locate the face in an image or video stream. This can be done using various computer vision algorithms and deep learning models. As presented in Fig. 2, the system automatically scans its surrounding and attempts to map each detected face with the already created image in the database. The comparison feature of the algorithm is initiated to compare all detected face until an accurate and exact facial recognition is attained (Fig. 4).

5.2 *Face Alignment*

The next step is to align the face so that it is properly positioned for recognition. This step is important as it helps to normalize the face for comparison and eliminate variations caused by head poses, proximity and facial expressions. The module and program for Facial alignment is presented in Fig. 5.

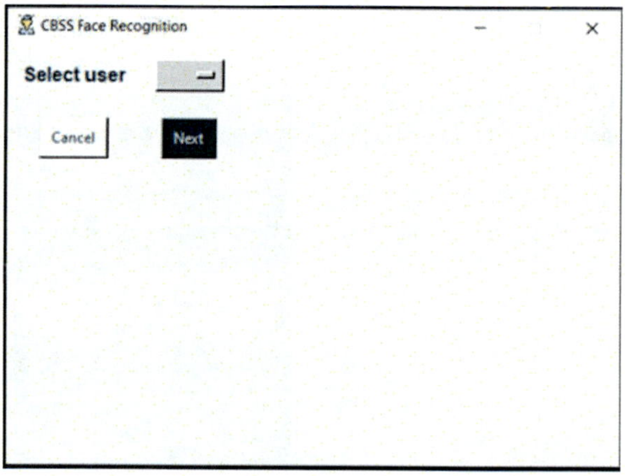

Fig. 4 CBSS user selection page

```
if confidence > 50:
    #if u want to print confidence level
        #confidence = 100 - int(confidence)
        pred += +1
        text = name.upper()
        font = cv2.FONT_HERSHEY_PLAIN
        frame = cv2.rectangle(frame, (x, y), (x + w, y + h), (0, 255, 0), 2)
        frame = cv2.putText(frame, text, (x, y-4), font, 1, (0, 255, 0), 1, cv2.LINE_AA)

else:
        pred += -1
        text = "UnknownFace"
        font = cv2.FONT_HERSHEY_PLAIN
        frame = cv2.rectangle(frame, (x, y), (x + w, y + h), (0, 0, 255), 2)
        frame = cv2.putText(frame, text, (x, y-4), font, 1, (0, 0,255), 1, cv2.LINE_AA)
```

Fig. 5 Facial alignment module

5.3 *Feature Extraction*

In this step, features such as the shape of the face, the eyes, nose, and mouth are extracted from the aligned face. These features are then used to create a unique representation of the face, known as a feature vector [4]. The module for creating the feature extraction is represented in Fig. 6.

5.4 *Comparison*

The feature vector is then compared to a database of known faces to determine the closest match. As seen in Fig. 7, the comparison process typically involves finding the euclidean distance between the feature vectors.

```
face_cascade = cv2.CascadeClassifier('./data/haarcascade_frontalface_default.xml')
recognizer = cv2.face.LBPHFaceRecognizer_create()
recognizer.read(f"./data/classifiers/{name}_classifier.xml")
cap = cv2.VideoCapture(0)
pred = 0
while True:
    ret, frame = cap.read()
    #default_img = cv2.cvtColor(frame, cv2.COLOR_BGR2RGB)
    gray = cv2.cvtColor(frame, cv2.COLOR_BGR2GRAY)
    faces = face_cascade.detectMultiScale(gray,1.3,5)
```

Fig. 6 Feature extraction module

```
roi_gray = gray[y:y+h,x:x+w]

id,confidence = recognizer.predict(roi_gray)
confidence = 100 - int(confidence)
pred = 0
if confidence > 50:
    #if u want to print confidence level
        #confidence = 100 - int(confidence)
        pred += +1
        text = name.upper()
        font = cv2.FONT_HERSHEY_PLAIN
        frame = cv2.rectangle(frame, (x, y), (x + w, y + h), (0, 255, 0), 2)
        frame = cv2.putText(frame, text, (x, y-4), font, 1, (0, 255, 0), 1, cv2.LINE_AA)

else:
        pred += -1
        text = "UnknownFace"
        font = cv2.FONT_HERSHEY_PLAIN
        frame = cv2.rectangle(frame, (x, y), (x + w, y + h), (0, 0, 255), 2)
        frame = cv2.putText(frame, text, (x, y-4), font, 1, (0, 0,255), 1, cv2.LINE_AA)
```

Fig. 7 Face comparison module

5.5 *Identification*

If a match is found, the system will identify the person as the one who corresponds to the closest match. If no match is found, the system may prompt the user to enter their information or offer the option to add their face to the database. It's important to note that the accuracy of facial recognition systems can be impacted by various factors, such as lighting, pose, and occlusions. In addition, there are also ethical and privacy concerns surrounding the use of facial recognition technology.

5.6 *Facial Detection Analysis*

In this subsection, we discuss the analysis of facial detection by the proposed CBSS with respect to distance, angular position and rate of detection.

5.6.1 Experimentation of Face Detection Distance and Time

One of the aims of our experiment is to establish how well our proposed model performed in terms of detection based on different distances. This is to ensure that no matter the range in proximity, the AI-based model can be able to efficiently detect and provide accurate result. Table 1 presents the results of detection for different distances and the rate at which the system established an accurate detection. The result of the experiment indicates that no matter the distance, the system was still

Table 1 Detection result according to distance and detection time

Distance (cm)	Status	Detection rate (s)
100	Detected	5.1
150	Detected	5.5
200	Detected	7.5
250	Detected	8.0
300	Detected	10.0
500	Detected	11.2
750	Detected	12.1
1000	Detected	15.0
1500	Detected	28.0

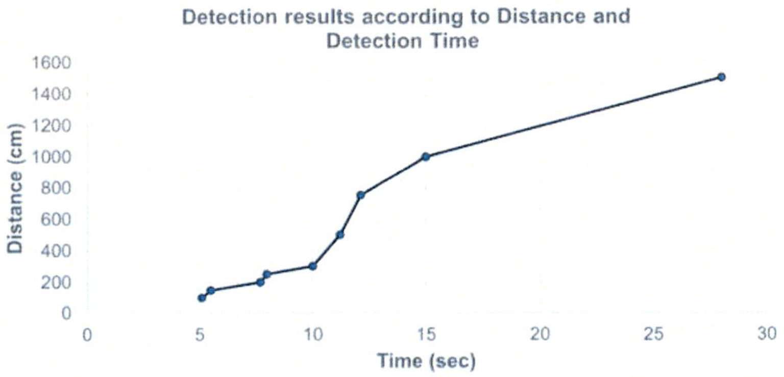

Fig. 8 Detection result with respect to distance and detection time

able to make accurate detection, however, there is a little variation in the detection times as the distances increases. Unlike previous techniques [16, 17] which only detected for a distance of <500 cm, our proposed system outperformed their research (Fig. 8).

5.6.2 Facial Positioning Experiment

In this experiment, we attempt to position the face in different angular positions in order to determine the accuracy in detection as illustrated in Table 2. The result shows that only when the face is directly facing the camera we can have a 100% detection, however, 80% and 70% detection rates can be achieve when the face is twisted at 15° and 30° respectively. This result shows that the missing person does not have to be facing the camera straight on for a near-accurate detection to be achieve. As long as there are some facial features detected by the AI-based camera, a true positive detection can be realized. On the other hand, the system have been programmed to

Table 2 Face detection result based on face position

Facial angular position (°)	Status	Detection rate (%)
Straight look	Detected	100
15 right twist	Detected	80
30 right twist	Detected	75
45 right twist	Not detected	0
15 left twist	Detected	80
30 left twist	Detected	75
45 left twist	Not detected	0
Upward	Not detected	0
Downward	Not detected	0

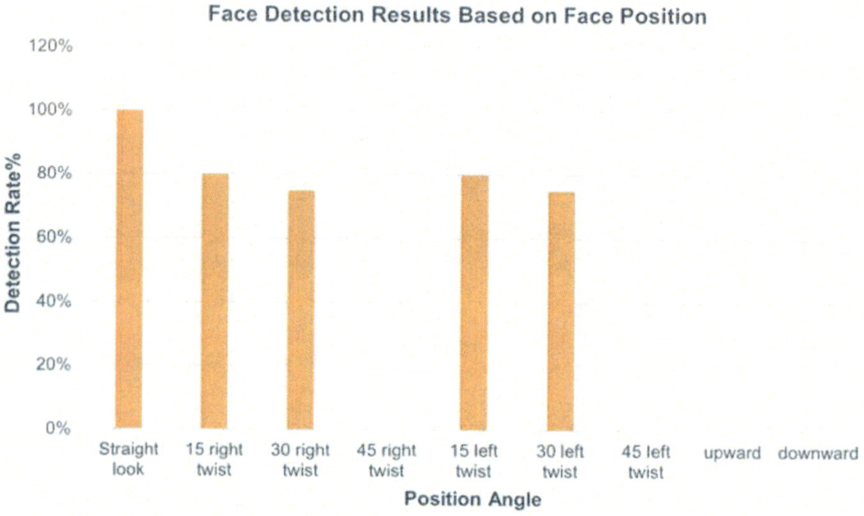

Fig. 9 Face detection result based on proximity

record a 0% detection whenever the face is twisted 45° or more. This is because, at this position that are limited feature to be captured for accuracy in detection, therefore, any detection capture at these angles is considered null (Fig. 9).

6 Conclusion

The proposed system is Implemented to present a type of facial recognition and bio-metric system, based on the face recognition, which can be deployed in some strategic areas in a community for locating missing persons who are captured by the

CBSS. The proposed software system is expected to show that the technology of face recognition is suitable in preventing and reducing the long time spent in search and finding of missing persons, due to the possibility of employing it in several implementations such as access control, surveillance, security system, etc.

References

1. Arulananth, T., Manjula, B., Baskar, M.: Human position tracking and detection using geometric active contours. In: 2020 Second International Conference on Inventive Research in Computing Applications (ICIRCA), pp. 509–512 (2020). IEEE
2. Baggio, D.L.: Mastering OpenCV with Practical Computer Vision Projects. Packt Publishing Ltd (2012)
3. Bencheriet, C.E.: New face features to detect multiple faces in complex background. Evol. Syst. **10**(2), 79–95 (2019)
4. Bendjillali, R.I., Beladgham, M., Merit, K., Taleb-Ahmed, A.: Illumination-robust face recognition based on deep convolutional neural networks architectures. Indones. J. Electr. Eng. Comput. Sci. **18**(2), 1015–1027 (2020)
5. Bradski, G., Kaehler, A.: Learning OpenCV: Computer Vision with the OpenCV Library. O'Reilly Media, Inc. (2008)
6. Gupta, A., Tiwari, R.: Face detection using modified Viola Jones algorithm. Int. J. Recent Res. Math. Comput. Sci. Inf. Technol. **1**(2), 59–66 (2014)
7. Hossen, A.M.A., Ogla, R.A.A., Ali, M.M.: Open CV Viola-Jones algorithm based on coding eyes. Iraqi J. Sci. **58**(2A), 735–745 (2017)
8. Joshi, P.: OpenCV with Python by Example. Packt Publishing Ltd. (2015)
9. McKinney, W.: Python for Data Analysis: Data Wrangling with Pandas, NumPy, and IPython. O'Reilly Media, Inc. (2012)
10. Raschka, S.: Python Machine Learning. Packt Publishing Ltd. (2015)
11. Minichino, J., Howse, J.: Learning OpenCV 3 Computer Vision with Python. Packt Publishing Ltd. (2015)
12. Castrillón, M., Déniz, O., Hernández, D., Lorenzo, J.: A comparison of face and facial feature detectors based on the Viola-Jones general object detection framework. Mach. Vis. Appl. **22**, 481–494 (2011)
13. Delac, K., Grgic, M., Grgic, S.: Image compression effects in face recognition systems. In: Face Recognition. Citeseer (2007)
14. Ephraim, T., Himmelman, T., Siddiqi, K.: Real-time Viola-Jones face detection in a web browser. In: 2009 Canadian Conference on Computer and Robot Vision, pp. 321–328 (2009). IEEE
15. Wilson, P.I., Fernandez, J.: Facial feature detection using Haar classifiers. J. Comput. Sci. Coll. **21**(4), 127–133 (2006)
16. Mittal, M., Saraswat, L.K., Iwendi, C., Anajemba, J.H.: A neuro-fuzzy approach for intrusion detection in energy efficient sensor routing. In: 2019 4th International Conference on Internet of Things: Smart Innovation and Usages (IoT-SIU), pp. 1–5 (2019). https://doi.org/10.1109/IoT-SIU.2019.8777501
17. Sackey, S.H., Chen, J., Henry, A.J., Zhang, X.: A clustering approach based on genetic algorithm for wireless sensor network localization. In: 2019 15th International Conference on Computational Intelligence and Security (CIS), pp. 54–58 (2019). https://doi.org/10.1109/CIS.2019.00020

Design and Implementation of a Secure Patient Recommender and Prediction System

Asmaa Ahmed, Amel Al Ali, Wedad Mousa, Haifa Alhosani, Joseph Anajemba, and Celestine Iwendi

Abstract Early disease prediction can help sick persons determine the severity of the disease and take quick action, thus, a healthcare recommended system is viewed as additional tools to help patients control and manage their ill-health. Medical recommended system which provides users with quick and optimal disease predictions has been in existence for a while; however, it is faced with several data security issues. Sometimes, patients confidential data which are stored of the archive after each recommendation may be accessed by unauthorized persons, and this can warrant a serve data breach and disclosure of private medical information. Thus, the focus of our project is to design a privacy-aware recommended system that not just makes facilitates quick and easy recommendation for sick persons but also securely protects stored medical information from unauthorized access. This system will be designed to support quick search, recommendation septimal confidentiality and integrity.

Keywords Recommendation system · Privacy · Smart healthcare web application

1 Introduction

A sort of information filtering system called a recommendation or recommender system aims to anticipate preferences using a model. By selecting important information fragments from a huge amount of dynamically created material based on the user's choices, interests, or observed behavior about the item, recommender systems

Amel Al Ali, Wedad Mousa, Haifa Alhosani, Joseph Anajemba and Celestine Iwendi—These authors contributed equally to this work.

A. Ahmed · A. Al Ali · W. Mousa · H. Alhosani · J. Anajemba (✉)
Information Security Engineering Technology, Abu Dhabi Polytechnic, MBZ, Abu Dhabi 111010, United Arab Emirates
e-mail: joseph.anajemba@adpoly.ac.ae

C. Iwendi
School of Creative Studies, University of Bolton, Bolton 10587, UK

C. Iwendi et al. (eds.), *Proceedings of ICACTCE'23—The International Conference on Advances in Communication Technology and Computer Engineering*, Lecture Notes in Networks and Systems 735, https://doi.org/10.1007/978-3-031-37164-6_19

are information filtering systems that address the issue of information overload. Based on the user's profile, a recommender system can determine if a specific user will favor an item or not. User information overload is addressed by recommender systems by offering users individualized, exclusive content and service recommendations [1]. Why the Recommender systems? Benefits users in finding items of their interest, help item providers in delivering their items to the right user, Identity products that are most relevant to users, personalized content, help websites improve user engagement [2]. Data can be used for various research activities and to generate new real-life data sets. Insurance companies can also make use of authentic data to bring down the fraudulent claims which on an average range between almost 10 to 15 [3]. The system is equipped with a prediction algorithm that works in the back end to predict the disease from the given symptoms. Prediction techniques have been used for several application domains such as the future box office movie collections, price of a house with given parameters or bankruptcy prediction, human face recognition systems and voice recognition, etc. [4, 5]. The ensemble approach [6, 7] has been proved to be very effective and more accurate [8] in many medical applications such as heart disease prediction [1, 2, 9], cancer prediction [8], hepatitis disease prediction [10], diabetes prediction [11] and Parkinson's disease prediction [12]. With this motivation, to predict plausible disease an ensemble approach is proposed in the present research by combining the features of three state-of-the-art classification algorithms using Logistic Regression. The selection of these algorithms follows from various research papers [5, 10, 13] which are focused on determining how the different classification algorithms behave with respect to the different data chunks and different applications. Following this, the ensemble approach is combined with the voting mechanism [7, 10, 14] to provide the best prediction of the disease.

The main contributions of this research are as follows:

- **Disease prediction**: This paper provides an algorithm to predict diseases based on the given symptoms using state-of-the-art machine learning classification algorithms.
- **Security and privacy**: Security: as we are dealing with sensitive data and confidential information we need to ensure the safety and security of our users. So, we have implemented multi-factor authentication with email OTP. This security measure provides an added layer of protection by requiring users to provide two or more factors of authentication to access the system. With email OTP, users receive a unique one-time password via email, which they must enter in addition to their usual login credentials [15].

According to Citrix (2021), email OTP is a secure and reliable authentication method that can be easily integrated with different systems. The unique password generated by email OTP is only valid for a short period, usually between 30 seconds to 10 minutes, which minimizes the risk of unauthorized access even if the password is intercepted. In addition, email OTP does not require the use of hardware tokens, making it a economical solution for small to medium-sized enterprises [15].

The use of multi-factor authentication with email OTP can greatly reduce the risk of cyber attacks, such as phishing and brute-force attacks, which are common methods used by hackers to gain unauthorized access to systems. By requiring users to provide multiple factors of authentication, we can ensure that only authorized users can access our system and protect it from potential security breaches [15].

In conclusion, the implementation of multi-factor authentication with email OTP is a crucial security measure that helps to ensure the safety and security. With its ease of use and reliable security features, email OTP is a cost-effective solution that can greatly reduce the risk of cyber attacks [15].

1.1 Organization of Paper

The remaining part of the paper is organized as follows; Section 2 is the project design where we have discussing the structure and process of our research. Related work is presented in Sect. 3 while the methodology of our research is presented in Sect. 4. In Sect. 5, we have elaborated all the experiments and results we achieved and the conclusion is presented in Sect. 6.

2 Project design

The following is a system for making recommendations. A user can request recommendations in any domain, such as music, movies, or in this case, diagnosis based on symptoms submitted by the user during the user interface phase. As a result, the user interface will access the database where the data related to the query is stored. Next comes the Top-N Recommendations are generated by the Recommendation Engine. The user will receive these recommendations. There are five phases to our recommended system, each of which is described below.

- Phase 1: As seen in Fig. 1, the user logs in and creates a new diagnosis request.
- Phase 2: The user will enter the create page and submit the symptoms he/she is experiencing, then ask for recommendations based on those symptoms. Upon entering the user's information, the system will automatically suggest a few more systems that are aligned with what was previously entered. The user's information will be stored and protected using encryption and will remain confidential.
- Phase 3: The information entered will be stored in a secure database.
- Phase 4: At this stage, the system analyses the inputs and makes recommendations for the users to choose from.
- Phase 5: Based on the user's symptoms, the system will be able to determine the most appropriate diagnosis.

It is possible to filter content using different techniques, some of which are more straightforward than others. Machine learning models are commonly used to build recommendation systems by programmers.

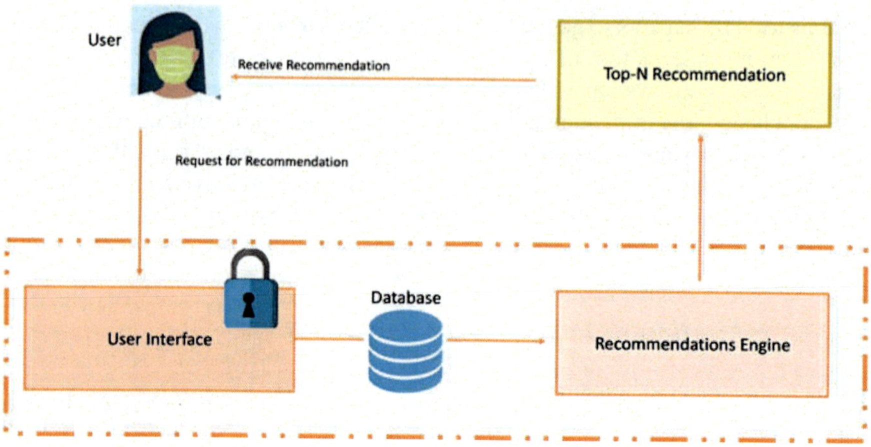

Fig. 1 Project design

3 Literature Review

3.1 Recommender Systems

A disease prediction system is an important aspect of a developing nation as part of a strong healthcare infrastructure. With this notion, the present research proposes a machine learning equipped web-based disease prediction system along with a recommended system to ensure a more strategic and robust healthcare environment. With deep insight into the existing e- healthcare system, it is evident that there are few diseases specific solutions available; but a diverse disease prediction system considering all common diseases is still missing (to the best of our knowledge). Kumar et al. [7] has focused on the prediction of cancer while Bashir et al. [6] proposed a solution to predict the heart disease for a patient. In similar approach, Dutta et al. [10] proposed an approach for hepatitis disease prediction, Rao et al. developed a Parkinson's prediction approach [12] and Pathirage et al. [11] offered a diabetic prediction system. All these methods are algorithmic approaches and do not offer user interfaces for end users. The absence of a GUI creates a gap between the system and the user. Following this, the present research focuses on common disease symptoms (including rare) to cater to the needs of a large group of people and offers a user-friendly web application.

3.2 *Algorithms of a Recommender System*

Logistic Regression is a type of machine learning algorithm, Machine Learning algorithms such as logistic regression belong to the Supervised Learning dependent variable.

A categorical dependent variable is predicted by logistic regression. There is therefore no choice but to have a categorical or discrete outcome. Rather than giving an exact value such as 0 and 1, the probabilistic values lie between 0 and 1. It can be either Yes or No, 0 or 1, true or false, etc.like this:

- During this step, we will preprocess/prepare the data so that we can use it efficiently in our code. Data pre-processing will be the same as before.
- Identify the dependent and independent variables in the dataset.
- The dataset is then divided into a training and a test set [15].

3.3 *Secured Recommended Systems*

The development of recommended systems is a multi-disciplinary effort that involves experts from various fields such as Artificial intelligence, Human- Computer Interaction, Data Mining, Statistics, Adaptive User Interfaces, Decision Support Systems, Marketing, or Consumer Behavior [4]. Presenting Logical Neural Networks (LNNs), a neuro-symbolic framework designed to simultaneously provide key properties of both neural nets (NNs) (learning) and symbolic logic (knowledge and reasoning) Riegel et al. [12] The vast majority of database-driven web applications perform, at a logical level, fundamentally simple INSERT/UPDATE/DELETE commands. In response to a user action on the browser, the web application executes a program that transitions the old state to a new state Kian Win Ong, Michalis Petropoulos.

Chaitrali and Dangare [14] The recommender system deals with a large volume of information present by filtering the most important information based on the data provided by a user and other factors that take care of the user's preference and interest Rohit Dwivedi [2]. According to recent studies 81% of U.S. adults use the Internet and 59% say they have looked online for health information regarding diseases, diagnoses and different treatments Wiesner M, Pfeifer D.

Ramasamy [9] Family medicine is an important medical discipline worldwide. Basically, the family is the smallest social unit that everyone has to be a member. Family medicine is a specific discipline that focuses on family. The discipline deals with the medical practice at the family level, which is the interface between individual and community medicine Wiwanitkit Viroj.

4 Methodology

To gain a deeper understanding of the system's nature. The part that follows is an examination of the proposed system's methodology.

4.1 Flow Chart

The proposed system's flowchart, which outlines the actions the user will follow when using the system, is shown in Fig. 2. The system will initially ask the user if he already has an account. If not, the user can create a new account by providing his name, phone number, address, user ID, and password. These details will be recorded in both the user registration form and the login details. Alternatively, if he already has an account, he can log in. He will be sent an OTP for authentication [16], and after entering it successfully, he will be logged in and able to create new diagnoses (Fig. 3).

Provide authentication: After the user enters the username and password, he will not be able to enter his account unless he signs in with his OTP (Fig. 4).

This will prevent and limit attackers from using the brute-force technique to hack or attack the account. One of the main benefits of using OTP is that it prevents many attackers from using identity theft, the methodology they tend to use is by making

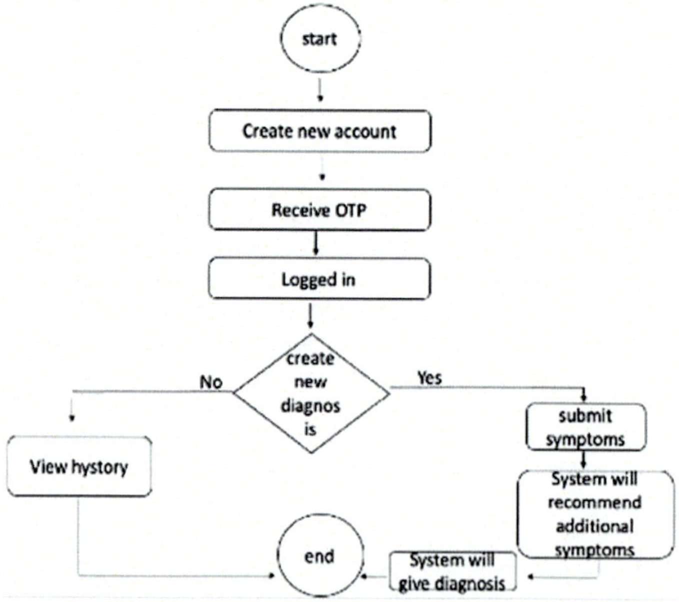

Fig. 2 Project flowchart

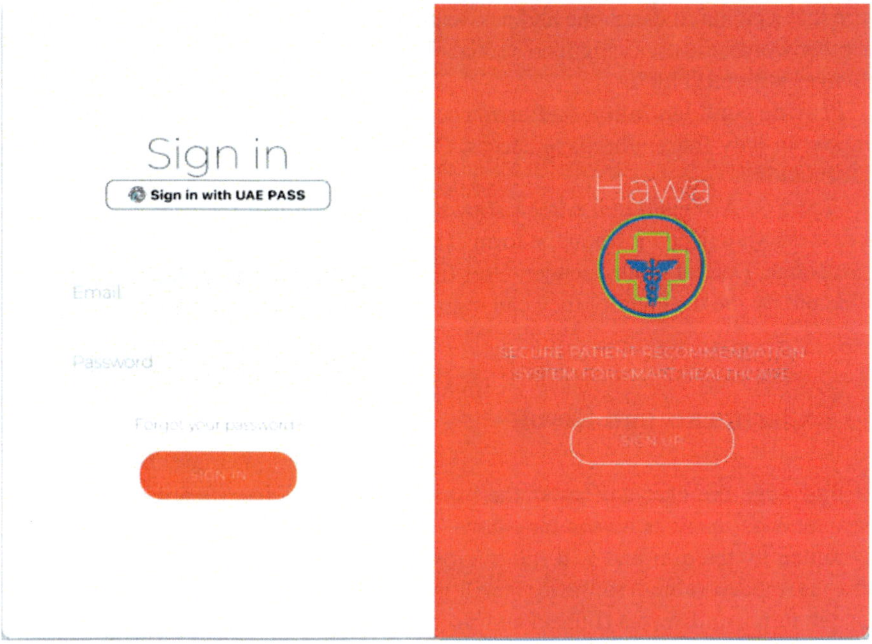

Fig. 3 Create new account

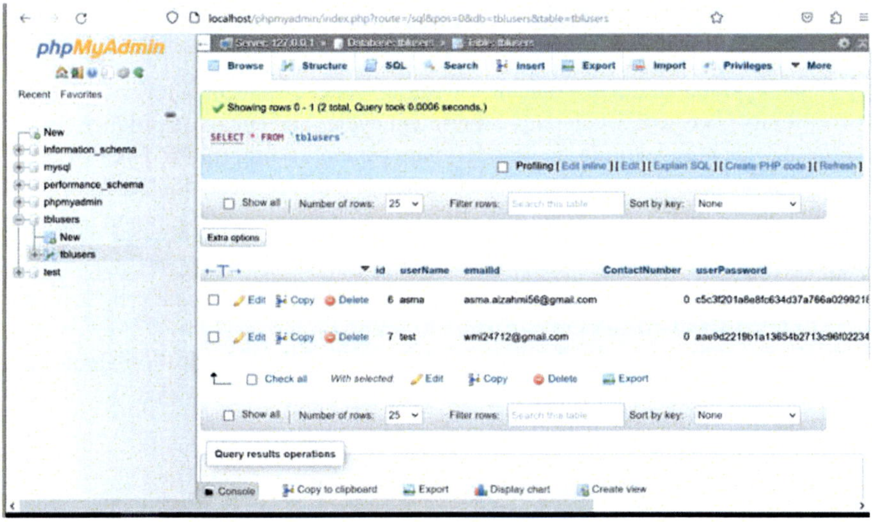

Fig. 4 Two-factor authentication: one time password

sure that a captured username and password that cannot be used a second time [16]. In other words, the user's login credentials stays the same, and the one-time password changes with each login.

Provide confidentiality: by Restrict access to data the user's information will be limited to only those who have access. Others will not be able to access sensitive information.

Figure 3. A registration screen shows login details that the user must enter to access the system as shown in the form. There are six fields in this form: First Name, Last Name, Date of Birth, Gender, User ID, Password, and Location, which must be filled out by the user before they can register.

5 Experiments and Results

At this stage the user will enter the user interface, the user now can enter there symptoms as many as he/she may have. The system will study the the symptoms submitted by the user and will run to cross match the best match corresponding to what have been submitted, lastly it will give a recommendation on what they may have in terms of diseases (Figs. 5 and 6).

5.1 Pycaret

In Python, PyCaret automates machine learning workflows using a low-code, open-source library. This tool accelerates experiment cycles and enhances productivity by integrating machine learning and model management. Furthermore, Pycart's users have access to many libraries that simplify coding. Users can also use Pycart to build simple websites and applications that allow them to demonstrate and preview their work.

- PyCaret is a Python open-source machine learning library designed to make performing standard tasks in a machine learning project easy.
- We use this library to quickly run and evaluate multiple classification algorithms and choose the best-performing one on our dataset.

5.2 Gradio

In this project we are using gradio to help us build mock up of the system. This Python library allows users to create easy-to-use, customized component demos for users machine learning models that you can use anywhere. Gradio works with Scikit-Learn, PyTorch, NumPy, Seaborn, Pandas, TensorFlow, and many other Python libraries.

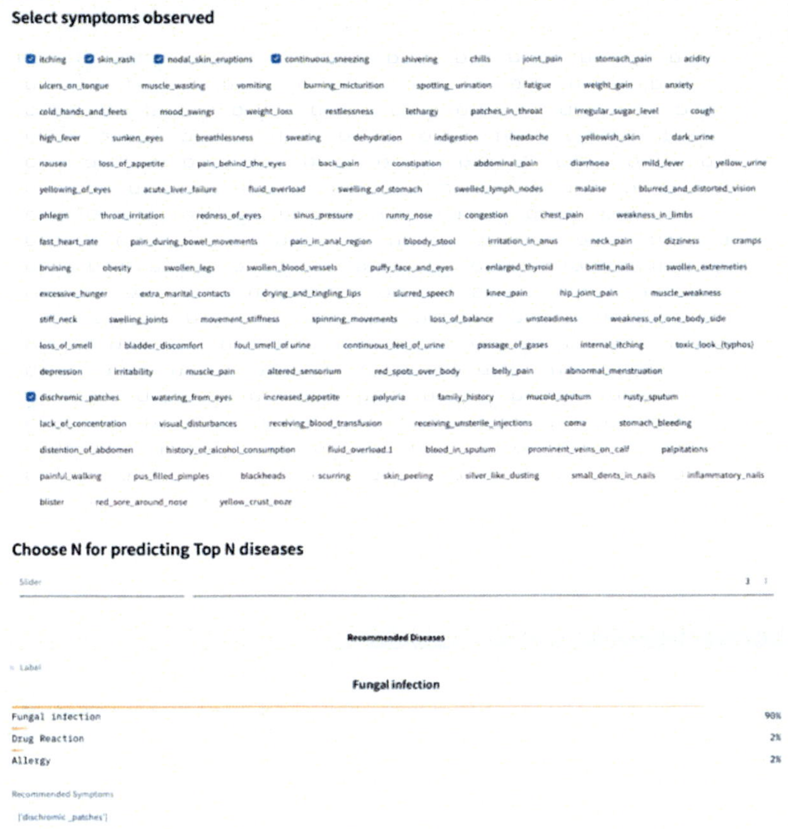

Fig. 5 User interface page: choosing symptoms

5.3 *Dataset*

Medical science knowledge is applied with the aim of facilitating doctors' work with this dataset. This dataset contains 132 parameters that can be used to predict 42 different types of disease. We train our model using a CSV file, then test it against another. Based on the information in the CSV files, these symptoms can be classified into 42 diseases. Each CSV file contains 133 columns. Each column represents a symptom experienced by the individual, with the last column representing a diagnosis [17, 18] (Fig. 7).

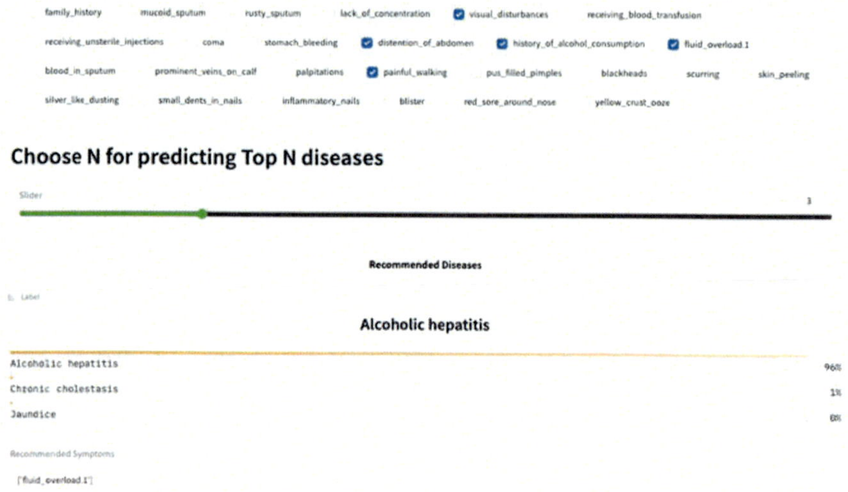

Fig. 6 User interface page: disease predictions

```
Successfully installed numpy-1.20.0

!pip install gradio

Requirement already satisfied: toolz in /usr/local/lib/python3.8/dist-packages (from altair>=
Requirement already satisfied: entrypoints in /usr/local/lib/python3.8/dist-packages (from al
Requirement already satisfied: jsonschema>=3.0 in /usr/local/lib/python3.8/dist-packages (frc
Collecting mdurl~=0.1
  Downloading mdurl-0.1.2-py3-none-any.whl (10.0 kB)
Collecting mdit-py-plugins
  Downloading mdit_py_plugins-0.3.3-py3-none-any.whl (50 kB)
     ━━━━━━━━━━━━━━━━━━━━━━━━━━━━━━━━━━━━━━━━ 50.5/50.5 KB 5.3 MB/s eta 0:00:00
Collecting linkify-it-py~=1.0
  Downloading linkify_it_py-1.0.3-py3-none-any.whl (19 kB)
Requirement already satisfied: pytz>=2017.3 in /usr/local/lib/python3.8/dist-packages (from p
Requirement already satisfied: python-dateutil>=2.7.3 in /usr/local/lib/python3.8/dist-packag
Requirement already satisfied: attrs>=17.3.0 in /usr/local/lib/python3.8/dist-packages (from
Requirement already satisfied: yarl<2.0,>=1.0 in /usr/local/lib/python3.8/dist-packages (from
Requirement already satisfied: aiosignal>=1.1.2 in /usr/local/lib/python3.8/dist-packages (fr
Requirement already satisfied: async-timeout<5.0,>=4.0.0a3 in /usr/local/lib/python3.8/dist-p
Requirement already satisfied: frozenlist>=1.1.1 in /usr/local/lib/python3.8/dist-packages (f
Requirement already satisfied: multidict<7.0,>=4.5 in /usr/local/lib/python3.8/dist-packages
Requirement already satisfied: charset-normalizer<3.0,>=2.0 in /usr/local/lib/python3.8/dist-
Collecting starlette<=0.23.0,>=0.22.0
```

Fig. 7 Gradio library import stage

5.4 *Training and Testing*

There are two types of datasets being used, first is training in this stage we are taring the machine and flirting the data to obtain the type of results that are useful to this project. On the other hand the testing file is used to test the code is working or not, for example two cases are given and our system must give the correct diagnoses, this file is impotent to make sure the system is working on the right path. in results the

▾ Load and View data

```
[2] #give appropriate file path
    train = pd.read_csv('Training.csv')
    test = pd.read_csv('Testing.csv')
```

```
train.head(3)
```

	itching	skin_rash	nodal_skin_eruptions	continuous_sneezing	shivering	chills	joint_pain	stomac
0	1	1	1	0	0	0	0	
1	0	1	1	0	0	0	0	
2	1	0	1	0	0	0	0	

3 rows × 134 columns

Fig. 8 Training and testing

system read the cases and gave correct diagnosis and it was proven that its working 99% on accuracy (Fig. 8).

5.5 *Classification*

Gradio is an open-source Python library that is used to build machine learning and data science demos and web applications.

- It allows us to develop an easy-to-use customized component demo for our machine learning model that anyone can use anywhere.
- We use this library to demo our disease and symptoms recommendation application.
- Classification is a Machine Learning technique which learns from training data to predict our target (in our project—prognosis) into categories.
- We use symptoms as input features to our classification model and predict the prognosis.
- We recommend top 3 or 5 diseases/prognosis on the basis of the confidence score generated by the classification algorithm.
- We recommend symptoms on the basis of the top disease recommended by the model. With that recommended disease, we look for other symptoms that our exhibited by the users in our training data. For example, if a user selects the symptoms as Skin Rash and Blackheads, the model will predict Acne as a top disease. Based on Acne, we look for other possible symptoms in our data like Scurrying, Pimples, etc. and recommend it.
- Pycarte has many classification in order to make sure we choose the right one we decided to select and run all classification, after that we start ruining and comparing

different models against each other to find out the best fit classification with the highest score, for our case it was logistic regulations in short IR. after that we start evaluating the model training file and lastly move to testing stage.

6 Conclusion

This project goal is to obtain recommendations for users based on submitted symptoms and to insure that the users information is kept secure at all times. First the registration page will be secured through one time password or by signing in using UAE Pass. second stage is submitting the symptoms into the system and the systems is going to give recommendations that have top 3 scores, next it will display more symptoms options for users to select if they were experiencing them, if not they can move on to the final stage of giving the final best fit recommendation.

References

1. Amruta Powar, V.G.: Survey on heart disease prediction system using data mining technique. Int. J. Innov. Eng. Res. Technol. **4** (2017)
2. Rao, M.V., Pandu Ranga Vittala, V.: Intelligent Parkinson disease prediction using machine learning algorithms. Int. J. Eng. Innov. Technol. (2013)
3. Mudaliar, V., Savaridaasan, P., Garg, S.: Disease prediction and drug rec-ommendation android application using data mining (virtual doctor). Int. J. Recent Technol. Eng. **8** (2019) 12 Rajora et al. Tools
4. Pathirage, N., Jinasena, K.K., De Silva, L.: Diabetic prediction system using data mining. In: 9th International Research Conference-KDU (2016)
5. Sethi, S.K.: Insurance frauds control act; an urgent need in india. https://www.businesstoday.in/opinion/columns/insurance-frauds-control-act-an-urgent-need-in-india-fraudulent-claims-indian-penal-code/story/400212.html (2020), [Online; Business Times] Security
6. Dutta, S., Bandyopadhyay, S.K.: A voting ensemble approach for hepatitis disease detection. Int. J. Res. Appl. Sci. Biotechnol. **7** (2020)
7. Barbozaa, F., Kimura, H., Altman, E.: Machine learning models and bankruptcy prediction. Expert Syst. Appl. (2017)
8. Paul, T.K., Iba, H.: Prediction of cancer class with majority voting genetic programming classifier using gene expression data. IEEE/ACM Trans. Comput. Biol. Bioinf. **6** (2009)
9. Ramasamy, S., Nirmala, K.: Disease prediction using decision tree based weighted voting system in data mining. Int. J. Res. Anal. Rev. (2018)
10. Hazra, A., Mandal, S.K., Gupta, A., Provo Mukherjee, A., Mukherjee A: Heart disease diagnosis and prediction using machine learning and data mining techniques: a review. Adv. Comput. Sci. Technol. **10** (2017)
11. Bashir, S., Qamar, U., Khan, F.H.: A multicriteria weighted vote-based classifier ensemble for heart disease prediction. Comput. Intell. **0** (2015)
12. Patil, R., Chopade, P., Mishra, A., Sane, B., Sargar, Y.: Disease prediction system using data mining hybrid approach. Commun. Appl. Electron. 4
13. Bashir, S., Qamar, U., Khan, F.H.: A novel ensemble for heart disease prediction bootstrap aggregation with multi-objective optimized voting. Aust. Coll. Phys. Sci. Eng. Med. (2015)
14. Miss. Chaitrali, S., Dangare, M.S.S.A.: A data mining approach for prediction of heart disease using neural networks. Int. J. Comput. Eng. Technol. (IJCET) **3** (2012)

15. Hastie, T., Friedman, J., Tisbshirani, R.: The elements of Statistical Learning: Data Mining, Inference, and prediction. Springer, New York (2017)
16. Email OTP authentication (no date) Email OTP authentication, Authentication, authorization, and auditing application traffic. https://docs.citrix.com/en-us/citrix-adc/current-release/aaa-tm/authentication-methods/email-otp.html. Accessed 16 Feb 2023
17. Iwendi, C., Khan, S., Anajemba, J.H., Bashir, A.K., Noor, F.: Realizing an efficient IoMT-assisted patient diet recommendation System through machine learning model. IEEE Access **8**, 28462–28474 (2020). https://doi.org/10.1109/ACCESS.2020.2968537
18. Ngabo, D., Wang, D., Iwendi, C., Anajemba, J.H., Ajao, L.A., Biamba, C.: Blockchain-based security mechanism for the medical data at fog computing architecture of internet of things. Electronics **10**(17), 2110 (2021)

SDD: An Efficient Stacked Deep Learning Technique for Social Distance Detection

Sundaravadivazhagan Balasubaramanian, Robin Cyriac, Sahana Roshan, Kulandaivel Maruthamuthu Paramasivam, and Boby Chellanthara Jose

Abstract The use of social distance as a tool throughout the battle with COVID-19 has shown promising results. In recent years, artificial intelligence (AI) and deep learning (DL) have emerged as a powerful resource for dealing with a wide range of practical challenges and delivering impressively positive outcomes. This article delves into the usage of Object Recognition, and Deep Learning to monitor personal and professional interactions between people at a distance. This research aims to aid in the pandemic fight by creating a technology that may be used as a social distance monitoring system. As evidence mounts in favour of social separation as a first-line, non-pharmaceutical intervention in the precautionary measure against the fastest spreading disease COVID-19, researchers feel that its rigorous observance should be encouraged. This proposal analyses live or recorded video to identify and evaluate correlation metric distances among individual people to verify to see if social distancing is preserved in crowded places.

Keywords Covid-19 · Faster R-CNN · Image recognition · Social distancing

1 Introduction

Since the COVID-19 pandemic's initial outbreak in the Chinese city of Wuhan in December 2019, it has substantially worsened as a global health issue [1]. An infection brought on by a virus that causes pneumonia and a severe acute pulmonary illness that can proceed to respiratory failure and death. In January 2020, it was discovered that the pathogenic microbe was a novel corona virus that was eventually designated severe acute respiratory coronavirus 2 [2]. The epidemic has caused many deaths and several serious illnesses in little than a year. The pandemic's dread and anxiety forced several changes in people's lives on a local, national, and international level.

S. Balasubaramanian (✉) · R. Cyriac · S. Roshan · K. M. Paramasivam · B. C. Jose
Department of Information Technology, University of Technology and Applied Sciences, Al Musanna, Oman
e-mail: sundaravadi@act.edu.om

C. Iwendi et al. (eds.), *Proceedings of ICACTCE'23—The International Conference on Advances in Communication Technology and Computer Engineering*, Lecture Notes in Networks and Systems 735, https://doi.org/10.1007/978-3-031-37164-6_20

Many people have suffered as a result of these changes, notably the most vulnerable groups like the homeless, elderly, crippled, imprisoned, and illegal immigrants, among others. Many people have lost their employment, and many households are experiencing decreased income [3]. Some people no longer have access to basic necessities like food, shelter, or clothing.

During the height of this pandemic, there was neither a treatment nor a vaccine available. Since the disease spreads so quickly through intimate contact with others, many non-medicinal preventative precautions were put into place in most parts of the world. Distancing oneself socially, tracking down and isolating potential contacts, preventing those infected or exposed from interacting with others, and outlawing most types of social gatherings were all part of the plan. The Centers for Disease Control and Prevention (CDC) identified social distancing, which is described as a minimum distance of six feet among people from different families, that was the most efficient method in reducing the speed of illness scattered. To this day, the vast majority of nations within and throughout countries have implemented this measure as part of their ongoing efforts to combat the pandemic. Several number of research studies [4–6] have shown that social isolation is, in fact, an essential component in preventing the further dissemination of the disease. It has been shown to be beneficial in preventing the spread of diseases during previous pandemics [7, 8], and it is still showing signs of being useful in the pandemic that is occurring right now.

Severe acute respiratory syndromes [9] have been reported all over the world due to social isolation because of COVID-19 since December of this year. Distancing oneself socially has been found to be an effective non-pharmaceutical technique in limiting the transmission of the COVID-19 virus. Separating yourself from other people by at least two metres is considered socially acceptable behaviour. Since the COVID-19 vaccines have begun to be distributed, and the issue does not appear to be one that will be resolved quickly, it may be vital for the society to have an automated system to track and analyse social isolation.

In respiratory pandemics like COVID-19, social isolation can reduce infection rates. Socially distant behaviour in cities can be best seen and measured at busy crossroads. Bird's-eye view video recordings of people crossing traffic crossings are used in the suggested and evaluated confidentiality social distancing analysis system (B-SDA) by the authors [10]. For the problem of spotting tiny objects or pedestrians in footage shot from above, the authors of [10, 11] have proposed a set of techniques for visual preprocessing, object detection, and tracking that build on existing computer-vision and DL methods but are tailored to this specific challenge. The authors suggest using pedestrian grouping as a means of identifying instances of inappropriate social distance. Using footage from before and after a pandemic, B-SDA analyses how people in a major city walk.

The process of object detection in computer vision comprises looking for and pinpointing instances of known semantic classes. The majority of cutting-edge object detectors use deep learning. R-CNN [12], Fast R-CNN [13], Faster R-CNN (FRC) [14], and Mask R-CNN [15] are just a few of the well-known methods that employ a two-stage framework for object detection, one that divides the process of feature map from that of classification.

Ren et al. [14] came up with the idea of using a FRC to lessen the amount of computing work involved in the production of recommendations. This strategy is broken up into two different parts or components. RST component is a fully convolutional network known as the Regional Proposal Network (RPN). It is responsible for generating object suggestions, which are then passed on to the final component. The final module is called the FRC detector (FRCD), and its job is to make the recommendations even more precise. The selective search technique that is utilised by FRC is in contrast to the framework such as Visual Geometry Group (VGG) and Inceptions that are utilised by FRC for the purpose of proposal generation. This is the primary distinction between the two versions of R-CNN. After that, the CNN features are distributed over the region via a network of regional proposals. After that, a sliding window is applied to the features, and the subsequent output contains projected bounding boxes as well as scores. Additionally, the FRC is quicker than the Fast R-CNN since it only computes the features one time, hence reducing the amount of processing that is required. Because of this, it is significantly quicker than the Fast R-CNN.

In the future, cutting-edge tools like Wi-Fi, networked computers, and AI will make social isolation easier to achieve and maintain [16]. This study examined new social distance theories, solutions, simulations, and future possibilities. In another paper [17], researchers categorised new methods into human-centric and smart-space categories and performed a SWOT analysis on each. A basic YOLOv3 and Deepsort technique [18] detected and followed people after measuring anti-social distance violation indication.

In another research work, the authors [19] have introduced a novel model using blockchain technology to ensure the Covid quarantine. The framework introduced in that model provided better accuracy. Haq et al. [20] proposed a hybrid CNN model to detect the social distancing in the crowd. The proposed smart social distancing model collected the location and information of infected person. The proposed framework [20] sent alerts to mobile device about the social distancing and infected person. Various deep learning techniques [21–23] have been used to detect the social distancing in public places such as retail malls, movie theatres and railway station.

2 SDD Technique

The proposed SSD (Social Distance Detector) technique is able to give assessment of distance observations in social framework with a high degree of precision whether in instantaneously or recorded video. This analysis may take performed whether in instantaneously or in a video feed that has been previously recorded. The model can be simplified by dividing it into two separate stages. The first one involves the detection and location of human beings, while the second involves the detection of social distance.

Figure 1 depicts the workflow of the proposed technique. The workflow given in the Fig. 1 is described as follows:

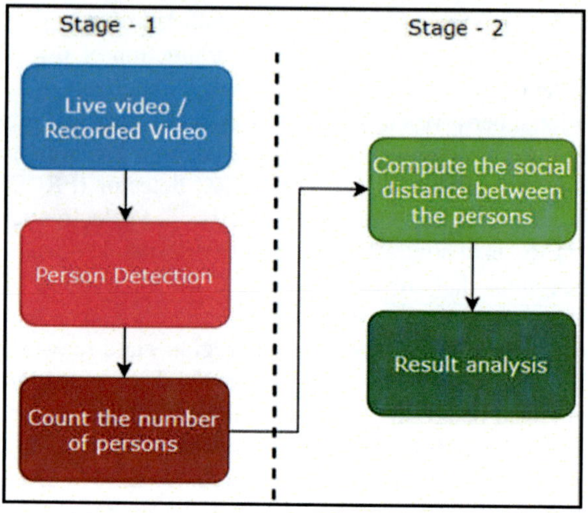

Fig. 1 Proposed SDD workflow

i. Get ready to receive photos or a live video feed from a camera that is focused on individuals.
ii. Using a more efficient version of the FRC to search for humans in the input.
iii. Count how many of persons who can be seen in the images or the live video.
iv. Determine the distance between the centres of the outlined objects that include the people who have been found using a distance calculator.
v. Finally, the algorithm will decide whether or not the social distancing protocol is being followed by considering the number of persons present as well as the distance that has been measured among the centroids of the outlined objects.

2.1 *Person Detector*

During this stage of the proposed work, it initially constructed a detector for a single class of objects that is capable of detecting with a high degree of true positives and accurateness. The detector will first locate the object. In this work, it detects person and draw an outline box around it. For instance, if it determines that there is more than one person, it will draw outline box for all the identified persons in the live or recorded video. If this is the case, the bounding box will contain all of the detected people.

To start with this object detection, the proposed technique is made with an advantage of the FRC. A conventional convolution layer with a kernel size of three by three comes first in the architecture of the FRC. Next comes a network of detectors, then a regional proposal network (RPN). Input images are scaled down such their smallest dimension is no less than 600 pixels and their largest dimension is no greater than 1000 pixels before RPN is applied. The image is then run through RPN. The RPN process will result in a significantly smaller feature mapping, making it easier to work with. A feature map with dimensions of 30 × 50 × 256 will be generated from an input image with dimensions of 600 × 1000 × 3. In the proposed work, the network stride is set to 9, since, 3 × 3 pixels is reduced. Before entering the RPN, a picture that is 1000 pixels wide by 600 pixels tall will be reduced to (1000/9 * 600/ 9) or a size feature map that is approximately 111 by 67.

The network is responsible for the generation of the greatest number of k-anchor boxes for each sliding window. The default value of k is nine, and there are three different scales and three different aspect ratios for each sliding place in the image. N is equal to W * H * k anchor boxes when referring to a W * H convolution feature map. These region recommendations were passed on to an intermediary layer, which consisted of 3 × 3 convolution, 1 padding and 256 (VGG-16) output networks. The output of this layer is sent to the 1 × 1 convolution layers that are used for classification and regression. The regression layer contains 4*N (W * H * (4*k)) output parameters (outlined object's coordinates) and the classification layer have 2*N.

After that, the results of the Region of Interest (RoI) pooling are flattened using some layers that are completely connected. The very last stage involves applying a softmax function to adjust the place of the outlined objects in order to perform classification.

2.2 *Distance Identifier*

In the second stage of the proposed work, it is tested whether or not the suggested tool is effective at analysing video stream in order to discover and compute the pairwise metric distances between persons. This stage also tests whether or not social distance is persistent. This phase of the experiment is designed to test whether or not social distancing is maintained. Calculations are made to determine the centroid of each set of coordinates that make up a bounding box. The Manhattan distance is then used to determine the distance between any two centroids. The sum of the absolute distances that exist between two points across all of the dimensions is the Manhattan Distance.

The following equation can be thought of as the generalised formula for an n-dimensional space:

$$D_m = \sum_{i=1}^{n} |p_i - q_i| \tag{1}$$

where, n represents the number of dimensions in the vfs (image), p_i and q_i are the data points in the image matrix.

3 Results and Discussion

The results of the losses, the amount of computational time needed while training, the dataset that was utilised for the study, and the findings are discussed in this section. This section summarises the findings acquired by utilising an FRC inception model. At this point in the work that is being proposed, the person detector stage, the procedure for constructing the person detection model is dissected in deeper detail.

The datasets that were used for this research came from a variety of different places. Both Google Images and the INRIA Person dataset [11] were utilised to compile the aforementioned datasets. For the purpose of this investigation, a total of 2916 photos were downloaded, with around 14% of the images coming from Google Images and approximately 86% coming from the INRIA Person collection. Each image was manually labelled with an outlined objects on the individuals present inside each frame by using the application label Img [11].

Train and test sets were created from the original dataset. The test dataset accounted for the 20% of the whole datasets, while the train datasets comprised 80% of the complete datasets. The selection of the split was done with the intention of facilitating the generalisation of detection to new datasets whose contents are unknown. A single class was used for training the model for a total of 10,726 steps before any hyper-parameters were adjusted. Figures 2 and 3 represent, respectively, the classification of losses and the overall accuracy sustained as a result of the training procedure.

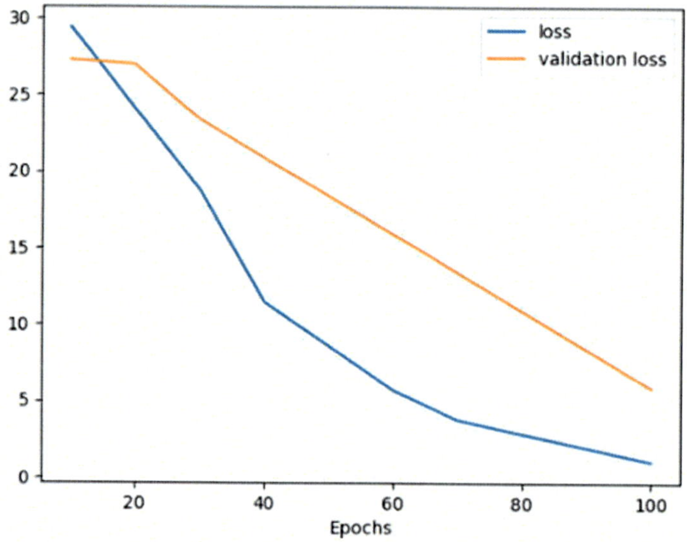

Fig. 2 Total loss

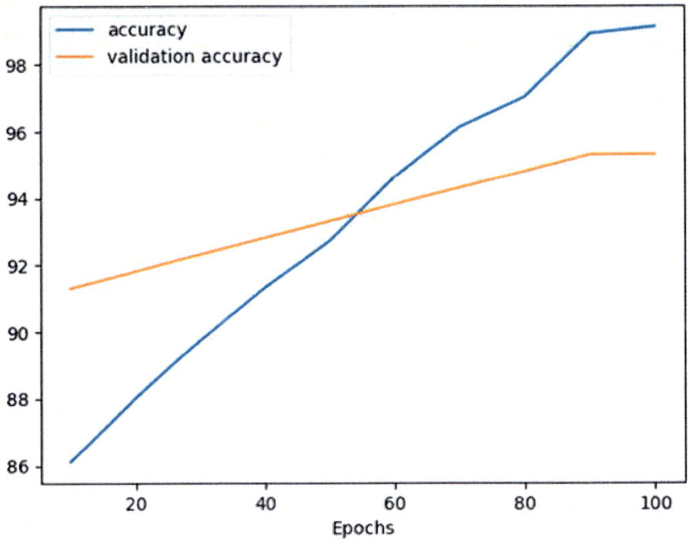

Fig. 3 Total accuracy

The essential hyper-parameters, if appropriate, were used during model training: The learning rate is 0.0002 and the total number of epochs is 100 with a batch size of 64.

Tables 1 and 2 represents the training loss and accuracy with respect to number of epochs.

Figure 2 depicts the training and validation loss of the proposed model. The x-axis that displays the number of training epochs, and y-axis that depicts the amount of loss that was experienced during model training. This is because the amount of loss that is reduced in the network is a function of the bias and weight that are present in it. This can be seen as an indication of how well the network learned.

Table 1 Loss: Training model on adam optimizer

Epoch	Loss	Validation loss
10	29.239	27.164
20	23.939	26.864
30	18.639	23.264
40	11.339	20.764
50	8.439	18.264
60	5.539	15.764
70	3.639	13.264
80	2.739	10.764
90	1.839	8.264
100	0.939	5.764

Table 2 Accuracy: Training model on adam optimizer

Epoch	Accuracy	Validation accuracy
10	86.29	91.6
20	88.19	92.1
30	89.89	92.6
40	91.49	93.1
50	92.89	93.6
60	94.79	94.1
70	96.29	94.6
80	97.19	95.1
90	99.09	95.6
100	99.29	95.6

Figure 3 represents the training and validation accuracy of the proposed model. The x-axis that displays the number of training epochs, and y-axis that depicts the amount of accuracy that was experienced during model training.

The comparison of the effectiveness of proposed algorithm makes use of a confusion matrix. This matrix is used to construct a variety of measures by combining the values of True Negative (TN), True Positive (TP), False Negative (FN), and False Positive (FP). The performance metrics used to assess models using the confusion matrix are listed below.

Accuracy is a measure of how well or closely a model's estimated value corresponds to its real or actual value, which corresponds to the proportion of all samples that are properly categorised. The model's accuracy is determined using the following formula:

$$Accuracy = \frac{TP + TN}{TP + TN + FP + FN} \tag{2}$$

Precision reveals the quantity of relevant occurrences among the chosen examples that are truly affirmative. To determine accuracy, apply the formula below:

$$Precision = \frac{TP}{TP + FP} \tag{3}$$

Table 3 Comparative analysis of proposed method with existing methods

	Precision	Recall	F-score	Accuracy
MobileNetV2	0.97659	0.94534	0.9607	0.93796
InceptionV2	0.98613	0.95457	0.97009	0.95168

The percentage of genuine positives that are accurately detected is determined by recall or true positive rate (TPR). The following equation is used to calculate recall:

$$Recall = \frac{TP}{TP + FN} \tag{4}$$

The harmonic mean of accuracy and memory, which sums the weighted average of precision and recall, is how the F1 score is often understood. The F1-score is calculated using the following formula:

$$F - Score = \frac{Precision * Recall}{Precision + Recall} \tag{5}$$

The proposed work is experimented using Ubuntu 21 computer, an Intel Core i7 processor, and Nvidia GTX 1080 GPU for the tests. Programming was carried out using Python 3.8. The pre-trained model is using 224 × 3 STIF frames.

In order to prevent the presentation of forecasts with a low level of accuracy, a confidence threshold has been defined. In addition, a relation known as "meter-to-pixel" has been established in order to accurately evaluate the distances. Both of these settings are done in conjunction with one another. The distance, in metres, that was subjectively calculated between each detected individual in the frame.

Table 3 represents the comparative results of proposed work with InceptionV2 and Mobilenetv2 based models. The mobilenetV2 uses the depth-wise separable CNN technique. For performance comparison mobilenetV2 performs less than inceptionv2 because of its nature. The smaller number of parameters affect the model's accuracy.

Figure 4 represents the comparative chart of the proposed model implemented with two different models such as MobileNetV2 and InceptionV2. The InceptionV2 based trained model outperforms well than the MobilNetV2 model.

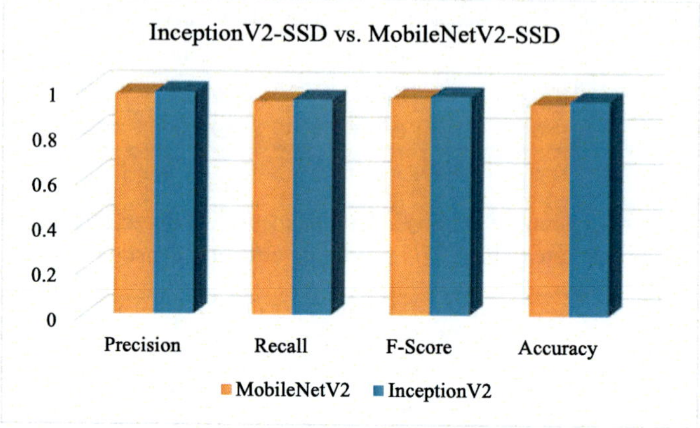

Fig. 4 Comparative chart of mobilenetV2 and InceptionV2 models

4 Conclusion

The proposed research work uses the FRC to identify the person detection in the public crowds and Manhattan Distance to calculate the social distancing. It is possible that the efficacy of the detection can be related to the RPN module. Because the RPN and the FRC detector module use the same convolutional layer, it is possible to add many convolutional layers to the RPN without placing any additional load on the network's computational resources. This is because both modules use the same convolutional layer. Taking into account one-of-a-kind patterns in addition to a variety of human positions can result in an even more significant improvement to the functionality. In addition, the proposed model is inspected with two different models such as Mobilenetv2 and Inceptionv2. InceptionV2 outperforms well than mobilenetv2. This research ignored the viewpoint that is provided by the camera position during estimating the distances between the locations. This will be considered for the future development.

References

1. Paules, C.I., Marston, H.D., Fauci, A.S.: Coronavirus infections more than just the common-cold. JAMA **323**(8), 707–708 (2020)
2. Saadat, S.I., Rawtani, D., Hussain, C.M.: Environmental perspective of covid-19. Sci. Total Environ. 138870 (2020)
3. Bodas, M., Peleg, K.: Self-isolation compliance in the COVID-19 era in influenced by compensation: findings from a recent survey in Israel: Public attitudes toward the COVID-19 outbreak and self-isolation: a cross sectional study of the adult population of Israel. Health A Airs **39**(6), 936–941 (2020)
4. Ting, D.S., Wei, L.C., Dzau, V., Wong, T.Y.: Digital technology and COVID-19. Nat. Med. **26**(4), 459–461 (2020)

5. Leung, N.H., Chu, D.K., Shiu, E.Y., Chan, K.H., McDevitt, J.J., Hau, B.J., Yen, H.L. et al.: Respiratory virus shedding in exhaled breath and efficacy of face masks. Nat. Med. **26**(5), 676–680 (2020)
6. Ather, B., Mirza, T.M., Edemekong, P.F.: Airborne Precautions. [Updated 2022 Aug 29]. In: StatPearls [Internet]. Treasure Island (FL): StatPearls Publishing (2022). https://www.ncbi.nlm.nih.gov/books/NBK531468/
7. Bae, S., Kim, M.C., Kim, J.Y., Cha, H.H., Lim, J.S., Jung, J., Kim, M.J. et al.: Effectiveness of surgical and cotton masks in blocking SARS–CoV-2: a controlled comparison in 4 patients. Ann. Intern. Med. **173**(1), W22-W23 (2020)
8. Tang, J.W., Li, Y., Eames Chan, P.K., Ridgway, G.L.: Factors involved in the aerosol transmission of infection and control of ventilation in healthcare premises. J. Hosp. Infect. **64**(2), 100–14 (2006)
9. Bharati, P., Pramanik, A.: Deep learning techniques—R-CNN to mask R-CNN: a survey. In: Computational Intelligence in Pattern Recognition, pp. 657–668. Springer, Berlin (2020)
10. Yang, Z., Sun, M., Ye, H., Xiong, Z., Zussman, G., Kostic, Z.: Bird's-eye view social distancing analysis system. In: 2022 IEEE International Conference on Communications Workshops (ICC Workshops), pp. 427–432 (2022)
11. Class-Peters, F., Adoni, W.Y.H., Nahhal, T., Byed, A.E., Krichen, M., Kimpolo, C., Kalala, F.M.: Post-COVID-19: deep image processing AI to analyze social distancing in a human community. In: Advances on Smart and Soft Computing. Advances in Intelligent Systems and Computing, 2022, vol. 1399. Springer, Singapore (2022)
12. Girshick, R., Donahue, J., Darrell, T., Malik, J.: Region-based convolutional networks for accurate object detection and segmentation. IEEE Trans. Pattern Anal. Mach. Intell. **38**(1), 142–158 (2016)
13. Girshick, R.: Fast r-cnn. In Proceedings of the IEEE International Conference on Computer Vision (ICCV), December 2015 (2015)
14. Ren, S., He, K., Girshick, R., Sun, J.: Faster r-cnn: towards real-time object detection with region proposal networks. IEEE Trans. Pattern Anal. Mach. Intell. **39**(6), 1137–1149 (2017)
15. He, K., Gkioxari, G., Dollar, P., Girshick. R.: Mask r-cnn. In Proceedings of the IEEE International Conference on Computer Vision (ICCV), Oct 2017 (2017)
16. Agarwal, S., Punn, N.S., Sonbhadra, S.K., Nagabhushan, P., Pandian, K., Saxena, P.: Unleashing the power of disruptive and emerging technologies amid covid 2019: a detailed review. arXiv: 2005.11507 (2020)
17. Punn, N.S., Sonbhadra, S.K., Agarwal, S.: Monitoring COVID-19 social distancing with person detection and tracking via fine-tuned Yolo v3 and Deepsort techniques. arXiv: https://arxiv.org/abs/2005.01385 (2020)
18. Krizhevsky, A., Sutskever, I., Hinton, G.E.: Imagenet classification with deep convolutionalneural networks. In: Advances in Neural Information Processing Systems, pp. 1097–1105 (2012)
19. Devi, M.S., Carmel Mary Belinda, M.J., Aruna, R., Ramesh, P.S., Sundaravadivazhagan, B.: A novel model for IoT blockchain assurance-based compliance to COVID quarantine. In: System Design for Epidemics Using Machine Learning and Deep Learning. Signals and Communication Technology. Springer, Cham (2023)
20. Haq, I.U., Du, X., Jan, H.: Implementation of smart social distancing for COVID-19 based on deep learning algorithm. Multimed. Tools Appl. **81**, 33569–33589 (2022)
21. Salagrama, S., Kumar, H.H., Nikitha, R., Prasanna, G., Sharma, K., Awasthi, S.: Real time social distance detection using deep learning. In: 2022 International Conference on Computational Intelligence and Sustainable Engineering Solutions (CISES), Greater Noida, India, pp. 541–544 (2022)
22. Vibhuti, Jindal, N., Singh, H. et al.: Face mask detection in COVID-19: a strategic review. Multimed. Tools Appl. **81**, 40013–40042 (2022)
23. Sangeetha, T., Miruthula, V., Kavimalar, C., Aakash, V.: Face mask detection and social distancing using machine learning with Haar cascade algorithm. Sentim. Anal. Deep Learn. Adv. Intell. Syst. Comput. **1432**. Springer, Singapore (2023)

Junk Mail Content Detection Using Logistic Regression Algorithm

Tao Hai, Shaoyi Li, Ezinne C. Maxwell-Chigozie, Chidera Eze, Zunhai Gao, Celestine Iwendi, and Zakaria Boulouard

Abstract In contemporary times, things have moved away from traditional method to sophisticated way of communication via social media. One of the common ways information is disseminated amongst people is in the use of emails. Emails are very effective, easy and less costly to use by the sender but invariably costly to the recipient. This is due to the effect unwarranted messages which are thrown in tons are being received daily. This paper focus is on developing an effective junk mail content detector to effectively detect the content of messages and properly classify them thereby eliminate spurious emails. Logistic regression and Random Forest algorithms were employed and the result showed thar our model Logistics regression proves a superior performance.

Keywords Junk mail content decision · Machine learning · Random forest · Logistic regression

T. Hai
School of Computer and Information, Qiannan Normal University for Nationalities, Duyun, Guizhou 558000, China

T. Hai · S. Li · Z. Gao
School of Information and Artificial Intelligence, Nanchang Institute of Science and Technology, Nanchang, China

E. C. Maxwell-Chigozie (✉) · C. Eze · C. Iwendi
School of Creative Technologies, University of Bolton, Bolton BL3 5AB, UK
e-mail: ezinne.maxwell-chigozie@ieee.org

Z. Boulouard
LIM, Hassan II University of Casablanca, Casablanca, Morocco

C. Iwendi et al. (eds.), *Proceedings of ICACTCE'23—The International Conference on Advances in Communication Technology and Computer Engineering*, Lecture Notes in Networks and Systems 735, https://doi.org/10.1007/978-3-031-37164-6_21

1 Introduction

1.1 An Overview

The world has grown to be a global village where information can be disseminated from one place to another in no time. There exist varied ways of communication, which include the use of emails. Emails are quick and easy to use but are also susceptible to spurious and unwanted content [1], most often shrouded in secrecy to cause harm or force unsolicited content on the receiver.

Spam or junk messages are unsolicited messages [2, 3] and are more often moved to a junk folder following the non-identification of their routing source [4]. A ham is an email with no malicious content. Sometimes good (or ham) emails may also drop into junk folders. Spam mail might cause one to miss vital information. Care is taken into consideration to ensure harmful or fraudulent emails are spotted early to prevent an unfavourable outcome. This study aims to develop a machine learning model to effectively detect the content of junk mail as either ham or spam. Machine learning techniques can effectively handle classification. In this study, the Logistic regression and Random Forest are implemented on a spam mail dataset from Kaggle in a python environment. The obtained result showed that Logistic regression performed better than Random Forest regressor. The remaining paper is sectioned as follows; Sect. 2 discusses previous work, Sect. 3 describes the dataset, its pre-processing, and the building of the model preceded performing exploratory data analysis. Section 4 shows the built models, their comparison, and the metrics used for their performance evaluation. Section 5 discusses the findings. Section 6 concludes the paper and advises on future works.

2 Literature Review

Authors in Wardani et al. [5] described junk mail as an electronic mail service with defaming and disruptive connotations. Accordingly, the work dealt with using metadata to uncover the URLs through which pornographic content was streamed; then the identified spam messages are released to spam mail. Spam or junk emails are unwanted and unsolicited emails sent by another person to be noticed or pass a malicious code. Bringing to the fore the thin line separating good or spurious emails was the focus of the author in Debnath et al. [2]. In their work, they employed a deep learning approach using BERT, LSTM and applied NLP method on Enrol email dataset, and got 99.14% accuracy. The aftermath of the ever-increasing unsolicited emails, motivated the researchers in Junnarkar et al. [1] to develop a model that yields high efficiency and performance. This system can filter spam using sematic based URL and NLP based on semantics-based test classification.

Hassan and Mtetwa [6] expounded on the usefulness of feature extraction and the need to classify emails based on their content. These were demonstrated by conducting performance evaluations on two supervised machine learning algorithms. In their work [7], an effective and optimal way of detecting phishing emails, was their primary focus. Accordingly, feature extraction and selection played more frequently between 2006 and 2022. Also, SVM detected phishing mail in conjunction with the NLP approach. The works of [8] iterate the importance of the extraction of keywords to the identification of message content. Accordingly, the authors utilised SVM and statistical features to extract the keyword content of Arabic words, with 0.58 recall and 0.77 precision.

The increase in spam or junk emails has been described as a logic gene [9]. The researchers opined that the appearance of unsolicited mail degrades the authenticity attributed to trusted mail. To help curb this menace, they used deep neural networks and the min-hash method to unveil the true content of emails. The result obtained an accuracy of 98% for the detection of spam emails. Much research has been conducted on the detection of junk mail, classifying them as either ham or spam [10]. Thus, for a more secure email, the detection of the content of emails and the initiator must be known. In the study, word count and stopwords were employed to classify emails received as either ham or spam. ANN, Linear SVM, Logistic regression, and Multinomial Naïve Bayes were utilised. Taloba and Ismail [4] demonstrated a more efficient spam email detection. In their work, a hybrid of genetic algorithm and decision tree for spam detection on email was proposed, with the application of the PCA method. The genetic algorithm optimises the decision tree while PCA helps in cleaning the data. Their work showed superior performance over earlier text classifiers.

In Al-Rawashdeh et al. [11], the challenge of proper email content classification and feature selection, were seen as impeding the performance of existing email content classifications. For an improved feature selection, the study employed water cycle meta-heuristic optimization and three ways simulated annealing hybridisation. In the study, used cross-validation was used for training and validation dataset and seven testing datasets. Their work gave an improved feature reduction and outperformed Genetic algorithm, particle swarm optimization and harmony search, with an accuracy of 96.3% with SVM having an f-measurement of 96.3% against earlier multiple classifiers; Naïve Bayesian, KNN and SVM. Krause et al. [12] tackled the detection of spam from meta data on email headers. The research work used the CSDMC2010 dataset with both auto-extracted features and static set engineered features. With a focus on the meta header, and not the body, they achieve a high classification speed with an accuracy above 99%.

The aftermath of the ever-increasing unsolicited emails, motivated the researchers in Junnarkar et al. [1] to develop a model that yields high efficiency and performance. This system can filter spam using sematic based URL and NLP based on semantics-based test classification. Peng et al. [13] proposed an enhanced spam filter using Naïve Bayes that takes note of any text change and accurately identifies it as either spam or ham. Keyword and semantic-based machine learning improved the Naive Bayes accuracy. The study showed that Bayesian Poisoning obstructs the detection of spam mail. Jain et al. [14] used a novel deep learning architecture based on CNN and LSTM to improve the precision of text detection. The research employed two datasets and yielded an effective F1 score and accuracy. In Dada et al. [3], spam filtering detection and prevention were juxtaposed with their strength and weakness, to find a lasting solution. The deep learning and adversarial learning as an effective technique in solving spam mail intrusion.

Spammers aim to spread fake information. These spurious data kick in a mistrust of legitimate content. In finding a solution to this issue, Rayan [15] developed a machine learning model that will correctly detect spam content and prevent it from achieving its aim. In the pre-processing stage, appropriate tasks such as tokenisation, etc., were done. The authors compared a hybrid of random forest and decision tree (J48) with a hybrid of both classifiers. The outcome presented an accuracy of 98%.

3 Materials and Methods

Dataset

The objective of this study is to use logistic regression to identify junk spam mail. To achieve this, the dataset for spam_ham.csv was used, which was gotten from Kaggle. The dataset consists of 5171 messages, of which 1499 spam emails make up 29.0% of the dataset, and 3672 legitimate/ham emails make up 71% of the dataset. The three columns in the dataset are label, text, and label num. The graph below displays the ratio of spam emails to ham emails present in the dataset (Fig. 1).

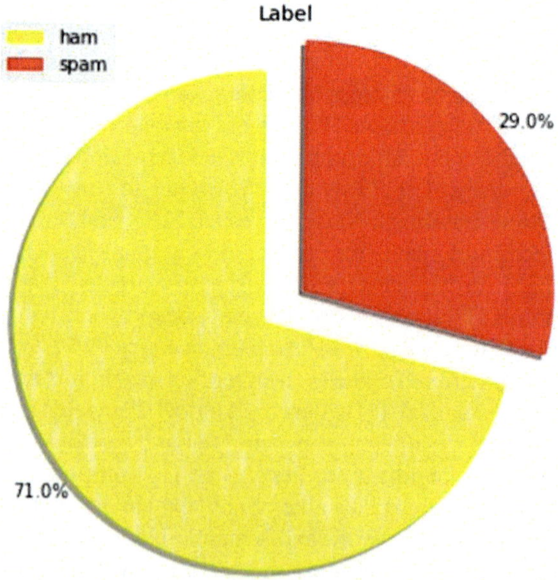

Fig. 1 Percentage of ham to spam emails

Packages

Python packages such as pandas, matplotlib, scipy, seaborn, and scikit-learn, were used to complete the analysis. The Python package "Pandas" is used to import the datasets into the Jupyter notebook IDE. It also used data manipulation, pre-processing, and analysis. The libraries "matplotlib" and "seaborn" were used for graph plotting which includes the confusion matrix result. The "scikit-learn" is a machine learning library that is built on numpy, scipy, and matplotlib. The imported "train test split" function is used to divide data into train and test datasets. The scikit-learn function "TfidfVectorizer" is used to convert text data into feature vectors. The "RandomForestClassifier" function was imported for the random forest classifier model while "LogisticRegression" was imported for the logistic regression model. For model evaluation, the "mean squared error" function was used to calculate the "root mean square error," and the accuracy score function was used, to calculate the accuracy of the models used.

4 Proposed Analysis Approach and Results

There are various processes that the dataset has to undergo for it to be used for the detection of junk mail. The processes used include data pre-processing, feature extraction, machine learning model (logistic regression and random forest classification), and classification into spam and ham. The diagram below shows the flow of the process involves (Fig. 2).

4.1 Data Pre-processing

Pre-processing is the process of removing text information from the dataset that is not required before text vectorization will be done. Text mining is at this crucial stage. Tokenization, stop word removal, and stemming are steps in this procedure. Tokenization process enables the removal of symbols such as "$","@" and" %" that are found inside the email text. In information retrieval (IR) and text mining, many commonly used English phrases are useless. These are known as "Stop words." Stop-words are frequent words that offer no information. They are functional terminology unique to a language (i.e., pronouns, prepositions, conjunctions). At this point, the most frequent words in a text—common terms—such as "we," "are," "is," and so on—are deleted. In the English language, stop words number between 400 and 500. The frequency of the terms provides the foundation for the stop word list. This approach will identify the terms that match the stop word lists by comparing the two lists. Removing these words and phrases makes room for document storage and cuts down on search time. The process of reducing a word to its root word, usually by

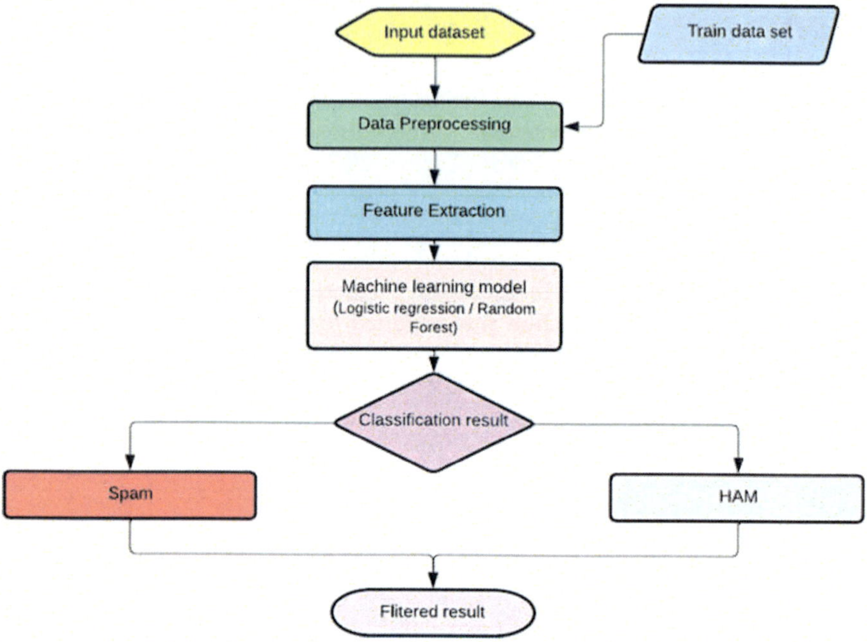

Fig. 2 Flow diagram for detecting spam

removing the suffix, is known as stemming. It reduces the vocabulary space, which speeds up the process. It's yet another way to normalise sentences for machines.

4.2 Feature Extraction

Feature Extraction attempts to decrease the number of features in a dataset by generating new features from the ones that already exist (and then discarding the original features). It entails turning random data, like text or photos, into numerical characteristics that may be used in machine learning.

4.2.1 Text Vectorization

The general technique of converting a group of text documents into numerical feature vectors is known as vectorization. The Bag of Words or "Bag of n-grams" is the name of this particular strategy. To achieve this vectorization, an sklearn.feature_extraction.text function called TfidfVectorizer was used. The below diagram shows the implementation of TfidfVectorizer in the dataset (Fig. 3).

```
from sklearn.feature_extraction.text import TfidfVectorizer

feature_extraction = TfidfVectorizer(min_df = 1, stop_words='english', lowercase='True')

X_train_features = feature_extraction.fit_transform(X_train)
X_test_features = feature_extraction.transform(X_test)

# convert Y_train and Y_test values as integers

y_train = y_train.astype('int')
y_test = y_test.astype('int')
```

Fig. 3 Vectorization of the dataset using TfidfVectorizer

4.3 Implementation of Machine Learning Models

The application of the machine learning models employed in this work begins after the dataset has undergone feature extraction and data pre-processing. The Random Forest Classifier and the Logistic Regression Model were applied in the course of this work. According to our research, the two models can be used to identify spam emails, but our goal is to compare the two to determine which is the best.

4.3.1 Logistic Regression and Evaluation Measures

Logistic Regression is a good predictive modelling approach for classifying issues with two or more probable outcomes. Logistic Regression is comparable to Linear Regression in that we receive an S-shaped line instead of a straight line to produce output in either 0's or 1's. Logistic Regression employs the sigmoid function to generate this S-shaped curve. The sigmoid function returns probabilities ranging from 0 to 1. This shows the logistic regression model used on the dataset with the accuracy and mean square error evaluation result (Fig. 4).

```
logistic_reg = LogisticRegression()
logistic_reg.fit(X_train_features, y_train)

LogisticRegression()

y_prediction_on_test_data = logistic_reg.predict(X_test_features)
accuracy_on_test_data = accuracy_score(y_test, y_prediction_on_test_data)

print('Accuracy on test data : ', accuracy_on_test_data)

Accuracy on test data :  0.9913043478260870

error = np.sqrt(mean_squared_error(y_test, y_prediction_on_test_data))

print('The mean square error : ', error)

The mean square error :  0.09325048082403138
```

Fig. 4 Logistic regression and evaluation metrics

```
random_forest_reg = RandomForestClassifier(random_state=2529)
random_forest_reg.fit(X_train_features, y_train)

RandomForestClassifier(random_state=2529)

y_pred_ram = random_forest_reg.predict(X_test_features)
accuracy_on_test_data = accuracy_score(y_test, y_pred_ram)

print('Accuracy on test data : ', accuracy_on_test_data)

Accuracy on test data :  0.9816425120772947

y_pred_ram.shape

(1035,)

error2 = np.sqrt(mean_squared_error(y_test, y_pred_ram))

print('The mean square error : ', error2)

The mean square error :  0.13548980744951006
```

Fig. 5 Random forest and evaluation metrics

4.3.2 Random Forest Classifier and Evaluation Measures

Random Forest is a meta-learner composed of several individual trees. Each tree votes on the overall categorization of the data set, and the random forest technique chooses the classification with the most votes. Replacement sampling is used to build each decision tree from a random portion of the training dataset [16]. This shows the Random Forest model used on the dataset with the accuracy and mean square error evaluation result (Fig. 5).

5 Discussion on Findings

We utilised the "spam_ham.csv" dataset in our analysis, which comprises 5171 emails, of which 1499 are spam and 3672 are ham. We got an accuracy score of 99.13 percent and 0.0932505 roots mean square error after gathering data, preparing it, balancing it, and then using the Logistic Regression method to classify messages as spam or ham. While with Random Forest Classifier method we got a 98.16 percent accuracy score and 0.13549 roots mean square error. Below shows the junk email prediction using the train logistic regression model showing that the sample email is a ham (Fig. 6).

```
input_mail = ["I've been searching for the right words to thank you for this breather. I promise i wont take your help for grante

# convert text to feature vectors
input_data_features = feature_extraction.transform(input_mail)

# making prediction

prediction = logistic_reg.predict(input_data_features)
print(prediction)


if (prediction[0]==1):
  print('Ham mail')

else:
  print('Spam mail')

[1]
Ham mail
```

Fig. 6 Checking an example email to know whether ham or spam

6 Conclusion and Future Scope

In this study, we looked at detection of junk mail content using a logistic regression model and then compared the results to a random forest regression model to see which one was better. The process entails reviewing publicly available papers on the related topic to gain a better understanding of what others have done to address the junk mail problem. The dataset obtained from the Kaggle website was pre-processed, which assisted in determining the number of ham and spam records. The logistic regression model and the random forest classifier can be used to detect junk mail in the future, but further research should focus on logistic regression due to its ability to achieve higher accuracy and lower error.

Acknowledgements This work was supported by Key Laboratory of Advanced Manufacturing Technology of the Ministry of Education of Guizhou University (GZUAMT2022KF[07]), the National Natural Science Foundation of China (No.61862051), the Science and Technology Foundation of Guizhou Province (No.[2019]1299, No.ZK[2022]449), the Top-notch Talent Program of Guizhou province (No.KY[2018]080), the Natural Science Foundation of Education of Guizhou province (No.[2019]203) and the Funds of Qiannan Normal University for Nationalities (No.qnsy2019rc09). The Educational Department of Guizhou under Grant NO.KY[2019]067.

References

1. Junnarkar, A., Adhikari, S., Fagania, J., Chimurkar, P., Karia, D.: E-Mail spam classification via machine learning and natural language processing. In: 2021 Third International Conference on Intelligent Communication Technologies and Virtual Mobile Networks (ICICV), Tirunelveli, India, pp. 693–699 (2021). https://doi.org/10.1109/ICICV50876.2021.9388530
2. Debnath, K., Kar, N.: Email spam detection using deep learning approach. In: 2022 International Conference on Machine Learning, Big Data, Cloud and Parallel Computing (COM-IT-CON), Faridabad, India, pp. 37–41 (2022). https://doi.org/10.1109/COM-IT-CON54601.2022.9850588

3. Dada, E.G., Bassi, J.S., Chiroma, H., Adetunmbi, A.O., Ajibuwa, O.E.: Machine learning for email spam filtering: review, approaches and open research problems. Heliyon **5**(6), e01802 (2019). ISSN 2405–8440, https://doi.org/10.1016/j.heliyon.2019.e01802
4. Taloba, A.I., Ismail, S.S.I.: An intelligent hybrid technique of decision tree and genetic algorithm for E-Mail spam detection. In: 2019 Ninth International Conference on Intelligent Computing and Information Systems (ICICIS), Cairo, Egypt, pp. 99–104 (2019). https://doi.org/10.1109/ICICIS46948.2019.9014756
5. Wardani, D., Siwi, R., Harjito, B., Marshallia, M.: Using metadata in detection spam email with pornography content. In: 2018 International Conference on Electrical Engineering and Computer Science (ICECOS), Pangkal, Indonesia, pp. 213–218 (2018). https://doi.org/10.1109/ICECOS.2018.8605220
6. Hassan, M.A., Mtetwa, N.: Feature extraction and classification of spam emails. In: 2018 5th International Conference on Soft Computing & Machine Intelligence (ISCMI), Nairobi, Kenya, pp. 93–98 (2018). https://doi.org/10.1109/ISCMI.2018.8703222
7. Salloum, S., Gaber, T., Vadera, S., Shaalan, K.: A systematic literature review on phishing email detection using natural language processing techniques. In: IEEE Access, vol. 10, pp. 65703–65727 (2022). https://doi.org/10.1109/ACCESS.3183083
8. Armouty, B., Tedmori, S.: Automated keyword extraction using support vector machine from arabic news documents. In: 2019 IEEE Jordan International Joint Conference on Electrical Engineering and Information Technology (JEEIT), Amman, Jordan, pp. 342–346 (2019). https://doi.org/10.1109/JEEIT.2019.8717420
9. Marza, N.H., Manaa, M.E., Lafta, H.A.: Classification of spam emails using deep learning. In: 2021 1st Babylon International Conference on Information Technology and Science (BICITS), Babil, Iraq, pp. 63–68 (2021). https://doi.org/10.1109/BICITS51482.2021.9509909
10. Sethi, M., Chandra, S., Chaudhary, V., Dahiya, Y.: Spam email detection using machine learning and neural networks. In: Shakya S., Balas V.E., Kamolphiwong S., Du K.L. (eds.) Sentimental Analysis and Deep Learning. Advances in Intelligent Systems and Computing, vol. 1408. Springer, Singapore (2022). https://doi.org/10.1007/978-981-16-5157-1_22
11. Al-Rawashdeh, G., Mamat, R., Hafhizah Binti Abd Rahim, N.: Hybrid water cycle optimization algorithm with simulated annealing for spam E-mail detection. IEEE Access **7**, 143721–143734 (2019). https://doi.org/10.1109/ACCESS.2019.2944089
12. Krause, T., Uetz, R., Kretschmann, T.: Recognizing email spam from meta data only. In: 2019 IEEE Conference on Communications and Network Security (CNS), Washington, DC, USA, pp. 178–186 (2019). https://doi.org/10.1109/CNS.2019.8802827
13. Peng, W., Huang, L., Jia, J., Ingram, E.: Enhancing the naive bayes spam filter through intelligent text modification detection. In: 2018 17th IEEE International Conference On Trust, Security And Privacy In Computing And Communications/ 12th IEEE International Conference On Big Data Science And Engineering (TrustCom/BigDataSE), New York, NY, USA, pp. 849–854 (2018). https://doi.org/10.1109/TrustCom/BigDataSE.2018.00122
14. Jain, G., Sharma, M., Agarwal, B.: Spam detection in social media using convolutional and long short term memory neural network. Ann. Math. Artif. Intell. **85**, 21–44 (2019). https://doi.org/10.1007/s10472-018-9612-z
15. Rayan, A.: Analysis of e-Mail spam detection using a novel machine learning-based hybrid bagging technique. Comput. Intell. Neurosci. **9**(2022), 2500772 (2022). https://doi.org/10.1155/2022/2500772.PMID:35983156;PMCID:PMC9381222
16. Nithesh, R.K., Vijayalakshmi, K.: Classification of spam messages using random forest algorithm (2021).https://doi.org/10.37896/jxu15.8/048

Predicting Fish Habitat in the Persian Gulf Using Artificial Intelligence

Tao Hai, Jincheng Zhou, Hoorieh Ahmadi, Ayibatonbo Ebiare Ekiye, Yangping Wei, Celestine Iwendi, and Zakaria Boulouard

Abstract The Persian Gulf is one of the most important habitats in the Middle East. It can be extremely beneficial to aquatic species' survival and environmental preservation to continuous monitoring and collect data about aquatic animals, their habitats, and behaviours. Finding a novel and suitable method to carry out accurate and automatic monitoring with low timing and low cost for monitoring aquatic species' behaviour in this high potential area is helpful. To predict fish habitat in Persian Gulf Convolutional Neural Network method and Naïve Bayes algorithm are used. Deep learning convolutional neural network technology is mostly used for data science classification and recognition because of its exceptional accuracy and to solve search and optimization issues, the Naïve Bayes algorithm is employed. Results indicate for predicting fish habitat in the Persian Gulf, the accuracy of the Convolutional Neural Network algorithm and the Naïve Bayes algorithms is 97.32% and 95.47%, respectively. With $p = 0.025$ (p0.05), there is a substantial difference between the Naïve Bayes method and the Convolutional Neural Network algorithm. Therefore, The Convolutional Neural Network method seems to be more accurate than the Naïve Bayes method at predicting fish habitat in the Persian Gulf.

Keywords Fish habitat · Convolutional neural network algorithm · Naïve bayes · Novel approach · Deep learning

T. Hai · J. Zhou · Y. Wei
School of Computer and Information, Qiannan Normal University for Nationalities, Duyun, Guizhou 558000, China

H. Ahmadi (✉) · A. E. Ekiye · C. Iwendi
School of Creative Technologies, University of Bolton, Bolton, UK
e-mail: ha7crt@bolton.ac.uk

Z. Boulouard
LIM, Hassan II University of Casablanca, Casablanca, Morocco

T. Hai
School of Information and Artificial Intelligence, Nanchang Institute of Science and Technology, Nanchang, China

C. Iwendi et al. (eds.), *Proceedings of ICACTCE'23—The International Conference on Advances in Communication Technology and Computer Engineering*, Lecture Notes in Networks and Systems 735, https://doi.org/10.1007/978-3-031-37164-6_22

1 Introduction

One of the most important aquatic habitats in Iran and the world is the Persian Gulf. This region has always been the focus of various scientists because of its unique nature, the diversity of different aquatic species, and the abundance of oil and gas resources it has [1]. Aquatics are very important for nutrition, pharmaceutical companies, industries, etc. (Selamoglu, 2021). Due to illegal fishing and sometimes pollution caused by oil extraction, some aquatic species may be at risk of extinction [3]. Constant observation and collection of information about aquatic animals, and their habitats, and investigating their behaviours can greatly help their survival and preserve the environment [4]. Considering the high potential of this region, finding a new and appropriate approach to perform precise and automatic monitoring in a short time and low cost is of great importance to the behaviour of aquatic animals in the region [5].

There are lots of cases that are dealing with uncertainties like medal diagnoses and weather forecasting and need to make decisions based on collecting data from the environment and uncertain-domain learning. All sorts of decision-making involve models created using ML[1] and DL,[2] so it is crucial to assess the dependability and effectiveness of AI[3] systems before putting them to use [6].

Multi-tasking cascaded CNN was used to accurately recognize the pattern by removing other history points. It has also been shown to be accurate in the position of features in detail. In addition, cleaning and analysis of data sets have been performed to better understand the features in a dataset. The planned approach has been checked and its accuracy, MAE,[4] and RMSE[5] were compared with other methods [7].

One of the popular classification methods is the Naïve Bayes algorithm (NB). It is based on the application of the Bayes theorem along with the strong assumption that all predictors are independent of each other. In simpler terms, it is assumed that the presence of a feature in a class is independent of the presence of any other feature in the same class. For example, a phone may be considered a smartphone if it has a touch screen, internet facilities, a good camera, etc. Although all these features are interdependent, they independently contribute to the possibility of the phone becoming smart [8].

The aim of this research is to provide a new method to predict the fish habitat based on the convolutional neural network technique by using the Naïve Bayes algorithm.

[1] Machine Learning.

[2] Deep Learning.

[3] Artificial Intelligence.

[4] Mean Absolute Error.

[5] Root Mean Squared Error.

2 Literature Review

One of the most important tasks in a nuclear power plant is regularly checking its function to be sure of its safety. This task requires a long and time-consuming process where staff must always carefully watch the videos of the power plants to identify cracks in the reactor body. These cracks are tiny, and then detecting them is a really hard task. Chen and Jahanshahi (2018) provide a framework based on a convolutional neural network (CNN) and a Naïve Bayes data fusion plan to check videos separately and discover cracks, whereas a unique data fusion strategy is suggested to combine the data taken from each video to improve the system's performance and reliability. Then they used CNN to discover cracks in each video, and the data fusion plan saved the spatial and temporal integrity of cracks in the frames of videos. In this framework, false positives are successfully discarded by the Naïve Bayes process. This framework improved the process of discovering cracks [9].

In the medical field, data mining can help to discover and find some precious information and important pattern. This data can be useful in diagnosing the disease and finding a suitable treatment. Repaka et al. [10] provided a prediction model based on Naïve Bayes to predict risk factors for heart disease. they presented a mobile web application, by analysing the received information and patterns. They collect information like age, sex, blood sugar, etc. in a standard form and use them as input for Naïve Bayes classification. They used 80% of the data for training and 20% for testing. After collecting information, the user should register and log in to the mobile application. Then Naïve Bayes classification and, secure data transmission and prediction are done using AES.[6] The results show the presented method is successful in predicting the potential risks of heart diseases [10].

Detecting fish by machine learning is an important technology and helps scientists in their research. Machine learning is a new method in the marine field that provides lots of benefits for scientists to gather their required data. Detecting different fish species when they look similar is a challenging task for the detecting system. There is more difficulty such as the background and the fish body angle in the image. Tan [11] provides a fish detection system based on CNN. This system used Network-in-Network (NIN) pattern to detect fish. At the end of the research, this model was analysed, and its accuracy was 83% [11].

Identification of fish is one of the tasks that can be useful for everybody. Machine learning can help to make a classification model to identify fish species. Ahmed et al. [12] made a classification model based on machine learning. This model helps everybody to identify all fish species even if they don't have any knowledge about the different types and species of extinct or existing fish. As fishes are a source of good vitamins and nutrition for people, this research can help create a suitable intake model. To achieve this goal, they used "Keras",[7] to create an improved VGG16 transition learning pattern combined with a CNN. Results show that using the CNN model is helpful, and the training accuracy and categorization accuracy are about

[6] Advanced Encryption Standard.

[7] "Keras" is an API designed for human beings, not machines.

97%. Accuracy for reliability is 99.92% using the CNN model and 99.76% for categorization with the VGG16 [12].

As marine scientists are looking for more efficient ways to understand and gather more information about the types of fish and their environments, Iqbal et al. [13] create an automated system to recognize and categorize fish species. This method is using convolutional neural networks and a lower-resolution AlexNet model that included 2 fully linked layers and 4 convolutional layers. They also provide a comparison between VGGNet and AlexNet. At the end of the research, this model with fewer layers has more test accuracy than primary AlexNet. They entered the dropout layer that increased the efficiency of their model [13].

One of the most important issues to manage the marine environment is analysing Acoustic-trawl. Explanation of the acoustic signals and assigning them to the appropriate species is an important challenge and to support this need some camera systems are developed. Analysing the images of these cameras helps to identify the species. Allken et al. [14] create and implement a deep-learning neural network to automatically categorize the species shown in photos taken by the cameras. To reduce training information, they used a new method by using a realistic simulation of the camera photos. The results show this method has high accuracy and efficiency and using synthetic data can improve performance [14].

Regularly checking the bridges to find any cracks or weaknesses in them is one of the most important tasks in their maintenance of them. The traditional method for detecting cracks has some weaknesses that can provide engineer needs. The deep learning method has grown and improved recently but still, there are lots of problems like the need for big model training and lots of parameters. Li et al. [15] create a new crack-detecting algorithm to classify cracks and noises automatically by using a new model by combining Naïve Bayes data fusion and convolutional network (NB-FCN). They gather bridge images in two categories of training and testing from different distances and situations. Then they trained noise and cracks using many repetitions. Then they identify some features of cracks like the width and length by digital distance measurement to find out the fault rate of their method. Results show their algorithm is more accurate in detecting cracks and have a shorter time to calculate and less fault rate [15].

One of the most known causes of death in today's world is cancer. This disease has various types, like colon cancer. Every year, many people die from colon cancer, and unfortunately, it does not have any obvious symptoms in the early stages. Although there are many distinctions between the ways that various cancers develop and their initial symptoms, accurately classifying this type of disease remains a significant difficulty. Researchers have repeatedly used machine learning to achieve a classification of cancer-related genes. Salmi et al. [16] used a Naïve Bayes model to classify the data of colon cancer patients. They present a prediction technique based on probabilistic and Naïve Bayes. Results show this model with less complexity has an acceptable accuracy [16].

Prediction analysis is a strategy that extrapolates potential outcomes from historical data. Tripathi et al. [17] present a model to predict Student Performance based on Naïve Bayes. This model has three stages including pre-processing, categorization, and character recognition. They executed this model in python and analysed the run time and accuracy. The results show this model has a high accuracy and low run time in comparison to other models [17].

3 Methods and Research Finding

CNN and NB algorithms are the two classifiers that are employed in this approach on two different with the 20-sample size. These algorithms were contrasted for greater accuracy scores and loss values in order to determine the best algorithm for predicting fish habitat. The Standard deviation has been calculated to be 0.27 for NB and 0.14 for CNN.

There are some cameras in the Persian Gulf whose duty is to collect images from the environment and these data are used as input for CNN and NB. These images were resized and the colour and the contrast were converted to a predefined standard. After Image segmentation, the futures would be extracted as a property of each fish and kept in a database as a final result (Fig. 1).

Moreover, the proposed architecture is shown in Fig. 2, which covers two modules. The first module is considered the training phase and the second one is prediction.

In the first phase, the images in the database should be processed and the features extract by ML to train the machine. In the next phase, the outcome would be anticipated for each given image after processing the image to identify the features based on the trained model.

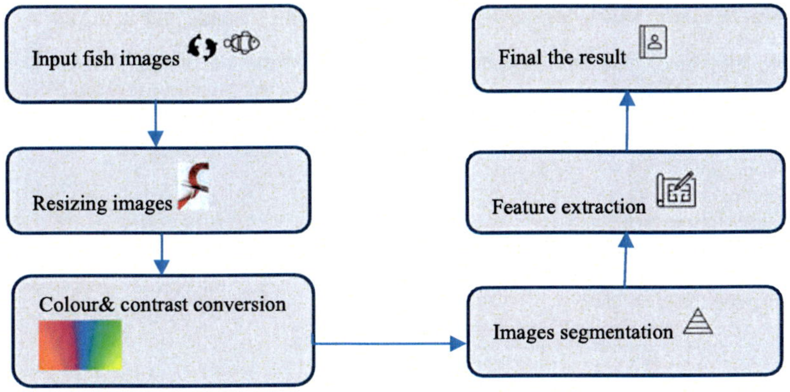

Fig. 1 Steps to implement the fish habit prediction (framework)

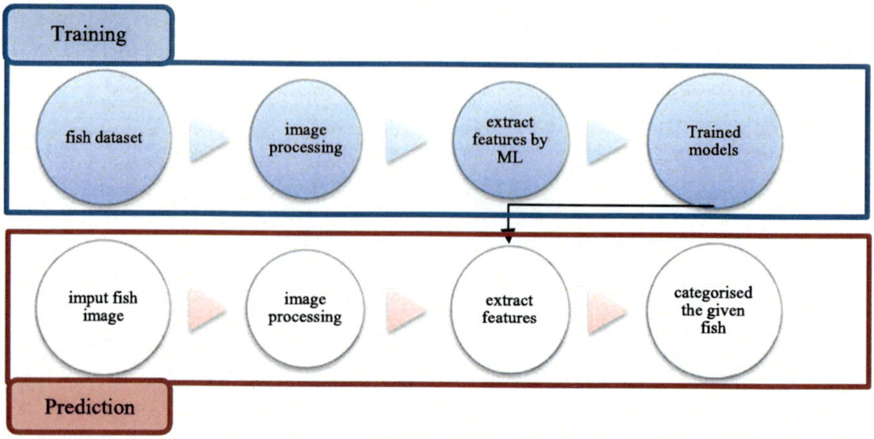

Fig. 2 Proposed architecture to implement the fish habit prediction

Algorithm

Convolutional Neural Network Algorithm

The classification and differentiation of input data kinds were done using CNN. A convolutional neural network, a type of Deep Learning system, can be used to prioritise and differentiate between distinct parts of an image after receiving it as input.

Pseudocode for CNN Algorithm

It is vital to understand The Nave Bayes pseudocode to create the required algorithm. The following contains the Naïve Bayes pseudocode.

```
Import Convolutional Network
Classifier Import Convolutional Neural Network as CNN
file_name, file_path = uigetfile("*.jpg, ""*.bmp, ""*.tif,""*.gif, "*.png, "".jpeg, ""*.gif")
'Load Image File';
if isequal(file_name,0)||isequal(file_path,0)
warndlg(' Press OK to continue', 'Warning');
Else
image aqa = imread([file_path file_name]);
imshow(image aqa);
title('Preprocess');
image aqa = imresize(image aqa);
Compare images and give accuracy as the results;
Plot the graph for accuracy;
Plot the graph for specificity;
Accuracy of the Convolutional Neural Network classifier;
```

Naïve Bayes Algorithm

This classification technique is based on the Bayes Theorem and the idea of predictor independence. Simply defined, a Naïve Bayes classifier holds that the existence of one attribute in a class has no bearing on the existence of any other features.

Pseudocode for Naïve Bayes

To write the needed algorithm it is necessary to know about the Naïve Bayes pseudocode. The Naïve Bayes pseudocode is given below.

Step 1: Import libraries
Step 2: Give input data (the training dataset T)
Step 3: Determine each class's standard division and mean of the predictor variable
Step 4: Repeat until all prediction variables (f1, f2, f3, ..., fn) have been determined. Use the gauss density equation to calculate the probability of fi in each class
Step 5: Calculate each class's probability
Step 6: Obtain the highest probability

4 Results

The testing setup has settings for both hardware and software configurations. The laptop has a solid-state drive, a 64-bit operating system, an × 64-based processor, an Intel Core i5 11th generation CPU, and 8 GB of RAM. The Windows 11 operating system is currently programmed using Python. When the program is complete, the accuracy value will appear. Procedure: Wi-Fi-connected laptop. Google Collaboration could be used as an environment to implement by python language as other environments like Jupiter Notebook. Execute the application. Make a folder for the file, and then upload it to the disc to store it. To log in, use the message ID Execute the code to create a graph and output the accuracy.

IBM SPSS version 28 was used to analyse the data, and it is a statistical software program. Ten iterations with a maximum of 20 samples were run for both the current algorithms and the proposed Innovative models to evaluate the accuracy of the Novel Approach. We recorded the projected accuracy for each iteration. The rounds findings of the Independent Sample T-test were applied. Numerous independent variables are needed to predict fish habitat.

Table 1 Group statistics results

Group statistics					
Accuracy	Group	N	Mean	Std division	Std. error mean
	CNN	20	97.45	0.14	0.04
	Naïve Bayes	20	95.30	0.27	0.09

Table 2 Independent samples T-test

Accuracy	Independent sample test									
	Levene's test for equality of variances					T-test for equality of means				
	F	Sig	t	df	Significance		Mean difference	Std error difference	95% Confidence interval of the difference	
					One sided p	Two sided p			Lower	Upper
Equal variance assumed	5.98	0.03	19.2	18	<0.001	<0.001	1.855	0.097	1.652	02.058
Equal variance not assumed			19.2	13.48	<0.001	<0.001	1.855	0.097	1.647	2.063

The predicted fish habitat is checked using images taken from the dataset.

The average accuracy for CNN is 97.45%, with a standard deviation of 0.14, whereas the average accuracy for NAÏVE BAYES is 95.30%, with a standard deviation of 0.27 (Table 1).

Results of the Independent Samples T-test show that CNN was significantly better than Naïve Bayes (Table 2).

In Fig. 3, there is a bar graph that compares the average accuracy of CNN and the Naïve Bayes algorithm. This graph shows Naïve Bayes' accuracy is 95.47%, while CNN's mean accuracy is 97.32%.

In Fig. 4, standard division and standard error are compared between CNN and Naïve Bayes. The results show that CNN has a better performance than Naïve Bayes.

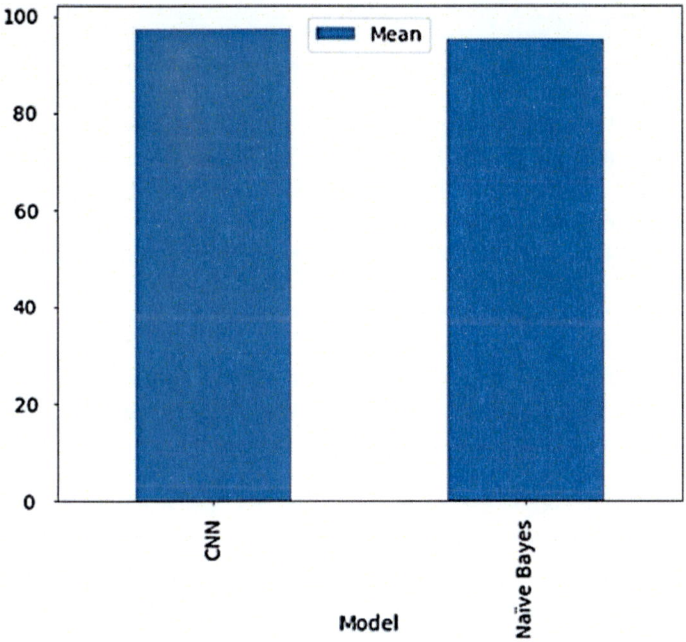

Fig. 3 Accuracy comparison between CNN and Naïve Bayes

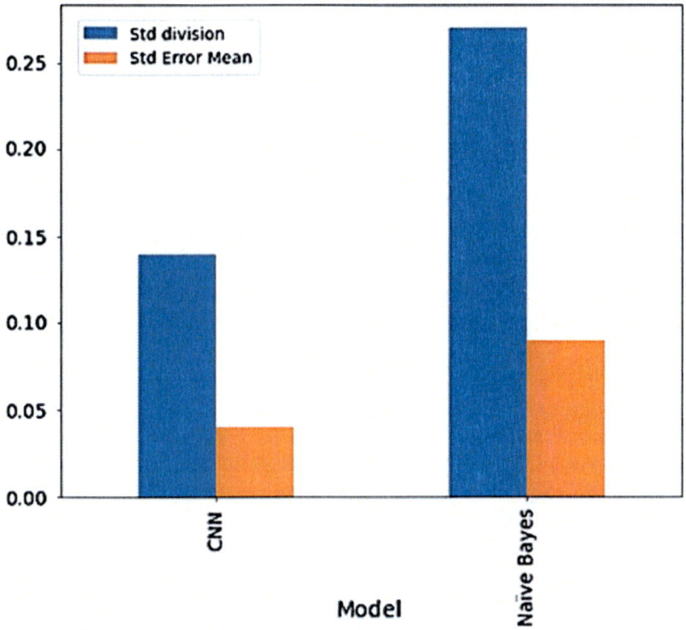

Fig. 4 Compare of standard division and standard error between CNN and Naïve Bayes

5 Conclusion

Continuously monitoring and observing fish's behaviour and their habitats can provide scientists with lots of precious information. This research purpose is to find an efficient model to predict fish habitat in the Persian Gulf. All sorts of prediction models used ML and DL. So, it is essential to evaluate the dependability and efficiency of AI systems before using them.

This study aims to give an appropriate approach for predicting the behaviour and habitat of fishes by comparing the results of two well-known prediction algorithms (CNN and NB) that use data collected from the environment.

The two classifiers used in this approach on two separate datasets with the 20-sample size are the CNN and NB algorithms. To choose the best algorithm for predicting fish habitat, these algorithms were compared for higher accuracy scores and loss values. For NB and CNN, the standard deviation has been determined to be 0.27 and 0.14, respectively.

Input for CNN and NB comes from cameras in the Persian Gulf that were responsible for gathering images of the environment. These images were resized, and a predefined standard was used to change the colour and contrast. Futures would be retrieved as a property of each fish following image segmentation, and as a result, they would be stored in a database. In this step there are two phases. In the first phase, based on the fish image dataset and processing the whole images, the features of each species are extracted and categorised to train an ML model. Then in the second phase, for each given image, after processing the image to identify the features based on the trained model, the result would be predicted.

In this research, the average accuracy of CNN and NB are compared. Results show Naïve Bayes' accuracy is 95.47%, while CNN's mean accuracy is 97.32%.

Standard division and standard error are also compared between CNN and Naïve Bayes. The results show that CNN has a better performance than Naïve Bayes.

To achieve better performance other criteria such as precision or MCC[8] can be considered to compare two algorithms or do this comparison on a larger scale to achieve more accurate results. There are many algorithms for predicting based on machine learning then it would be a good idea to check other models and compare them with CNN and NB to find the best algorithm for predicting fish habitat.

Acknowledgements This work was supported by Key Laboratory of Advanced Manufacturing Technology of the Ministry of Education of Guizhou University (GZUAMT2022KF[07]), the National Natural Science Foundation of China (No.61862051), the Science and Technology Foundation of Guizhou Province (No.[2019]1299, No.ZK[2022]449), the Top-notch Talent Program of Guizhou province (No.KY[2018]080), the Natural Science Foundation of Education of Guizhou province(No.[2019]203) and the Funds of Qiannan Normal University for Nationalities (No. qnsy2019rc09). The Educational Department of Guizhou under Grant NO. KY[2019]067.

[8] Matthews Correlation Coefficient.

References

1. Sharifinia, M., Daliri, M., Kamrani, E.: Estuaries and coastal zones in the northern Persian gulf (Iran). In Coasts and Estuaries, pp. 57–68. Elsevier (2019)
2. Selamoglu, M.: Importance of the cold chain logistics in the marketing process of aquatic products: an update study. J. Surv. Fisheries Sci. 25–29 (2023)
3. Duarte, C.M., Agusti, S., Barbier, E., Britten, G.L., Castilla, J.C., Gattuso, J.P., Fulweiler, R.W., Hughes, T.P., Knowlton, N., Lovelock, C.E., Lotze, H.K.: Rebuilding marine life. Nature **580**(7801), 39–51 (2020)
4. Saleh, A., Sheaves, M., Rahimi Azghadi, M.: Computer vision and deep learning for fish classification in underwater habitats: a survey. Fish Fish. **23**(4), 977–999 (2022)
5. Jo, J., Jo, B., Kim, J., Kim, S., Han, W.: Development of an IoT-based indoor air quality monitoring platform. J. Sens. **2020**, 1–14 (2020)
6. Abdar, M., Pourpanah, F., Hussain, S., Rezazadegan, D., Liu, L., Ghavamzadeh, M., Fieguth, P., Cao, X., Khosravi, A., Acharya, U.R., Makarenkov, V.: A review of uncertainty quantification in deep learning: techniques, applications and challenges. Inf. Fusion **76**, 243–297 (2021)
7. Gadekallu, T.R., Iwendi, C., Wei, C., Xin, Q.: Identification of malnutrition and prediction of BMI from facial images using real-time image processing and machine learning. IET Image Process **16**, 647–658 (2021)
8. Chen, S., Webb, G.I., Liu, L., Ma, X.: A novel selective naïve Bayes algorithm. Knowl. Based Syst. **192**, 105361 (2020)
9. Chen, F.C., Jahanshahi, M.R.: NB-CNN: Deep learning-based crack detection using convolutional neural network and Naïve Bayes data fusion. IEEE Trans. Industr. Electron. **65**(5), 4392–4400 (2017)
10. Repaka, A.N., Ravikanti, S.D. and Franklin, R.G.: Design and implementing heart disease prediction using naives Bayesian. In 2019 3rd International conference on trends in electronics and informatics (ICOEI), pp. 292–297. IEEE (2019)
11. Tan, Y.Y.: Fish species recognition using convolutional neural network/Tan Ying Ying. Doctoral dissertation, University of Malaya (2018)
12. Ahmed, F., Basak, B., Chakraborty, S., Karmokar, T., Reza, A.W., Imam, O.T., Arefin, M.S.: Developing a classification CNN model to classify different types of fish. In: Intelligent Computing & Optimization: Proceedings of the 5th International Conference on Intelligent Computing and Optimization 2022 (ICO2022), pp. 529–539. Springer International Publishing, Cham (2022)
13. Iqbal, M.A., Wang, Z., Ali, Z.A., Riaz, S.: Automatic fish species classification using deep convolutional neural networks. Wireless Pers. Commun. **116**, 1043–1053 (2021)
14. Allken, V., Handegard, N.O., Rosen, S., Schreyeck, T., Mahiout, T., Malde, K.: Fish species identification using a convolutional neural network trained on synthetic data. ICES J. Mar. Sci. **76**(1), 342–349 (2019)
15. Li, G., Liu, Q., Zhao, S., Qiao, W., Ren, X.: Automatic crack recognition for concrete bridges using a fully convolutional neural network and naive Bayes data fusion based on a visual detection system. Meas. Sci. Technol. **31**(7), 075403 (2020)
16. Salmi, N., Rustam, Z.: Naïve Bayes classifier models for predicting the colon cancer. In: IOP Conference Series: Materials Science and Engineering, vol. 546, no. 5, p. 052068. IOP Publishing (2019)
17. Tripathi, A., Yadav, S., Rajan, R.: Naive Bayes classification model for the student performance prediction. In: 2019 2nd International Conference on Intelligent Computing, Instrumentation and Control Technologies (ICICICT), vol. 1, pp. 1548–1553. IEEE (2019)

Evaluation of Text Classification Using Support Vector Machine Compare with Naive Bayes, Random Forest Decision Tree and K-NN

Tao Hai, Jincheng Zhou, Shirin Abolfath Zadeh, Oluwabukola A. Adetiloye, Mingjiang Li, Ikpenmosa Uhumuavbi, and Celestine Iwendi

Abstract This paper aims to find the boost model which brings the best accuracy in text classification by using Support Vector Machine in comparison with other models namely Naive Bayes, Random Forest Decision Tree and K-NN. For the text classification and processing, the planned system will have to apply with the Support Vector Machine and the result is decided by major roles. Based on the Machine Learning algorithms used for the implementation of the research- the BBC news dataset- illustrates that the Support Vector Machine has better accuracy and result.

Keywords Text processing · Text classification · Machine learning · Support vector machine · SVM · Language processing · Classification · Semi-supervised learning · Naive bayes · Random forest · Decision tree · K-NN

T. Hai · J. Zhou · M. Li
School of Computer and Information, Qiannan Normal University for Nationalities, Duyun, Guizhou 558000, China
e-mail: haitao@bjwlxy.edu.cn

J. Zhou
e-mail: guideaaa@126.com

M. Li
e-mail: 511399831@qq.com

S. A. Zadeh (✉) · O. A. Adetiloye · I. Uhumuavbi · C. Iwendi
School of Creative Technologies, University of Bolton, Bolton BL3 5AB, UK
e-mail: Shirin.abolfathzadeh@ieee.org

T. Hai
School of Information and Artificial Intelligence, Nanchang Institute of Science and Technology, Nanchang, China

C. Iwendi et al. (eds.), *Proceedings of ICACTCE'23—The International Conference on Advances in Communication Technology and Computer Engineering*, Lecture Notes in Networks and Systems 735, https://doi.org/10.1007/978-3-031-37164-6_23

1 Introduction

Text classification is a common problem in Natural Language Processing (NLP), and several algorithms can be used for this task, including Support Vector Machine (SVM), Naive Bayes, Random Forest, and k-Nearest Neighbours (k-NN).

SVM is a linear or non-linear model that is effective in high-dimensional spaces. It works well when the data is separable but can struggle with noisy or unbalanced datasets.

Naive Bayes is a probabilistic model based on Bayes' theorem. It is fast, simple, and works well for high-dimensional datasets, but its assumptions of independence between features can limit its performance.

Random Forest is an ensemble method that combines multiple decision trees to make a prediction. It's robust to overfitting and can handle non-linear relationships and missing data well but can be slow and require more computational resources than other algorithms.

K-NN is a non-parametric algorithm that makes predictions based on the class of the nearest neighbours in the feature space. It's simple, fast, and requires little training data, but can be sensitive to the choice of k and can struggle with high dimensional datasets.

The aim of text classification and processing is achieved presumably through natural language processing and machine learning. With the help of all solutions, with the help of all particles affected by the management of all sources related to memory creation, all sentiment analysis of all activities before creating support vectors create. Because in the field of speech awareness, there is evidence that the substance of absolute speech is used to interpret what is said, rather than linearly [1–18].

The research in this study is primarily about the proposed system. While our existing system made a change and found high accuracy through research work, different areas are affected by text processing. Text classification, in the field of information processing, maps vast sets of documents to predefined text attributes for archiving a task that transforms hypertext files into understandable content by machine. This is to classify them into some predefined categories accordingly. Problems arise today in many areas, such as email filtering, web page searching, office automation, topic indexing, news article classification, and more. Finding requested information is typically time-consuming and tedious for the average Internet user. To retrieve the information, users often need to connect to multiple servers and search through numerous documents to find them. The most natural and direct way for humans to access such tasks is by asking questions in natural language and eliciting responses to gain clarity. All these require text processing to discover the relevant questions.

2 Related Works

Identifying the words in the text and classifying and normalizing them is a pivotal part of the task in text processing and converting the text to speech in any language. While there are different ways algorithms are applied to concepts, this article proposes the idea of employing supportive machine learning to categorise the words in a text. As part of normalizing the text after the single-word tokenized text, finding out standard words including types of acronyms and abbreviations is an issue. Therefore, to convert them to normal words based on the pre-defined rules, Moattar et al. [1] this paper proposes the idea of using Machine learning methods to identify the Persian Non-standard words by using the Support vector machine and classification and regression tree. This method uses specific rules to group the data and split words into the tree and evaluate the cost of any subtrees that were divided based on the base tree.

While the mass data need to be improved and organized, a handy method would not be effective. As a result, some studies [2] sought to compare the impact of linear and non-linear kernels of SVM classification in text classification by using two experiments to show the effectiveness and efficiency of each method. They argued that the RBF kernel achieved above 95% accuracy in solving issues in text classification. In addition, they argued that this result could be affected by the given datasets to find out which linear or non-linear SVM would bring the better result.

Moreover, the first step in the answering system is always to find out the classification of the question. Lin et al. [3] proposed the One-against-One SVM algorithm to extract the features in question in the Chinese language, especially in internet searches. They provided the classifier, based on SVM to support high accuracy by segmenting the words automatically including keyword extraction. However, there are issues in the web area, including filtering emails, searching for a subject and indexing. Another more problematic case for researchers is in categorizing the text and documents. In Hao et al. [4] the researchers assumed a new text-processing approach based on the Naive Bayes SVM. This was considered more reliable and efficient compared to the traditional algorithm. Notably, algorithms improve classification precision and speed by removing unrelated texts and dispensing with the need for testing related texts. Such an approach is counter-productive as it affects the recall rate thereby creating additional classification problems. This aspect of the experiment needs further improvement.

To find the fastest and shortest approach in text classification in the era of massive information, and bring excellent accuracy in results, (like the feature mapping in neural networks), Wang et al. [5] proposed a new module to map the features and classify the text by showing text as a vector using SVM on Cross-connected GRU Kernel. This method would help to make short the text that is not valuable and change the non-standard and structured information into an acceptable format. In the case of semi-supervised learning and classifying text and making them labelled, Zhang et al. [6] proposed the idea of learning by looking at the unlabelled data and improving the 1-DNF approach and splitting the data. In this idea, the machine would learn by adopting the Roc-SVM algorithm and Co-EM SVM to classify the text whilst using unlabelled examples. The method includes 4 steps: splitting features in text, producing trustable examples, applying the algorithm, and then selecting the high-accuracy output.

Categorizing the grammar, symbols, and characters in the range of different languages is an issue that [7] focused on and improved. The researchers used the TF-IDF weighting algorithm while at the same time adopting statistical methods to identify features of Tibetan words in the text. This approach enhanced the TF-IDF method to find the high-effect classification. They used the vector space model to express the Tibetan characters and symbols by finding the importance of words in the text based on the weight of vectors. Another idea of two-way Bi-directional Long Short-Term Memory was suggested by Brownlee [8]. The SVM data analysis approach was used to handle the sources of memory information with the help of all solutions. Padmanabha Reddy [9] initially established the use of two-dimensional sequencing in text-to-speech recognition.

This study is essential because it identifies and introduces new classifications to achieve accuracy and reliability in processing. In text processing, there are numerous research papers on the case of text classification. SVM is a fully operational network model that adopts width spacing in a hyperplane and is mapped in the ML algorithm. The SVM has a high accuracy rate. Manoharan et al. [10] argued that the two-way LSTM shows good accuracy of the new short-term correct classifier, and the Bi-LSTM shows different ways of working than the SVM [10].

3 Materials and Methods

Activities related to research were done at the processing laboratory of the University of Bolton. They are identified as two formations, using a sample size of 26 per group [11]. The Data set was about BBC News with more than 10 million text-based records. All process of cleaning the data was done on the data before applying the models and converting the mass data to clear one. This study uses 7 steps to investigate issues. The list of procedures for the suggested models is shown in Fig. 1.

Since the SVM is proposed from the best hyperplane in the linearly detachable facts [11], the main problem faced by the SVM is how to find the highest hyperplane in the nonlinear cases. SVM solves the nonlinear case by modifying the input space into a higher dimensional feature space by the nonlinear map and linearly decomposed input sample text in the dimensional feature space [11].

In the type and frequency concept, there are some words which are out of the standard in text processing in each language compared to English. As an example, as [1] and [3] argued, in Persian and Chines languages there are some words which do not follow the standards, as some of them are listed in Table 1 [8, 13].

The Dataset for text categorization was obtained from Kaggle.com, and this dataset is used to train five different classification algorithms SVM. The Data set

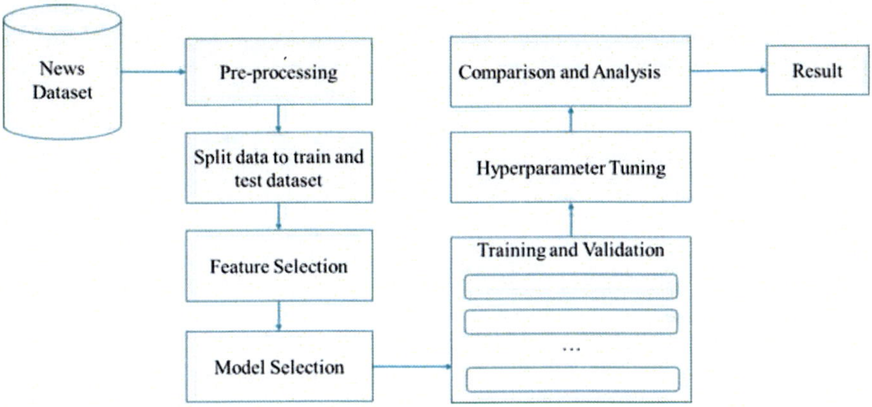

Fig. 1 Analysis method

Table 1 List of the non-standard word in different languages

Row	Non-standard word
1	Acronym
2	Abbreviation
3	Special keyboard character

contained 2 main columns as features, the 'Text' of news and the 'Category' which defines the category of the news (see Fig. 2).

The following steps applied in this paper, as our algorithm, are listed below to prepare the data to process the text (see Figs. 3 and 4):

- Removing any Null Values in the 'Text' Column
- Create a new column, based on the string value of the 'Text', which was lowercase.
- Tokenization- each entry in the corpus will be broken into a set of words.

	category	text
0	tech	tv future in the hands of viewers with home th...
1	business	worldcom boss left books alone former worldc...
2	sport	tigers wary of farrell gamble leicester say ...
3	sport	yeading face newcastle in fa cup premiership s...
4	entertainment	ocean s twelve raids box office ocean s twelve...

Fig. 2 Selected feature based on the CNN-news dataset

```
# 1. Removing Blank Spaces
Corpus['text'].dropna(inplace=True)
# 2. Changing all text to lowercase
Corpus['text_original'] =str( Corpus['text'])
Corpus['text'] = [entry.lower() for entry in Corpus['text']]
# 3. Tokenization-In this each entry in the corpus will be broken into set of words
Corpus['text']= [word_tokenize(entry) for entry in Corpus['text']]
# 4. Remove Stop words, Non-Numeric and perfoming Word Stemming/Lemmenting.
# WordNetLemmatizer requires Pos tags to understand if the word is noun
# or verb or adjective etc. By default it is set to Noun
tag_map = defaultdict(lambda : wn.NOUN)
tag_map['J'] = wn.ADJ
tag_map['V'] = wn.VERB
tag_map['R'] = wn.ADV

Corpus.head()
```

Fig. 3 Cleaning the 'Text' column and removing stop words

```
for index,entry in enumerate(Corpus['text']):
    # Declaring Empty List to store the words that follow the rules for this step
    Final_words = []
    # Initializing WordNetLemmatizer()
    word_Lemmatized = WordNetLemmatizer()
    # pos_tag function below will provide the 'tag' i.e if the word is Noun(N) or Verb(V) or something else.
    for word, tag in pos_tag(entry):
        # Below condition is to check for Stop words and consider only alphabets
        if word not in stopwords.words('english') and word.isalpha():
            word_Final = word_Lemmatized.lemmatize(word,tag_map[tag[0]])
            Final_words.append(word_Final)
    # The final processed set of words for each iteration will be stored in 'text_final'
    Corpus.loc[index,'text_final'] = str(Final_words)
```

Fig. 4 Preparing the 'Text_Final' column

- Remove Stop words, non-Numeric and perform Word Stemming/Lamenting. The word requires tags to understand the type of word (noun - verb – adjective) By default, it is set to a Noun.
- for each index and entry in the 'Text' Column, the 'WordNetLemmatizer' method was initialised.
- 'pos_tag' function provided the 'tag', for example, finding out the word is a Noun(N) or Verb(V) or something else.
- processed set of words for each iteration and saved in the new column named 'text_final'.

4 Results

Python programming language in the Jupyter Notebook environment was used in this paper to find standard deviation, mean, mean of standard error, and mean difference. The following algorithms were applied to the dataset to find the best model:

The k-nearest Neighbours Algorithm, called KNN and k-NN, is a supervised classifier, which uses proximity to make classifications about the data point [12]. As

```
# Training the K-NN model on the Training set
from sklearn.neighbors import KNeighborsClassifier
classifier_KNN = KNeighborsClassifier(n_neighbors = 5, metric = 'minkowski', p = 2)
classifier_KNN.fit(Train_X_Tfidf,Train_Y)
# predict the labels on validation dataset
predictions_KNN = classifier_KNN.predict(Test_X_Tfidf)
# Use accuracy_score function to get the accuracy
print("KNN Accuracy Score -> ",accuracy_score(predictions_KNN, Test_Y)*100)
score.append(Set_score("KNN",predictions_KNN))
```

```
KNN Accuracy Score ->  94.76047904191617
```

Fig. 5 Training dataset on K-NN classifier

it is shown in Fig. 5 the K-NN with n_neighbors equal 5 as a parameter reached 94.76% accuracy.

Support Vector Machine (SVM) is a supervised learning method used for the classification, regression, and detection of outliers[14]. the Linear-SVM recorded 98.35% accuracy which is shown in Fig. 6.

The Naive Bayes method is a supervised algorithm cantered on applying Bayes' theory with the "naive" assumption of restrictive individuality between pair of features. Gaussian Naive Bayes called GaussianNB implements the Gaussian Naive Bayes algorithm for classification [14]. the Naive Bayes algorithm registered 97.30% accuracy (see Fig. 7).

```
# Classifier - Algorithm - SVM
# fit the training dataset on the classifier
SVM = svm.SVC(C=1.0, kernel='linear', degree=3, gamma='auto')
SVM.fit(Train_X_Tfidf,Train_Y)
# predict the labels on validation dataset
predictions_SVM = SVM.predict(Test_X_Tfidf)
# Use accuracy_score function to get the accuracy
print("SVM Accuracy Score -> ",accuracy_score(predictions_SVM, Test_Y)*100)
score.append(Set_score("SVM",predictions_SVM))
```

```
SVM Accuracy Score ->  98.35329341317365
```

Fig. 6 Training dataset on SVM classifier

```
# fit the training dataset on the NB classifier
Naive = naive_bayes.MultinomialNB()
Naive.fit(Train_X_Tfidf,Train_Y)
# predict the labels on validation dataset
predictions_NB = Naive.predict(Test_X_Tfidf)
# Use accuracy_score function to get the accuracy
print("Naive Bayes Accuracy Score -> ",accuracy_score(predictions_NB, Test_Y)*100)
score.append(Set_score("Naive Bayes",predictions_NB))
```

```
Naive Bayes Accuracy Score ->  97.30538922155688
```

Fig. 7 Training dataset on the NB classifier

```
# fit the training dataset on the Decision Tree classifier
from sklearn.tree import DecisionTreeClassifier
classifier_tr = DecisionTreeClassifier(criterion = 'entropy', random_state = 0)
classifier_tr.fit(Train_X_Tfidf,Train_Y)
y_pred_dt = classifier_tr.predict(Test_X_Tfidf)
print("DecisionTree Accuracy Score -> ",accuracy_score(y_pred_dt, Test_Y)*100)
score.append(Set_score("DecisionTree",y_pred_dt))

DecisionTree Accuracy Score ->  81.2874251497006
```

Fig. 8 Training dataset on decision tree classifier

```
# fit the training dataset on the Random Forest classifier
from sklearn.ensemble import RandomForestClassifier
classifier_rf = RandomForestClassifier(n_estimators = 10, criterion = 'entropy'
                                       , random_state = 0)
classifier_rf.fit(Train_X_Tfidf,Train_Y)
y_pred_RF = classifier_rf.predict(Test_X_Tfidf)
print("Random Forest Accuracy Score -> ",accuracy_score(y_pred_RF, Test_Y)*100)
score.append(Set_score("Random Forest",y_pred_RF))

Random Forest Accuracy Score ->  92.51497005988024
```

Fig. 9 Training dataset on random forest classifier

Decision Tree (DT) is another supervised method helpful in classification and regression. this algorithm has aim to develop a model which guesses the value of a purpose variable by realising straightforward instructions based on features [16]. As depicted in Fig. 8, the Decision tree classifier displayed 81% accuracy, the lowest result.

The Random Forest comprises decision trees which drive together. Each separate tree in the random forest spits out a prediction and the most votes become the result prediction [14]. The result of accuracy based on this algorithm was registered 92.51% (see Fig. 9).

The data table of accuracy for all five Machine learning model has shown in Table 2 which illustrate the difference of analysis by Naive Bayes, SVM, Decision Tree, Random Forest and K-NN. The Accuracy Score of the Decision Tree was the lowest and The SVM reached the high rank. In addition, the mean absolute error followed the same pattern with 0.4 by Decision Tree as a heist and 0.3 as the lowest one by SVM.

In addition, these results could be illustrated in a bar chart as is shown in Fig. 10. This bar chart was generated by plotting the accuracy score and mean absolute error on the Y-axis and the classification models on the X-axis. The graph clearly shows that SVM is more accurate than other algorithms. The error rate of SVM is lower than that of others.

Table 2 Compare the result of the modelling

Model	Accuracy_score	MAE	MSE	RMSE
Naïve Bayes	0.973	0.058	0.151	0.389
SVM	0.984	0.034	0.085	0.292
Decision tree	0.813	0.401	1.054	1.027
Random forest	0.925	0.166	0.478	0.691
KNN	0.948	0.117	0.311	0.558

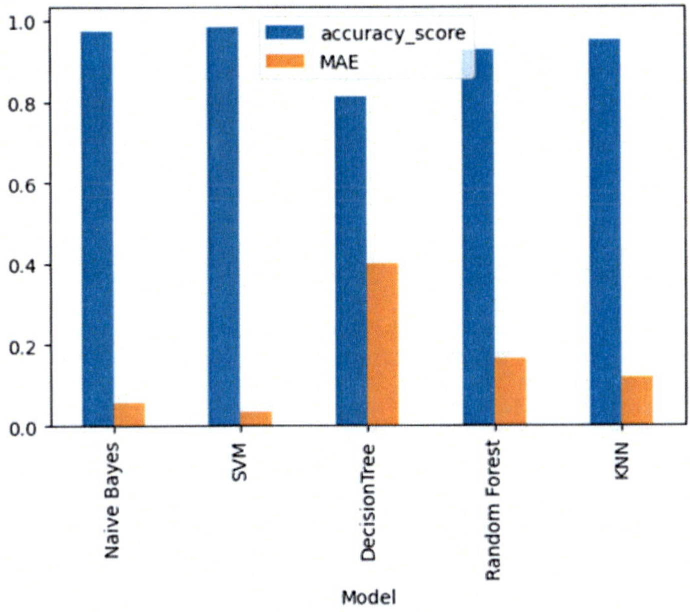

Fig. 10 Depict the comparison of the model results

5 Conclusion

For accurate text classification, the proposed SVM had greater accuracy than the others. Also, Naïve Bayes performs all the methods and activities very well based on the comparison of Support Vector Machines.

In addition, SVM is a fully operating network model that assumes a gap of the width in a hyperplane and is mapped into the ML algorithm. Since the SVM has a high accuracy rate [16]. The study aimed to classify the accurate classification in Long Short-Term Memory to improve the accuracy rate using machine learning [17].

Linear-SVM will be the major contact with all methods to get a good processing model that helps to get an acceptable and higher accuracy than the others. The main factors affecting the algorithm are pre-processing, feature extraction, normalization, and deep belief networks. The limitations of SVM are hard to train a bit more accurately and make speech recognition clear to the estimation of natural language processing [16]. The future scope is to maintain all the models and fields based on the different Languages, especially which contain the not common characters, and aim to explore all the sequential problems by natural language processing that will be solved by the Machin Learning algorithm.

Acknowledgements This work was supported by the Key Laboratory of Advanced Manufacturing Technology of the Ministry of Education of Guizhou University (GZUAMT2022KF[07]), the National Natural Science Foundation of China (No.61862051), the Science and Technology Foundation of Guizhou Province (No. [2019]1299, No.ZK[2022]449), the Top-notch Talent Program of Guizhou province (No.KY[2018]080), the Natural Science Foundation of Education of Guizhou province (No. [2019]203) and the Funds of Qiannan Normal University for Nationalities (No. qnsy2019rc09). The Educational Department of Guizhou under Grant No. KY[2019]067.

References

1. Moattar, M.H., Homayounpour, M.M., Zabihzadeh, D.: Persian text normalization using classification tree and support vector machine. In: 2006 2nd International Conference on Information & Communication Technologies, Damascus, Syria, pp. 1308–1311 (2006). https://doi.org/10.1109/ICTTA.2006.1684569
2. Gao, Y., Sun, S.: An empirical evaluation of linear and nonlinear kernels for text classification using Support Vector Machines. In: 2010 Seventh International Conference on Fuzzy Systems and Knowledge Discovery, Yantai, China, pp. 1502–1505 (2010). https://doi.org/10.1109/FSKD.2010.5569327
3. Lin, X.-d., Peng, H., Liu, B.: Support vector machines for text categorization in chinese question classification. In: 2006 IEEE/WIC/ACM International Conference on Web Intelligence (WI 2006 Main Conference Proceedings)(WI'06), Hong Kong, China, pp. 334–337 (2006). https://doi.org/10.1109/WI.2006.163
4. Hao, P., Ying, D., Longyuan, T.: Application for web text categorization based on support vector machine. In: 2009 International Forum on Computer Science-Technology and Applications, Chongqing, China, pp. 42–45 (2009).https://doi.org/10.1109/IFCSTA.2009.132
5. Wang, Q., Liu, Z., Zhang, T., Li, Y.: Short text classification based on cross-connected GRU kernel mapping support vector machine. In: 2021 2nd Asia Symposium on Signal Processing (ASSP), Beijing, China, pp. 201–207 (2021).https://doi.org/10.1109/ASSP54407.2021.00038
6. Zhang, B.-z., Zuo, W.-l.: Co-EM support vector machine based text classification from positive and unlabeled examples. In: 2008 First International Conference on Intelligent Networks and Intelligent Systems, Wuhan, China, pp. 745–748 (2008). https://doi.org/10.1109/ICINIS.2008.29
7. Cao, H., Jia, H.: Tibetan text classification based on the feature of position weight. In: 2013 International Conference on Asian Language Processing, Urumqi, China, pp. 220–223 (2013). https://doi.org/10.1109/IALP.2013.63
8. Brownlee, J.: Long Short-Term Memory Networks With Python: Develop Sequence Prediction Models with Deep Learning. Machine Learning Mastery (2017)
9. Dr Padmanabha Reddy, V., Padmanabha Reddy, V.: Human Cognitive State Classification Using Support Vector Machine. JARDCS (2020)
10. Manoharan, P., Walia, R., Iwendi, C., Ahanger, T., Balaji, S., Kamruzzaman, M.M., Bourouis, S., Alhakami, W., Hamdi, M.: SVM-based generative adverserial networks for federated

learning and edge computing attack model and outpoising. Expert. Syst. (2022). https://doi.org/10.1111/exsy.13072

11. Murty, M.N., Rashmi.: Support Vector Machines and Perceptrons: Learning, Optimization, Classification, and Application to Social Networks. Springer (2016)
12. Raschka, S., Mirjalili, V.: Machine Learning and Deep Learning with Python, scikit-learn, and TensorFlow (2017)
13. Dhiman, P., Kukreja, V., Manoharan, P., Kaur, A., Kamruzzaman, M.M., Dhaou, I.B., Iwendi, C.: A Novel deep learning model for detection of severity level of the disease in citrus fruits. Electronics **11**(3), 495. https://doi.org/10.3390/electronics11030495
14. Burkov, A.: The Hundred-Page Machine Learning Book. ISBN: 978–1999579500 (2019)
15. Iwendi, C., Khan, S., Anajemba, J.H., Mittal, M., Alenezi, M., Alazab, M.: The use of ensemble models for multiple class and binary class classification for improving intrusion detection systems. Sensors (Basel). **20**(9), 2559 (2020). https://doi.org/10.3390/s20092559.PMID:32365937;PMCID:PMC7249012
16. Manoharan, P., Walia, R., Iwendi, C., Ahanger, T.A., Suganthi, S.T., Kamruzzaman, M.M., Bourouis, S., Alhakami, W., Hamdi, M.: SVM-based generative adverserial networks for federated learning and edge computing attack model and outpoising. Expert. Syst. (2022). https://doi.org/10.1111/exsy.13072
17. Lv, G., Ganyun, L.V., Wang, X., Jin, Y.: Short-term load forecasting in power system using least squares support vector machine. In: Computational Intelligence, Theory and Applications
18. Mairittha, T., Mairittha, N., Inoue, S.: Integrating a spoken dialogue system, nursing records, and activity data collection based on smartphones. Comput. Method. Programs Biomed. **210**, 106364 (2021)

Emotion AI

P. Sowmya and Saroja Devi

Abstract Facial expressions are a type of emotion expressed by humans to convey different meanings in different contexts. And it is also a non verbal way of communication. Due to the rigorous development in the area of Face detection and emotion classification, many applications like real time health monitoring system, living assistance system, surveillance systems, access control systems and many more are using this technology. As mouth and eye are the parts that are mainly focused for emotion detection, extracting the gestures and categorizing them are very crucial task. Using Machine Learning and Neural network concepts, good results are obtained in emotion detection and categorization. So, in this paper we have combined the Facial Key-point Detection and Emotion Detection models together to get a better Emotion Classification model.

Keywords Facial key-point · Facial emotion · Convolutional neural networks · Haar cascade · Residual neural networks

1 Introduction

Emotions are conscious mental reactions which includes human thoughts, behaviors, feelings, actions etc. Facial expressions are a type of non-verbal communication and also a natural form of expressing human emotions. The seven emotional expressions of happiness, anger, sadness, fear, disgust, surprise, and neutral can be distinguished from the facial expressions. Emotions are so delicate that detecting the difference in different expressions is very challenging as the movement of facial muscles are very minute.

P. Sowmya (✉) · S. Devi
Nitte Meenakshi Institute of Technology, Bengaluru, India
e-mail: sowmya.p@nmit.ac.in

S. Devi
e-mail: sarojadevi.n@nmit.ac.in

C. Iwendi et al. (eds.), *Proceedings of ICACTCE'23—The International Conference on Advances in Communication Technology and Computer Engineering*, Lecture Notes in Networks and Systems 735, https://doi.org/10.1007/978-3-031-37164-6_24

The facial expressions of every human being may differ for same emotion depending on what occurs around a person and so emotions are said to be context dependent. The main focus for emotion detection targets around eyes and mouth and so separating and classifying these expressions is very crucial. Neural network models and Machine learning techniques help in getting best results in emotion detection and categorization.

One of the neural network deep learning techniques is the convolutional neural network (CNN), which helps in analyzing visual images using image detection methodologies. The input image is assigned biases and weights based on different features in the images to differentiate one image from another. The pre processing required is less compared to other image processing algorithms or natural processing algorithms.

Residual Neural Networks (ResNet) are a type of neural network algorithms constructed from pyramidal neurons in cerebral cortex which are primary excitation unit. In this network, skip connections are used which skip layers in between by connecting activation of a layer to further layer. And it also includes ReLu which is an activation function and batch normalization. The skip connections used in the model help in maintaining a low error rate much deeper in the network.

2 Related Work

Facial Emotion Recognition of Students using Convolutional Neural Network [1], in this paper, the authors have proposed a model that consists of Convolutional, maximum pooling, and completely linked layers are the different types of layers. Convolutional layer extract features from an input image. Max pooling helps in reducing dimensionality of each feature map by retaining the important data. And fully connected layer helps in classifying input image into various classes. The model uses Haar Cascades to recognize facial features from students' input images and classifies them into seven facial emotions that are happy, sad, angry, fear, surprise, disgust and neutral. The model achieved an accuracy rate of 70%.

An improved method to recognize hand over face gesture based facial motion using convolutional neural network [2], in this paper, an improvised version of hand and face gesture based emotion identification method is used to recognize emotions like confident, fear, ashamed and okay along with normal emotions like happy, sad, angry etc. To increase the accuracy of the proposed system, two major components are integrated namely a coding model with huge count of hand and face gestures and CNN model to obtain required features and to categorize the emotions.

Emotional stability detection using convolutional neural networks [3], here the authors have proposed a CNN model that recognizes facial emotions from three different emotions on the spectrum namely positive emotion, negative emotion and neutral emotion. The proposed work uses residual blocks and convolutions that are separable to decrease the number of training features. And the system achieved an overall accuracy of 81%.

Modified CNN Architecture Analysis for Facial Emotion Recognition [4], in this paper, Convolutional neural networks and deep neural networks are employed by the authors to construct a emotion classification model that classifies facial emotions into seven types of emotions. They have also compared their proposed model namely Venturi model with two existing deep neural network models namely modified triangular model and rectangular model in terms of accuracy and loss. Two set of layers of feature maps are used to obtain the required characteristics from the images and passed into six hidden layers of deep neural network.

3 Implementation

The datasets used were obtained from Kaggle. Two datasets are used-one dataset for detection of facial key points and the second dataset for facial emotion detection. Figure 1 shows the seven universal facial emotions collected from the second dataset.

The facial keypoint detection dataset contains 2,140 images of 96*96 pixels. After augmentation, 5,136 images are used as training data to train the model. Each image in the dataset has fifteen significant key locations' x and y coordinates, as well as a row-by-row pixel list used to order image data. A total of 1284 images are used as test images. The second dataset contains images of size 48*48 pixels. A total of 28,709 images belonging to seven different categories of emotions are used for training the model and 7,178 images are used for testing and validation.

The implementation of the model is divided into mainly two phases.

- Facial key point detection phase
- Facial emotion detection phase

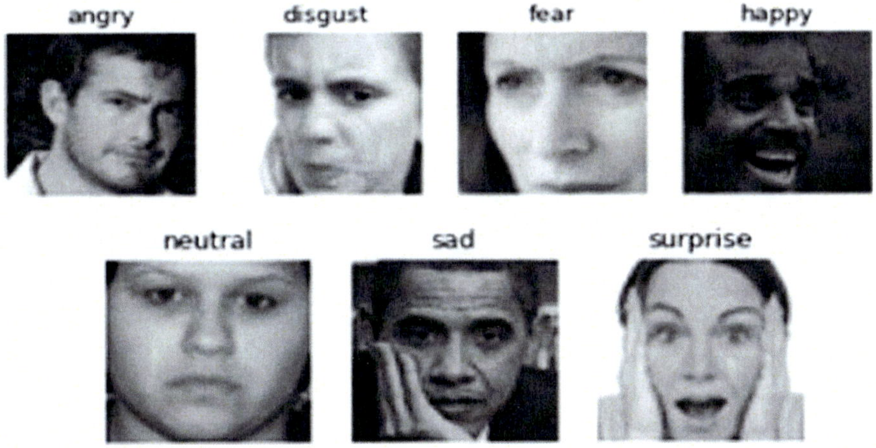

Fig. 1 Seven universal facial emotions

3.1 Facial Key Point Detection Phase

Facial Key Point is used to extract the required features from the face. Figure 2 shows the steps involved in Facial Key Point Detection Model.

3.1.1 Importing Libraries and Dataset

All the required libraries used in the implementation are imported. Also the datasets which are used in building the model are uploaded from Kaggle. 90% of the images from the dataset are used for training the model and the rest 10% are used for testing the model.

Fig. 2 Facial key point detection model

3.1.2 Visualizing the Image and Image Augmentation

Images are visualized in order to identify several face important points. Image augmentation is used to artificially increase the size of the training dataset. We receive updated versions of the photographs through this procedure, and the dataset size also grows [5].

3.2 Normalization of Image and Preparation of Training Data

After augmentation, from 6,420 images 80% of the images are used in order to train the model, for key point detection and rest 20% are used for testing. During normalization, the images are resized so that the original range of value is between 0 and 1.

3.2.1 Building Facial Key Point Detection Model Using Residual Neural Network

In Residual Neural Networks (ResNets), skip connections are used to connect the activation of one layer to a subsequent layer without experiencing vanishing gradient problems. Identity mappings are added to CNN to make ResNets function. The steps involved are as follows.

- **Zero Padding**: By adding pixel border to the edge of the image, which includes only zero values, the original size of the input image is recreated in this case.
- **Conv2D**: Convolution technique uses kernel to extract features from an input image. In this layer a convolution kernel is developed which convolves input layer with the kernal to produce tensor of output.
- **BatchNorm**: Batch Normalization helps in stabilizing the deep learning model during training phase. It is standard practice to normalize input data to zero mean and unit variance while inputting data to deep learning models. So in this layer batch normalization is applied to adjust mean value to zero and standard deviation value to nearly one [6].
- **ReLU**: The Rectified Linear Unit or ReLU is an activation unit and a linear piecewise function that transforms its inputs into outputs that have a certain range. The ReLU function, for negative input data, returns a value 0 and for any positive input x, that value will be returned back.
- **Res Block**: Residual blocks allow the information to flow from initial to last layers in spite of absence of gates in their skip connections.
- **Average Pooling 2d**: In average pooling, the images received from previous layers are downsampled along its spatial dimensions, that is along its height and width by averaging the values for each input channel over an input window.

- **Flatten**: Flattening is a process by which the resultant two dimensional arrays from the Average pooling phase are converted into single dimension which acts as input for next layer.
- **Dense Layer**: The Dense layer is fully connected and the neurons of the dense layer receives the neurons from the previous layers which makes them fully connected. It performs dot product of all input values for getting the output.
- **Drop out**: The Dropout layer helps in reducing overfitting in neural models. The purpose of this layer is to eliminate noise present in the input of neurons. The input values are set randomly to zero at each step by using frequency of rate to reduce over-fitting.

3.2.2 Compiling the Data and Training the Model for Facial Keypoint Detection

After undergoing all the above phases, the model is made ready to train and use.

3.3 Facial Emotion Detection

3.3.1 Loading the Emotion Dataset

The facial emotion detection dataset consists of total of 28,709 images. The images are of 7 types of emotions. They are classified as happy, sad, angry, fear, disgust, surprise and neutral. Figure 3 graphically represents the number of images of each emotion in the facial emotion detection dataset.

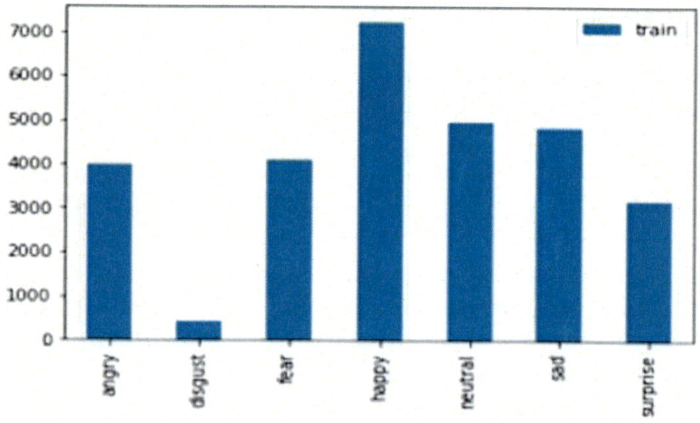

Fig. 3 Class distribution

3.3.2 Perform Data Preparation

The data is cleaned to remove unwanted noise from the images and structuring the data so that it can be sent to next phase to train the model.

3.3.3 Building and Training the Model for Emotion Detection

The model employed for facial keypoint recognition is also utilised to detect emotions, but this time with the help of emotion images as shown in Fig. 4.

A flask web application is used to serve combined predictions on real time input (video or webcam) as shown in Fig. 5. It loads a video stored locally and uses OpenCV to get frames of images and feeds it to the combined model to make predictions [7]. A camera.py file is used to get images frame by frame from the video and convert it to grayscale using OpenCV and detects the face drawing bounding boxes using HaarCascade. A model.py file loads the model architecture from a json file and loads the weights from the previously trained model. Finally, a main.py file runs the html file for the web application on local host and serves the predictions from the model.

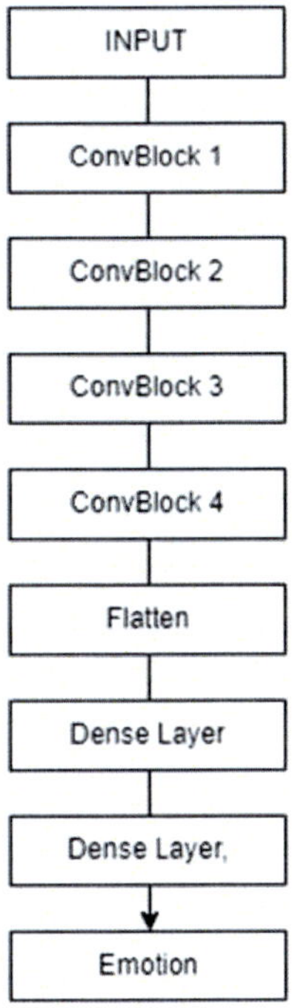

Fig. 4 Facial emotion detection model

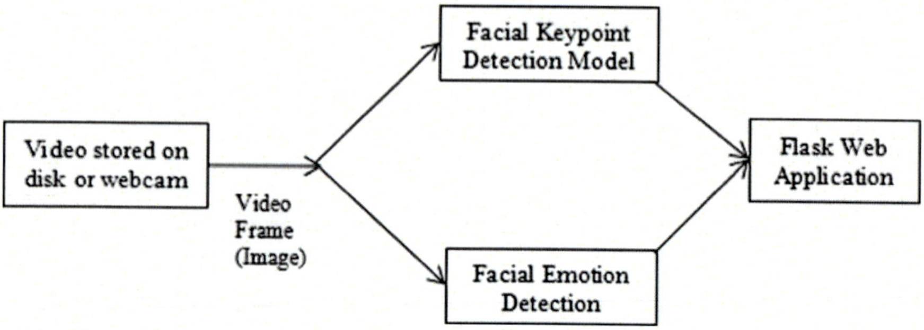

Fig. 5 Serving predictions on web application

4 Results

In the facial keypoints detection model, 15 epochs are used to train the model. The model performance is shown in the Table 1

In the facial emotion detection model, 60 epochs are used to train the model. The model performance is shown in the Table 2.

Both the models combined together gives a final model which detects different types of emotions along with which it extracts 15 key facial points.

Table 3 shows the performance index with respect to precision, recall and F-score for seven different emotions.

Figure 6 shows the combined model output where 7 different types of emotions are detected.

Table 1 Facial key point detection performance analysis

	Accuracy (%)
Training	89
Testing	85

Table 2 Facial emotion detection performance analysis

	Accuracy (%)
Training	85
Testing	87

Table 3 Performance index

Emotion	Recall	Precision	F-score
Happy	0.94	0.92	0.93
Sad	0.85	0.88	0.87
Anger	0.8	0.83	0.81
Fear	0.77	0.68	0.72
Surprise	0.95	0.88	0.92
Disgust	0.93	0.67	0.78
Neutral	0.76	0.69	0.73

Fig. 6 Combined model predictions

5 Conclusion

In this paper we have used CNN and ResNet to detect facial emotions. Facial keypoint detection mainly focuses on eye and mouth part to extract features. The model is able to detect 15 facial keypoints in the input image and is able to classify the facial emotions into seven different emotion classes. The accuracy of the model is around 85%.

In future, along with eye and mouth features, other features can also be inculcated in the model to get improved results. The facial key points detection mode can also be used in biometrics and face recognition applications. In institutions, emotion recognition models help in understanding the emotions of students and provide them guidelines. Facial recognition also serves as efficient verification method in security systems. Emotion recognition also helps in mental diseases diagnosis.

References

1. Lasri, Solh, A.R., Belkacemi, M. E.: Facial Emotion Recognition of Students Using Convolutional Neural Network. In: 3rd International Conference on Intelligent Computing in Data Sciences (ICDS), pp. 1–6. IEEE, Marrakech, Morocco (2019)
2. Naik, N., Mehta, M.A.: An Improved Method to Recognize Hand-over-Face Gesture based Facial Emotion using Convolutional Neural Network. In: 6th International Conference on Electronics, Computing and Communication Technologies (CONECCT), pp. 1–6. IEEE, Bangalore, India (2020)
3. Hussein, E.S., Qidwai, U., Al-Meer, M.: Emotional Stability Detection Using Convolutional Neural Networks. In: International Conference on Informatics, IoT and Enabling Technologies (ICIoT), pp. 136–140. IEEE, Doha, Qatar (2020)
4. Verma, A., Singh, P., Rani Alex, J.S.: Modified Convolutional Neural Network Architecture Analysis for Facial Emotion Recognition. In: International Conference on Systems, Signals and Image Processing (IWSSIP), pp. 169–173. IEEE, Osijek, Croatia (2019)
5. Sooch, S.K., Anand, D.: Emotion Classification and Facial Key Point Detection Using AI. In: 2nd International Conference on Advances in Computing, Communication, Embedded and Secure Systems (ACCESS). IEEE, Ernakulam, India (2021)
6. Asad, U., Kashyap, N., Singh, S.N.: Recent Advancements in Facial Expression Recognition Systems a Survey. In: International Conference on Computing Communication and Automation (ICCCA), pp. 1203–1208. IEEE, Greater Noida, India (2017)
7. Baskar, A., Gireesh, K.T. (eds.): Facial Expression Classification Using Machine Learning Approach-A Review. In: Data Engineering and Intelligent Computing, vol. 542, pp. 337–345. Springer, Singapore (2017)

Accuracy Prediction of Rainfall Using Decision Tree Algorithm and Random Forest

Dan Wang, Tao Hai, Doyinsola Ayandiran, Chijioke Victor Uzochukwu, Xiaofeng Ding, Celestine Iwendi, and Zakaria Boulouard

Abstract Climate change has made accurate rainfall forecasting more difficult than ever. In this paper, the decision tree algorithm and Random Forest is used to predict the rainfall accuracy based on historical climate data. The classification and regression tree (CART) approach is employed to this result, producing a better accuracy rate. The algorithm can determine the probabilities of rain on any given day, making it an ideal choice for various applications involving large datasets.

Keywords Rainfall prediction · Decision tree · Classification · Random forest

1 Introduction

The fluctuation in rainfall is mostly due to its relationship with Humidity, Temperature, Pressure, Wind Speed, and Dew Point, among other factors therefore, climate change has made accurate rainfall forecasting more difficult than ever. Several efforts in this topic have been completed throughout the last few decades. An accurate

D. Wang
School of Mathematics and Statistics, Qiannan Normal University for Nationalities, Duyun, Guizhou 558000, China

D. Ayandiran (✉) · C. V. Uzochukwu · C. Iwendi
School of Creative Technologies, University of Bolton, Bolton BL3 5AB, UK
e-mail: doyinsola.aa@ieee.org

Z. Boulouard
LIM, Hassan II University of Casablanca, Casablanca, Morocco

T. Hai
School of Computer and Information, Qiannan Normal University for Nationalities, Duyun, Guizhou 558000, China

T. Hai · X. Ding
School of Information and Artificial Intelligence, Nanchang Institute of Science and Technology, Nanchang, China

C. Iwendi et al. (eds.), *Proceedings of ICACTCE'23—The International Conference on Advances in Communication Technology and Computer Engineering*, Lecture Notes in Networks and Systems 735, https://doi.org/10.1007/978-3-031-37164-6_25

prediction of rainfall occurrences can help with appropriate financial planning for the nation's economy. Natural calamities such as floods and droughts have an impact not only on a country's economy but also on the lifestyle of rural residents. Data mining is a popular strategy for extracting hidden information from large databases and making judgements based on knowledge mining hence this helps predict necessary patterns for rainfall prediction accurately [7]. There are several hardware devices available for forecasting rainfall based on weather data including temperature, humidity, and pressure. Classification algorithm has been found to be efficient in the accurate prediction of rainfall. Some classification algorithms are Decision Tree, Extreme Gradient Boosting, Random Forest, Multilayer Perception [1].

2 Related Works

According to [2] after comparing kNN (k nearest neighbor) and Decision tree, the decision tree technique is an effective algorithm in creating decision trees and rules for categorising meteorological information such as minimum temperature, maximum temperature, humidity, and wind speed. Through the analysis of the experimental trends affecting the appropriate data throughout time, noteworthy variances were identified which pointed to a shift in climate patterns. The decision tree technique has been used in many applications from forecasting stock prices to predicting rainfall. In particular, its effectiveness in rainfall prediction has been widely recognised. It is capable of determining the probabilities of rain on any given day, making it a valuable tool for short-term weather forecasts. The decision tree technique is a powerful tool to analyse data and develop accurate predictions. Its ability to detect changes in weather patterns over time makes it a useful instrument in meteorology and other related fields.

Also [5] mentioned that the decision tree algorithm is a non-parametric technique that uses a tree-like structure to classify the input dataset instances. It works in a top-down manner, assigning each sample a class label when it reaches the end of the branches. Decision tree classification is easy to understand and use, but pruning the tree using cost complexity method can be quite time consuming due to its intensive calculations. Since decision trees are data-driven algorithms, they can be applied on a vast amount of data and thus can provide accurate results. For instance, in the context of rainfall prediction, decision trees can analyze large datasets of temperature and humidity records to predict rainfall probabilities accurately. The ability of this algorithm to classify data with high accuracy makes it an ideal choice for various applications involving large datasets. Furthermore, it can also be used for feature selection as it helps in selecting the best subset of variables which are important for making accurate predictions.

Prasad [4] considered attributes like temperature, dewpoint, humidity, windspeed for classification and predicting rainfall using Decision tree technique. In rainfall prediction, accuracy and scalability plays an important role. Decision tree algorithm provides accurate rainfall prediction based on historical climate data while Random

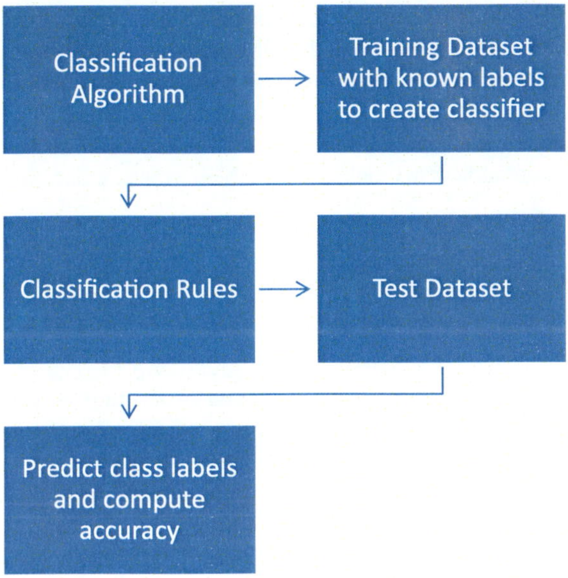

Fig. 1 Classification process

Tree has more accuracy when compared. In other word, Random Forest technique gives maximum accuracy with less learning time and scalability.

According to [6] the accuracy of a classification algorithm is crucial in predicting trends from historical and current data. Through the process of classification, discrete and unordered class labels are analyzed and classified for future data trends. To begin this process, learning algorithms are applied to training datasets, resulting in classification rules that are used to compute the accuracy of the algorithm on test data. This method has proven effective in various fields, including predicting rainfall patterns. By applying these classification algorithms to historical rainfall data, scientists can forecast future patterns with a high degree of accuracy. In essence, the classification algorithm serves as a powerful tool for decision-making and trend analysis across different industries. As a skilled digital marketer, my ability to understand and utilize these techniques allows me to create compelling content that resonates with audiences while also being optimized for conversation. The classification process is explained in Fig. 1 below.

3 Methods

Generally, there are 2 accepted types of classifiers, the Decision Tree and Random Forest algorithms. In this research, Decision tree uses a sample size of 2500 dataset, with each having its own accuracy and precision score. The same dataset was used for the random forest accuracy prediction algorithm. The Decision Tree Method is a practical tool often used for categorization problems, and is structured as a flowchart, with each internal node defining a condition or test on an attribute. Each branch then depends on the result of that test, and a class label is assigned to the leaf node once all attributes have been determined. The distance from the root to the leaf denotes the novel categorization rule, and the capability to work with dependent variables is remarkable. This makes it possible to make a decision in a tree-like format [2].

4 Methodology

i. Data Preprocessing: The initial step of data preprocessing is to acquire and generate rainfall data. An automated tool is developed to obtain and generate information from an online resource. After the data acquisition and formation, the raw data is divided into three sections: cleaning, interpolation, and splitting. In the raw datasets, any empty entries and duplicates are eliminated. To tackle the vast number of vacant values in the dataset, data interpolation is used to determine the projected value by taking the average from its adjacent chronological entries. To meet the needs of training and testing in machine learning models, the dataset is split with an 80/20 ratio.
ii. Data Splitting: Train/Test is a technique used to assess the reliability of a model. The term Train/Test originates from the fact that the data set is divided into two components: a training set and a testing set. The training set is used to construct the model and the testing set to evaluate it. Building the model indicates constructing the model, and evaluating the model signifies determining the model's accuracy. The training set should consist of a random sample of 80% of the original data, with the remaining 20% to be used for the testing set. During this process, the independent (X) and dependent variables (Y) are identified.
iii. Decision Tree & Random Forest Technique prediction: The decision tree technique accurately predicts rainfall and is solely dependent on historical data. To improve accuracy, the random forest technique is used to this result, producing a better accuracy rate. To build decision trees, the Classification and Regression Trees (CART) approach is employed. The dataset is divided into training and testing samples, with the training sample being the largest. The system is subsequently put through its paces using the testing sample. This way, the rate of inaccuracy is less [3].

5 Data Analysis

The dataset was gotten from Kaggle in order to conduct an accuracy study. In addition, the accuracy forecast was recorded. Calculations were performed to ascertain the outcome of the Independent Sample. Rain tomorrow (label) is a dependent factor, whereas humidity, temperature, windspeed, rainfall and pressure are independent

Table 1 Parameters in the dataset

Parameter	Data type
Date	Text
Min temperature	Numerical
Max temperature	Numerical
Windspeed	Numerical
Average temperature	Numerical
Rainfall (mm)	Numerical
Date	Numerical
Humidity	Numerical
Pressure	Numerical

variables. These figures have been thoroughly investigated in order to get the accuracy of the rainfall forecast.

Has been cleaned and all other preprocesses done, the accuracy prediction was carried out on google collab using both Decision Tree classifier and Random Forest classifier. Table 1 below shows the parameters in the dataset used for predicting rainfall. Figure 2 below shows an overview of the dataset and the libraries imported.

A decision tree and random forest algorithm is fitted to training and test set: `X _train, X_test, y_train, y_test = train_`

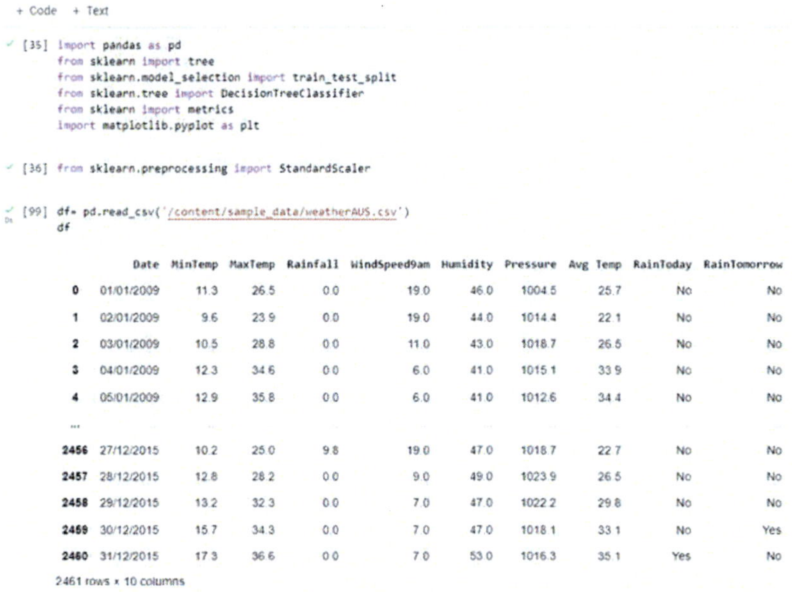

+ Code + Text

```
[35] import pandas as pd
     from sklearn import tree
     from sklearn.model_selection import train_test_split
     from sklearn.tree import DecisionTreeClassifier
     from sklearn import metrics
     import matplotlib.pyplot as plt

[36] from sklearn.preprocessing import StandardScaler

[99] df= pd.read_csv('/content/sample_data/weatherAUS.csv')
     df
```

	Date	MinTemp	MaxTemp	Rainfall	WindSpeed9am	Humidity	Pressure	Avg Temp	RainToday	RainTomorrow
0	01/01/2009	11.3	26.5	0.0	19.0	46.0	1004.5	25.7	No	No
1	02/01/2009	9.6	23.9	0.0	19.0	44.0	1014.4	22.1	No	No
2	03/01/2009	10.5	28.8	0.0	11.0	43.0	1018.7	26.5	No	No
3	04/01/2009	12.3	34.6	0.0	6.0	41.0	1015.1	33.9	No	No
4	05/01/2009	12.9	35.8	0.0	6.0	41.0	1012.6	34.4	No	No
...	...	...	...	...	...	...	...	...	...	...
2456	27/12/2015	10.2	25.0	9.8	19.0	47.0	1018.7	22.7	No	No
2457	28/12/2015	12.8	28.2	0.0	9.0	49.0	1023.9	26.5	No	No
2458	29/12/2015	13.2	32.3	0.0	7.0	47.0	1022.2	29.8	No	No
2459	30/12/2015	15.7	34.3	0.0	7.0	47.0	1018.1	33.1	No	Yes
2460	31/12/2015	17.3	36.6	0.0	7.0	53.0	1016.3	35.1	Yes	No

2461 rows × 10 columns

Fig. 2 Overview of dataset and the libraries

```
test_split ( X, Y, test_size = 0.2, random_state =
1)# 80% training and 20% tes t .
```

6 Results

After using Decision tree classification model, the prediction accuracy was 0.6815415821501014 just as seen in Fig. 3 below while the Random Forest Classification model accuracy prediction was 0.7606490872210954 as seen in Fig. 4 below. Therefore, Random Forest Classifier has the maximum prediction accuracy.

```
+ Code   + Text

[188] X=df[['MinTemp','MaxTemp','Rainfall','WindSpeed9am','Humidity','Avg Temp']]
      Y=df['RainTomorrow']

[189] X_train, X_test, y_train, y_test = train_test_split(X, Y, test_size=0.2, random_state=1)# 70% training and 30% test

[190] scaler=StandardScaler()
      X_train = scaler.fit_transform(X_train)
      X_test = scaler.transform(X_test)

      dtree = DecisionTreeClassifier(random_state=0)
      dtree.fit(X_train,y_train)
      y_predict=dtree.predict(X_test)
      print("Accuracy:", metrics.accuracy_score(y_test, y_predict))

      Accuracy: 0.6815415821501014

[163] Y.unique()

      array([0, 1])
```

Fig. 3 Decision tree model and accuracy prediction result

```
[198] from sklearn.ensemble import RandomForestClassifier

[199] rf = RandomForestClassifier(random_state=0)
      rf.fit(X_train,y_train)
      y_predict=rf.predict(X_test)
      print("Accuracy:", metrics.accuracy_score(y_test, y_predict))

      Accuracy: 0.7606490872210954
```

Fig. 4 Random forest model and accuracy prediction result

7 Discussion

The decision tree classification model was used to predict the accuracy of 0.6815415821501014, as seen in Fig. 3. On the other hand, the random forest classification model was used to predict the accuracy of 0.7606490872210954, as seen in Fig. 4. From the analysis of both the models, it is clear that Random Forest Classifier has a higher prediction accuracy than Decision Tree Classifier. Therefore, this model should be adopted for future predictions and analysis. Further research should focus on improving the accuracy of the Random Forest Classifier by using different techniques such as hyperparameter tuning and feature selection to get better results. Additionally, increasing the number of trees in the forest may also enhance the predictive power of this model since a large number of decision trees provide more reliable predictions. The performance of Random Forest Classifier can also be improved by utilizing new data pre-processing techniques for better data representation and cleaner features. Lastly, ensembling techniques such as bagging and boosting can be used to further improve the efficiency and accuracy of this model.

Acknowledgements This work was supported by Key Laboratory of Advanced Manufacturing Technology of the Ministry of Education of Guizhou University (GZUAMT2022KF[07]), the National Natural Science Foundation of China (No.61862051), the Science and Technology Foundation of Guizhou Province (No.[2019]1299, No.ZK[2022]449), the Top-notch Talent Program of Guizhou province (No.KY[2018]080), the Natural Science Foundation of Education of Guizhou province (No.[2019]203) and the Funds of Qiannan Normal University for Nationalities (No. qnsy2019rc09). The Educational Department of Guizhou under Grant NO. KY[2019]067.

References

1. Appiah-Badu, N.K.A., Missah, Y.M., Amekudzi, L.K., Ussiph, N., Frimpong, T., Ahene, E.: Rainfall prediction using machine learning algorithms for the various ecological zones of Ghana. IEEE Access **10**, 5069–5082 (2022). https://doi.org/10.1109/ACCESS.2021.3139312
2. Biruntha, S., Sowmiya, B.S., Subashri, R., Vasanth, M.: Rainfall Prediction Using kNN and Decision Tree. In: Proceedings of the International Conference on Electronics and Renewable Systems, pp. 1757−1763. ICEARS. https://doi.org/10.1109/ICEARS53579.2022.9752220
3. Prasad, N., Kumar, P., Naidu, M.M.: An Approach to Prediction of Precipitation Using Gini Index in SLIQ Decision tree. In: Proceedings-International Conference on Intelligent Systems, Modelling and Simulation, pp. 56–60. ISMS. https://doi.org/10.1109/ISMS.2013.27
4. Prasad, N., Reddy, P.K., Naidu, M.M.: A Novel Decision Tree Approach for the Prediction of Precipitation Using Entropy in SLIQ. In: Proceedings-UKSim 15th International Conference on Computer Modelling and Simulation, pp. 209–216. UKSim 2013 (2013). https://doi.org/10.1109/UKSim.2013.74
5. Shukla, B., Khatri, S. K., Kapur, P. K., Amity University, Amity University. Amity Institute of Information Technology, Computer Society of India., Institute of Electrical and Electronics Engineers. Uttar Pradesh Section, & Institute of Electrical and Electronics Engineers.: 2015 4th International Conference on Reliability, Infocom Technologies and Optimization (ICRITO) (Trends and Future Directions) : September 2–4, 2015. venue, Amity University Uttar Pradesh, Noida, India (n.d.-a).

6. Shukla, B., Khatri, S. K., Kapur, P. K., Amity University, Amity University. Amity Institute of Information Technology, Computer Society of India., Institute of Electrical and Electronics Engineers. Uttar Pradesh Section, & Institute of Electrical and Electronics Engineers.: 2015 4th International Conference on Reliability, Infocom Technologies and Optimization (ICRITO) (Trends and Future Directions) : September 2–4, 2015. venue, Amity University Uttar Pradesh, Noida, India (n.d.-b).
7. Tang, F., Wang, X., Wang, Z., Xiao, M., Ma, Y., Wan, Y.: Displacement Prediction of Rainfall Induced Landslide Based on AdaBoost BP Neural Network. In: 2022 3rd International Conference on Geology, Mapping and Remote Sensing, pp. 794–799. ICGMRS 2022 (2022). https://doi.org/10.1109/ICGMRS55602.2022.9849366

Fraud Detection Using Decision Tree Algorithm to Curb Identity Theft

Tao Hai, Jincheng Zhou, Oluwakemi A. Ajoboh, Timothy Olatunji, Xiaoshan Zhou, Celestine Iwendi, and Boluwatife Oyesola

Abstract Identity theft is a growing concern that can cause significant financial and emotional harm to individuals. One way to detect and prevent identity theft is by using machine learning algorithms, such as decision tree. In this study, we investigate the effectiveness of using a decision tree algorithm in detecting and preventing identity theft. A dataset consisting of personal information, as well as information on suspicious activity, was collected from a financial institution. The dataset included a total of 284,807 rows of data and 30 columns. The decision tree algorithm was implemented using the Python programming language and the scikit-learn library. The algorithm was trained on the training set and used to classify new cases as either fraudulent or non-fraudulent. The performance of the decision tree algorithm was evaluated using several performance metrics such as accuracy, precision, recall and F1-score. Results showed that the decision tree algorithm was effective in detecting and preventing identity theft, with an overall accuracy of 99%. These findings demonstrate the potential of using decision tree algorithms in detecting and preventing identity theft, which can help to curb the increasing problem of identity theft and protect individuals from financial and emotional harm.

Keywords F1-score · Recall · Precision · Identity theft · Decision tree algorithm · Fraud detection

---

T. Hai · J. Zhou
School of Computer and Information, Qiannan Normal University for Nationalities, Duyun, Guizhou 558000, China

O. A. Ajoboh (✉) · T. Olatunji · C. Iwendi · B. Oyesola
School of Creative Technologies, University of Bolton, Bolton, UK
e-mail: aaliyahajoboh@ieee.org

T. Hai · X. Zhou
School of Information and Artificial Intelligence, Nanchang Institute of Science and Technology, Nanchang, China

C. Iwendi et al. (eds.), *Proceedings of ICACTCE'23—The International Conference on Advances in Communication Technology and Computer Engineering*, Lecture Notes in Networks and Systems 735, https://doi.org/10.1007/978-3-031-37164-6_26

1 Introduction

The prevalence of fraudulent activities such as identity theft, financial scams, and cyber-attacks has increased dramatically in recent years [1–3]. These activities have resulted in significant financial losses to both individuals and businesses, and the trend shows no signs of slowing down [2]. To combat this problem, there is a need to develop effective and reliable methods to detect and prevent fraudulent activities from occurring. Traditional methods of fraud detection, such as manual inspection and rule-based systems, are often time-consuming, resource-intensive, and not very effective in detecting modern-day fraudsters who use sophisticated methods and tools to perpetrate their crimes. Fortunately, advances in machine learning and artificial intelligence (AI) have provided new opportunities for fraud detection and prevention. Machine learning techniques are developed from statistical learning and data mining techniques and have shown great potential in performing complex tasks such as fraud detection [4, 5]. One effective way to combat this problem is through the use of decision trees, which are a machine learning algorithm used to make decisions based on a series of data points [4]. Decision trees are a powerful tool that can be used to identify patterns in data and determine the most likely outcome of a given situation. In the context of fraud detection, decision trees can be used to identify potential fraudulent activities, such as identity theft, by analysing data and detecting unusual patterns [5]. The decision tree algorithm works by dividing the dataset into smaller subsets based on the most relevant features. It then creates a tree-like structure, where each node represents a feature, and each branch represents a possible outcome of that feature. The algorithm continues to split the data until it reaches a point where it can no longer create a useful split, and this forms the leaf nodes of the tree. The decision tree algorithm has several advantages over traditional rule-based systems. First, it can handle both categorical and numerical data, making it more versatile. Second, it is easy to interpret and visualize, making it easy to understand how the algorithm makes decisions. Third, it is computationally efficient and can handle large datasets with ease. In this paper, we propose a fraud detection system based on machine learning techniques using decision trees. The proposed system leverages a combination of supervised and unsupervised learning techniques to identify potential fraudulent activities. The system works by first pre-processing the data and then training a decision tree model using a labelled dataset. Once the model is trained, it is used to classify new instances of data as either fraudulent or not.

To evaluate the effectiveness of the proposed method, we conducted experiments on publicly available datasets. We compared the performance of the decision tree model against other machine learning models and traditional rule-based systems. Our results show that the decision tree model outperforms other models in terms of accuracy and computational efficiency. Ultimately, the proposed fraud detection system based on machine learning techniques using decision trees is a viable solution for detecting and preventing fraudulent activities. The system is easy to implement, computationally efficient, and can handle both categorical and numerical data. The results of our experiments demonstrate the effectiveness of the proposed method in detecting and preventing fraud, providing a valuable tool for businesses and individuals seeking to combat fraudulent activities.

2 Objective of the Study

The main objective of this study is to investigate the effectiveness of a decision tree algorithm for fraud detection. In particular, we aim to demonstrate the effectiveness of this algorithm in identifying potential fraudulent activities and preventing them from occurring. To achieve this objective, we will conduct experiments on publicly available datasets, evaluate the performance of the algorithm, and compare it against other machine learning models and traditional rule-based systems.

3 Related Works

There have been many attempts in the literature to address the early detection of credit card fraud [6–8]. However, the existing methods suffer from several drawbacks such as the inability to deal with time-varying features, a large number of false alerts, and low accuracy [9]. Recently, neural networks have received much attention due to their high computational efficiency and ability to cope with high-dimensional data [10]. Deep learning techniques have also been used for various types of fraud including banking fraud and insurance fraud [11, 12].

Abdelhalim [13, 14], proposed an approach for the automatic identification of valid customers based on analysing the frequency and duration of their credit card transactions, which can be used to prevent fraudulent behaviour. Similarly, [15] developed a deep learning model called Deep Identity Network to identify fraudulent users by analysing their transaction records. However, the performance of these approaches is limited by the fact that they require access to transaction records to train the model, which can potentially compromise the privacy of the users.

Berezkin [16] Used behavioural biometrics for online credit card transactions to detect fraudulent activities. The proposed approach relies on individuals' at the point of sale and classification of user identity based on their biometric data such as fingerprints or hand gestures. The objective of this technique is to protect businesses from the rising rate of online credit card fraud and protect the privacy of individuals at the same time. However, the approach is susceptible to spoofing attacks since attackers can easily mimic the fingerprint of the legitimate user in order to gain access to the system.

Liu [17, 18] Proposed a behavioural pattern recognition system to identify potential fraudsters. The system uses features extracted from credit card transaction logs to analyse the purchasing patterns of a user and classify him as a fraudster or a non-fraudster. The model employs a supervised learning technique based on the Random Forest classifier to classify transactions as either fraudulent or legitimate.

Furthermore, the model can also be adapted to perform anomaly detection to identify anomalous transactions for further analysis.

According to [19, 20] different technologies such as machine learning, neural networks, fuzzy logic, support vector machines, and so on, have been explored for the early detection of credit card fraud. Among these technologies, the support vector machine has proven to be effective in the early detection of various frauds including credit card fraud. The main advantage of this technique is that it does not need any prior information about the transaction pattern and hence it can be used independently without any prior knowledge about the characteristics of the transactions. However, this method is not robust to external noises and hence it is not able to identify fraud without involving any additional techniques.

Razak [21] Proposed the use of statistical learning techniques for detecting and preventing credit card fraud. More specifically, they made use of linear classifiers such as support vector machines and logistic regression to analyse the purchasing behaviour of cardholders in order to detect suspicious transactions. In addition, they also incorporated temporal-difference learning models to classify customer behaviour. However, this approach suffers three major limitations: it is not suitable for online transactions as it involves long training times; it also suffers from high computational complexity and requires a high memory storage; and finally, it is very difficult to adapt this method for financial institutions that do not use a standard format for recording card transactions.

Prakash [22] The use of a hierarchical probabilistic approach to detect various types of credit card fraud. In particular, they used a Bayesian network to construct a hierarchy of probabilistic rules which identifies different types of frauds and assigns them to the appropriate category. They also used clustering techniques to assign the discovered fraudulent transactions into pre-defined categories and then applied sophisticated data mining algorithms to perform further classification. As a result, this approach is able to accurately identify various types of credit card fraud and distinguish them from legitimate transactions.

These approaches have usually relied on either handcrafted rules or machine learning algorithms such as neural networks. However, due to the dynamic nature of credit card transactions and the high dimensionality of the feature space, these methods often encounter a high error rate when deployed in real applications. Despite these works, none of them consider the problem of detecting multiple types of fraud at the same time.

3.1 Identity Theft Prevention with Decision Tree Algorithm

The decision tree algorithm is a machine learning technique that can be used to make decisions based on a series of data points. The algorithm is a powerful tool that can be used to identify patterns in data and determine the most likely outcome of a given situation. It has been used in several applications, such as fraud detection. In fraud detection, the decision tree algorithm can be used to identify patterns in the data.

By doing this, the decision tree algorithm can help to prevent identity theft from happening.

There are a number of different algorithms that can be used to create a decision tree. The most common algorithm is the training algorithms. The training algorithm is used to determine the accuracy of the decision tree. The training algorithm is used to determine the best model for the decision tree. The accuracy of the decision tree is determined by how well it can identify different types of fraud. The other important factor that is used to determine the accuracy of the decision tree is how well it can predict the outcome of a situation.

This paper proposes a new approach to the problem of identifying customers who are at high risk of identity theft. Specifically, we use a decision tree algorithm to learn this risk and apply it during subsequent interactions with these customers to identify potentially fraudulent transactions.

Using a dataset publicly obtained from Kraggle, we first extract several features from the credit card transaction records using a feature extraction algorithm. Next, we train our decision tree classifier using the extracted features and test it on a test dataset containing real credit card transactions. Finally, we apply the algorithm to a customer's new transaction record and identify high-risk transactions based on the output produced by the decision tree classifier. Our approach has several advantages over existing methods. First, it does not require a large set of hand-crafted rules or complex machine-learning algorithms. Second, it is applicable to a wide variety of customer behaviours including regular purchases as well as fraudulent transactions. Finally, it can be used in real time to provide real-time fraud prevention.

4 Materials and Methods

The structure of the proposed system consists of two major components: an ontology-based feature extraction mechanism and a decision tree-based classification model. In this study, we aimed to investigate the effectiveness of using a decision tree algorithm in detecting and preventing identity theft. The following methods were used to accomplish this aim:

I. ***Data Collection***: A dataset consisting of 284,807 rows of data and 30 columns. Each transaction is characterized by a set of attributes describing various aspects of the transaction, such as the cardholder's name, the payment method used, the purchase price, and so on Table 1.

II. ***Feature Extraction***: We extracted the features from each transaction record and then transformed them into a common format using decision tree. These features represent various properties of the transaction such as the customer name, transaction amount, time of purchase, etc. We use these features to train a decision tree classifier and identify high-risk transactions.

Table 1 Data dictionary

S/N	*Strings*	*Description*
1	trans_date_trans_time	The date and time of the transaction
2	cc_num	Credit card number
3	merchant	Merchant who was getting paid
4	category	In what area does that merchant deal
5	amt	Amount of money in American Dollars
6	first	First name of the card holder
7	last	Last name of the card holder
8	Gender of the cardholder	Just male and female
9	street	Street of card holder residence
10	city	City of card holder residence
11	state	State of card holder residence
12	zip	ZIP code of card holder residence
13	lat	Latitude of card holder
14	long	Longitude of card holder
15	city_pop	Population of the city
16	job	Trade of the card holder
17	dob	Date of birth of the card holder
18	trans_num	Transaction ID
19	unix_time	Unix time which is the time calculated since 1970 to today
20	merch_lat	latitude of the merchant
21	merch_long	Longitude of the merchant
22	is_fraud	Whether the transaction is fraud(1) or not(0)

4.1 Feature Extraction Algorithm

We use a combination of genetic programming and traditional machine learning techniques to extract features from each transaction record. Genetic programming uses an evolutionary algorithm to create programs based on a set of given specifications while traditional machine learning methods use techniques such as support vector machines to create a decision function based on a training dataset. Our feature set consists of a string representing the name of the customer, a vector representing the transaction amount, and a vector of other features describing other characteristics of the transaction such as the date of the transaction and the zip code of the customer's billing address. Each record contains information about the customer who made the purchase, the date and time of the transaction, the type of transaction performed, the merchant category code associated with the transaction, the items purchased, and the total amount spent by the customer during the transaction.

4.2 Extracting Features from Transaction Records

i. We retrieve the records from a database that contains information about all credit card transactions performed by our customers over a period of several months. The database contains the following information:

   (a) Customer information such as name and address;
   (b) Date-time stamps for the first and last date of transaction;
   (c) Details of the transaction such as amount of the transaction, whether the purchase was a swipe or manual entry, etc.; and
   (d) Other details about the purchase such as the items purchased, merchant category code, etc.

ii. For each transaction record in the database, we retrieve the following features:

   (a) Customer name;
   (b) Date and time of transaction;
   (c) Type of transaction - whether it is a manual entry or a swipe;
   (d) Merchant category code;
   (e) Item purchased; and
   (f) Total amount of money spent for the transaction.

III. ***Data Anonymization***: We anonymize the data in the dataset using PCA Dimensionality Reduction Technique, which reduces the data to a lower dimension without losing any of the original information. This is done to prevent any identifiable information from being identified. V1-V28 are the new columns gotten as a result of the PCA Dimensionality reduction to protect user identities and other sensitive features.

IV. ***Data Pre-processing***: The collected data is pre-processed to remove any missing or irrelevant information. The data is randomly split into a training set (70%) and a test set (30%).

V. ***Decision Tree Algorithm***: The decision tree algorithm is implemented using the Python programming language and the scikit-learn library. The algorithm is trained on the training set and used to classify new cases as either fraudulent or non-fraudulent. A fraud score is assigned to each case based on the classifier output. If a case is classified as fraudulent, it is added to the set of fraudulent cases. Otherwise, it is classified as a non-fraudulent case.

VI. ***Classification Accuracy***: The percentage of correct classifications in the test set is used to calculate the classification accuracy of the model. The model accuracy can be as high as 100% if the number of cases classified as fraudulent is equal to the number of cases classified as non-fraudulent. Otherwise, the model has a lower accuracy percentage.

VII. ***Evaluation***: The performance of the decision tree algorithm is evaluated using several performance metrics such as accuracy, precision, recall and F1-score. These metrics help us determine the effectiveness of the model in identifying fraudulent cases.

Table 2 Model results

	Results (%)
Accuracy score using decision tree classifier	99
Precision score	85.12
Recall	75
F1-score	78.6

VIII. ***Comparison***: The results of the decision tree algorithm is compared to a baseline model, such as logistic regression, and to the results of other state-of-the-art algorithms used in the literature for the same task. This comparison allows us to determine the effectiveness of the decision tree algorithm compared to other approaches.

5 Results

I. ***Model Performance***: The performance of our model in terms of accuracy, precision, recall, and F1-score is shown in Table 2 below. As you can see, our decision tree algorithm achieves high classification accuracy with a value of 99%. This suggests that our model identifies most fraudulent cases correctly. It also achieves a very high precision with a value of 85.12%. This means that in most cases where our model predicts that a case is fraudulent, it is actually fraudulent. On the other hand, our model has a relatively low recall with a value of 75%. This means that in many cases where our model predicts that a case is not fraudulent, it actually is not fraudulent. Our model also achieves a low F1-score with a value of 78.6%. This is mainly because our system makes many incorrect predictions in the test set. Finally, these results suggest that the decision tree algorithm is an effective approach for fraud detection.

II. ***Analysis of Fraudulent Cases***: We analyse the fraudulent cases identified in our model in order to identify the patterns used by fraudsters to defraud their victims. In the results section, we categorize each fraudulent case based on the types of fraud used by the fraudsters. In total, we found a total of 492 fraudulent cases which is 0.17% in the dataset.

6 Discussion

Identity theft is the most common type of fraud, as it involves stealing a person's personal information to commit fraudulent activities such as opening new bank accounts or purchasing expensive items using the victim's credit cards or loans. It is therefore critical to prevent this type of fraud by detecting it early through automated analysis. In this study, we showed that a decision tree algorithm can be used to detect cases of identity theft accurately. By comparing the results of the baseline model and the model created using the decision tree algorithm, we found that the performance of the decision tree algorithm was significantly better compared to that of the baseline model. This highlights the effectiveness of the decision tree algorithm in detecting cases of identity theft.

7 Conclusion

Identity theft is a growing concern in today's digital age, and fraud detection is crucial in preventing it. This study has shown that decision tree algorithms can be an effective tool for identifying identity theft with high accuracy. Future research should continue to explore the use of decision tree algorithms for identity theft prevention, as well as other methods for protecting personal information. One such tool is decision tree algorithms. This study has shown that these algorithms can be an effective means of identifying identity theft with high accuracy. The decision tree works by splitting the data into subsets based on the most important features that can differentiate fraudulent cases from legitimate ones. This enables the algorithm to identify the patterns and relationships in the data that are indicative of identity theft.

Despite the promising results, there is still room for improvement. Future research should continue to explore the use of decision tree algorithms for identity theft prevention and investigate other methods that can help protect personal information. This may include developing new algorithms, improving the quality of the data used, or incorporating additional data sources.

In conclusion, the study highlights the importance of addressing the issue of identity theft and the potential of decision tree algorithms as a means of achieving this goal. With the continued advancement of technology and the increasing need for effective fraud detection, it is crucial to explore new and innovative ways to protect personal information.

Acknowledgements This work was supported by Key Laboratory of Advanced Manufacturing Technology of the Ministry of Education of Guizhou University (GZUAMT2022KF[07]), the National Natural Science Foundation of China (No.61862051), the Science and Technology Foundation of Guizhou Province (No.[2019]1299, No.ZK[2022]449), the Top-notch Talent Program of Guizhou province (No.KY[2018]080), the Natural Science Foundation of Education of Guizhou province(No.[2019]203) and the Funds of Qiannan Normal University for Nationalities (No. qnsy2019rc09). The Educational Department of Guizhou under Grant NO. KY[2019]067.

References

1. Behind identity theft and fraud in cyberspace: the current landscape of phishing vectors. Int. J. Recent. Trends Eng. Res. **4**(3) (2018)
2. Dev, Rajput, S.: Fraudulent credit card transactions classification using randomized search CV with XGB classifier. Int. J. Eng. Tech. **6**(5) (2020)
3. Manning, P.: A novel understanding of fraudsters. Emerald (2022)
4. Subudhi, S., Panigrahi, S.: Detection of automobile insurance fraud using feature selection and data mining techniques. Int. J. Rough Sets Data Anal. **5**(3) (2018)
5. Gayathri, R., Malathi, A.: Investigation of data mining techniques in fraud detection: credit card. Int. J. Comput. Appl. **82**(9) (2013)
6. Ozili, P.: Forensic accounting and fraud: a review of literature and policy implications. Int. J. Account. Econ. Stud. **3**(1) (2015)
7. Save, P., Tiwarekar, P., Jain, K. N., Mahyavanshi, N.: A novel idea for credit card fraud detection using decision tree. Int. J. Comput. Appl. **161**(13) (2017)
8. Performance Analysis of machine learning based algorithms for detecting fraudulent credit card transactions. J. Xidian Univ. **14**(4) (2020)
9. Rahmani, F., Valmohammadi, C., Fathi, K.: Detecting fraudulent transactions in banking cards using scalefree graphs. Concurr. Comput: Pract. Exp. **34**(19) (2022)
10. Shi, Y., Liu, J., Qi, Z., Wang, B.: Learning from label proportions on high-dimensional data. Neural Netw. **103**, 9–18 (2018)
11. Voican, O.: Credit card fraud detection using deep learning techniques. Inform. Econ. **25**(1) (2021).
12. Samy, N., Mohamed Mohamed, S.: Credit card fraud detection using machine learning techniques, Futur. Comput. Inform. J. **7**(1) (2022)
13. Abdelhalim, A., Traore, I.: Identity Application Fraud Detection using Web Mining and Rule-based Decision Tree. [Online]. https://www.uvic.ca/ecs/ece/isot/assets/docs/ijcns-v0.2.pdf
14. Abdelhalim, A., Traore, I., Nakkabi, Y.: Creating decision trees from rules using RBDT-1. Comput. Intell. **3**(2) (2014)
15. Chikontwe, P., Lee, H.J.: Deep multi-task network for learning person identity and attributes. IEEE Access **6**, 60801–60811 (2018)
16. Berezkin, D.V., Runfang, S., Tengjiao, L.: Applying and Comparing Multiple Machine Learning Techniques to Detect Fraudulent Credit Card Transactions. Publishing house "Radiotekhnika" (2021)
17. Liu, G., Guo, J., Zuo, Y., Wu, J., Guo, R.: Fraud detection via behavioral sequence embedding. Knowl. Inf. Syst. **62**(7) (2020)
18. Xie, Y, Liu, G., Yan, C., Jiang, C., Zhou, M., Li, M.: Learning Transactional Behavioral Representations for Credit Card Fraud Detection. Institute of Electrical and Electronics Engineers (IEEE) (2022)
19. Design of a Model in Machine Learning For Credit Card Fraud Detection. International Institute for Science, Technology and Education (2022)
20. Decision Tree Algorithm for Credit Card FraudDetection. Granthaalayah Publications and Printers (2021)
21. Razak, A., Ahmed, N.: Detecting credit card fraud using data mining techniques-meta-learning. Indian J. Sci. Technol. **8**(28) (2015)
22. Prakash, A., Chandrasekar, C.: A novel hidden markov model for credit card fraud detection. Int. J. Comput. Appl. **59**(3) (2012)

Academic Performance Prediction Using Machine Learning Algorithms

Tao Hai, Jincheng Zhou, Shirin Abolfath Zadeh, Afolake O. Adedayo, S. F. Gan, Celestine Iwendi, and Zakaria Boulouard

Abstract The objective of the study is to use a method to predict student performance during the semesters and to compare accuracy perceptron for a dataset of student performance. In this regard, Machine Learning techniques were applied to the student performance dataset provided by the Kaggle.com website. Multilayer Perceptron, Random Forest, SVM, Naïve Bayes, Decision tree and K-NN algorithms were used to predict the Grade result of students as a factor of performance. The Student Performance dataset is used to forecast how well students will perform in their tests. As a result, Random Forest with 94.9% accuracy was the best prediction algorithm.

Keywords Academic performance · Prediction · Multilayer perceptron · Random forest · SVM · Naïve bayes · Decision tree · K-NN · Machine learning

1 Introduction

Academic observation is crucial and has been regularly used these days. Although this type of observation is thought to be necessary, it would be challenging to carry out on every student in a class, especially if the class has many pupils. Observing students in the classroom enables us to recognise their behaviour, which enables us to give

T. Hai · J. Zhou · S. F. Gan
School of Computer and Information, Qiannan Normal University for Nationalities, Duyun, Guizhou 558000, China

S. Abolfath Zadeh (✉) · A. O. Adedayo · C. Iwendi
School of Creative Technologies, University of Bolton, Bolton BL3 5AB, UK
e-mail: Shirin.abolfathzadeh@ieee.org

Z. Boulouard
LIM, Hassan II University of Casablanca, Casablanca, Morocco

T. Hai
School of Information and Artificial Intelligence, Nanchang Institute of Science and Technology, Nanchang, China

C. Iwendi et al. (eds.), *Proceedings of ICACTCE'23—The International Conference on Advances in Communication Technology and Computer Engineering*, Lecture Notes in Networks and Systems 735, https://doi.org/10.1007/978-3-031-37164-6_27

them the appropriate direction and also contributes to improving student behaviour (Mindell 1974). The COVID-19 pandemic has increased demand for online learning strategies and Learning Management Systems (LMS) and its developing field of Learning Analytics (LA) has grown in significance as a result. These tools track and assess students' activities to boost teaching and make decisions in the educational system. Because of its accuracy and speed, the Online Examination System is now thought of as a rapidly growing examination technique. Almost every organisation uses testing systems. Exams also enable organisations to easily monitor students' advancement over time. Consequently, the result may be computed more quickly [1].

Both teaching and learning profit from the prediction of student academic success [1, 2]. To be proactive, educators can utilise the anticipated results to determine how many students will perform well, averagely, or poorly in a class. For instance, teachers may think about adopting proactive actions to help those students perform better in the semester if the expected results reveal that some students in the class would be "academically at risk." Representative examples of preventative approaches involve expanding recitation sessions, increasing office hours, enhancing student problem-solving skills with computer simulations and animations, implementing a range of active and interactive learning methodologies, and so forth.

2 Related Works

As noted in [1], educational institutions can enhance student performance using Educational Data Mining (EDM) techniques. This involves the analysis and visualization of data related to the education system, including domain modelling and student learning behaviour. The paper in question applied linear regression and multilayer perceptron algorithms to this end, ultimately finding that the multilayer perceptron provided more accurate predictions of students' final examination results compared to linear regression.

Kiu [2] performed analysis to find out the significant impact of student background, and their coursework achievement in predicting their result and reaching high performance. The prediction was applied to the 2-level class (pass/fail) and 5-level class (A / B / C / D / F) on the final grade. The paper used Naïve Bayes, Multilayer perception, J48 and random forest and approved that J48 was the best classifier by more than 90% in Accuracy of 2-Level Classification and 80% in Accuracy of 5-Level Classification.

In addition, [3] develop an understanding of academic performance to measure student learning improvement on 2,151 students in the Dataset about the Engineering Dynamics course. They applied multivariate linear regression (MLR), multilayer perceptron (MLP) neural networks, radial basis function (RBF) neural networks, and support vector machines (SVMs) and the result shows support vector machine model with an accuracy higher than 90% the best.

In [4] multilayer perceptron and decision tree were used to predict the student grades from the repositories in which MLP (Multilayer Perceptron), while data has been taken from 480 students in offline processed by Weka (3.8). in addition, the aim was to find the method based on the best accuracy and short-time answer model. The result shows that MLP brings 93% accuracy.

Covid- 19 pandemic affection on the online education system was studied by [5] which presents two regression models based on Multilayer Perceptron (MLP) neural network and Random Forest (RF) regressor to analyse and predict Jordan students' performance by using Python on the JUSPH dataset with 775 records. Results presented that the proposed MLP models achieved the best accuracy score of 99.94%.

As [6] explore some physical and medical situation affect academic performance. The study exposed that, by using online education, the prevalence of sadness and anxiety decreased, which shows that medical students' overall mental health had improved [7].

Moreover, as [8–11] argued there are different significant features which could affect Academic performance, as it is shown in Fig. 1, and needs to be considered in the data to have a better prediction (see Fig. 1).

To identify which Students are at a high risk of dropping out of the study, [12, 13] offered two prediction models: the input–output hidden Markov model (IOHMM) and the logistic regression with the addition of a regularisation term. Consequently, this gives the course instructors the information they need to make prompt and effective interventions. When compared to ML models that served as a baseline, this model's accuracy score for identifying students who were likely to drop out was 84%.

Fig. 1 Features effect academic performance

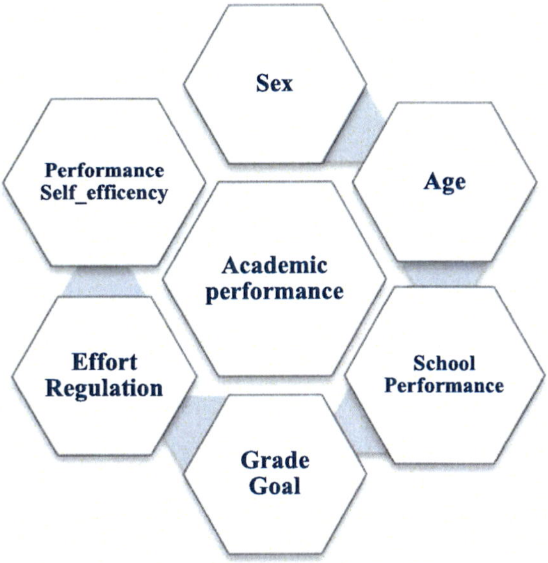

3 Materials and Methods

The proposed plan was studied in the Computer Laboratory in the School of Creative Technologies, University of Bolton. The dataset used is Students Performance which was collected from the Kaggle.com website. The database consists of 10,000 rows, about gender, past education, jobs, facilities, health situations and exam preparations of students. Jupyter notebook was the platform used to implement the algorithms. All process of cleaning the data was done on the data before applying the models and converting the mass data to clear one [14].

This study uses some steps to investigate issues and anticipate how education would affect students' academic achievement and feature usage. The list of procedures for the suggested models is shown in Fig. 2.

To check the performance of educational activities by students, in the dataset, there was a column about the Grade result of students classified into five different categories of the Grade Class, as it is shown in Fig. 3.

The following steps applied in this paper to find the best model are listed below:

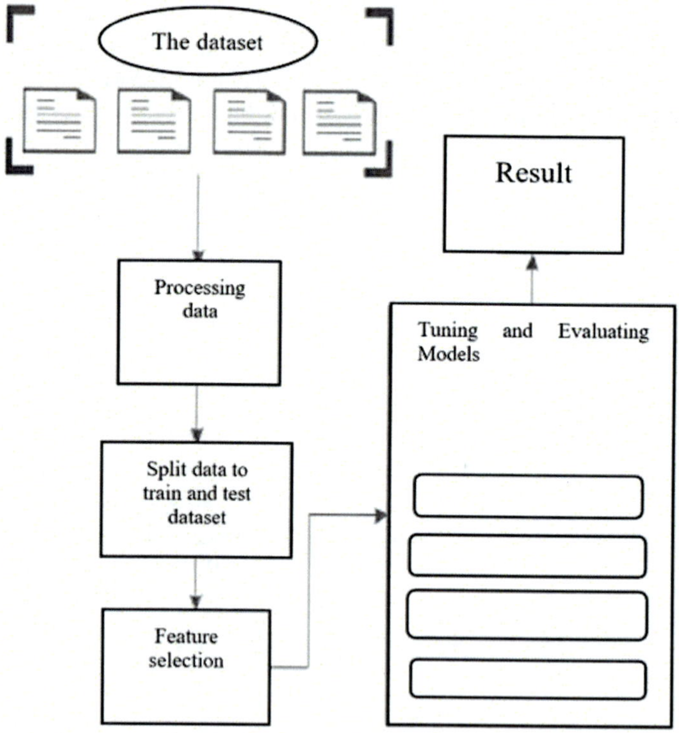

Fig. 2 Steps to predict the students' academic performance

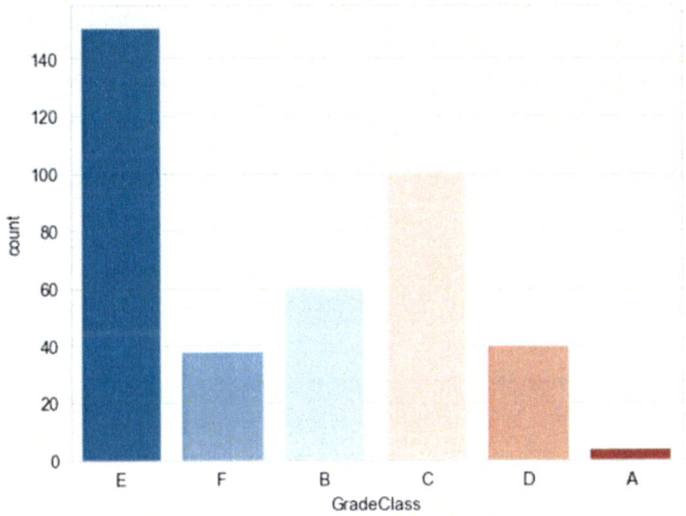

Fig. 3 Depicting the destiny of the grade

Pseudocode of algorithm:

1. Begin the Program
2. Import the relevant libraries.
3. read the CSV-formatted dataset file.
4. Cleaning the data.
5. Encoding the categorical columns
6. Feature selection by SelectKBest
7. loading the data to train and test set.
8. Normalizing(Scaling) the Numerical Column values
9. Use different Models.
10. Evaluate Models
11. Tune Models to find the best parameters.
12. Print the accuracy scores to find the best model.
13. Finish the programme.

This implementation can be done in Python using various libraries such as scikit-learn, TensorFlow, PyTorch, etc. The steps involved in this process include:

- **Data collection**: Collecting the relevant data such as students' grades, attendance, demographic information, etc.
- **Data pre-processing**: Cleaning and transforming the data to make it suitable for training the model.

- **Feature engineering**: Creating new features from the existing data to help the model learn patterns better.
- **Model selection**: Selecting the appropriate machine learning algorithm based on the type of problem (regression or classification), the size of the data, and the computing resources available.
- **Model training**: Using the training data to train the model to predict the academic performance of students.
- **Model evaluation**: Evaluating the performance of the model using metrics such as accuracy, precision, recall, F1-score, etc.
- **Model deployment**: Deploying the trained model in a production environment to make predictions on new data.

Overall, the implementation of academic performance prediction using machine learning algorithms in Python requires a good understanding of both machine learning concepts and the Python programming language.

The Data set contained 35 columns which provided 13 main columns as the features. To find the effective features SelectKBest technique was used, and the result of the best features are shown in Fig. 4.

In the Encoding section, the "LabelEncoding" was applied to the categorical columns and in the section of Normalisation data was normalised with standard Scaling. In addition, the train and test dataset was split as 80% training part and 20% test part (see Fig. 5).

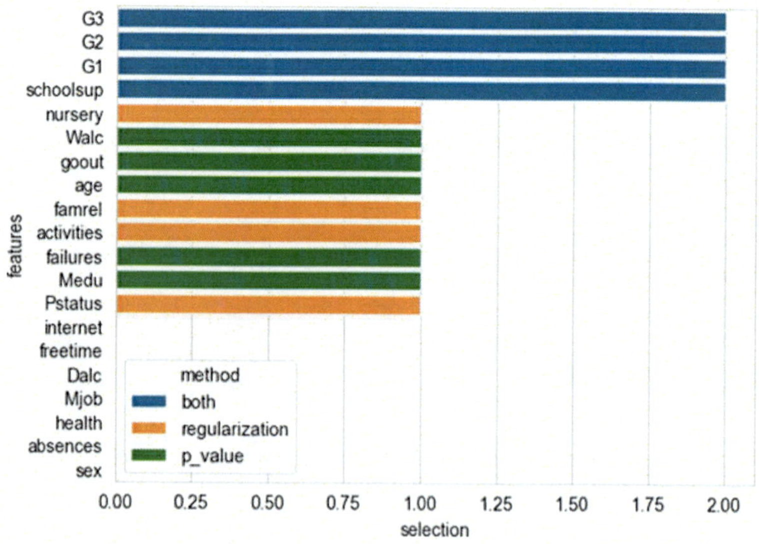

Fig. 4 Selecting the best features

```
norm_data_random = df_selected.sample(frac=1 )
# Spliting data
X=norm_data_random[['G1','G2','G3','nursery', 'Walc', 'goout','schoolsup','age'
                    ,'famrel','failures', 'Medu','activities','Pstatus']]
y=norm_data_random['GradeClass']

# Import train_test_split function
from sklearn.model_selection import train_test_split

# Split dataset into training set and test set
X_train, X_test, y_train, y_test= train_test_split(X, y, test_size=0.2, random_state=0)  # 80% training and 20% test
```

Fig. 5 Splitting data to test and train datasets

```
# fit the training dataset on the Random Forest classifier
from sklearn.ensemble import RandomForestClassifier
classifier_rf = RandomForestClassifier(n_estimators = 10, criterion = 'entropy'
                                       , random_state = 0)
classifier_rf.fit(X_train,y_train)
y_pred_RF = classifier_rf.predict(X_test)
print("Random Forest Accuracy Score -> ",accuracy_score(y_test,y_pred_RF)*100)
score.append(Set_score("Random Forest",y_pred_RF))

Random Forest Accuracy Score ->  94.9367088607595
```

Fig. 6 Fit the training dataset on the random forest classifier

To train and evaluate the models, the following Algorithms [15–17] were applied to the data and tuned to find the best parameters:

- Multilayer Perceptron
- Random Forest
- SVM
- Naïve Bayes
- Decision
- K-NN algorithms

Additionally, for each model, the training model was fitted on the classifier models, as the exact step can be seen in Fig. 6, based on the Random Forest Classifier.

As the actual and predicted data for the random forest classifier is depicted in the scatter plot, while this model can predict with 94.94% accuracy, there are some blue points, which show the wrong prediction by the model (see Fig. 7).

4 Results

Table 1 shows descriptive statistics of accuracy for all Algorithms. This table also shows which algorithms give the low Mean Absolute Error (MAE), Mean Square Error (MSE) and Round Mean Square Error (RMSE). As it is depicted the Accuracy Score of the Random Forest with 94.9% was the best. In addition, the mean absolute

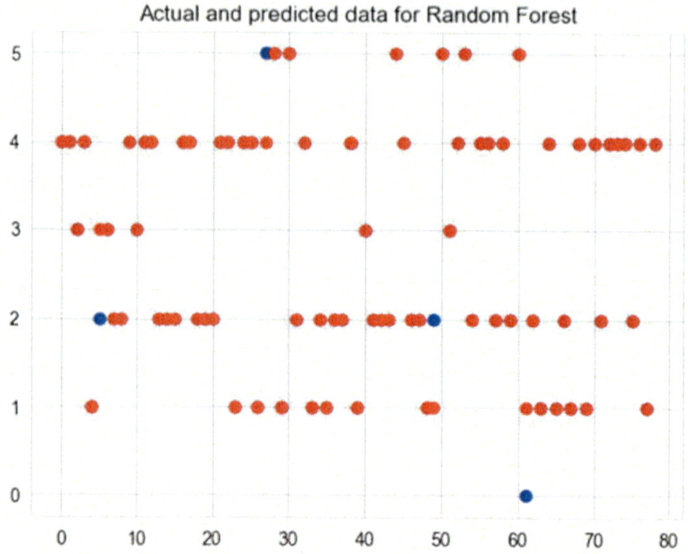

Fig. 7 Scatter plot to show the actual and predicted data for random forest classifier

error followed the same pattern with 0.05 by Random Forest as the lowest and 0.29 as the highest one by K-NN.

In addition, these results could be illustrated in a bar chart as is shown in Fig. 8. This bar chart was generated by plotting the accuracy score and mean absolute error on the Y-axis and the classification models on the X-axis. The graph clearly shows that Random forest is more accurate than other algorithms. The error rate of Random forest is lower than others.

In addition, Table 2 displays the findings of the independent test analysis and it shows a significance of 0.889. From the outcomes, it is perceptible that the Random Forest performs better when compared to other models.

Table 1 List of ML Algorithms that were applied to data to predict the result

	Model	accuracy_score	MAE	MSE	RMSE
1	MLP classifier	0.911	0.089	0.089	0.298
2	Random forest	0.949	0.051	0.051	0.225
3	SVC-RBF	0.911	0.089	0.089	0.298
4	SVC-linear	0.899	0.101	0.101	0.318
5	GaussianNB	0.886	0.114	0.114	0.338
6	Decision tree classifier	0.937	0.063	0.063	0.252
7	KNeighbors classifier	0.772	0.266	0.342	0.585

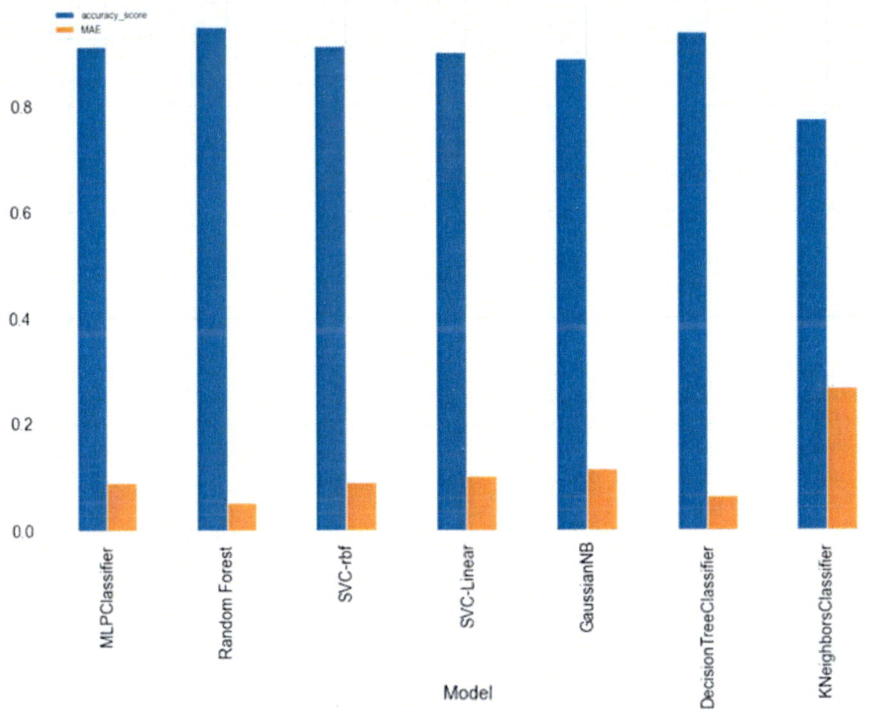

Fig. 8 Depict the comparison of the model results

Table 2 List statistical significance and standard error determined using the independent sample test

Equal variance	Levene's test for equality of variance		T-test for equality of means						
	F	Sig	t	df	Sig (2-tailed)	Mean Dif	Std error diff	95% confidence interval of the difference	
								Lower	Upper
Assumed	0.020	0.889	4.950	18	0.035	6.900	1.394	3.971	9.829
Not assumed			4.950	17.9	0.035	6.900	1.394	3.971	9.829

5 Conclusion

The proposed model approaches the students' performance in examination using Machin learning algorithms. The datasets from different students and various semester marks helped in improving the accuracy percentage.

After validating the models using the Performance Students datasets, it was noticed that, for an accurate result, the proposed Random Forest had a greater accuracy the student performance in exams was predicted using six different algorithms. The Random Forest algorithm demonstrates improved performance than others with an accuracy of 94.94%. the results achieved after all iterations on the dataset showed a constant 94.94% accuracy.

6 Future Work

There is a lot of scope for future work in academic performance prediction using machine learning algorithms, some of which are:

- Incorporating additional data sources: Collecting and incorporating additional data sources such as student behaviour, learning style, socio-economic status, etc. can improve the accuracy of the model.
- Exploring advanced algorithms: There is a lot of room to experiment with advanced machine learning algorithms such as decision trees, random forests, gradient boosting, neural networks, etc. to see if they perform better than simple linear regression.
- Ensemble methods: Combining the predictions from multiple models can often lead to better results. Ensemble methods such as stacking, bagging, and boosting can be used to improve the performance of the model.
- Hyperparameter tuning: Fine-tuning the hyperparameters of the model can often lead to improved performance. Grid search and random search can be used to find the optimal hyperparameters.
- Model interpretability: Interpreting the results of complex machine learning models can be challenging. Techniques such as feature importance, partial dependence plots, and local interpretable model-agnostic explanations (LIME) can be used to make the results of the model more interpretable.
- Incorporating domain knowledge: Incorporating domain knowledge into the model can often lead to improved performance. This can be done by incorporating domain-specific features, using domain-specific algorithms, or incorporating expert knowledge into the model.

Acknowledgements This work was supported by the Key Laboratory of Advanced Manufacturing Technology of the Ministry of Education of Guizhou University (GZUAMT2022KF[07]), the National Natural Science Foundation of China (No.61862051), the Science and Technology Foundation of Guizhou Province (No.[2019]1299, No.ZK[2022]449), the Top-notch Talent Program of Guizhou province (No.KY[2018]080), the Natural Science Foundation of Education of Guizhou province(No.[2019]203) and the Funds of Qiannan Normal University for Nationalities (No. qnsy2019rc09). The Educational Department of Guizhou under Grant NO. KY[2019]067.

References

1. Widyahastuti, F., Tjhin, V.U.: Predicting Students' Performance in Final examination Using Linear Regression and Multilayer Perceptron. In: 2017 10th International Conference on Human System Interactions (HSI), pp. 188–192. Ulsan, Korea (South) (2017). https://doi.org/10.1109/HSI.2017.8005026
2. Kiu, C-C.: Data Mining Analysis on Student's Academic Performance through Exploration of Student's Background and Social Activities. In: 2018 Fourth International Conference on Advances in Computing, Communication & Automation (ICACCA), pp. 1–5. Subang Jaya, Malaysia (2018). https://doi.org/10.1109/ICACCAF.2018.8776809.
3. Huang, S., Fang, N.: Work in Progress—Prediction of Students' Academic Performance in an Introductory Engineering Course. In: 2011 Frontiers in Education Conference (FIE), pp. S4D-1-S4D-3. Rapid City, SD, USA (2011). https://doi.org/10.1109/FIE.2011.6142729.
4. Jain, A., Shah, K., Chaturvedi, P., Tambe, A.: Prediction and Analysis of Student Performance using Hybrid Model of Multilayer Perceptron and Random Forest. In: 2018 International Conference on Advanced Computation and Telecommunication (ICACAT), pp. 1–7. Bhopal, India (2018). https://doi.org/10.1109/ICACAT.2018.8933580
5. Alkilani, A.H., Nusir, M.I.: Students' Performance-Prediction-Model Based on Physical and Physiological Constraints. In: 2022 19th International Multi-Conference on Systems, Signals & Devices (SSD), pp. 47–52, Sétif, Algeria (2022). https://doi.org/10.1109/SSD54932.2022.9955809
6. Bolatov, A.K., Seisembekov, T.Z., Askarova, A.Z., Baikanova, R.K., Smailova, D.S., Fabbro, E.: Online learning due to COVID-19 improved mental health among medical students. Med. Sci. Educ. **31**(1), 183–192 (2021)
7. Zong, J., Cui, C., Ma, Y., Yao, L., Chen, M., Yin, Y.: Behavior-Driven Student Performance Prediction with Tri-Branch Convolutional Neural Network. In: Proceedings of the 29th ACM International Conference on Information & Knowledge Management (2020). https://doi.org/10.1145/3340531.3412110
8. Fairbrother, E.: Students' examination performance. Electron. Power (1978). https://doi.org/10.1049/ep.1978.0077
9. Lee, S., Choeh, J.Y.: Predicting the helpfulness of online reviews using multilayer perceptron neural networks. Expert Syst. Appl. (2014). https://doi.org/10.1016/j.eswa.2013.10.034
10. Mindell, M.G.: The Relationship between Student Performance on Objective Exams and Student Performance in Class Activities as Reflected by Assigned Grades in a Basic Speech Course (1974).
11. Sander, P., Putwain, D., de la Fuente, J.: Using Structural Equation Modelling to Understand Predictors of Undergraduate Students' Academic Performance. In: Theory and Method in Higher Education Research, vol. 9, pp. 219–241. International Perspectives on Higher Education Research, Emerald Group Publishing Limited, Bingley. https://doi.org/10.1108/S1479-3628(2013)0000009015
12. Jain, A., Shah, K., Chaturvedi, P., Tambe, A.: Prediction and Analysis of Student Performance Using Hybrid Model of Multilayer Perceptron and Random Forest. In: 2018 International Conference on Advanced Computation and Telecommunication (ICACAT) (2018). https://doi.org/10.1109/icacat.2018.8933580
13. Mubarak, A.A, Cao, H., Zhang, W.: Prediction of students' early dropout based on their interaction logs in the online learning environment. Interact. Learn. Environ. 1–20 (2020)
14. Harris, L.: An examination of 8th grade students' perceptions of learning environment in relation to their academic performance (n.d.). https://doi.org/10.17760/d20003393.
15. Krishnan, R., Jeba, S., James, M., Al Badi, A., Kumari, S., Anajemba, J.H., Iwendi, C., Harinda, E.: Case Study-Based Predictive Linear Regression Model to Measure Anxiety and Depression as the Impact of COVID-19 among Students in Higher Education (2022)

16. Krishnan, R., Nair, S., Saamuel, B.S., Justin, S., Iwendi, C., Biamba, C., Ibeke, E.: Smart analysis of learners performance using learning analytics for improving academic progression: a case study model. Sustain. **14**(6), 3378 (2022).
17. Tarasov, D., Buevich, A., Shichkin, A., Subbotina, I., Tyagunov, A., Baglaeva, E.: Chromium Distribution Forecasting Using Multilayer Perceptron Neural Network and Multilayer Perceptron Residual Kriging (2018). https://doi.org/10.1063/1.5044048.

Accuracy of Water Quality Prediction Using Random Forest Regression and Artificial Neural Network

Tao Hai, Jincheng Zhou, Chijioke Victor Uzochukwu, Doyinsola Ayandiran, Yi Liu, and Celestine Iwendi

Abstract Water is life, human and all activities of living things needs water for its survival. There are growing concerns about the quality of water available for human consumption and other daily human activities. The pollution of water are traceable to economic activities, industrialization, urbanization, agricultural activities and other human activities. These activities also affect human life and water quality. This concern has necessitated the studies on prediction of water quality and an efficient water monitoring system. The most common method of ensuring good quality of water is by testing them in the laboratory, this current method wastes time, not economical, wastes human power. Many studies on the use of Artificial neural network and machine learning models to predict the quality of water has been on the increase, this study creates an Artificial neural network and random forest regression machine learning model using parameters such as ph., Hardness, Conductivity, Solids, Conductivity, Turbidity, Portability and does a comparative analysis of the best fit model and it was determined that the random forest model had the best fit model with a lower mean squared error of 2.325103283606211 when compared to the Artificial neural network with a mean squared error of 3.3122911 by determining the model with the least mean squared error. This study postulates that random forest method is the best fit model for prediction of water quality when compared to the Artificial neural network.

Keywords Artificial neural network · Random forest · Water quality · Machine learning · Prediction

T. Hai · J. Zhou · Y. Liu
School of Computer and Information, Qiannan Normal University for Nationalities, Duyun, Guizhou 558000, China

C. V. Uzochukwu · D. Ayandiran (✉) · C. Iwendi
School of Creative Technologies, University of Bolton, Bolton, UK
e-mail: doyinsola.aa@ieee.org

T. Hai
School of Information and Artificial Intelligence, Nanchang Institute of Science and Technology, Nanchang, China

C. Iwendi et al. (eds.), *Proceedings of ICACTCE'23—The International Conference on Advances in Communication Technology and Computer Engineering*, Lecture Notes in Networks and Systems 735, https://doi.org/10.1007/978-3-031-37164-6_28

1 Introduction

Over the years, the demand for water has been on high, this increasing demand for quality water is because of the importance and the use of water to human daily life and activities. There is also a rising industrialization, urbanization, agricultural developments and other actions of man on the environment and water bodies, these activities have effect on the quality of water index, the quality of water continues to reduce on daily basis dues to these activities. To ensure high quality of the water, there is an increasing research and studies on predicting and monitoring the quality of water. There is a need to predict the accuracy of the quality of water, this will help in effective management, conservation and preservation of our bodies and its quality [1].

Many studies have tried to predict the accuracy of the quality of water using various methods and arrived at varying degrees of accuracy. Prediction is finding an unknown characteristic of an element in a system when it is subjected to other different available elements. The formulae that are used for the prediction is gotten from already solved cases of historical data. During the use of the formula the new value of the element will be defined by different conditions [2].

Therefore, in this study a machine learning model like decision tree regression and Artificial neural network model will be created and a comparison made to determine the model that has the best prediction for the quality of water. This model can be applied in some areas like modelling of hydrology, water engineering etc. and can help people predict the quality of water used in those areas and consequently lead to better treatment and preservation of water bodies.

2 Related Works

Many studies have been done to ensure that the quality of water id of good grade. A study on the Predicting the quality of water, the study Investigated and estimated the quality grade of water [3]. This examination and prediction was be done using a grade prediction model that was be based on Artificial neural network, to carry out the modelling the crawler technology was deployed to obtain monitor historical data water quality, this collected data was subjected to further analysis and a model is created from the trained data. The created Artificial neural network model was trained by repeatedly changing the weight and biased parameters till the model is properly trained. The newly trained model was therefore deployed in the prediction of the quality of the water grade. The testing of the new model continues till an accuracy of over 96% is achieved, the model achieves a good generalization ability and can help in the future prediction of water quality level [3].

Further study on the indicator analysis of lake water quality based on fuzzy neural network and spectrophotometry [4]. Applied the use of fuzzy neural network model and spectrophotometry for the prediction of the quality of Lake water and its indicator analysis, this prediction made use of the principle of spectrophotometry, flexible temperature and Ph meter were used for checking the quality indicators of lake water. Data was collected for the purposes of data training and the correlation coefficient method was used in checking the quality of the water. 73 of the collected data was used for training of the model for prediction while 16 was used for the prediction. When this model was created, it was used to predict the chemical oxygen demand (COD), the COD prediction was within 0.8–0.95 correlation coefficient and relative error was within 15% of the data used for the training. From the results it shows that the Fuzzy neural network has a good accuracy in prediction and the generalization ability of water. The created fuzzy neural network helps in the prediction of the quality of water.

Another study on water prediction did some investigations on the Prediction of the quality of water using machine learning. Ragi [5] machine learning methodology used the Levenberg–Marquardt algorithm which uses already established parameters like PH, the electrical conductivity, TDS for the testing for the values of parameters like Alkalinity, chloride and sulphate using etc. the results gave an accuracy of 83.94%,87.9%,81.736%,79.48% respectively. This model will help to save time, it is financially economical and save wastage of human strength. It also offers the solution to chemical method of testing water.

A Mariculture Fish Mortality Prediction Using Machine Learning Based its analysis on Water Quality Monitoring. Develop a risk mortality Prediction of mariculture fish using machine learning to prediction of the analysis of quality of water monitoring [2]. How: There is an unstable production of fishes among fish farmers in Gondol, due to the high mortality rate of this mariculture, Predicted the quality of water using the Method Based on LSTM Neural Network, [6] this prediction was built on the creation of long term and the short-term memory neural network (LSTM NN), using 6 a year data collected from Tiahu late between 2000–2006 for the training of the data model, to enhance the prediction capability of the model some stimulations were carried out and some parameters selected before doing a comparison of the two methos. One of the methods was based on back propagation of neural network while the second one is based on online sequential extreme learning machine. This model is highly accurate and generalized.

This model was used in the solving of the time series prediction problem, which could not be solves using the traditional neural network and solved problems like water pollution.

3 Materials and Methods

3.1 Data Acquisition

The dataset for this study was downloaded from Kaggle website with name Water portability, the contents of the dataset are historical data collected from 3276 different water bodies. The dataset had 10 columns namely ph., Hardness, Solids, Chloramines, Sulfate, Conductivity, Organic carbon, Trihalomethanes, Turbidity, Potability (Table 1).

Pre-processing operations were carried out on the data, and it was divided into two different parts a training data which is 70% of the data and a testing data which is 30% of the dataset. When the training of the model is completed a random portion of about 0.04% will be taken out of the data to test the precision of the model sample.

The models were created using the python panda's library in the google colab, the study tested.

This study was carried out in the school of creative technologies, university of Bolton. For the purposes of the study, t 30 samples were selected and the alpha value of 0.07 and the beta value of 0.3. The pretest power of the study is 81% and the certainty stretch value is 96% [7]. This paper will be using the Artificial neural network and Decision tree classifier as the algorithm for the research and will report on the findings (Fig. 1).

3.2 Artificial Neural Network

The first proposal of Neural network mode was in 1943 by Warren McCulloh a psychologists and Walter pits a mathematician who used mathematical methods to form a biological neural network [7]. The essential unit of the neutral network is

Table 1 Dataset showing water features

Water features	Datatype	Count
Ph	Float 64	3276 non-null
Hardness	Float 64	3276 non-null
Solids	Float 64	3276 non-null
Chloramines	Float 64	3276 non-null
Sulfate	Float 64	3276 non-null
Conductivity	Float 64	3276 non-null
Organic carbon	Float 64	3276 non-null
Trihalomethanes	Float 64	3276 non-null
Turbidity	Float 64	3276 non-null
Portability	int64	3276 non-null

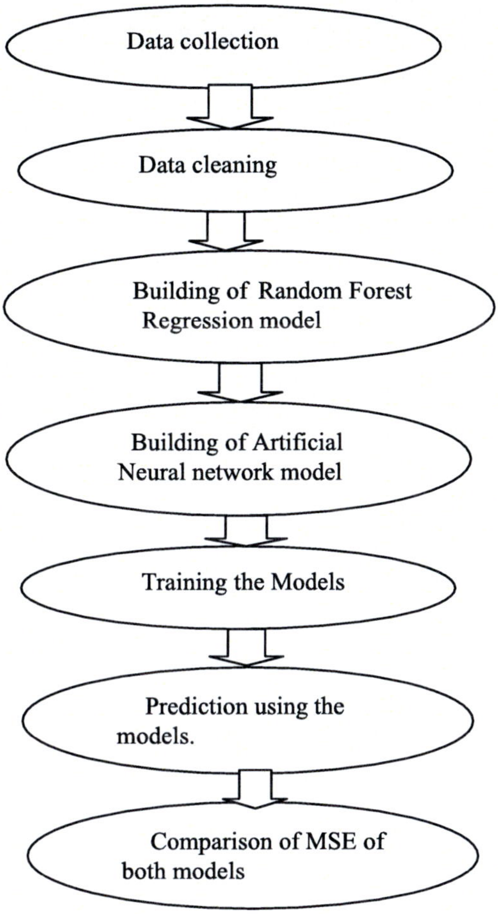

Fig. 1 System application flowchart

called neuron, the neuron has many inputs points, and the outputs are in tandem with a weight coefficient. The formula for obtaining the weighted results can be seen in the formula. Artificial neural networks are designed to learn and will not be reprogrammed [8].

When the model is applied it is used for testing water quality index which an instrument that is used for measuring the quality of water using different parameters that can be used in testing for water drinkability [9]. That uses nodes and can accept input and give out output.

3.3 Random Forest Regression

Random Forest is an aggregation of some of tree structure classifiers. It possesses many good qualities that makes it suitable for classification, prediction and can also be applicable for (Regression) purposes. These features make it very favorable when compared to some other traditional machine learning models [10–13].

4 Results

To test for the quality of the water index, two models were built to predict the PH value of the water. For the training of the model 6 features were chosen for the training which include Hardness, Conductivity, Solids, Conductivity, Turbidity, Potability, and this is represented by the X_train while we test for the PH value of the model. Then a comparative analysis of the two models were carried by comparing the mean squared error of the Artificial neural network and Random Forest Regression. The mean squared error of the Artificial neural network is 3.31229 while the mean squared error of the random forest regression is 2.32510. The model with the mean squared error is the best fit for the prediction of the ph. value of the water which is the measure for determining the quality of the water (Figs. 2, 3, 4, 5, 6, and 7).

```
[ ]  from sklearn.ensemble import RandomForestRegressor

[ ]  rf_model = RandomForestRegressor(max_depth=2, random_state=0)

[ ]  rf_model.fit(X_train, Y_train)

     RandomForestRegressor(max_depth=2, random_state=0)

[ ]  rf_model_y_pred = rf_model.predict(X_test)
```

Fig. 2 Random forest regression training model

```
# mean squared error for the Random Forest  y_prediction and Y_test
mse(Y_test, rf_model_y_pred).numpy()

2.325103283606211
```

Fig. 3 Mean squared error for random forest regression model

```
from keras.models import Sequential
from keras.layers import Dense


model = Sequential([
    Dense(24, activation='relu', input_shape=(6,)),
    Dense(32, activation='relu'),
    Dense(32, activation='relu'),
    Dense(32, activation='relu'),
    Dense(1),
])


model.compile(optimizer=keras.optimizers.Adam(learning_rate=1e-3),
              loss='mse',
              metrics=['mse'])


hist = model.fit(X_train, Y_train, epochs=100,
          validation_split=0.1)
```

Fig. 4 Artificial neural network training model

```
hist = model.fit(X_train, Y_train, epochs=100,
          validation_split=0.1)

Epoch 1/100
65/65 [==============================] - 2s 15ms/step - loss: 37125.9141 - mse: 37125.9141 - val_loss: 743.5979 - val_mse: 743.5979
Epoch 2/100
65/65 [==============================] - 1s 8ms/step - loss: 345.4397 - mse: 345.4397 - val_loss: 320.9673 - val_mse: 320.9673
Epoch 3/100
65/65 [==============================] - 1s 8ms/step - loss: 181.4945 - mse: 181.4945 - val_loss: 108.2182 - val_mse: 108.2182
Epoch 4/100
65/65 [==============================] - 0s 6ms/step - loss: 77.1977 - mse: 77.1977 - val_loss: 51.4418 - val_mse: 51.4418
Epoch 5/100
65/65 [==============================] - 0s 5ms/step - loss: 41.9702 - mse: 41.9702 - val_loss: 27.8911 - val_mse: 27.8911
Epoch 6/100
65/65 [==============================] - 0s 4ms/step - loss: 19.8343 - mse: 19.8343 - val_loss: 12.1801 - val_mse: 12.1801
Epoch 7/100
65/65 [==============================] - 0s 6ms/step - loss: 12.2126 - mse: 12.2126 - val_loss: 8.0751 - val_mse: 8.0751
Epoch 8/100
65/65 [==============================] - 0s 7ms/step - loss: 8.0052 - mse: 8.0052 - val_loss: 10.4161 - val_mse: 10.4161
Epoch 9/100
65/65 [==============================] - 0s 7ms/step - loss: 7.4450 - mse: 7.4450 - val_loss: 10.5734 - val_mse: 10.5734
Epoch 10/100
65/65 [==============================] - 0s 4ms/step - loss: 7.7130 - mse: 7.7130 - val_loss: 7.5686 - val_mse: 7.5686
```

Fig. 5 Sample of training epochs

5 Discussion

This figure shows the comparison of the mean squared error Artificial neural network and random forest. The mean squared error of the random forest machine learning model is lower than the mean squared error of the Artificial neural network The Artificial neural network and random forest are represented on the On the X-axis, while the Mean squared error is represented on the Y-axis. From the findings of this research, the Artificial neural network has a lower mean squared error, this model can be used to predict the quality of water in real time. Artificial neural network

```
model.evaluate(X_test, Y_test)

31/31 [==============================] - 0s 2ms/step - loss: 3.2329 - mse: 3.2329
[3.2329328060150146, 3.2329328060150146]

y_pred = model.predict(X_test)

31/31 [==============================] - 0s 1ms/step

# mean squared error for the y_prediction and Y_test
mse = keras.losses.MeanSquaredError()
mse(Y_test, y_pred).numpy()

3.3122911
```

Fig. 6 Model evaluation and mean square error of artificial neural network

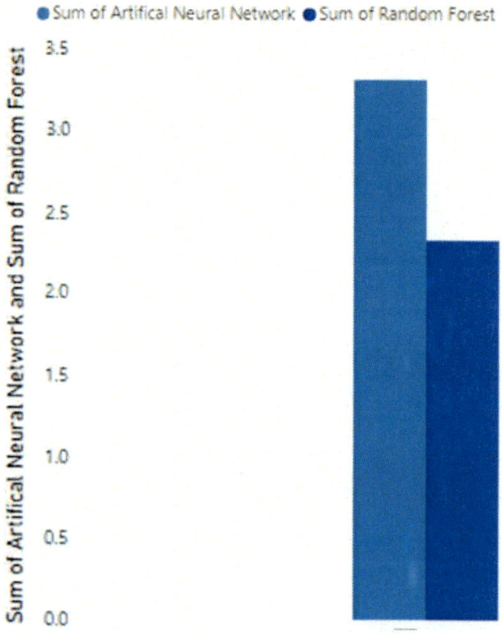

Fig. 7 Comparison of the mean square error of artificial neural network and random forest regression model

also has possibilities to be enhanced further by add more deeper layers, this makes it more versatile than the random forest model.

However random forest has churned out lower WQI predictions when compared to the ones produced by Artificial neural network.

6 Conclusion

From the results of the study conducted, the observations show that the decision tree classifier is considerably more suitable model because it shows as a better accuracy for prediction of water quality than the random forest model.

References

1. Ying, Z., Qianqian, G.: Comprehensive prediction model of water quality based on grey model and fuzzy neural network. Chin. J. Environ. Eng. **9**(2), 537–545 (2015)
2. Saville, R., Hatanaka, K., Fujiwara, A., Wada, M., Puspasari, R., Albasri, H., Muzaki, A.: A Mariculture Fish Mortality Prediction Using Machine Learning Based Analysis of Water Quality Monitoring. In: OCEANS 2022, Hampton Roads, pp. 1–4. IEEE (2022)
3. Hu, Z., Zhang, Y., Zhao, Y., Xie, M., Zhong, J., Tu, Z., Liu, J.: A water quality prediction method based on the deep LSTM network considering correlation in smart mariculture. Sensors **19**(6), 1420 (2019)
4. Yu, Y., Qu, Y., Zhang, H., Jiang, L., Shao, M., Wei, D., Zhang, D.: Indicator Analysis of Lake Water Quality Based on Fuzzy Neural Network and Spectrophotometry. In: 2022 IEEE 10th Joint International Information Technology and Artificial Intelligence Conference (ITAIC), vol. 10, pp. 1543–1546. IEEE (2022)
5. Ragi, N.M., Holla, R., Manju, G.: Predicting Water Quality Parameters Using Machine Learning. In: 2019 4th International Conference on Recent Trends on Electronics, Information, Communication & Technology (RTEICT), pp. 1109–1112. IEEE (2019)
6. Wang, Y., Zhou, J., Chen, K., Wang, Y., Liu, L.: Water Quality Prediction Method Based on LSTM Neural Network. In: 2017 12th international conference on intelligent systems and knowledge engineering (ISKE), pp. 1–5. IEEE (2017)
7. Zhao, H., Wang, S., Jiao, L., Yang, S., Liu, W.: Characteristics of temporal and spatial distribution of different forms of phosphorus in the sediments of Erhai Lake. Res. Environ. Sci. **26**(3), 227–234 (2013)
8. Xu, L., Liu, S.: Study of short-term water quality prediction model based on wavelet neural network. Math. Comput. Model. **58**(3–4), 807–813 (2013)
9. Custodio, M.: A review of water quality indices used to assess the health status of high mountain wetlands. Open J. Ecol. **9**(3), 66–83 (2019)
10. Iwendi, C., Bashir, A.K., Peshkar, A., Sujatha, R., Chatterjee, J.M., Pasupuleti, S., Jo, O.: COVID-19 patient health prediction using boosted random forest algorithm. Front. Public Health **8**, 357 (2020)
11. Kim, B.G., Goswami, K.: Basic Prediction Techniques in Modern Video Coding Standards. Springer (2016)
12. Liu, Y., Wang, Y., Zhang, J.: New Machine Learning Algorithm: Random Forest. In: Information Computing and Applications: Third International Conference, ICICA 2012, Chengde, China, September 14–16, 2012 Proceedings 3, pp. 246−252. Springer, Berlin, Heidelberg
13. Salami, E.S., Ehteshami, M.: Simulation, evaluation and prediction modeling of river water quality properties (case study: Ireland Rivers). Int. J. Environ. Sci. Technol. **12**, 3235–3242 (2015)

Indian Sign Language Recognition System for Interrogative Words Using Deep Learning

Sagar Rawat, Dahlak Daniel Solomon, Kushal Kanwar, Sonia Garg, Karan Kumar, Maad M. Mijwil, and Eleonora Beňova

Abstract With around 63 million population of deaf/mute society in India, an effective way of communicating with them is needed in order to understand them as deaf/mute people use hand shape, facial expressions, and body movements for communication. Developing an effective, accessible, and reliable way of communication is vital research in progress in machine translation. In this paper, proposed methodology considers isolated dynamic Indian Sign Language (ISL) gestures to recognize human actions on interrogative words. A self-recorded dataset of such gestures is made to contour the unavailability and variation of such dataset. Video frames are extracted from these dynamic gestures which are fed into proposed CNN architecture for training followed by prediction using Single-Frame CNN method on input videos. The proposed method has been tested with different optimizers and has achieved highest accuracy of 99.6% with Adam optimizer.

Keywords Indian sign language · Convolutional neural network · Deep learning · Dynamic gestures · Single-frame CNN method

S. Rawat · D. D. Solomon · K. Kanwar · S. Garg
Yogananda School of Artificial Intelligence, Computers and Data Science, Solan, Himachal Pradesh 173229, India

K. Kumar (✉)
Electronics and Communication Engineering Department, Maharishi Markandeshwar Engineering College, Maharishi Markandeshwar (Deemed to be University), Mullana, Ambala 133207, India
e-mail: karan.170987@gmail.com

M. M. Mijwil
Computer Techniques Engineering Department, Baghdad College of Economic Sciences University, Baghdad, Iraq

E. Beňova
Faculty of Management, Comenius University in Bratislava, 25 Odbojárov 10, 82005 Bratislava, Slovakia

C. Iwendi et al. (eds.), *Proceedings of ICACTCE'23—The International Conference on Advances in Communication Technology and Computer Engineering*, Lecture Notes in Networks and Systems 735, https://doi.org/10.1007/978-3-031-37164-6_29

1 Introduction

Sign Language (SL) is a medium by which people with hearing disability can share their thoughts, opinions, and ideas with normal person. Visual actions are performed in Sign language which is different from any spoken language. Both spoken language and sign language makes use of rules to produce clear messages with complex grammar [1]. SL gestures involve gestures for whole words instead of finger spelling for the sake of tempo of spoken conversation [2]. These gestures include hand movement, face expressions and body movement for depicting words or sentences. Sign languages have distinct lexicon and grammar which makes them complete natural languages in their own manner. Different SLs such as Indian Sign Language (ISL), American Sign Language (ASL), Australian Sign Language, British Sign Language (BSL), French Sign Language (FSL), and many others have been developed for deaf society. These SL's had many similarities but are not mutually understandable and universal [3]. Therefore, this complicates interaction between the hearing and Deaf community. Alternatives of written communication is difficult, because in writing a spoken language Deaf community is generally less skilled [4]. Each sign language has many thousands of gestures (sign), each different from the next in hand motion, position, shape, or context which makes it complex as any spoken language [5]. Spoken and Sign Language have different grammar, these differences are multiple channels used to give parallel information, different word order, using space and direction to explain connections between things [6].

Deaf community usually uses interpreters for communicating with the world but have drawback because of high hourly costs and privacy concerns like at a lawyer's, doctors or psychologist's office. Also, specialized translators for Deaf people are expensive and are difficult to find [7].

To establish a proper communication interface among the Deaf community and normal people an automatic recognition system and its translation is required [3]. The unfulfilled demand for good translating services can be met by such computer automated recognition system. Deaf society who are deaf due to sickness or injury but have unaffected manual ability can be helped through such systems. A crucial role in sign language recognition is played by such sign language recognition systems (SLRS) [8]. With the introduction of deep learning a considerable amount of applied research in computer vision field is being carried out. But research based on recognition of Indian Sign Language (ISL) dynamic gestures has plenty of work to improve on. So, with the motivation to work and enhancing the same, our paper focuses on establishing criteria for classifying and predicting the dynamic gestures of ISL by applying proposed methodology.

Recognition system for sign gestures is a hard job as the input fed into the system is high dimensional patio-temporal data, models are required that understand sign videos which involves how signer interact and move in given space. The model needs to interpret what these aspects mean when combined [6]. With that ISL uses both one handed and two-handed method for making sign gestures. These gestures are way more complex and identical to each other as compared to gestures in other Sign

Languages due to which it is more challenging to work on such gestures. The performances of traditional neural networks have significantly improved with the use of deep learning techniques, particularly for image classification and prediction problems. These techniques utilize huge amount of data to automatically retrieve features with same power as human brain to observe, analyze, learn, and make an interpretation [3]. Interpreting and shapes and motion trajectory are the primary challenges in sign language recognition. But it is difficult to obtain comprehensive result for SLR (Sign Language Recognition) due to the inconsistency and occlusion of hands and body joints. Although plenty research has been done to fix these issues as it's a complex problem to integrate the trajectory and hand shape features together [9]. Therefore sign language recognition has become an ideal field for computer vision and human language modeling algorithms for gesture recognition using computer intelligence. Sign language theory has well defined rules for hand body interaction as the manual and non-manual parts. Recognition system needs to be capable of adjusting with various input sequences and execution pace for videos which represents a time series of frames [10]. Amongst all available techniques for gesture recognition the Convolutional Neural Network (CNN) which is multi-layered Artificial Neural Network (ANN) to give high accuracy in medical image analysis, voice and gesture recognition [3].

We have taken gestures of interrogative words from ISL into consideration to make it easy to understand what a person with hearing disability is trying to ask. These words include 10 dynamic gestures of interrogative words taken from ISL dictionary. This paper focuses on vision-based technique for hand sign recognition and CNN technique was used to recognize and classify dynamic gestures of ISL.

Such systems which show skin color hand tracking, only single-handed sign language is considered in some of these systems. Most of them were done with real gestures done by non-native signers [11].

The major purposes of this paper are as follows.

- An improvised method to dynamic gesture detection with high accuracy rate.
- This system recognizes hand gestures of what, why, when, who, which, how many which will eventually help in better understanding.

The paper is organized in five sections—Sect. 2 describes the literature survey while Sect. 3 gives the steps involved in pre-processing. Section 4 evaluate the result of proposed system and lastly Sect. 5 concludes the paper and gives the scope for future work.

2 Literature Survey

Various approaches have been implemented by researchers to make a sign language recognition system. This mainly comes under two approaches which are Sensor based and vision-based methods as shown in Fig. 1. Sensor based approach uses glove fitted with motion sensors which provide good accuracy as compared to other

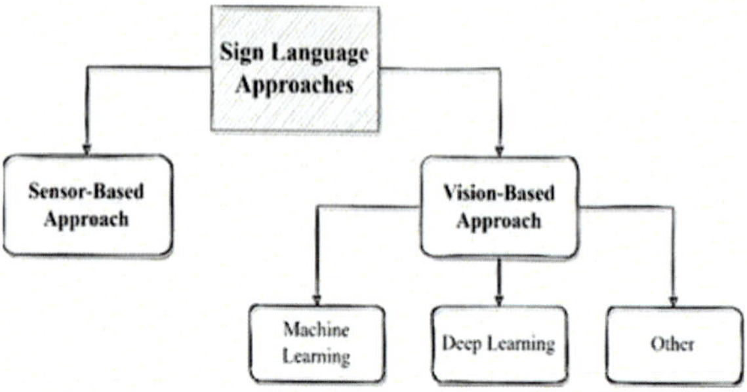

Fig. 1 Classification of sign language approach

approach but have a demerit of carrying and maintaining the hardware device. Unlike vision-based approaches, which detect movements from pictures or videos based on features generated using image processing method, vision-based methods identify gestures from images or videos. When it comes to overcoming the challenge of gesture recognition, there are essentially two options. The following categories can be used to classify them.

2.1 Sensor Based

Suri et al. [12] have implemented a CNN model which uses the dataset of Inertial Measurement Unit (IMU) signals collected using a GY-80 and Arduino UNO board containing a microcontroller. Caps Net architecture was used for recognizing above signals. Optimization was done by using Adam optimizer with Amstrad gradient optimization and learning rate of 0.001. Highest accuracy values for the model with 3 routings were observed as 99.72% in training and 94.00% in validation. The above model was performed on 20 sentences gathered with each sentence containing two to four signs1s apart. Suri et al. [13] proposed a method which uses Wearable Inertial Measurement Units (IMUs) for acquiring data. Each network contains nine layers which are convolutional, pooling, flattening and fully connected layers. Accuracy rate of 99.30% and 95.00% during training and validation was observed for CNN using question subset, and 98.42% and 94.20% average accuracy rate during training and validation for CNN using sentence subset. Abraham et al. [14] proposed a system which recognizes ISL dynamic gestures by making use of a glove with flix sensors on it to observe the movement of each finger. IMU was used to observe the orientation of the hand and collect data of the motion. Long Short-Term Memory (LSTM) networks were implemented for classification of above collected data and corresponding speech is given as output. An accuracy of 98% was observed for 26 gestures from the proposed model. Singh et al. [15] proposed a system that employs machine learning to identify the wearer's gesture and then the audio clip that corresponds to it. Fifty-six dynamic ISL gestures were built and evaluated in the system using three different machine learning algorithms where SVM based classifier achieved highest accuracy of 93% from the suggested method.

2.2 *Vision Based*

2.2.1 Machine Learning

Raheja et al. [14] proposed a technique for pre-processing video that requires transforming it to HSV color space followed by segmentation of skin pixels. Developed system was tested using both a camera and a Microsoft Kinect. Kumar et al. [16] proposed a method to detect the letters a-z in American Sign Language. To distinguish between fingertips and defect points between fingers contour was computed using Centre of Gravity (CoG) with standard formulae. The distance they move was used to find if a gesture is static or dynamic. The accuracy of static gesture was 93%, and 100% for four dynamic gestures. Rekha et al. [17] used a method that uses a YCbCr skin color model reference to segment and detect the hand region. Each hand posture was classified using SVM. Recognition rate of 86.3% was observed when Dynamic Time Warping (DTW) with the trajectory feature vector was used for classification of dynamic actions.

Rao et al. [18] have proposed a continuous ISL recognition application that runs in real time using a selfie camera. Data was collected using the camera from the signer in the proposed system. The suggested method was put to the test on real-world sentences. The performance of the proposed system was measured by using Word Matching Score (WMS). An average WMS of 85.88% was observed from MDC for classification.

Bashab et al. [19] proposed a system which uses Direction histogram because of its appeal for orientation variance illumination, it has been classified. Euclidean distance and KNN metrics are two alternative techniques used for recognition. The proposed system showed a comprehensive result for above metrics for dynamic gestures. Athira et al. [20] proposed a method for identifying dynamic gestures as well as a static gesture. Proposed method uses algorithm for removing co-articulation between static and dynamic gesture combinations. With 91 % accuracy, the system detected finger spelled alphabets and single-handed dynamic words with 89 % accuracy.

2.2.2 Deep Learning

Rao et al. [21] used a technique that uses two-dimensional video shot with a smartphone camera to detect Indian sign language signs. A dataset of 200 ISL widely used signs done at a rate of 30fps by five ISL users in five various camera angles (user dependent angles). Stochastic pooling was used to construct the CNN architecture. An average accuracy rate of 92.88 % was observed by suggested CNN model. Bhagat et al. [22] introduced two different models for detecting static and dynamic

ISL words. A Microsoft RGB-D camera was applied in the proposed experiment to gather depth information from sign photos. An accuracy of 98.81% was observed for depth + segmented static model. A CNN model was proposed for recognizing static gesture-based words and LSTM model was used to achieve an accuracy of 99.08% for dynamic gestures of words like mall, win, machine, Wi-Fi, lock, low, local and airplane. Jayadeep et al. [23] introduced a recognition system for classifying ISL words that are commonly used every day and in banks. The suggested model achieved 85 % of accuracy for words used in banks and 92 % for gesture of words commonly used every day. Dudhal et al. [24] used a CNN model in which convolutional layers were classified into five types. Two convolutional layers were present in each group. The training accuracy for CNN with adaptive thresholding was 97.58 %, while the validation accuracy was 91.84%. CNN with hybrid SIFT produced an increase in training accuracy to 98.83 % and validation accuracy to 92.72%. Salian et al. [25] proposed a system which uses RGB images with segmented hands. The system uses a CNN model consisting of two set of convolutional layers Training of the model was done on overall ASL dataset consisting of 2515 images of 36 signs. A validation accuracy of 98% was observed for 200 iterations and batch size of 50. Kishore et al. [26] used a CNN model in which layers were created using various filter windows with widths of 32 × 32, 16 × 16, 9 × 9 and 5 × 5. Three data sets were used in the training (i.e., 180,000 video frames). Recognition rate of 91.33 and 89.84% was produced when max pooling and mean pooling was applied. The average recognition rate of a stochastic pooling was 92.88%.

2.2.3 Other

Marrippan et al. [27] used raw videos with dynamic background as an input to the proposed SLRS which recognized the words of ISL with an accuracy of 75%.

Hore et al. [28] created a system after using multiple combined techniques like NNEA, NN-PSO and NN-GA. Nanivadekar et al. [29] create a recognition system based on self-captured ISL movies of the English alphabets A–Z, digits 1–10, and a few words helpful in normal conversation and in an emergency. To minimize the influence of uneven lighting in a picture, frames were recovered from these films after skin color segmentation in YCbCr color space. To determine the bounds of the hand movement, thresholding and edge detection were used. A skin color-based model was included in the system for improved feature extraction. Sahoo et al. [30] discussed an ISL recognition system which identifies alphabets, numerals, and words using static and dynamic gesture. To determine the performance of SLRS, the proposed method was implemented to several kinds of data independently. The suggested study demonstrated that the neural network classifier outperforms other classifiers in terms of accuracy. Joy et al. [31] implemented Sign Quiz. A low-cost web-based sign learning application forISL that relies on automated sign language recognition. By analyzing the literature survey, it is found that amongst all the techniques used by researchers CNN's have optimum accuracy and error rate for gesture recognition. It is because its system do feature extraction on its own, it utilizes image and filter convolution to produce consistent features that are passed on to the following layer. The features in the following layer are twisted with different filters to yield more consistent and basic features, also the process is repeated until the final output is occlusion-free. Also, CNNs can retrieve relevant characteristics from pictures, removing the use for manual image data processing methods that other approaches cannot achieve.

3 Methodology

3.1 Dataset Used

For training our proposed model we created our own ISL dataset as ideal ISL data is not easily available. We collected 10 samples each of 10 dynamic gestures of interrogative words from 3 subjects. All the volunteers were non-signers (had not learned and practiced sign gestures in past). These videos were taken from smart phone camera with resolution of 1280 × 720 and 30 fps. All videos were taken in white background with natural lighting. Figure 2 shows the extracted random frames from their respective class. Frames with no hand sign are either starting or end frame of the video. Though as stated by [18] the camera and the detection system must be placed at a distance from the signer, which is a key challenge with camera-based detection systems. but sensor based has its own downside in terms of economical, maintenance, and comfortableness which makes researchers to take vision-based approach as it was more flexible and economical. Several techniques have been followed funder this approach which includes, mobile camera [18, 20, 23, 26, 31], Microsoft Kinect RGBD camera [22, 32], web camera [20, 24, 31] and digital camera [27, 29–31, 33].

Researchers have been focusing more on the CNN in deep learning applications to increase performance and make the system more appropriate for real-time circumstances. [32]. Most computer vision problems, such as object detection and classification, can be solved with CNN [26]. CNNs have seen an impressive results as it can handle a variety of tasks related to processing data [25]. The proposed methodology uses sequential model consisting of convolutional layers which uses image frames from videos for processing tasks. Features extracted from these frames are used to classify these frames into their respective classes and later used for prediction. Classification and feature extraction of data was done by proposed CNN architecture. Lastly Single-Frame CNN Method is used to predict the videos in real time. All computation was done in python with OpenCV and Keras library. Figure 3 gives the working diagram of proposed system.

Fig. 2 Sample image frames from their respective class

3.2 *Frame Extraction from Input Videos*

To extract frames from the input videos we used video capture module cv2 from OpenCV (Open-Source Computer Vision Library) [25, 26]. These frames stored into their classes are later used for training and feature extraction purpose.

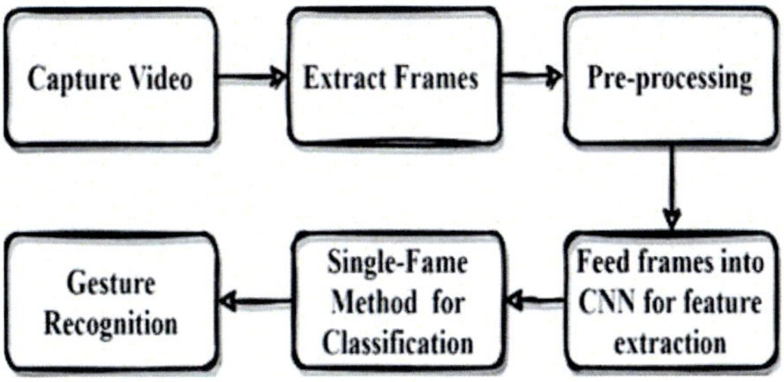

Fig. 3 Proposed methodology block diagram

4 Preprocessing

To establish a base size for all images fed into our network resizing was done followed by normalizing to make computation efficient by reducing values between 0 and 1. For that reason, these steps are also followed by [14, 17, 19, 22, 24, 26–28, 30]. Unwanted noise was removed from the videos before extracting and resizing the image frames from the videos. Frames extracted are converted to fixed size of 64 × 64 and normalized later for converting each pixel value in range 0–1 before storing them to their corresponding classes.

4.1 Feature Extraction

To extract features from the collected frames, a CNN sequential model was created. This proposed model uses convolutional layers as input and hidden layer followed by dense layer as output layer. The frames extracted from frame extraction method are passed through series of these layers. Dimension of the first input image is 64 × 64 which is converted to dimension of 62 × 62 × 64 after convolving it with 64 filters of size 3 × 3 which is then followed by pooling layer reducing the image dimension to 31 × 31 × 32. The second and third layer consists of same metrices as first. Dropout layer is added to avoid overfitting and better generalization of training data. Flattening is done before moving to fourth layer which is a fully connected convolutional layer with 256 filters each of size 3 × 3. Each neuron of this layer is connected to the 30,752 (31 × 31 × 32) neurons from the previous layers. The final fifth layer is a SoftMax output layer for classification of 10 classes. The CNN architecture is displayed in Table 1.

Table 1 CNN architecture summary

Layer type	Output shape	No. of parameters
Conv2d (Conv2D)	(None, 64, 64, 64)	1792
Max_pooling2d (MaxPooling2D)	(None, 32, 32, 64)	0
Conv2d_1 (Conv2D)	(None, 32, 32, 64)	36,928
Max_pooling2d_1 (MaxPooling2D)	(None, 16, 16, 64)	0
Conv2d_2 (Conv2D)	(None, 16, 16, 64)	36,928
Max_pooling2d_2 (MaxPooling2D)	(None, 8, 8, 64)	0
Dropout (Dropout)	(None, 8, 8, 64)	0
Flatten (Flatten)	(None, 4096)	0
Dense (Dense)	(None, 256)	1,048,832
Dense_1 (Dense)	(None, 10)	2570
Total params: 1,127,050 Trainable params: 1,127,050 Non-trainable params: 0		

4.2 *Single Frame-CNN Method*

Smoothing of the predictions via averaging, results in a superior video classifier. Using rolling prediction averaging (RPA), a list of the last K predictions from the CNN is preserved. Last K predictions are taken, label is selected after averaging them, selected label has the largest probability, and classification of frame is done [32]. In Single Frame-CNN (SFCNN) method 'n' frames from the entire video are taken for classification and predictions. The predictions of those n frames are averaged in the end to give us the final classification for input video [34]. SFCNN method is used in this paper for predicting gestures by creating a function which uses SFCNN method and displays the average probability of all the classes and arranges them in decreasing order giving the best class for the input video.

The areas where cybercrime is most likely to occur [13]. AI can find suspicious and criminal conduct by searching through the disorganized data that the investigators have gathered [14]. AI provides cognitive-data analytics, allowing people to review and consume data quickly. Additionally, it might make it easier for investigators to review criminal convictions and find potential offenders [16]. AI can help in detecting specific elements in the images and movies under investigation. AI also helps in spotting patterns in communication, setting, and timing. Detectives can utilize this knowledge to find the next crime's scene.

5 Training and Results

This section describes the results collected by training and testing the models described in the previous section. The dataset consists of 10 videos each of 10 dynamic gestures of interrogative words used in ISL which was divided in 80:20 ratio for the training and testing set, after collecting the data captured from all five volunteers by a smart phone camera. All models were trained on NVIDIA GEFORCE 940MX GPU. Different optimizers were implemented for the same training set to calculate the best model out from the same. Same hyper-parameters were followed in all models for the training:

- Epochs: 10
- Batch Size: 10
- Loss Function: CategoricalCrossentropy
- Learning rate: 0.01.

Table 2 gives results obtained after training a model with different optimizers. Accuracy and Error graphs of the different optimizers are shown in Figs. 2, 3 and 4 respectively. Results show by Adam optimizer fits best for the model by achieving the highest accuracy amongst the optimizers. Once the model is created, other metrices like confusion matrix, accuracy, recall and f1score are also calculated for model to better distinguish between and come up with the best optimizer for the proposed CNN architecture. Table 3 and Fig. 5 are the metrices calculated for the model with Adam optimizer.

Table 2 Training results

SN	Optimizer	Training accuracy (%)	Testing accuracy (%)
1	Adam	99.75	99.60
2	RMSProp	99.25	99.00
3	SGD	97.50	96.30

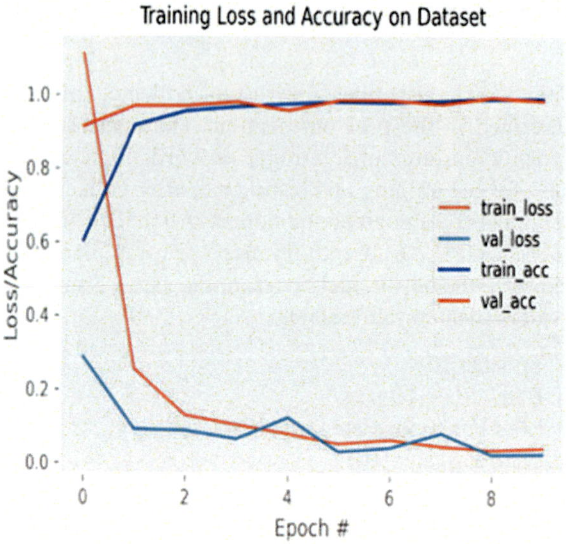

Fig. 4 Accuracy and loss error graph of adam optimizer

Table 3 Classification report of model

Class	Precision	Recall	f1-score	Support
0	0.98	1.00	0.99	108
1	0.99	1.00	1.00	107
2	1.00	0.98	0.99	95
3	1.00	0.99	0.99	93
4	1.00	1.00	1.00	96
5	0.99	1.00	1.00	103
6	1.00	1.00	1.00	102
7	1.00	1.00	1.00	95
8	1.00	1.00	1.00	104
9	1.00	0.99	0.99	97
Accuracy			1.00	1000
Macro avg	1.00	1.00	1.00	1000
Weighted avg	1.00	1.00	1.00	1000

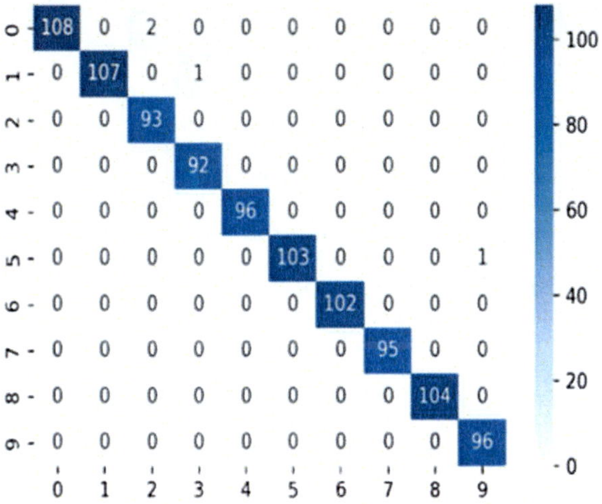

Fig. 5 Confusion matrix of model

6 Conclusion

The results show that the suggested system is capable of correctly understanding dynamic ISL gestures of interrogative words. Its real-world application can be deployed in public places like airports, railways, banks, and malls where normal human usually needs help with directions, places, shops, counter, or checkouts. This will help people with hearing disability in communicating with normal people by understanding what they are asking while having a normal conversation which usually involves asking and answering questions. Since it is economically friendly because of image -based approach therefore can be deployed in an application. It can be well applied to know what people with hearing disability are trying to ask thus helping in better understanding.

Further enhancements can be done by extending the dataset to add more dynamic gestures which can help initial conversation with a person with hearing disability. Other techniques can be applied in well created dataset which covers real world scenario of video recording i.e., with moving background, low lighting, and complex gestures.

Acknowledgements "This work was supported by the Slovak Research and Development Agency under the contract No. APVV-19-0581"

References

1. Sahoo, A.K., Mishra, G.S., Ravulakollu, K.K.: Sign language recognition: state of the art. ARPN J. Eng. Appl. Sci. **9**, 116–134 (2014)
2. Starner, T., Pentland, A.: Real-time American sign language recognition from video using hidden Markov models. In: Motion-based Recognition. Springer, pp. 227–243 (1997)
3. Rahman, M.M., Islam, M.S., Rahman, M.H., Sassi, R., Rivolta, M.W., Aktaruzzaman, M.: A new benchmark on american sign language recognition using convolutional neural network. In: 2019 International Conference on Sustainable Technologies for Industry 4.0 (STI) (2019)
4. Pigou, L., Dieleman, S., Kindermans, P.-J., Schrauwen, B.: Sign language recognition using convolutional neural networks. In: European Conference on Computer Vision (2014)
5. Cooper, H., Holt, B., Bowden, R.: Sign language recognition. In: Visual Analysis of Humans. Springer, pp. 539–562 (2011)
6. Camgoz, N.C., Koller, O., Hadfield, S., Bowden, R.: Sign language transformers: joint end-to-end sign language recognition and translation. In: Proceedings of the IEEE/CVF Conference on Computer Vision and Pattern Recognition (2020)
7. Brashear, H., Starner, T., Lukowicz, P., Junker, H.: Using multiple sensors for mobile sign language recognition (2003)
8. Zheng, L., Liang, B., Jiang, A.: Recent advances of deep learning for sign language recognition. In: 2017 International Conference on Digital Image Computing: Techniques and Applications (DICTA) (2017)
9. Huang, J., Zhou, W., Li, H., Li, W.: Sign language recognition using 3d convolutional neural networks. In: 2015 IEEE International Conference on Multimedia and Expo (ICME) (2015)
10. Koller, O., Zargaran, S., Ney, H., Bowden, R.: Deep sign: enabling robust statistical continuous sign language recognition via hybrid CNN-HMMs. Int. J. Comput. Vision **126**, 1311–1325 (2018)
11. Imagawa, K., Lu, S., Igi, S.: Color-based hands tracking system for sign language recognition. In: Proceedings Third IEEE International Conference on Automatic Face and Gesture Recognition (1998)
12. Suri, K., Gupta, R.: Continuous sign language recognition from wearable IMUs using deep capsule networks and game theory. Comput. Electr. Eng. **78**, 493–503 (2019)
13. Suri, K., Gupta, R.: Classification of hand gestures from wearable IMUs using deep neural network. In: 2018 Second International Conference on Inventive Communication and Computational Technologies (ICICCT) (2018)
14. Abraham, E., Nayak, A., Iqbal, A.: Real-time translation of Indian sign language using LSTM. In: 2019 Global Conference for Advancement in Technology (GCAT) (2019)
15. Singh, A.K., John, B.P., Subramanian, S.V., Kumar, A.S., Nair, B.B.: A low-cost wearable Indian sign language interpretation system. In: 2016 International Conference on Robotics and Automation for Humanitarian Applications (RAHA) (2016)
16. Kumar, A., Thankachan, K., Dominic, M.M.: Sign language recognition. In: 2016 3rd International Conference on Recent Advances in Information Technology (RAIT) (2016)
17. Rekha, J., Bhattacharya, J., Majumder, S.: Shape, texture and local movement hand gesture features for indian sign language recognition. In: 3rd International Conference on Trendz in Information Sciences & Computing (TISC2011) (2011)
18. Rao, G.A., Kishore, P.V.V.: Selfie video based continuous Indian sign language recognition system. Ain Shams Eng. J. **9**, 1929–1939 (2018)
19. Bashab, A., Ibrahim, A.O., Tarigo Hashem, I.A., Aggarwal, K., Mukhlif, F., et al.: Optimization techniques in university timetabling problem: constraints, methodologies, benchmarks, and open issues. Comput., Mater. Continua **74**(3), 6461–6484 (2023)
20. Athira, P.K., Sruthi, C.J., Lijiya, A.: A signer independent sign language recognition with co-articulation elimination from live videos: an Indian scenario. J. King Saud Univ.-Comput. Inf. Sci. (2019)

21. Rao, G.A., Syamala, K., Kishore, P.V.V., Sastry, A.S.C.S.: Deep convolutional neural networks for sign language recognition. In: 2018 Conference on Signal Processing And Communication Engineering Systems (SPACES) (2018)
22. Bhagat, N.K., Vishnusai, Y., Rathna, G.N.: Indian sign language gesture recognition using image processing and deep learning. In: 2019 Digital Image Computing: Techniques and Applications (DICTA) (2019)
23. Bhakuni, M., Kumar, K., Iwendi, C., Singh, A.: Evolution and evaluation: sarcasm analysis for twitter data using sentiment analysis. J. Sens. **2022**, Article ID 6287559, 10 (2022). https://doi.org/10.1155/2022/6287559
24. Dudhal, A., Mathkar, H., Jain, A., Kadam, O., Shirole, M.: Hybrid SIFT feature extraction approach for indian sign language recognition system based on CNN. In: International Conference on ISMAC in Computational Vision and Bio-Engineering (2018)
25. Salian, S., Dokare, I., Serai, D., Suresh, A., Ganorkar, P.: Proposed system for sign language recognition. In: 2017 International Conference on Computation of Power, Energy Information and Communication (ICCPEIC) (2017)
26. Kishore, P.V.V., Kumar, D.A., Sastry, A.C.S., Kumar, E.K.: Motionlets matching with adaptive kernels for 3-d indian sign language recognition. IEEE Sens. J. **18**, 3327–3337 (2018)
27. Mariappan, H.M., Gomathi, V.: Real-time recognition of Indian sign language. In: 2019 International Conference on Computational Intelligence in Data Science (ICCIDS) (2019)
28. Hore, S., Chatterjee, S., Santhi, V., Dey, N., Ashour, A.S., Balas, V.E., Shi, F.: Indian sign language recognition using optimized neural networks. In: Information Technology and Intelligent Transportation Systems. Springer, pp. 553–563 (2017)
29. Nanivadekar, P.A., Kulkarni, V.: Indian sign language recognition: database creation, hand tracking and segmentation. In: 2014 International Conference on Circuits, Systems, Communication and Information Technology Applications (CSCITA) (2014)
30. Sahoo, A.K., Sarangi, P.K., Goyal, P.: Indian sign language recognition using soft computing techniques. Mach. Vision Inspection Syst.: Image Process., Concepts, Methodologies Appl. **1**, 37–65 (2020)
31. Joy, J., Balakrishnan, K., Sreeraj, M.: SignQuiz: a quiz based tool for learning fingerspelled signs in Indian sign language using ASLR. IEEE Access **7**, 28363–28371 (2019)
32. Raheja, J.L., Mishra, A., Chaudhary, A.: Indian sign language recognition using SVM. Pattern Recognit Image Anal. **26**, 434–441 (2016)
33. Aggarwal, K., Mijwil, M.M., Al-Mistarehi, A.H., Alomari, S., Gök, M., Alaabdin, A.M.Z., Abdulrhman, S.H.: Has the future started? The current growth of artificial intelligence, machine learning, and deep learning. Iraqi J. Comput. Sci. Math. **3**(1), 115–123 (2022)
34. Anwar, T.: Introduction to video classification and human activity recognition, LearnOpenCV. https://learnopencv.com/introduction-to-video-classification-and-human-activity-recognition/ Accessed 20 May 2022
35. Sharma, V., Jaiswal, M., Sharma, A., Tomar, R.: A study of cnn architectures over two hand Indian sign language dataset. In: International Conference on Advanced Communication and Networking (2019)

New Approach of Artificial Intelligence in Digital Forensic Investigation: A Literature Review

Raghav Verma, Sonia Garg, Karan Kumar, Gaurav Gupta, Waleed Salehi, Piyush Kumar Pareek, and Jaroslava Kniežova

Abstract A Digital Forensic is a subfield of forensic science known as "digital forensic science" that focuses on the recovery and examination of data from digital devices that are connected to cybercrime. Computer forensics was the original meaning of the word "digital forensics." it has broadened to include any device that could store digital data. National policies on digital forensics didn't start to arise until the beginning of the twenty-first century. The process of locating, protecting, analyzing, and documenting digital evidence is known as digital forensics. When necessary, this is done so that evidence can be presented in court. AI in Digital forensics is one of the technologies that is developing the fastest, and it has had a significant impact on the methods and tools used to examine, monitor, and visualize crime scenes as well as develop effective strategies for dealing with impending threats and attacks on the internet and cyberspace. Modern technology has replaced human labor with machine-oriented designed work that has the capability and ability to minimize error and increase quality, reducing human effort and achieving maximum results with the fewest possible errors. It can influence the outcome and analyze the evidence in the field of digital forensics in a more effective manner to track the outcomes.

Keywords Digital forensic · Artificial intelligence

R. Verma · S. Garg · G. Gupta · W. Salehi
Yogananda School of Artificial Intelligence, Computers and Data Science, Solan, Himachal Pradesh 173229, India

K. Kumar (✉)
Electronics and Communication Engineering Department, Maharishi Markandeshwar Engineering College, Maharishi Markandeshwar (Deemed to be University), Mullana, Ambala 133207, India
e-mail: karan.170987@gmail.com

P. K. Pareek
Department of Artificial Intelligence and Machine Learning, Nitte Meenakshi Institute of Technology, Bengaluru, India

J. Kniežova
Faculty of Management, Comenius University in Bratislava, 25 Odbojárov 10, 82005 Bratislava, Slovakia

C. Iwendi et al. (eds.), *Proceedings of ICACTCE'23—The International Conference on Advances in Communication Technology and Computer Engineering*, Lecture Notes in Networks and Systems 735, https://doi.org/10.1007/978-3-031-37164-6_30

1 Introduction

Digital Forensic In terms of process models that the steps of the inquiry, digital forensic science is the study of the recovery of evidence from digital devices. In order [1] to determine the reason for the assault and which area of the system was targeted, digital forensics also deals with gathering and analysing information. Digital forensics is a learning tool that will enhance the system's security and analytical capacity to monitor information and computer risk. New security users and end users may gain access [1] of technical advancements in information creation and results. Algorithms used by various researchers and practitioners for the analysis of data that is accessible on the cloud and other devices are highly helpful in the detection of hazards as well as the prevention and prediction of criminal activities. AI is [2] used to quickly analyse the various factors that are taken into account by conventional algorithms, allowing the pertinent data to be quickly recognised and the insights to be categorised in the proper category. It is well recognised that the usage of sophisticated algorithms is necessary for AI. These algorithms can assist one in locating, observing, analysing [7], and forecasting various potential outcomes depending on time, place, and previous events. The detection of risk factors and their solutions depend heavily on a number of AI-related sub-fields. Three types of artificial intelligence are categorised: super, general, and narrow. They are utilized to resolve issues that human intelligence is unable to resolve, and as a result, they are superior to human intelligence [3] supervised and unsupervised learning, machine learning, optimization, deep learning, hybrid learning, and reinforcement learning are some of the most important AI approaches. Additionally, case-based reasoning (CBR), an efficient method for solving issues with various conventional systems such as symbolic rule-based systems, is used in digital forensics. A condition that the system has never encountered before can be handled by Case Based Reasoning (CBR) (Fig. 1).

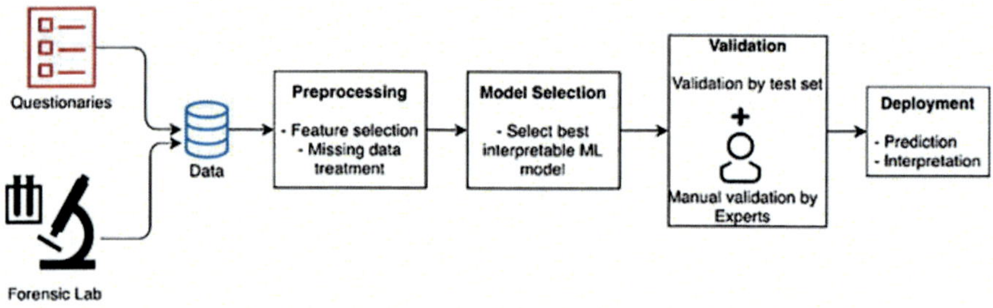

Fig. 1 Block diagram of digital forensic investigation

1.1 Artificial Intelligence and Its Implementation in Digital Forensics

Artificial intelligence is used today to manage systems autonomously, using proper inputs from past outcomes, and by analysing the consequent values for the future course. Artificial intelligence reduces the workload of input to acquire maximum results. Artificial intelligence is concerned with managing resources, analysing, and adapting resources and usage patterns. Virtually every field of technology and every application has been touched by digital forensics. Artificial intelligence can assist systems in analysing the pre-deterministic strategy to address current and impending faults, as well as security breaches and attacks [2]. Artificial intelligence can also assist us in anticipating potential problem-solving approaches so that security is not jeopardised. By incorporating the procedures and tactics employed by the features of artificial intelligence to deliver the best results for the evidence that has been analysed, artificial intelligence can also aid the field of digital forensics. Artificial intelligence is required in the field of digital forensics in two ways: first, to analyse data, and second, to forecast the potential vulnerabilities of computers [3]. Artificial intelligence may use any forensic tool created to prevent an intrusive party from concealing evidence wherever inside the system that conventional forensics tools are unable to analyse. When an attacker breaks into a system, they leave behind some evidence. If they are aware of the sort of forensics tool that will be used, they will hide their information in memory blocks where the tool will not be able to find it. Hence, no recoverable evidence was produced [4]. But if the forensic tool is correctly constructed, it can also analyse the leftover memory block and analyse the evidence.

1.2 Artificial Intelligence Methods that also Impact Digital Forensic

Information representation: This has to do with what a computer programme needs to know to do tasks intelligently, as well as how computation methods may feed that knowledge to the programme, which can then be used to develop better defences against cyberattacks.

Pattern recognition is a technique that recognises specific groups of data in an inquiry. It can be used to identify spam emails, suspicious file folders on hard drives, and the contents of pictures. When used in conjunction with knowledge discovery, it can help analysts find patterns in massive amounts of data.

2 Literature Review

2.1 *Digital Forensics*

The preservation, removal, examination, and recording of a (II. LITERATURE REVIEW, n.d.) for legitimate purposes is known as digital forensics. Finding specific traces of evidence on digital media is a science. It offers a variety of forensic techniques to handle and resolve complex cases using digital technology. It is also described as a procedure that entails a variety of jobs and processes that are involved in an inquiry rather than just one. Over a hundred different tasks and procedures must be completed in various digital forensic examinations. While some of these activities and procedures concentrate on data collecting technologies, others concentrate on the study field of data analysis. Many of these procedures were developed to handle the unique innovation used in the evaluated device; if the target device's underlying technology evolves, new procedures must be developed. These instances contain several processes that are utilised and are referred to as standard practises in digital forensic methodologies. However, occasionally there are some inconsistencies between them.

2.2 *Artificial Intelligence*

It is difficult to define artificial intelligence. AI is not well-defined in a single way. The major (Representation of Knowledge, n.d.) definitions of AI that do exist typically refer to "developing a computer process that responds intelligently" (but what is intelligence?). or "developing a computer process that can mirror human behaviour" (do people always act rationally, and what if a machine can typically perform more effectively than a person?). Other definitions mention "rational behaviour" but what exactly is rational behaviour? These terms are equally useless in this topic, such as "performing tasks that are hard for a computer to accomplish" (does this imply that once an AI system has been designed to perform the task, [5] it is no longer AI?) [6]. Therefore, in order to make the task more manageable for the purposes of this article, a practical approach is taken, AI is defined as "creating a computer process that acts in a manner that an ordinary person would deem intelligent," and consideration is given to some of the various types of AI and AI technologies that might be of interest to people in the community of digital forensics [2]. Therefore, methods like robotics will not be taken into consideration.

2.3 *Machine Learning*

Machine learning is a method for finding answers and reinventing risk management. The existing methods and data demonstrate the significant benefits of machine learning approaches in risk management. Machine learning is used in a variety of ways in AI systems that are used for risk management. The management of credit is one area in which machine learning is used. It is the financial loss incurred as a (II. LITERATURE REVIEW, n.d.) result of the counterparties' failure to fulfil their contractual obligations or from the increased transaction time risk. Regression models are typically employed to solve such issues. Institutions are becoming more intrigued by employing artificial intelligence and machine learning to enhance risk management procedures because some traditional risk management strategies are unsuccessful.

2.4 Supervised Learning

In artificial intelligence, supervised learning refers to a process where the system is given labelled input data with an anticipated output. It's a method where the computer algorithm is trained to receive specific outputs from given data [7, 8]. This model is trained until it finds the underlying relationships, patterns, and similarities between the input and output data that allow precise labelling. The risk management of regression and classification processes, such as determining an article's category or forecasting future sales volume, benefits greatly from supervised learning [6]. By monitoring a model's outputs frequently to approach the system's desired accuracy, supervised learning is made better in neural networks of algorithms [9]. The availability of labelled data and the algorithm used both affect how accurate results are.

2.5 Some AI Techniques that Can Help in Digital Forensics

Living forensics: These enable one to plan how to deal with the threat by immediately identifying, limiting, and eliminating threats. While doing this, it's crucial to search through criminal histories to identify anyone who might be at fault.

Data restoration: Data that was deleted or lost can now be recovered. This method of recovering prospective evidence for use in digital forensics is efficient.

Recovery of passwords: When investigating involving password-protected files, this is useful. It permits

Understand file filtering: This makes it possible for investigators to find relevant fields.

Timeline evaluation: This provides investigators with the sequence of events leading up to the incident they are looking into.

2.6 *Machine Learning Forensics*

The application of machine learning in the field of Machine learning is a new field that forensics has established. forensics, which can identify criminal activity patterns and predictions, such as where and whenever crimes are most likely to occur. In this case, digital Forensics must be conducted using a framework that meets the capacity to monitor and analyse servers, whether online or as well as numerous other forms of data, via a wireless connection. For segmenting, visualising, and link association illegal activity gathering [10]. There are numerous methods and machine learning forensics technologies.

Incident and crime clustering: Incident and crime clustering One can create groups using software for clustering. Based on spoken and written human behaviour. When this cluster is a forensic machine learning technique using the data, the software is independently constructed [11]. To isolate, a certain type of exploratory analysis is used [12]. Abnormalities in the data that may exist, such as an avoidance for instance, conduct that may result in criminal behaviour accessing a computer network using a dubious port of possible evidence of money laundering.

Attacks and crimes prognosis: A significant number of assaults and crimes take place annually, making it challenging, time-consuming, and costly to determine whether a crime was committed, use crime analysts by the exact same thief. An actual crime or an online assault typically include a person choosing a target, a time, and analytical context for their criminal behaviour techniques are employed to identify crime hotspots. Future scenarios through statistical projections.

Fraud detection in machine learning forensics: The most significant source that the data that machine learning forensics investigators use as they save their information concerning customers, tax paying citizens, clients, online users, and many others other things. We can use these data repositories to develop guidelines and models that are based on the daily encounters of people with websites and organizations around the world [13]. These resulting models may be applied sequentially. to both find fraud and help organisations get better techniques that may not be well thought out or even detect any relationships that the data may have obscured, which can impact and boost effectiveness while diminishing the rate of potential revenue loss because of crimes and invasions.

3 Use of AI in Crime Investigation

Forensic investigation calls for thorough research, gathering information from various sources, and synthesising it in order to reach reasonable conclusions. While acquiring such knowledge from unidentified sources Despite the fact that dealing with enormous volumes of data is occasionally chaotic and confounding, it may be a rewarding and exciting vocation. During an investigation, AI can assist forensic specialists in managing data efficiently and doing multiple layers of meta-analysis.

This might potentially save forensic investigators a significant amount of time and give them more time to focus on other crucial responsibilities! The information acquired must also be accessible to all parties without restriction so that it can be accessed as needed. AI technology can be used to create data repositories that can be periodically stored in your systems and accessed as needed. All technology increases the possibility of discovering and looking into crimes. This enables forensic experts to identify the problem's core cause swiftly and effectively [10]. Al helps investigators rapidly solve a crime and helps them save a lot of money in the process. As a result, they will be able to concentrate more on the areas where cybercrime is most likely to occur [14]. Al can find suspicious and criminal conduct by searching through the disorganised data that the investigators have gathered [1]. Al provides cognitive-data analytics, allowing people to review and consume data quickly. Additionally, it might make it easier for investigators to review criminal convictions and find potential offenders [15]. Al can help in detecting specific elements in the images and movies under investigation. Al also helps in spotting patterns in communication, setting, and timing. Detectives can utilise this knowledge to find the next crime's scene.

3.1 AI Can Be a Great Tool in Many Aspects Like

- Analysis of the data and its accessibility to support the investigation.
- Addressing challenges with a clear scope and case technique.
- Recognizing patterns
- Clearly articulating the thought process.
- Forensic science frequently involves lowering the proportion of false-positive or false-negative results during analysis.
- Formally organising knowledge representation, which will aid the legal community in finding prompt and accurate solutions [16].
- Having a performance evaluation that is well-structured.
- Knowledge discovery and data mining.

3.2 Steps to Involve in Digital Forensic Investigation

- Collection
- Examination
- Analysis
- Reporting.

3.3 Artificial Intelligence in Crime Detection

As governments work to support their country's rapid expansion, infrastructure in every region is becoming more sophisticated. Government authorities are receiving real-time data from the more intelligent and networked infrastructure of nations [17]. Together, real-time data and AI can aid in the early detection of criminal activity.

Examining crime scenes: Every crime scene is unique. Some might take place indoors, some outside, and some virtually. Investigators must come up with a unique investigative plan for each case, depending on the crime scene. One investigative approach cannot be used for all crimes because they are all unique. Nevertheless, there are a few fundamental procedures that crime scene investigators always follow. AI can be utilized by police departments to reduce costs in addition to helping them solve crimes. Police agencies all across the world are looking for ways to cut costs, and many of them are turning to reducing the number of officers and investigators they have on staff. This is because police budgets are being cut worldwide. Cities can benefit from artificial intelligence by using it to [18] perform jobs that would otherwise need to be completed by humans. In the past, catching suspects required spending endless hours reviewing evidence files, security camera footage, the crime scene, and other materials. All of this can be completed quicker, more affordably, and more effectively thanks to recent advancements in AI.

Recognizing Suspects: It might be exceedingly challenging to recognise one person among a large group of people. Furthermore, although it could be extremely simple for the person to make a mistake, it might be challenging to show in court [19]. However, facial recognition technology has made all of this simpler. Law enforcement organisations all over the world are using this new technology to identify persons both offline and online. For instance, if you have images of the suspect, you may feed them into machine learning software so that it can learn to recognise the suspects on millions of web pages. Consider the difficulty of manually searching all of these websites for the person you are looking for. This is especially helpful for looking into the 'Dark Web,' where many criminals hide and unlawful activity occurs.

6.3 Making Use of Police Databases' Power: As part of this technique, machine learning algorithms are trained to find every type of particle the researchers are interested in. As a result, the computer will need to be able to discriminate between particles left at the crime scene by officers, photographers, and pets. Medical experts must annotate the data in this, which might be expensive and time-consuming. Having said that, if such technology could be developed to the point where it could recognise every type of particle left at a crime scene, it would greatly benefit in the investigation of crimes.

4 Methodology

4.1 Recognize Pattern

To recognize patterns, AI technology may help with tasks like identifying the many elements of a single image, spotting patterns in emails and messages, and comparing new information with various types of pre-existing data in system databases [20]. Additionally, it can help investigators make connections between suspect information and already-available criminal records and tell them of any prior criminal activity the suspect in question may have been associated with.

4.2 Rendering Legal Responses

Scientific approaches to dealing with evidence in the legal system are provided by forensic statistics. AI can offer the legal community speedy solutions, when necessary, by using more complex and extensive information databases. Access the electronic version.

4.3 Advancement of Communication

Communication between forensic statisticians, lawyers, criminal investigators, and other experts is necessary for forensic investigation. Miscommunication between these parties could lead to inaccurate data interpretation or conclusions, which would delay or render justice incorrectly. AI helps narrow the communication gap between the many industry actors.

4.4 Data-Based Evidence

In forensic science, the narrative and arguments are supported by solid statistical evidence. To help with the development of scenarios and case studies, AI can design graphical frameworks. In order to help the law, make better decisions, it can also be utilised to produce graphical representations of the facts that can be used to support or contradict assertions.

4.5 Creating Repositories

AI can also help with the establishment of an online repository that would hold all information gathered during digital forensic investigations. It is becoming more challenging for forensic science investigators to store and analyse all this data due to the exponential development of storage capacity, including USB, hard drives, optical media, and flash drives. AI has the potential to be an effective tool for storing, analysing, and using this data legally.

5 Conclusion

The application of artificial intelligence is soon overtaking all other fields of applied science in importance. If our system does not completely rely on it, it is also advantageous to the forensic industry. People are working hard to understand AI through digital science, which is now accessible to everyone, as more and more people become aware of its significance in all aspects of life. Although it can facilitate their work, technology will never be able to fully replace people. Because forensic science requires a high level of specialisation, AI can only be used as a support tool in this field. Artificial intelligence will assist digital forensics tools in analyzing the evidence and make it simpler for forensics experts to examine the data and draw accurate conclusions in order to locate the scene of the crime. Additionally, keeping threat recordings to update the system for future usage and pre-analysing current security threats considering historical threats may be helpful. This study has demonstrated how several AI algorithms are now being employed in various branches of digital forensics.

6 Future Scope

The earlier sections have demonstrated that there is a substantial existing effort on using AI to facets of digital forensics. This section examines general challenges and untapped opportunities in the field of existing and developing AI. Techniques are still being used. These technologies can process and analyse data within minutes making the process of eDiscovery and investigations more agile. AI and deep learning in digital forensics are transforming the investigative environment, empowering organisations with the ability to access evidence faster and uncover more relevant findings.

Acknowledgements "This work was supported by the Slovak Research and Development Agency under the contract No. APVV-19-0581".

References

1. Irons, A., Lallie, H.: Digital forensics to intelligent forensics. Future Internet **6**(3), 584–596 (2014). https://doi.org/10.3390/fi6030584
2. Ali, M.R.: Digital forensics and artificial intelligence a study (2020). www.ijisrt.com651
3. 'II. LITERATURE REVIEW'
4. 'Representation of knowledge'. http://archive.ics.uci.edu/ml/
5. Jeong, D.: Artificial intelligence security threat, crime, and forensics: taxonomy and open issues. IEEE Access **8**, 184560–184574 (2020). https://doi.org/10.1109/ACCESS.2020.3029280
6. Hall, S.W., Sakzad, A., Choo, K.R.: Explainable artificial intelligence for digital forensics. WIREs Forensic Sci. **4**(2) (2022). https://doi.org/10.1002/wfs2.1434
7. Du, X., et al.: SoK: exploring the state of the art and the future potential of artificial intelligence in digital forensic investigation. ACM Int. Conf. Proc. Ser. (2020). https://doi.org/10.1145/3407023.3407068
8. Ukwen, D.O., Karabatak, M.: Review of NLP-based systems in digital forensics and cybersecurity. In: 9th International Symposium on Digital Forensics and Security, ISDFS 2021 (2021). https://doi.org/10.1109/ISDFS52919.2021.9486354
9. Karie, N.M., Kebande, V.R., Venter, H.S.: Diverging deep learning cognitive computing techniques into cyber forensics. Forensic Sci. Int. **1**, 61–67 (2019). https://doi.org/10.1016/j.fsisyn.2019.03.006
10. Solanke, A.A., Biasiotti, M.A.: Digital forensics AI: evaluating, standardizing and optimizing digital evidence mining techniques. KI - KunstlicheIntelligenz (2022). https://doi.org/10.1007/s13218-022-00763-9
11. Rizvi, S., Scanlon, M., McGibney, J., Sheppard, J.: Application of artificial intelligence to network forensics: survey, challenges and future directions. IEEE Access (2022). https://doi.org/10.1109/ACCESS.2022.3214506
12. Bhatt, P.: Machine learning forensics: a new branch of digital forensics. Int. J. Adv. Res. Comput. Sci. **8**(8), 217–222 (2017). https://doi.org/10.26483/ijarcs.v8i8.4613
13. Gupta, S., Sharma, M.V., Johri, P.: Artificial intelligence in forensic science. Int. Res. J. Eng. Technol. (2020). www.irjet.net
14. Beebe, N.: 'IFIP AICT 306 - Digital Forensic Research: The Good, the Bad and the Unaddressed'
15. Quick, D., Choo, K.K.R.: Impacts of increasing volume of digital forensic data: a survey and future research challenges. Digital Invest. **11**(4), 273–294 (2014). https://doi.org/10.1016/j.diin.2014.09.002
16. Kim, H., Kim, I.S., Kim, K.: AIBFT: artificial intelligence browser forensic toolkit. Forensic Sci. Int.: Digital Invest. **36** (2021). https://doi.org/10.1016/j.fsidi.2020.301091
17. Mitchell, F.R.: An overview of artificial intelligence based pattern matching in a security and digital forensic context. In: Cyberpatterns: Unifying Design Patterns with Security and Attack Patterns, vol. 9783319044477, Springer International Publishing, pp. 215–222 (2014). https://doi.org/10.1007/978-3-319-04447-7_17
18. 'isda19-ss4'
19. Venter, H.S., Coetzee, M., Loock, M.: Institute of electrical and electronics engineers. In: 2013 Information Security for South Africa: Proceedings of the ISSA 2013 Conference: 14–16 August 2013: Radisson Blu Gautrain Hotel, Sandton, Johannesburg, South Africa. IEEE (2013)
20. Solanke, A.A.: Explainable digital forensics AI: towards mitigating distrust in AI-based digital forensics analysis using interpretable models. Forensic Sci. Int.: Digital Invest. **42**, 301403 (2022). https://doi.org/10.1016/j.fsidi.2022.301403

Comparative Analysis of Machine Learning Classifiers for Plant Leaf Disease Detection

Sachin Dahiya, Tarun Gulati, Dushyant Gupta, Karan Kumar, B. D. Parameshachari, and Rastislav Kulhánek

Abstract In this decade artificial intelligence is making a tremendous growth in every field. Machine learning which is a subset of AI is founds application in various fields. Plant leaves disease detection and classification is one such area in which the use of AI and ML techniques is growing exponentially. Manual inspection of crops for disease detection is sometimes not possible because of large cultivation area. Use machine learning techniques facilitates the rapid detection of disease at early stages. In this study SVM, KNN and Tree classification algorithm of supervised machine learning techniques are studied on three dataset plant village dataset, Binary class plant village dataset and Taiwan dataset. Cubic SVM classifier shows highest accuracy among all the classifiers with all the datasets. Results also indicates that classifiers performance degrades with increased number of classes.

Keywords Artificial intelligence · Machine learning · Classification · Plant village dataset

S. Dahiya · T. Gulati · K. Kumar (✉)
ECE Department, M. M. Engineering College, M. M. Deemed to Be University, Mullana, Ambala, India
e-mail: karan.170987@gmail.com

D. Gupta
Institute of Integrated and Hons, Studies, Kurukshetra University, Kurukshetra, Haryana, India

B. D. Parameshachari
Department of Electronics and Communication Engineering, Nitte Meenakshi Institute of Technology, Bengaluru, India

R. Kulhánek
Faculty of Management, Comenius University in Bratislava, 25 Odbojárov 10, 82005 Bratislava, Slovakia

C. Iwendi et al. (eds.), *Proceedings of ICACTCE'23—The International Conference on Advances in Communication Technology and Computer Engineering*, Lecture Notes in Networks and Systems 735, https://doi.org/10.1007/978-3-031-37164-6_31

1 Introduction

Economy of any country largely depends upon the agriculture. Human livelihood largely depends on the quality and quantity of crop production. Farmers across the globe tries to give their best efforts by selecting good seeds and by selecting the types of crops according to the environmental condition. Plant disease severely affects the crops and decrease the amount of production as well as the quality of the production. To save the crops against various types of diseases and to achieve the targets of production to meet out the growing demand of the food it is now demand of hour to adopt new technologies for monitoring the crops for disease detection. Earlier the farmers inspect the crops manually by visiting the cultivation area which is very time consuming and sometimes inadequate in early detection of disease. In the era of technology need may be felt to design an automated system that can detect the plant disease at an early stage before being spread over the whole cultivation area. Early detection not only save the crops from disease but also helps in reducing the use of fertilizers and overuse of fertilizers. Lesser use of pesticides in the crop production ensures high quality food items and is safe for human consumption. Use of machine learning very beneficial in agriculture [1].

With the advancement in the image acquisition techniques, image processing and in artificial intelligence it is now possible to design a system that capture the images of the plants, process them on the computers and may inform the farmers through IoT technologies [2]. Deep learning [3] which is a form of machine learning was also utilized in the agriculture image analysis.

Automatic disease detection system includes image acquisition system, image processing techniques to extract features of interest and analysis methods to categorize various disease of plants. The tremendous growth in artificial intelligence makes it possible to devise such methods that makes machines intelligent enough to identify the various disease at the time of first occurrence. Use of machine intelligence helps in reducing the human errors and avoids the excessive use of harmful chemicals over the crops. In his paper various machine learning algorithms are studied and tested over three datasets. The accuracy of these classifiers is compared for different disease classes.

1.1 Related Work

In [4], Gergantopolus et al. used AlexNet architecture for tomato disease detection. In this work a 10 class classifiers is trained with 1 healthy and 9 diseased classes and in the second case 9 different two class classifiers are trained with healthy and unhealthy classes. It is also pointed out in the study that some of the classes in the dataset are recoded in different condition and can not be separable by using a single metric.

In Dahiya et al. [5] compared the different deep learning architectures using plant village dataset. A smaller version of plant village dataset is used, which contains 20,640 images of 15 classes of three species of pepped, potato and tomato. It is found that ResNet 50 and ResNet 101 performs better than all other deep learning architecture.

Maniyath [6] make use of random forest classifier for differentiating the healthy and the diseased leaves. HOG technique is used for feature extraction. Proposed algorithm give approximate 70% of accuracy.

Sujatha et al. in [7] compare the performance of machine learning models like Stochastic gradient decent, random forest and SVM and Deep learning models like inception v3, VGG-16 and Vgg19 are compared for citrus plant disease detection. Highest 89.5% accuracy is achieved with VGG-16 whereas Random Forest gives only 76.8% accuracy.

Lamba et al. in [8] authors propose a Parkinson's disease detection system using speech signals. A hybrid MIRFE feature selection approach is used and the results of proposed feature selection approach is compared with the 5 standard feature selection methods by XGBoost classifier. The proposed MIRFE approach selects 40 features out of 754 features with a feature reduction ratio of 94.69%. An accuracy of 93.88% and area under curve (AUC) of 0.978 is obtained by the proposed system.

2 Machine Learning Techniques

Machine learning is a technique by which the computers/machine can be made to learn, just like the human brain tries to learn i.e. learning from experience. There are various computational methods by which machine learning algorithm learn directly from the data set without any need of predefined equations. The accuracy of the model can be increased if the size of the data set is large. Machine learning techniques basically divided into two types, that is supervised learning and unsupervised learning. In supervised machine learning, models are trained on known inputs and outputs data so that the future outputs can be predicted. Unsupervised learning methods tries to find the hidden pattern in the given data set. The categorization of various machine learning algorithm shown in Fig. 1.

2.1 Supervised Learning Methods

Classification Learning Technique

In the classification techniques of supervised learning models are developed that can categorize the data in different classes. For example, plant leaf is characterized into healthy and unhealthy classes. These methods are used in various fields, such as in plant leaf disease detection, speech recognition, and in medical imaging.

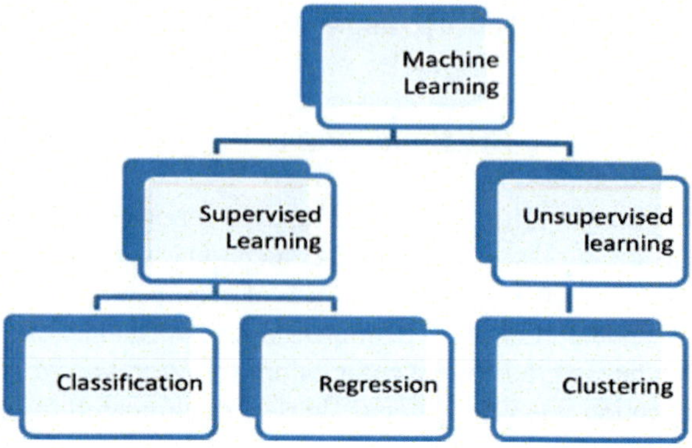

Fig. 1 Types of machine learning techniques [9]

Regression Learning Technique

These techniques are used for predicting responses which are continuous in nature. The application area includes electricity load forecasting and variation in temperature.

2.2 Unsupervised Learning Technique

This technique tries to find the hidden pattern and make inferences from the data set which contains unlabeled data. Clustering is a supervised technique and is utilized in object recognition and in market research.

3 Types of Classifiers

In this paper various classifiers are analyzed for the classification of various plant leaf disease. The classifiers used in the study are given below:

- K-Nearest Neighbor.
- Support Vector Machine
- Decision Tree.

3.1 K-Nearest Neighbor

Machine learning algorithm and falls under the category of supervised learning techniques. KNN [10] calculate the Euclidean distance between the data points. In this algorithm, firstly the number K is selected of the neighbors. Then Euclidian distance will be calculated among the K number of neighbors. Next, based upon the Euclidean distance, k neighbors will be selected. Within these K neighbors, data point corresponding to each category will be selected. Finally, the new data points will be assigned to those categories for which number of members maximum. KNN classifier is used for different values of K as mentioned in Table 1.

3.2 SVM (Support Vector Machine)

Support vector machine [11] is a technique that classify data by determining the most appropriate hyperplane that that separate out all the data points from one class to the data points from the other class. When the dataset points can be separated linearly than we can use linear SVM. Linear SVM separates the data points in to two classes by using a single straight line (Table 2).

3.3 Decision Tree. Classifiers

Decision trees are constructed in an algorithmic approach that defined the difference ways to divide a dataset depending upon certain condition. Decision Tree is a supervisor training method and is widely used in many applications. It is a non-parametric supervised learning method and can be used for classification as well as regression problems. Decision Tree is a very useful technique as it can be utilized in solving the various. Problems. Decision Tree algorithm is very simple and easy to understand (Table 3).

Table 1 Types of KNN

Classifier type	Prediction speed	Model flexibility
Fine KNN	Medium	Finely detailed distinctions between classes. The number of neighbours is set to 1
Medium KNN	Medium	Medium distinctions between classes. The number of neighbours is set to 10
Coarse KNN	Medium	Coarse distinctions between classes. The number of neighbours is set to 100
Cosine KNN	Medium	Medium distinctions between classes, using a Cosine distance metric. The number of neighbours is set to 10
Cubic KNN	Slow	Medium distinctions between classes, using a cubic distance metric. The number of neighbours is set to 10
Weighted KNN	Medium	Medium distinctions between classes, using a distance weight. The number of neighbours is set to 10

Table 2 SVM classifiers

Classifier type	Speed	Model flexibilty
Linear SVM	Binary: fast Multiclass: medium	Low Makes a simple linear separation between classes
Quadratic SVM	Binary: fast Multiclass: slow	Medium
Cubic SVM	Binary: fast Multiclass: slow	Medium
Fine Gaussian SVM	Binary: Fast Multiclass: slow	High—decreases with kernel scale setting Makes finely detailed distinctions between classes, with kernel scale set to sqrt(P)/4
Medium Gaussian SVM	Binary: fast Multiclass: slow	Medium Medium distinctions, with kernel scale set to sqrt(P)
Coarse Gaussian SVM	Binary: fast Multiclass: slow	Low Makes coarse distinctions between classes, with kernel scale set to sqrt(P)*4, where P is the number of predictors

Table 3 Tress classifiers

Classifiers type	Speed	Model flexibility
Coarse tree	Fast	Low Few leaves to make coarse distinctions between classes (maximum number of splits is 4)
Medium tree	Fast	Medium Medium number of leaves for finer distinctions between classes (maximum number of splits is 20)
Fine tree	Fast	High Many leaves to make many fine distinctions between classes (maximum number of splits is 100)

4 Experimental Study

In this experimental Study 3 different types of data set are used. First data set is a version of Plant Village [12] data set which contains. 15 classes 20,640 images as shown in Fig. 2. This plant village data set is converted into a new data set by converting the plant village data set in two classes. One is healthy and other is unhealthy to study the performance of various classifiers in case of binary class. Images in plant village data sets are taken into control condition. Therefore, to study the performance of classifiers on images taken under natural condition another dataset

Fig. 2 Plant village dataset (full version) [13]

is used which contains 4976 images (containing natural as well multiple leaves images) and having Six Classes:—01 Healthy and 05 unhealthy and is known as Taiwan dataset. Details of six classes are given below:

- Bacterial spot,
- Black leaf mold
- Gray leaf spot,
- Healthy
- Late blight
- Powdery mildew.

MATLAB tool is used for the study (Table 4).

4.1 Comparison of KNN Classifiers

K-Nearest Neighbor algorithm is analyzed for plant village dataset having 15 classes and its modified version having only binary classes. The results are summarized

Table 4 Detailed description of plant village dataset

S. no	Species	Class
1	Pepper bell	Bacterial spot
2	Pepper bell	Healthy
3	Potato	Early blight
4	Potato	Healthy
5	Potato	Late blight
6	Tomato	Target spot
7	Tomato	Tomato mosaic virus
8	Tomato	Yellow leaf curl virus
9	Tomato	Bacterial spot
10	Tomato	Early blight
11	Tomato	Healthy
12	Tomato	Late blight
13	Tomato	Leaf mold
14	Tomato	Septoria leaf spot
15	Tomato	Spider mites Two spotted spider mite

in Fig. 3. Results shows that for binary class coarse KNN has the highest learning accuracy whereas for multiple classes (15) cosine KNN performs better than other KNNs. However, for multiple classes the training accuracy of Coarse KNN is poor as compared to other KNN. It means cosine KNN is most suitable for plant leaf disease detection.

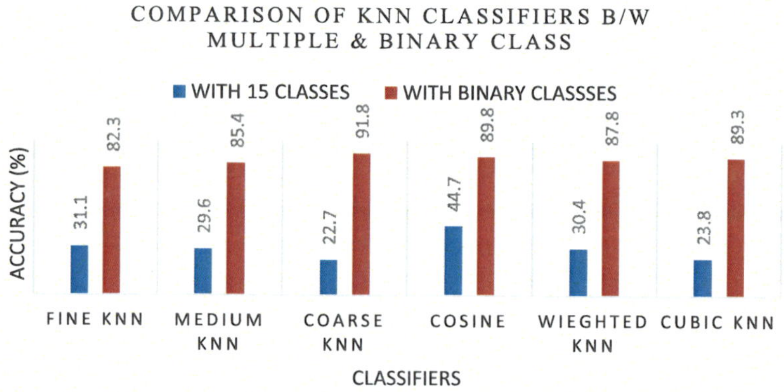

Fig. 3 Comparison of KNN classifiers for binary and multiple classes

4.2 *Comparison of SVM Classifiers*

Support vector machine is tested in the same set of condition as in case of KNN and results are summarized in Fig. 3. Cubic support vector machine outperforms all other SVM versions both in case of binary and multiple classes. Fine gaussian performs worst in case of multiple classes. All the above models perform well in case of binary classes whereas performance deteriorate in case of multiple classes.

4.3 *Comparison of Tree Classifiers*

Three types of tree classifiers are compared on the above two datasets. Their performance is summarized in Fig. 4. Results indicates that performance of fine, medium and coarse tree is comparable for binary class whereas fine tree performs somewhat better than other tree in multiple class datasets. All the above results indicates that the SVM, KNN and decision tree classifiers performs good when the number of classes is less i.e. in binary class case where only two classes are considered. Accuracy of all the classifiers degrades as we increase the number of classes to 15 (Fig. 5).

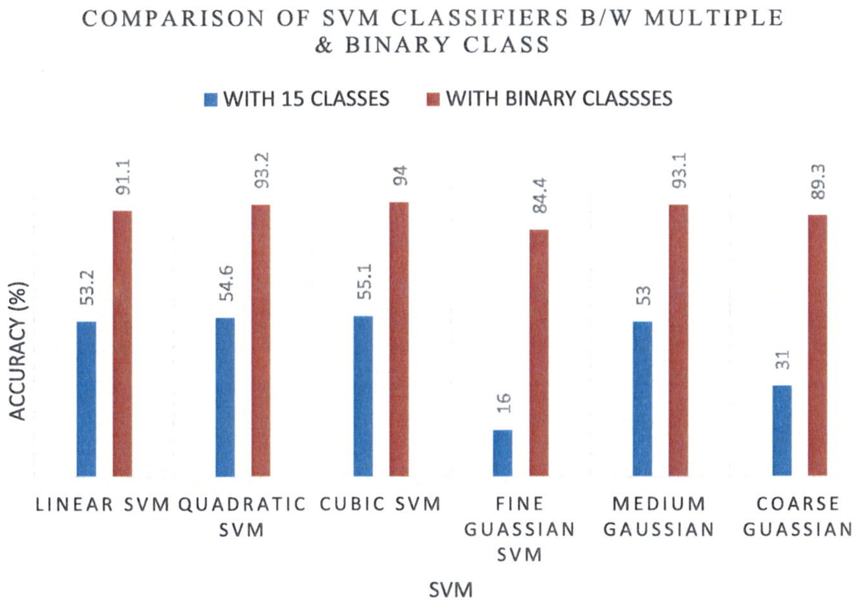

Fig. 4 Comparison of SVM classifiers for binary and multiple classes

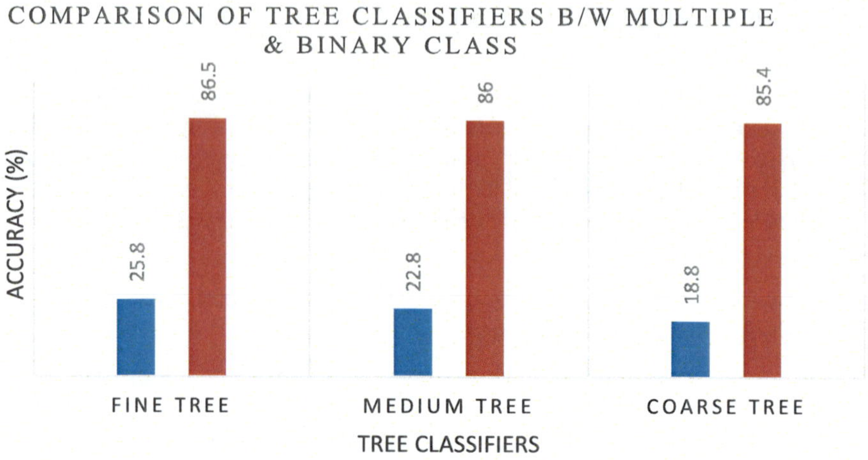

Fig. 5 Comparison of Tree classifiers for binary and multiple classes

4.4 *Comparison of SVM, CNN and KNN for Binary Plant Village Dataset Versus Taiwan Dataset*

In this set of experiments performance of all the classifiers is analyzed for Plant Village dataset and for Taiwan dataset. Taiwan dataset set contains six classes. Images in this dataset contains images of leaf with branches also. This dataset contains natural images as compared to the plant village dataset that contains images of leaves only and are taken under controlled conditions. Performances of tree classifiers is shown in Fig. 6.

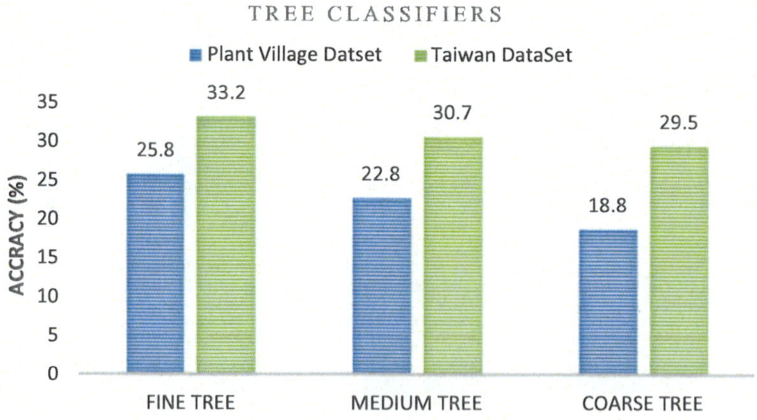

Fig. 6 Tree classifiers (plant village vs. Taiwan dataset)

Results indicate that the trees classifiers perform better in case of Taiwan dataset because it contains a smaller number of classes. SVM classifiers is also tested on the above-mentioned dataset, and it has been observed that cubic SVM performs better than other SVM and its accuracy is more in case of Taiwan dataset.

Similar study is also carried out for KNN classifiers which also confirms that the classifiers work better with dataset having lesser number of classes. Fine KNN gives highest accuracy with Taiwan dataset (Figs. 7 and 8).

Performance of best classifiers are summarized for all the three datasets is summarized in Table 5.

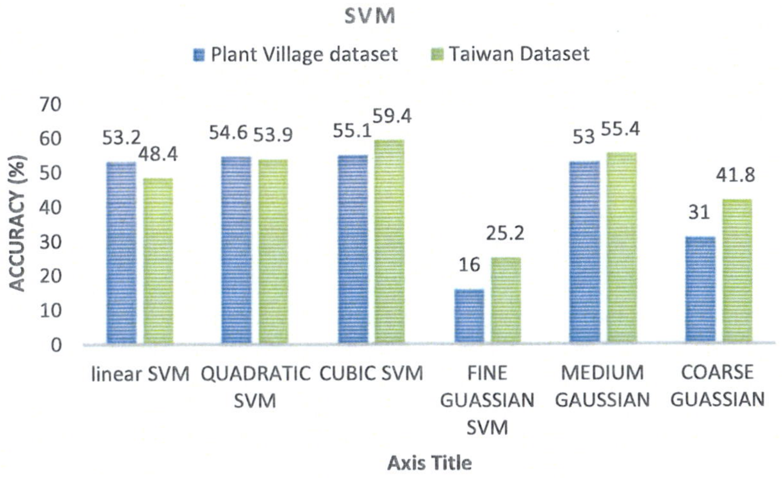

Fig. 7 SVM (plant village vs. Taiwan dataset)

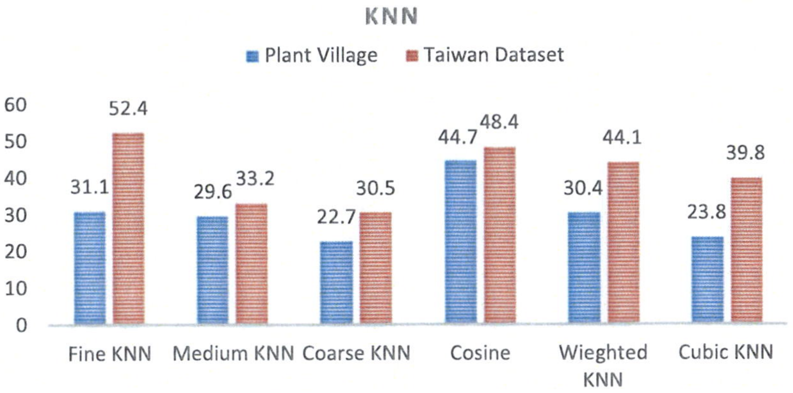

Fig. 8 KNN (plant village vs. Taiwan dataset)

Table 5 Best classifiers

DataSet	Best tree type classifiers accuracy	Best SVM Accuracy	Best KNN Accuracy
Plant village dataset with binary class	Fine tree (86.5%)	Cubic SVM (94%)	Coarse KNN (91.8%)
Plant village dataset with 15 classes	Fine tree (25.8%)	Cubic SVM (55.1%)	Cosine KNN (44.7%)
Taiwan dataset	Fine tree (33.2%)	Cubic SVM (59.4%)	Fine KNN (52.4%)

Results in above table indicates that cubic SVM has achieved the highest accuracy with all the datasets among all the classifiers. It performs better than tree and KNN classifiers. Hence it will be utilized for future work in plant leaves disease detection classification problem.

5 Conclusion

Machine learning techniques are efficiently utilized in various fields by the researcher as it reduces the human efforts and human errors. In this paper various machine learning algorithms for classification like Decision tree, K-Nearest Neighbor and Support Vector Machine are implemented for plant leaf disease detection and classification. A version of Plant Village data set contains 15 classes is utilized. This dataset is converted into new dataset by merging the various classes in two classes i.e. healthy and unhealthy, which is a binary class case. Finally, a new dataset known as Taiwan dataset is also used for study which contains six classes and having images with branches along with the leaf. Results indicates that performance of classifiers degrades with increase in number of classes in the dataset. Most of the classifiers performs well with binary classes, but performance severely degrades with multiple classes. Cubic SVM classifiers performs better among all the other classifiers with the Taiwan dataset, binary class dataset and Plant village dataset. Cubic SVM is selected as a classifier for future work in Plant Leaves disease detection and classification.

Acknowledgements "This work was supported by the Slovak Research and Development Agency under the contract No. APVV-19-0581".

References

1. Benos, L., Tagarakis, A.C., Dolias, G., Berruto, R., Kateris, D., Bochtis, D.: Machine learning in agriculture: a comprehensive updated review. Sensors (Basel). **21**(11), 3758 (2021). https://doi.org/10.3390/s21113758.PMID:34071553;PMCID:PMC8198852
2. Khanna, A., Kaur, S.: Evolution of internet of things and its significant impact in the field of precision agriculture. J. Comput. Electron. Agric. **157**, 218–231 (2019)
3. Lecun, Y., Bengio, Y., Hinton, G.: Deep learning. Nature **521**, 436–444 (2015)
4. Georgantopoulos, P.S., Constantinopoulos, C., Kosmopoulos, D.: Tomato disease classification using AlexNet. Mosaic **299**, 74
5. Dahiya, S., Gulati, T., Gupta, D.: Performance analysis of deep learning architectures for plant leaves disease detection. Measure.: Sens. **24**, 100581 (2022). ISSN 2665-9174
6. Maniyath, R.S., Vinod, P.V., Niveditha, M., Pooja, R., Prasad, N., Shashank, N., Hebbar, R.: Plant disease detection using machine learning, 41–45 (2018). https://doi.org/10.1109/ICDI3C.2018.00017
7. Sujatha, R., Chatterjee, J.M., Jhanjhi, N.Z., Brohi, S.N.: Performance of deep learning vs machine learning in plant leaf disease detection. Microprocessors Microsyst. **80**, 103615 (2021). ISSN 0141-9331
8. Lamba, R., Gulati, T., Jain, A.: An intelligent system for Parkinson's diagnosis using hybrid feature selection approach. Int. J. Softw. Innovation (IJSI) **10**(1), 1–13 (2022). https://doi.org/10.4018/IJSI.292027
9. Machine Learning with MATLAB - MATLAB & Simulink (mathworks.com)
10. Alsharef, A., Aggarwal, K., Sonia, et al.: Review of ML and AutoML solutions to forecast time-series data. Arch. Comput. Methods Eng. **29**, 5297–5311 (2022). https://doi.org/10.1007/s11831-022-09765-0
11. Alsharef, A., Kumar, K., Iwendi, C.: Time series data modeling using advanced machine learning and AutoML. Sustainability **14**(22), 15292. https://doi.org/10.3390/su142215292
12. Mohanty, S.P., Hughes, D.P., Salathé, M.: Using deep learning for image-based plant disease detection. Front. Plant Sci. **7**, 1419 (2016). https://www.kaggle.com/emmarex/plantdisease
13. https://www.frontiersin.org/files/Articles/215232/fpls-07-01419-g001.jpg (964×824) (frontiersin.org)

Effective House Price Prediction Using Machine Learning

Jincheng Zhou, Tao Hai, Ezinne C. Maxwell-Chigozie, Afolake Adedayo, Ying Chen, Celestine Iwendi, and Zakaria Boulouard

Abstract In recent times, there have been a surge in the housing business, such that prediction of houses is of utmost important both for the seller and the potential buyer. This has been influenced by several key indices. Many approaches have been used to tackle the issue of predicting house prices to help the house owners and real estate agents maximise their profit while the prospective buyers make better informed decision. This study focuses on building an effective model for the prediction of house prices. Since price is a continuous variable, it was expedient we used regression models. Some regression models like linear regression, Random Forest regressor (RF), Extreme Gradient Boosting Regressor (XGBoost), Support Vector Machine (SVM) regressor, K-Nearest Neighbor (KNN) and Linear regression were employed. The result showed that Random Forest Regressor showed a superior performance having an R2 score of 99.97% while SVM regressor performed poorly with an R2 score of −4.11%. The result proved that Random Forest regressor as an effective machine learning model to predicting house prices.

Keywords House price prediction · Regression algorithm · Machine learning

J. Zhou · T. Hai
School of Computer and Information, Qiannan Normal University for Nationalities, Duyun, Guizhou 558000, China

E. C. Maxwell-Chigozie (✉) · A. Adedayo · C. Iwendi
School of Creative Technologies, University of Bolton, Bolton BL3 5AB, UK
e-mail: ezinne.maxwell-chigozie@ieee.org

T. Hai · Y. Chen
School of Information and Artificial Intelligence, Nanchang Institute of Science and Technology, Nanchang, China

Z. Boulouard
LIM, Hassan II University of Casablanca, Casablanca, Morocco

C. Iwendi et al. (eds.), *Proceedings of ICACTCE'23—The International Conference on Advances in Communication Technology and Computer Engineering*, Lecture Notes in Networks and Systems 735, https://doi.org/10.1007/978-3-031-37164-6_32

1 Introduction

Housing is essential to everyone for shelter and warmth. Over the years, there have been a continuous surge in the housing sector, such that the prediction of houses is of utmost importance both for the seller and the potential buyer. This has been influenced by several key indices. Many approaches have been used to tackle the issue of predicting house prices to help house owners and real estate agents maximise their profit while prospective buyers also make better-informed decisions. According to contemporary studies on the erratic nature of house pricing, several key indices affect the prediction of house prices. These factors are but not limited to both either macro and micro in nature such as budget, hedonic, catchment area, closeness to water, prevailing interest rate, economic situation, etc. [1–4]. To effectively predict housing pricing, machine learning algorithms can be applied [1–3, 5]. Machine learning is a technique that enables computers to learn from data for correct prediction [6]. It helps to get the best model to accurately predict prices [3]. Regression models notably predict prices considering varied indices [5]. The Random Forest regressor, Gradient Boosting regressor (Gboost), Extra Gradient Boosting regressor (XGboost), Linear regression and KNN were considered. This study focuses on effectively predicting house prices using machine learning. These models were to confirm the best model for the predicting house prices. Our dataset was house prediction collected from the trustworthy and open source Kaggle [7]. The dataset consists of 545 records and was run with Python, a good programming language for running machine learning. The Random Forest regressor had highest R2 score of 99.990158%. The remaining of the paper is sectioned thus: Sect. 2 discusses previous related work, Sect. 3 describes the dataset, its pre-processing and the building of the model preceded performing exploratory data analysis. Section 4 shows the built models, their comparison, and the metrics used for their performance evaluation. Section 5 gives the result of the models and discusses them and Sect. 6 then concludes the work while offering ideas on further improvement of the model.

2 Literature Review

The prediction of house prices has continued to garner several studies among researchers. These studies reflect the underpinning factors influencing the erratic prices and then offer machine learning methods to solve them. [3] discussed influences on the pricing of houses from the primary level such as the size and context. SVM, regression, and random forest were considered and over R2 was gotten. This result proved that specific factors influence the prediction of house prices. Considering the fluctuating nature of prices, especially in housing [5], came up with the task of developing a model to discover factors influencing house pricing. Accordingly, several regression models, such as XGBoost, Linear regression, SVR, and Lasso and Ridge regression techniques, were evaluated for the best-fit model. To help the less affluent predict house pricing while factoring their budget, [2] analysed the performance of several optimal regression methods such as Decision tree regression, XGBoost, Support vector regression, and Random Forest, in an attempt to get the best predictive model. The models were also checked for accuracy using error metrics. These models were considered and the accuracy was determined. XGBoost produced a stronger classifier while SVM handled the house prediction with R2 score of over 0.8. The daunting misclassification of house price prediction was the focus of the authors [8]. In their paper, many models were compared and arrived at a favourable solution for which house sellers and buyers alike can mitigate exposure to the fluctuating housing price. The authors used logistic regression, decision tree, random forest, and naïve bayes.

While there is an inefficient model assessment for house price prediction, authors [1] compared and analysed five deep learning and machine learning models namely deep neural network, backpropagation neural network, Linear regression, and Support Vector Machine, for effective prediction. The outcome demonstrated that SVM performed best for the predicting prices of houses. The authors in [9] focused on the creation of proper predictions. Accordingly, the researchers used the Boston dataset and applied ridge regression, polynomial regression, lasso regression, and simple linear regression models. R2, RMSE, and Cross-validation were the performance metrics used. The result showed that Lasso and Ridge performed best by applying regulation techniques to reduce overfitting. Lasso regression performed best with an R2 of 88.79%, while simple linear regression performed the least. The author in [10] compared selected regression models in a bid to arrive at a model with better performance for house price prediction. For the evaluation of the models' performance, shuffle-split and k-fold cross-validation and score functions were applied. Score selection, parameter search, and grid search were used to improve the models' performance. The RF shows overfitting and performed poorly with a score of 0.1 while SVM with RBF kernel function performed best with an R2 score of 0.7. The work yielded positive. [11] focused on the prediction of house prices. In their work, Random Forest and multiple linear regression were compared. The work showed that multiple regression performed better than random forest. To make an informed cost-effective decision for aspiring tenants, motivated the authors in [12] to develop a model to predict house prices. In their paper, various housing features like the number of floors, the area of the house, etc. were analysed using random forest regression, KNN regression, Decision tree, and Catboost. The Catboost regression algorithms gave a superior performance with an R2 of 0.82.

The above review gives an overview of different ways used in prediction of house prices, the machine learning algorithms and techniques employed in obtaining results and the evaluation of the models' performance.

3 Materials and Method

The housing prices dataset provided by Kaggle [7] was used in this work. The dataset consists of 545 housing data records with 13 features, having both categorical and numerical data. The housing features are namely price, area, bedrooms, bathrooms, etc., with price as the target variable or feature. These features represent the relevant information of each house and are useful in building machine learning models to predict house prices.

The dataset was searched for any missing values and outliers were also checked to ensure the prevention of incorrect model performance. The target variable, price, was visualised (see Fig. 1) and checked for skewness. It was discovered that the target variable, was highly skewed-tailed to the right with a value of 1.2 (see Fig. 2). The skewness of a variable shows its distribution. The acceptable skewness is between -1 and $+1/2$. The data was then normalised from 1.2 to 0.14 (see Fig. 3). Following this was feature selection. Here, we employed Spearman's coefficient (see Fig. 4) which checks the dependence of the variables depending given their ranking. The features, area, and bathroom, showed a strong correlation to price. The Heatmap method helped find the correlation of the numerical variables, with area and bathroom showing as major influencing factors to pricing.

The categorical features were label encoded, which is useful step in fitting the data to build an effective model. The categorical encoding simply involved the conversion of the non-numeric features or variables to their numeric representation. This can be in the form of a one-hot encoder or label encoder. For this work, a label encoder was used and this was gotten from the sklearn library.

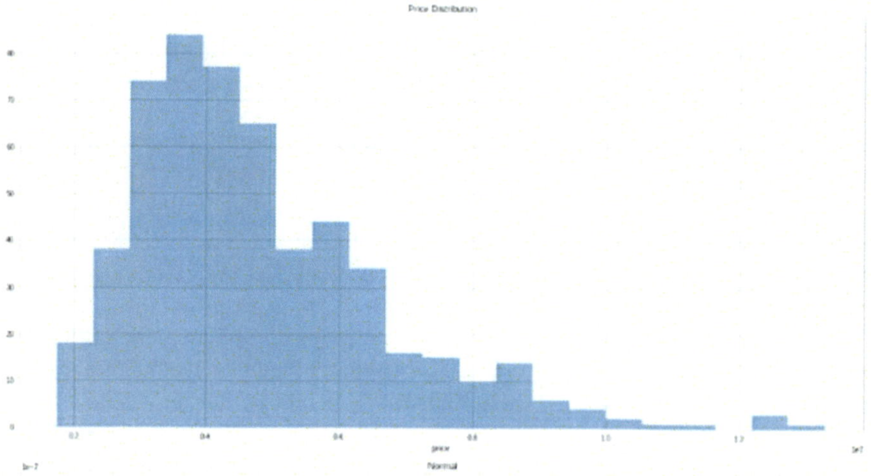

Fig. 1 Price distribution

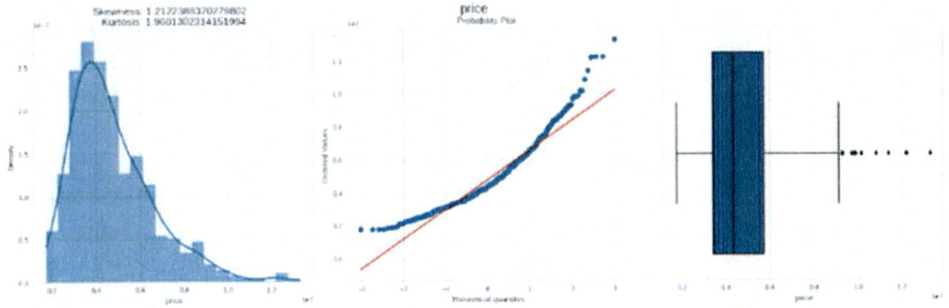

Fig. 2 Skewness of the target variable price

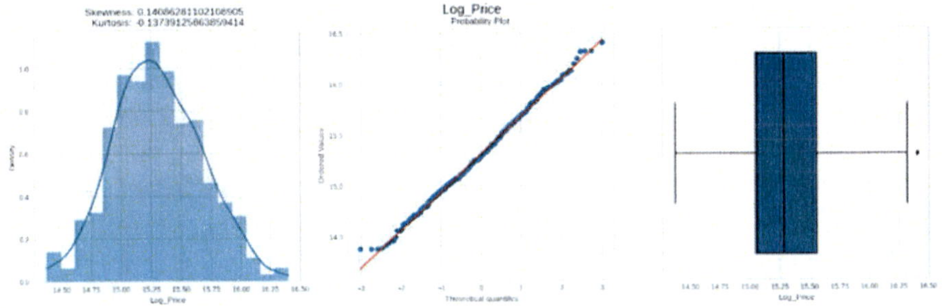

Fig. 3 Normalised distribution (skewness) of the target variable price

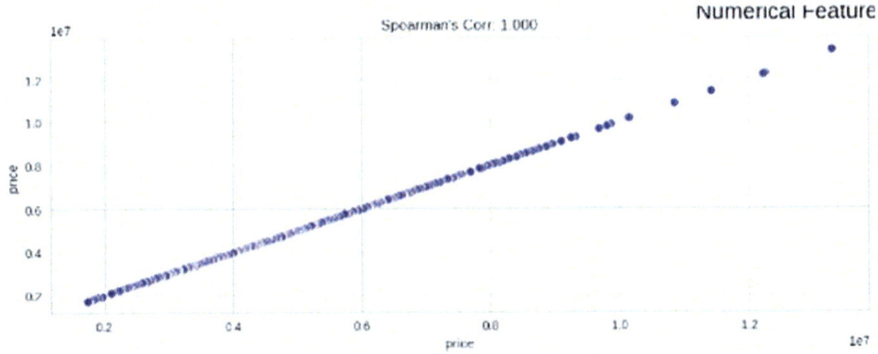

Fig. 4 Price correlation using Spearman's coefficient

4 Modelling and Evaluation

After performing the necessary data pre-processing, the dataset was split. This was done using the train_test_split method from the sklearn library. The dataset was split into training data and testing data in the ratio of 80:20 respectively. There are several ways of splitting the data. These include 80 and 20%, 75 and 25%, 70 and 30%, etc.

Though there are no stamped rules to the proportion to split, an 80 and 20% data split handout gives a better model performance. Thus, in this study, the 80 and 20% split rule for training and testing data respectively were employed. The first 80% of the data was used for the model training while the 20% was in the testing of the data. It was necessary to test the trained data to ensure the model can be useful in prediction.

Since the target feature, 'price' is a continuous variable, various regression models were implemented and evaluated. These models are Extra gradient boosting (XGBoost), Gradient Boosting Regressor, Random Forest Regressor, SVM Regressor, and Linear regression. After performing the necessary data pre-processing on the dataset, various models were implemented and evaluated. These models are Extra gradient boosting (XGBoost), Gradient Boosting Regressor, Random Forest Regressor, SVM Regressor, and Linear regression.

Random Forest Regressor

Random Forest Regressor (RF) is an ensemble machine learning algorithm used for regression problems. It combines the predictions of multiple decision trees to make a final prediction. It is a simple regressor that can capture the relationships between various features and target features. It gives a robust prediction and ensures a minimised overfitting occurs.

The equation for a Random Forest Regressor can be represented as follows:

$$f(x) = 1/N * \Sigma\_i(f\_i(x)) \tag{1}$$

where:

f(x) is the final prediction for a given instance x
f_i(x) is the prediction of the i-th decision tree in the ensemble for
N is the number of trees in the ensemble
Σ_i represents the sum over all trees in the ensemble.

Linear Regression

Linear Regression is a simple regression model used for regression problems. This model the relationship between a dependent variable (the target) and one or more independent variables (the features) useful for prediction. It assumes a linear relationship between the target and other features and finds the best line that fits the data. It is easy and efficient and but it is limited to computational resources.

The equation for a simple Linear Regression model can be represented as follows:

$$y = \beta\_0 + \beta\_1x\_1 + \beta\_2x\_2 + \ldots + \beta\_px\_p + \varepsilon \tag{2}$$

where:

y is the target value
x_1, x_2, ..., x_p are the feature values
β_0, β_1, β_2, ..., β_p are the coefficients (weights) of the model
ε is the residual or error term, representing the difference between the true target value and the predicted value.

K-Nearest Neighbors

K-Nearest Neighbors (KNN) is a non-parametric supervised learning algorithm used for prediction in regression and classification problems [12]. It predicts a given query by finding its K nearest neighbors (data points) and applying the average values of the data points or the majority vote for regression and classification purposes, respectively.

Gradient Boosting Regressor (GBoost)

This is an ensemble machine learning algorithm used for regression problems. It can handle both linear and non-linear relationships between the target and other features, making it a suitable algorithm for house price prediction that involve complex relationship between variables. It also captures interactions between various features and improves the model's predictive accuracy.

The equation for a Gradient Boosting Regressor can be represented as follows:

$$f(x) = f0(x) + \Sigma\_i(g\_i(x)) \tag{3}$$

where:

f(x) is the final prediction for a given instance x

f0(x) is the initial prediction, which is typically the mean of the target values for all instances in the training data

g_i(x) is the contribution of the i-th decision tree in the ensemble to the final prediction for instance x

Σ_i represents the sum over all trees in the ensemble.

Extra Gradient Boosting Regressor (XGBoost)

The Extra gradient boosting regressor is a machine learning algorithm widely used for classification and regression problems. It is highly efficient and solves real-world problems, like house price prediction.

The basic equation for prediction in XGBoost is as follows:

$$f(x) = \Sigma(w\_i * h\_i(x)) \tag{4}$$

where:

f(x) is the final prediction for a given instance x

w_i is the weight assigned to each decision tree h_i(x) in the ensemble

h_i(x) is the prediction of the i-th decision tree in the ensemble for instance x.

SVM Regressor

Support Vector Machine (SVM) Regressor is a machine learning algorithm for regression problems. The SVM regressor follows the classification principle of Support Vector Machines but finds the best line or hyperplane to approximate the target values.

SVM Regressor can handle linear and non-linear relationships between the target and the other features. It also handles high-dimensional feature spaces effectively. It also makes use of kernel functions.

The equation for a Support Vector Machine (SVM) Regressor can be represented as follows:

$$f(x) = w^{Tx} + b \tag{5}$$

where:

f(x) is the predicted target value for a given instance x
w is the weight vector
x is the feature vector
b is the bias term
T represents the transpose operation.

Evaluation Metrics

This research used Root squared (R2) and Root mean square error (RMSE) for the evaluation metrics of the models' performance.

Root squared (R2) score measures the linear relationship between X and y. It is the proportion of the variance in the dependent variable that is predictable from the independent variable. It is used in regression to check how best a model fit data. R2 score ranges from 0 to 1 with 1 being the best fit while 0 is the worse fit. It can also be used to confirm the performance of different models. The R2 also work with other evaluation metrics like RMSE.

The RMSE is used in evaluating the performance of models. It measures the deviation of the predicted value from the actual value. The lower an RMSE is the more accurate the model [12].

Residual: Residual in regression shows the variation between the actual value and the predicted value. A regression line tries to minimise the residual value. A residual value of 0 shows a better prediction.

Algorithm for the Models

Input *Data for Both Training and Evaluation Are Inputs.*

Output *Root Mean Square Error and R2 Score Are the Result*

Step 1. From sklearn import libraries
Step 2. Import dataset
Step 3. Pre-process the data
Step 4 Split the data into train dataset and test dataset.
Step 5: Use the training and test datasets to fine-tune the algorithm
Step 6. Train the algorithm
Step 7. Predict the result
Step 8. Calculate the root mean square error and the R2 score.

5 Result and Discussion

Table 1 displays the results of the evaluation of the various models using R2 and RMSE performance metrics.

The models are compared in terms R2 Score (R2) and Root Mean Square Error (RMSE) (see Figs. 5 and 6). Table 1 shows that Random Forest (RF) has the highest R2 value and the lowest RMSE among the five models. This means that RF with an R2 score of 99.990158% which is closest to 1, fits better the data, and also its RMSE of 17,726.68 shows RF has a closer predictive value to the actual or target value than other models. The Gboost, XGboost, Linear regression and KNN also show good performance in house price prediction as their R2 score is close to 1. The Linear regression model and KNN almost have the same R2 and RMSE results. The SVM gave an R2 score of −4.107443 and an RMSE of 18,232.06. For SVM, though its low RMSE value indicates it can give an accurate model it's considered the worst performing model as it has a negative R2 value. We can see that the RMSE gave a rather high range of value and as such deem unfit to measure the model's performance. Therefore, we focus only on using the R2 score to measure performance of the model. Random Forest with the best R2 value proved a better prediction between the actual and the predicted value (see Fig. 7). It has a residual value hovering on 0 (see Fig. 8). When the residual value was compared with price, it showed a minimized error tending to 0 (see Fig. 9). RF is considered the most effective house price prediction model among the 5 models.

Table 1 Evaluation of the performance of the models

Model	R-squared score (%)	RMSE (test)
Random forest	99.990158	17,726.68
Gboost	99.987087	20,305.64
XGboost	99.971619	30,102.80
Linear regression	94.293882	42,684.02
KNN	94.293875	42,684.04
SVM	-4.107443	18,232.06

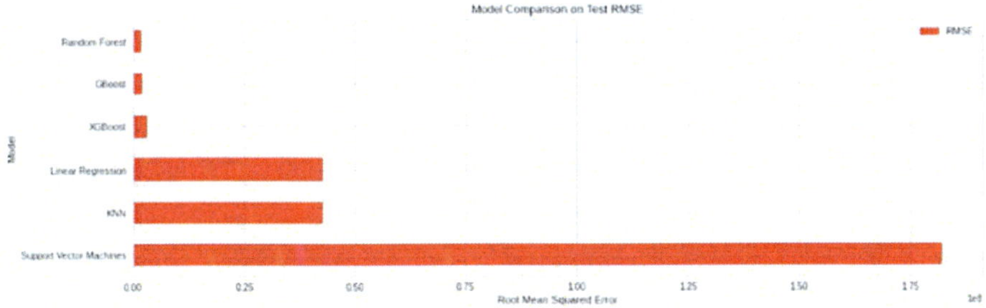

Fig. 5 The performance of the models

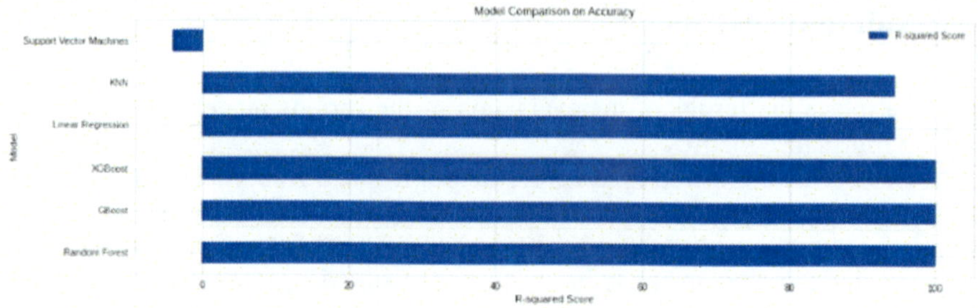

Fig. 6 The accuracy of the models

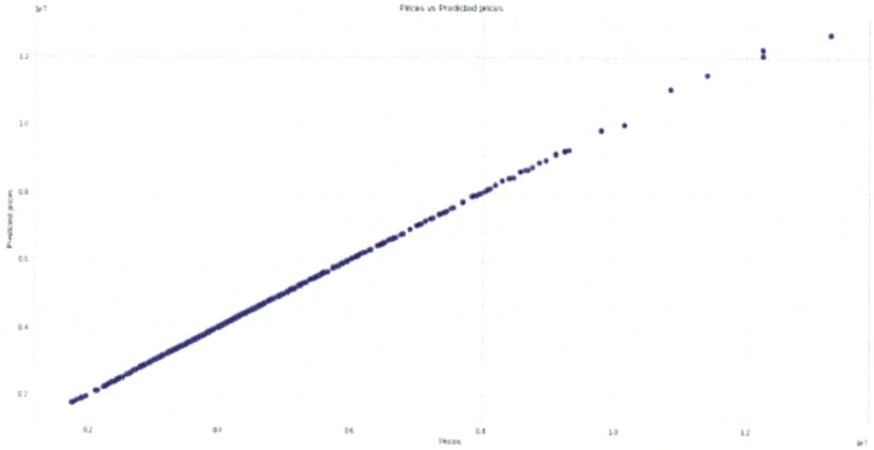

Fig. 7 Random forest showing difference between actual and predicted price

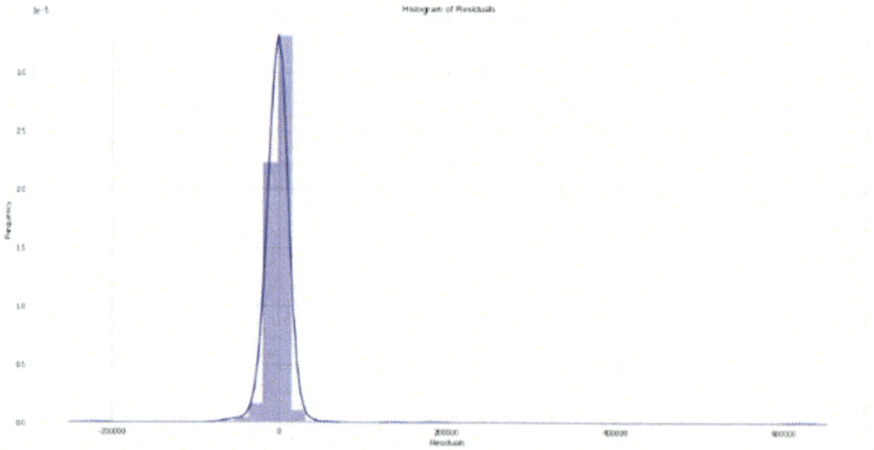

Fig. 8 Random forest showing histogram of residuals

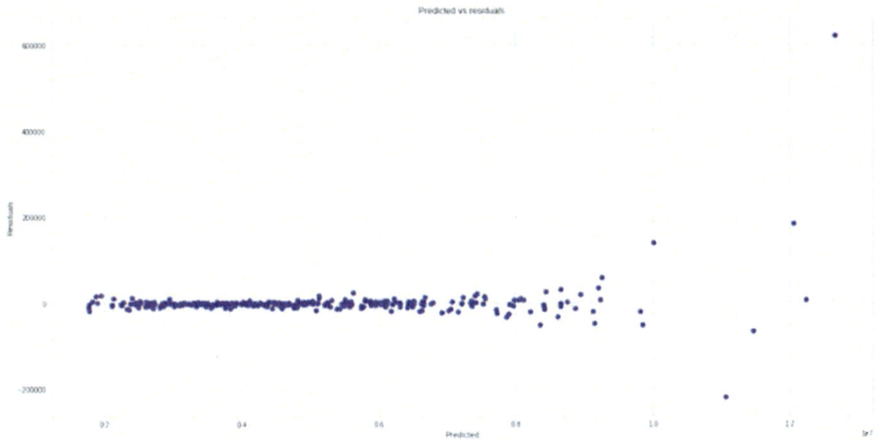

Fig. 9 Random forest showing predicted versus residuals

6 Conclusion

The application of machine learning is important in the prediction of house prices, for both sellers and customers, to maximise profit and make better informed decision respectively. Several factors have affected the pricing of houses. This research gave an effective house prediction model than was shown in [12] which implemented CatBoost with R2 of 0.82 on a different dataset. We employed various machine learning models to build an effective house price prediction model. The R2 and RMSE evaluation metrics were used to train the models to ascertain their performance. While XGboost, Gboost, Linear regression model and KNN show promising predictive house price performance with their R2 closer to 1, Random Forest regressor outperformed them with R2 score of 99.990158% (0.99990158). The SVM gave the worst performance with a negative R2 score. Thus, the Random Forest regressor is considered an effective predictive model for house prices. For improved performance, other advanced regression methods may be considered.

Acknowledgements This work was supported by Key Laboratory of Advanced Manufacturing Technology of the Ministry of Education of Guizhou University (GZUAMT2022KF[07]), the National Natural Science Foundation of China (No.61862051), the Science and Technology Foundation of Guizhou Province (No.[2019]1299, No.ZK[2022]449), the Top-notch Talent Program of Guizhou province (No.KY[2018]080), the Natural Science Foundation of Education of Guizhou province(No.[2019]203) and the Funds of Qiannan Normal University for Nationalities (No. qnsy2019rc09). The Educational Department of Guizhou under Grant NO. KY[2019]067.

References

1. Chen, Y., Xue, R., Zhang, Y.: House price prediction based on machine learning and deep learning methods. In: 2021 International Conference on Electronic Information Engineering and Computer Science (EIECS), Changchun, China, pp. 699–702 (2021). https://doi.org/10.1109/EIECS53707.2021.9587907
2. Rana, V.S., Mondal, J., Sharma, A., Kashyap, I.: House price prediction using optimal regression techniques. In: 2020 2nd International Conference on Advances in Computing, Communication Control and Networking (ICACCCN), Greater Noida, India, pp. 203–208 (2020). https://doi.org/10.1109/ICACCCN51052.2020.9362864
3. Luo, Y.: Residential asset pricing prediction using machine learning. In: 2019 International Conference on Economic Management and Model Engineering (ICEMME), Malacca, Malaysia, pp. 193–198 (2019). https://doi.org/10.1109/ICEMME49371.2019.00046
4. Yu, D., Wang, Z., Wei, W.: House price prediction based on a machine learning model. In: 2021 2nd International Seminar on Artificial Intelligence, Networking and Information Technology (AINIT), Shanghai, China, pp. 391–395 (2021). https://doi.org/10.1109/AINIT54228.2021.00082
5. Manasa, J., Gupta, R., Narahari, N.S.: Machine learning based predicting house prices using regression techniques. In: 2020 2nd International Conference on Innovative Mechanisms for Industry Applications (ICIMIA), Bangalore, India, pp. 624–630 (2020). https://doi.org/10.1109/ICIMIA48430.2020.9074952
6. Iwendi, C., Huescas, C.G.Y., Chakraborty, C., Mohan, S.: COVID-19 health analysis and prediction using machine learning algorithms for Mexico and Brazil patients. J. Exp. Theor. Artif. Intell. (2022). https://doi.org/10.1080/0952813X.2022.2058097
7. Housing Prices Dataset. https://www.kaggle.com/datasets/yasserh/housing-prices-dataset?select=Housing.csv
8. Sawant, R., Jangid, Y., Tiwari, T., Jain, S., Gupta, A.: Comprehensive analysis of housing price prediction in Pune using multi-featured random forest approach. In: 2018 Fourth International Conference on Computing Communication Control and Automation (ICCUBEA), Pune, India, pp. 1–5 (2018). https://doi.org/10.1109/ICCUBEA.2018.8697402
9. Sanyal, S., Kumar Biswas, S., Das, D., Chakraborty, M., Purkayastha, B.: Boston house price prediction using regression models. In: 2022 2nd International Conference on Intelligent Technologies (CONIT), Hubli, India, pp. 1–6 (2022). https://doi.org/10.1109/CONIT55038.2022.9848309
10. Wang, Y.: The comparison of six prediction models in machine learning: based on the house prices prediction. In: 2022 International Conference on Machine Learning and Intelligent Systems Engineering (MLISE), Guangzhou, China, pp. 446–451 (2022). https://doi.org/10.1109/MLISE57402.2022.00095
11. Durganjali, P., Pujitha, M.V.: House resale price prediction using classification algorithms. In: 2019 International Conference on Smart Structures and Systems (ICSSS), Chennai, India, 2019, pp. 1–4 (2019). https://doi.org/10.1109/ICSSS.2019.8882842
12. Kumar, G.K., Rani, D.M., Koppula, N., Ashraf, S.: Prediction of house price using machine learning algorithms. In: 2021 5th International Conference on Trends in Electronics and Informatics (ICOEI), Tirunelveli, India, pp. 1268–1271 (2021). https://doi.org/10.1109/ICOEI51242.2021.9452820
13. Madhuri, C.R., Anuradha, G., Pujitha, M.V.: House price prediction using regression techniques: a comparative study. In: 2019 International Conference on Smart Structures and Systems (ICSSS), Chennai, India, 2019, pp. 1–5 (2019). https://doi.org/10.1109/ICSSS.2019.8882834

Colour Trend Analysis Using Machine Learning and Histogram

Shirin Abolfath Zadeh, Celestine Iwendi, Ikpenmosa Uhumuavbi, and Zakaria Boulouard

Abstract A significant aspect of colour prediction is the process of detecting a colour palette that represents a collection at a fashion show. People do this manually, but often with too many images, it becomes a hard task. An automated machine-learning method has been developed to generate colour palettes for fashion shows. the model was obtained by fashion pictures dataset, representing each of 48 images of her from a particular fashion show. This work can be extended to analyse millions of images from social media feeds and provide data-driven insights for colour prediction.

Keywords Fashion · Colour · Trend · Machine learning · Big data · K-Means · Clustering · Predict · Histogram · Python

1 Introduction

Machine learning and artificial intelligence are advanced research topics motivated by real-world dilemmas. Machine learning (ML) helps computers to predict based on provided data sets or past experience. ML techniques -supervised and unsupervised -have the power to handle large amounts of data and solve classification problems. In past years, big data has become progressively critical in the fashion era. It is increasingly used in trend forecasting, supply chain management, and analysing customer behaviour. While the needs of customers are constantly changing, people like clothes based on their own style. As result, fashion companies pay a lot of money due to excess inventory becoming obsolete due to changing trends.

Big data helps the fashion industry in many ways. Madsen [1] declare that the fashion industry is able to make the power of Big Data to distinguish markets, analyse trends, and enhance cross-selling and sustainability. Trend forecasting is important in

S. A. Zadeh (✉) · C. Iwendi · I. Uhumuavbi
School of Creative Technologies, University of Bolton, Bolton BL3 5AB, UK
e-mail: Shirin.abolfathzadeh@ieee.org

Z. Boulouard
LIM, Hassan II University of Casablanca, Casablanca, Morocco

C. Iwendi et al. (eds.), *Proceedings of ICACTCE'23—The International Conference on Advances in Communication Technology and Computer Engineering*, Lecture Notes in Networks and Systems 735, https://doi.org/10.1007/978-3-031-37164-6_33

fashion and Big Data helps it by plotting periodic variations and giving an indication of general trends that could materialize. In addition, Big Data like Google Trends is able to enhance if overburdened properly [2]. There are so many opportunities waiting to be tapped as fashion brands start to get involved and take advantage of big data [3]. Extracting attributes like comments on social media is a facility to measure customer feedback on their purchase and is valuable for companies [1].

2 Literature Review

2.1 Power of Big Data in Fashion

As [4] argued, at one glance, the significant impact of big data in the fashion industry could be listed in Fig. 1. one of the most important of them is "Trend Forecasting", which is the process of analysing and making predictions about the future purchasing habits of the customer [1].

In addition, ML has motivated many aspects of the fashion industry. Trend cycles in fashion are noticeably short, but with the help of ML, trends can be analysed immediately and quantitatively with trustable results. As a result, big data or ML has become necessary to analyse trends [4, 5].

Big Data can be used to study purchaser behaviour across multiple variables. purchaser engagement and shopping time could be tracked on social media and can rate them. Big data can catch more than basic customer data. A wide range of purchaser behaviours can be collected and evaluated, providing insights not only about who the customer is, but who they are as well [6].

Picking suitable colours were one of the most substantial requests of big data. Designers may get sell data from previous gatherings. They could rapidly assess

Fig. 1 Big data in fashion

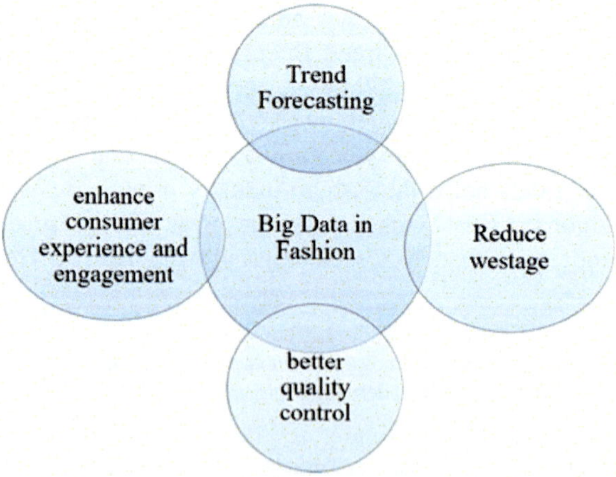

which colours the client's interests. Designers need to be knowledgeable about existing trends, however, it's even more vital for them to know which colours their consumers favour [7]. Fashion designers and producers may use big data to identify the acceptance of colours and adjust their designs to meet their end users' demands [6].

Moreover, selecting the right colour in fashion design is one of the important applications of Big Data. Designers can pull market data from past collections and determine what colour customers want. However, they need to have info about current fashion trends. Fashion designers and manufacturers can use big data to determine the popularity of colours and tailor their designs to meet customer needs. Big data can also be used to help price clothes [6].

2.2 *Big Data's Influence on the Fashion Industry*

Fashion attribute detection has earned growing interest in the computer vision community, which is essential for experts in the fashion industry, especially trend forecasting. Fashion forecasting companies, namely WGSN, discover info about fashion attributes from images through fashion shows, and sales. They evaluate info through experience, observations, media analysis, interviews, and exposure to novel places. Such a process of information analysis is called abstraction, which notes the analogy or contrast between all garments and collections [3].

Fashion Trend Analysis Based on Fashion Attributes Early efforts on trend analysis broke down images from New York fashion shows to identify style trends in high fashion. Recent advances in A.I. Algorithms have empowered more work in this area [8]. Numerous recent methods have used AI. algorithm to extract clothing attributes from images and create visual integration of clothing style groups to study worldwide clothing fashion trends. However, it is not entirely clear whether the trends detected in these studies are accurate [3].

Therefore, all data related to a fashion product is called fashion data. This data can be used for trend analysis, client behaviour analysis, forecasting, and more. The fashion industry generates many different data sources. All this data comes in various forms like words, images, etc. As this is the era of fast fashion, data is growing and developing rapidly. The era of fashion data could be depicted in Fig. 2, as it is shown the highlighted part is about colours, which is studied in this report [8].

2.3 *Colour as Fashion Attribute*

An important attribute in a fashion that benefits its trends is the "Colour". Colour is a capable sales tool, mainly in the fashion industry, where products are aimed at visually inspiring consumers. Colour cycle analysis studies periodic cycles of trends [9]. Traditional fashion colour cycle search and prediction is done by observing and

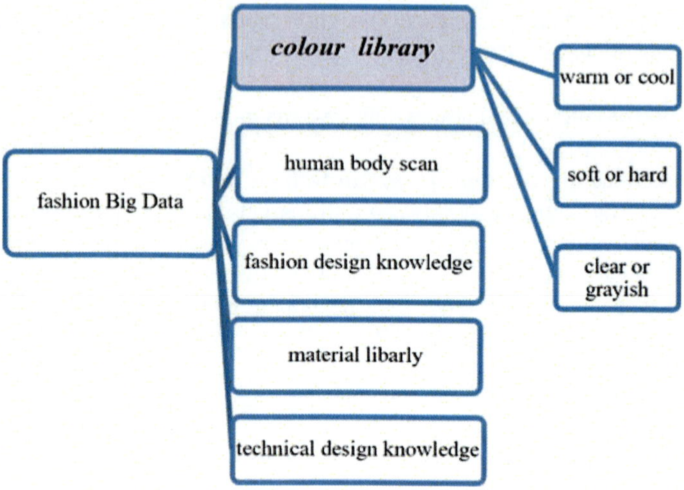

Fig. 2 Data in fashion

concluding from trends evident in fashion shows. Widespread access to the Internet and digital content has democratized access to enormous amounts of fashion data, and the growing speed of digital information has accelerated the cycle of fashion trends. Therefore, there is a need for a quick and capable way to comprehensively summarize fashion show trends [10].

Colour prediction advises manufacturers to offer new products while avoiding loss of gain from products that customers are not into them. Colour forecast has a durable directional signal when several identical trends are observed over the same period [11]. Colour prediction service customers include companies in the apparel supply chain that sell to the region. These customers need information on colour trends to clearly identify consumer flavours [4].

Analysis of image data by computer vision is a common topic in fashion informatics. In this industry, visual-based content and promotional images are generated repeatedly and there is a need to develop potentially substantial amounts of data for image analysis. computer vision researchers are increasingly excited about the application of image data analysis in the fashion while lots of Previous studies have approved this power of computer vision analysis [4].

However, expert opinion plays a vital role in the fashion industry, but trends have a direct impact on business and therefore need to be requested more objectively. Quantitative analytics is becoming increasingly essential due to the improvement of social media. Therefore, it is obligatory to lead a data analysis method in this industry that can act in diverse ways using ML [4].

To detect fashion attributes by AI, the algorithm, called "training", learns to determine the theory of mode description from a dataset of images annotated with certain mode attributes. Then A.I. algorithm can start to predict the properties of the clothes in the new image [3].

3 Materials and Methods

3.1 Methodology of Analysing Fashion Trend

This study employs a quantitative research design and experimental study. This study used 6 steps as a methodology to investigate the issues. The list of the procedures for the suggested models is shown in Fig. 3.

To describe the fashion attributes, the two popular machine learning algorithms called 'clustering' and 'descriptive' can be used. Clustering is a technique which is employed in the field of pattern recognition, information retrieval and image analysis. In this report, by looking at the idea of clustering is used to evaluate modal attributes on images and find similar attributes in the same group (cluster). By analysing clusters, common properties are detected. moreover, this method can be used to analyse fashion properties and trends that are visually correlated in terms of colour perspective [3]. In this algorithm, any fashion images, based on the AI clustering Model, properties like colour and styles would be detected (see Fig. 4) and the focus is just on detecting and using the colour attributes.

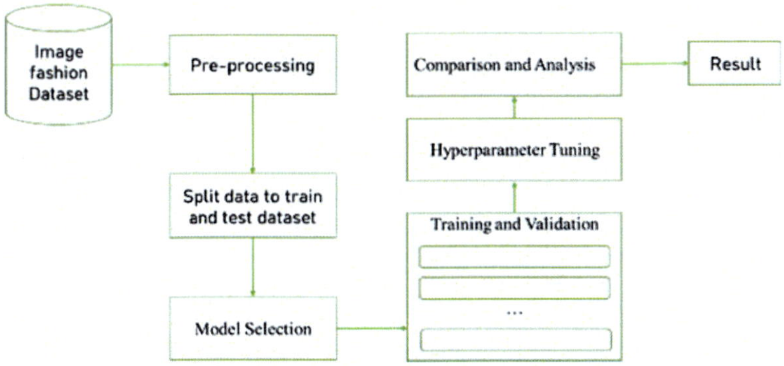

Fig. 3 Methodology of fashion trend analysis

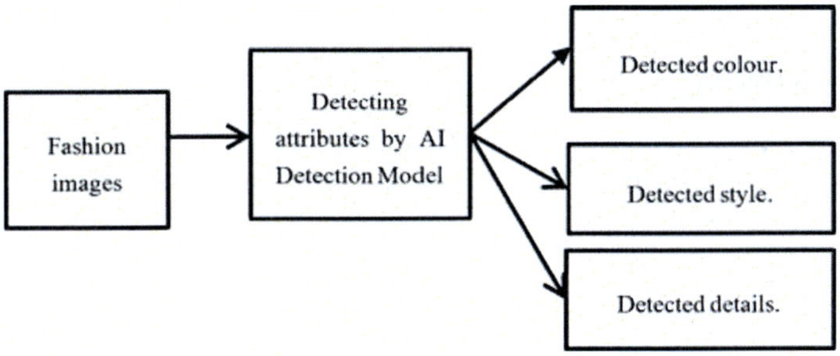

Fig. 4 Detecting attributes, especially colour, from fashion image

Fig. 5 Sample of the result of colour processing by AI

One viewpoint of detecting colour, and analysing it is about using AI to explore the colour data in digital photos of ready-to-wear luxury brands from websites to generate insights about colour trends. In this report, AI is used to analyse and depict the colours based on images as a sample as shown in Fig. 5 [12].

3.2 *Depicting Fashion Trends by Histogram*

A histogram plans a visual depiction of the distribution of data; additionally, it helps to envisage whether the distribution is symmetrical or skewed left or right. In addition, if it is unimodal, bimodal, or multimodal. It can also act outlier or gaps in the data [13].

Colour Clustering

So far, this report has considered models that depend upon training data in the form of features. It is time to show how a Histogram can help to depict colour trends in an advanced way. Based on the concept of clustering methods, data can be put as each point in a group. In different words, data can be categorized as a set of entities, based on their properties, automatically (Unsupervised Learning). Here, based on the most popular and error-proof k-means-based colour clustering method colours were clustered and enumeration. The k-means-based method may have different resolutions, but it can be adjusted to capture the most important colours based on results; Colours are grouped first and then enumerated. Colour names and numbers in parentheses demonstrate colour clusters and colour numbers [12, 5, 14, 15].

K-Means Clustering

K-Means clustering is a vector quantization approach that originated in signal processing and is often used in data mining cluster analysis. K-means clustering intends to divide n experience into k clusters [12]. Each inspection belongs to the cluster with the closest mean and serves as a cluster prototype [16]. This divides the data space into Voronoi cells (see Fig. 6). Based on this algorithm offered the following steps:

1. Get random k points (called centroids) from X.

```
def exact_color_KMeans(img_url):
    img=cv2.imread(img_url)
    img=cv2.cvtColor(img,cv2.COLOR_BGR2RGB)
    img=img.reshape((img.shape[1]*img.shape[0],3))
    kmeans=KMeans(n_clusters=5)
    s=kmeans.fit(img)
    labels=kmeans.labels_
    labels=list(labels)
    centroid=kmeans.cluster_centers_
    percent=[]
    for i in range(len(centroid)):
      j=labels.count(i)
      j=j/(len(labels))
      percent.append(j)
    plt.pie(percent,colors=np.array(centroid/255),labels=np.arange(len(centroid))
            , autopct='%1.1f%%', explode=[0.1,0,0,0,0], shadow=True, startangle=90)
    plt.legend(loc="upper left")
    plt.show()
    Img_Colot_to_histogram(img_url)
```

Fig. 6 Sample code to classify the colour of the image in python based on Kmean

2. Match each point to the nearest centroid. The lately formed bundle of points is called a cluster.
3. Find new centroids for each cluster by computing new centres from the points.
4. Redo steps 2 and 3 until the centre of gravity does not change.

3.3 *Python Implementation*

The basic python implementation of this algorithm is the below code (see Fig. 7) when a file is passed to the function of "exact-colour" to detect the colours and "img_colour_to_histogram" to depict the result. First, in "exact-colour" the palette of the image was extracted and shown in a donate chart. Then, need to get the points, to cluster them by the 'K_means' ML algorithm in the second function. In this research the cluster was set to 5, to find the 5 most used colours in each image. next, in each cluster, the RGB code is calculated and returned. As a result, the list of colours of the image would be returned as a histogram [10].

As a result of the code that was shown in Fig. 7, firstly, the colour can be categorized based on the k-means model. In addition, to draw a Histogram it is better to find the weight of each colour, which could be calculated by the percentages of continence, as implemented by python code and shown in Fig. 8. After the calculation of the weight, all the colours collected are normalized as a percentage, as it is depicted in Fig. 9.

```
def exact_color(input_image, resize, tolerance, zoom):
    #background
    bg = 'bg.png'
    fig, ax = plt.subplots(figsize=(192,108),dpi=10)
    fig.set_facecolor('white')
    plt.savefig(bg)
    plt.close(fig)
    #resize
    output_width = resize
    img = Image.open(input_image)
    if img.size[0] >= resize:
        wpercent = (output_width/float(img.size[0]))
        hsize = int((float(img.size[1])*float(wpercent)))
        img = img.resize((output_width,hsize), Image.ANTIALIAS)
        resize_name =  input_image
        img.save(resize_name)
    else:
        resize_name = input_image

    #crate dataframe
    img_url = resize_name
    colors_x = extcolors.extract_from_path(img_url, tolerance = tolerance, limit = 13)
    df_color = color_to_df(colors_x)

    #annotate text
    list_color = list(df_color['c_code'])
    list_precent = [int(i) for i in list(df_color['occurence'])]
    text_c = [c + ' ' + str(round(p*100/sum(list_precent),1)) +'%'
             for c, p in zip(list_color, list_precent)]
    fig, (ax1, ax2) = plt.subplots(1, 2, figsize=(140,100), dpi = 10)

    #donut plot
    wedges, text = ax1.pie(list_precent,
                           labels= text_c,
                           labeldistance= 1.05,
                           colors = list_color,
                           textprops={'fontsize': 130, 'color':'black'})
    plt.setp(wedges, width=0.3)

    #add image in the center of donut plot
    img = mpimg.imread(resize_name)
    imagebox = OffsetImage(img, zoom=zoom)
    ab = AnnotationBbox(imagebox, (0, 0))
    ax1.add_artist(ab)

    #color palette
    x_posi, y_posi, y_posi2 = 160, -170, -170
    for c in list_color:
        if list_color.index(c) <= 5:
            y_posi += 180
            rect = patches.Rectangle((x_posi, y_posi), 360, 160, facecolor = c)
            ax2.add_patch(rect)
            ax2.text(x = x_posi+400, y = y_posi+100, s = c, fontdict={'fontsize': 190})
        else:
            y_posi2 += 180
            rect = patches.Rectangle((x_posi + 1000, y_posi2), 360, 160, facecolor = c)
            ax2.add_artist(rect)
            ax2.text(x = x_posi+1400, y = y_posi2+100, s = c, fontdict={'fontsize': 190})

    fig.set_facecolor('white')
    ax2.axis('off')
    bg = plt.imread('bg.png')
    plt.imshow(bg)
    plt.tight_layout()
    Img_Colot_to_histogram(img_url)

    return plt.show()
```

Fig. 7 The implementation of the algorithm to pars the image to detect the colour

```
# construct the argument parser and parse the arguments
def Img_Color_to_histogram(img_url):
    # load the image and convert it from BGR to RGB so that
    # we can dispaly it with matplotlib
    image = cv2.imread(img_url)
    image = cv2.cvtColor(image, cv2.COLOR_BGR2RGB)
    image = image.reshape((image.shape[0] * image.shape[1], 3))
    # cluster the pixel intensities
    clt = KMeans(n_clusters =5)
    clt.fit(image)
    # build a histogram of clusters and then create a figure
    # representing the number of pixels labeled to each color
    hist = centroid_histogram(clt)
    bar = plot_colors(hist, clt.cluster_centers_)
    # show our color bart
    plt.figure()
    plt.axis("off")
    plt.imshow(bar)
    plt.show()
```

Fig. 8 Python code for clustering colours of the image by histogram

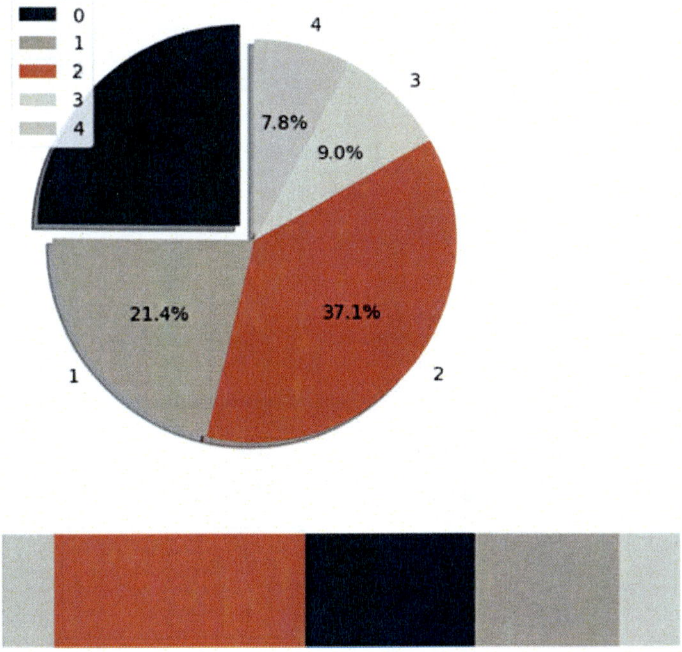

Fig. 9 Analysing colour data by K-Means

3.4 Colour Visualization

First, explain each node in V to show the enumeration colour group and their proportions in the form (colour) for each colour that is displayed on the object to construct a histogram of the object colour data. For simplicity, Hear Nodes contain only colour information, but they are capable of describing ratio, texture, and other features. The geo-location of the gathered colours was then displayed on the graph's edges [12].

The V and E as two matrices can be preserved in the straightforward below constructed to draw a graph, as they are shown in Fig. 8 [12], to make the display and maintenance of the extracted colour information simple. V displays the colour code weights, while E describes how the colours in an image relate to one another. As a result, as is depicted in the right part of Fig. 10, each colour of the image could be shown based on their frequency and their relationship.

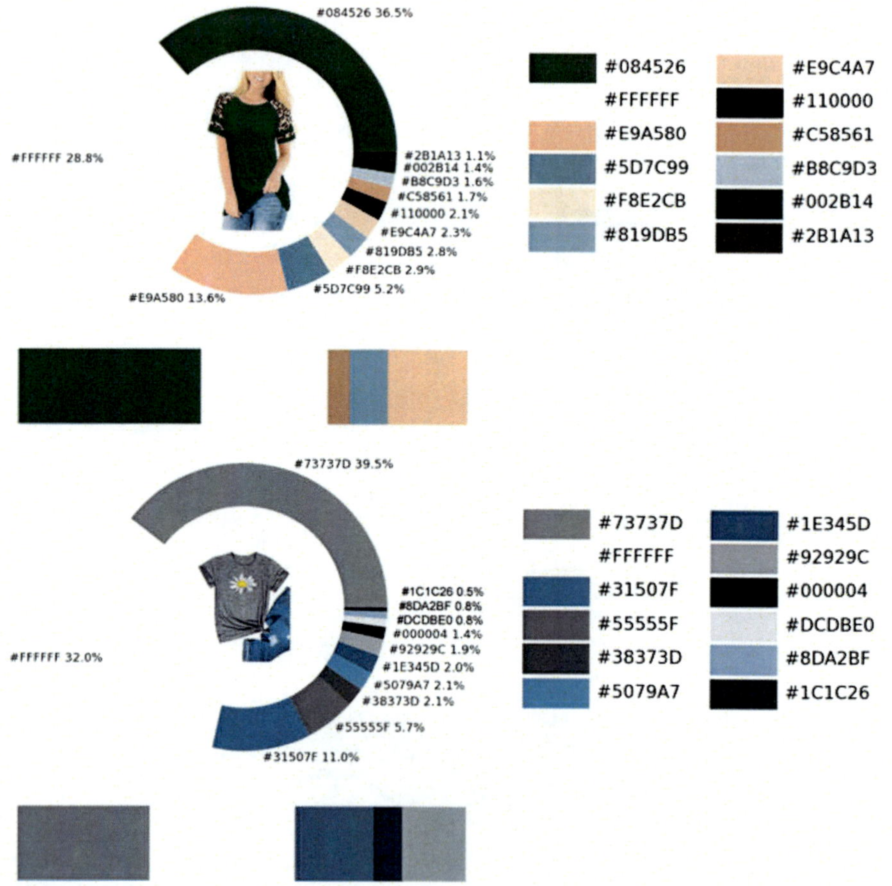

Fig. 10 Depicting the colour processing by donate chart and histogram

Fig. 11 Colour detection by ML in depicting in histogram

The final result is also shown below (see Fig. 11).

Histogram-based colour detection algorithms use a histogram to represent the distribution of pixel values in an image. The histogram of an image is a graphical representation of the number of pixels in the image that have a particular colour or intensity. Machine learning algorithms, such as Multi-Layer Perceptron (MLP), Random Forest (RF), Support Vector Regression (SVR), and Multiple Linear Regression (MLR)-M5, was used to analyse the histogram data and detect colours in an image.

The detection of colours using machine learning is a popular technique that has various applications in image processing, computer vision, and object recognition. Among the different algorithms used for this purpose, MLP is a type of feedforward neural network that has gained wide acceptance. The training of MLP algorithm involves feeding it with a dataset of labelled colour images, enabling it to learn and identify different colour clusters by analysing the input histogram data. Other popular algorithms for colour detection include Random Forest and Support Vector Regression, which employ decision trees and statistical models to analyse the histogram data and identify various colour s in an image. MLR-M5 is another algorithm that uses a combination of linear regression and decision trees to detect colour s in an image. Machine learning algorithms are efficient in detecting colour s in an image using histogram-based techniques. The comparison of the four machine learning algorithms' results is presented in Fig. 12, showing that MLP provided the best model, achieving an accuracy rate of approximately 0.93%.

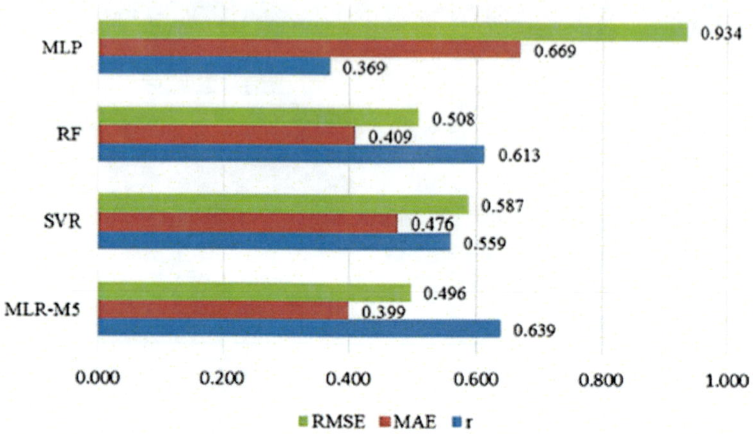

Fig. 12 The comparison of the results of the four machine learning algorithms

4 Conclusion

While big data has a significant effect on any industry, the key role of the on market cannot be denied and the fashion industry is not exempted from this concept. Fashion and mode, like other markets, need to improve, not only by modern design and novel patterns but also by taking advantage of technology. Technology is not only software to use in designing websites to increase the selling rate. big data, especially analysing fashion data and improving fashion design based on data and Machine learning are other advantages of technology. While trends are so important nowadays, finding ways to classify and follow trends could help designers and fashion companies to economical approaches. As discussed in this report, trend colour could be analysed by Machine learning. Clustering is the best way to classify the colour attributes of any image on social media to identify the colour trend. Histograms on the other hand, add value to that through their capacity to depict it and make it more efficient.

References

1. Silva, E.S., Hassani, H., Madsen, D.Ø.: Big data in fashion: transforming the retail sector (2019)
2. Giri, C., Thomassey, S., Zeng, X.: Customer analytics in fashion retail industry (2018)
3. Mengyun, S., Van, D.L.: Using artificial intelligence to analyse fashion trends (2020)
4. Han, A., Kim, J., Ahn, J.: Colour trend analysis using machine learning with fashion collection images (2022)
5. Mittal, M., Iwendi, C., Khan, S., Rehman Javed, A.: Analysis of security and energy efficiency for shortest route discovery in low-energy adaptive clustering hierarchy protocol using Levenberg-Marquardt neural network and gated recurrent unit for intrusion detection system. Trans Emerging Tel Tech. **32**, e3997 (2021). https://doi.org/10.1002/ett.3997

6. Jitender, K., Simran, S.: Big data in fashion industry (2022)
7. Oh, K.: The roles of data analytics in the fashion industry (2020)
8. Hyosun, A., Kim, S., Choi, Y.: Sportive fashion trend reports: a hybrid style analysis based on deep learning techniques (2021)
9. Jain, S., Bruniaux, J., Zeng, X., Bruniaux, P.: Big data in fashion industry (2017)
10. Wong, M.Y., Zhou, Y., Xu, H.: Big data in fashion industry: colour cycle mining from runway data (2016)
11. Lai, P., Westland, S.: Machine learning for colour palette extraction from fashion runway images (2020)
12. Chen, Y., Dai, Y., Li, L., Ma, C., Liu, X.: A graph-based representation method for fashion colour (2020)
13. Kolossváry, M., Karády, J., Kikuchi, Y., Ivanov, A., Schlett, C.L., Lu, M.T., Foldyna, B., Merkely, B., Aerts, H.J., Hoffmann, U., Maurovich-Horvat, P.: Radiomics versus visual and histogram-based assessment to identify atheromatous lesions at coronary CT angiography: an ex vivo study. Radiology **293**(1), 89–96 (2019)
14. Awotunde, J.B., Misra, S., Obagwu, D., Florez, H.: Multiple colour detection of RGB images using machine learning algorithm. In: Applied Informatics: 5th International Conference, ICAI 2022, Arequipa, Peru, October 27–29, 2022, Proceedings. Springer International Publishing, Cham, pp. 60–74 (2022)
15. Si, J., Yu, L., Wang, Y., Huang, Z., Homewood, K., Gao, Y.: Colour centre controlled formation of stable sub-nanometer transition metal clusters on TiO2 nanosheet for highly efficient H2 production. Appl. Surf. Sci. **511**, 145577 (2020)
16. Kamarudin, N.D., Ooi, C.Y., Kawanabe, T., Odaguchi, H., Kobayashi, F.: A fast SVM-based tongue's colour classification aided by k-means clustering identifiers and colour attributes as computer-assisted tool for tongue diagnosis. J. Healthcare Eng. (2017)

A New AI-Based Approach for Rental Tax Evasion Management in Iran (Ethical Consideration)

Shirin Abolfath Zadeh, Celestine Iwendi, Ikpenmosa Uhumuavbi, and Zakaria Boulouard

Abstract In any nation, some parts of society are unsatisfied with the idea of paying taxes. Some authorities have tried to fill the gaps in law by adopting techniques to discourage tax evasion and tax fraud. In this case, technology, especially big data have been used to enable tax collection and regulation. This paper studied the pivotal role of big data in rental tax evasion management in different countries. In order to solve the issues, this paper investigates tax evasion in the rental era and then applies the GIS approach in describing geo-behaviours, social connections, and the interaction between taxpayers and properties. The idea is to reduce evasion and fraud in Tax Management in Iran by applying the 'National Licensing Schema' for landlords, using Tax Profiling System combined with GIS and GraphDB. This is to identify the landlord's and tenants' relationship.

Keywords Tax · Ethical issues · Rental income tax · Tax evasion · Tax fraud · Big data · Machine learning · Iran · Tax administration · Artificial intelligence (AI)

1 Introduction

Paying tax is one responsibility of anyone in society which is based on income, wealth, or transaction as a society membership fee. According to this paper, a big part of ethical issues in taxation is related to fraud, evasion, and avoidance of tax. A group of people are unsatisfied with the service that they gain after paying their taxes, but some more are not satisfied because of the "unfair paying of tax". This subject is about paying different amounts of tax by people, which are not balanced with their income or breaking the tax laws. Therefore, they know governments and authorities as the responsible agents to avoid these problems [1–3].

S. A. Zadeh (✉) · C. Iwendi · I. Uhumuavbi
School of Creative Technologies, University of Bolton, Bolton Postcode: BL3 5AB, UK
e-mail: Shirin.abolfathzadeh@ieee.org

Z. Boulouard
LIM, Hassan II University of Casablanca, Casablanca, Morocco

C. Iwendi et al. (eds.), *Proceedings of ICACTCE'23—The International Conference on Advances in Communication Technology and Computer Engineering*, Lecture Notes in Networks and Systems 735, https://doi.org/10.1007/978-3-031-37164-6_34

In the case of rental tax, why do some landlords have to pay their exact tax amount, while some avoid or hide it? Why do some individual landlords of small properties pay their taxes but some big letting companies are able to use gaps in laws or hide their earnings?

The solutions lay in the IT-based service and facilities which can be used by governments. Naming these kinds of taxpayers based on the new techniques is one solution which is used by different countries. Using AI, supplying national licensing for owners and using GIS are common these days in the case of identifying and detecting rental tax issues.

2 Literature Review

2.1 Rental Tax

People and businesses who have any revenue, are involved to pay their taxes to authorities to support governmental activities, namely construction, Security, and Medicare. It is significant to consider there are numerous taxes from a financial point of view, involving payroll taxes, income taxes, wealth taxes, and sales taxes. In Iran, and some other countries, rental income is subject to taxation.

An income tax is a tax earned on a taxpayer concerning the income or profit collected by the taxpayer. Generally calculated as the product of tax rate and taxable income. A special division of income tax is the rent of real estate. The money that property owners collect from properties is preserved as unearned revenue. Proprietors are involved to prepare (file) a self-assessment tax return and must pay their taxes on the turnover made from their houses or flats [1].

Moreover, Rental income is the earn that landlords catch from renters. This forms any spending for the use of equipment, and charges for added help like cleaning, heating, and in-need repairs. This payment depends on how much profit landlords make and their personal circumstances. Managing this specific type of tax requires controlling and improving the performance of the housing rental market, not only real estate, but also landlords' information must be in control [3].

Governments need to keep track of property finances to manage better rental taxation and predict how much funds they would have based on this type of tax for the next years. however, there are some people or businesses that avoid paying their taxes by hiding the exact amount of their earnings or not filling out the self-assessment. These actions have effects on different areas such as government funds and social services, therefore, it is Tax Authorities to find and address these ways and taxpayers [4].

However, taxation on rental house or flats are regular across the globe, but guidelines are different from country to country. For an instance, in Iran, it is different as the tax change on the amount of rent earned and the size of the estate, and the type of family relationship between the landlord(s) and tenants. In each country, these roles would be in different situations. As an example, in UK and Iran, if someone has more than one estate to let others, the revenues and costs from those would be included together. The Czech Republic has a sizeable supportive housing part or companies, as it was seen as a personal rental estate, and the rental earnings after the subtraction of all charges would be taxed [5].

In some nations, this type of income is added to other sources and is taxed in total, but in Iran, Greece, Hungary, Ireland, Italy, Norway, Romania, and Sweden as well as in Belgium and Switzerland, rental income is taxed distinctly [1].

2.2 *Tax Avoidance, Evasion and Fraud*

Most Issues in Tax Administration (TA) are connected to Evasion, Avoidance and Fraud. it is an argument for the TA in any nation and has an impression on the in-service market that pays their tax fairly [6]. In addition, surges the cost of production and, as a result, increases the price of products and services contrasted to those of non-paying participants' Taxes and Duties.

- Tax evasion is criminal; when people or companies intentionally fail to announce or record what they owe, including hidden economy, when a person completely conceals a taxable activity.
- Tax avoidance associates breaking the taxation rules to obtain a tax advantage that Parliament never envisaged. It's about obeying the letter, but not the spirit of the law.
- Tax Fraud happens when the taxpayer files incorrect data on their self-assessment tax returns, which can be including preparing records to overstate their expenses or abandoning to paper complete earnings. At one glance, when a taxpayer knowingly or intentionally gives wrong info, it might reflect tax fraud [7].

While many landlords pay their taxes, Tax evasion is rampant around the world by some people or businesses. When it comes to rental tax, some property owners must pay an accurate amount, while others escape or ignore it. It is an important question for people, why do some let companies take advantage of legal loopholes or hide their income? And the response, all the time, is related to governments [2, 8].

2.3 *Ethical Issues Raised*

Abstaining taxes by the "Turing point" of the tax system rules and taking advantage of legal loopholes sometimes is not criminal, but many consider it more literal than legal. Any big or small business is supposed to pay constitutional taxation, while this matter falls into the realm of ethics as they can select how to take taxation laws and can extend to where they pay their taxes [7].

As [8] argued the gigantic winners in this negative competition are rich people, and multinational corporations, who abuse position and impact to take advantage of legal loopholes, who have to pay the most taxation alternately boost earnings by paying less as can or covering parts. And as some tax avoidance applies are unlawful, many exploit weaknesses in nowadays tax system—with no interrupting the laws.

On the other hand, based on the [9], according to companies -bigger taxpayers-tax ethical issues can be clearer. It is an idea for businesses to have a transparent announcement about their tax payment, especially for those that due to achieve according to a set of ethical values. Part of businesses handles taxes as a concern or engagement. The majority simply state that they are committed to elusive banded evasion and observing the legal obligations of the nations in which they run.

In addition, one of the major issues facing TAs is the ethical disagreement between the request to gather and custom data to recover performance and client service and the responsibility to regard the privacy of those involved [9]. It is the TAs responsibility to society to detect it. But the basis of any technique for identifying or detecting is Data. TAs need to prepare data or gain access to data that is in the systems of the agents, like councils, banks, real estate, markets and so on. This kind of access brings some worries in society about the privacy of data and "Data ethical issues".

Jonathan [10] claimed that data ethics can be described by 2 core principles that lead to the formation of data criteria:

- Privacy and Private do not necessarily mean classified. Certifying privacy is an element of defining and inducing information rules, not only about data collection but also about data usage and detainment. People need to control the flow of personal information through large third-party analytics systems.
- Shared private info can remain Secret. It is unrealistic to think of info as classified or shared, wholly public, or secret. Data is shared or activated with trusted services, only because data is developed and shared, it does not mean that it is allowed to do everything.

2.4 Solutions—Strategic Transitions

The commercial concept of tax evasion states that people and businesses only pay taxes because they are forced to do so. This is because they will be prosecuted if they do not. Pay their taxes. If so, the likelihood of being arrested for tax evasion and the scale of penalties for being charged by crime must be high enough to prevent tax evasion [11].

Tax evasion is detailed as knowingly failing to pay taxes by failing to disclose or declare income, by charging prohibitive costs to diminish taxable earnings, or by claiming tax credits to which one is not entitled. This is irregular and punishable by law. Individuals may evade taxes due to high rates, complex tax regimes, or bad practices. Tax avoidance allows taxpayers to organize their finances to pay less. This is done entirely by exploiting existing tax law loopholes [12].

These days, not only the loopholes in the existing tax legislation but also the lack of using technology to implement the high-tech approach in identifying taxpayers might lead to Evasion or Fraud. Therefore, TAs are responsible to use technologies like Data Mining to advance the use of their Data [13].

Solution 1: National Licensing Scheme for Landlords

In case of evasion in Rental Tax, one part of the problem is recognizing who is the landlord. a potential solution could lie in the UK landlord approach. Scotland, Wales, Northern Ireland, and England already have landlord registration systems. This will allow HMRC to verify who the landlord is and expeditiously verify if the landowner has declared property earnings on the tax return or filed the tax return [7].

To see how this works in process, Newham in London is a good case. Newham Council introduced a mandatory landlord licensing scheme across the UK in 2013. Before this program, the UK predicted 30,000 rental landscapes. But after that, they found nearly 50% of the total, had not signed up to self-report their rental earnings, and HMRC said £115 m in unpaid tax has been recovered in cooperation with the county. Therefore, in England, from 2018, any property-owner who lets a house or flat to at least 5 people from at least 2 households has to be licensed by local authorities. It helps the government to identify high-risk property owners [7].

This solution solves the problem of not filing the tax return owner because by this certificate they were forced to sign up for their self-report rental earnings and property owners would be identified by TAs. However, even if the governments have no intention of introducing a national landlord licensing scheme, there is much more that could be done. First, TAs can use already available data more effectively. A lot of data is previously available—land registers, real estate company reports, rent guarantee schemes—but these are not kept in a single record and are only referenced during individual HMRC enquiries [7].

Solution 2: Geographic Information System (GIS)

The rental income data, like most other datasets, has a geographical component. The use of Geographic Information Systems (GIS) can therefore help provide the info needed to support decision-making on a variety of tax issues. With the latest developments in geospatial technology, it is possible to link data from different fields and sources and use the resulting database to query tax evaders by address, it is now possible to view, analyse and even find [12].

The first action in identifying tax fraud is recognizing the exact address of them. These locations can be clearly identified on the map by their coordinates. By using the data produced by the database, queries can identify individuals who paid their taxes on time and those who did not. For example, if an organization's name appears in the registrar's database but not the TA's database, the organization most likely does not pay taxes [12].

Solution 3: AI for Evasion Detect

AI has the power to change the work of tax gathering and auditing. Tax gathering and control are two assumed tasks by TAs that could be accomplished by AI. The depletion of AI in TA is powerful to add value to TA, grow tax proportions, identify tax avoidance and evasion, and inspire taxpayers to fill out their tax papers [14]. AI can be used in different models to boost services for IT-based taxpayers. AI in the TA system could support a joined database and build harmony with organisations or institutions, better than other consistent parties. For example, the core TA system, which could aid TAs and taxpayers by including AI [15].

Try out to apply AI in TA has been directed in several nations. AI has been deployed by China as a robot to help with tax routines. This robot could confirm a taxpayer's identity, including providing insightful replies to taxpayer queries. In addition, India has started integrating AI into its tax-management system to execute repeated file treating, gather data and decide if taxpayers' tax returns are accurate [15].

Analysis of Ethical Solutions to Be Provided

In some countries like Iran where governments trust people and asked them to declare their income to calculate their taxes, problems related to tax fraud and tax evasion have happened. Sometimes people or companies do not sign up their-self as taxpayers, and sometimes they hide the information. In case of whom do not sign up for their info, the solution of the landlord's national certificate would help to find them. But there are some ways to escape while tenants accept to sign a letting-rental contract with not certified the owner paying less rent. This ethical issue only is solved when real states, tenants, and banks, collaborate to avoid uncertified landlords [1, 8, 15].

On the other hand, by looking at AI-based solutions, financial services are integral to our daily lives and ethics are needed more urgently. Privacy, data safeguarding, and ethical issues have been debated for years in social research and related fields, and approaches to these issues have been developed in the process. BD is no longer a new phenomenon; descriptions of its potential and applications are progressively prevalent in emerging fields such as computational social sciences. The procreation of personal and business-like data is challenging conventional beliefs about privacy and autonomy, but the debate on such ethical challenges appears to be lagging [16].

A major issue in the principled debate around BD is the invasion of privacy. It is often questioned that the rampant exploitation of personal data for profit and government surveillance threaten freedom and self-rule. However, while generic ethical attention to the use of BDs is broad, research ethics issues are more granular and specific and thus diverge significantly from general ethical attention in the field. One logic for this is the prerogative of science. The individual's right to absolute privacy runs counter to the public concern for good science and well-founded scientific knowledge. However, the literature on distinct topics of research ethics and BD is codified in ethical advice for scientific research and is therefore less common [16, 17].

Fraud Detection is a bold issue for any authority. This has been done through advances, involving direct and indirect ways. In the direct mode, the items are specifically shown. They are the surest way to match the info that already exists with the info expressed by the taxpayer [18].

By applying algorithms that illustrate guides and exceptions data analysts TAs named fraud. This discovery could be inaccessible in the condition of statistical data analysis or AI. Data study methods like computation of factors, regression analysis, prediction models, and data comparison. AI skills contain use Data Mining by sorting and segmenting data to display transactions and find forms of fraud. Machine Learning routinely cabs separate features that seem fake, also, model detection release patterns of doubtable actions [2, 3, 19, 20].

3 Creative Problem-Solving Methods and Solutions

Aligning the problem statement is the first step to structured problem-solving. After clearing the statement of the problem, it is time to illustrate the novel solution.

TAs in all countries have incorporated recent technologies to improve their structure. They demand technology to ensure operational transparency, adaptability, and better response to the needs of governmental authorities and taxpayers. Application of modern technologies by TAs differs from developing countries to developed ones. Regard to some studies, about 160 UN member states (out of 193) use ICT systems for tax administration [21]. AI is not yet well used in Iran and especially in the tax industry. AI could be used to revamp Iran's taxation core, making technical support more adept for all involved.

3.1 Proposed Solution

AI-based approaches can help tax administration in several ways, some are listed below:

- **Data analysis**: AI can quickly and accurately analyse large amounts of data, helping identify tax compliance issues and potential fraud.
- **Automation of repetitive tasks**: AI can automate routine tasks, freeing up staff time to focus on more complex and strategic issues.
- **Predictive analytics**: AI can use historical data to identify trends and patterns, allowing for more informed decision-making and improved revenue forecasting.
- **Enhanced compliance**: AI can assist in identifying tax compliance issues and suggest remedies, thereby increasing voluntary compliance.
- **Improved accuracy**: AI can help reduce human error and improve the accuracy of tax assessments and collections.
- **Real-time processing**: AI can process data in real time, providing up-to-date information for tax administrators.

Overall, an AI-based approach in tax administration can lead to increased efficiency, improved accuracy, and enhanced compliance, resulting in increased revenue collection and improved taxpayer satisfaction.

An AI-based approach in tax administration can bring numerous benefits, such as:

- Improved accuracy: AI can analyse large amounts of data and identify potential errors or fraud more effectively than manual processes.
- Enhanced compliance: AI can help identify tax compliance issues and suggest remedies, thereby increasing voluntary compliance.
- Streamlined processes: AI can automate many repetitive tasks, freeing up staff time to focus on more complex and strategic issues.
- Increased efficiency: AI can process data in real time, reducing the time it takes to complete tasks and provide results.
- Predictive analytics: AI can use historical data to identify trends and patterns that can be used to predict future behaviour and tax liabilities.

These benefits can ultimately lead to increased revenue collection and improved taxpayer satisfaction. However, it is important to note that AI implementation in tax administration must be carefully planned and managed, with appropriate safeguards and privacy controls in place.

One new AI-based approach in tax administration is the use of machine learning algorithms to automate tax compliance processes. This approach involves the following steps:

- **Data collection**: Tax administrators collect data from various sources, such as tax returns, financial statements, and other relevant documents.
- **Data processing**: The collected data is processed and organized into a format suitable for machine learning algorithms.
- **Model training**: Machine learning algorithms are trained using the processed data, allowing them to identify patterns and make predictions.
- **Automation of compliance processes**: The trained algorithms are used to automate routine compliance tasks, such as the identification of potential tax fraud and the calculation of tax liabilities.
- **Monitoring and improvement**: The performance of the algorithms is continuously monitored and improved over time, ensuring accurate and efficient compliance processes.

This approach can help TAs make more informed decisions, improve the accuracy of tax assessments and collections, and reduce the time and resources required for tax compliance processes. It also has the potential to increase voluntary compliance and improve taxpayer satisfaction.

Another approach in tax administration is the use of natural language processing (NLP) for improving communication between taxpayers and tax administrators. This approach involves the following steps:

- Development of a conversational AI system: A conversational AI system, such as a chatbot, is developed to communicate with taxpayers in natural language.
- Integration with existing systems: The chatbot is integrated with existing tax administration systems, such as databases and processing systems.

- Communication with taxpayers: Taxpayers can use the chatbot to ask questions, receive answers, and access information about their tax obligations and entitlements.
- Data analysis: The chatbot can also collect and analyse data about taxpayer interactions, providing insights into taxpayer needs and preferences.
- Continuous improvement: The performance of the chatbot is continuously monitored and improved over time, ensuring accurate and effective communication with taxpayers.

This approach can improve communication between taxpayers and tax administrators, reducing misunderstandings and increasing taxpayer satisfaction. It can also provide valuable data insights, which can be used to inform tax administration policy and improve service delivery.

The proposed solution in this paper is a combination of mentioned approaches above, but a new localized version. Based on the solutions which were mentioned in the earlier section, a new AI-based profile approach, combined with the following methods is offered.

National Licensing Scheme for Landlords

The Iranian government has adjusted TA by digitizing services, starting with the running of electronic tax filing. The employment of E-Services needs taxpayers to file their tax returns electronically. These services include enrolment, online dispatch, and payment. The government has launched an application skilled in mixing e-services such as e-filing and e-billing into a single online website.

Iran's TA proves its commitment to modernizing itself by continuously improving techno-based services. This is also in the use of technology or the delivery of technology-based services (service digitization), that can cut down the operating costs of collection. This usage has the potential to improve efficiency. The Iranian government is currently updating its basic TA core system to improve its previously owned technology system.

To be skilled in using AI in the Iranian tax system, it is necessary to revise databases and systems. This concerns the scope of legislation that may boost the expansion of AI for TA and the necessary database to use AI in supporting the work of TA in Iran. An insight into the concept of an unstructured database is required because for many purposes it may be more useful to use this type of database and will also be necessary.

However, in the face of IT-based changes, to avoid rental tax evasion, some business-based changes are more obligatory. Like the UK approach [7]. The "National licensing for landlords" is a solution to classifying landowners in Iran. It is time for the government to trust the IT industry and enact legislation to support them. This type of support includes requesting any organization, agent, or group to ask for the national ownership license. This helps TA recognise owners nationwide, not only in major cities but also in other regions. Under this law, all real estate is needed to register the rental information of licensees only. Moreover, it helps tenants recognise trustworthy landlords, while also paying huge sums of money to bail in Iran.

Here is a simple pseudocode algorithm for using AI in a national licensing scheme for rental tax:

---

1. Create a national licensing scheme for rental properties
2. Require landlords to register their rental properties with the licensing scheme
3. Assign unique licenses to each rental property
4. Collect rental tax based on the rental income of each licensed property
5. Load the tax and licensing data into the system
6. Clean and pre-process the data
7. Train a machine learning model to predict the likelihood of compliance with the licensing scheme and rental tax obligations
8. Apply the model to the tax and licensing data to identify instances of potential non-compliance
9. Further investigation of those instances to confirm or dismiss the potential non-compliance
10. Take appropriate enforcement action, such as fines or legal proceedings, against confirmed non-compliant landlords
11. Continuously review and improve the machine learning model to ensure its effectiveness in predicting non-compliance and promoting compliance with the licensing scheme and rental tax obligations

---

This algorithm demonstrates the basic steps for using AI in a national licensing scheme for rental tax. The specific details of the algorithm, such as the choice of machine learning model, will depend on the nature of the tax and licensing data and the goals of the tax administration. However, this algorithm provides a general framework for using AI to promote compliance and collect rental tax in a systematic and automated manner.

Profiling and Central Database

Based on this idea, the owner will be identified, classified, and observed by TA. In this regard, TA can use AI to leverage data in enhanced ways. The use of AI-based digital technology for TA provides the ability to guide conformity between owners and taxpayers while decreasing the administrative burden on tax administration. AI assists this process of analysing substantial amounts of data and even helps TA analyse errors that taxpayers may make [15]. While this data is necessary for the TA, it may be shared as an electronic service by other organizations such as councils, since in the case of council taxes or other concepts there is a relationship tight between them.

Profiling is an approach for the automatic processing of personal and non-personal data, marked at developing predictive knowledge from the data in the form of profiling, as an example of discovering not common patterns. Transform data into data sets that could then be activated as a base for decision-making. profiling also is known as a method of data mining and be seen as an automatic process for examining data sets to generate classes or feature categories, which can be used to make profiles of individuals, groups, or anything to generate foreboding information to predict eventual trends and predict behaviour, processes to check the risks or opportunities. Financial and tax "big data mining", which involves the extensive collection of tax

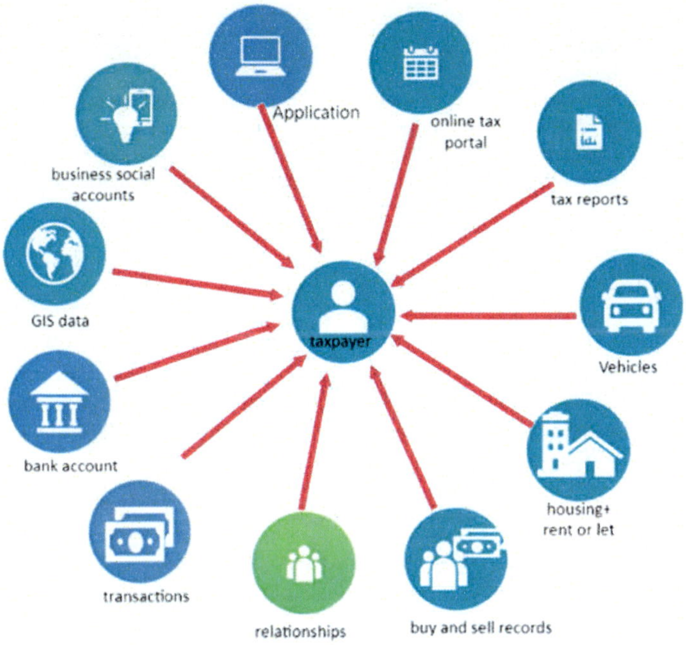

Fig. 1 Taxpayer profiling

and financial data and the processing of that data with intelligent algorithms to figure out detailed profiles or automated decision-making raises concerns about privacy, data protection, and other crucial rights.

As the Fig. 1 depicts, in taxpayer profiling, all data related to each taxpayer is collected, classified, and reorganized as an integrated profile. In this approach personal data and non-personal data, both, complete the profile. For example, bank accounts and transactions, the vehicles or houses that are on their own or any record about the taxation must be organized as a taxpayer profile. Therefore, AI helps TAs to predict fraud based on this dataset.

Undoubtedly, the appearance of BD has extremely contributed to the approach of profiling with data mining. BD analytics, detailed primarily using ML algorithms and a large Dataset of new or often reused data types, is becoming progressively popular. Changes in businesses and society that benefit consumers. BD analytics also encourages the public side to provide more active and efficient services and conclude complete results that boost everyone's life aspect. Moreover, the national licensing would use as a source in the profiling.

Here is a simple pseudocode algorithm for profiling in taxation:

1. Load tax data into the system
2. Clean and pre-processing the data
3. Create tax profiles for each taxpayer based on their tax data and demographic information
4. Analyse the tax profiles to identify patterns and trends
5. Identify instances where the tax profile deviates from the expected pattern
6. Further investigation of those instances to confirm or dismiss the potential fraud or non-compliance
7. If confirmed, take appropriate action to address the issue

This algorithm demonstrates the basic steps for profiling in taxation. The specific details of the algorithm, such as the choice of analysis method and the definition of expected patterns, will depend on the nature of the tax data and the goals of the tax administration. However, this algorithm provides a general framework for using profiling to identify potential tax fraud and non-compliance in a systematic and automated manner.

GIS Information to Landlords' Properties

The need is to implement a system that concedes TA to describe all taxable entities and determine the amount of tax to be paid. This data could be associated with a Geographic Information System (GIS) and Accounting-System that will generate invoices, check receipts, and brief the demand for action on liabilities [12].

The implementation of the GIS-based tax system will begin once the programmed map base and automatic payment database are in place. This system should be able to oversee all revenue activities, such as creating a new property in the system, tax invoices, collecting taxes, controlling debt, and splitting and/or merging two land units.

As it is shown in Fig. 2, this new tax-based GIS system, which is integrated with the profiling system, mentioned in the earlier part, all tax-related locations, like rent/let houses, companies and shops must be marked on the GIS system. This data would be used as a source in the profiling of taxpayers.

Here is a simple pseudocode algorithm for using GIS in taxation:

1. Load tax data and geographic information into the system
2. Clean and pre-process the data
3. Join the tax data with the geographic information based on the location
4. Create a spatial representation of the tax data using GIS techniques
5. Analyse the spatial representation to identify patterns and trends
6. Identify instances where the tax data deviates from the expected pattern in a specific location
7. Further investigation of those instances to confirm or dismiss the potential fraud or non-compliance
8. If confirmed, take appropriate action to address the issue

This algorithm demonstrates the basic steps for using GIS in taxation. The specific details of the algorithm, such as the choice of analysis method and the definition of expected patterns, will depend on the nature of the tax data and the goals of the tax administration. However, this algorithm provides a general framework for using GIS

Fig. 2 Tax-based GIS

to identify potential tax fraud and non-compliance in a systematic and automated manner.

GraphDB and Graph (Show Relationships)

Tax evasion, a serious dilemma for governments, could be addressed with modern technology. Criminals can now set up shell companies and make them look like proper entities. Funds flow back and forth between multiple accounts, following devious and consciously baffling paths. Tracing the track is a daunting task, but a Graph database can help to explain the structure of Shell's business units, bring visualization tools to aid hand-operated inquiry, and helps to Identify doubtful patterns, uncover tortuous paths, and ultimately return to corrupt individuals or organizations.

From a different view of tax evasion, graph technology also reveals buried traits and wages that people try to hide. As it is shown in Fig. 3, there are relationships between taxpayers, with their commercial actions, like buying or selling properties or cars, transferring money to other bank accounts, or signing contracts with other people. As a case, a person who earnings from various sources, such as rental properties, royalties, cooperation, and real estate, especially when multiple people have ownership of the property. All these kinds of relations or collaboration can be depicted by a graph to illustrate any fraud or hiding information. Graphs can be used to signify these assets and stakeholders, making the relationships between them (and the money they owe) much clearer. Finding at-risk people graphically, gaining more insight into places, and finding potential criminals by applying graph-based algorithms such as PageRank and Centrality would be easier to do.

Graph databases, such as Neo4j or Amazon Neptune, can be used in taxation to manage and analyse tax-related data more intuitively and efficiently. Here's a simple pseudocode algorithm for using a graph database in taxation:

Fig. 3 Graphing the tax data

---

1. Load tax data into the graph database
2. Define the relationships between the tax entities, such as taxpayers, properties, and tax obligations
3. Store the tax data and relationships in the graph database
4. Run graph queries to analyse the tax data and relationships
5. Visualize the tax data and relationships using graph visualization tools
6. Identify instances where the tax data deviates from the expected relationships
7. Further investigation of those instances to confirm or dismiss the potential fraud or non-compliance
8. If confirmed, take appropriate action to address the issue

---

This algorithm provides a general framework for using a graph database to manage and analyse tax-related data more intuitively and efficiently, and to identify potential tax fraud and non-compliance in a systematic and automated manner.

A simple outline of the framework which can help to implement this approach is:

- Data collection and processing: Tax data is collected from various sources, such as tax returns and financial statements, and processed into a format suitable for machine learning algorithms.
- Model development: Machine learning algorithms are developed and trained using the processed data to automate tax compliance processes.
- Implementation and integration: The developed algorithms are integrated with existing tax administration systems and processes.
- Automation of compliance processes: The trained algorithms are used to automate routine compliance tasks, such as the identification of potential tax fraud and the calculation of tax liabilities.

- Monitoring and improvement: The performance of the algorithms is continuously monitored and improved over time, ensuring accurate and efficient compliance processes.
- Communication with taxpayers: A conversational AI system, such as a chatbot, is developed to communicate with taxpayers in natural language, providing answers to questions and information about their tax obligations and entitlements.
- Data analysis and insights: The chatbot collects and analyses data about taxpayer interactions, providing valuable insights into taxpayer needs and preferences.

This outline demonstrates how AI-based approaches in tax administration can support the automation of compliance processes, improve communication with taxpayers, and provide valuable data insights.

3.2 Ethical Contextual Analysis of the Solution Proposed

Financial and tax data mining has involved collecting significant personal tax and commercial data and processing that data with intelligent algorithms to build records [13]. Detailed personal records or automated decision-making can raise concerns about privacy, data protection, and other crucial rights.

Tax secrecy is considered one of the most ethical issues regarding the use of BD in taxation. Tax secrecy concerning the violation of privacy when tax info is revealed to the public or given to other agencies for use or sent for re-uses to other experts. Economic privacy is often understood as the right of individuals to regulate what commercial info about them should be known to others. In contrast, the right to protect data, as provided in the protection law of data, not only concerns the clandestineness of the data collected and exchanged but also gives the data subject more rights. With data protection principles in mind, tax and financial confidentiality today have a broader meaning, including the negative privacy impacts of extensive collection by experts and companies. As result, privacy experts have frequently pointed out that financial data is considered one of the conscious forms of personal info, such as information on income, expenses, savings, hiring status and healthiness.

Although a load of private and particularly sensitive, non-financial information that makes up financial and tax information is large, recognizing certain details as private does not necessarily mean that collecting them is not fair. Concerns are not always greater than other causes. Instead, they should be treated against these aspects. Indeed, as other researchers explain, the right to tax and financial secrecy is not complete because absolute secrecy would render any modern tax system impracticable. However, it can be assumed that it is the responsibility of the authorities to give compelling reasons why privacy should be violated, rather than the nation to point out why privacy should be respected. In the tax context, individuals have only procedural guarantees.

In addition, while fairness is another issue for taxpayers, the proposed solution supports equity in the rental tax. When all property owners have the same responsibility to society by paying their exact tax, and no one can fraud or evasion while all landlords must sign for the national certificated, fairness would be identified.

From a fairness point of view, implementing a GraphDB helps the TA to recognize high-risk taxpayers who might have fraud or evasion based on the relationship of tenants, real estate agents and landlords.

However, GIS-based systems are used in tax organisations and councils, they can be beneficial in the case of recognizing the zones which have more unknown landlords as risk-able zones, which need to be labelled and managed by authorities.

Moreover, there are other ethical issues associated with using AI in tax administration, including:

- **Bias and discrimination**: AI algorithms can perpetuate and amplify biases in the data they are trained on. This can result in discriminatory outcomes, such as unequal enforcement of tax laws or unequal treatment of taxpayers.
- **Lack of transparency and accountability**: AI algorithms can be opaque, making it difficult for taxpayers and tax administrators to understand how decisions are being made. This can lead to questions about the accountability of the tax administration for decisions made by AI algorithms.
- **Privacy concerns**: The use of AI in tax administration can involve the processing of sensitive personal and financial information, raising privacy concerns and the need for robust data protection measures.
- **Job loss and economic impacts**: The automation of tax administration processes can result in job losses, particularly for lower-skilled workers, and have broader economic impacts.
- **Algorithm error**: AI algorithms can make mistakes, and those mistakes can have significant consequences for taxpayers and the tax administration.

To address these ethical issues, it is important for TAs to carefully consider the design, implementation, and oversight of AI systems, and to ensure that these systems are transparent, accountable, and consistent with ethical standards. This can include regular independent audits, providing clear explanations of AI-based decisions, and providing avenues for taxpayers to challenge those decisions.

4 Conclusion

While tax evasion and fraud are two of the issues of TA around the world, especially in the rental tax concept studied in this paper, AI brings solutions to tackle these societal issues. Countries use different methods to decrease fraud in tax and detect evasion, such as forcing to use of national landlord licensing, using GIS to identify the properties and their owners and machine learning algorithms to detect high-risk taxpayers [4].

The study reaches that, overall, landlords' opinion of rental tax, and the tax office was adverse, which strongly affected their disobedience. They do not look conscientiously at the operation of taxation and believe that the tax is not being used wisely. So, they enrol taxes in crucial conditions and are willing to use any means to avoid or underpay taxes. Knowledge of tax policy for rental income is also still low. This strongly affects non-compliance; However, many homeowners have a basic consideration of how taxes are calculated and the tax filing process.

In addition, this paper suggests that Iran TA must go on public alertness movement to instruct the public on their part and task in taxation compared to pending the matter from an allowed possibility perspective. It will generate a sense of duty for fulfilment rather than the horror of refusal.

Applying a tax-base system based on profiling and GIS info supplies TAs productivity in checking auditing activities as control approved in real-time. After implementing profiling and GraphDB, the TAs do not request to set time to check normal taxpayers because they can moderate arguments and in addition, everything could be remotely done.

Moreover, implementation of the tax-based data-integrated system would lead to the theory of improvement in Tax Compliance, and its execution will realise the 3 elements of this submission concept, namely law enforcement, provision of facilities, and trust. The application of a tax-based data-integrated system, in addition, affects costs by TAs and taxpayers. However, the power to develop the system and realize this kind of system is not low-cost and simple.

References

1. Strovolidis, G.: A comparative study on real estate property taxation in Europe (2017)
2. López, J.J.: A quantitative theory of tax evasion. J. Macroecon. **53**, 107–126 (2017)
3. Huang, Z.: The impact of replacing business tax with VALUE-ADDED tax based on big data technology on enterprise financial management. In: The 2021 International Conference on Machine Learning and Big Data Analytics for IoT Security and Privacy: SPIoT-2021, Vol. 1, pp. 1136–1142. Springer International Publishing (2022)
4. Tian, F., Lan, T., Chao, K.M., Godwin, N., Zheng, Q., Shah, N., Zhang, F.: Mining suspicious tax evasion groups in big data. IEEE Trans. Knowl. Data Eng. **28**(10), 2651–2664 (2016)
5. Lunde, J.: How taxation varies between owner-occupation, private renting and other housing tenures in European countries (2021)
6. Baesens, B., Vlasselaer, V.V., Verbeke, W.: Fraud Analytics Using Descriptive, Predictive, and Social Network Techniques. Wiley (2017)
7. HM revenue and customs.: Tackling tax avoidance, evasion, and other forms of noncompliance (2019). ISBN 978-1-912809-45-5
8. Atanasijević, J.: Using big data analytics to improve efficiency of tax collection in the tax administration of the republic of Serbia (2018)
9. Politou, E., Alepis, E., Patsakis, C.: Profiling tax and financial behaviour with big data under the GDPR Art (2019)
10. H. King, J., Richards, N.M.: What's Up with Big Data Ethics? (2015)
11. Gërxhani, K., Wintrobe, R.: Understanding tax evasion: combining the public choice and new institutionalist perspectives (2021)

12. Karanja, F., Kiema, J.: (2017) The Spatial Approach in the Identification of Tax Evasion on Rental Income: Case study of Umoja and Kileleshwa Estates in Nairobi, Kenya
13. Walker-Munro, B.: Use of big data analytics by tax authorities (2020)
14. Srinivasan, S.: Guide to Big Data Applications. Springer International Publishing (2017).
15. Saragih, A.H., Reyhani, Q., Seplian, M.: The potential of an artifcial intelligence (AI) application for the tax administration system's modernization: the case of Indonesia (2022)
16. Weinhardt, M.: Ethical issues in the use of big data for social research (2020)
17. OECD: Advanced Analytics for Better Tax Administration. OECD Publishing (2016)
18. Malaszczyk, K., Purcell, B. M.: Big data analytics in tax fraud detection. Financ. Account. **23** (2016)
19. Mekonnen, E.: Data Mining for Detection of Tax Evasion: The Case of Taxpayers in Addis Ababa (2021)
20. Lowman, M.: A Practical Guide to Analytics for Governments-Using Big Data for Good. Wiley (2017)
21. Scarcella, L.: Tax compliance and privacy rights in profiling and automated decision making. Internet Policy Rev. (2019). https://doi.org/10.14763/2019.4.1422
22. Badshah, A., Iwendi, C., Jalal, A., Hasan, S.S.U., Said, G., Band, S.S., Chang, A.: Use of regional computing to minimize the social big data effects. Comput. Ind. Eng. 171(C) (2022). https://doi.org/10.1016/j.cie.2022.108433

Bibliometric Analysis of Scientific Literature on Mental Health Research in Africa

Clara Egwuogu, Ebuka Ibeke, Priyanka Chaurasia, Celestine Iwendi, and Zakaria Boulouard

Abstract This bibliometric study presents a comprehensive summary of literature published on mental health research in Africa. The region has a large number of scientific studies conducted in this area. The purpose of this study was to investigate the contributions of African researchers to global mental research. It also investigated the quantity of research and publications about Africa. Bibliographic information for the analysis was retrieved from the Web of Science database. Over 11,960 and 1,144 articles from 1900 to the present were culled and examined respectively. Using scientific mapping tools and performance analysis methods, this study pinpointed the top countries, institutions, collaboration patterns, prolific authors, and developing themes in this research field. Although the results showed that African scholars contributed significantly to worldwide research and publication on mental health, the number of publications that are exclusively about Africa is only 0.09% of the global output. A closer look at the data showed that South Africa really outperformed all other countries in terms of research output. Anxiety, depression, and Covid-19 were the three most common terms used by authors. The results are noteworthy for the academic community because they provided a historical context for mental health studies.

Keywords Bibliometric analysis · Science mapping · Citation analysis · Co-occurrence keywords analysis · Performance analysis Co-authorship analysis · Mental health · Co-citation analysis

---

C. Egwuogu · E. Ibeke (✉)
School of Creative and Cultural Business, Robert Gordon University, Garthdee Road, Garthdee, Aberdeen AB10 7AQ, UK
e-mail: e.ibeke@rgu.ac.uk

P. Chaurasia
School of Computing, Eng Intel. Sys, Ulster University, Northlands Road, Londonderry BT48 7JL, UK

C. Iwendi
School of Creative Technologies, University of Bolton, Deane Rd, Bolton BL3 5AB, UK

Z. Boulouard
LIM, Hassan II University of Casablanca, Casablanca, Morocco

C. Iwendi et al. (eds.), *Proceedings of ICACTCE'23—The International Conference on Advances in Communication Technology and Computer Engineering*, Lecture Notes in Networks and Systems 735, https://doi.org/10.1007/978-3-031-37164-6_35

1 Introduction

According to the World Health Organization (WHO), a person is considered to be mentally healthy if they have a positive outlook on life, are confident in their own skills, are actively contributing members of society through their profession, and are not experiencing any serious mental health problems [1]. Felman defined mental health as the presence of a harmonious blend of cognitive ability, behavioural flexibility, and emotional security [2]. A report by the WHO stated that mental health is often given less attention than physical health in most countries [3]. According to this study, a large percentage of the world's population experiences difficulties relating to their mental health. While these statistics are certainly a cause for concern, the report claims that governments around the world have consistently and publicly ignored and undervalued the issue of mental health (MH). Further, millions of individuals all across the world suffer from mental health issues, as highlighted by the United Nations paper [4]. According to the report, one in four people will suffer from mental illness at some point in their lives. Other researchers have stressed the global nature of mental health issues and the consequential effects they have [5].

1.1 An Overview of Studies Done in Africa on Mental Health

Africa is the second most populous continent after Asia, with over a billion inhabitants. It is a diverse continent with a vast range of faiths, races, and cultures, as well as urban, suburban, and rural locations. Conflict and various forms of political turmoil, gender inequality as well as economic stagnation, impact several regions of the continent [6], and a study [7] suggests that these are the most prevalent social indicators of mental illness in Africa. Consequently, many scholars have reviewed the available material and concluded that mental health has been severely disregarded by successive leaders in developing countries [8]. Developing countries have very poor mental health policies and research funding. Many countries in Africa have only one psychiatrist for every one million citizens. Misconceptions abound about mental illness in Africa [9–11]. Other studies have presented overviews of the fundamental fields of research into mental health in Africa, particularly mental illnesses [12] and rates of mental illness in young people from Sub-Saharan Africa [13]. In contrast, bibliometric analysis [14, 15] is used in this study to look at Africa's overall impact on the domain of mental health research from 1900 to 2022. The volume of these articles that are focused on the continent of Africa was studied to meet the following goals:

- To assess the contributions of African academics in this field by looking into the number of publications on mental health research.
- To look into the amount of research done exclusively on the subject of mental health in Africa by African authors.

- Using the citations analysis parameter, to examine the number of citations these indicated publications have received to assess their influence.
- To use keyword analysis to determine the research themes and frequently used terms that have been investigated on the subject of mental health in Africa, as well as to gain an understanding of the conceptual framework of the body of knowledge and new developments in the field.
- To evaluate African authors', institutions', and nations' contributions and partnerships in the field of mental health research.

There are several reasons why it is essential to undertake bibliographic studies of research in the field of mental health in Africa. To begin, a bibliometric study would be helpful in that it would reveal previously unnoticed patterns in the success of publications and journals in this field. It would shed light on the collaborative patterns and key players in the field of African mental health research. A bibliometric study of regional publications in the field of mental health would also look into their theoretical foundations and highlight knowledge gaps in the most important research areas. Researchers, doctors, policymakers, government officials, academic institutions, and anyone interested in learning more about the research trends and intellectual framework in this topic will find the results of this study useful [16, 17].

Bibliographic analysis has strong empirical backing in the health sciences. However, no bibliometric analysis of journal articles devoted to research on mental health in Africa has been published as of yet. A few records were found, and they concerned studies of mental health all around the world and their connections to the virus Covid-19 and other illnesses [18–20]. This study examined the theoretical foundations and evolving trends in research articles on mental health in African countries to gain a better understanding of the topic at hand. The study used bibliometric analysis to cover a wider range of publication dates and concentrated on works authored by researchers with ties to African universities, filling a knowledge vacuum that had previously been overlooked.

2 Methodology

Bibliometric methods like performance analysis and science mapping were used to examine the impact and visibility of groups of scientific authors. Science mapping was used to assess the relevance of published papers to studies of mental health in Africa. Specifically, the project analysed the citation and co- citation patterns and the co-occurrence of words and the co-authorship in the selected academic publications.

Web of Science (WoS), Scopus, and Google Scholar are just a few of the databases that may be utilised to conduct a bibliometrics search and extract data [21]. A bibliometric data search, however, requires careful consideration when deciding which source to use. The WoS database was utilised for data search because, compared to Pub Med, Scopus and Google Scholar (three other sources), it was deemed the most appropriate index for use in this bibliometric analysis. WoS was selected because it is

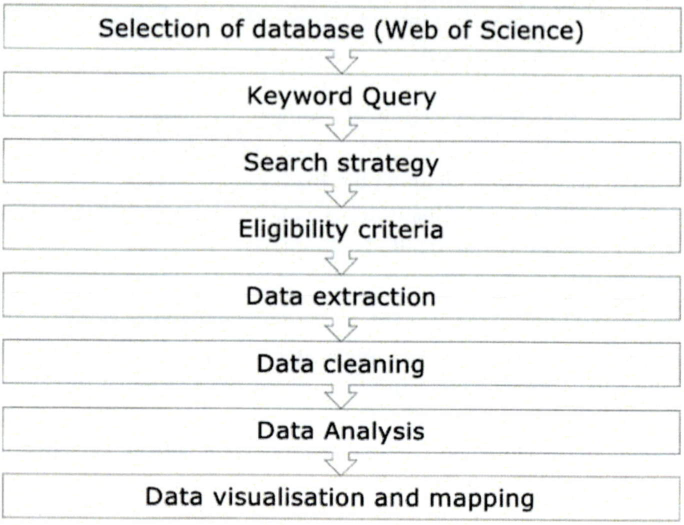

Fig. 1 The flowchart of the proposed methodology

widely recognised as a top resource for scholarly article citations and abstracts [22]. The database is said to index roughly 8500 of the most prestigious and influential scientific papers in the world [23] In addition, the WoS database is widely acknowledged as the most relevant bibliometric data source from which information may be retrieved for comparative analysis of research production across nations, authors, and institutions [24].

The methodology was based on a statistical evaluation of relevant papers from the WoS. The abstract, topic, and title of the identified literature served as criteria for selecting these works. The study's step-by-step procedures are depicted in Fig. 1.

During the data-gathering stage of this research, the team used two strategies. The first method retrieved records about research on mental health by African writers for Africa, while the second method retrieved records about research on mental health by African scholars in general. The mesh word *mental health* and Africa were linked using the Boolean operator "AND". This method was used to collect papers on the topic of mental health in Africa. Using this information, we looked at how many scholarly articles had been written by African authors and about Africa. To assess the contribution of African academics to global mental health research, we gathered articles that were not narrowly focused on the continent using the second keyword, "mental health" and this was used for comparative analysis. The keywords used in the search were derived from an analysis of twenty-five articles from a variety of sources that dealt with the research topic.

Using mesh terms from queries, a thorough search was performed in the WoS database between August 10 and 18, 2022, with one scenario focusing on "mental health*" and Africa and the other focusing on "mental health*" alone. The research utilised asterisks (*) as wildcards in both search techniques to account for variations in keyword spelling to retrieve any publications that shared similarities with the keywords [25]. Publications were obtained on a variety of mental health-related themes, including mental healthcare, mental health care, and mental illness. The quotation marks symbols were used to ensure that only results that had the precise phrase "mental health*" would be returned by the search engine.

2.1 *Eligibility*

The only parameters employed in the initial search were the article title, abstract, and keywords. To enhance the search, the following inclusion criteria were established because the WoS database allows users to do a variety of searches, including those based on publication year, country/region, source type, and document type:

Date: 1900 to August 2022. There was no time range included in the search. All years of publications in WoS were taken into account, as long as they were relevant to the predetermined search terms.

Document Types: To narrow the search, we disregarded documents that merely corrected previous mistakes or reported recent events. This was achieved by employing WoS's "refine" and "limit" functions. We restricted the scope of the research to scholarly publications such as peer-reviewed articles, early-access journals, review articles, book chapters, book reviews, conference papers, and letters. The majority of the documents obtained for analysis were journal papers, which are regarded as original and novel works [26]. While articles and reviews made up the bulk of the documents extracted, the other publication types listed in Table 3 were also retrieved because each piece contains original research and provides insight into the state of the domain. **Languages**: Articles written in English, French, and Afrikaans Country affiliation to authors: Publication locations were also used as a criterion for the data. Only articles written by African researchers were included in the analysis. The "Countries/Regions" section on WoS was used to manually pick and review African countries and regions.

2.2 *Data Extraction and Cleaning*

After taking into account the aforementioned criteria, the final search yielded 11,960 records that were analysed to determine the extent to which African experts have contributed to global mental health research. While the second keyword search query generated a total of 1,144 papers utilised to examine the number of research by African writers and for Africa specifically.

The bibliometric indicators obtained from the extracted data were publication year, keywords, abstract, author, nations, regions, institutions, topic field, document type, journal, and citation rate. Many researchers from different parts of the world have studied mental health extensively over the years. According to WoS database records, however, the earliest or first registered publication on mental health research that largely focused on Africa was published in 1924, while the most recent is expected to be published in 2022. In contrast, the other keyword ("mental health*") search query revealed that mental health research on the continent dates back to 1903 when the oldest publication was published as noted in WoS.

The identified publications were exported as a tab-delimited file which contained the Full Record Content and Cited References. A total of 27 separate files were downloaded for the keyword search of "mental health*". While 3 separate files were downloaded for the second keyword search query "mental health*" and Africa.

Due to a limitation in WoS, large quantities of bibliometric data could only be exported in increments of 500 records. In addition, comma-separated values files with the Wos's Citation Reports for each document returned by the search were created. The significance of publications in the field of mental health research in Africa was evaluated using this exported data set. It is important to remember that, as a dynamic process, citation might cause the citation rate of the retrieved publications to rise and fall. Accordingly, all citation data was accurate at the time this study was completed.

During the data-cleaning phase of the investigation, two methods were used. To begin, the dataset was manually inspected to rule out the possibility of any major missing records, outliers, or coding discrepancies. Second, we made sure that the field content was consistent with the field titles and that the coverage of the content was the same as stated in the field names. Several glaring inaccuracies were uncovered, including misspelt words, inaccurate publishing dates, and the use of anonymous authors. The end result of this was the deletion of all data that had been input erroneously. Data identified through citation reviews were selected, filtered, extracted, and saved in a new file format following data cleansing.

2.3 Data Analysis

The data analysis phase made use of two quite different kinds of software to perform the analysis and display the findings. A descriptive analysis of the retrieved records was performed first, using Microsoft Excel. In this study, we analysed the number of scholarly articles published on mental health in Africa, as well as the characteristics of the research journals themselves and the number of citations to those articles. Data, such as highly cited journals, prominent research institutes and nations in the domain, and the pattern of publications, were analysed, rated, and tallied in MS Excel.

Second, we utilised the VOSviewer, a bibliometric visualisation network tool, to create a network map of the retrieved data. VOSviewer may display data in three distinct ways: through a network, an overlay, and via density [26]. The co-authorship, co-citation, and citation networks were visualised by network visualisation in this study, while the co-occurrence keywords network was revealed through overlay visualisation. In network visualisation, the inter-actions and interconnections between entities and between entities and themes published in a given field are depicted graphically. These connections provide graphical depictions of how each node in the network functions [15] and that the distance between two circles is representative of the depth of the connection between the ideas [24]. We compared the relative strength of connections by measuring line thickness. This was employed as a metric of author and institution collaboration. It was believed that greater relative link strength led to greater collaboration. For the most part, a larger circle size indicated increased output or citations. In addition, the VOSviewer's colour-coded data helps users see the trajectory and association between various term components. The colour of the connecting phrases grows more prominent as the "phrase" is used or mentioned in more works of literature. However, a tenuous bond is indicated when the colour is pale.

No other sources' keywords were used in the co-occurrence analysis. Using the link strength measurement, the research determined the frequency with which two or more terms co-occur in a single publication. A visual depiction of the keyword analysis is provided below.

The co-authorship analysis aimed to do the same thing by looking into the level of cooperation between the most influential authors in the field. The number of documents written by two or more related institutions served as the analytical unit, while an examination of the total link strength provided insight into the depth of the co-authorship ties between all the countries associated with a certain nation. Therefore, authors without a relationship were omitted from the analysis. Excluding these authors from the analysis would not significantly alter the outcome because they only had one publication and made minor contributions to the study. This research treated university departments and research centres within a single country or region as discrete units for the purposes of its institutional analysis. It should be emphasised, however, that when such information was missing, the university was regarded as a single entity. This research used the author's institution's home nation as the basis for its country analysis. The purpose of the journal analysis was to determine which journal is the most prominent in this field of study. Journal characteristics were analysed to see how representative the journals were of the articles that were retrieved. For this journal evaluation, we employed a cutoff of 10 articles per journal. Journals were sorted by their publication output and citation frequency.

3 Results

3.1 Description and Publications Growth

Only 5,168 (or 42.8%) of the 11,960 data used for analysis were research publications. Also, 11,645 (or 97.3%) of the documents retrieved and analysed were written in English. There were 17,771 authors from 1,164 different institutions, and the papers appeared in 2,654 journals. WoS claims that the first related paper appeared in the Archives De Psychologie in 1903. Figure 2 shows the total number of academic

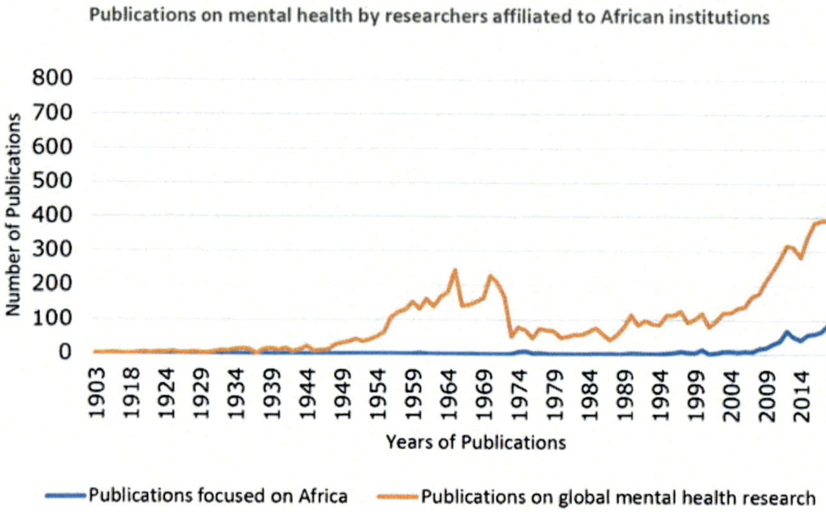

Fig. 2 Chronological analysis of Global versus African focused publications

papers published between the years 1903 and 2022. It can be seen from the graph that between 1904 and 1959, there was a negligible increase in the number of research publications in this sector, with the number fluctuating between zero and ten per year. Slight increases of around 30–50 publications per year were seen between 1961 and 1970, but the number of publications has been on a steady decline since then. After a lengthy period of gradual growth, however, there was a sudden rise between 2003 and 2022. According to the graph's tendency, there was a maximum of 762 records of papers authored by African scholars in the year 2021. From 2011 to 2021, there was a significant increase of 485 total publications. This highlighted the increasing attention paid to studies of mental health in Africa, both by academics and publishing houses.

In contrast, the current research combined the two data sets and cross-referenced the results to count the number of articles that specifically addressed Africa. As the investigation progressed, it became clear that research publications likewise made up the bulk of published content. On the other hand, the data showed that only 0.09% of all studies on mental health around the world included Africa as a setting. According to the figure, between 1924 and 1959, there was essentially no change in the number of publications. Two publications were cited as being available in 1960. The total number of published papers fluctuated between one and eight between 1961 and 1990. In 2021, a total of 160 articles were produced, making it the most prolific year.

3.2 *Institution and Country Analysis*

As part of our study, we tallied up the number of scholarly works published by each African country to evaluate their relative impact and research output. According to the data, as shown in Table 1, South Africa had the most articles published and link popularity (59,821). Nigeria published the second-highest records (464) and link strength of 31,154; while Ethiopia came in third place (24,543). Table 1 summarised the five leading countries in research on mental health on the continent, while Fig. 3 shows the strength of the connections between these countries in terms of publications.

The research also investigated the region's top research institution for mental health publications. Based on their link counts, the analysis revealed which institutions and organisations were the most active. After analysing the published research on mental health in Africa, researchers discovered 1,164 affiliated institutions. Table 2 below lists the top 10 of these organisations. With a total of 62,412 links and 307 articles, the University of Cape Town topped the charts in this section. Additionally, the university (4,573) obtained the highest number of citations, solidifying its

Table 1 Top five countries with the most publications

Countries	No. of publications	Total link strength
South Africa	1,543	59,821
Nigeria	464	31,154
Ethiopia	285	24,543
Ghana	120	12,977
Egypt	171	9,130

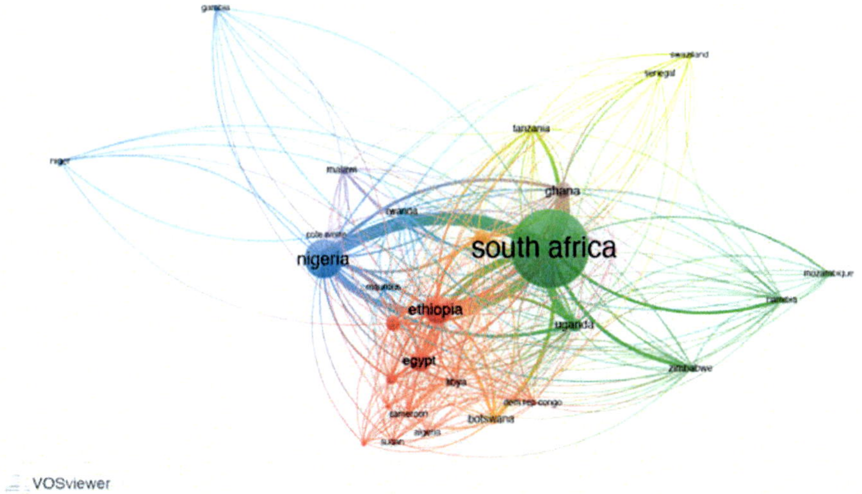

Fig. 3 The link strength of countries by publication count

Table 2 Affiliations and publication count

Affiliations	Publication count	Total links	Location
University of Cape Town	307	62,412	South Africa
University of KwaZulu Natal	271	47,698	South Africa
University of Witwatersrand	250	33,095	South Africa
Stellenbosch University	237	28,226	South Africa
University of Pretoria	160	22,781	South Africa
North-West University	153	21,713	South Africa
South African Medical Research Council	139	21,248	South Africa
University of Ibadan	125	15,599	Nigeria
University of Gondar	125	15,373	Ethiopia
University of Ghana	82	13,159	Ghana

Table 3 The 5 most frequently referenced publications

Article title	Cited reference count
"Handbook for the Study of Mental Health: Social Contexts, Theories, and Systems, 3rd Edition"	4003
"The handbook of sociology and human rights preface"	2392
"Movements for human rights locally and globally introduction"	2390
"Peace Psychology: A Comprehensive Introduction"	1641
"Child Welfare Challenge: Policy, Practice, and Research"	1379

position as one of the continent's leading centres for mental health research. The inter-institutional network of collaboration in Fig. 4 shows a visualisation of these connections. It was found that over 270 different institutions collaborated with one another across African regions, as evidenced by the presence of more than 50 links between them.

3.3 *Analysis of Country/Region Collaboration*

Furthermore, the publication collaboration pattern by country was shown in Fig. 5. VOSviewer is based on the general tenet that a stronger relationship exists between any two countries that are significantly closer to one another in the application's user interface and that the thickness of the line between them reflects the degree of

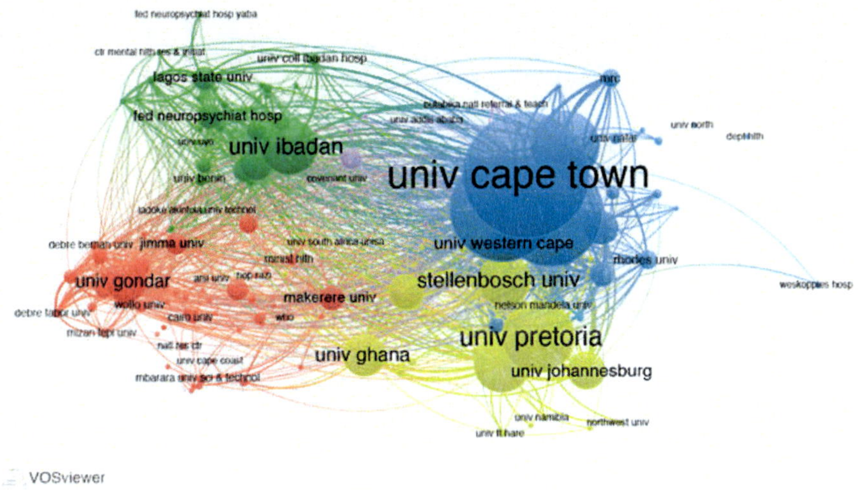

Fig. 4 The institutional interaction network

this relationship [27]. According to the results of the co-authorship analysis, South Africa is associated with 19 other countries and/or territories through a total of 1,546 co-authored works. The number of documents written in collaboration by authors from Nigeria (464 total) placed them in second place. Despite having more scholarly articles and citations than Kenya and Uganda, Egypt had a lower regional interaction profile. Kenya is the third most linked country in international collaborations, despite having only published 88 documents with 442 citations. With a total of 25 links, 120 co-authored papers, and 904 citations, Ghana is certainly an active research hub in the continent. The research also found that 41 of the identified countries, or 2/3 of them, had an inter-regional collaboration with fewer than 2 nations. However, researchers from Libya, Algeria, Senegal, Cote d'Ivoire, and Madagascar were not affiliated with any other countries in the publication of articles on mental health research.

3.4 Analysis of Scholarly Journals

In total, 2,654 different journals from all over the continent were found to have published articles related to mental health. Based on a comprehensive literature assessment, the study identified the top ten journals in the field (shown in Fig. 6) as measured by the total number of articles published. The American Journal of Psychiatry, the American Journal of Orthopsychiatry, and the Mental Hygiene Journal were the top three journals in this category.

Several scholarly journals, including the Lancet and the Australian and New Zealand journal of psychiatry, have played important roles in disseminating research

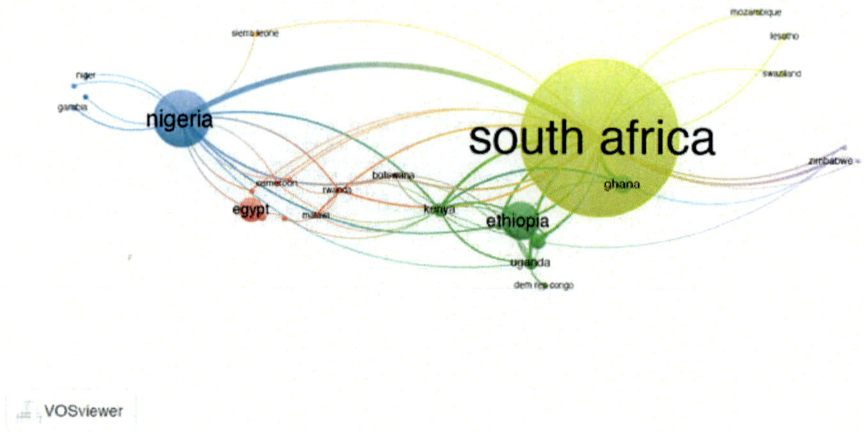

Fig. 5 Country collaboration network analysis based on co-authorship

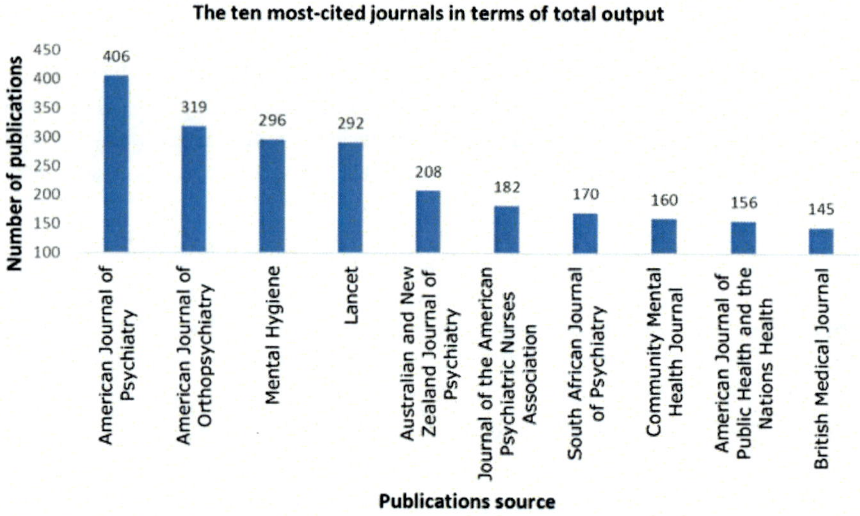

Fig. 6 The top ten journals in the publications of Africa's global mental health research output

findings in this area. In Fig. 6, we can see that most studies that analysed the total contributions of African scholars to global mental health research appeared in peer-reviewed international periodicals. This proved that academics from different regions worked together to advance global mental health research.

Moreover, we looked at the volume of citations received by each publication to determine its influence in the field of mental health research in the region. While the American Journal of Psychiatry may have produced more papers in the domain, VOSviewer's examination of citation data shows that the Journal of Paediatrics is

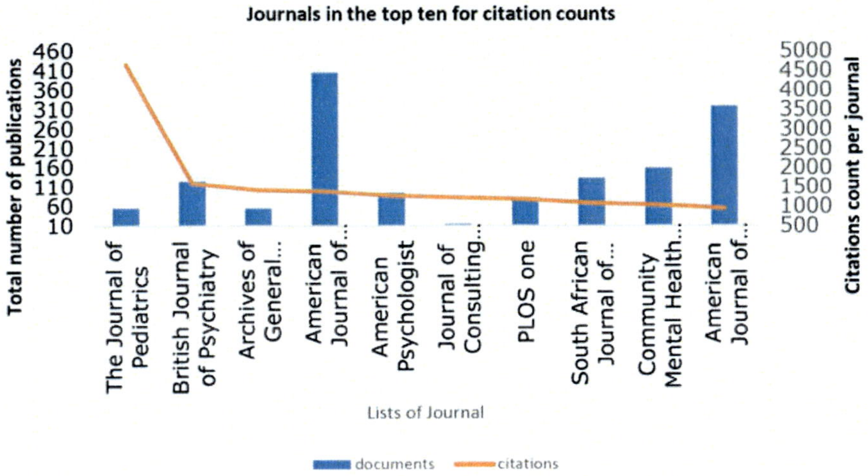

Fig. 7 Journals in the top ten for citation counts

more frequently cited. As shown in Fig. 7, the latter received 4,659 citations. The British Journal of Psychiatry and Archives of General Psychiatry came next.

3.5 *Analysis of Citations*

The purpose of the citation study was to look at how citations have changed over time and to provide a descriptive analysis of the most cited publication in the field. Figure 8 analyses the total number of citations per year in relation to the number of publications from 1900 to 2022, thereby illuminating the citation pattern. A study of the fluctuation of citation counts over a five-year period is presented. The total number of citations showed a gradual increase from 1930 to 1958, as depicted by the graph. The graph showed that the total number of citations rose steadily between 1959 and 1975. During this time span, there was a rise in citations from 72 in 1959 to 486 in 1975. Still, between the years 1990 and 1993, there was a significant drop. The subsequent years saw a dramatic increase in citations, from 370 to 1063. The number of citations began rising again in 2010, reaching a high of 8,276 in 2021.

The retrieved documents also produced a total of 63,867 citations, with citation counts ranging from 0 to 8,276. The overall h-index is 88 and the average number of citations per publication is 12.1. The h-index measures an author's overall contribution to their field and serves as a measure of their success, as stated by [28]. The high h-index of the identified records indicates widespread interest from both authors and readers. Among academic works released in 2017, "Handbook for the Study of Mental Health: Social Contexts, Theories, and Systems, 3rd Edition" was the most often cited. Understanding mental health and mental disease in the context of mental

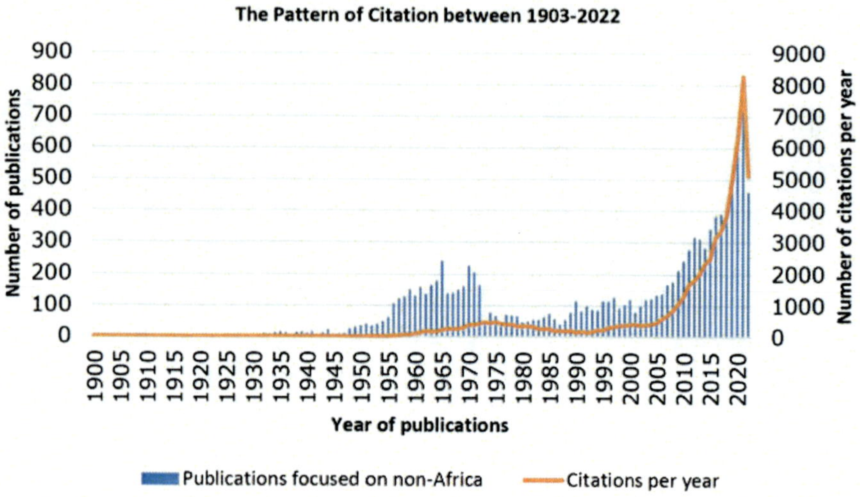

Fig. 8 The pattern of citation over time

health delivery systems was the focus of this document. Table 3 displays the 20 most-cited papers. There was a total of ten published books, four editorial pieces/chapters, and three scholarly works included on the list. However, most of these articles need payment to see their content. Articles published in open-access journals are more likely to be referenced by other researchers.

In addition, a citation analysis of African institutions was conducted to assess the research output of a specific institution by looking at the number of citations generated by that institution's academic activity. Figure 9 depicts the citation network of various educational institutions. The study revealed that the University of Cape Town, the University of Stellenbosch, the University of KwaZulu Natal, the University of Witwatersrand, and the University of Ibadan were the top five most referenced universities in the study of mental health on the continent.

3.6 *Analysis of the Authors' Contributions*

A count of publications was used in an authorship analysis to determine which individuals had the greatest impact on the field. Authors with at least 10 works published on the topic were considered for the analysis. According to the data, 11,960 articles collected from WoS were written by 17,771 different individuals. The threshold was reached at 82 authors. Table 4 displays the results of the study, which analysed only the publications of the top ten most prolific authors. Without a shadow of a doubt, these writers are the most prominent on the continent in the study of mental health around the globe.

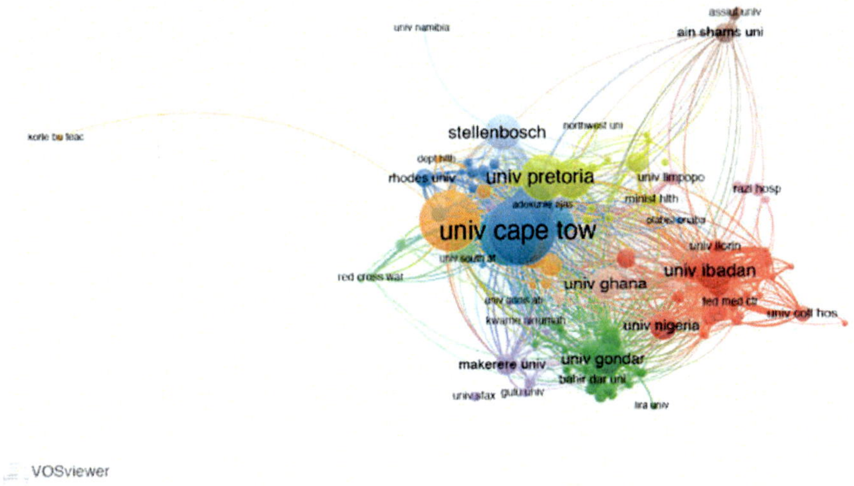

Fig. 9 The citation network of institutions

Table 4 List of the ten most prolific researchers in the subject of global mental health

Author	Documents	Citations	Affiliations
Stein, DJ	60	1217	University of Cape Town
Lund, C	55	1283	University of Cape Town
Swartz, L	47	439	Stellenbosch University
Petersen, I	43	1131	University of KwaZulu Natal
Seedat, S	41	1013	Stellenbosch University
Gureje, O	35	944	University College Ibadan
Kagee, A	30	463	Stellenbosch University
Atilola, O	26	207	University College Ibadan
Pillay, AL	26	185	University of KwaZulu Natal
Rothmann, S	25	206	North West University

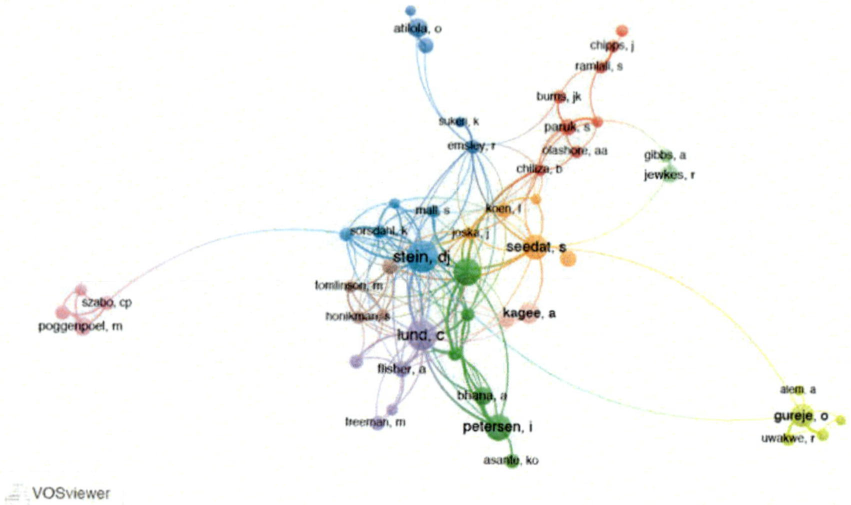

Fig. 10 Co-authorship network of collaboration among African researchers

These authors were found to have ties to the top 25 universities. The research found that DJ Stein, with 60 publications and 1,217 citations, had the largest impact in this area. With 55 records and 1,283 citations, Lund C is the most often cited journal, placing it in second place. Table 4 below provides a visual representation of these results.

It was also found that African researchers do engage in some type of collaboration. Only 86 authors out of a total of 17,771 authors satisfied the criterion, according to the co-authorship study. Table 4 shows that Lund C. is the most linked author, followed by Stein DJ (77 links), and then Swartz I. (47 links). Figure 10 depicts a typical example of collaboration between African scholars. There are a minimum of 10 publications by each author with a connection strength of at least 2. The network visualisation reveals six unique groups of co-authorship.

3.7 *Analysis of Co-occurrence Keywords*

The most used words and phrases by writers reflect the primary areas of attention and interest in the scientific community. To understand the changing trend in the research field, we used VOSviewer to conduct a co-occurrence analysis of the authors' keywords. The analysis's findings were presented in a graphical overlay that showed how often specific keywords occurred annually. The information was analysed so that the study's fourth goal could be achieved. The investigation found 6,408 unique writers' keywords. The analysis was refined by excluding some keywords in order to create a keyword map that reflected a more thorough grasp of the concepts

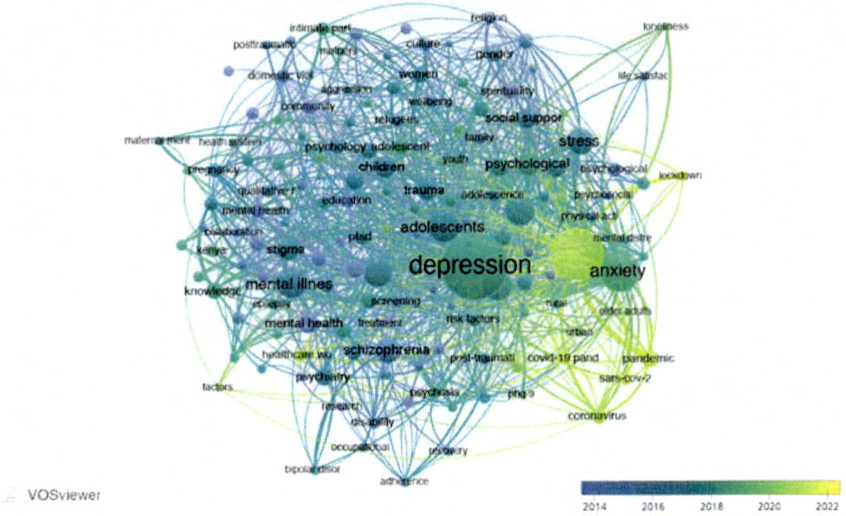

Fig. 11 Analysis of the time-related co-occurrence of keywords

involved. In total, 157 of the authors' keywords that appeared at least 10 times or more across the works were used for analysis. Network visualisation of the analysis of co-occurrence keywords is shown in Fig. 11.

With 333 co-occurrences, the term "depression" has the highest overall link strength, followed by Covid-19 (221), anxiety (141), teenagers (138), and mental illness (114). Furthermore, the selected keywords were divided into six categories as illustrated in Fig. 11. The six groups on the map represented various subjects connected to African mental health research. In general, these findings are consistent with prior findings on the intellectual structure of mental health-related knowledge during the Covid-19 pandemic [24].

4 Discussion

Good mental health is a given, but when it declines to the point that it interferes with daily life, it is considered a disability. The review demonstrates a rise in research focused on mental well-being in recent years. Following the devastating Covid-19 outbreak [29–31], this curiosity skyrocketed. This study evaluated the trajectory of African authors' explorations of mental health by analysing bibliometric and visualisation data. As part of this investigation, the current state of affairs and the characteristics of these scholarly works were analysed. Researchers in this study also analysed the impact of scholarly authors in the field. According to the findings, there has been a significant rise in the number of academic journals during the past five years. The research team hypothesised that poverty, conflict, social exclusion, refugees, discrimination, poor health policies, the rising HIV prevalence in Africa, violence against women, stigmatisation, and, most importantly, the impact of the Covid-19 pandemic were to blame for this rise in the occurrence of mental health problems like schizophrenia, anxiety, depression, and suicide.

It seemed that the most commonly reported issues in this area were depression, Covid-19, and anxiety [32, 33]. This supported [34] claim that mental health and depressive illnesses accounted for a sizable proportion of the world's disease burden in the year 2000. Similarly, [35] foresaw a significant shift in the mental health requirements of the world's population over the next two decades as a result of the increasing incidence of mental diseases such as depression, dementia, alcoholism, and schizophrenia. By 2030, depression is expected to overtake cancer as the biggest cause of death worldwide, according to a United Nations assessment published document [36].

Furthermore, it was not unexpected that African scholars were also interested in Covid-19. For all human history, the year 2020 will go down as a turning point of monumental significance. The consequences of Covid-19 were obvious all throughout. This year was marked by increased security measures, strained healthcare services, corpse bags, deserted streets, and a general sense of hopelessness. Modern society was profoundly affected by the pandemic. In addition, despite having the longest history of wars, Africa ranked last in the number of country-specific studies published in this area of study.

Journals in psychiatry, psychology, public and occupational health, internal medicine, social work, nursing, and rehabilitation were mined for research articles on mental health. Mental health has become more of a mainstream research topic in the medical and public health fields thanks in large part to the efforts of specialised journals like Mental Hygiene, Community Mental Health Journal, American Journal of Psychiatry, British Medical Journal, and American Journal of Orthopsychiatry. There was also evidence of increased international interest in the publication of mental health research through the inclusion of papers authored in languages other than English. According to the findings, researchers studying mental health who are connected to institutions in Africa have made significant contributions. However, different regions of the continent had varying degrees of involvement and attention paid by these publications. According to the findings, South Africa is the leading nation on the continent in this research area. The analysis reveals and places Nigeria as the continent's number two nation in the domain. There have been major contributions from other countries, such as Ethiopia, Ghana, and Egypt. Each of the four nations represented a different region of Africa: West Africa, East Africa, and North Africa. Despite making such a significant impact, the analysis showed that only 0.09% of the world's research in this sector was conducted in Africa. Because mental health was not seen as a public problem in the regions under study until relatively recently, the research output of African-specific publications was limited. A lack of resources to carry out the study and publish findings could be another contributing factor. Figure 2 displays an upward trend in the number of articles published in this sub-set between 2018 and 2021, which may be indicative of a growing interest in research with an African focus. Contrarily, the research team opines that there is not enough demand for the service. There will likely be more papers released in the near future.

Furthermore, universities were also identified as the most crucial organisational support structures for scientific research in the region. The University of Cape Town in South Africa is the organisation with the most impact. Universities in the South, West, and East African regions clearly exhibit geographical differences when it comes to organisational collaboration. Universities were also identified as the most crucial organisational support structures for scientific research in the same way. The University of Cape Town in South Africa is the most significant school there. When comparing universities in South Africa, West Africa, and East Africa, there are clear geographical differences. Moreover, the analysis conducted using VOSviewer reveals that South African institutions have the most bibliographic couplings and links with institutions from other countries. Based on this analysis, it is clear that South African institutions are at the forefront of this field of study in Africa. Articles made up roughly 42.3% of all retrieved documents when analysed by type. This indicated that non-periodical publications such as books, reviews, book reviews, letters, and editorials also contributed significantly to the area. Co-occurrence keyword analysis found that major central topics within the domain included depression, Covid-19, anxiety, mental disease, HIV, teenagers, stress, and quality of life. Figure 11's use of colour to represent the development of concepts across time is very instructive. The keywords that were more than a few years old were shown in a dark purple colour, while the more recent ones were shown in a brilliant yellow. Moreover, the image showed six separate clusters related to studies on mental health in Africa.

5 Conclusion and Limitations

Findings from this study dispute the belief that African researchers have made no contributions to this field of study. The data showed that during the past five years, there has been an increasing rate of research being published. Publication growth has been strong, and it is expected to climb even further as governments around the world (African countries/regions) continue to fight the consequences of the Covid-19 pandemic and other economic, and sociopolitical challenges.

Conclusions for future researchers on the subject of mental health include suggestions for reading material, authors, and topics to consider. Interesting findings were found through the bibliometric analysis, but there are still questions that need answering. To begin, the research spanned a huge time frame, from 1900 to 2022. The most influential academic journals in this area during the past decade could be the focus of follow-up studies. Second, it's possible that the interdisciplinary approach used produced conflicting results. Reviewers in the future may wish to pay special attention to the literature concerning mental health disorders as a whole in order to address this issue. Last but not least, because the data utilised in the analysis came from the WoS, there is always the possibility that some significant papers from other databases were left out, making the data suspect. It's possible that in the future, researchers will use data from additional bibliometric resources.

References

1. Word Health Organization, et al.: World mental health report: transforming mental health for all (2022)
2. Felman, A., Tee-Melegrito, R.A.: Mental health: Definition, common disorders, early signs, and more (2022). https://www.medicalnewstoday.com/articles/154543. Accessed 22 Dec 2022
3. ESA, U.: World Population Prospects: The 2004 Revision, Volume 3 Analytical Report. UN Population Division, New York (2006)
4. UN: World population prospects, the 2004 revision: population database. United Nations Department of Economic and Social Affairs, Population Division, New York, N.Y. (2005)
5. Peltzer, K., Phaswana-Mafuya, N.: Depression and associated factors in older adults in South Africa. Glob. Health Action **6**(1), 18871 (2013). https://doi.org/10.3402/gha.v6i0.18871
6. Mars, B., et al.: Suicidal behaviour across the African continent: a review of the literature. BMC Public Health **14**(1) (2014). https://doi.org/10.1186/1471-2458-14-606
7. Patel, V., Kleinman, A.: Poverty and common mental disorders in developing countries. Bull. World Health Organ. **81**, 609–615 (2003)
8. Eaton, J., Westcott, N., Okolo, S.: Mental Health in Africa: Innovation and Investment. Mental Health Innovation Network (2019). https://www.mhinnovation.net/blog/2019/oct/10/mental-health-africa-innovation-and-investment
9. Mosanya, T.J., Adelufosi, A.O., Adebowale, O.T., Ogunwale, A., Adebayo, O.K.: Self-stigma, quality of life and schizophrenia: An outpatient clinic survey in Nigeria. Int. J. Soc. Psychiatry **60**(4), 377–386 (2014)
10. Kabir, M., Iliyasu, Z., Abubakar, I.S., Aliyu, M.H.: Perception and beliefs about mental illness among adults in karfi village, northern nigeria. BMC Int. Health Hum. Rights **4**(1), 1–5 (2004)
11. Cortina, M.A., Sodha, A., Fazel, M., Ramchandani, P.G.: Prevalence of child mental health problems in sub-saharan africa: a systematic review. Arch. Pediatr. Adolesc. Med. **166**(3), 276–281 (2012)
12. Hendler, R., Kidia, K., Machando, D., Crooks, M., Mangezi, W., Abas, M., Katz, C., Thornicroft, G., Semrau, M., Jack, H.: "we are not really marketing mental health": mental health advocacy in Zimbabwe. PLoS ONE **11**(9), 0161860 (2016)
13. Daar, A.S., Jacobs, M., Wall, S., Groenewald, J., Eaton, J., Patel, V., Dos Santos, P., Kagee, A., Gevers, A., Sunkel, C., et al.: Declaration on mental health in Africa: moving to implementation. Glob. Health Action **7**(1), 24589 (2014)
14. Donthu, N., Kumar, S., Mukherjee, D., Pandey, N., Lim, W.M.: How to conduct a bibliometric analysis: an overview and guidelines. J. Bus. Res. **133**, 285–296 (2021)
15. Aria, M., Cuccurullo, C.: Bibliometrix: an r-tool for comprehensive science mapping analysis. J. Informet. **11**(4), 959–975 (2017)
16. Ferreira, M.P., Santos, J.C., de Almeida, M.I.R., Reis, N.R.: Mergers & acquisitions research: a bibliometric study of top strategy and international business journals, 1980–2010. J. Bus. Res. **67**(12), 2550–2558 (2014)
17. Mas-Tur, A., Kraus, S., Brandtner, M., Ewert, R., Ku¨rsten, W.: Advances in management research: a bibliometric overview of the review of man-agerial science. Rev. Manag. Sci. **14**, 933–958 (2020)
18. Akintunde, T.Y., Musa, T.H., Musa, H.H., Musa, I.H., Chen, S., Ibrahim, E., Tassang, A.E., Helmy, M.S.E.D.M.: Bibliometric analysis of global sci- entific literature on effects of covid-19 pandemic on mental health. Asian J. Psychiatry **63**, 102753 (2021)
19. Larivi'ere, V., Grant, J.: Bibliometric analysis of mental health research (2016)

20. Zeinoun, P., Akl, E.A., Maalouf, F.T., Meho, L.I.: The Arab region's contribution to global mental health research (2009–2018): a bibliometric analysis. Front. Psych. **11**, 182 (2020)
21. Mart´ın-Mart´ın, A., Thelwall, M., Orduna-Malea, E., Delgado L´opez- C´ozar, E.: Google scholar, microsoft academic, scopus, dimensions, web of science, and opencitations' coci: a multidisciplinary comparison of coverage via citations. Scientometrics **126**(1), 871–906 (2021)
22. Ramlal, A., Ahmad, S., Kumar, L., Khan, F.N., Chongtham, R.: From molecules to patients: the clinical applications of biological databases and electronic health records. In: Translational Bioinformatics in Healthcare and Medicine, pp. 107–125. Elsevier (2021)
23. Ronda-Pupo, G.A.: Research evaluation of author's citation-based performance through the relative author superiority index. Transinforma¸c˜ao **29**, 191–201 (2017)
24. Chen, Y., Zhang, X., Chen, S., Zhang, Y., Wang, Y., Lu, Q., Zhao, Y.: Bibliometric analysis of mental health during the covid-19 pandemic. Asian J. Psychiatr. **65**, 102846 (2021)
25. Kolkailah, A.A., Fugar, S., Vondee, N., Hirji, S.A., Okoh, A.K., Ayoub, A., Al-Ogaili, A., Rios, L.H.P., Kumar, S.K., Camacho, M.T., et al.: Bibliometric analysis of the top 100 most cited articles in the first 50 years of heart transplantation. Am. J. Cardiol. **123**(1), 175–186 (2019)
26. Sweileh, W.M., Al-Jabi, S.W., Sawalha, A.F., AbuTaha, A.S., Zyoud, S.H.: Bibliometric analysis of medicine-related publications on poverty (2005–2015). Springerplus **5**(1), 1–13 (2016)
27. Khudzari, J.M., Kurian, J., Tartakovsky, B., Raghavan, G.V.: Bibliometric analysis of global research trends on microbial fuel cells using scopus database. Biochem. Eng. J. **136**, 51–60 (2018)
28. Zhang, C.-T.: The h'-index, effectively improving the h-index based on the citation distribution. PLoS ONE **8**(4), 59912 (2013)
29. Ugochukwu-Ibe, I.M., Ibeke, E.: E-learning and covid-19: the Nigerian experience: challenges of teaching technical courses in tertiary institutions. CEUR Workshop Proceedings (2021)
30. Iwendi, C., Mohan, S., Ibeke, E., Ahmadian, A., Ciano, T., et al.: Covid-19 fake news sentiment analysis. Comput. Electr. Eng. **101**, 107967 (2022)
31. Ngabo, D., Dong, W., Ibeke, E., Iwendi, C., Masabo, E.: Tackling pandemics in smart cities using machine learning architecture. Math. Biosci. Eng. **18**(6) (2021)
32. Kumar, R.L., Khan, F., Din, S., Band, S.S., Mosavi, A., Ibeke, E.: Recurrent neural network and reinforcement learning model for covid-19 prediction. Front. Public Health **1437** (2021)
33. Amujo, O., Ibeke, E., Fuzi, R., Ogara, U., Iwendi, C.: Sentiment computation of UK-originated covid-19 vaccine tweets: a chronological analysis and news effect. Sustainability **15**(4), 3212 (2023)
34. Meyer, C.: Depressive disorders were the fourth leading cause of global disease burden in the year 2000. Evid. Based Ment. Health **7**(4), 123 (2004)
35. Murray, C.J., Lopez, A.D., Organization, W.H., et al.: The Global Burden of Disease: a Comprehensive Assessment of Mortality and Disability from Diseases, Injuries, and Risk Factors in 1990 and Projected to 2020: Summary. World Health Organization (1996)
36. U¨stun, T.B., Ayuso-Mateos, J.L., Chatterji, S., Mathers, C., Murray, C.J.: Global burden of depressive disorders in the year 2000. Br. J. Psychiatry **184**(5), 386–392 (2004)

Multi-modal Sentiment Analysis Using Text and Audio for Customer Support Centers

Hardik Srivastava, Sneha Sunil, K. Shantha Kumari, and P. Kanmani

Abstract Customer service has become crucial for every organization looking to grow and improve its clientele in today's cutthroat marketplace. Companies cannot afford to fall short of consumer expectations. With the recent progress in Artificial Intelligence, companies have adopted AI techniques like sentiment analysis to measure customer satisfaction. Chatbots are the foundation for most AI applications in call centers trained for either question-answering tasks or calculating sentiment out of user feedback via surveys. These existing expert systems only utilize text mining techniques to classify sentiment as positive or negative. In this paper, we propose a Multimodal learning framework for tackling the sentiment classification task, employing acoustic and linguistic modalities from real-world conversations between support representatives and customers and survey feedback data using the decision-level fusion technique. Leveraging the classification abilities and feature representations from both modalities, our model achieves the best results amongst all implementations and enables us to better analyze the sentiment expressed by the user than using a single modality like text or audio.

Keywords Sentiment analysis · Multimodal learning · Text mining · BERT · Speech representation learning

H. Srivastava · K. Shantha Kumari · P. Kanmani (✉)
Data Science and Business Systems, SRM Institute of Science and Technology, Chennai, India
e-mail: kanmanip@srmist.edu.in

S. Sunil
Networking and Communications, SRM Institute of Science and Technology, Chennai, India

C. Iwendi et al. (eds.), *Proceedings of ICACTCE'23 – The International Conference on Advances in Communication Technology and Computer Engineering*, Lecture Notes in Networks and Systems 735, https://doi.org/10.1007/978-3-031-37164-6_36

1 Introduction

The internet is an excellent platform for people to express their views on various products and services [1]. Using feedback or communication channels, people produce a substantial amount of multimodal data each day that is rich in sentiment information. As people have the ability to comprehend and react and be transparent at the same time, it becomes a necessity for organizations to analyze human emotions regarding a product or a service. A positive opinion could increase customer retention, while a negative opinion can deteriorate possible customer relationships. By analyzing the customer emotions behind these opinions, Artificial Intelligence can help service providers achieve their desired goals, enabling them to scale up their businesses.

Existing research in sentiment analysis and opinion mining has mostly concentrated on interpreting sentiment toward movies, goods, services, acts, etc., which only includes a single entity's perspective [2]. With the introduction of Voice Assistants, Interactive Voice Response (IVR) Systems, and Chatbots, researchers have begun to develop affinity-based conversation systems to enhance the entire experience by adjusting to users' emotions [3]. However, when we consider an instance of a Human-Human Interaction like a Customer-Agent conversation in a support center, it becomes essential for the service providers to ensure customer satisfaction through quality delivery of services. Customer sentiment analysis has gained widespread importance in customer helpline sectors. Support centers are a natural resource of real-world speech data. Every day, a large number of calls are recorded in order to evaluate how well customer support agents (CSRs) and clients communicate. Sentiment analysis is carried out by analyzing the calls made between consumers and representatives.

As opposed to the artificial datasets collected in controlled environments that have traditionally been utilized in multimodal sentiment research, the real-world data-service calls-that we target and employ in this paper provide additional difficulties, such as noise and natural disruptions. The four phases of sentiment analysis are data collection, data engineering, data analysis, and interpretation [4]. However, it has been fraught with many challenges. For instance, if the opinion was neutral or contained an idiom, the trained model could have trouble identifying the sentiment. This problem arises while performing sentiment analysis using conventional techniques considering only text as an input source [5].

In this research, we provide a Multi-Modal method for tackling the sentiment analysis problem, making use of the sentiment data that is present in both the audio modality (vocal intonations) and the text modality (feedback) for a more precise analysis. Since call records as a whole are not consistent and usually contain a mix of negative and positive utterances, sentiment analysis is addressed at the phrase level. The primary objective here is to determine the customer's emotional tone during the call, capture the sentiment from the shared feedback, and classify them as positive or negative.

The paper is structured as follows. The introduction is followed by Sect. 2 which describes the various Sentiment Analysis methods that have been studied in the previous papers, including text-based, speech-based, and multimodal approaches, their shortcomings, and the novel method we have proposed for our area of application. Section 3 provides a detailed description of our proposed methodology and system implementation, along with the various speaker recognition and speech processing techniques we have employed. Later in this section, we talk about the various experiments that were carried out. Section 4 talks about our results and discussions. The potential research directions to advance the field are covered in Sect. 5.

2 Related Work

2.1 Text-Based Sentiment Analysis

Text Sentiment Analysis is the process of identifying and analyzing the polarity expressed in a text sequence to find whether it has a positive or negative sentiment. There has been a lot of research in this domain focusing on using algorithms like Naive Bayesian, Decision Trees, and Maximum Entropy [6–8]. Vimali et al. [9] discussed a methodology for text sentiment classification using a BiLSTM RNN addressing text categorization and cross-lingual issues. They depicted how social media platforms like Twitter allow people to pool many positive and negative opinions, and a deep learning model with multiple layers is required for sentimental analysis. They took the input data from an E-Commerce platform for their research and displayed a visual representation of the positive and negative data obtained after classification. Haifeng et al. [10] achieved better accuracy in a similar task by utilizing the pre-trained BERT model for text classification from the text comments obtained from social media and obtained an accuracy rate of 85.83%. They even considered categories such as uncivilized and questionable comments while classifying the data, which was a drawback in the latter. Movie reviews are a real-world example and a good source for sentiment analysis to predict if people liked or disliked the movie. Zhongxiang et al. [11] showed how it is manually impossible to go through each review to guess the audience's emotions as the review data scales. They built an Albert model classifier [12] and used the dataset provided by Stanford University that contained a tremendous amount of movie reviews for training. Compared to the analysis performed earlier using traditional LSTMs, the accuracy of the Albert model was improved by 3% over the 89.05% accuracy rate.

2.2 Audio-Based Sentiment Analysis

In text-based classification systems, we cannot model acoustic features like emotions. Voice's tonal qualities need to be extracted and represented in a form to orderly classify emotions and sentiments when dealing with audio data. For feature extraction from speech data, several researchers have used MFCCs (Mel-Frequency Cepstrum Coefficients) [13] and spectral centroid features. Christine et al. [14] developed a deep-learning model to analyze emotions in a human voice. They developed a CNN deep learning model for emotion analysis and classified it into four categories: happiness, sorrow, neutral, and anger. Their approach showed how sentiment analysis is needed in today's world in developing smart cities that can accelerate economic growth. Idioms and proverbs used in social media communication are figurative and can be very hard to detect the polarity. Bashar et al. [15] proposed a method that uses the automated annotation of idioms that do not require human intervention to enrich the performance of the sentiment classifier. They depicted how their model did not require any fine-tuning for BERT.

2.3 Multimodal Sentiment Analysis

Multi-Modal machine learning is a very recent research topic in the industry [16]. It is gaining popularity among several researchers because of the unlimited multimodality source of information on the internet, like text, audio, and video. Multimodal sentiment analysis systems use the multimodal fusion technique that merges the modalities together. In order to get emotional features via deep multimodal learning, Roy et al. [17] used a Deep 2D-CNN with multi-scale coupled to dense DNN. Their techniques to model the Speech Emotion Recognition [18] task also solve the problem of data sparsity. They discussed several methods to improve the accuracy of existing models by performing data augmentation, like pitch modification and noise insertion. A more effective multimodal strategy for sentiment analysis is suggested by Goecke et al. [19]. The paper's main objective was to categorize emotions into two categories: positive and negative. The voice mood, age, or gender of the speaker was identified for a better user experience. Cheong et al. [20] proposed using 2-D CNNs and DNNs for encoding each segment into a vector of fixed length by merging the activations together.

The variability and noise that occur in the actual world must be handled by a strong expert system performing sentiment analysis. Our proposed system implementation collects real-world recorded calls that contain ambient noise as well as a variety of person-to-person communication patterns, in contrast to earlier proposed methodologies on multimodal sentiment analysis that used audio data recorded in laboratory settings. To properly extract usable data from various sources, these two circumstances create challenges that must be resolved.

We aim to train a Multi-Modal Sentiment Analysis model to leverage this information from speech data and text to measure customer satisfaction by directly analyzing the customer-agent interactions recorded in an uncontrolled environment and content extracted from survey and feedback forms.

3 Methodology

3.1 Dataset

We have used two datasets for our experiments, one for audio and another for text. The speech dataset is collected from Summa Linguae Technologies, which provided us with actual Call Center calls in the US English language. The dataset contains 150 audio files of call center conversations sampled at 8000 Hz with a bit rate of 64 kb/s. The speech data between support agents and customers is recorded in noisy real-world conditions and not in a controlled environment like a laboratory, giving our model the benefit of generalizing to real-world data. The audio dataset contains male and female speakers selected randomly. All calls are in wave format and recorded for an average of 13 min. We propose leveraging audio labeling techniques to effectively annotate the speech data in a format that is subsequently compatible with our deep learning model. Our entire dataset, on average, amounts to approximately 117,000 s of speech data, which due to infrastructure constraints, is challenging to process. Hence we consider taking a sample size of 60 audio files from the entire dataset, out of which we prepared 11,700 chunks of audio samples which were, on an average of 4 s duration each. These utterances are labeled manually with Positive (value of 1) and Negative (value of 0) based on the sentiment of the dialogue in that particular audio sample. All the audio segments are transcribed into text dynamically using Python scripting. The annotations are then exported into JSON format and saved for our model to ingest and perform further analysis. Table 1 shows an example annotation of the first five utterances of one of the calls from our audio dataset. We labeled the utterances with either positive or negative sentiment according to the emotion expressed in the speech. The transcription for the utterance with Segment ID *4* is shown in Table 2.

The data pipeline also consists of speaker diarization [21] and automatic tagging which is discussed later in this section.

For our text modality, we utilize the Multi-Domain Sentiment Dataset [22], which contains product reviews and customer feedback on multiple product and service types collected from an e-commerce website. The dataset was categorized into 4 product types, with each review consisting of a customer rating (0–5) and some meta-data.

Table 1 Representation of a sample annotation of the first five utterances of one of the calls from our audio dataset

Segment ID	Start	End	Total duration	Sentiment
1	0:00	0:04	0:04	1
2	0:04	0:06	0:02	1
3	0:06	0:08	0:02	1
4	0:08	0:22	0:14	0
5	0:22	0:36	0:14	0

Table 2 Representation of the transcription for utterance with Segment ID 4, as discussed in Sect. 3.1

Segment ID	Transcription
4	Oh not so good. I found out that #ah I'm not able to find any locations that are offering a rental car to throw in my One hundred dollars as bank balance I have ((Pennzoil)) I keep searching every single time What is it? But there's no results

Reviews with ratings ≥3 were labeled as positive, those with ratings <3 as negative, and the other reviews were disregarded because of their unclear polarity. Following this conversion, we got a balanced number of 1000 positive and 1000 negative reviews for each domain as in the polarity dataset [23]. We made sure that there were equal amounts of good and negative examples because we were able to collect labels for all of the reviews. After applying various pre-processing techniques like removing null data, filtering out feedback in languages other than English, and removing special characters and punctuation, we obtained 2000 rows of processed data that were suitable for fine-tuning the BERT model for our Sentiment Analysis task.

3.2 Problem Formulation

Now we formally propose our formulation of the problem. Our problem consists of three parts. The first part deals with extracting features from our speech data. Given a speech sequence X_S that is sampled from a conversation between a user and a support agent, its audio features are denoted as $X_S = [S_1, S_2, S_3, \ldots, S_n]$. Similarly, for our second part, consider a text sequence of word-piece tokens generated from the feedback shared by a user after his interaction with the agent. Let us assume the sequence to be $T = [T_1, T_2, T_4, \ldots, T_n]$, where n is the maximum length of the sequence. Since we are using BERT to generate the text embeddings, the embedding layer will add an additional embedding for the special [CLS] token. Hence, the output of the last encoder layer becomes $X_T = [E_{[CLS]}, E_1, E_2, \ldots, E_n]$. Now that we have X_S features and X_T embeddings representing our audio and text modalities, we utilize the decision-level fusion technique to merge these modalities for our Multi-Modal Sentiment Analysis model. We utilize the interaction between X_S and X_T and pass the merged embedding to the Cross-Modal BERT [24] network to perform the sentiment analysis task. Our goal is to create a model that can learn from these multi-modal training data in a way that, given an audio sample $x_1 \in X_S$ and a text sequence $x_2 \in X_T$, that has not yet been encountered, we can correctly estimate a sentiment value based on the ongoing conversation between a user and a support center agent and the feedback shared by him to efficiently and effectively analyze customer satisfaction.

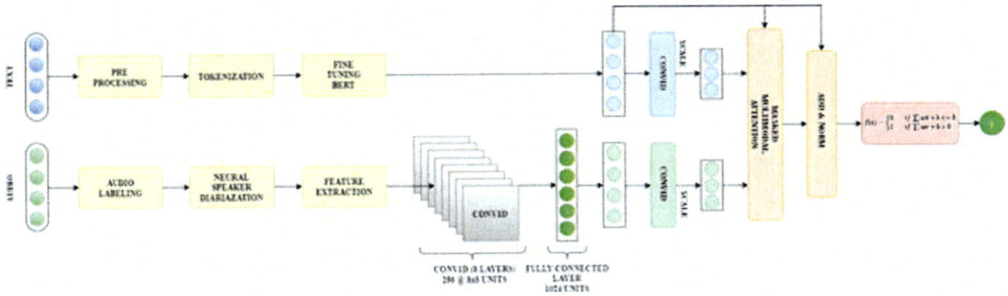

Fig. 1 Representation of our overall system implementation

3.3 System Implementation

In this paper, we propose a Multi-Modal Sentiment Analysis technique that aims to utilize the information available in different modalities to predict customer satisfaction during the customer-agent interaction using sentiment prediction. In our model, we use features extracted from speech signals from conversations between support agents and customers and text data from feedback and survey forms to build a learning framework that leverages the classification ability of each modality. Figure 1 represents our overall system implementation.

We begin with our modeling phase by initially annotating our audio files to extract utterances from the speech samples separated by speaker identifiers. These segments are manually labeled by us in accordance with the output type of our binary classification model. We perform Neural Speaker Diarization [21] or the Speaker Recognition process to fetch the speakers involved in the conversation using PyTorch's API implementation of the Pyannote Audio Toolkit [25]. After generating the chunks of data, we automatically tag them with unique speaker identifiers so that it becomes easy to transcribe their respective data. The tagging algorithm works in an unsupervised fashion, i.e., it first finds whether the chunks are from the same speaker or a different speaker. In addition to transcribing the chunks to text, we also tag them with necessary meta-data like the total running time, the start and end time, the primary language used, etc. This annotated data in JSON format now serves as a segmented and well-labeled representation of the entire audio file with separate information for each speaker's speech output as they converse in the call. Table 3 shows audio annotations for five utterances of an audio file in a tabular format.

After generating our audio annotations, we generate their speech signals using the Librosa framework [26] in Python. Since humans can hear sounds with a comparatively small range of frequencies and loudness, we also produced an identical signal for time-frequency analysis of the various frequencies and amplitudes of our signal over time. The frequency and amplitude are calculated which helps us plot the spectrogram for analysis, showing decibels of the various frequencies over time.

Following that, we extract the Mel Frequency Cepstral Coefficients [13] from each signal wave. These coefficients are used for the feature extraction process when dealing with speech data. They represent a sound's power spectrum across a short

Table 3 Representation of a sample annotation of the first five utterances of one of the calls from our audio dataset

Segment ID	Primary type	Locale	Speaker ID	Start	End	Total duration	Sentiment
1	Speech	*en_US*	*en_US*_1000	0:00	0:04	0:04	1
2	Speech	*en_US*	*en_US*_1001	0:04	0:06	0:02	1
3	Speech	*en_US*	*en_US*_1000	0:06	0:08	0:02	1
4	Speech	*en_US*	*en_US*_1001	0:08	0:22	0:14	0
5	Speech	*en_US*	*en_US*_1000	0:22	0:36	0:14	0

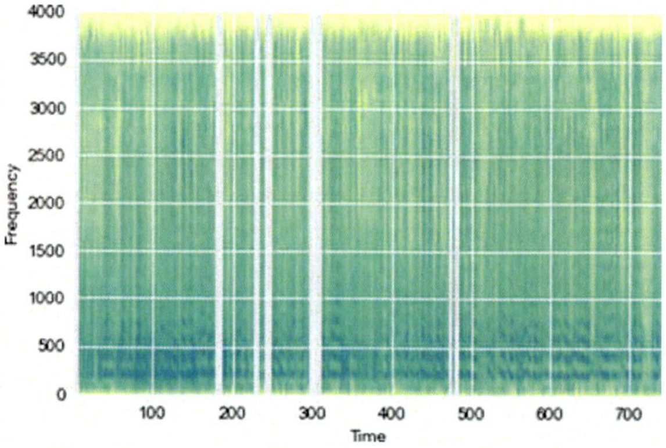

Fig. 2 Representation of an MFCC spectrogram of an audio file in the time-frequency domain

time span [27]. These MFCC values help in further making our speaker diarization system more robust. Figure 2 shows an MFCC spectrogram plotted for one of the audio files.

Since speech signals are sequence data, it is recommended to use an Encoder-Decoder framework, as RNNs are noted for being effective in audio modeling tasks. However, we propose using a Feed-Forward Convolutional Neural Network to model the speech data in our system implementation, as recent studies prove that CNNs tend to perform better than RNNs in various scenarios. The network architecture consists of an input layer, eight convolutional layers (Conv1D) for feature extraction, and one fully connected layer with 1024 hidden neurons. A Max Pooling or an Average Pooling Layer, and a Dropout layer with a dropout value of 0.5 was placed after every Conv1D Layer. Convolutional and Fully connected layers employ Rectified Linear Units (ReLU) [28] as activation functions to introduce non-linearity. The number of kernels is set to eight for all convolutional layers, respectively. Cross-entropy was used to calculate the loss while training and validation, and the Stochastic Gradient Descent [29] algorithm was employed as the optimizer over RMSProp to minimize the loss function over the mini-batches of the training data. We use a learning rate, best suited for our task, of 1e–4 and a batch size of 32 to train our model for 100 epochs over our call center audio data for the sentiment analysis task.

For our text modeling task, we leveraged the Transformers API from Hugging Face to implement the BERT Language Model [30] in TensorFlow and tweaked it with additional training data to make it perform our Sentiment Analysis task.

The multimodal machine learning technique with decision-level fusion (late fusion) was utilized to merge the audio data from our tests with the BERT results in the last phase of our research. We mixed the aforementioned models by fusing feature representations from both our CNN network and the BERT model using the Cross-Modal BERT (CM-BERT) [24] architecture into a single output representation.

3.3.1 Neural Speaker Diarization

Speaker diarization [21] is a method of segmenting a human voice audio stream into homogenous parts based on the identification of each speaker. The readability of the speech transcription is improved by annotating the speech signal by segmenting the audio stream into speaker turns and identifying the speaker's real identity when used in conjunction with speaker identification technologies. NSD answers the question of "Who spoke when?" Speaker segmentation and speaker clustering are used to create speaker diarization. One of the most popular techniques to implement NSD is Gaussian mixture models. We leverage Pyannote's Audio API, an open-source toolkit in Python, to perform our diarization task. After all the speakers participating in the conversation are identified, we assign them unique identifiers and tag them with their respective transcribed text. This helps us generate our annotated call center audio data. Table 3 represents an example annotation showing the first five utterances of one of our audio files.

3.3.2 Mel Frequency Cepstral Coefficients (MFCCs)

Any automatic speech recognition system starts with the process of extracting features, which involves recognizing the parts of the audio signal that are good for detecting the linguistic content and ignoring anything else that contains information such as background noise, emotion, etc. We use MFCCs [13] to divide the frequency band of the signal into sub-bands using the MEL scale and then extract the Cepstral Coefficients using a linear cosine transform. A pure tone's perceived frequency or pitch, and its actual measured frequency is related using the Mel scale. Using this scale makes our features more similar to human hearing. A spectrogram showing the distribution of frequency versus time of several pronounced words is shown in Fig. 2. Equation 1 shows the relation between frequency and the Mel scale.

$$M(f) = 1125 \ln (1 + f/700) \tag{1}$$

3.3.3 Audio Feature Extraction

Raw audio signals contain a lot of noise and disturbance, even when recorded in a controlled environment. However, our call center audio dataset contains recordings of actual conversations from an organization's support center that took place in an uncontrolled environment. Using noisy disturbances and real-world data helps us model the actual speech features which our algorithm will be evaluated on when deployed in a real-world scenario. Because of the noise and disturbance, these signals cannot be fed directly as input to our network. It is observed that for using the base model's input, features extracted from the audio signal rather than the raw audio signal alone would result in significantly improved performance. Hence, MFCCs are used to extract characteristics from audio signals, and therefore we have used them to be our input feature. Loading audio data and converting it to MFCC format can be quickly done by the Python package Librosa.

3.3.4 Text Encoder

The text input X^t for our model, extracted from customer feedback, is first tokenized by the BERT Tokenizer using WordPiece [31]. We mark the beginning of sentences using the [CLS] special tokens respectively. Adding the [CLS] token is necessary for BERT to generate our word embeddings. These tokens are then converted to token ids using BERTTokenizer's fixed-size vocabulary. Assuming the token ids to be X^{tk}, the encoded feature representation in the form of word embeddings is generated and given as output by BERT. This computation is formally represented as $E^t = BERT(X^{tk})$.

3.3.5 Fusion

Fusion is the technique to merge multiple representations from different classifiers into a single representation. There are two ways to perform this task: decision-level fusion (late fusion) and feature-level fusion (early fusion). We used decision-level fusion in our experiments to merge both our modalities into one single representation. The benefits of decision-level fusion are numerous [32]. We employed classifiers for text and audio characteristics independently, which is one of the major advantages when using late fusion. A further advantage is the method's processing speed and high-performance capabilities.

3.4 Experiments

3.4.1 Data Pre-processing

The methods of pre-processing while extracting features from our audio and text modalities are as below:
Text: The reviews are tokenized using WordPiece, and the encoded text representations are generated in accordance with the BERT's format of input embeddings.
Audio: We annotate our call center audio data to prepare chunks of shorter audio clips segregated by respective speaker ids. Annotating the calls helps us to label them with respect to the sentiment expressed in the dialogue in accordance with our Sentiment Analysis task. We use MFCCs to extract the audio features for our model and further analyze the respective speech signals.

After several minor changes and tweaks in hyperparameter values like the optimizer, learning rate schedule, and even our network architecture, we discovered that our CNN model performs better generalization if we introduced augmented audio samples to our network. Hence, we explored some audio augmentation methods to increase the variability in our audio data. We explored techniques like Pitch Tuning and Shifting, which shift the pitch of the speech signal up or down depending on the pitch interval that is chosen beforehand.

3.4.2 Fine-Tuning

We leverage the BERT Language Model [30] and tweak it with additional training data to make it perform our Sentiment Analysis task for our text modality. We express the input embeddings as the sum of token, segmentation, and position embeddings to fine-tune the sentiment analysis model on our multi-domain sentiment dataset [33]. The pre-training head of the model is removed and replaced with a classification head. When fine-tuning, only the classification layer weights $W \in R^{N \times H}$, where N represents the number of output labels, are added as additional parameters. We use a batch size of 16 and fine-tune for 100 epochs over the data. We choose the best fine-tuning learning rate of 2e–5 for each task on the validation set. During the evaluation, our model achieves an accuracy of 77% and a weighted F1-score of 0.76 on the test set.

4 Results and Discussions

We use accuracy and the weighted F1-score, which is the weighted harmonic mean of precision and recall, as evaluation metrics. The likelihood of delivering accurate values is known as precision. The likelihood that the algorithm would provide meaningful values is known as recall. Both precision and recall for our positive and negative classes are also considered for evaluation.

Table 4 Representation of the results from the sentiment classification task on our test data. Audio and text are treated as individual and separate modalities here

	Audio			Text			
Metrics	CNN	CNN (Aug)	RNN	MNB	SVM	LSTM	BERT
Accuracy	0.69	0.83	0.77	0.74	0.58	0.68	0.77
F1 (w)	0.68	0.81	0.81	0.75	0.30	0.69	0.76
Precision (+)	0.59	0.79	0.83	0.74	0.94	0.57	0.77
Precision (−)	0.55	0.77	0.80	0.76	0.55	0.81	0.78
Recall (+)	0.78	0.85	0.79	0.77	0.18	0.88	0.76
Recall (−)	0.78	0.76	0.83	0.73	0.99	0.66	0.73

Note F1 (w) is the macro-weighted F1-score. Precision (+) and Recall (+) denote the precision and recall values for our positive class and Precision (−) and Recall (−) denote the precision and recall values for our negative class

When we ran our experiments and recorded our results, as tabulated in Table 4, we observed that CNNs performed better than RNNs in modeling the audio features. Our CNN model when evaluated over our data, initially achieved an accuracy of 69% and a weighted F1-score of 0.68, after applying augmentation techniques, scaled its performance by 17% to reach an accuracy of 83% and a weighted F1-score of 0.81. These findings were better than what our RNN network produced. For our text modality, we ran experiments on machine learning-based classifiers to get an analytical overview. Multinomial Naïve Bayes [34, 35] and Support Vector Machines [36] achieved overall weighted F1-scores of 0.75 and 0.30. Long Short-Term Memory Network (LSTM) [37] performs better than these classifiers by reaching an accuracy of 68% and a weighted F1-score of 0.69. When we choose to fine-tune BERT over our multi-domain sentiment dataset, our model achieves acceptable results, like an accuracy of 77%, considering that our dataset contains some instances of transcribed examples and some grammatically incorrect data points that are insignificant and vague.

In order to achieve better results, we merge the two modalities (audio and text) together using the CM-BERT (Cross-Modal BERT) architecture. The multimodal model shows better results than the CNN model (with augmented data) for audio and the classifiers employed for text sentiment analysis. Table 5 represents the tabulated results for our Multi-Modal model when evaluated on both modalities. Our Multi-Modal Sentiment Analysis model achieved an overall accuracy of 89.41% and a weighted F1-score of 0.88 on our call center audio and customer sentiment datasets. Our results prove that leveraging both linguistic and acoustic modalities enable us to better analyze the sentiment expressed by the user than using a single modality like text or audio. The audio features help us to capture the customer's tonality and emotion when engaged in a dialogue by analyzing the respective speech signals, which is not possible when considering only text.

Table 5 Representation of the results from the sentiment classification task on both audio and text modalities

Metrics	Audio and text Multi-modal fusion model
Accuracy	0.89
F1 (w)	0.88
Precision (+)	0.86
Precision (−)	0.84
Recall (+)	0.90
Recall (−)	0.88

5 Conclusion

Insights gained from customer sentiment analysis are rich and valuable, allowing organizations to build customer relationships and enhance loyalty. In this paper, we proposed a system that performs the sentiment analysis task using a multi-modal representation of both audio and text. We benchmark the Call Center Dataset by Summa Linguae Technologies for audio modality and Multi-Domain Sentiment Analysis Dataset for text modality with four modeling approaches, including both machine learning and deep learning techniques. Our Multi-Modal Sentiment Analysis model demonstrates that fusing the audio and text representations to measure customer sentiment in support center areas is better than adopting a single text representation, delivering the best results among all implementations.

Our future research will focus on broadening the existing network architecture, incorporating new features, and increasing the data domain for our model, especially for the audio modality. We are seeking more effective ways to analyze acoustic and text features and how they can help drive important decisions. Exploring additional modalities like video and their role in assessing customer sentiment will open new areas of research. Furthermore, we would like to improve our multi-modal fusion process to achieve better results.

References

1. Mishra, P., Rajnish, R., Kumar, P.: Sentiment analysis of Twitter data: case study on digital India. In: 2016 International Conference on Information Technology (InCITe)—The Next Generation IT Summit on the Theme—Internet of Things: Connect your Worlds, Noida, India, pp. 148–153 (2016). https://doi.org/10.1109/INCITE.2016.7857607
2. Malini, R., Sunitha, M.R.: Opinion mining on movie reviews. In: 2019 1st International Conference on Advances in Information Technology (ICAIT), Chikmagalur, India, pp. 282–286 (2019). https://doi.org/10.1109/ICAIT47043.2019.8987366

3. Sehgal, R.R., Agarwal, S., Raj, G.: Interactive voice response using sentiment analysis in automatic speech recognition systems. In: 2018 International Conference on Advances in Computing and Communication Engineering (ICACCE), Paris, France, pp. 213–218 (2018). https://doi.org/10.1109/ICACCE.2018.8441741
4. Nair, A.J., Veena, G., Vinayak, A.: Comparative study of Twitter sentiment on COVID-19 tweets. In: 2021 5th International Conference on Computing Methodologies and Communication (ICCMC), Erode, India, pp. 1773–1778 (2021). https://doi.org/10.1109/ICCMC51019.2021.9418320
5. Chen, Y., Zhang, Z.: Research on text sentiment analysis based on CNNs and SVM. In: 2018 13th IEEE Conference on Industrial Electronics and Applications (ICIEA), Wuhan, China, pp. 2731–2734 (2018). https://doi.org/10.1109/ICIEA.2018.8398173
6. Pang, B., Lee, L.: A sentimental education: sentiment analysis using subjectivity summarization based on minimum cuts. In: Proceedings of the 42nd Annual Meeting on Association for Computational Linguistics, p. 271. Association for Computational Linguistics (2004)
7. Pang, B., Lee, L.: Seeing stars: exploiting class relationships for sentiment categorization with respect to rating scales. In: Proceedings of the 43rd Annual Meeting on Association for Computational Linguistics, pp. 115–124. Association for Computational Linguistics (2005)
8. Pang, B., Lee, L., Vaithyanathan, S.: Thumbs up?: Sentiment classification using machine learning techniques. In: Proceedings of the ACL-02 Conference on Empirical Methods in Natural Language Processing, vol. 10, pp. 79–86. Association for Computational Linguistics (2002)
9. Vimali, J.S., Murugan, S.: A text based sentiment analysis model using bi-directional LSTM networks. In: 2021 6th International Conference on Communication and Electronics Systems (ICCES), Coimbatore, India, pp. 1652–1658 (2021). https://doi.org/10.1109/ICCES51350.2021.9489129
10. Liang, H., Tang, B., Cao, S.: Sentiment analysis of comment texts on converged media platforms based on BERT model. In: 2021 International Conference on Culture-Oriented Science and Technology (ICCST), Beijing, China, pp. 552–555 (2021). https://doi.org/10.1109/ICCST53801.2021.00120
11. Ding, Z., Qi, Y., Lin, D.: Albert-based sentiment analysis of movie review. In: 2021 4th International Conference on Advanced Electronic Materials, Computers and Software Engineering (AEMCSE), Changsha, China, pp. 1243–1246 (2021). https://doi.org/10.1109/AEMCSE51986.2021.00254
12. Lan, Z., Chen, M., Goodman, S., Gimpel, K., Sharma, P., Soricut, R.: ALBERT: A Lite BERT for Self-supervised Learning of Language Representations (2019). https://doi.org/10.48550/arXiv.1909.11942
13. Hossan, M.A., Memon, S., Gregory, M.A.: A novel approach for MFCC feature extraction. In: 2010 4th International Conference on Signal Processing and Communication Systems, Gold Coast, QLD, Australia, pp. 1–5 (2010). https://doi.org/10.1109/ICSPCS.2010.5709752
14. Sora, C.J., Alkhatib, M.: Speech sentiment analysis for citizen's engagement in smart cities' events. In: 2022 7th International Conference on Smart and Sustainable Technologies (SpliTech), Split/Bol, Croatia, pp. 1–5 (2022). https://doi.org/10.23919/SpliTech55088.2022.9854309
15. Tahayna, B.M.A., Ayyasamy, R.K., Akbar, R.: Automatic sentiment annotation of idiomatic expressions for sentiment analysis task. IEEE Access **10**, 122234–122242 (2022). https://doi.org/10.1109/ACCESS.2022.3222233
16. Ngiam, J., Khosla, A., Kim, M., Nam, J., Lee, H., Ng, A.: Multimodal deep learning. In: Proceedings of the 28th International Conference on Machine Learning, ICML, pp. 689–696 (2011)
17. Roy, S., Ghoshal, S., Basak, R., Basu, P., Roy, N.: Multimodal sentiment analysis of human speech using deep learning. In: Interdisciplinary Research in Technology and Management (IRTM), Kolkata, India, pp. 1–4 (2022). https://doi.org/10.1109/IRTM54583.2022.9791745
18. Aouani, H., Ayed, Y.B.: Speech emotion recognition with deep learning. Procedia Comput. Sci. **176**, 251–260 (2020). https://doi.org/10.1016/j.procs.2020.08.027

19. Dhall, A., Goecke, R., Joshi, J., Wagner, M., Gedeon, T.: Emotion recognition in the wild challenge 2013. In: Proceedings of the 15th ACM on International Conference on Multimodal Interaction (ICMI'13), pp. 509–516. Association for Computing Machinery, New York, NY, USA (2013). https://doi.org/10.1145/2522848.2531739
20. Cheong, S., Oh, S.H., Lee, S.-Y.: Support Vector Machines with Binary Tree Architecture for Multi-class Classification
21. Anguera, X., Bozonnet, S., Evans, N., Fredouille, C., Friedland, G., Vinyals, O.: Speaker diarization: a review of recent research. IEEE Trans. Audio Speech Lang. Process. **20**(2), 356–370 (2012). https://doi.org/10.1109/TASL.2011.2125954
22. Blitzer, J., Dredze, M., Pereira, F.: Biographies, Bollywood, boom-boxes and blenders: domain adaptation for sentiment classification. In: Proceedings of the 45th Annual Meeting of the Association of Computational Linguistics, pp. 440–447, Prague, Czech Republic. Association for Computational Linguistics (2007)
23. Yuan, Z., Wu, S., Wu, F., Liu, J., Huang, Y.: Domain attention model for multi-domain sentiment classification. Knowl. Based Syst. **155**, 1–10 (2018). https://doi.org/10.1016/j.knosys.2018.05.004
24. Yang, K., Xu, H., Gao, K.: CM-BERT: cross-modal BERT for text-audio sentiment analysis. In: Proceedings of the 28th ACM International Conference on Multimedia (MM'20), pp. 521–528. Association for Computing Machinery, New York, NY, USA (2020). https://doi.org/10.1145/3394171.3413690
25. Bredin H., et al.: Pyannote.Audio: neural building blocks for speaker diarization. In: ICASSP 2020 IEEE International Conference on Acoustics, Speech and Signal Processing (ICASSP), Barcelona, Spain, pp. 7124–7128 (2020). https://doi.org/10.1109/ICASSP40776.2020.9052974
26. McFee, B., Raffel, C., Liang, D., Ellis, D., Mcvicar, M., Battenberg, E., Nieto, O.: Librosa: Audio and Music Signal Analysis in Python, pp. 18–24 (2015). https://doi.org/10.25080/Majora-7b98e3ed-003
27. Wanli, Z., Guoxin, L.: The research of feature extraction based on MFCC for speaker recognition. In: Proceedings of 2013 3rd International Conference on Computer Science and Network Technology, Dalian, China, pp. 1074–1077 (2013). https://doi.org/10.1109/ICCSNT.2013.6967289
28. Agarap, A.F.: Deep Learning using Rectified Linear Units (ReLU) (2018). https://doi.org/10.48550/arXiv.1803.08375
29. Ruder, S.: An Overview of Gradient Descent Optimization Algorithms (2016). arXiv:1609.04747
30. Devlin, J., Chang, M., Lee, K., Toutanova, K.: BERT: Pre-training of Deep Bidirectional Transformers for Language Understanding (2018). https://doi.org/10.48550/arXiv.1810.04805
31. Wu, Y., Schuster, M., Chen, Z., Le, Q. V., Norouzi, M., Macherey, W., Krikun, M., Cao, Y., Gao, Q., Macherey, K., Klingner, J., Shah, A., Johnson, M., Liu, X., Kaiser, Ł., Gouws, S., Kato, Y., Kudo, T., Kazawa, H., Dean, J.: Google's Neural Machine Translation System: Bridging the Gap between Human and Machine Translation (2016). https://doi.org/10.48550/arXiv.1609.08144
32. Roheda, S., Krim, H., Luo, Z.-Q., Wu, T.: Decision level fusion: an event driven approach. In: 26th European Signal Processing Conference (EUSIPCO), Rome, Italy, pp. 2598–2602 (2018). https://doi.org/10.23919/EUSIPCO.2018.8553412
33. Guven, Z.A.: Comparison of BERT models and machine learning methods for sentiment analysis on Turkish tweets. In: 2021 6th International Conference on Computer Science and Engineering (UBMK), Ankara, Turkey, pp. 98–101 (2021). https://doi.org/10.1109/UBMK52708.2021.9559014
34. Xu, S., Li, Y., Zheng, W.: Bayesian Multinomial Na ïve Bayes Classifier to Text Classification, pp. 347–352 (2017). https://doi.org/10.1007/978-981-10-5041-1_57
35. Syrien, A., Hanumanthappa, P., Balasubaramanian, S.: Sentiment analysis of tweets using Nave Bayes algorithm through R programming. Int. J. Comput. Sci. Eng. **6**, 884–889 (2018). https://doi.org/10.26438/ijcse/v6i11.884889

36. Evgeniou, T., Pontil, M.: Support Vector Machines: Theory and Applications, vol. 2049, pp. 249–257 (2001). https://doi.org/10.1007/3-540-44673-7_12
37. Hochreiter, S., Schmidhuber, J.: Long short-term memory. Neural Comput. **9**, 1735–80 (1997). https://doi.org/10.1162/neco.1997.9.8.1735

Comparative Analysis on the Use of Teleconsultation Using Support

Vector Regression and Decision Tree Regression to Predict Patient Satisfaction

Chijioke Victor Uzochukwu, Oluwabukola A. Adetiloye, Afolake O. Adedayo, and Celestine Iwendi

Abstract Teleconsultation is the use of electronic information and communication technology to assist and provide medical care to patients who are unable to go to a healthcare facility for treatment. Globally, teleconsultation is used to provide medical care in a variety of specialisations, for different ailments, and in a variety of ways. Over the past years, significant advancement in technology has improved the accessibility and standard of care that is received by patients through teleconsultation. Over time, researchers have examined the benefits and drawbacks of teleconsultation in comparison to conventional patients visit but still note that the benefits of teleconsultation outweigh its drawbacks as patients can easily have access to quality medical care attention remotely, easily, and timely. Recently, researchers have documented the use of machine-learning techniques to predict diseases and patient experiences like satisfaction, however, there are few papers on the prediction of experiences compared to the prediction of diseases. Therefore, this paper adopts the use of Supervised Machine Learning techniques in training and testing patients' dataset and specifically used regression to predict patient satisfaction. Comparing the results gotten for Support Vector Regression (SVR) Model (using radial basis function kernel) with Decision Tree Regression, it is evident that SVR Model is the best-fit model for the dataset because it has a lower mean squared error of 1.061899561612261 for test data compared to the mean squared error value of DTR Model which is 2.6202531645569622. Hence, this paper concludes that SVR Model is the best-fit model to be used with teleconsultation in predicting patient satisfaction. However, the paper recommends that more machine learning algorithms should be explored and implemented with teleconsultation in treating patients and improving other healthcare services.

Keywords Comparative analysis · Teleconsultation · Support vector regression · Decision tree regression · Patient satisfaction prediction

C. V. Uzochukwu (✉) · O. A. Adetiloye · A. O. Adedayo · C. Iwendi
School of Creative Technologies, University of Bolton, Bolton, UK
e-mail: uzochukwuchijioke@ieee.org

C. Iwendi et al. (eds.), *Proceedings of ICACTCE'23—The International Conference on Advances in Communication Technology and Computer Engineering*, Lecture Notes in Networks and Systems 735, https://doi.org/10.1007/978-3-031-37164-6_37

1 Introduction

Teleconsultation is the use of electronic information and communication technology to assist and provide medical care to patients who are unable to go to a healthcare facility for treatment. It includes a broad range of technological innovations, including videoconferencing, wearable technology, internet platforms, mobile applications, streaming media, etc. [1]. In teleconsultation, patients' data and all healthcare-related data are stored in the cloud for easy access, sharing and security. Cloud storage is advantageous because it enhances speedy implementation and effective/reliable data management [2]. Globally, teleconsultation is used to provide medical care in a variety of specialisations, for different ailments, and in a variety of ways. Numerous studies suggest that teleconsultation-delivered treatment can sometimes produce better results than face-to-face care in terms of safety and effectiveness. In a variety of situations and scenarios, teleconsultation services can also be economically advantageous [3]. Over the past years, significant advancement in technology has improved the accessibility and standard of care that is received by patients through teleconsultation. Over time, researchers have examined the benefits and drawbacks of teleconsultation in comparison to conventional patients visit but still note that the benefits of teleconsultation outweigh its drawbacks as patients can easily have access to quality medical care attention remotely, easily, and timely. During the Covid-19 pandemic, teleconsultation significantly increased patients' access to high-quality and reasonably priced medical care remotely which assisted in abiding by the precautionary measure of social distancing for both patients' and medical personnel's safety. In addition to teleconsultation, patients and medical personnel can communicate through text messages, emails, mobile applications, and wearable devices [4]. Most hospitals in developed countries use teleconsultation in providing effective and efficient medical care services to patients. Physicians also use teleconsultation in radiology where a remote diagnosis is carried out through cloud-based imaging systems, however, this communication is largely between physicians and not patients [5].

2 Related Works

2.1 Use of Machine Learning Techniques in Healthcare

Recently, Artificial Intelligence (AI) has prominently been used in the healthcare sector. Machine Learning (ML) is a subset of AI described as the capability of a computer to learn on its own, generate, and provide reliable predictions [6]. Prediction of outcomes is a major area of machine learning. Research has shown that the use of predictive models based on machine learning techniques is widely adopted for use in the healthcare sector these days as doctors need accurate and credible predictions for the outcomes of patients' ailments which in turn guides in making timely treatment decisions [7]. ML is widely used in healthcare for predictive analytics, natural language processing, automated image interpretation, extraction of data from electronic health records, electrocardiograms, early and accurate detection of patients' diseases/ailments, early treatment decisions and modalities, remote healthcare monitoring, medication development, prediction of mortality or complications after surgical procedures, and cost reduction among others [6, 8, 9]. Although, the implementation of machine intelligence algorithms is expensive, nevertheless, recent improvements in computational power have made it easy to incorporate machine intelligence algorithms with healthcare applications [8, 10]. Furthermore, healthcare management uses machine intelligence to predict patient wait times in urgent care waiting areas. The ML models calculate wait times based on patients' information, their level of discomfort, charts of the busiest departments, etc. The adoption and use of ML in healthcare is therefore beneficial to medical professionals in terms of timely and accurate delivery of medical services, and to patients in terms of cost reduction, accuracy, reliability, and satisfaction [11].

2.2 *Review of Past Studies*

Over the years, numerous machine learning techniques have been applied in making teleconsultation smarter in making predictions [12]. This paper discusses some of the ML techniques that have been explored. The implementation and use of a Wireless Body Area Network (WBAN) with the incorporation of a Decision Tree Algorithm (which is one of the machine learning algorithms) have helped patients to learn about their blood sugar levels, blood pressure, and blood viscosity [2]. Decision Support Systems using Decision Tree and Support Vector Algorithms are used in investigating patients' diseases to provide background information about patients' specific health issues and guidance on when to commence treatment and how to administer drugs for the identified diseases [11]. Support vector machines and decision tree are also used in making thyroid disease predictions. Majali et al. used decision tree and frequent pattern to predict breast cancer, in which decision tree gives an accuracy of 94% [11, 13]. Support Vector Machine has been used to predict kidney failure [12]. Support vector algorithm with different kernel functions have also been used to predict and detect diabetes [14]. Over the years, the combination of support vector machines and artificial neural networks with medical infrastructures has been effective and efficient in predicting human diseases [15]. It is known from previous studies that naive Bayes has been used to diagnose heart disease with an accuracy of 86% and Support Vector Machine with an accuracy of 96.40% for breast diagnosis. Hence, machine learning algorithms have been used to predict a range of illnesses [11]. Furthermore, research has shown that machine learning techniques have also been used to predict patients' experiences like satisfaction and some of this research includes Paper [3] which discusses how patient satisfaction is predicted using traditional statistical methods specifically binomial correlation and linear regression in predicting patient satisfaction with linear regression having the highest accuracy of

88%. Paper [16] also explores patients' satisfaction using teleconsultation otherwise known as telemedicine in Bangladesh using structural equation modelling and artificial neural network approaches, the findings of this paper show the determinants of patients' satisfaction with teleconsultation which in turn help medical personnel, academics, and Information Technology experts to make wise decisions in their subsequent application of teleconsultation. The authors of the paper recommended future research directions. Although there has been numerous research in the use and incorporation of machine learning algorithms with teleconsultation in healthcare, with the identified machine learning algorithms used being effective and efficient, still there is a need for more evaluation and improvements of these algorithms and also the addition of other machine learning algorithms that have not been widely used to detect human diseases. Through a systematic review of past studies, it has been discovered that most ML algorithms used are for the prediction of specific human diseases in teleconsultation and few works on the use of machine learning techniques for the prediction of patient satisfaction using teleconsultation. Hence this paper improves on these gaps by making comparisons between two supervised machine learning techniques (support vector regression and decision tree regression) in predicting patient satisfaction using teleconsultation.

3 Materials and Methods

The tools and libraries used for this paper include Matplotlib, NumPy and pandas, Scikit Learn, this study made use of Google Colaboratory as a web Integrated Environment (IDE) using the CPU runtime model: 79, CPU Family: 6, model name: Intel(R) HP(R) CPU @ 2.20 GHz and cache size: 56,320 KB. The storage used is Google Drive. The paper creates two models and makes comparisons to determine the best-fit model.

3.1 Research Design

The questionnaire used has 24 items, and the questions of the questionnaire are sourced from previous studies that are related to the subject of this paper. The questionnaire focuses on getting the patients' experiences while using teleconsultation and telehealth, some of the questions are adapted from [17–22]. Furthermore, the items in the questionnaires were developed by other researchers, e.g., perceived usability, perceived ease of use, patient satisfaction, future use, perceived quality interaction, and perceived effectiveness. The questions that are adapted from other questionnaire items have undergone the test for reliability and construct validity in previous studies (Cronbach's Alpha 0.80-0.90). A face and content validity test are conducted on the questionnaire. This method employs one of the most effective techniques for confirming the questionnaire's content and the ability of people to understand it [23].

3.2 *Data Understanding*

This paper used primary data, which are collected using a convenient sampling method. A questionnaire containing 22- close-ended questions are prepared using google forms a free web-based platform and effectively dispersed online to residents in North Cyprus to explore the impact and how the patients feel while using teleconsultation in providing health care services. This application is used to prepare the questions, collect the data from the questions, investigate the responses from the participants from the survey, and provide a good representation of the data in charts and tables [24, 25]. There are 462 respondents drawn from a population of 326,000 with an error margin of 5% and a confidence level of 95%. After the pre-processing of the data collected, the number of accepted responses is 391, which is adjudged to be valid and complete. Consequently, the sample space is (n = 391). This pegs the response rate at 84.63%. In the Questionnaire, question 1 seeks to know the patient's frequency of use of teleconsultation services, questions 2–4 captures the demographic information about the participant which include Age, Gender, Educational qualifications, questions 5 and 6 examines the type of digital technology used and the services offered to the patient. While questions 7, 8, and 9 investigate the perceived usefulness of teleconsultation services. Question 10-12 investigates the ease of use of teleconsultations services by the patients, question 10, 12, 14, 17 of the questionnaire investigates the patient's satisfaction and future use of teleconsultations services, and question 15 tests the perceived quality of communication or interaction, while question 19–21 test for the perceived effectiveness of treatment received while using teleconsultation services. The questionnaire used a 5-point Likert scale (where 5 = strongly disagree; 4 = disagree; 3 = neutral; 2 = agree; 1 = strongly agree). The 5-point Likert scale is very effective in capturing the best feeling of the respondents. This Likert scaling method was applied to all the questions in the questionnaire except for questions 19, and 20 which are open questions, where the participants are expected to type in their suggestions and responses, question 5 contains multiple-choice questions where respondents are allowed to tick all the answers that apply to the questions being asked, also questions 4 and 6 contains multiple-choice questions, where the participants are expected to choose the options best suited for them [9]. (i) perceived usefulness, (ii) perceived ease of use, (iii) patient satisfaction and future use, (iv) perceived quality interaction, and (v) perceived effectiveness. Table 1 shows the summary of the dataset used.

Table 1 Summary of dataset

Columns	Descriptions	Values (for categorical variables)		Data Type
Q1	Frequency of the use of teleconsultation	Daily	5	Int
		1–2 times weekly	4	
		1–2 times monthly	3	
		1–2 times yearly	2	
		Other	1	
Gender	Patients gender	Male	1	Int
		Female	2	
Age	Respondents age	18–30	1	Int
		31–40	2	
		41–50	3	
		51–60	4	
		61-above	5	
Education	Educational status of respondents	PhD Masters BSc High school Other		Int
Health service	Health care services received using teleconsultation	Diagnosis Prescription Consultation Other		Int
PU	Perceived usefulness of teleconsultation	Strongly disagree Disagree Neutral Agree Strongly agree		Int
EU	Ease of use of teleconsultation for receiving healthcare	Strongly disagree Disagree Neutral Agree Strongly agree		Int
CQ	Quality of communication using teleconsultation	Strongly disagree Disagree Neutral Agree Strongly agree		Int

(continued)

Table 1 (continued)

Columns	Descriptions	Values (for categorical variables)	Data Type
PS	Patient satisfaction with teleconsultation	Strongly disagree Disagree Neutral Agree Strongly agree	Int
EF	Effectiveness of teleconsultation	Strongly disagree Disagree Neutral Agree Strongly agree	Int
Challenges	Challenges faced using Teleconsultation	Challenges faced using teleconsultation	String

3.2.1 Data Visualizations

Figures 1, 2 and 3 show the visualizations of some of the responses of the respondents (patients). Figure 1 shows that 454 responses were gotten, however, out of the 454 respondents, 54.4% disclosed their gender as male and 45.2% disclosed their gender as female. Hence, from the responses, there are more males than females. Figure 2 shows patient satisfaction rate in terms of using the services provided by teleconsultation. From the result gotten 42.5% are satisfied with the services provided by teleconsultation. Figure 3 shows that 39% of patients are unsure if they will continue to use teleconsultation while 28.3% are willing to keep using teleconsultation services.

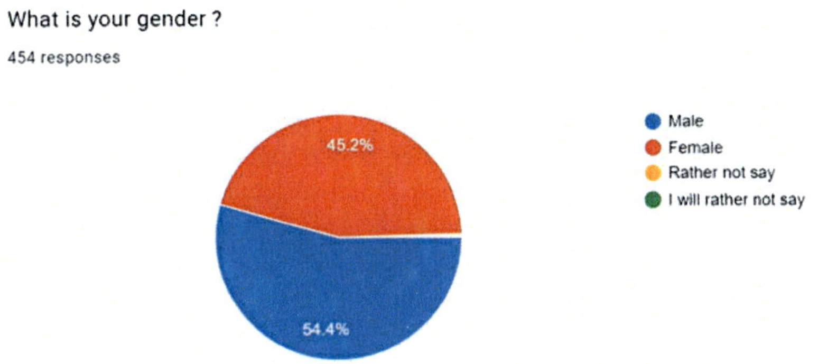

Fig. 1 Gender percentage

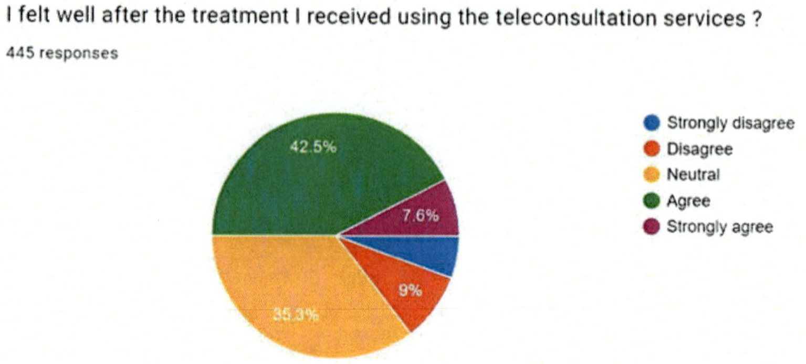

Fig. 2 Patient satisfaction using teleconsultation services

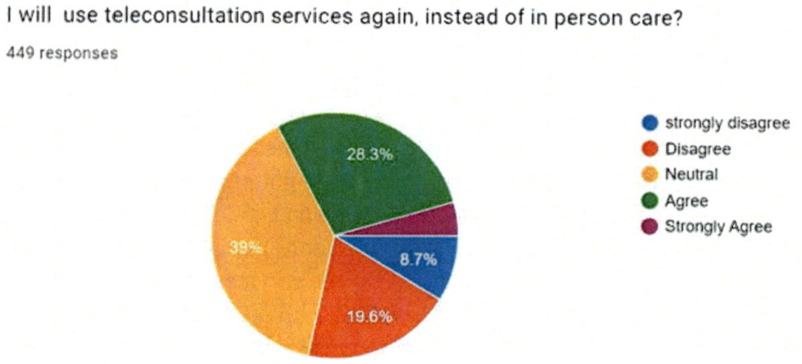

Fig. 3 Patients' willingness to continue using teleconsultation services

4 Machine Learning Models and Results

Support Vector Regression and Decision Tree Regression models are created to predict patient satisfaction using the dataset. 80% of the dataset is for training and 20% for testing. The dataset is cleaned and normalised before applying machine learning techniques. Figures 4, 5, 6, 7 and 8 show the lines of codes written to create the models.

5 Discussion

This paper adopts Supervised Machine Learning techniques in training and testing patients' dataset and specifically used regression because the aim of the study is to predict patient satisfaction. Comparing the results gotten for Support Vector Regression (SVR) Model (using only rbf kernel) with Decision Tree Regression, it is evident

```
X=df [['PU1','PU2','PU3','EU1','EU2','EU3','CQ1','CQ2','CQ3','EF1','EF2','EF3',]]
Y=df['PS1']

X_train, X_test, y_train, y_test = train_test_split(X, Y, test_size=0.2, random_state=1) # 80% training and 20% test

scaler  = StandardScaler()
X_train = scaler.fit_transform(X_train)
x_test = scaler.transform(X_test)
# x_train = X_train.values
# x_test = X_test.values
```

Fig. 4 Splitting dataset into test and train data (80% training and 20% test)

```
#SUPPORT VECTOR REGRESSION
from sklearn.svm import SVR
model= SVR (kernel='rbf')
model.fit(X_train, y_train)

SVR()

y_predict = model.predict(X_test)
y_predict_train = model.predict(X_train)

/usr/local/lib/python3.8/dist-packages/sklearn/base.py:443: UserWarning: X has feature names, but SVR was fitted without feature names
  warnings.warn(
```

Fig. 5 Support vector regression model using radial basis function (RBF) kernel

```
from sklearn import metrics
import numpy as np
#Train Accuracy
print("\nFor train Data: ")
print ('Mean Squared Error: ', metrics.mean_squared_error(y_train, y_predict_train))
print("\nFor test Data: ")
#Test Accuracy
print ('Mean Squared Error: ', metrics.mean_squared_error(y_test, y_predict))



For train Data:
Mean Squared Error:  0.020266326267136973

For test Data:
Mean Squared Error:  1.061899561612261
```

Fig. 6 Support vector regression test and train result

```
# DECISION TREE REGRESSION
from sklearn.tree import DecisionTreeRegressor


dc_tree_model = DecisionTreeRegressor(random_state=0)
dc_tree_model.fit(X_train, y_train)

DecisionTreeRegressor(random_state=0)

y_dc_tree_predict= dc_tree_model.predict(X_test)
y_predict_train = dc_tree_model.predict(X_train)

/usr/local/lib/python3.8/dist-packages/sklearn/base.py:443: UserWarning: X has feature names, but DecisionTreeRegressor was fitted without feature names
  warnings.warn(
```

Fig. 7 Decision tree regression model

```
from sklearn import metrics
import numpy as np
#Train Accuracy
print("\nFor train Data: ")
print ('Mean Squared Error: ', metrics.mean_squared_error(y_train, y_predict_train))
print("\nFor test Data: ")
#Test Accuracy
print ('Mean Squared Error: ', metrics.mean_squared_error(y_test,y_dc_tree_predict))
```

```
For train Data:
Mean Squared Error:  0.0

For test Data:
Mean Squared Error:  2.6202531645569622
```

Fig. 8 Decision tree regression test and train result

that SVR Model is the best fit model for the dataset because it has a lower mean squared error of 1.061899561612261 for test data compared to the mean squared error value of DTR Model which is 2.6202531645569622 for test data. Hence, this paper concludes that SVR Model is the best-fit model to be used with teleconsultation in predicting patient satisfaction.

6 Conclusion

The continuous advancement in teleconsultation has improved patient treatment over the years. Hence, the use of machine learning techniques with teleconsultation has made the system much better by providing the ability to predict variables like patient satisfaction. However, this work proposes that more machine learning algorithms should be explored and implemented with teleconsultation in treating patients and providing other health care services.

References

1. Oliveira Hashiguchi, T.: Bringing health care to the patient: an overview of the use of telemedicine in OECD countries. OECD Health Working Papers, No. 116. OECD Publishing, Paris (2020) https://doi.org/10.1787/8e56ede7-en
2. Saleem, T.J., Chishti, M.A.: Exploring the applications of machine learning in healthcare. Int. J. Sens. Wirel. Commun. Control **10**(2019). https://doi.org/10.2174/2210327910666191220103417
3. Sabarmathi, G., Chinnaiyan, R.: Reliable machine learning approach to predict patient satisfaction for optimal decision making and quality health care. In: International Conference on Communication and Electronics Systems (ICCES), pp. 1489–1493. Coimbatore, India, (2019). https://doi.org/10.1109/ICCES45898.2019.9002593

4. Barbosa, M.T., Sousa, C.S., Morais-Almeida, M., Simões, M.J., Mendes, P.: Telemedicine in COPD: an overview by topics. COPD: J. Chronic Obstr. Pulm. Dis. **17**(5), 601–617 (2020). https://doi.org/10.1080/15412555.2020.1815182
5. Kichloo, A., Albosta, M., Dettloff, K., Wani, F., El-Amir, Z., Singh, J., Aljadah, M., Chakinala, R.C., Kanugula, A.K., Solanki, S., Chugh, S.: Telemedicine, the current COVID-19 pandemic and the future: A narrative review and perspectives moving forward in the USA. Fam. Med. Community Health **8**(3), e000530 (2020). https://doi.org/10.1136/fmch-2020-000530
6. Alanazi, H.O., Abdullah, A.H., Qureshi, K.N.: A critical review for developing accurate and dynamic predictive models using machine learning methods in medicine and health care. J. Med. Syst. **41**(4) (2017). https://doi.org/10.1007/s10916-017-0715-6
7. Shailaja, K., Seetharamulu, B., Jabbar, M.A.: Machine learning in healthcare: a review. In: Second International Conference on Electronics, Communication and Aerospace Technology (ICECA), pp. 910–914. Coimbatore, India (2018). https://doi.org/10.1109/ICECA.2018.8474918
8. Kilic, A.: Artificial intelligence and machine learning in cardiovascular healthcare. Ann. Thorac. Surg. (2019). https://doi.org/10.1016/j.athoracsur.2019.09.042
9. Thandapani, S., Mahaboob, M.I., Iwendi, C., Selvaraj, D., Dumka, A., Rashid, M., Mohan, S.: IoMT with Deep CNN: ai-based intelligent support system for pandemic diseases. Electronics **12**(2), 424 (2023). https://doi.org/10.3390/electronics12020424
10. Iwendi, C., Huescas, C.G.Y., Chakraborty, C., Mohan, S.: COVID-19 health analysis and prediction using machine learning algorithms for Mexico and Brazil patients. J. Exp. Theor. Artif. Intell. 1–21 (2022). https://doi.org/10.1080/0952813x.2022.2058097
11. Loeb, A.E., Rao, S.S., Ficke, J.R., Morris, C.D., Riley, L.H., Levin, A.S. (2020). Departmental experience and lessons learned with accelerated introduction of telemedicine during the COVID-19 crisis. J. Am. Acad. Orthop. Surg. **1**. https://doi.org/10.5435/jaaos-d-20-00380
12. Majali, J., Niranjan, R., Phatak, V., Tadakhe, O.: Data mining techniques for diagnosis and prognosis of cancer. IJARCCE **4**(3), 613–615(2015). https://doi.org/10.17148/ijarcce.2015.43147
13. Latha, R., Vetrivelan, P., Geetha, S.: Telemedicine setup using wireless body area network over cloud. Procedia Comput. Sci. [online] **165**, 285–291 (2019). https://doi.org/10.1016/j.procs.2020.01.074
14. Sharma, N., Dalwal, S.: A novel approach to predict kidney detection using support vector machine. Int. J. Sci. Res. (IJSR) **8**(4), 1984–1991 (2019). https://doi.org/10.21275/19031902
15. Gheorghe, M.: A support vector machine approach for developing telemedicine solutions: Medical diagnosis. Netw. Intell. Stud. **3**(05), 43–48 (2015). https://www.ceeol.com/search/article-detail?id=537388
16. Aiken, L.H., Clarke, S.P., Sloane, D.M., Sochalski, J., Silber, J.H.: Dotación de personal de enfermería hospitalaria y mortalidad de los pacientes, agotamiento de las enfermeras e insatisfacción laboral. J. Am. Med. Assoc. **288**(16), 1987–1993 (2002)
17. Amez, S., Vujić, S., Soffers, P., Baert, S.: Yawning while scrolling? Examining gender differences in the association between smartphone use and sleep quality. J. Sleep Res. **29**(6), 1–13 (2020). https://doi.org/10.1111/jsr.12971
18. Ayatollahi, H., Langarizadeh, M.: Clinicians ' Knowledge and Perception of Telemedicine Technology. (2019)
19. Balasopoulou, A., Kokkinos, P., Pagoulatos, D., Plotas, P., Makri, O.E., Georgakopoulos, C.D., Vantarakis, A., Li, Y., Liu, J.J., Qi, P., Rapoport, Y., Wayman, L.L., Chomsky, A.S., Joshi, R.S., Press, D., Rung, L., Ademola-popoola, D., Africa, S., Article, O., Loukovaara, S.: Symposium recent advances and challenges in the management of retinoblastoma globe-saving treatments. BMC Ophthalmol. **17**(1), 1 (2017). https://doi.org/10.4103/ijo.IJO
20. Becevic, M., Boren, S., Mutrux, R., Shah, Z., Banerjee, S.: User satisfaction with telehealth: study of patients, providers, and coordinators. Health Care Manag. **34**(4), 337–349 (2015). https://doi.org/10.1097/HCM.0000000000000081
21. Glaser, J.R., Glaser, E.M. Microbrightfield review. **20**, 1–12 (2000). papers2://publication/uuid/D3B808FD-66A3-4307-B242-9DABEFB2A398

22. Nayak, M., Narayan, K.A.: Strengths and weakness of online surveys. IOSR J. Hum.Ities Soc. Sci. (IOSR-JHSS) **24**(5), 31–38 (2019). https://doi.org/10.9790/0837-2405053138
23. Alaboudi, A., Atkins, A., Sharp, B., Balkhair, A., Alzahrani, M., Sunbul, T.: Barriers and challenges in adopting Saudi telemedicine network: the perceptions of decision makers of healthcare facilities in Saudi Arabia. J. Infect. Public Health **9**(6), 725–733 (2016). https://doi.org/10.1016/j.jiph.2016.09.001
24. Taherdoost, H.: What is the best response scale for survey and questionnaire design: Review of Different lengths of rating scale/attitude scale/Likert scale by Hamed Taherdoost . SSRN **8**(1), 1–10 (2019). https://papers.ssrn.com/sol3/papers.cfm?abstract_id=3588604
25. Pillai, R., Oza, P., Sharma, P.: Review of machine learning techniques in health care. In: Singh, P., Kar, A., Singh, Y., Kolekar, M., Tanwar, S. (eds) Proceedings of ICRIC 2019. Lecture Notes in Electrical Engineering, vol. 597. Springer, Cham (2020). https://doi.org/10.1007/978-3-030-29407-6_9

COVID-19 in the UK: Sentiment and Emotion Analysis of Tweets Over Time

Olasoji Amujo, Ebuka Ibeke, Celestine Iwendi, and Zakaria Boulouard

Abstract We performed an analysis of tweets concerning the COVID-19 pandemic in the UK over a two-year period, selecting fifteen timelines. Over 110,000 tweets were obtained from Twitter and analysed using BERT and Text2Emotions for sentiment and emotion analysis, respectively. The most common emotions expressed on Twitter about COVID-19 in the UK appeared to be surprise and fear. This is not unusual, given the unprecedented nature of the pandemic. However, as time passed, there was a notable shift in sentiment towards other emotions, such as sad- ness and happiness. Moreover, more positive than negative sentiments were observed over the fifteen timelines studied: eight positive sentiments to seven negative ones. Further, results indicated that confirmed cases, deaths, and government policy heavily influenced public sentiment. This study sheds light on the collective state of mind surrounding the pandemic and provides insight into how people reacted emotionally over time to COVID-19. The results provide valuable insights for policymakers and other stakeholders looking to understand how people respond in times of crisis. Furthermore, it illustrates how sentiment analysis can be used effectively to gain deeper insights into public perception over time. As such, this study is a valuable contribution to understanding the human emotional response, demonstrating how sentiment and emotion can be used to better comprehend a situation and react accordingly.

Keywords COVID-19 · Sentiment analysis · Emotion analysis · Text2Emotion · BERT model

O. Amujo · E. Ibeke (✉)
School of Creative and Cultural Business, Robert Gordon University, Garthdee Road, Garthdee, Aberdeen AB10 7AQ, UK
e-mail: e.ibeke@rgu.ac.uk

C. Iwendi
School of Creative Technologies, University of Bolton, Deane Rd, Bolton BL3 5AB, UK

Z. Boulouard
LIM, Hassan II University of Casablanca, Casablanca, Morocco

C. Iwendi et al. (eds.), *Proceedings of ICACTCE'23—The International Conference on Advances in Communication Technology and Computer Engineering*, Lecture Notes in Networks and Systems 735, https://doi.org/10.1007/978-3-031-37164-6_38

1 Introduction

COVID-19, also known as the coronavirus, was first discovered in Wuhan, China, over two years ago [1]. Its effects were so devastating that the World Health Organisation declared it a public health emergency on January 30, 2020, barely a month after its discovery in December 2019 [2]. Within three months of its discovery, the virus had spread to almost 38 nations, causing chaos around the world and alarm and uncertainty among many [3, 4]. Hospitals were crowded at this time, and governments found it difficult to provide basic healthcare for their people. The COVID-19 virus was formally classified as a global pandemic by Dr Tedros Adhanom Ghebreyesus, Director-General of the World Health Organization, on March 11, 2020. One of the nations most severely affected by the virus was the United Kingdom, which as of September 2022 had over 23 million confirmed cases and more than 192,000 deaths [5].

1.1 The COVID-19 Pandemic in the UK

The COVID-19 pandemic has no doubt been one of the most devastating and disruptive events to happen to humanity in recent memory. No facet of life has been spared from the harsh effects of the pandemic. The outbreak has significantly impacted how people interact in all contexts, including work, school, travelling, and even social gatherings [6]. The UK was one of the European countries with the highest number of COVID-19 cases, with the first confirmed case being reported on January 31, 2020, [7]. Barely nine months later, in October 2020, the UK recorded over 400,000 confirmed cases of COVID-19 and over 42,000 deaths [5]. To combat the pandemic and mitigate its effects, the UK government put several measures in place, such as imposing lockdown restrictions, implementing social distancing guidelines, and using a contact tracing app. In a bid to stop the spread of the virus, the lockdown restrictions were announced in 2020 and lasted for a year [8]. These control measures have been met with much controversy from the citizens. Some argue that they are too restrictive and negatively impact mental health. Others, however, say that they are essential to stop the virus from spreading. As of November 2020, the virus had spread across 180 countries [9]. Since its discovery till this time, neither the UK nor most countries have developed a vaccine to combat the virus and prevent it from killing so many people. Following MHRA clearance, however, the United Kingdom allowed COVID-19 vaccine use in emergencies for the first time on December 2nd, 2020.

1.2 Benefits of Understanding People's Sentiment and Emotion of COVID-19 (During and Post-covid)

As of July 2022, there were over 23 million confirmed cases and 180,000 deaths from the pandemic in the UK. Therefore, it is imperative to study the sentiments and emotions of UK citizens during the pandemic and consider their impact on people's lives. Carrying out a sentiment and emotion analysis is an effective way to study people's feelings and emotions during the pandemic. This analysis will help us discover and understand how people's sentiments and emotions evolved from the news of the first COVID-19 death in the UK to two years after that.

1.3 Emotional Analysis of Tweets

The outbreak of COVID-19 has resulted in a global health crisis. As the events of the outbreak unfolded, many people worldwide turned to social media as an avenue to share information and express their opinions about the outbreak [10]. Twitter is a social media site where users can communicate with one another by sharing brief messages known as "tweets." Twitter has become one of the most widely used social media platforms, with over 229 million daily active users worldwide [11]. Twitter serves as a platform for users to share their thoughts and feelings about current events as well as a source of real-time information about these events [12]. As of January 2022, there were an estimated 18.4 million Twitter users in the United Kingdom. This is a significant increase compared to the 13.7 million users reported in 2019. The vast majority of UK Twitter users are between 18 and 49, with nearly two-thirds falling into the 18–34 age bracket [13]. According to 2020 survey data, the UK has an active Twitter audience in addition to a massive number of users since roughly 87% of users aged 15 and over check in to their Twitter accounts at least once a week, and 59% use it every day [14]. This makes Twitter a valuable source of data for sentiment analysis.

In this paper, we will use sentiment and emotion analysis to critically analyse how the UK public has felt about the coronavirus over time through social media, specifically Twitter. Sentiment analysis, also known as opinion mining, is the use of natural language processing, text analysis, and computational linguistics to identify and extract subjective information from source materials. It can be used to determine the emotional tone of a piece of text, such as a social media post or a product review, and can be used in various applications such as marketing, customer service, and politics [15, 16]. Twitter is a popular microblogging site that allows users to share short messages, or "tweets", of up to 280 characters. Twitter has developed into a valuable data source for sentiment and emotion analysis in relation to the COVID-19 discourse due to its popularity and global reach. This paper used a dataset of over 118,000 tweets about COVID-19 in the UK, collected between March 2020 and January 2022. We will be using this dataset to examine how the sentiment and emotion of tweets about COVID-19 have changed over time, adopting the BERT model for sentiment analysis and Text2Emotion for emotion analysis. The contribution of this paper is divided into seven folds.

- Extract tweets based on different incidents across various timelines of the pandemic.
- Examine the general sentiment about COVID-19 and people's perceptions.
- Explore the regularity of sentiment evolution by examining the tweets from (1).
- Identify and explore the evolution of people's emotions over time based on the tweets.
- Analyse and compare emotions in the tweets captured for the time frames.
- Discuss the impact of government regulations and announcements on overall emotions.
- Make recommendations for further research.

Although research on the sentiment analysis of COVID-19 on people's emotions has been conducted, previous researchers have been unable to analyze tweets within the timelines following the lockdown. There were also a lot of studies that did not focus on a specific country. People's lives and the economies of countries, including the UK, are still being impacted by the COVID-19 pandemic. Although the number of reported deaths or confirmed cases has significantly decreased over time, the fight to eradicate the virus is far from over.

Therefore, the main objective of this research is to chronologically study and understand the sentiments and emotions of UK citizens on Twitter across different timelines from the first COVID-19 death until two years later. We expect that the sentiment and emotions of tweets about COVID-19 will change over time, in line with the changing situation of the pandemic. In the early stages of the outbreak, when information about the virus emerged, we expected to see a lot of fear and uncertainty. As the situation progresses and more information becomes available, we expect to see a decrease in negative sentiments and emotions as people become more accustomed to the idea of the virus and are better able to cope with it.

2 Literature Review

Since the inception of the COVID-19 pandemic, a lot of research has been done worldwide. This section looks into previous research efforts that have been made concerning the COVID-19 pandemic. It is crucial to conduct a critical analysis of earlier studies in order to show the originality of this research work. In their investigation of the emotions and psychology of Twitter users during the COVID-19 lockdown, Priyadarshini et al. [17] made the case that the rapid spread of the virus may have impacted people's psychology. The study also draws attention to the potential link between increased multimedia device use and psychological disorders due to government emergency protocols. The study conducts a sentiment analysis of Twitter data during the pandemic lockdown, covering only the second and fourth weeks after the lockdown was imposed. However, the study covers tweets from all over the world without considering that lockdowns were not imposed at the same time across the different regions. Although the study claims to have covered the whole world, it didn't consider people's sentiments before the lockdown.

During the COVID-19 pandemic, Mathur et al. [18] made an effort to research the feelings that Twitter users around the world were experiencing. The study's goal was to divide tweets into positive and negative categories before further subdividing them into the six fundamental emotions Ekman suggested: surprise, joy, sadness, anger, fear, and disgust [19]. Their findings indicate that the number of positive and negative tweets was nearly equal. The emotion classification reveals that a higher percentage of tweets are connected to trust, demonstrating that people were optimistic about fighting COVID-19 and the measures put in place by governments around the world. The majority of other tweets are filled with fear as COVID-19 cases continue to rise and spread quickly around the globe. At the time of the research, there was no effective vaccination, which could have accounted for the fear in the tweets. The research, however, did not show a discernible shift in people's emotions at particular timelines (for example, during the lockdown in various countries). Also, since government regulations would be different in various regions of the world, the research could have attempted to understand the emotions in each country based on their government regulations.

Dubey et al. [20] carried out a Twitter sentiment study for specific countries, including France, Germany, Belgium, Switzerland, Australia, China, India, and the United States of America, during the COVID-19 pandemic. The study takes the eight fundamental emotions—fear, pleasure, anticipation, anger, disgust, sorrow, surprise, and trust—into account. Most tweets about anger and anticipation come from France and Germany, respectively. While Switzerland has more tweets about fear, the United States reports the most tweets of disgust. The country with the highest percentage of joy was India, while Switzerland had the highest percentage of sadness. The majority of tweets that express surprise and trust come from Belgium. The shortcoming of this study is that it only summarizes the feelings rather than segmenting them into distinct timelines for each nation.

An infoveillance analysis showing the top concerns of tweeters during the COVID-19 pandemic was published by Abd-Alrazaq et al. in [21]. The study uncovered specific themes, such as the virus's origins, modes of trans- mission, effects on economies, societies, and people, as well as methods of prevention. The majority of the tweets had positive sentiments. The negative tweets focused on COVID-19-related mortality and a spike in racism. On average, tweets about economic loss were more common than tweets about travel restrictions and warnings. In this study, only the COVID-19 pandemic concerns were taken into consideration when the analysis was done, not the public's psychology after the lockdown.

In 2020, Dhar et al. [22] presented a study on the impact of COVID-19 on the psychology of more than 15,000 university students. Anxiety levels have been found to be correlated with stressors associated with epidemics. The effects of the pandemic on the economy, daily life, education, and social support are the main sources of stress. The psychological effects of the COVID-19 lock-down on law enforcement personnel were examined by Varalakshmi and Swetha [23]. The execution of the legislation resulted in a huge population adhering to the regulations and restrictions during emergency situations, as well as quarantining those infected with the disease. Amujo et al. [24] performed chronological sentiment analysis of UK-originated vaccine-related tweets. These studies had two drawbacks: they were restricted to a small population and neglected to account for how emotions change over time.

3 Methodology

According to Fig. 1, the study focuses on sentiment analysis and emotion detection of tweets relating to COVID-19 over a predetermined timeline (limited to tweets in the English language posted from the UK). The timeline was chosen based on the announcement of different measures by the government to control the spread of the virus. BERT and Text2emotions were used for sentiment analysis and emotion analysis, respectively, to precisely analyze the sentiments and emotions expressed in tweets about COVID-19. BERT was selected for this study because it has consistently outperformed a number of established NLP techniques [25]. The majority of BERT-based text classification studies produced positive outcomes across a range of languages [26].

Text2emotions is relatively fast, and the results are simple to comprehend. The same emotion classifier was also used in some earlier literature [20]. It begins by determining each word's grammatical function before identifying any adjectives and adverbs that may be present. The library then specifies the emotion connected to the word by comparing the adjectives with the pre-trained database. This section explains the methodology that was used in the study.

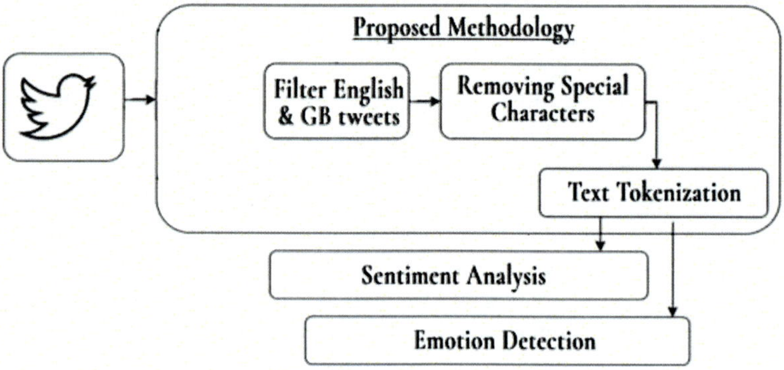

Fig. 1 The proposed methodology

3.1 Data Collection

The application programming interface (API) for retrieving tweets on Twitter was used to extract the data. However, in order to access Twitter's API, a developer account is required. In order to ensure that the data obtained is used for legitimate or research purposes, Twitter validated the justifications for requesting an API key during this process. Tweets from the Twitter API were gathered using Tweepy, a Python library for extracting tweets from the Twitter API. This study uses fifteen timelines that are based on the crucial moments between the first COVID-19 death and two years later. The following keywords were used to extract tweets centred on this research: "Covid", "COVID19", "Covid-19", "COVID 19", "Coronavirus", "COVID19Pandamic", "Lockdown", "StayHome" and "StayHomeSaveLives". The first step is to collect data from Twitter. Several pieces of metadata were also retrieved, such as the time and date of creation and the tweet's text. As a result, no personally identifiable information was gathered. Table 1 below shows the timelines selected for this study.

Table 1 COVID-19 event timeline in the UK

S/N	Timeline	Date of event
1	The first death from COVID-19 in the UK was confirmed	5 March 2020
2	Government Advised self-isolation	12 March 2020
3	The Prime Minister ordered all cafés, pubs, and restaurants to close	20 March 2020
4	Death tolls reached 10,000 amidst the lockdown in the UK	12 April 2020
5	Lockdown lifted, but social distancing advised	13 May 2020
6	Health Secretary Matt Hancock announces hope for COVID vaccine	13 June 2020
7	The majority of schools in the UK reopened	1 September 2020
8	The UK reaches a million COVID-19 cases	31 October 2020
9	The UK hits 50,000 COVID-related deaths	11 November 2020
10	First COVID-19 Vaccine	8 December 2020
11	COVID variants in the UK	14 December 2020
12	Death toll reaches 100,000	15 January 2021
13	A year since the UK recorded its first domestic cases of COVID-19	30 January 2021
14	A third of UK adults are now fully vaccinated against COVID-19	9 May 2021
15	Two years since the UK recorded its first domestic cases of COVID-19	30 January 2022

3.2 Data Preparation

Texts derived from tweets are inherently noisy; cleaning the texts before further processing helps to produce higher quality and reduce computational complexity [27]. After extracting the tweets, we put the raw data through four preprocessing stages.

- Punctuation removal
- Blank space removal
- Removal of Twitter hashtags, usernames, retweets, Unicode characters, word lengthening, URLs, emoticons and internet shorthand/slang
- Text tokenization

3.3 Sentiment Detection Using BERT

In order to understand how generally positive, negative, or neutral the tweets were, this phase entails analyzing the polarity of each tweet. BERT's model architecture (shown in Fig. 2 consists of a multi-layer bidirectional trans- former encoder, BERT, and a new language representation model that has achieved state-of-the-art performance on the majority of NLP tasks to which it has been applied (Hadjer 2021). It is pre-trained on a huge corpus of unlabeled text, which includes the entire English Wikipedia (2,500 million words) and the Book Corpus (800 million words). This process is used to train deep bidirectional representations from these texts in order to understand language context as best as possible. "Bidirectional" refers to the simultaneous consideration of both left and right contexts (previous and posterior tokens) in all layers [28].

3.4 Emotion Detection Using Text2Emotion

Text2Emotion is a Python package created to identify appropriate emotions embedded in text data [29]. Humans use their emotions appropriately when communicating and choose the right words to convey those emotions. Text2Emotion analyzes textual data for embedded emotions and generates an output dictionary. The top five emotion categories for tweets that are most effective are fear, sadness, anger, happiness, and surprise. The emotions are based on the fundamental emotions that leading studies on emotional theory have identified [30].

The user can access this data through a Python dictionary that has five keys for the five emotions that this program can identify. These values indicate how frequently each emotion appears in the text. The values of the emotions must be aggregated to a specific value, which must be either 1 or 0, with 0 only appearing in cases where no emotions were present in the text (such as for strings of characters that make no

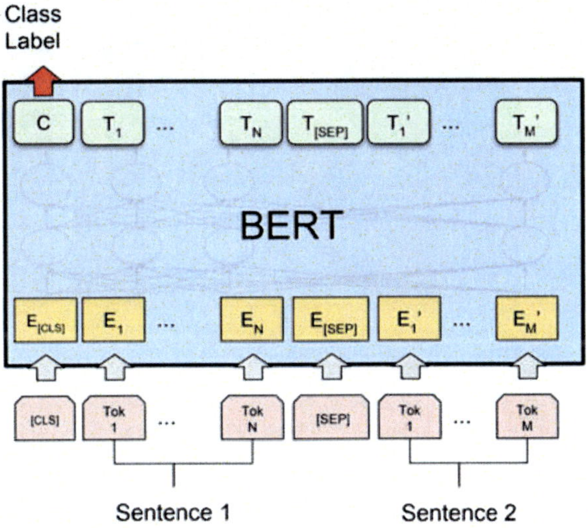

Fig. 2 The BERT model architecture

sense). So, for example, a passage of text that only expresses anger would have an "angry" value of 1 and a value of 0 for every other emotion.

4 Result

This section presents the findings of the analysis, discussed in terms of sentiments and emotions over time. Fear, sadness, and surprise were the most frequently expressed feelings in the tweets, which is understandable given the situation. Fear is a natural reaction to something potentially dangerous, sadness is often associated with loss, and surprise is an unexpected or astonishing event. Other emotions expressed included happiness and anger.

The results from each timeline are shown in the section below. The report represents the sentiments around COVID-19 in the UK. Stars represent the sentiment of a tweet, with 1 being the most negative and 5 being the most positive. In the following sections, we discussed the first five timelines and provide a summary of the sentiment polarity of the 15 timelines in the discussion section.

4.1 Timeline 1: The First Death from COVID-19 in the UK was Confirmed (5 March–11 March 2020)

On 5th March 2020, the UK had its first COVID-19 death. The first death showed the magnitude of the threat COVID-19 poses to human lives, prompting a flood of

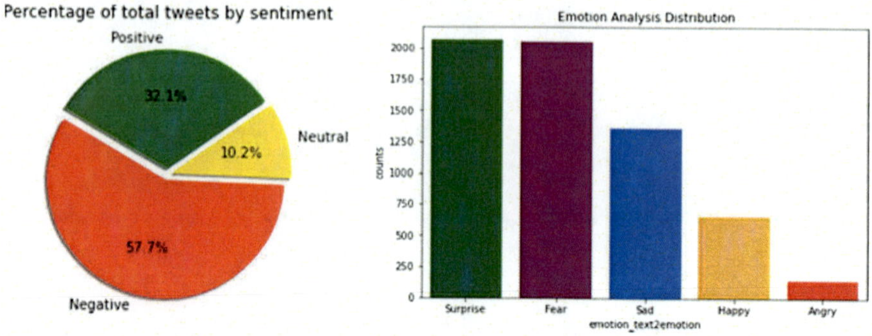

(a) Sentiment Polarity (b) Emotion Analysis

Fig. 3 Timeline 1—sentiment and emotion analysis

negative tweets. Figure 3a shows that public opinion about the pandemic quickly turned negative after the first death was recorded. Negative sentiments, therefore, made up more than 57.7% (n = 3628) of all tweets, while neutral and positive sentiments made up 10.2% (n = 641) and 32.1% (n = 2017), respectively.

Figure 3b demonstrates that "surprise" and "fear," which together accounted for more than 60% of the emotions expressed in this timeline, were the most frequently expressed feelings. Given that there were more deaths during this time, this was to be expected. Due to the nature of the pandemic, the rate of spread, and the fatality rate, the predominant emotions of surprise, fear, and sadness were expected.

4.2 Timeline 2: Government Advised Self-isolation (12 March–18 March 2020)

During this timeline, Prime Minister Boris Johnson confirmed that COVID-19 is not the festive flu, as many presumed, but a more lethal virus that can kill many people. Figure 4a shows that 52.4% (n = 11768) of people had negative sentiments toward self-isolation. 37.5% (n = 8427) had positive sentiments, while 10.1% (n = 2269) had neutral sentiments about the total tweet.

Figure 4b shows that "surprise" and "fear" were the most common responses of people to the imposed self-isolation. Although the sentiments for this time-line were a little bit more positive than the earlier timeline, there is a change in the dominant emotion from surprise to fear compared to Fig. 3b.

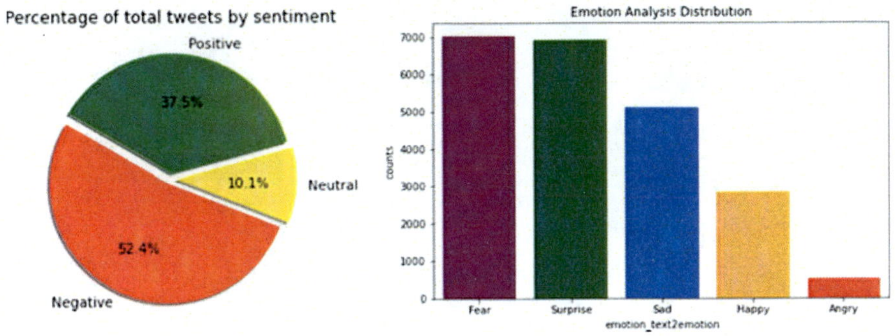

(a) Sentiment Polarity (b) Emotion Analysis

Fig. 4 Timeline 2—sentiment and emotion analysis

4.3 Timeline 3: Prime Minister Orders All Caf´es, Pubs, and Restaurants to Close (20 March–26 March 2020)

During this timeline, Aberdeen FC players had a night out at a coronavirus hotspot just days before two players tested positive (Daily Record 2020). This period saw a huge increase in positive tweets as shown in Fig. 5a. Here, we see that 47.8% (n = 16,404) of tweets are positive, while 43.0% (n = 14,765) and 9.2% (n = 3145) are negative and neutral, respectively. This increase may be due to people's concern about the spread of the virus, and hence, the close of the entertainment and leisure centres seems to be received more positively.

Compared to timeline 2, this timeline meant safety and benefit for some people, as the Nursing and Midwifery Council confirmed that over 5,600 former nurses have offered to help combat the virus's spread, which correlated with a rise in positive sentiments in this timeline. Also, we can see in Fig. 5b that people became happier and less afraid.

4.4 Timeline 4: Death Tolls Reached 10,000 Amidst the Lockdown in the UK (12 April–18 April 2020)

Figure 6a shows that many tweets were positive amidst the sad news. This could be because of Capt. Tom Moore's NHS fundraiser. During Tom Moore's 100 laps of his garden, about a million people contributed to his JustGiving page, raising over £25 million for the NHS [31]. The percentage of positivity is 52.8% (n = 11166), the percentage of negativity is 39.0% (n = 8234), and the percentage of neutrality is 8.2% (n = 1736). As illustrated in Fig. 6b, the high death toll amounted to 10,000 people during the lockdown. In an interview with BBC News, Health Secretary Matt

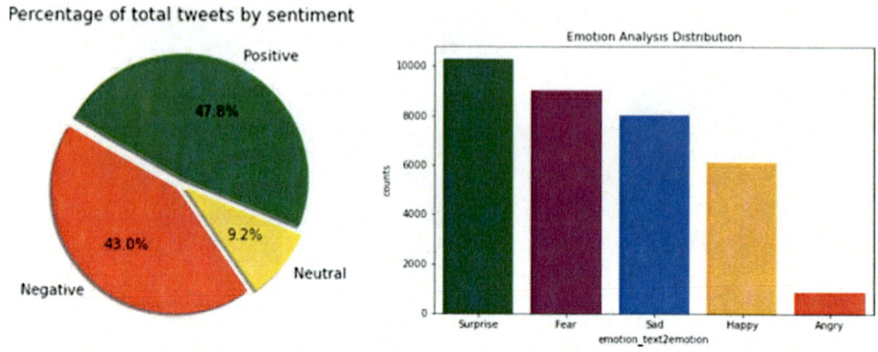

(a) Sentiment Polarity (b) Emotion Analysis

Fig. 5 Timeline 3—sentiment and emotion analysis

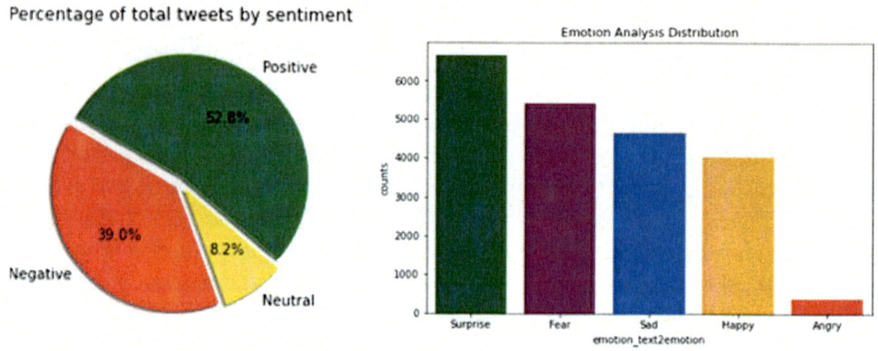

(a) Sentiment Polarity (b) Emotion Analysis

Fig. 6 Timeline 4—sentiment and emotion analysis

Hancock stated that a decrease in the number of new cases and deaths all add to the effectiveness of the lockdown [32]. Here, we also see that the anger levels dropped.

4.5 Timeline 5: Lockdown Lifted, But Social Distancing Advised (13 May–19 May 2020)

During this timeline, Chancellor Rishi Sunak announced a recession in the UK economy. According to the data, 49.7% (n = 4658) of tweets were positive, 41.8% (n = 3912) were negative, and 8.5% (n = 796) were neutral. Positive tweets have decreased compared to timeline four because of the decline in the UK economy and the possibility of a noteworthy recession, as stated by Chancellor Rishi Sunak, Chancellor of the Exchequer.

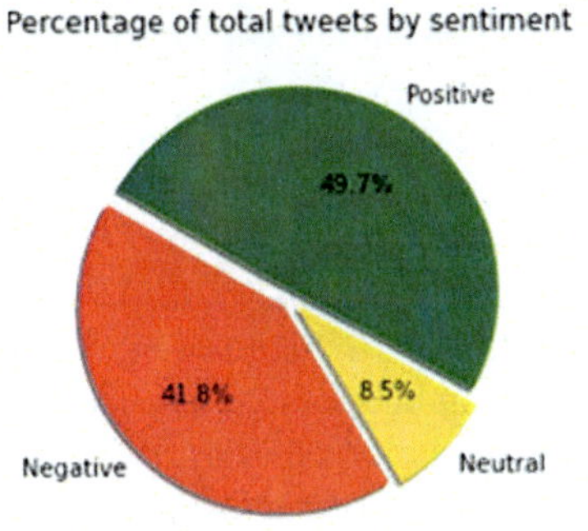

(a) Sentiment Polarity

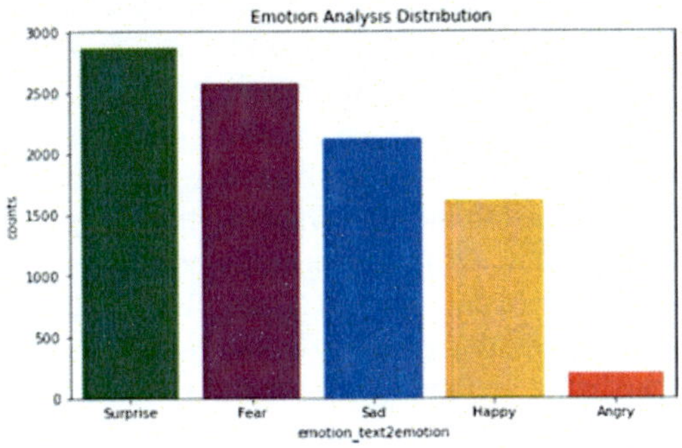

(b) Emotion Analysis

Fig. 7 Timeline 5—sentiment and emotion analysis

9371 tweets were examined, and approximately 3000 tweets showed surprise at the lift in lockdown and recommendations on social distancing (see Fig. 7b). We observed an increase in the level of fear in the UK community after government scientists confirmed that the COVID-19 infection had progressed and could spread spontaneously.

5 Discussion

This study examined how the COVID-19 pandemic was perceived by UK citizens. It analysed tweets that were posted between the first case confirmed in the UK and two years after the initial case was reported. The data suggested a relationship between the various timelines and their impacts on the sentiment surrounding the timelines, including the emotions expressed towards an event or governmental regulations. The sentiment of the tweets around particular timelines (events happening at specific times) and the evolution of emotions in the tweets were the two main sections we extracted from the 15 timelines of this study.

The sentiments surrounding each timeline showed slight variations, as shown in Table 2 below. The perspectives of the general public on COVID-19, governmental rules, and health-related information surrounding the timelines are reflected in these changes. The news about the first death from COVID-19 in the UK had a negative sentiment of 57.7%, only surpassed by the timeline 15: two years since the UK recorded its first domestic cases of COVID-19. In contrast, the positive sentiment was only 32% from the first timeline. This gradually increased as the government's efforts to stop the spread of the virus by closing all cafés, bars, and restaurants grew more intense. People's sentiments appear to be positively influenced by governmental regulation and positive health information. On the other hand, a rise in negative sentiments resulted from the virus's spread and the development of COVID variants. Only a small portion of tweets in each timeline was neutral.

Then, we examined the emotions expressed in the tweets toward each time-line using the five primary emotions of surprise, fear, sadness, happiness, and anger. A bar chart was used to represent these emotions graphically.

In contrast to the time frame of the study by Priyadarshini et al. [17], the tweets analyzed in this study covered the period from March 5, 2020, to February 6, 2022, and they were only limited to the UK to obtain more streamlined results as opposed to the methodology used by Imran et al. [33]. According to the hypothesis previously stated in the literature review for this study, the emotions most commonly expressed

Table 2 Overview of sentiment in each of the timeline

S/N	Timeline	Negative	Neutral	Positive
1	The first death from COVID-19 in the UK was confirmed	3628 (57.7%)	641 (10.2%)	2017 (32.1%)
2	Government Advised self-isolation	11,768 (52.4%)	2269 (10.1%)	8427 (37.5%)
3	Prime Minister orders all Cafés, pubs, and restaurants to close	14,765 (43.0%)	3145 (9.2%)	16,404 (47.8%)
4	Death tolls reached 10,000 amidst the lock-down in UK	8234 (39.0%)	1736 (8.2%)	11,166 (52.8%)
5	Lockdown lifted, but social distancing advised	3912 (41.8%)	796 (8.5%)	4658 (49.7%)
6	Health Secretary Matt Hancock announces hope for COVID vaccine	1930 (42.1%)	403 (8.8%)	2253 (49.1%)
7	The majority of schools in the UK reopened	885 (49.0%)	184 (10.2%)	738 (40.8%)
8	The UK reaches a million COVID-19 cases	2444 (53.5%)	337 (7.4%)	1787 (39.1%)
9	The UK hits 50,000 COVID-related deaths	1256 (45.0%)	220 (7.9%)	1316 (47.1%)
10	First COVID-19 Vaccine	807 (45.0%)	158 (8.8%)	830 (46.2%)
11	COVID variants in the UK	1746 (53.6%)	325 (10%)	1188 (36.5%)
12	Death tolls reach 100,000	1325 (42.5%)	302 (9.7%)	1491 (47.8%)
13	A year since the UK recorded its first domestic cases of COVID-19	1108 (43.9%)	218 (8.6%)	1197 (47.4%)
14	A third of UK adults are now fully vaccinated against COVID-19	437 (47.3%)	73 (7.9%)	414 (44.8%)
15	Two years since the UK recorded its first domestic cases of COVID-19	191 (58.1%)	21 (6.4%)	117 (35.6%)

at the beginning of the pandemic were surprise and fear, which gradually subsided post-pandemic.

According to the study's findings, we observed a consistent rise in the overall number of tweets extracted from the first timeline to the third timeline. The total number of tweets extracted decreased from the fourth to the seventh timeline. One thing to note is that there was a significant drop in tweets from timeline four to timeline five, totaling a difference of 11,765 tweets. At the eighth timeline, the number of tweets increased once more. This rise can be attributed to the increase in COVID-19 cases discovered during England's second lockdown on November 5, 2022. At timelines 9 and 10, the number of tweets started to decline. However, these timelines still showed a lot of positive sentiments. In addition, many tweets expressed surprise, fear, and sadness because the vaccine was known to have some side effects, while only a small number of people were happy about the development. When the COVID variants were found in the UK, many people began tweeting about it and expressing their emotions, primarily surprise and fear, with a lot of negative sentiment. As a result, more tweets were extracted at timeline 11. The number of tweets decreased from timeline 12 to timeline 15.

6 Conclusion

Online activity increased due to the global COVID-19 outbreak. People, including those in the UK, used social media platforms to share and receive information about the virus as well as to express their emotions. In order to understand how emotions have changed over time, we examined more than 100,000 tweets from the UK across 15 timelines for this study.

According to our findings, the most frequent emotions discussed in tweets about COVID-19, particularly in the early stages of the outbreak, were surprise and fear. As the pandemic spread, it became more common to see sad and happy emotions. This might be because lockdown procedures were successfully implemented in the UK and helped to contain the virus's spread. However, the decline in tweets indicated that people are likely less worried about the virus. It was also observed that positive sentiments increased steadily across the timelines. However, it is noteworthy that the 15th timeline, two years since the UK recorded its first domestic cases of COVID-19, had a negative sentiment of 58.1%.

Regardless of what the future holds, it is evident that governmental restrictions and policy modifications significantly impact people's emotions and how they express them on social media. Future research could expand the analysis to other languages and nations, as this study was only able to look at tweets in English and the United Kingdom. The use of irony and sarcasm in tweets, where the written literal meaning is frequently the opposite of the intended meaning, was also not taken into account in this study. Finally, there is the problem of nontext data in sentiment analysis. Emojis and other non- text data were not taken into account in the study. Despite these limitations, this study sheds light on how UK citizens have felt about COVID-19 over time. The findings can guide future policy choices and assist decision-makers in comprehending how their choices may affect people's emotions.

References

1. Huang, C., et al.: Clinical features of patients infected with 2019 novel coronavirus in Wuhan, china. Lancet **395**(10223), 497–506 (2020)
2. WHO: COVID-19 Public Health Emergency of International Con- cern (PHEIC) Global research and innovation forum (2020). https://www.who.int/publications/m/item/covid-19-public-health-emergency-of-international-concern-(pheic)-global-research-and-innovation-forum Accessed 31 July 2022
3. Ugochukwu-Ibe, I.M., Ibeke, E.: E-learning and covid-19: the nigerian experience: challenges of teaching technical courses in tertiary institutions. (2021). CEUR Workshop Proceedings
4. Iwendi, C., Mohan, S., Ibeke, E., Ahmadian, A., Ciano, T., et al.: Covid- 19 fake news sentiment analysis. Comput. Electr. Eng. **101**, 107967 (2022)
5. Worldometers: COVID-19 Coronavirus Pandemic. https://www.worldometers.info/coronavirus/ (2022). Assessed 1 Aug 2022
6. Kumar, R.L., Khan, F., Din, S., Band, S.S., Mosavi, A., Ibeke, E.: Recurrent neural network and reinforcement learning model for covid-19 prediction. Front. Public Health **1437** (2021)
7. BBC: Coronavirus: Two Cases Confirmed in UK. https://www.bbc.co.uk/news/health-51325192. Accessed 2 July 2022
8. Ngabo, D., Dong, W., Ibeke, E., Iwendi, C., Masabo, E.: Tackling pandemics in smart cities using machine learning architecture. Math. Biosci. Eng. **18**(6) (2021)
9. BBC: Covid-19 Vaccine: First Person Receives Pfizer Jab in UK. https://www.bbc.co.uk/news/uk-55227325. Accessed 2 July 2022
10. Kumar, D.A., Chinnalagu, A.: Sentiment and emotion in social media covid-19 conversations: Sablstm approach. In: 2020 9th International Conference System Modeling and Advancement in Research Trends (SMART), pp. 463–467 (2020). IEEE
11. Statista: Twitter usage frequency in the United Kingdom.https://www.statista.com/statistics/1176756/frequency-of-use-among-twitter-users-in-the-united-kingdom/ (2020), (2021). Accessed 1 Aug 2022
12. Mansoor, M., Gurumurthy, K., Prasad, V.R.: Global sentiment analysis of covid-19 tweets over time. arXiv preprint (2020) arXiv:2010.14234
13. Statista: Number of monetizable daily active Twitter users (mDAU) worldwide from 1st quarter 2017 to 1st quarter 2022. https://www.statista.com/statistics/970920/monetizable-daily-active-twitter-users-worldwide/#:text=Twitter%3A%20number%20of%20monetizable%20daily%20active%20users% 20worldwide%202017%2D2022&text=In%20the%20last%20reported%20quarter,mDAU%20in%20the%20previous%20quarter (2022). Accessed 1 Aug 2022
14. Statista: Leading countries based on number of Twitter users as of January. https://www.statista.com/statistics/242606/number-of-active-twitter-users-in-selected-countries/ (2022). Accessed 1 Aug 2022
15. Ibeke, E., Lin, C., Wyner, A., Barawi, M.H.: A unified latent variable model for contrastive opinion mining. Front. Comp. Sci. **14**, 404–416 (2020)
16. Ibeke, E., Lin, C., Wyner, A., Barawi, M.H.: Extracting and understanding contrastive opinion through topic relevant sentences. In: Proceedings of the Eighth International Joint Conference on Natural Language Processing (Vol. 2: Short Papers), pp. 395–400 (2017)
17. Priyadarshini, I., Mohanty, P., Kumar, R., Sharma, R., Puri, V., Singh, P.: A study on the sentiments and psychology of twitter users during the covid-19 lockdown period. Multimed. Tools Appl. 1–23 (2021)
18. Mathur, A., Kubde, P., Vaidya, S.: Emotional analysis using twitter data during pandemic situation: Covid-19. In: 2020 5th International Conference on Communication and Electronics Systems (ICCES), pp. 845–848 (2020). IEEE

19. Tocoglu, M.A., Ozturkmenoglu, O., Alpkocak, A.: Emotion analysis from Turkish tweets using deep neural networks. IEEE Access **7**, 183061–183069 (2019). https://doi.org/10.1109/ACCESS.2019.2960113
20. Dubey, A.D.: Twitter Sentiment Analysis during COVID-19 Out- break. SSRN: https://ssrn.com/abstract=3572023 or https://doi.org/10.2139/ssrn.3572023 (2020). https://ssrn.com/abstract=3572023
21. Abd-Alrazaq, A., Alhuwail, D.-A., Househ, M., Hamdi, M., Shah, Z.: Top concerns of tweeters during the covid-19 pandemic: infoveillance study. JMIR Public Health Surveill. **6**(4), 19016 (2020)
22. Dhar, B.K., Ayittey, F.K., Sarkar, S.M.: Impact of covid-19 on psychology among the university students. Global Chall. **4**(11), 2000038 (2020)
23. Varalakshmi, R., Swetha, R.: Covid-19 lock down: people's psychology due to law enforcement. Asian J. Psychiatr. **51**, 102102 (2020)
24. Amujo, O., Ibeke, E., Fuzi, R., Ogara, U., Iwendi, C.: Sentiment computation of UK-originated covid-19 vaccine tweets: a chronological analysis and news effect. Sustainability **15**(4), 3212 (2023)
25. Gonzalez-Carvajal, S., Garrido-Merch'AN, E.C.: Comparing bert against traditional machine learning text classification. ArXiv preprint (2021) arXiv:2005.13012
26. Al-Garadi, M.A., Yang, Y.-C., Cai, H., Ruan, Y., O'Connor, K., Graciela, G.-H., et al.: Text classification models for the automatic detection of nonmedical prescription medication use from social media. BMC Med. Inform. Decis. Mak. **21**, 21–27 (2021)
27. Lagrari, F.-E., Elkettani, Y.: Customized Bert with convolution model: a new heuristic enabled encoder for twitter sentiment analysis (2020)
28. Hadjer, M., Ounadi, I., Benkhelifa, Y.: Tweets categorization using fine-tuned Bert model (2021)
29. Aslam, N., Rustam, F., Lee, E., Washington, P.B., Ashraf, I.: Sentiment analysis and emotion detection on cryptocurrency related tweets using ensemble Lstm-Gru model (2022)
30. Hakak, N., Mohd, M., Kirmani, M., Mohd, M.: Emotion analysis: a survey. In: IEEE International Conference on Computational Intelligence and Computing Research, pp. 397–402 (2017). https://doi.org/10.1109/COMPTELIX.2017.8004002
31. BBC: Capt Tom Moore's NHS Fundraiser Hits £17m. https://www.bbc.co.uk/news/uk-england-beds-bucks-herts-52303859. Accessed 15 Sept 2022
32. BBC: Students 'scared and Confused' as Halls Lock Down. https://www.bbc.co.uk/news/uk-54308329. Accessed 15 Sept 2022
33. Imran, A.S., Daudpota, S.M., Kastrati, Z., Batra, R.: Cross-cultural polarity and emotion detection using sentiment analysis and deep learning on covid-19 related tweets. IEEE Access **8**, 181074–181090 (2020)

Deep Learning-Based Sentiment Analysis of Customer Reviews on Hotels Through Tweets

M. Srisankar

Abstract Tweets are continually produced and short. Sentiments play a vital role in comprehending personal feelings. Multitudes show their opinions on any subject or object on social media. The public's sentiment is split into different categories namely, positive, negative and neutral. In this paper, data is acquired from hotel reviews on Twitter using Python's Tweepy library and pre-processed using Python scripts. Inconsistent data or noise, retweets, tags, URLs along with hashtag symbols and duplicate entries are removed. The tweets are upsampled and split by using scikit-learn in Python. The textual data is converted into vectors using Keras Tokenizer in Python. Bi-Sense Emoji Embedding (BSEE) is used for performing SA. The sentiments are categorized using Support Vector Machine (SVM) and Random forest (RF), and performance is compared with BSEE based on Accuracy, Recall, F-measure, Precision and Time period. It is seen that the proposed classifier offers better results.

Keywords Random forest (RF) · Support vector machine (SVM) · Bi-sense emoji embedding (BSEE)

1 Introduction

Sentiment detection and classification are currently the emphasis of social analytics on the online platform. Nowadays, Twitter has turned into a significant tool for procurement of peoples' feelings. Sentiment Analysis (SA) of tweets becomes a captivating exploration in the area of Artificial Intelligence (AI) [1]. SA is widely utilized in several fields like marketing, education, e-commerce, so on. The goal of SA is to analyse the online reviews and determine the scores for sentiments. Categorizing sentiments primarily depends on labelled data. English has more labelled texts in contrast to other languages [2]. SA is necessary for information alignment

M. Srisankar (✉)
Department of Computer Science, Government Arts College, Udumalpet, Bharathiar University, Udumalaipettai, Tamil Nadu, India
e-mail: Shankar276@gmail.com

C. Iwendi et al. (eds.), *Proceedings of ICACTCE'23—The International Conference on Advances in Communication Technology and Computer Engineering*, Lecture Notes in Networks and Systems 735, https://doi.org/10.1007/978-3-031-37164-6_39

in social media as user engagement rates are significantly higher on Twitter [3]. Twitter has evolved into an important tool to collect information about people's emotions. Emotion recognition of tweets turns into a fascinating AI experiment. Several methods are formulated for evaluating the SA of Twitter users, yet some improvisation is needed to improve accuracy and execution time. An efficient algorithm is needed to extract and classify tweet sentiments in Twitter [4]. A dataset for SA is a group of tweets about customer perceptions. Every tweet is given a score demonstrating whether it is positive, negative or neutral. For each tweet, SA techniques determine the sentiment. Using the sentiment score in the dataset, the calculated sentiment is then assessed as right or wrong. Accuracy identifies tweets that are accurately evaluated by algorithm [5]. Micro-blogging sites like Twitter involve infinite source of diverse types of information. People post opinions on varied topics, contend about present challenges, criticize as well as express emotions. Day-by-day, the number of users of platforms in addition to social networks gradually increase [6–8]. Data got from sources are employed in opinion mining, and sentiment along with emotion analysis. Sentiments, opinions besides emotions are included in opinion mining and SA. Data on social media are noisy involving errors in spelling, punctuation and grammar. Emoji is the general form of communication [9]. They are visual representations of emotions, symbols or objects. Ideograms as well as smileys are used in electronic messages in addition to web pages. They are of varying kinds comprising of facial expressions, weather, universal objects, places, animals and birds [10]. They look like emoticons, but Emojis are pictures and not typo-graphics. Emoticons are typographic exhibits of emotion, while Emojis are images that express facial expressions, emotions, activities, objects and animals [11, 12]. Emojis and Emoticons are used in present applications for communication. They include text messaging and applications involved in social networking like Instagram, Facebook, Twitter and Snapchat [12]. As SA plays a major role in social media, in this paper, Bi-Sense Emoji Embedding (BSEE) is proposed for performing SA of hotel reviews. Data is gathered using Python's Tweepy library and pre-processed using Python scripts [13]. Noise, duplicate, tags, URLs along with hashtag symbols are removed. Tweets are upsampled and split using scikit-learn in Python. It is seen that the proposed classification model offers improved results.

2 Related Work

Salampasis et al. [14] have used Twitter and also a system for gathering user produced data for a particular duration. An algorithm based on unsupervised and lexicon-based approach is used to perform SA to automatically handle the conveyed emotion of microblogs using ternary classification. It offers better performance in a variety of environments. Howells and Ertugan [15] have focussed on developing a model which analyses content of microblog and is capable of analysing customer feedback or perception. The model forms basis of application which an organisation uses for marketers for timely as well as strong illustration of customers' insights on any product as well as service. Initially, a Fuzzy Logic (FL)-based model is designed to convert linguistic variables into membership functions which define variables in a fuzzy manner. Every membership function signifies linguistic variable and forms Fuzzy Set (FS) which is related to specific factor. Once FS is formed, the ensuing step is defining if-then rules which describe FSs interaction. Operators are included to form connections amid FSs namely max (and) and min (or). Once connections are formed, propositions are combined into final FS. FS releases fuzzy number and is formed into crisp value using diverse methods. Al-Otaibi et al. [16] have used Twitter data for gaining understanding from public opinion in data. SVM is employed for classifying sentiments obtained from tweets and unigram is applied for extracting features. Decision-making in Organizations as well as individuals by offering investigation of information about products, customer opinions and product reviews found in social media are handled. It aids in providing details about competitors and assess products available in market. This saves customer's time enabling them to know about the products and stimulate production of many products by knowing the customer viewpoints and other competitors' products. Textual data of products as well as services are considered in Twitter and features like age, education, followers, gender are considered. Experiments are carried out for large training datasets and it is seen that the algorithm offers improved accuracy. Smetanin and Komarov [17] offers a method for SA of Russian product reviews using Convolutional Neural Networks (CNNs). Vectors pre-trained with Word2Vec are given as inputs to NNs. This scheme does not use any hand-crafted attributes or lexicons. Training dataset is formed from reviews on high rated goods from main e-commerce sites, where user-rated scores are used as labels for classes. The proposed scheme offers improved F-measure. The training dataset as well as Word Embeddings (WEs) can be accessed by research community. Moussa [18] have introduced an emoji-based metric for observing emotions of consumers available on social media related to brands. Initial findings from study tests the metric based on 720 consumer tweets of 18 principal universal brands demonstrating 6 product groups or markets. This metric can be easily implemented and interpreted by every researcher as well as manager. Researchers and managers compute it using information that can be freely got by mining social media account. Huge amount of information seen in emoji-based User-Generated Content (UGC) regardless of language can be analysed. It is highly useful for researchers who perform multi-cultural study of emotions of consumers as well as managers of universal brands who are interested in observing consumers' feelings. Behera et al. [19] have proposed a hybrid approach involving CNN as well as Long Short Term Memory (LSTM) for classification of sentiments in reviews related to numerous domains. Deep CNNs are highly useful in selection of local features, while Recurrent Neural Networks (RNNs) yield improved outcomes in sequential analysis of longer text. The propounded Co-LSTM model is highly flexible in observing huge volume of social data, managing scalability, and unlike conventional Machine Learning (ML) schemes, it is not restricted to any specific domain. The experiment is applied on 4 review datasets from varied domains to train model that can deal with every dependency which arises in post. The outcomes show that propounded ensemble model outdoes other ML schemes based on Accuracy and other factors. Classify emotions functions makes use of Naïve Bayes algorithm for classifying the emotions into seven categories such as anger, disgust, fear, joy, sadness, surprise, and best fit. Classify polarity function receives two arguments, cleaned tweets and Naïve Bayes algorithm for classifying the polarity into positive sentiment and negative sentiment. Naïve Bayes, Maximum entropy, SVM, and ensemble classifier are the other data mining techniques available for classification of tweets in sentiment analysis, though Naïve Bayes is simple classifier it gives better precision and accuracy compared to other classifiers and vastly used in the sentiment analysis [20]. Machine/deep learning methods with different set features that are derived from URL, content, and server-based information of the web page to detect or classify the malicious web pages. Most attacks, including drive-by downloads, phishing, and injections, conceals the maliciously intended code or link in the content of legitimate or purportedly legitimate webpage. Many current research work uses the URL based features due to its risk-free processing. Only less attention is given to the content-based features. That hides the vital area of features to better classification. To address this issue, this paper focus on web content-based features for malicious web page classification [21]. Deep neural networks are trained regularly to be able to read, classify, diagnose, and make appropriate decisions with high accuracy. Deep learning depends on data representation and provides advantages in many applications and learns from the amount of data. The more data increases, the more the ability of deep learning in execution will grow, unlike other machine learning techniques. In addition, deep learning has the advantage of building very deep structures to learn more abstract information. The quality of deep learning techniques is that they learn feature representations automatically, thus increasing performance speed in a shorter time [22]. Deep learning is another booming field that is mostly used in computer applications. This work proposes an integrated environment of a blockchain-deep learning environment for analyzing the Electronic Health Records (EHR). The EHR is the medical documentation of a patient which can be shared among hospitals and other public health organizations. The proposed work enables a deep learning algorithm act as an agent to analyze the EHR data which is stored in the blockchain [23].

3 Classification Models

Classification models incorporating Random Forest (RF) and SVM are discussed in detail. The proposed Bi-Sense Emoji Embedding (BSEE) for performing SA is detailed in the next section.

3.1 Random Forest (RF)

RF classifier, a supervised learning algorithm may be used for regression besides classification. It is a popular ML algorithm as it is highly flexible and easy to implement. It includes several Decision Trees (DTs) like a forest. Randomness is used for improving Accuracy and fight overfitting that may be an issue. DTs are created depending on arbitrary choice of data samples and obtain predictions from each tree. Best feasible solutions are obtained through votes. It is used in fraud detection, disease prediction etc., Let the amount of features in dataset be 'm'. RF chooses 'k' features arbitrarily, where $k < m$. The algorithm determines root node among 'k' features by choosing the node with increased information gain. The node is split into children and the process is repeated 'n' times. The results of DTs are combined to get the final result. This ensemble algorithm produces separate DTs by selecting attributes. The votes from every tree are taken and most prevalent class is taken as the final outcome. In a regression problem, the average of tree outputs is computed which is the end outcome.

3.2 Support Vector Machine (SVM)

SVM is also supervised learning algorithm that is used for classification in addition to regression problems. It is employed for classification in ML. It is used for generating decision boundary which can classify n-dimensional space into categories such that data point is put in right category. Hyperplane is the decision boundary. It selects extreme points or vectors which aid in producing hyperplane. The extreme cases are known as SVs. SVM determines a hyperplane which generates a boundary amid different kinds of data. Every data item in dataset is plotted in n-Dimensional space for 'n' features in data. Ideal hyperplanes are used to separate data. SVM performs binary classification. Nevertheless, several schemes that can be used for multi-class problems are available. By applying SVM to multi-class problems, a binary classifier can be created for every class of data.

4 Proposed Methodology

The propounded method involves the ensuing steps.

- **Data acquisition**: Python's Tweepy library is employed for collecting tweets related to customer reviews on hotels in CSV format during the period 01 April 2022 to 30 May 2022. The tweets include features like Tweet ID, text and date and time when it was posted. An investigative analysis shows occurrence of tweets of diverse polarities in dataset.

- **Data Pre-processing**: Pre-processing is performed using Python scripts. Raw data from Twitter is pre-processed to remove inconsistent data or noise before feeding to model. The dimensionality of data reduces which improves performance in SA. By using Python's Emoji module, Emojis are transformed to Common Locale Data Repository Project (CLDR) short names. Text is transformed to lowercase, followed by elimination of retweets, URLs, tags, and hashtag symbols. Incomplete words are rewritten. Duplicate entries are removed from text.
- **Data Upsampling along with Splitting**: As the data is not balanced, most of the tweets fall under positive class. Neutral as well as negative tweets are upsampled in addition to split by using scikit-learn in Python. Hence, they are resampled and split into training (80%), validation and testing (20%).
- **Tokenisation, Padding and WEs**: The textual data is converted to vectors called WEs using Keras Tokenizer in Python. The length of the sentence is fixed to 150. Hence, WEs of sentences with lesser lengths are padded with zeros.

4.1 Bi-sense Embedding

Word-Guide Attention-based LSTM (WGA-LSTM) as well as Multi-Level Attention-based LSTM (MLA-LSTM) is proposed. Bi-Sense Emoji Embedding (BSEE) is used for performing SA. The proposed system includes the ensuing steps:

- Initialization of BEE
- Generation of Senti-Emoji Embedding (SEE) depending on self-chosen attention
- Classification of sentiments through attention-dependent LSTM networks.

Currently, WEs like word2vec and fasttext play a major role [24, 25]. Fasttext aids initialisation of Emoji embeddings by considering emoji as word along with WEs. In contrast to traditional schemes where every emoji corresponds to an embedding vector which is Word-Emoji Embedding (WEE), emoji is embedded into 2 different vectors (BSEE). Two separate tokens are assigned to every emoji wherein, one is for a specific emoji employed in positive sentiments, and other is used in negative sentiments. Text sentiments are initialized using Vader [26]. Fasttext training helps in embedding tokens into diverse vectors. This aids in obtaining positive in addition to negative-sense embeddings for each emoji. Word2vec is based on skip-gram model that is used for maximising log probability determined by totalling probabilities of current word incidences for a set of adjoining words. In case of fasttext model, the problem is formulated as binary classification for predicting incidence of every Context Word (CW) with negative samples being arbitrarily chosen from absent CWs. For a word structure $\{w_1, w_2 \ldots w_T\}$ given as input, and CW set $\left(w_{c_t}\right)$ and collection of negative samples $\left(w_{n_t}\right)$ of present word (w_t), objective function depends on Binary Logistic Loss (BLL).

$$\sum_{t=1}^{T}\left[\sum_{w_c \in w_{c_t}} \underset{\cdot}{\overline{L}}(S(w_t, w_c)) + \sum_{w_c \in w_{n_t}} \underset{\cdot}{\overline{L}}(S(w_t, w_n))\right] \tag{1}$$

where,

$\underset{\cdot}{\overline{L}}(s(\cdot\ ,\ \cdot))$—Score function's logistic loss

$s(\cdot\ ,\ \cdot)$—Determined by summing scalar products amid *n*-gram embeddings of present word and CW embedding that varies from word2vec wherein score is scalar product amid embeddings Fast text is chosen as it is computationally efficient. The models offer improved performance.

4.2 Word-Guide Attention-Based LSTM (WGA-LSTM)

LSTM is widely used for encoding textual contents. Fundamental encoder model includes text embedding, LSTM and Fully-Connected (FC) layers for text classification depending on encoded feature. Functions in LSTM for time step (t) are shown below.

$$I_t = \sigma(S_i \cdot X_t + T_i \cdot H_{t-1} + B_i) \tag{2}$$

$$F_t = \sigma\left(S_f \cdot X_t + T_f \cdot H_{t-1} + B_f\right) \tag{3}$$

$$O_t = \sigma(S_o \cdot X_t + T_o \cdot H_{t-1} + B_o) \tag{4}$$

$$G_t = \tanh(S_c \cdot X_t + T_c \cdot H_{t-1} + B_c) \tag{5}$$

$$C_t = F_t \odot C_{t-1} + I_t \odot G_t \tag{6}$$

$$H_t = O_t \odot \tanh(C_t) \tag{7}$$

where,

H, H_{t-1}—Present and former hidden states
X_t—Present LSTM input
S and T—Weight matrices
σ—Sigmoid function.

To make use of the benefits of BSEE, the input layer is modified into LSTM units. SEE is got as weighted average of BSEE depending on self-chosen attention scheme.

Let $e_{\mathrm{t}}^{i}, i \in (1, m)$ signifies 'ith' sense EE (e_{t}) ($m = 2$ in BSEE) and $f_{\mathrm{att}}(\cdot, w_{\mathrm{t}})$ represents attention function on present WE. The attention weight (ρ_i) and SEE vector (V_{t}) are shown below.

$$u_{\mathrm{t}}^{i} = f_{\mathrm{att}}\left(e_{\mathrm{t}}^{i}, W_{\mathrm{t}}\right) \tag{8}$$

$$\rho_{\mathrm{t}}^{i} = \frac{\exp(u_{\mathrm{t}}, i)}{\sum_{i=1}^{n} \exp(u_{\mathrm{t}}, i)} \tag{9}$$

$$v_{\mathrm{t}} = \sum_{i=1}^{m} \left(\rho_{\mathrm{t}}^{i}, e_{\mathrm{t}}^{i}\right) \tag{10}$$

FC layer along with ReLU activation is used as attention function, and attention vector (V_{t}) is integrated with WE to LSTM. 'X_{t}' in Eqs. (2), (3), (4), (5), (6) and (7) becomes [W_{t}, V_{t}]. The LSTM unit's output is fed into FC layer with sigmoid activation to return sentiment. Binary Cross-Entropy (BCE) loss is employed as objection function with 'n' samples. The inspiration behind the proposed model is that every CW guides 'ρ_i' to impose model to choose embedding sense it attends. This model is denoted as WGA-LSTM with BSEE (WGA-BSEE-LSTM).

$$L(\theta) = -\frac{1}{n} \sum_{i=1}^{n} X_i \cdot \log(P_i) + (1 - X_i) \cdot \log(1 - P_i) \tag{11}$$

4.3 *Multi-level Attention-Based LSTM (MLA-LSTM)*

The attention scheme is formulated in a different way, where 'V_{t}' indicates how image information (emoji) is dispersed in CWs as seen in [27, 28]. The adapted SEE vector (V) is at tweet rather than word-level in Eqs. (8), (9) and (10) by substituting 'W_{t}' with final state vector (H) received as output from final LSTM unit.

$$\rho_i = \frac{\exp(f_{\mathrm{att}}(e_i, h))}{\sum_{i=1}^{m} \exp(f_{\mathrm{att}}(e_i, h))} \tag{12}$$

$$V' = \sum_{i=1}^{m} (\rho_i, e_i) \tag{13}$$

The derived SEE (V') is used for determining an added layer of attention. For sequence $\{W_1, W_2, \ldots W_T\}$, attention weight $\rho_t' \in (1, T)$ on SEE is expressed as follows:

$$\rho_t' = \frac{\exp\left(F_{att}\left(W_t, V'\right)\right)}{\sum_{i=1}^{m} \exp(F_{att}(W_t, V'))} \tag{14}$$

In the MLA-LSTM with BSEE (MLA-BSEE-LSTM) model, new input (u_t) for every LSTM unit is constructed by integrating original WE and 'V_t' in Eq. (15) to allocate SE information to every step. BCE is selected as loss function with comparable network configuration with WGA-BSEE-LSTM.

$$u = \left[W_t, \rho_t', V'\right] \tag{15}$$

5 Results and Discussion

Data acquired from hotel reviews available on Twitter by using Python's Tweepy library. Data pre-processing is performed using Python scripts to remove inconsistent data, noise, retweets, tags, URLs, hashtag symbols and identical entries. Upsampling and splitting of samples is carried out by using scikit-learn. The textual data is converted into vectors using Keras Tokenizer. SA is performed using BSEE. Sentiments are categorized using SVM and RF. It is seen that the proposed scheme outperforms the existing schemes in terms of Accuracy, Recall, F-measure, Precision and Time period. BSEE offers 5.7% and 1.5% better Accuracy in contrast to SVM and RF respectively (Fig. 1).

BSEE offers 3.7% and 2.2% better Recall when compared to SVM and RF respectively (Fig. 2).

BSEE offers 2.6% and 1.3% better F-Measure in contrast to SVM and RF respectively (Fig. 3).

BSEE offers 3.8% and 2.1% better Precision when compared to SVM and RF respectively (Fig. 4).

BSEE offers 1.66 and 1.26 times reduced Time period in contrast to SVM and RF respectively (Fig. 5).

Fig. 1 Accuracy

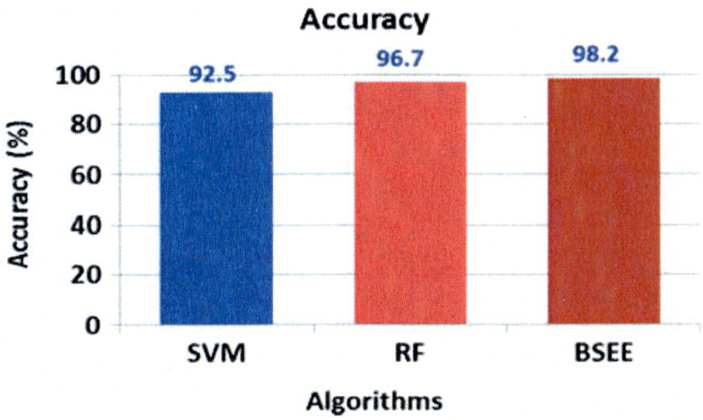

Fig. 2 Recall

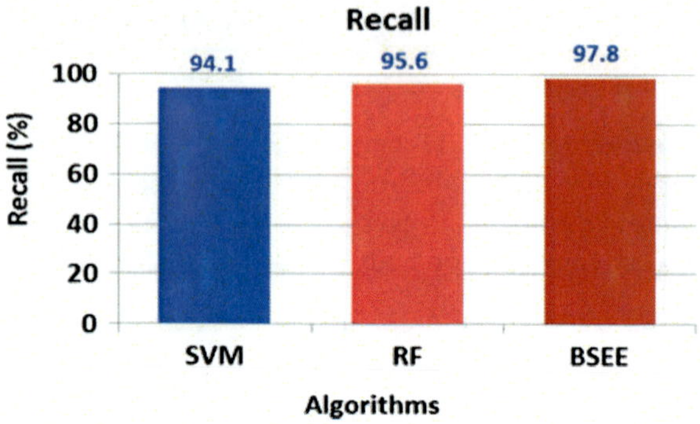

Fig. 3 F-measure

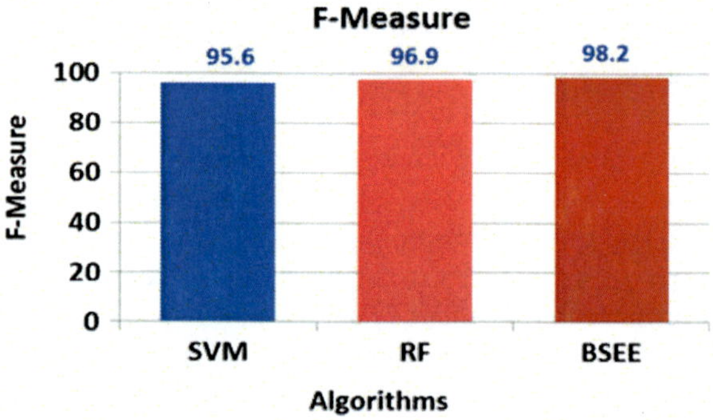

Fig. 4 Precision

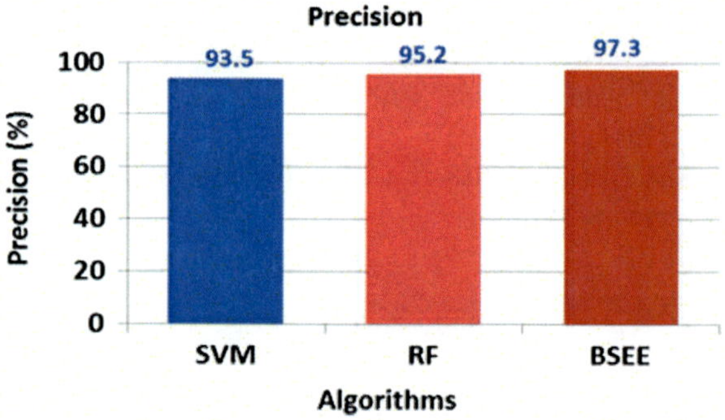

Fig. 5 Time period

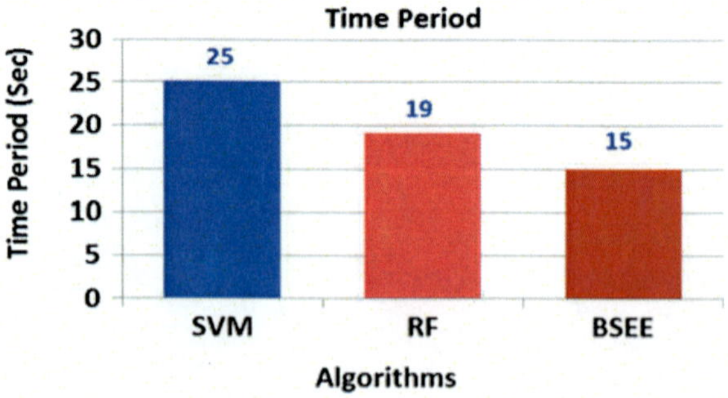

6 Conclusion

In this paper, Bi-Sense Emoji Embedding (BSEE) is used for performing SA. Sentiments are classified using RF and SVM and performance is compared with BSEE in terms of Accuracy, Recall, Precision, Time period and F-measure. Python's Tweepy library is used in data acquisition and Python scripts are used in pre-processing. Inconsistent data, noise, URLs, retweets, tags, hashtag symbols and duplicate entries are removed. Scikit-learn in Python is employed in upsampling and splitting of tweets. Keras Tokenizer is used for converting textual data into vectors. It is seen that the proposed classifier outperforms other models.

References

1. Giachanou, A., Crestani, F.: Like it or not: a survey of twitter sentiment analysis methods. ACM Comput. Surv. (CSUR) **49**(2), 1–41 (2016)
2. Sarlan, A., Nadam, C., Basri, S.: Twitter sentiment analysis. In: Proceedings of the 6th International conference on Information Technology and Multimedia, pp. 212–216. IEEE (2014, November)
3. Zimbra, D., Abbasi, A., Zeng, D., Chen, H.: The state-of-the-art in Twitter sentiment analysis: a review and benchmark evaluation. ACM Trans. Manag. Inf. Syst. (TMIS) **9**(2), 1–29 (2018)
4. Sahayak, V., Shete, V., Pathan, A.: Sentiment analysis on twitter data. Int. J. Innov. Res. Adv. Eng. (IJIRAE) **2**(1), 178–183 (2015)
5. Neethu, M.S., Rajasree, R.: Sentiment analysis in twitter using machine learning techniques. In: 2013 Fourth International Conference on Computing, Communications and Networking Technologies (ICCCNT), pp. 1–5. IEEE (2013, July)
6. Sebastian, T.: Sentiment analysis for Twitter. Doctoral dissertation (2012)
7. Martínez-Cámara, E., Martín-Valdivia, M.T., Urena-López, L.A., Montejo-Ráez, A.R.: Sentiment analysis in Twitter. Nat. Lang. Eng. **20**(1), 1–28 (2014)
8. Ramadhani, A.M., Goo, H.S.: Twitter sentiment analysis using deep learning methods. In: 2017 7th International annual engineering seminar (InAES), pp. 1–4. IEEE (2017, August)
9. Shiha, M., Ayvaz, S.: The effects of emoji in sentiment analysis. Int. J. Comput. Electr. Eng. (IJCEE) **9**(1), 360–369 (2017)
10. Fernández-Gavilanes, M., Costa-Montenegro, E., García-Méndez, S., González-Castaño, F.J., Juncal-Martínez, J.: Evaluation of online emoji description resources for sentiment analysis purposes. Expert Syst. Appl. **184**, 115279 (2021)
11. Guibon, G., Ochs, M., Bellot, P.: From emojis to sentiment analysis. In: WACAI 2016 (2016, June)
12. Lou, Y., Zhang, Y., Li, F., Qian, T., Ji, D.: Emoji-based sentiment analysis using attention networks. ACM Trans. Asian Low-Resour. Lang. Inf. Process. (TALLIP) **19**(5), 1–13 (2020)
13. Grover, V.: Exploiting emojis in sentiment analysis: a survey. J. Inst. Eng. (India): Ser. B **103**(1), 259–272 (2022)
14. Salampasis, M., Paltoglou, G., Giachanou, A.: Using social media for continuous monitoring and mining of consumer behaviour. Int. J. Electron. Bus. **11**(1), 85–96 (2014)
15. Howells, K., Ertugan, A.: Applying fuzzy logic for sentiment analysis of social media network data in marketing. Procedia Comput. Sci. **120**, 664–670 (2017)
16. Al-Otaibi, S., Alnassar, A., Alshahrani, A., Al-Mubarak, A., Albugami, S., Almutiri, N., Albugami, A.: Customer satisfaction measurement using sentiment analysis. Int. J. Adv. Comput. Sci. Appl. **9**(2) (2018)

17. Smetanin, S., Komarov, M.: Sentiment analysis of product reviews in Russian using convolutional neural networks. In: 21st IEEE Conference on Business Informatics (CBI), vol. 1, pp. 482–486 (2019, July)
18. Moussa, S.: An emoji-based metric for monitoring consumers' emotions toward brands on social media. Mark. Intell. Plan. (2019)
19. Behera, R.K., Jena, M., Rath, S.K., Misra, S.: Co-LSTM: convolutional LSTM model for sentiment analysis in social big data. Inf. Process. Manag. **58**(1), 102435 (2021)
20. Syrien, A., Hanumanthappa, M., Sundaravadivazhagan, B.: Sentiment analysis of tweets using Naïve Bayes algorithm through R programming. IJCSE 884–889 (2018)
21. Saleem Raja, A., Sundarvadivazhagan, B., Vijayarangan, R., Veeramani, S.: Malicious webpage classification based on web content features using machine learning and deep learning. In: 2022 International Conference on Green Energy, Computing and Sustainable Technology (GECOST), Miri Sarawak, Malaysia, pp. 314–319 (2022)
22. Mijwil, M.M., Aggarwal, K., Doshi, R., Hiran, K.K., Sundaravadivazhagan, B.: Deep learning techniques for COVID-19 detection based on chest X-ray and CT-scan images: a short review and future perspective. Asian J. Appl. Sci (2022)
23. Mantey, E.A., Zhou, C., Srividhya, S.R., Jain, S.K., Sundaravadivazhagan, B.: Integrated blockchain-deep learning approach for analyzing the electronic health records recommender system. Front. Public Health (2022)
24. Tomas, M., Kai, C., Greg, C., Jeffrey, D.: Efficient estimation of word representations in vector space. CoRR (2013). arXiv preprint arXiv:1301.3781
25. Piotr, B., Edouard, G., Armand, J., Tomas, M.: Enriching word vectors with subword information. TACL **5**(2017), 135–146 (2017)
26. Clayton, J.H., Eric, G.: VADER: a parsimonious rule-based model for sentiment analysis of social media text. In: 8th International Conference on Weblogs and Social Media (ICWSM 2014) (2014)
27. Tianlang, C., Yuxiao, C., Han, G., Jiebo, L.: When E-commerce meets social media: identifying business on WeChat moment using bilateral-attention LSTM. In: 27th International Conference on World Wide Web Companion, pp. 343–350 (2018)
28. Yequan, W., Minlie, H., Li, Z., Xiaoyan, Z.: Attention-based lstm for aspect-level sentiment classification. In: Conference on Empirical Methods in Natural Language Processing, 606–615 (2016)

Diagnosis of Diabetes Type Using Random Forest Algorithm and SVM for Improving Accuracy

Tao Hai, Jincheng Zhou, Timothy A. Olatunji, Oluwakemi A. Ajoboh, Lee Chen, Celestine Iwendi, Nkechi Omeoga, and Anurag Sinha

Abstract To improve the accuracy in the detection of diabetes-type diagnosis using Random Forest Algorithm and Support Vector Method. A dataset on diabetes was collected on kaggle.com. Support vector method analysis was performed on the data using kernel as linear, poly, and rbf with random state = 42 and test size = 0.2. The Random Forest Algorithm with n_estimators = 100 has the same accuracy as the support vector method with the kernel as rbf. When it comes to type diagnosis for a patient having diabetes, the Random Forest Algorithm and Support Vector Method with kernel rbf can be used, unlike another kernel of the Support Vector Method. The dataset was also visualized using power.

T. Hai · J. Zhou
School of Computer and Information, Qiannan Normal University for Nationalities, Duyun 558000, Guizhou, China

T. Hai
Key Laboratory of Advanced Manufacturing Technology of the Ministry of Education, Guizhou University, Guiyang 550025, Guizhou, China

T. Hai · L. Chen
School of Information and Artificial Intelligence, Nanchang Institute of Science and Technology, Nanchang, China

J. Zhou
Key Laboratory of Complex Systems and Intelligent Optimization of Guizhou Province, Duyun 558000, China

Key Laboratory of Complex Systems and Intelligent Optimization of Qiannan, Duyun 558000, China

T. A. Olatunji (✉) · O. A. Ajoboh · C. Iwendi
School of Creative Technologies, University of Bolton, Bolton BL3 5AB, UK
e-mail: timothyolatunji@ieee.org

N. Omeoga
Manchester Metropolitan University, Manchester, UK

A. Sinha
Department of Computer Science and Information Technology, IGNOU, New Delhi, India

C. Iwendi et al. (eds.), *Proceedings of ICACTCE'23—The International Conference on Advances in Communication Technology and Computer Engineering*, Lecture Notes in Networks and Systems 735, https://doi.org/10.1007/978-3-031-37164-6_40

Keywords Diabetes diagnosis · Support vector method · Machine learning · Regression · Random forest algorithm · Power BI

1 Introduction

In the health sector, diabetes is one of the most chronic diseases in the world. Diabetes resulting from obesity or excessive blood glucose levels, etc. It alters the hormone insulin, leading to an irregular crab metabolism, and reduces blood sugar levels. Diabetes happens when the body produces insufficient insulin. According to the World Health Organization, approximately 422 million people suffer from diabetes, particularly in countries with low or no income. And by 2030, this number could reach 490 billion [1]. Existing diabetes screening methods utilize medical and laboratory tests such as fasting blood glucose and oral glucose tolerance. This procedure is, however, time-consuming. Diabetes is classified into various types [2].

Type-1 Diabetes Mellitus (DM) is sometimes called Insulin-Dependent Diabetes Mellitus (IDDM) [2–4]. The inability of the human body to produce sufficient insulin is the cause of this kind of DM; hence, insulin injections are required [5]. Type-2 diabetes is sometimes known as non-insulin-dependent diabetes (NIDDM). This kind of Diabetes occurs when cells are unable to correctly utilize insulin [5, 6]. This kind of gestational diabetes is caused by an increase in blood sugar in pregnant women whose diabetes was not recognized earlier. Long-term problems are related to diabetes [7, 8].

Additionally, a diabetic is at great risk for a variety of health issues [8–10]. A methodology known as Predictive Analysis employs a range of machine learning algorithms, data mining techniques, and statistical methodologies to discover information and forecast future events using current and historical data [11, 12]. By applying predictive analysis to healthcare data, it is possible to make major decisions and predictions [13].

Using machine learning and regression, it is possible to perform predictive analytics [14, 15]. The objective of predictive analytics is to accurately diagnose disease, improve patient care, optimize resources, and enhance clinical outcomes [16]. Machine learning is regarded as one of the most essential characteristics of artificial intelligence since it enables the construction of computer systems that can acquire information from prior experiences without the need for explicit programming [17, 18]. Machine learning is regarded as an absolute necessity in the current environment to eliminate human labor by enabling automation with few errors [19].

1.1 Problem Statement

Various health problems cumulate to knowing the diabetes type a patient may be experiencing. Do these health problems duly contribute to diabetes type? Which health problem has an adverse impact and can be rapidly used in testing for diabetes type in patients? Does smoking constitute diabetes type? Which model is best for analyzing diabetes type? This problem statement is what this research looks into and provides answers to.

2 Literature Review

Diabetes is one of the chronic diseases affected by the majority and is liable to obesity, age, and lots more. Prediction is expensive and detection is not very easy, hence the proposition for machine learning methods for detection [20]. HbA1c and the estimate of plasma glucose (using either an FPG or an OGTT) are the standard benchmarks for diagnosing diabetes mellitus [21, 22]. The relationship between FPG or HbA1c and retinopathy is used to estimate the glucose and HbA1c cut off values [23]. Diagnosing diabetes mellitus requires a fasting plasma glucose level of 126 mg/dL (7.0 mmol/L), a plasma glucose level of 200 mg/dL (11.1 mmol/L) following a 2-hour oral glucose tolerance test (OGTT), a HbA1c level of 6.5% (48 mmol/mol), or the presence of symptoms of hyperglycemia [24]. Diabetes invents varieties of disorders such as urinary organ disease, coronary failure, and many more. It takes time to detect these disorders and how badly patients have been affected but machine learning methods were proposed to help detect other inventory of diabetes in the patient system [25]. In several nations of the world, diabetes is a known and growing disease and these nations are working hard to curb this disease at an early stage with various techniques employed [26]. Symptoms of diabetes at an early stage are not evident at early stage hence resulting in late diagnosis and a machine learning method has been adopted to improve early diagnosis of diabetes [1].

3 Materials and Methods

A diabetes dataset consisting of various determining variables for the diabetes binary index. The dataset is cleaned by checking for null values, in which none existed but there was a duplicated copy of data which was removed by keeping the first duplicate copy. After cleaning the data, a comparative study of the two groups of algorithms is made for improving accuracy in the diagnosis of diabetes type. The data collection is done through kaggle.com. The cleaned dataset consists of 69,046 rows and 22 columns. The dataset is then tested and trained for the process of diagnosing diabetes type.

3.1 Random Forest Algorithm

In the diagnosis of diabetes type, a Random Forest Algorithm (RFA) is sometimes utilized to forecast and predict diabetes type. The program, therefore, checks various determining variables such as age, income, blood pressure, cholesterol, and other 17 variables in the patient to determine the diabetes type. This algorithm involves randomly selecting data points from the data set. From that decision, the tree is constructed using the data points. Class of testing and training dataset is produced.

3.2 Support Vector Method

Regression problems are also solved by the support vector method. The support vector method is most frequently used to solve regression problems [27]. The support vector method has various kernels for solving regression methods such as 'linear', 'poly' and 'rbf' and this paper uses the three kernels.

4 Results

The accuracy value and statistics for RFA and SVM are displayed in Table 1. It shows that RFA and RBF SVM have the same statistical result and another kernel of SVM has a different statistical result. The proposed RF algorithm and RBF SVM perform better than the Poly-SVM and Linear-SVM.

5 Discussion

In this modeling, the RF algorithm and RBF SVM outperform POLY SVM and LINEAR SVM at diagnosing diabetes type, with a statistical mean squared error of 0.2587842292238949 for RFA, 0.25884157143302583 for POLY SVM, 0.25880521361325687 for LINEAR SVM, and 0.2587842292238949 for RBF SVM as shown in Table 1.

This shows that both the RBF SVM and RF algorithms can be used interchangeably to diagnose diabetes type for patients with other kernels of SVM less effective concerning giving datasets.

Figure 1 shows the correlation of the sum of every medical condition and how they affect the diabetes type of each patient. It shows that being a smoker does not impact being diabetic, but BMI greatly affects the diabetes type a patient may have.

Table 1 RFA and SVM statistics result

Group statistics					
	Groups	N	Mean squared error	Mean absolute error	Root mean squared error
Accuracy	RFA	100	0.2587842292238949	0.34882418908024637	0.5087083931132794
	POLY SVM	42	0.25884157143302583	0.3487962304605996	0.5087647505802911
	LINEAR SVM	42	0.25880521361325687	0.3488052136134473	0.5087290178604489
	RBF SVM	42	0.2587842292238949	0.34882418908024637	0.5087083931132794

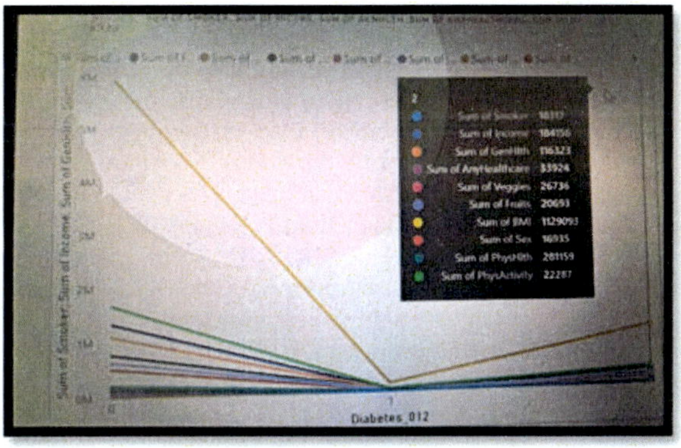

Fig. 1 Visualization of dataset

6 Conclusion

The health sector in recent times has been stressed coupled with a shortage in staff and this affects the sector and humans at large. Diabetes as a disease is not quite easy to detect in humans at an early stage and it claims the lives of patients. And to reduce death and increase early diagnosis of diabetes type, machine learning techniques are suggested and a Random Forest Algorithm with different kernels of Support Vector Method is employed in this research to help improve accuracy.

In this paper, diagnosis of diabetes type was performed using the RFA and different SVM kernels. The accuracy value of the RFA and RBF SVM are the same for the dataset used for analysis and are better than other kernels of SVM which is LINEAR and POLY. Further analysis can be performed using other datasets on the SVM kernels as the RBF SVM being the best model might just imply the dataset is not linear with its regression. And other kernels of SVM might be the best models in other research. It is also discovered that smoking does not contribute to the detection of diabetes type but BMI has a higher effect in discovering the diabetes type of patient.

Acknowledgements This work was supported by: Key Laboratory of Advanced Manufacturing Technology of the Ministry of Education of Guizhou University (GZUAMT2022KF[07]). The National Natural Science Foundation of China (No. 61862051). The Science and Technology Foundation of Guizhou Province (No. [2019]1299, No. ZK[2022]449). The Top-notch Talent Program of Guizhou province (No. KY[2018]080). The Natural Science Foundation of Education of Guizhou province (No. [2019]203). The Funds of Qiannan Normal University for Nationalities (No. qnsy2019rc09). The Educational Department of Guizhou under Grant No. KY[2019]067.

References

1. Hao, J., Luo, S., Pan, L.: Rule extraction from biased random forest and fuzzy support vector machine for early diagnosis of diabetes. Sci. Rep. **12**(1) (2022)
2. National Research Council, Division of Behavioral and Social Sciences and Education, Center for Education, and Committee on Scientific Principles for Education Research: Scientific Research in Education. National Academies Press (2002)
3. Radbruch, A.: Flow Cytometry and Cell Sorting. Springer Science & Business Media (2013)
4. Bhatia, A., Kedia, V., Shroff, A., Kumar, M., Shah, B.K., Aryan: Fake currency detection with machine learning algorithm and image processing. In: 2021 5th International Conference on Intelligent Computing and Control Systems (ICICCS) (2021). https://doi.org/10.1109/iciccs51141.2021.9432274
5. Iwendi, C., Bashir, A.K., Peshkar, A., Sujatha, R., Chatterjee, J.M., Pasupuleti, S., Mishra, R., Pillai, S., Jo, O.: Covid-19 patient health prediction using boosted random forest algorithm. Frontiers (2020). https://doi.org/10.3389/fpubh.2020.00357/full. Accessed 15 Jan 2023
6. Bartram, J., Ballance, R.: Water Quality Monitoring: A Practical Guide to the Design and Implementation of Freshwater Quality Studies and Monitoring Programmes. CRC Press (2020)
7. United States. Secret Service: The Detection of Counterfeit Currency: A Law Enforcement Officer's Guide (1975)
8. Shu, K., Liu, H.: Detecting Fake News on Social Media. Morgan & Claypool (2019)
9. Ángel, G.A.D., Del Ángel G.A.: Cash and dash: how ATMs and computers changed banking. By Bernardo Bátiz-Lazo. Oxford: Oxford University Press, 2018. Pp. xiv, 324 pp. $75.00, hardcover. J. Econ. Hist. **79**(4), 1208–1210 (2019). https://doi.org/10.1017/s0022050719000706
10. Bleay, S.M., Croxton, R.S., De Puit, M.: Fingerprint Development Techniques: Theory and Application. John Wiley & Sons (2018)
11. Covid-19 health analysis and prediction using machine learning algorithms for Mexico and Brazil patients. Taylor & Francis. https://doi.org/10.1080/0952813X.2022.2058097. Accessed 15 Jan 2023
12. Hai, T., Zhou, J., Padmavathy, T.V., Md, A.Q., Jawawi, D.N.A., Aksoy, M.: Design and validation of lifetime extension low latency MAC protocol (LELLMAC) for wireless sensor networks using a hybrid algorithm. Sustainability **14**(23), 15547 (2022). https://doi.org/10.3390/su142315547
13. Bhavikatti, S.K., et al.: Investigating the antioxidant and cytocompatibility of *Mimusops elengi* Linn extract over human gingival fibroblast cells. Int. J. Environ. Res. Public Health **18**(13) (2021). https://doi.org/10.3390/ijerph18137162
14. Mehta, P., Pandya, S., Kotecha, K.: Harvesting social media sentiment analysis to enhance stock market prediction using deep learning. PeerJ Comput. Sci. **7**, e476 (2021)
15. Ireton, C., Posetti, J.: Journalism, Fake News and Disinformation: Handbook for Journalism Education and Training. UNESCO Publishing (2018)
16. Kalyankar, G.D., Poojara, S.R., Dharwadkar, N.V.: Predictive analysis of diabetic patient data using machine learning and Hadoop. In: International Conference on I-SMAC (2017). 978-1-5090-3243-3

17. Preethi, K.A., Auxzilia Preethi, K., Lakshmanan, G., Sekar, D.: Antagomir technology in the treatment of different types of cancer. Epigenomics **13**(7), 481–484 (2021). https://doi.org/10.2217/epi-2020-0439
18. Tapscott, D., Iansiti, M., Lakhani, K.R.: Blockchain: The Insights You Need from Harvard Business Review. Harvard Business Press (2019)
19. Mujumdar, A., Vaidehi, V.: Diabetes prediction using machine learning algorithms. Procedia Comput. Sci. **165**, 292–299 (2019)
20. Palimkar, P., Shaw, R.N., Ghosh, A.: Machine learning technique to prognosis diabetes disease: random Forest classifier approach. In: Advanced Computing and Intelligent Technologies, pp. 219–244 (2021)
21. Karobari, M.I., et al.: An in vitro stereomicroscopic evaluation of bioactivity between Neo MTA Plus, Pro Root MTA, BIODENTINE and glass ionomer cement using dye penetration method. Materials **14**(12) (2021). https://doi.org/10.3390/ma14123159
22. Shanmugam, V., et al.: Circular economy in biocomposite development: state-of-the-art, challenges and emerging trends. Compos. Part C: Open Access **5**, 100138 (2021)
23. Sawant, K., et al.: Dentinal microcracks after root canal instrumentation using instruments manufactured with different NiTi alloys and the SAF system: a systematic review. NATO Adv. Sci. Inst. Ser. E Appl. Sci. **11**(11), 4984 (2021)
24. Muthukrishnan, L.: Nanotechnology for cleaner leather production: a review. Environ. Chem. Lett. **19**(3), 2527–2549 (2021)
25. Sonar, P., JayaMalini K.: Diabetes prediction using different machine learning approaches. In: 2019 3rd International Conference on Computing Methodologies and Communication (ICCMC), Erode, India, pp. 367–371 (2019). https://doi.org/10.1109/ICCMC.2019.8819841
26. Kandhasamy, J.P., Balamurali, S.: Performance analysis of classifier models to predict diabetes mellitus. Procedia Comput. Sci. **47**, 45–51 (2015)
27. Irizarry, R.A.: Introduction to Data Science: Data Analysis and Prediction Algorithms with R. Chapman & Hall/CRC Data Science Series (2019)

Biometric Recognition Systems: A Short Survey

Souhir Sghaier, Moez Krichen, Abir Osman Elfakki, Mutiq Almutiq, Mariya Ouaissa, and Mariyam Ouaissa

Abstract By describing the various biometric techniques used to identify people, we concentrate on the broad context of biometrics in this work. The biometric system, its architecture, and the metrics used to assess it are then presented. We also provide an outline of the many issues that biometric systems have run into. We then highlight the various industries that use biometrics. Then, we define the position that the face recognition biometric system holds in the market when compared to other biometric systems. Additionally, we focus on the biometric facial recognition system. We show the system's general diagram. We conclude by summarizing research on methods that utilize the biometric face recognition systems in 2D and 3D.

Keywords Biometrics · Face recognition · 2D · 3D · Architecture · Survey

S. Sghaier · A. O. Elfakki
Department of Science and Technology, University College of Ranyah, Taif University, Street, Taif 21944, Saudi Arabia
e-mail: sdsghaier@tu.edu.sa

A. O. Elfakki
e-mail: aoelfakki@tu.edu.sa

M. Krichen
Department of Information Technology, FCSIT, AlBaha University, AlBaha 65528, Saudi Arabia
e-mail: moez.krichen@redcad.org

ReDCAD Laboratory, Sfax University, Sfax 3029, Tunisia

M. Almutiq
Department of Management Information Systems and Production Management, College of Business and Economics, Qassim University, 6633, Buraidah 51452, Saudi Arabia
e-mail: mmatk@qu.edu.sa

M. Ouaissa · M. Ouaissa (✉)
Moulay Ismail University, Marjane 2, Meknes 50050, Morocco
e-mail: mariyam.ouaissa@edu.umi.ac.ma

M. Ouaissa
e-mail: mariya.ouaissa@edu.umi.ac.ma

C. Iwendi et al. (eds.), *Proceedings of ICACTCE'23 – The International Conference on Advances in Communication Technology and Computer Engineering*, Lecture Notes in Networks and Systems 735, https://doi.org/10.1007/978-3-031-37164-6_41

1 Introduction

Traditional identity verification methods for access controls, such as passports or identification cards, passwords or secret codes, are readily falsifiable, forgettable, etc. The use of biometrics [1] can mitigate some of these problems, although its uses remain limited to police work. Given its potential benefits, this technology is currently being evaluated for a significantly broader array of applications [2]. In fact, the use of biometrics to assure security in a variety of industries is growing [3]. Nowadays, the use of biometrics [4] is the only option to guarantee the efficiency of rigorous access control because it is based on taking into account unique morphological aspects. In addition, biometrics is a method of identification that is difficult to steal, lose, and fabricate, as well as impossible to borrow or forget [5].

Today's businesses recognize the benefits of adopting biometric security systems [6] to safeguard work PCs, server rooms, and other corporate assets. In a business setting, it is of the utmost significance that unauthorized users do not have access to secure systems and networks. In addition, for compliance purposes, it is vital to guarantee that workflow protocols are followed appropriately and that only designated workers have access to particular files. Because they can be shared by coworkers, passwords are not a suitable option. Instead, businesses can use fingerprint biometrics to control access to a computer or server room, as it will only grant access to individuals whose fingerprints match. There are numerous biometric systems [7] available today, including fingerprint and iris recognition. These two methods are the most reliable, but their intrusive character is frequently rejected by users. This explains why their use has been restricted to professional applications for the time being (access control of airport personnel for example).

Hence, we are interested in this article in studying and providing a short survey on biometric recognition systems [8, 9] in numerous sectors, including authentication, access control, and video surveillance. The remaining part of the paper is structured as follows:

- Section 2 gives a short definition of biometrics.
- Section 3 describes the architecture of a biometric system.
- Section 4 provides a performance evaluation of biometric systems.
- Section 5 provides a comparison with other biometric techniques.
- Section 6 concludes the paper.

2 Definition of Biometrics

Biometrics is the measure of life (bio = life, metrics = measurement), which is the technology for measuring and analyzing biological data (saliva, blood, DNA, etc.), morphology (veins, Hand, fingerprints, voice, retina, teeth, iris, hair, thermography, face, etc.) or behavioral (signature, writing, typing dynamics, etc.) of the human being.

Indeed, this term is generally used in the meaning of automatic authentication/identification of people in an irrefutable way based on its distinctive and measurable characteristics. These characteristics must have the seven basic criteria to be qualified as biometric modalities:

- Universality: The characteristic is found in each individual.
- Uniqueness: The characteristic must be distinct between individuals.
- Permanence: The characteristic must resist among problems and any other interference or noise over time.
- Collection: The characteristic must be easy to acquire for analysis.
- Representation: The characteristic must be accurate, fast and robust to use.
- Acceptability: The characteristic must have an acceptable or even high degree of acceptance.
- Checking: The characteristic must be flexible with other substitutes.

Person recognition is a preliminary and critical step for the construction of all intelligent machines in interaction with the person. It is based on biometric modalities that are measurable and invariable over a certain period of time. Figure 1 presents some examples of the most permanent biometric modalities used in the biometric market and in industry.

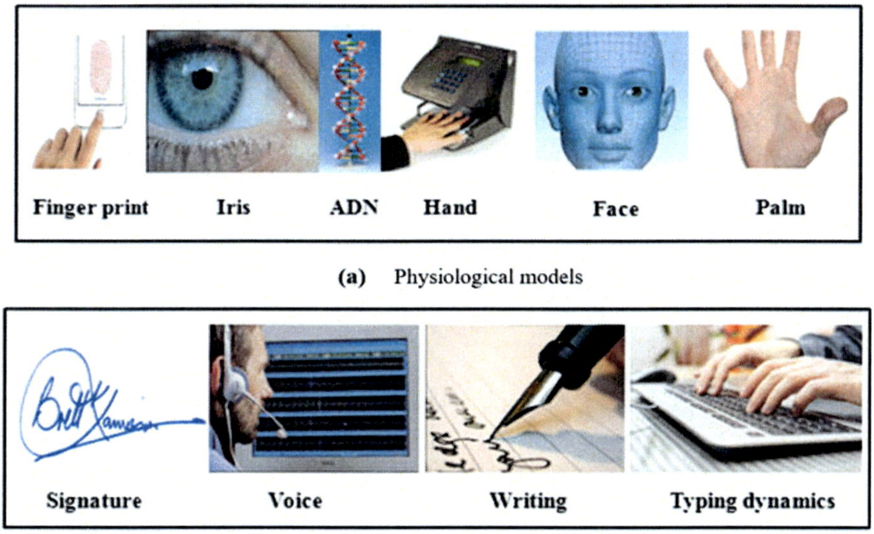

(b) Behavioral models

Fig. 1 Different biometric models

2.1 Multimodal Biometrics

Multimodal biometry is the combination of two or more biometric systems in the same application. The use of multimodal biometrics can solve some problems of unimodal systems such as problems of the data acquisition or intentional fraud and subsequently improve recognition performance [10]. The various types of multimodal biometric systems are discussed below:

- Multi-sensors: These systems use two or more different sensors to capture the same instance of a biometric feature. For example, capture the image of the face using a visible light camera and a fixed infrared camera with a particular frequency.
- Multi-instances: These systems use two or more different samples of the same biometric characteristic. An example of this could be a system for capturing images of several digital fingerprints of the same finger.
- Multi-algorithm: These systems take a single biometric sample from a single sensor and then process it using two or more algorithms.

2.2 Soft Biometrics

Soft biometrics is the features that provide weak information about an individual, devoid of distinctive character and permanence which is insufficient to properly identify a person. However, gentle modalities can be used to improve the performance of a biometric system (e.g. face recognition, fingerprints, and hand geometry). Soft biometrics have the advantage of being able to be acquired by remote surveillance cameras without any user cooperation, unlike traditional biometric features that require the cooperation of people [10]. We can cite as an example, facial measurements, face shapes, gender, age, height, weight, and occlusion.

3 Architecture of a Biometric System

A biometric system is essentially a system of automatic recognition of individuals based on their biometric characteristics. The biometric system consists of two main modules; Learning and recognition (identification or authentication).

3.1 Learning Module

During this phase (also called enrollment), the biometric system will acquire the biometric information necessary and insufficient for the recognition of the person in question. Indeed, the biometric characteristics extracted are processed by a biometric sensor, then represented in digital form (signatures), and finally recorded in the database. The treatments made during the learning process are performed offline.

3.2 *Recognition Module*

The operation of the recognition phase differs according to the operating mode of the system: identification or authentication:

- In identification mode: The system must recognize the person. He answers a question of type: "Who am I?" So the system will compare the processed signal with all the models stored in the database (the problem of one to all type).
- In authentication (or verification) mode: The system must authenticate the individual by answering the question of type: "Am I the person I claim to be?" Indeed, the system must validate that the identity of the person is the one proposed. It is, therefore, sufficient to compare the incoming signal with a single sample recorded in the database (one-to-one problem).

Figure 2 shows the main modules of a biometric system as well as the different modes. A brief description of the functioning of each module is presented as follows:

- Biometric sensor module: Measurement of the physiological, behavioral or biological characteristics of the individual;
- Data extraction module: Extraction of the discriminating attributes from the biometric data, for example, images of faces or regions of interest;

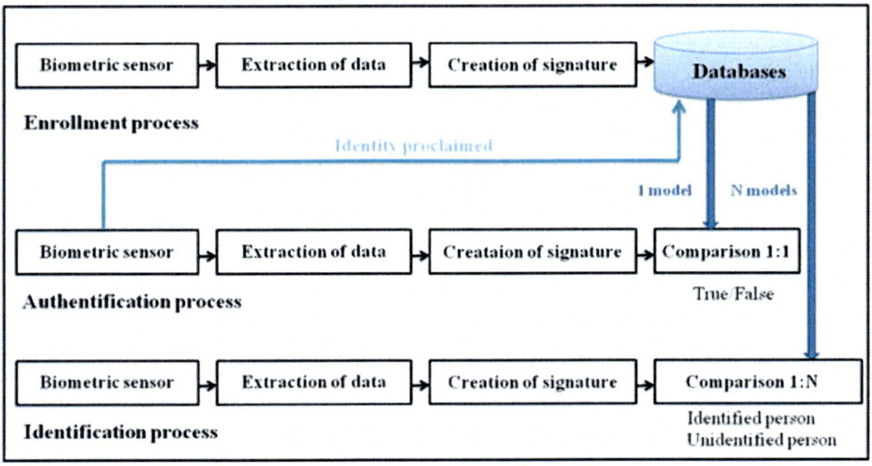

Fig. 2 Principle of the functioning of a biometric system

- Module creation of a signature: Construction of a numerical model to represent acquired biometric information. This model, also called signature, will be stored in a database;
- Comparison module: Comparison of the biometric characteristics of the individual subject to control with the registered signatures. This module works either in verification mode (for a proclaimed identity) or in identification mode (for a searched identity).
- Database module: Registration of biometric references for enrolled persons.

4 Performance Evaluation of Biometric Systems

Numerous studies have been made to evaluate the performance of biometric systems. Mainly performance evaluation criteria are Intrusive, cost, amount of information recorded for each person, efficiency, reliability, and accuracy. The biometric system can operate according to the context of the application either in authentication (verification) mode or in identification mode.

4.1 Evaluation of Authentication

When the system works in the context of verification it can accept users' impostors, we speak therefore of false acceptance. As it can reject legitimate users in this case we speak of false rejection. The performance of this system is thus measured by the two fundamental rates: False Acceptance Rate (FAR), False Rejection Rate (FRR) presented below:

- False rejection rate: This is the e proportion of individuals who are supposed to be recognized yet are rejected by the system.
- False acceptance rate: This is the proportion of applicants who are supposed to be rejected but who are accepted by the system.

In addition, the most efficient criterion for determining the performance of an algorithm in verification mode is to calculate the Equal Error Rate (EER): This point corresponds to the location where FAR = FRR. It is the best compromise between false rejections and false acceptances.

The Receiver Operating Characteristic (ROC) curve shown in Fig. 3 plots the false rejection rate as a function of the false acceptance rate. Moreover, this curve tends to conform to the shape of the reference frame. If the system is efficient, the recognition rate is more preferment. This curve presents graphical measures of the performance of a verification system. We can see on the ROC curve that if the FAR decreases, the FRR increases and vice versa.

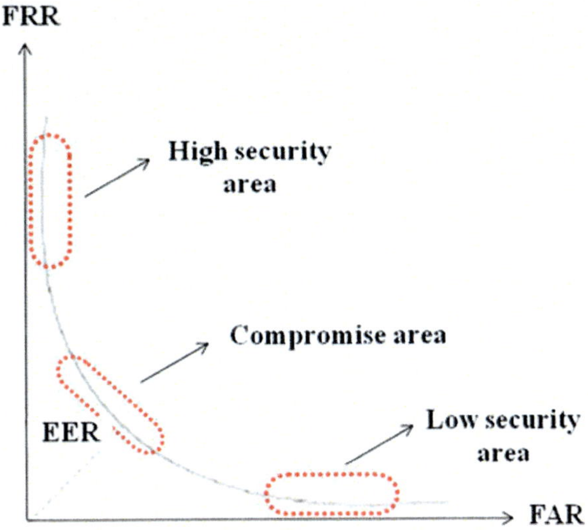

Fig. 3 ROC curve

4.2 Evaluation of Identification

In the identification mode, it is important to use the Cumulative Match Characteristics (CMC) curve. This curve plots the probability that the right choice is the first N. Indeed, it presents the rate of identification of persons recognized according to the rank, as illustrated in Fig. 4. We say that a system recognizes at rank 1 when it chooses the nearest image as a result of the recognition. We say that a system recognizes in rank 2 when it chooses, from two images, the one that best corresponds to the input image, etc.

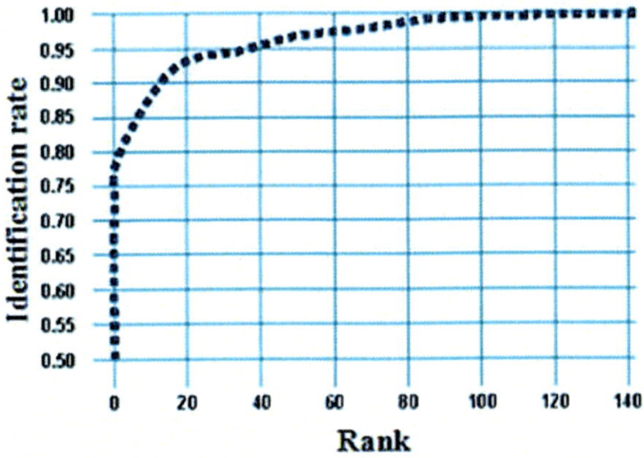

Fig. 4 CMC curve

5 Comparison with Other Biometric Techniques

Each biometric system can have various problems that can affect the performance rate. These problems are generally due to:

- Noise introduced by the sensor: It is the noise generated during the acquisition of the scans; it is generally due to a faulty sensor or poorly maintained sensor.
- Lack of individuality: It is the lack of uniqueness of characteristics between individuals. Indeed, we can find a great similarity between different people because of the genetic factors (true twins, father, and son) which negatively affect the recognition rate of the biometric system.
- Lack of invariant representation: It is the presence of intra-class variations between the images of the same individual. These variations may be due to poor user interaction with the sensor (e.g., pose and facial expression changes when the user is standing in front of a camera), or the use of different sensors during the enrollment and verification, also may be due to changes in environmental conditions (e.g., changes in illumination for a facial recognition system).
- Sensitivity to attacks: It is the possibility of using the usurped biometric modalities of the individuals.

Presently, the facial recognition system occupies a prominent position in the market and in the industry [11]. This form of the biometric system has various advantages over fingerprint, iris, and other biometric techniques, etc. Facial recognition is user-friendly, has cheap deployment costs, is natural, and is not invasive. This technique does not require the assistance of humans for recognition. Due to these factors, the scientific community is becoming increasingly interested in facial recognition [12].

Figure 5 demonstrates the scientific interest and degree of the use of this technique in comparison with other techniques such as vein recognition, voice recognition, and hand geometry. Although the fingerprint is the biometric characteristic the most commonly used, it is less accepted by the public than face recognition; also it obliged the intervention of the person.

The biometric system of face recognition has several advantages over other biometric modalities. It is well accepted by users and has attracted researchers from different fields: model identification, psychology, computer vision, neural networks, and computer graphics.

A recognition system generally involves the following steps: acquisition, preprocessing, detection, and extraction of characteristics followed by a decision phase (Fig. 6):

- Acquisition Phase: Take different face scans of an individual from a sensor to make a database.
- Detection phase: Detect the face as a region of interest.
- Pre-treatment phase: Process scans of unwanted noises that affect recognition.
- Characterization phase: Extract the most relevant and discriminating primitives.
- Learning Phase: Learn the different faces of the system.

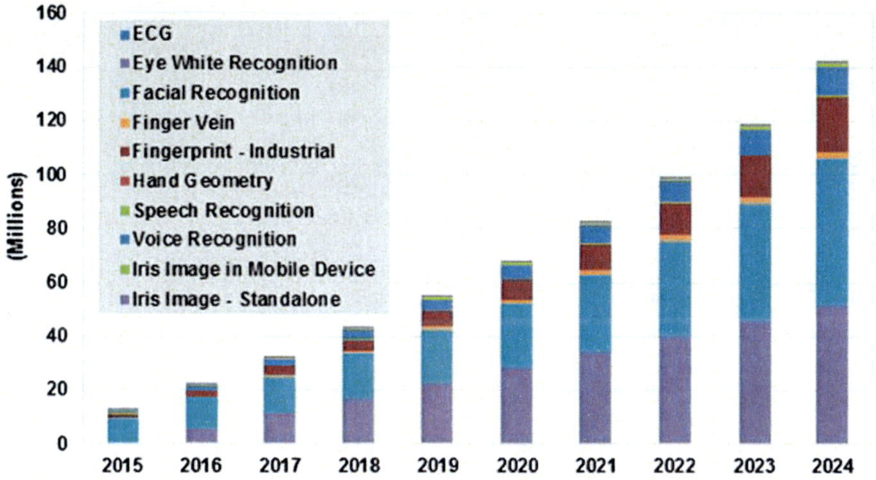

Fig. 5 Distribution of biometric techniques in the market

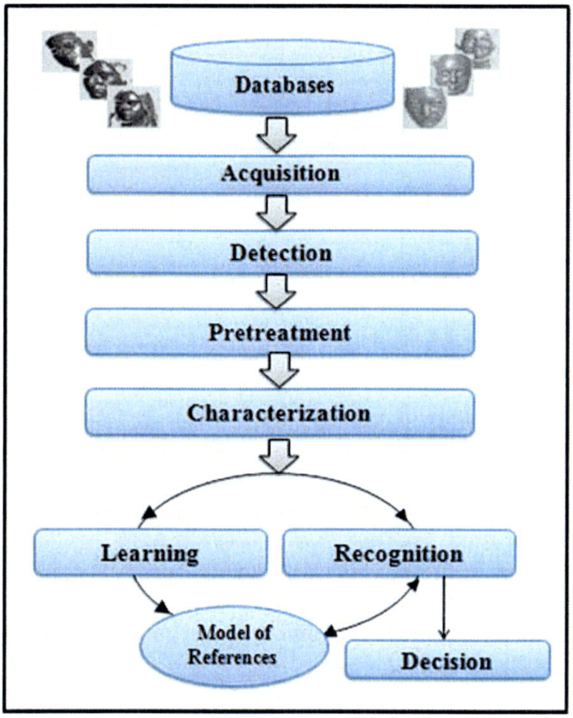

Fig. 6 General diagram of a facial recognition system [13]

Table 1 Selected works on 2D face recognition

Reference	Methods	Databases	Results
[14]	• Principal component analysis	ORL	TR = 84.37%
[15]	• The embedded image kernel • Support vector machine	ORL FEI	TR = 98.23% TR = 97.5%
[16]	• Local binary pattern • Gaussian filter	ORL	TSR = 93.33%

Table 2 Selected works on 3D face recognition

Reference	Methods	Databases	Results
[17]	• Cropping filter • Median filter • Laplacian filter • Delaunay triangulation • Anthropometric measurements • Euclidean distance • SVM	GAVAB Frav 3D	RR = 98.36% RR = 100%
[18]	• DL network Sur3dNet-face • PointNet • Gaussian process morphable model	FRGC v2.0 Bosphorus	TR = 98.85 TR = 99.33%
[19]	• Laplacian pyramid algorithm	BU-3DFE	TR = 93.60%
[20]	• Streamed attention network • Convolutional neural networks • 3D local binary pattern	FRGCv2.0 Bosphorus EURECOM Kinect face IIIT-D RGB-D face MICC-HR/LR 3D face	TR = 84.62% TR = 86% TR = 59.31% TR = 86.67%

- Classification/decision phase: Organize a characterization set in homogeneous and contrasting classes, generally based on the learning basis, and associate a symbolic description with the object, based on its characteristics.

Traditional methods of recognition ensure the safety of individuals. They are unreliable, and easy to borrow, steal, forget or falsify. Table 1 gives a selection of works on 2D face recognition, specifying the techniques used, databases, and the results recorded. 2D face recognition systems suffer from several problems such as changes in lighting, pose, and facial expression. Table 2 presents a selection of literature work on 3D facial recognition systems, specifying the techniques, databases used and the recorded results.

6 Conclusion

In this study, we have focused on the overall context of biometrics by describing the various biometric identification methods. Then, we introduced the biometric system, its architecture, and evaluation metrics. We also provided an outline of the many issues that biometric systems face. We have since emphasized the various applications of biometrics. Then, we clarified the market position of the face recognition biometric system in comparison to other biometric systems. In addition, we have focused on the face recognition biometric system. This system's general schematic has been given. Lastly, we provided a summary of a literature review of 2D and 3D biometric facial recognition system-based methodologies.

References

1. Labati, R.D., Piuri, V., Rundo, F., Scotti, F.: Photoplethysmographic biometrics: a comprehensive survey. Pattern Recognit. Lett. (2022)
2. Palma, D., Montessoro, P.L.: Biometric-based human recognition systems: an overview. Recent Adv. Biom. 1–21 (2022)
3. Liu, S., Shao, W., Li, T., Xu, W., Song, L.: Recent advances in biometrics-based user authentication for wearable devices: a contemporary survey. Digit. Signal Process. **125**, 103120 (2022)
4. Wang, M., Yin, X., Zhu, Y., Hu, J.: Representation learning and pattern recognition in cognitive biometrics: a survey. Sensors **22**(14), 5111 (2022)
5. Minaee, S., Abdolrashidi, A., Su, H., Bennamoun, M., Zhang, D.: Biometrics recognition using deep learning: a survey. Artif. Intell. Rev. 1–49 (2023)
6. Rayani, P.K., Changder, S.: Continuous user authentication on smartphone via behavioral biometrics: a survey. Multimed. Tools Appl. **82**(2), 1633–1667 (2023)
7. Maiorana, E.: A survey on biometric recognition using wearable devices. Pattern Recogn. Lett. **156**, 29–37 (2022)
8. Smith, M., Miller, S.: The ethical application of biometric facial recognition technology. In: AI & Society, pp. 1–9 (2022)
9. Yu, Z., Dong, Y., Cheng, J., Sun, M., Su, F.: Research on face recognition classification based on improved Googlenet. Secur. Commun. Netw. 1–9 (2022)
10. Ghalleb, A.E.K., Sghaier, S., Amara, N.E.B.: Face recognition improvement using soft biometrics. In: 10th International Multi-Conferences on Systems, Signals & Devices 2013 (SSD13), pp. 1–6. IEEE (2013)
11. Khan, A.A., Shaikh, A.A., Shaikh, Z.A., Laghari, A.A., Karim, S.: IPM-model: AI and metaheuristic-enabled face recognition using image partial matching for multimedia forensics investigation with genetic algorithm. Multimed. Tools Appl. **81**(17), 23533–23549 (2022)
12. Sun, Y., Ren, Z., Zheng, W.: Research on face recognition algorithm based on image processing. Comput. Intell. Neurosci. **2022** (2022)

13. Sghaier, S., Krichen, M., Elfaki, A.O., Abu Al-Haija, Q.: Efficient machine-learning based 3d face identification system under large pose variation. In: Advances in Computational Collective Intelligence: 14th International Conference, ICCCI 2022, Hammamet, Tunisia, September 28–30, 2022, Proceedings, pp. 273–285. Springer International Publishing, Cham, (2022)
14. Bazatbekov, B., Turan, C., Kadyrov, S., Aitimov, A.: 2d face recognition using PCA and triplet similarity embedding. Bull. Electr. Eng. Inf. **12**(1), 580–586 (2023)
15. Abdulhussain, S.H., Mahmmod, B.M., AlGhadhban, A., Flusser, J.: Face recognition algorithm based on fast computation of orthogonal moments. Mathematics **10**(15), 2721 (2022)
16. Bhavikatti, S., Bhairannawar, S.: Efficient reconfigurable architecture to extract image features using local binary pattern for face recognition (2023)
17. Sghaier, S., Farhat, W., Souani, C.: Novel technique for 3d face recognition using anthropometric methodology. Int. J. Ambient Comput. Intell. (IJACI) **9**(1), 60–77 (2018)
18. Yu, Y., Da, F., Zhang, Z.: Few-data guided learning upon end-to-end point cloud network for 3d face recognition. Multimed. Tools Appl. **81**(9), 12795–12814 (2022)
19. Kong, W., You, Z., Lv, X.: 3d face recognition algorithm based on deep Laplacian pyramid under the normalization of epidemic control. Comput. Commun. **199**, 30–41 (2023)
20. Neto, J.B.C., Ferrari, C., Marana, A.N., Berretti, S., Del Bimbo, A.: Learning streamed attention network from descriptor images for cross-resolution 3d face recognition. ACM Trans. Multimed. Comput. Commun. Appl. **19**(1s), 1–20 (2023)

Lung Histopathological Detection Using Image Classification

K. Ishwarya, J. Christy Jackson, Abdul Quadir Md, Senthilkumar Mohan, and Celestine Iwendi

Abstract Early detection of lung histopathology has become crucial and essential for humans. Rapid recognition gives many patients the greatest chance of recovery. Histopathological graphics of biopsy samples tissue from possibly infected areas of the lungs are used by doctors to best solution. The multiple type of lung disease is frequently misdiagnosed and prolonged to detect. The characteristics used to detect Lung Histopathology are extracted from Computed Tomography (CT scan) images. Deep Learning (CL) is a novel method that enables us to improve result's precision. In this paper, we create DL model to determine the type of lung cancers from Computed Tomography images. Convolutional Neural Networks (CNN) which recognize and categorize lung-cancer type within improved efficiency and less amount of time, that is crucial to deciding on the best treatment approach for patents and their risk of mortality. This paper proposes a tri-category classification which applies to images of lung cancer. Benign, adenocarcinoma and squamous cell carcinoma is performed by utilizing VL, VGG-16, and Le-Net to process an image of lung tissue and obtain functionalities effective for diagnostic techniques. Further, the paper analyzes how hand-crafted characteristics can be extracted from raw pictures after various processing steps, and finally, the Python framework (Django) is used to deploy the model. The purpose of using this technique is to obtain some characteristics primarily relevant to lung histopathology forecasting.

Keywords Le-Net · VGG-16 · Histopathological images · CNN · Lung cancer · Deep learning · Tensor flow · Keras · Computed tomography

---

K. Ishwarya · J. C. Jackson (✉) · A. Q. Md
School of Computer Science and Engineering, Vellore Institute of Technology, Chennai, India
e-mail: christyjackson.j@vit.ac.in

S. Mohan
School of Information Technology and Engineering, Vellore Institute of Technology, Vellore 632014, India

C. Iwendi
School of Creative Technologies, University of Bolton, Bolton BL3 5AB, UK

C. Iwendi et al. (eds.), *Proceedings of ICACTCE'23—The International Conference on Advances in Communication Technology and Computer Engineering*, Lecture Notes in Networks and Systems 735, https://doi.org/10.1007/978-3-031-37164-6_42

1 Introduction

1.1 An Overview

In the United States, lung carcinoma is both one of the most common diagnoses and the main factor in lung carcinoma deaths, accounting for about 25% among all cancer-related deaths. Lung carcinoma has a high mortality rate, in part because symptoms don't show up until the disease has progressed. For screening in densely populated areas, Low-Dose Computed Tomography (CT) has already been recommended as a safe and effective tool. Comparing yearly chest radiography to yearly CT monitoring after 7 years could reduce the incidence of lung cancer by at least 20% [1]. After breast cancer, lung carcinoma is the second-leading cause of death in Chinese women, and it has the highest mortality rates. Despite the low smoking rates among women, tumour rates are still high due to the ease with which women can receive second—hand tobacco, indoor smoke, pollution from the environment, as well as other factors [2]. Typically, the precise detection of Lung nodule in Computed tomography is the foundation for an initial cancer detection. A tumor or other malignant structure with a diameter with about 3 mm is called a Lung tumor. Radiologists read chest CT slices by slices for this purpose. However, this procedure is not only time-consuming but also prone to errors [3]. As the amount of CT images of lung nodules increases. Despite the fact that many [2–4] methods of tumor segmentation have been suggested in recent years, it remains difficult to achieve acceptable detection rate due to a variety of Lung tumor visible on Computed tomography. Unfortunately, in the early stages of lung cancer, there are often no symptoms. Symptoms appear later in the disease's progression. Hardly 17.4% of Americans survive for five years without being diagnosed, and also the percentage is lower in emerging regions.

Contrarily, promptly identifying and diagnosing cases of malignancy leads to a quicker rate of recovery, less complex and costly treatment, and—most importantly—a higher proportion of disease remission [5]. Carcinogenesis is divided into three stages. The initial stage is called the Initiation phase, during which any alterations brought on by gene variant in the tissue could cause a change in expression of genes and in some contexts, the deactivation of a section of DNA. The unit with genetic alterations stays consistent if such variations skip the conceptual phase of the cell growth. The altered cell begins to proliferate during the Publicity phase, which is the second stage. In the final phase, the Progressive phase, cells begin to proliferate aggressively in terms of number, size, and formation of primary tumours. Cells become obtrusive and metastatic at this stage. The name of a form of cancer is determined by the skin organ or type of cell from which the cancer develops. So far, and over 100 different types of cancer have been discovered. Cancer comes in many forms, including breast, brain, lung, and colon cancer. Cancer is a non-communicable disease (NCD), accounting for 71% of all mortality globally (World Health Organization) [6]. Cancer, also referred to as Lung Carcinoma, becomes a malignant tumor marked by unchecked tumor growth in Lung Tissue [7]. Treatment is necessary to prevent metastases from being used to scatter this accumulation towards other internal organs [8, 9]. A high proportion of carcinomas that begin inside the bronchi are carcinomas; the two primary kinds are carcinoma of the Small Cells of the lung and Non-Small Cells of the lung [9]. Everlasting smoking is the main cause of malignancy in 88% of cases. That almost 25% of cancer deaths are Lung disease, which affects all Men and Women equally. In about 80% of contexts, taking drugs seems to be the main factor contributing to lung cancer death. Carcinoma may occur in non-smokers who are sensitive to radon, second-hand tobacco, indoor air, or other elements like asbestosis at work, diesel exhaust, or even other compounds [10].

2 Literature Review

Ozdemir et al. In this paper introducing a new low-dose CT scan with computer-aided detection for screening for lung cancer… Presenting a Deep 3DCNN based end-to-end probability—based detection approach for lung cancer (CNNs)… Additionally, it has been divided into two categories: (i) a computer-aided detection (CADe) device which identifies and classifies suspected lung nodules, but instead (ii) a computer-aided diagnosis (CADx) device which identifies and classifies malignancies at its patient as well as nodule levels, respectively [1]. They To identify the type of lung carcinoma from Computed tomography for patients in a hospital database, they proposed a DL model. First use DCNN (DenseNet) to classify malignant tumor from image obtained from, followed by image rotation, translation, and modification technique to enlarge and equipoise our data for training. Finally, Adaptive Boosting (Adaboost) method is implemented to aggregate multiple classifiers to enhance classification performance. This gives pathologists an effective, non-invasive diagnosis tool for determining the type of lung cancer [11]. Sahu et al., The outcome of this experiment, they developed a Novel Multi-section Convolution Neural Network for categorizing pleura tumor and determining their likelihood of being cancerous. The outcomes of this study demonstrated that our suggested model outperformed a number of cutting-edge classification techniques. Lightweight, Multi-section Convolution Neural Network method seems to be portable to mobile gadgets such as smartphones and embedded devices and doesn't permit the time-consuming manual spatial annotation… picture a hybrid approach that makes use of cloud computing power to recognize the causes behind model performance [12].

In this study, the topic itself suggest that they are developing TSCNN to diagnose the bronchi lump… As we know by obtaining of CT images it is hard to see the diff. between lung nodule and non-lung nodules… In the present study, they propose a TSCNN to address the issues raised in the preceding disparate Computed tomography. An applicant tumor diagnosis stage of a system architecture would be based on the enhanced U-Net, and also the false-positive reducing stage is based on 3DCNN [13]. Alzubaidi et al. This outlines a thorough and contrasting regional and worldwide extraction of features for Computed tomography screening for bronchi tumor. Information gathering, worldwide training and local testing and training are the three primary stages with this approach. The above approach assesses the efficiency of 10 various processed image feature types. Additionally, it contrasts six frequently used ML Algorithms: DT, RF, K-nearest neighbors, support vector machine and naive Bayes. Developing any such approach is designed to assist with to identify the kind of Feature-extractor and understanding technique that will produce a most precise detection model [5]. This lung nodule malignancy prediction [14] is a early diagnosis of lung cancer… they build a own module i.e. multi-view divide and rule… it learns Nodule malignant prediction from both authentic and vague annotation… With the help of combining reliable and ambiguous, it's easy to find the predictions of the lung nodule malignancy and also performing some of the deep learning modules. They are dividing three nodule they are Consistent and Reliable Set CR set, Inconsistent IC set and Low-Reliable LR set… Each Set is opted for training and also each is annotated by some parameters. The above module performs as an architecture network for the predictions i.e. CF-net, LR-net and prediction network PRD-net… It also aims to boost the prediction ability of networks respectively (Figs. 1, 2 and 3)…

Lu et al. [8] As the title itself suggest that it's kind of screening trail in lungs organ… This model is fully depending on mortality prediction in lung screening trail… It focuses on deep learning concepts and it detects the severe place of human body that leads to mortality… So, to find that severe place on lung we are implementing this module… This will help the doctors to identify affected parts that leads to mortality… To avoid the malignancy also this model would be helpful for the patients… So doctors can easily start the treatment and it also reduces time for identifying the part… This paper proposes about disease diagnosis and this have been found with the help of probability analysis it's kind of predicting the disease diagnosis and also to improve predictions accuracy rate. We are enhancing us to lead in medical method which helps easy to find the particular diagnosis that is non-small cell lung cancer… The above study seeks to develop a model that takes into account not only statistical analysis and also planning to determine the disease prognosis with aid of findings… This even divides the likelihood of NSLC transitions into 4 phases… [9].

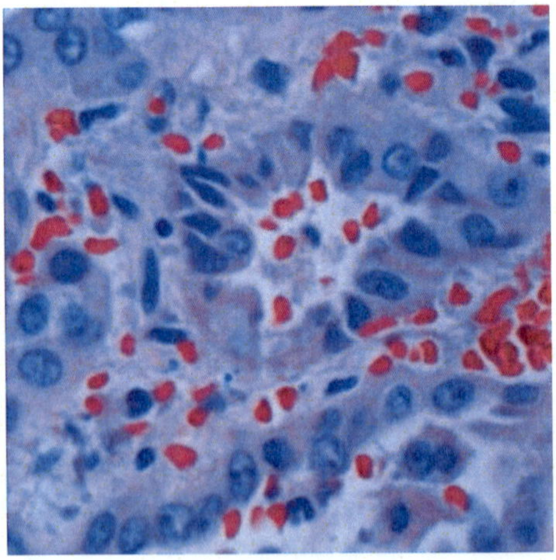

Fig. 1 Adenocarcinoma

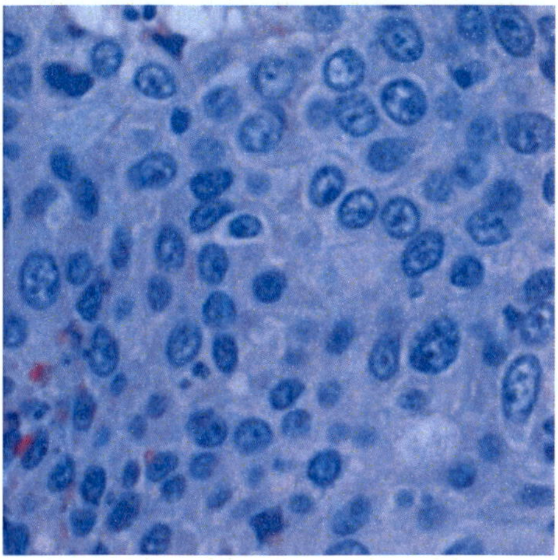

Fig. 2 Squamous cell carcinoma

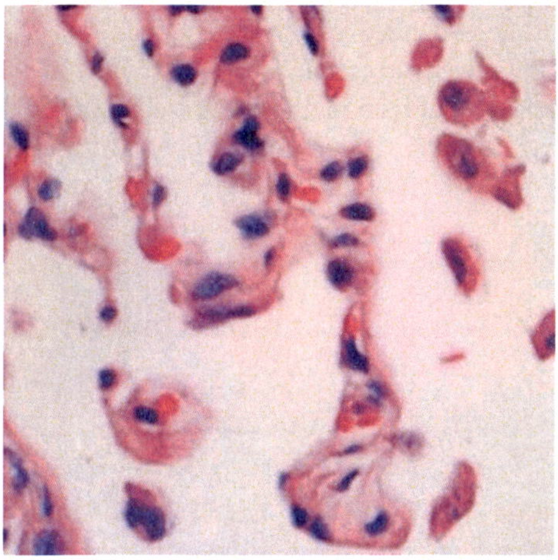

Fig. 3 Benign

According this paper, yes lung cancer [15] plays a vital role for humans and there are so many techniques to find the lung cancer by implementing some parameters. But in this model Image processing techniques like machine learning, image restoration, convolution etc.... have been extensively used for cancer predictions and also by performing some of the machine learning algorithms it detects the lung cancer... Liu et al. [16]. The prediction of cancer survival may offer guidance for smarter clinical choices and personalized medicine. Massive amounts of datasets are available from the SEER programme for ML analysis. The SEER dataset is used to identify and label SPLC cases, and the collected data is then enhanced with IECFS. Inter–Intra-class dispersions as well as the ranking criteria are used in the IECFS method. The approach produces the greatest efficacy by varying the parameter's value as well as the amount of feature chosen. In [17–43] the authors proposes different models based on QoS parameters.

3 Materials and Methods

Dataset Partitioning

There are typically two categories of information: learning information and test data. The training set includes a known output, and the prototype learns from it in order to later generalize to other data. The dataset is divided into a 7:3 ratio (i.e., 70% schooling and 30% testing). It has the set of data (or set of items) in order to run our designs, and it will accomplish this in Python and used the Tensor flow library and the Keras process. To categorise the lung cancer visuals, the set of data is acquired from the LC25000 datasets, that contains enhanced data with 5000 images—all the 3 categories of lung cancer image set. For our work, we consider 3 categories: Benign Tissue, Adenocarcinoma and Squamous Cell Carcinoma of the lungs, each consists of 5000 Histopathology Images. Before the model is constructed and tested, the above set of data is first processed using Python tools, as well as the feature are then extracted using CNN techniques. To make a comparison and classify the images, different CNN methodologies are used.

4 Proposed Analysis Approach and Results

4.1 Data Pre-processing

Pre-processing of data is a crucial step in improving image quality. It includes the data preparation, normalisation, cleaning, and formatting. Data preparation facilitates data transformation by customising it into the suitable form. Data normalisation converts a different image layout into a frequent format in which all images are uniform, whereas data transformation compresses the data. In order to reduce internet traffic and computation cost, image pre-processing is necessary. Resizing, rotation, horizontal_flip and vertical_flip and the Image Data Generator library from Keras are some of the data augmentation techniques used. Better capability is used to limit the input images between 0 and 1, which aids in the reduction of computational complexity. Our image was randomly rotated with both zero- and one-degree courses using rotation range = 1 degree. Automatically resize 1./255 is used to transform all pixel values between [0, 255] and [0, 1]. Grayscale pictures are converted to RGB image using Python's flow from directory and the colour mode function. A simple and fast method to enhance images is to use the Keras Image_Data_Generator class to resize the input_image from 512*512 to CNN input_size. While the neural net modelling is trying to train on the training examples, the image files are read from

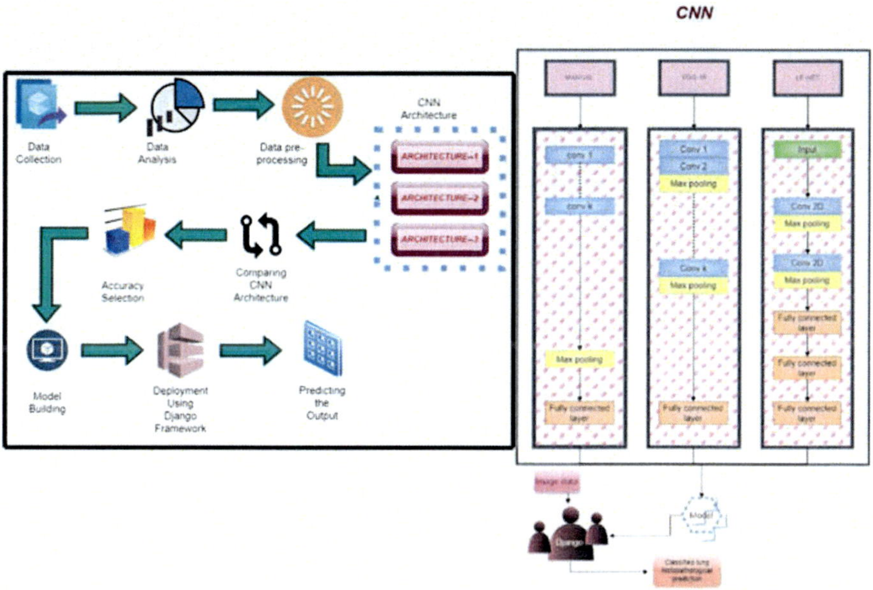

Fig. 4 Architecture diagram of the proposed work shows the number of convolutional layer, maximum pooling and densely connected layers sequentially

database and enhanced using the Image_Data_Generator flow_from_directory (). The target_size parameter, which specifies the dimensions source images, is crucial. It is a feature that resizes image according to the input size. In our instance, the traget_size attribute values were set using an image with a size of 250*250 (Fig. 4).

4.2 *Model Training, Testing and Prediction*

Development, Evaluation, and Prediction of Models A liner stack of layers was used to create the CNN or ConvNets for image detection and categorization. Training and test images were processed using convolution layers of kernel filters, pooling layer, and convolutional layers. The Softmax was utilised to categorise the provided object. The model was tested and trained using Jupyter Notebook (anaconda). A neural network with three hidden layers, one input node, and one fully connected layer was implemented for this mission. Images are divided in a ratio of 70:30 for training and validation. Input image with image size were sent to the input layer (250, 250). (3, 3) Kernel size with activation function $ReLu(x) = max(0, x)$. A max pool size of (2, 2) has been implemented in order to reduce a next convolutional layers data-processing parameters. Using the activation function of SoftMax and a dense value of three, the output data classes were determined. An Adam optimizer was used to determine the learning rates for various parameters. A Loss function is important for adjusting the network's weights in order for a machine learning algorithm to fit better while training neural networks. During deep network while training, the loss function penalises the model if the label it has predicted differs from the real target label. The Loss Function computes a difference between an input vector's for an expected and labelled output (Figs. 5 and 6).

$$\text{Loss} = \frac{1}{2}(y - \hat{y})$$

For this task, classification cross-entropy (CE) was employed as a algorithm, which would be estimated as:

$$CE = -\log\left(\frac{e^{s_p}}{\sum_j^C e^{s_j}}\right)$$

C denotes the no of output layers, S_p denotes CNN scores of provided true Positive, and S_j denotes net scores for each class C.

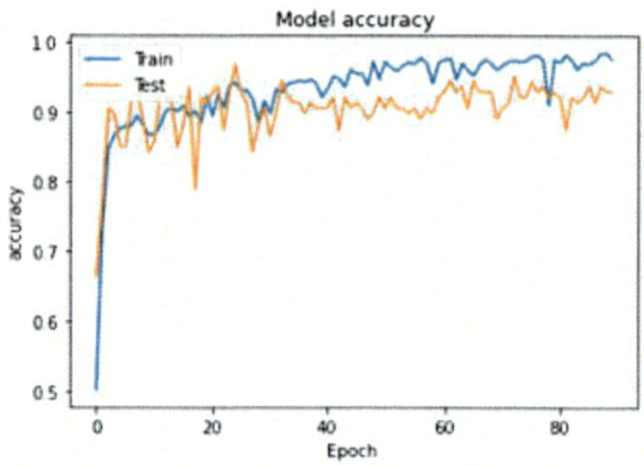

(a) VL architecture

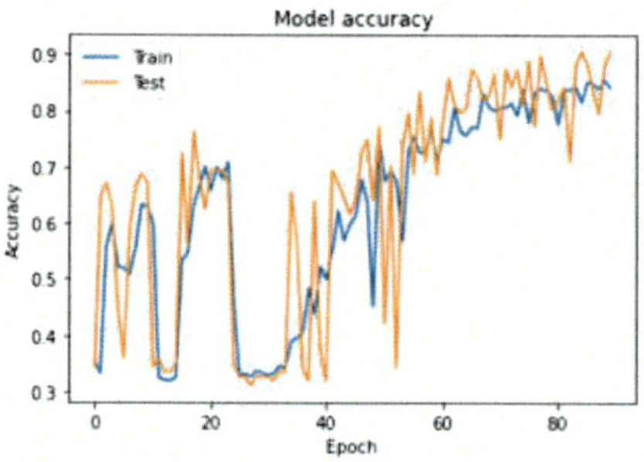

(b) VGG-16

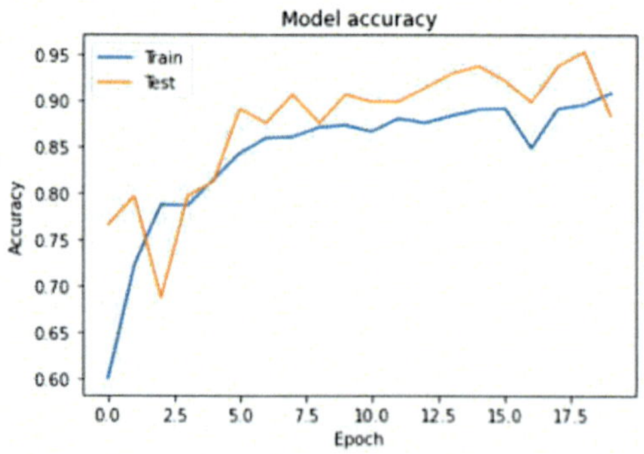

(c) LENET

Fig. 5 **a–c** Illustrates model accuracy for both test and training images with epoch

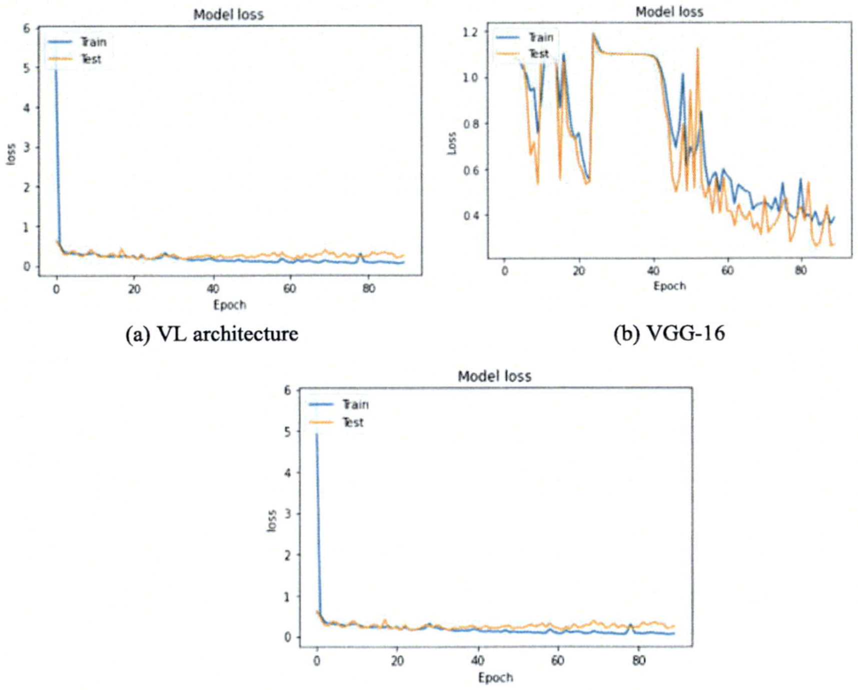

(a) VL architecture (b) VGG-16

(c) LENET

Fig. 6 **a–c** Illustrates model loss for both test and training images with epoch

4.3 Convolutional Neural Network

CNN design's three main layers are convolutional, pooling, and fully connected. The most popular architectures for CNN classification and pattern classification are Le-Net, Alex-Net, VGG-Net, Res-Net, ZF-NET, and Squeeze-Net.

Because of how easy it is to calculate and use, the back propagation belongs to the best learning algorithms. Back propagation makes a distinction between both the algorithms forecast accuracy and the labels assigned to the regression coefficients determined by the error function throughout training. Equation (1) states the cross-shape entropy. The total amount of training instances is given by the symbol n. In the below equation, training images are represented by the input parameters, while directly connected output neurons (j) have already been represented as input data in the output nodes ($a\ l\ j$).

The calculation is as follows:

$$C = -\frac{1}{n}\Sigma \times \sum_j \left(y_j a_j^l\right)\left(1 + y_j\right)\ln\left(1 - a_j^l\right) \quad (1)$$

4.3.1 Image Data Generator

In addition to rescaling the appearance, it zoom in on the image and rotates it horizontally. All image configurations are included in this image_data_generator.

4.3.2 Training Process

The function train_datagen_flow from directory is used to begin preparing information from the train dataset directory. The image's target size is specified by target size. Test_datagen_flow from directory is used to organise test data for the model, and everything is the same as described above. Fit generator is utilized to fit the data into the algorithm created above, and steps per epochs tell us how many times the design will implement for the training phase.

4.3.3 Convolutional Layer

Thus every subset of features is made up of a collection of neurons that work together to form a feature map. Equation (2) calculates the output of a convolutional layer. A kernel's (filter's) size, Feature maps, Bias and Weight is denoted by $F, \ m, \ B$ and W_j. Conv Layers produce y_i, where I is an index indicating the ith extracted features in a layer called l (2).

$$y_i^l = B_i^l + \sum_{j=1}^{ml(l-1)} F_{i,j}l \times W_j^{l-1} \tag{2}$$

4.3.4 Pooling Layers

Convolutional layers and pooling layers are frequently combined; as a result, subsampling shrinks the amount of information in all depth portions, assisting in preventing over fitting through learning the network. The depth dimension is unaffected by the pooling process because the input volume is reduced. As illustrated in Fig. 3, the most frequently used pooling technique is maximum and average pooling. In (3), two Equations determine the full scope and length of the pooled-layer performance.

$$B_2 = \left(\frac{B_1 + F}{S}\right) + 1 \tag{3}$$

$$H_2 = \left(\frac{H_1 + F}{S}\right) + 1$$

4.3.5 Flatten Layer

The flatten function is used to straighten the input. It serves to flatten the components of an image that has been convolved.

Flatten makes the following claim:

keras.layers.Flatten(data_format = None)

When switching between data_formats, the selectable argument file system is used to keep weight accumulating. It will accept either the first or last value, channels. Channels initially needs to recognize the information model as (batch_size, channel), wherein channels last identifies it as (channels last) (batch_size, channel).

4.3.6 Dense Layer

The operation is carried out by Dense as follows: output = activation(dot(input, kernel) + bias), in which activation seems to be the element-wise activation function carried as the activation argument, kernel is indeed a weights matrix constructed by the layer, as well as bias is a state vector constructed by the layer.

4.3.7 ReLU Activation

CNN determines neuron output using a non-linear activation function rather than a local connection. Using the operational sigmoid and hyperbolic tangent model parameters is advantageous due to enhanced results, faster acquiring knowledge, and simple structure (Fig. 4). The ReLU function is depicted in Eq. (4). If y is less than zero, the shading of the ReLU function is zero; alternatively, it is one (2).

$$f(y) = max(0, y) \tag{4}$$

If z is below zero, $f(z)$ is 0, and when $z \geq 0$, $f(z)$ is equivalent to z.

$$f(z) = \begin{cases} \text{zero}, & z < 0 \\ z, & z \geq 0 \end{cases}$$

4.3.8 SoftMax Activation

SoftMax is employed to determine the likelihood of the each ground truth label, producing an output that is between zero and 1, and then converting the outcomes to noticeable values Eq. (5). $f(z)$ [44] employs the SoftMax function to convert random variables (z) into relevant values ranging from zero and 1.

$$S(y)_i = \frac{\exp(y_i)}{\sum_{j=1}^{n} \exp(y_j)} \tag{5}$$

where y is a Softmax function to a state vector, s. It comprised of n components for different classifiers.

The i-th component of the input sequence is represented by y_i It can be any number between $-\infty$ and $+\infty$.

The standard exponential function, $\exp(y_i)$, is applied to y_i.

The term $\sum\limits_{j=1}^{n} \exp(y_i)$ is a standardisation term. It helps to ensure that the attributes of the output vector $S(y)$ isume to 1 for the i-th class and that they are all between 0 and 1, forming a valid probabilistic model.

n refers to the number of classes.

4.3.9 Epochs

An epoch is referred to the amount of repetitions of all the training_set examples of one process having trained the ML models when all of the training examples is used at once. An epoch can also be defined as the amount of passes a training set makes around an algorithm. When the set of data has made both ahead and reverse passes, only one pass is counted. The quantity of epochs is referred to as a hyperparameter. It specifies how many times the complete data set must be processed by the learning algorithm (Table 1).

Table 1 Comparative analyses in terms of accuracy among the existing and proposed models

References	Year	Dataset	Accuracy (%)
Xie and Yang [45]	2019	LUNA-16	86.42
Qin and Wang [46]	2020	Henan provincial people's hospital	72
Sahu and Yu [12]	2019	LIDC-IDRI	93.18
Afshar and Oikonomou [47]	2020	NIH's chest X-ray dataset, obtained from Kaggle	73
Bharati and Podder [48]	2020	LIDC-IDRI	83
Proposed (VL)	**2020**	**Histopathology dataset**	**98**

4.4 *Architecture of CNN*

4.4.1 Sequential Model in Keras

We can create models sequentially, layer upon layer. However, this doesn't enable us to construct models with a large number of inputs as well as outputs. It functions well for straightforward layers stacks with a single input and output tensor. This paradigm is useless if any tier with in stack has a large number of inputs or outputs. Although in the case of non-linear topology, it is not appropriate.

4.4.2 Model Subclassing in Keras

The sequence does not give you much leeway in creating your models. Functional API also allows for only a limited amount of customization. In Keras, however, you can start creating your own fully customizable models. Subclassing the Model class and trying to implement a call method accomplishes.

Input() is titled to create a Keras tensor.

A Keras tensor is a representational tensor-like object that we augment with characteristics that enable us to create a Keras model simply by knowing the model's inputs and outputs.

For example, if a, b, and c are Keras tensors, then it is possible to do: model = Model([a, b] input, [c] output).

4.4.3 Kernels

Each convolutional layer employs a series of filtering known as convolutional kernels. A subgroup of the input image pixels is subjected to the filtering, which is a matrices of integer a same dimensions as the kernel. A grid cell, which is comparable to a pixel with in output channel/feature map, is represented by multiplying each pixel by the kernel's real value. Those linear modifications are called convolutions, which are a kind of affine function (Fig. 7).

4.4.4 Padding

Convolutional neural systems are referred to as having "padding" since it describes the total pixels that are decided to be added to with an input prior to being processed by the CNN kernel. Each pixel contribution made will be 0, for instance, if indeed the padding in a CNN was set to zero. Just one border with a megapixel value of 0 would be added to the image, and if indeed the 0 padding is set to one.

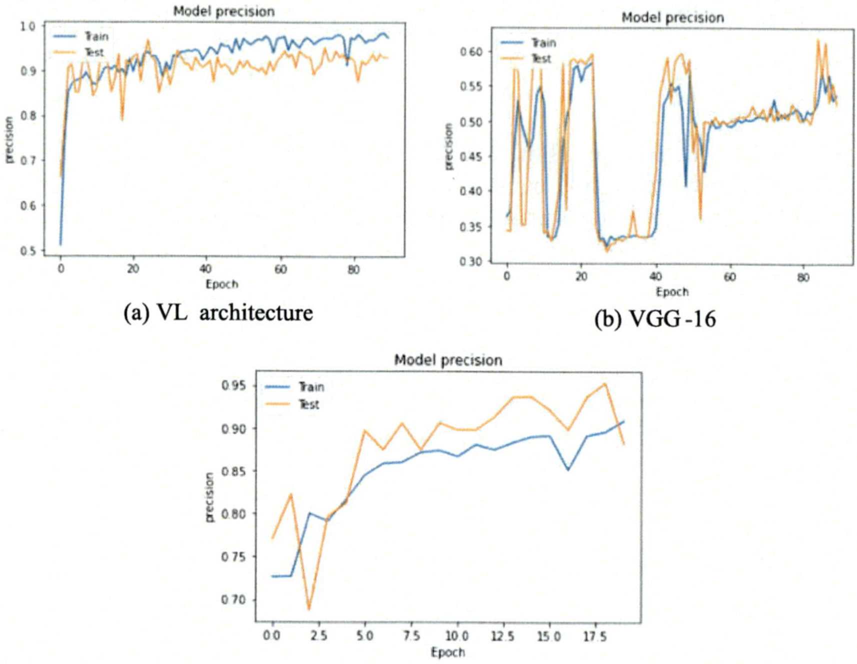

(a) VL architecture

(b) VGG-16

(c) LENET

Fig. 7 **a–c** Illustrates model precision for both test and training images with epoch

4.4.5 Model and Evaluation Metrics

For image detection and classification, a CNN is built with a stack of layers. Following parameters are measured to assess the model's performance:

Accuracy: Accuracy represents the accurately classified data over the total amount of information instances. The formula for calculating accuracy is represented by Eq. (6). However, accuracy alone might not be a great indicator of a models performance.

Precision: It's used to determine how many observations were correctly predictive. It means for the proportion of accurately guessed positive observations in the total number of correctly predicted observations. The precision is calculated using Eq. (7). A fewer detection rate is associated with high precision (Fig. 8).

Recall (Sensitivity): The percentage of correctly predicted good observation across all hugely positive observations is known as recall Eq. (8). A real positive rate or responsiveness are other names for it.

$$\text{Accuracy} = \frac{\text{True\_positive} + \text{True\_Negative}}{\text{True\_Postive} + \text{True\_Negative} + \text{False\_Positive} + \text{False\_Negative}} \tag{6}$$

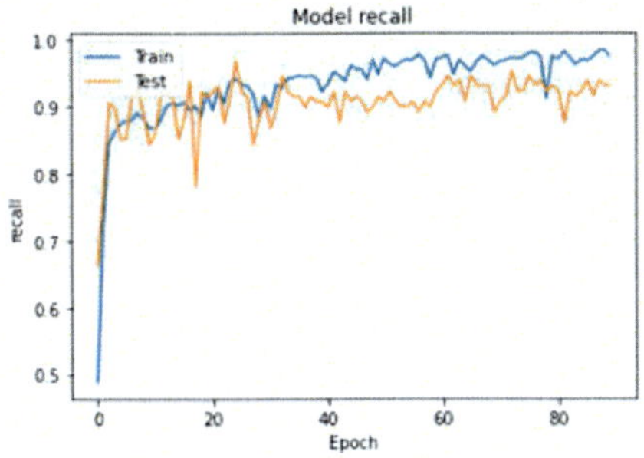

(a) VL architecture

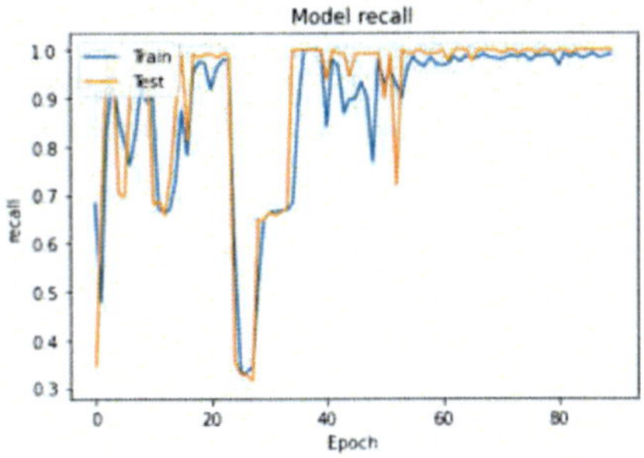

(b) VGG -16

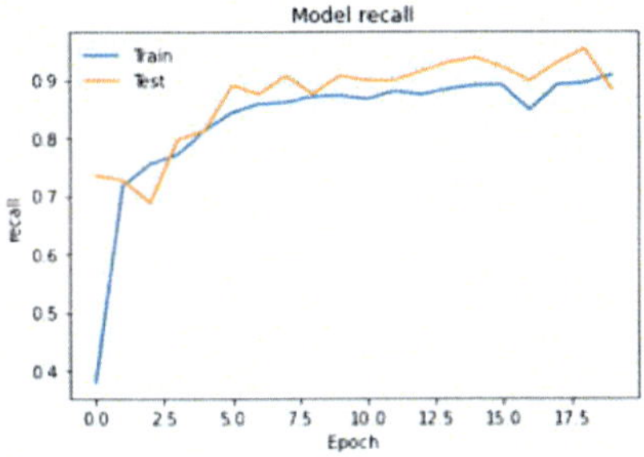

(c) LENET

Fig. 8 **a–c** Illustrates model recall for both test and training images with epoch

$$\text{Precision} = \frac{\text{True\_Positive}}{\text{True\_Positive} + \text{False\_Positive}} \tag{7}$$

$$\text{Recall} = \frac{\text{True\_Positive}}{\text{True\_Positive} + \text{False\_Negative}} \tag{8}$$

Such measurements, denoted as True_Positive, False_Positive, False_Negative, and True_Negative, depict the output measures for the validation and training images of the models as real positive, false positive, negative result, and true negative values. The trained model weight values were ended up saving in the h5 file format and then loaded into the model architecture to predict the future (Fig. 9 and Table 2).

4.5 Pretrained Models

Using the lung histopathology dataset, we used the considerations models for lung cancer detection and classification.

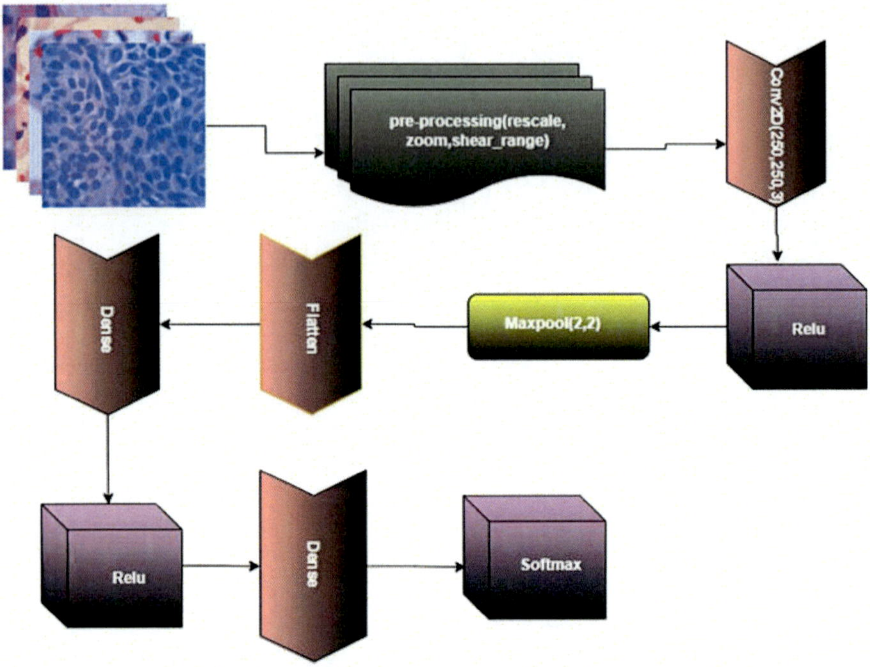

Fig. 9 Proposed architecture for the classification lung histopathology detection

Table 2 Shows the classification matrix of different models using histopathology dataset

Methods	Accuracy	Loss	Precision	Recall
VL architecture (proposed)	0.98	0.04	0.98	0.98
VGG-16	0.84	0.38	0.53	0.99
Le-Net	0.90	0.38	0.90	0.90

4.5.1 VGG-16

The VGG model, which exists for the Visual Geometry Group, is a well-known CNN architecture that demonstrated that it was possible to achieve efficient image recognition using a deep network and small convolutional filters. It is regarded as one of the finest vision model architectures available today. The most distinctive aspect of VGG16 is that it focused upon using convolution layers of a 3×3 filter with stride 1 instead of a lot of hyper-parameters and consistently uses the same padding and max pooling layer of a 2×2 filter with stride 2. The architecture keeps the layers of convolutional and max pooling in this arrangement. Two FC as well as a Softmax are included as its final features. VGG16 contains 16 weighted layers, as indicated by the number 16. The network has 138 million (estimated) parameters, making is fairly huge.

Step by step implementation of VGG-16:

Step 1: START
Step 2: import all libraries
INPUT:
Data Augmentation: ImageDataGenerator ()
Parameter: rescale-1. /255, shear_range-0.2, zoom_range-0.2, horizontal_flip- True
Step 3: Dataset
training_set ← train_datagen
test_set ← test_datagen
Parameter: target_size(224,244), batch_size-32, class_mode-categorical
Step 4: to building a sequential model,
model←Sequential()
Layers: C1: Conv1_2, 64 depth, 3x3
pool_size:2x2
C2: Conv2_2, 128 depth, 3x3
pool_size:2x2
C3: Conv3_3, 256 depth, 3x3
pool_size:2x2
C4: Conv4_3, 512 depth, 3x3
pool_size:2x2
C5: Conv5_3, 512 depth, 3x3
pool_size:2x2
Add ReLu activation to every layer to prevent all negative values from being passed onto subsequent layers.
Step 5: send data to the fully connected layer in order to flatten the vector,
F1: 256→relu
F2: 128→relu
F3: 3→sigmoid
Step 6: to compile the model,
Parameter: optimizer→SGD, loss→categorical_crossentropy,
metrics→accuracy, loss, precision, recall
Step 7: to create a ModelCheckpoint(),
mc←ModelCheckpoint()
Parameter: monitor, verbose
Step 8: to fit the model,
history← model.fit(steps_per_epoch-training_set.samples,
validation_steps-test_set.samples)
Step 9: observe train/test Accuracy, Loss, Precision & Recall using matplotlib
Step 10: OUTPUT
end

4.5.2 LENET

LeNet is an abbreviation for LeNet-5, which is a simple CNN that can respond to a subset of the neighbouring cells in the coverage areas and accomplish well in large-scale image analysis. Lenet-5 was one of the first pre-trained models, and its popularity stemmed from its simple and straightforward architectural style. It is an image classification multi-layer convolution neural network. The network is named Lenet-5 because it has five layers with learnable parameters. The LeNet-5 CNN architecture has seven layers. Three convolutional layers, two subsampling layers, and two fully connected layers make up the layer composition.

Step by step implementation of LENET:

Step 1: START
Step 2: import all libraries
INPUT:
Data Augmentation: ImageDataGenerator ()
Parameter: rescale-1. /255, shear_range-0.2, zoom_range-0.2, horizontal_flip-True
Step 3: Dataset
training_set ← train_datagen
test_set ← test_datagen
Parameter: target_size(224,244), batch_size-32, class_mode-categorical
Step 4: to building a sequential model,
Classifier←Sequential()
Layers: C1: Conv1_1, 32 depth, 3x3
pool_size:2x2
C2: Conv2_1, 128 depth, 3x3
pool_size:2x2
Add ReLu activation to every layer to prevent all negative values from being passed onto subsequent layers.
Step 5: send data to the fully connected layer in order to flatten the vector,
F1: 256→relu
F2: 3→softmax
Step 6: to compile the model,
Parameter: optimizer→Adam, loss→categorical_crossentropy,
metrics→accuracy, loss, precision, recall
Classifier←summary ()
Step 7: to create a ModelCheckpoint(),
callbacks←ModelCheckpoint()
Parameter: model_path, monitor, verbose
Step 8: to fit the model,
history← Classifier.fit(steps_per_epoch-training_set.samples,
validation_steps-test_set.samples)
Step 9: observe train/test Accuracy, Loss, Precision & Recall using matplotlib
Step 10: OUTPUT
End

4.5.3 Implementation of VL Architecture

The proposed model is built on the Jupyter Notebook platform, and it makes use of the Python libraries Tensorflow and Keras. Implementation of proposed architecture Image_details function is used as data analysis that contains image_count, min_width, min_height, max_width, max_height. All these arguments show the max width and height, min width and height and also total number of images in the dataset. Plot_image is used to plot the images which is related to the Image_details function. Image data generator includes all possible orientation of the image i.e., rescale = 1./255, shear_range = 0.2, zoom_range = 0.2, horizontal_flip = True.

The manual architecture is summarized as follows:

The classifier is the name of the architecture and here the model is a sequential model which allows us to create models layer by layer in sequential order. The first convolution layer with kernel_size = 3, fliter = 32, input_shape = 250, channels = 3, acyivaion = relu, maxpooling = (2,2), batch_size = 32, Flatten is used to flatten the dimensions of the image obtained after convolving it. Dense = 3, activation = softmax. To compile the model the arguments are optimizer = adam, loss = categorical_crossentropy.

Step by step implementation of VL architecture:

Step 1: START
Step 2: import all libraries
INPUT:
def Images_details_print_data ()
Printing each key & values
For k, v in data.items():
Print (k,v)
Step 3: def Images_details(path)
files← take all images store in var
data={} // empty dictionary
which contains image_count, max_width, max_height, min_width, min_height
for f in files
im←to open the image
Step 4: def plot_images()
//prints 10 images of each category along with image_count, min_width, max_width, min_height, max_height
Step 5: Data Augmentation: ImageDataGenerator ()
Parameter: rescale-1. /255, shear_range-0.2, zoom_range-0.2, horizontal_flip- True
Step 6: Dataset
training_set ← train_datagen
test_set ← test_datagen
Parameter: target_size(224,244), batch_size-32, class_mode-categorical
Step 7: to building a sequential model,
Classifier←Sequential()
Layers: C1: Conv1_1, 32 depth, 3x3
pool_size:2x2
Add ReLu activation to the conv layer to prevent all negative values from being passed onto subsequent layers.

Step 8: send data to the fully connected layer in order to flatten the vector,
F1: 38→relu
F2: 3→softmax
Step 9: to compile the model,
Parameter: optimizer→Adam, loss→categorical_crossentropy,
metrics→accuracy, loss, precision, recall
Classifier←summary()
Step 10: to create a ModelCheckpoint(),
mc←ModelCheckpoint()
Parameter: model_path, monitor, verbose
Step 11: to fit the model,
history← Classifier.fit(steps_per_epoch-training_set.samples,
validation_steps-test_set.samples)
Step 12: observe train/test Accuracy, Loss, Precision & Recall using matplotlib

Step 13: OUTPUT
end

5 Discussion on Findings

5.1 Libraries Required

Tensorflow: TensorFlow is a Google accessible library primarily designed for applications in deep learning. TensorFlow was originally designed for large numerical calculations rather than deep learning. Even so, it was also very helpful for deep learning development. Tensors are multi-dimensional arrays of extra dimensionality that TensorFlow accepts.

$$syntax: pip\ install\ tensorflow == 2.6.0$$

keras: An advanced deep learning API that incorporating NN is Google's Keras. It was created by utilising Python to write minimal NN framework. Moreover, it entails permitting the computing of several backend neural network.

$$syntax: pip\ install\ keras == 2.6.0$$

matplotlib: A popular Python package for visualisation methods is called Matplotlib. It is a platform-agnostic tool for 2D plotting array data. Python is used to create Matplotlib, which makes use of NumPy, Python's extensions for numerical mathematics.

os: The OS module in Python includes a number of functions that allow designers to communicate with the operating system they are working on. The Python OS module allows for the establishment of relationship between the user and the operating system. To connect the file system and read the image from our machines' train and test directories.

Django Framework: Django is an elevated framework that enables the quick creation of safe and sustainable internet sites. Django, which has been generated by application experts, handles much of the tedium of web design, so you can focus on composing your plugin rather than trying to reinvent the motor. It is a free and open-source framework that accelerate development of a web-applications which is written in python. Django tends to favor the reuse of parts, also renowned as Don't Repeat Yourself and involve willing features such as a user account, authentication process, and CRUD functions like create then read, updates or deleted.

$$syntax: pip\ install\ django == 3.2.7$$

6 Conclusion and Future Scope

The proposed work concentrated on how an image from a testing set (trained dataset) and a previous data set were used to estimate the sequence of Lung Histopathology using a CNN model. This provides some of the succeeding insights into the prediction of Lung Histopathology. The ability to automatically classify images is the primary advantage of the CNN classification framework. The Lung Histopathology primarily contributes to face misshape and is frequently irreversible because patients are diagnosed with the diseases too late. We have discussed an overview of methodologies for detecting abnormalities in Lung Histopathology images in this study, which includes the collection of Lung Histopathology image data sets, pre-processing methodologies, feature extraction techniques, and classification schemes. In future, we will deploy the cloud platform and also create a mobile based application.

References

1. Ozdemir, O., Russell, R.L., Berlin, A.A.: A 3D probabilistic deep learning system for detection and diagnosis of lung cancer using low-dose CT scans. IEEE Trans. Med. Imaging **39**(5), 1419–1429 (2019)
2. Zia ur Rehman, M., Javaid, M., Shah, S.I.A., Gilani, S.O., Jamil, M., Butt, S.I.: An appraisal of nodules detection techniques for lung cancer in CT images. Biomed. Signal Process. Control **41**, 140–151 (2018)
3. Zhang, J., Xia, Y., Cui, H., Zhang, Y.: Pulmonary nodule detection in medical images: a survey. Biomed. Signal Process. Control **43**, 138–147 (2018)
4. Zhang, W., Wang, X., Li, X., Chen, J.: 3D skeletonization feature based computer-aided detection system for pulmonary nodules in CT datasets. Comput. Biol. Med. **92**, 64–72 (2018)
5. Alzubaidi, M.A., Otoom, M., Jaradat, H.: Comprehensive and comparative global and local feature extraction framework for lung cancer detection using CT scan images. IEEE Access **9**, 158140–158154 (2021)
6. Baranwal, N., Doravari, P., Kachhoria, R.: Classification of histopathology images of lung cancer using convolutional neural network (CNN). arXiv preprint arXiv:2112.13553 (2021)

7. Coudray, N., Ocampo, P.S., Sakellaropoulos, T., Narula, N., Snuderl, M., Fenyö, D., Moreira, A.L., Razavian, N., Tsirigos, A.: Classification and mutation prediction from non-small cell lung cancer histopathology images using deep learning. Nat. Med. **24**(10), 1559–1567 (2018)
8. Lu, Y., Aslani, S., Emberton, M., Alexander, D.C., Jacob, J.: Deep learning-based long term mortality prediction in the national lung screening trial. IEEE Access **10**, 34369–34378 (2022)
9. Wu, J., Guan, P., Tan, Y.: Diagnosis and data probability decision based on non-small cell lung cancer in medical system. IEEE Access **7**, 44851–44861 (2019)
10. Hatuwal, B.K., Thapa, H.C.: Lung cancer detection using convolutional neural network on histopathological images. Int. J. Comput. Trends Technol. **68**(10), 21–24 (2020)
11. Pang, S., Zhang, Y., Ding, M., Wang, X., Xie, X.: A deep model for lung cancer type identification by densely connected convolutional networks and adaptive boosting. IEEE Access **8**, 4799–4805 (2019)
12. Sahu, P., Yu, D., Dasari, M., Hou, F., Qin, H.: A lightweight multi-section CNN for lung nodule classification and malignancy estimation. IEEE J. Biomed. Health Inform. **23**(3), 960–968 (2018)
13. Cao, H., Liu, H., Song, E., Ma, G., Xu, X., Jin, R., et al.: A two-stage convolutional neural networks for lung nodule detection. IEEE J. Biomed. Health Inform. **24**(7), 2006–2015 (2020)
14. Liao, Z., Xie, Y., Hu, S., Xia, Y.: Learning from ambiguous labels for lung nodule malignancy prediction. IEEE Trans. Med. Imaging (2022)
15. Jena, S. R., George, T., Ponraj, N.: Texture analysis-based feature extraction and classification of lung cancer. In: 2019 IEEE International Conference on Electrical, Computer and Communication Technologies (ICECCT), pp. 1–5. IEEE (2019, February)
16. Liu, P., Jin, K., Jiao, Y., He, M., Fei, S.: Prediction of second primary lung cancer patient's survivability based on improved eigenvector centrality-based feature selection. IEEE Access **9**, 55663–55672 (2021)
17. Priyadarshini, R., Quadir Md, A., Rajendran, N., Neelanarayanan, V., Sabireen, H.: An enhanced encryption-based security framework in the cps cloud. J. Cloud Comput. **11**(1), 64 (2022)
18. Md, A.Q., Varadarajan, V., Mandal, K.: Correction to: efficient algorithm for identification and cache based discovery of cloud services. Mob. Netw. Appl. **24**(4), 1198–1198 (2019)
19. Panja, A., Jackson, C., Md, A.Q.: An approach to skin cancer detection using Keras and Tensorflow. J. Phys.: Conf. Ser. **1911**(1), 012032 (2021)
20. Md, A.Q., Vijayakumar, V.: Combined preference ranking algorithm for comparing and initial ranking of cloud services. Recent Adv. Electr. Electron. Eng. (Formerly Recent Patents on Electrical & Electronic Engineering) **13**(2), 260–275 (2020)
21. Prassanna, J., Md, A.Q., Christy Jackson, J., Prabakaran, R.: SecrecyProtector: a novel data analytics based credit score management system. Int. J. Sci. Technol. Res. **9**(03) (2020)
22. Matthew, K.M., Quadir, A.: Comput. Netw. Inf. Secur. **2**, 36–42 (2017)
23. Md, A.Q., Vijayakumar, V.: Dynamic ranking of cloud services for web-based cloud communities: efficient algorithm for rating-based discovery and multi-level ranking of cloud services. Int. J. Web Based Communities **15**(3), 248–270 (2019)
24. Mathew, S.A., Md, A.Q.: Evaluation of blockchain in capital market use-cases. Int. J. Web Portals **10**(1), 54–76 (2018)
25. Md, A.Q., Kapoor, S., AV, C.J., Sivaraman, A.K., Tee, K.F., Sabireen, H., Janakiraman, N.: Novel optimization approach for stock price forecasting using multi-layered sequential LSTM. Appl. Soft Comput. **134**, 109830 (2023)
26. Basit, Z., Tabassum, M., Sharma, T., Furqan, M., Quadir Md, A.: Performance analysis of OSPF and EIGRP convergence through IPsec tunnel using multi-homing BGP connection. Mater. Today: Proc. (2022)
27. Md, A.Q., Jaiswal, D., Daftari, J., Haneef, S., Iwendi, C., Jain, S.K.: Efficient dynamic phishing safeguard system using neural boost phishing protection. Electronics **11**(19), 3133 (2022)
28. Matthew, K.M., Muhammed, A.Q., Varadarajan, V.: An improved key management scheme in cloud storage. Int. J. Adv. Intell. Parad. **14**(3/4), 197 (2019)

29. Matthew, K.M., Quadir Md, A.: Analysis framework for logs in communication devices. Int. J. Web Portals **10**(1), 15–26 (2018)
30. Srinivasan, A., Quadir, M.A., Vijayakumar, V.: Hybrid cloud for educational sector. Procedia Comput. Sci. **50**, 37–41 (2015)
31. Quadir, M.A.: An efficient algorithm to detect DDoS amplification attacks. J. Intell. Fuzzy Syst. **39**(6), 8565–8572 (2020)
32. Md, A.Q., Narayanan, S.S.S., Sabireen, H., Sivaraman, A.K., Tee, K.F.: A novel approach to detect fraud in Ethereum transactions using stacking. Expert Syst. e13255
33. Gupta, B., Madan, G., Quadir Md, A.: A smart agriculture framework for IoT based plant decay detection using smart croft algorithm. Mater. Today: Proc. (2022)
34. Sathish, S., Ashwin, S., Quadir, M.A., Pavithra, L.K.: Analysis of convolutional neural networks on Indian food detection and estimation of calories. Mater. Today: Proc. (2022)
35. Quadir, M.A., Prassanna, J., Christy Jackson, J., Sabireen, H., Gupta, G.: Efficient algorithm for CSP selection based on three-level architecture. Artif. Intell. Technol. 515–531 (2021)
36. Md, A.Q., Agrawal, D., Mehta, M., Sivaraman, A.K., Tee, K.F.: Time optimization of unmanned aerial vehicles using an augmented path. Futur. Internet **13**(12), 308 (2021)
37. Rajakumaran, G., Venkataraman, N., Quadir, A.: Early detection of LDoS attack using SNMP MIBs. ITM Web Conf. **37**, 01025 (2021)
38. Quadir, A., Jha, K., Haneef, S., Sivaraman, A.K., Tee, K.F.: A review on data-driven quality prediction in the production process with machine learning for industry 4.0. Processes **10**(10), 1966 (2022)
39. Joshua, C.J., Jayachandran, P., Md, A.Q., Sivaraman, A.K., Tee, K.F.: Clustering, routing, scheduling, and challenges in bio-inspired parameter tuning of vehicular Ad Hoc networks for environmental sustainability. Sustainability **15**(6), 4767 (2023)
40. Hai, T., Zhou, J., Padmavathy, T.V., Md, A.Q., Jawawi, D.N., Aksoy, M.: Design and validation of lifetime extension low latency MAC protocol (LELLMAC) for wireless sensor networks using a hybrid algorithm. Sustainability **14**(23), 15547 (2022)
41. Sivakumar, J., Md, A.Q., Vigneswaran, T., Punitha, K., Sivaraman, A.K.: Reliability of smart-wearables using PSO-GA optimized algorithm in terms of data analysis. In: 2022 Third International Conference on Intelligent Computing Instrumentation and Control Technologies (ICICICT), pp. 1614–1620. IEEE (2022, August)
42. Jojo, A., Sunil, G.K., Md, A.Q., Vigneswaran, T., Punitha, K., Sivaraman, A.K.: Intruder detection system using IoT with adaptive face monitoring and motion sensing algorithm. In: 2022 Third International Conference on Intelligent Computing Instrumentation and Control Technologies (ICICICT), pp. 1628–1634. IEEE (2022, August)
43. Setia, P., Singh, Y., Md, A.Q., Vigneswaran, T., Punitha, K., Sivaraman, A.K.: An efficient dynamic K-Papa architecture for communicating with the things using collision-free algorithm. In: 2022 Third International Conference on Intelligent Computing Instrumentation and Control Technologies (ICICICT), pp. 1640–1646. IEEE (2022, August)
44. Bouchard, G.: Efficient bounds for the Softmax function and applications to approximate inference in hybrid models. In: Proceedings of the Presentation at the Workshop For Approximate Bayesian Inference in Continuous/Hybrid Systems at Neural Information Processing Systems (NIPS), Meylan, France, vol. 31 (2008, May)
45. Xie, H., Yang, D., Sun, N., Chen, Z., Zhang, Y.: Automated pulmonary nodule detection in CT images using deep convolutional neural networks. Pattern Recognit. **85**, 109–119 (2019)
46. Qin, R., Wang, Z., Jiang, L., Qiao, K., Hai, J., Chen, J., et al.: Fine-grained lung cancer classification from PET and CT images based on multidimensional attention mechanism. Complexity
47. Afshar, P., Oikonomou, A., Naderkhani, F., Tyrrell, P.N., Plataniotis, K.N., Farahani, K., et al.: 3D-MCN: a 3D multi-scale capsule network for lung nodule malignancy prediction. Sci. Rep. **10**(1), 1–11 (2020)
48. Bharati, S., Podder, P., Mondal, M.R.H.: Hybrid deep learning for detecting lung diseases from X-ray images. Inform. Med. Unlocked **20**, 100391 (2020)

A Novel Approach of Machine Learning Application in Astrophysics: Morphological Feature Wrapping Based Ensemble Method for Galaxy Shape Classification Using GAMA Dataset

Anurag Sinha, Mohammad Shahid, Amritesh Nandan, Celestine Iwendi, Abhijeet Kumar Giri, and Saprit Anand

Abstract The numerous strategies for the automated morphological categorization of galaxies, which uses a variety of supervised machine learning techniques, have not been well examined or compared. As the majority of star galaxy classifiers in use today use condensed summary data from catalogues, rigorous feature extraction and selection are required. With the aid of Deep Convolutional Neural Networks (CNN), a development in machine learning, it may automate the process of feature detection from data by a computer, therefore lowering the demand for qualified human input. Low-level artificial classification has made great progress. While this is the case, Deep Learning consistently outperforms traditional computers. analyzing large datasets while learning. We examine three machine learning techniques for categorizing morphological galaxies: Support Vector Machines (SVM), Random Forests (RF), and Naive Bayes (NB). We examine the efficacy of several machine learning algorithms on each feature representation of a galaxy using a collection of morphological features produced by image analysis as well as the raw image pixel data compressed using PCA (Principal Component Analysis) into PCA features. According to our experiments, RF outperformed SVM and NB. The remainder of

A. Sinha
Department of Computer Science and Information Technology, IGNOU, New Delhi, India

M. Shahid
Department of Computer Science and Information Technology, Amity University Jharkhand, Ranchi, India

A. Nandan
Department of Computer Science and Information Technology, Graphic Era Deemed to be University, Dehradun, India

C. Iwendi (✉)
School of Creative Technologies, University of Bolton, Bolton BL3 5AB, UK
e-mail: Celestine.iwendi@ieee.org

A. K. Giri · S. Anand
Department of Computer Science and Engineering, KIIT, Bhubaneswar, India

C. Iwendi et al. (eds.), *Proceedings of ICACTCE'23—The International Conference on Advances in Communication Technology and Computer Engineering*, Lecture Notes in Networks and Systems 735, https://doi.org/10.1007/978-3-031-37164-6_43

the time, morphological features outperformed our PCA features in performance. Thus, the current mechanism is not extremely scalable. A probabilistic classifier that can scale, is based on source data, and requires the least amount of human interaction is essential to resolving these problems.

Keywords Machine learning · GAMA dataset · Supervised classifiers · Galaxy classification · Morphological features

1 Introduction

Astronomy has always placed a high value on galaxy classification. Galaxies appear in a wide variety of forms, sizes, and hues. As a result, categorizing them helps scientists understand how the cosmos formed and evolved. The categorization of galaxies is based on their form, structure, metallicity, brightness, and other features. There are several strategies to achieve this goal. Galaxy classification is the first step towards better comprehending the physics of galaxies. The availability of several digitized sky studies at various wavelengths has lately transformed astronomy into an exceedingly data-rich science. These pictures are normally categorized by visually examining photographic plates. However, because it requires knowledge and is not an easy task. The evolution of astronomy in the modern period has greatly mirrored technological advancements in computers. Due to the enormous amount of data available to us now and the greatly enhanced efficiency of computer performance, sophisticated computational approaches like machine learning models are now considerably more effective. A large number of seasoned individuals used to classify galaxies by hand long ago, and they would then analyze the findings using a cross-validation procedure. I'm going to expose to a Galaxy Classification Model with Machine Learning in light of this motivation. The SDSS (Sloan Digital Sky Surveys), one of the largest sky surveys, has produced a database of countless millions of galaxies. In order to categorize astronomical photos since it is difficult to manually evaluate them, the Galaxy Zoo Project was founded in 2007. The project was completed successfully and considerably sooner than anticipated. Due to advancements in computer vision and machine learning, particularly Neural Networks, which can finish the process relatively quickly in Fig. 1, it is more feasible to use these automated techniques, provided they have excellent accuracy.

The objective of the study: Astronomy relies on the study of morphological galaxy classification because it provides details about the frequency and spatial distribution of galaxies that aid in our comprehension of how they originate and change over time.

Scope of the study: When used to classify morphological galaxies, Support Vector Machine (SVM), Random Forests (RF), and Decision Tree classification results are compared. We assess the effectiveness of the various Machine Learning methods on each feature representation of a galaxy using both sets of morphological features produced from image analysis and the direct image pixel data compressed by PCA

Fig. 1 Galaxy classification sensor-based system

(Principal Component Analysis) into PCA features. The experiment demonstrates that Random Forest (RF) outperformed Support Vector Machine (SVM) and Decision Tree in terms of performance. Furthermore, morphological features outperformed PCA features in terms of effectiveness.

In line with our earlier findings and the studies from the literature stated above, we discover that the used categorization scheme is not entirely supported, even when taking into account additional catalogs and a different classifier. The Little Blue Spheroids galaxy class, which was recently proposed in [1], interestingly, remains the most prominent, also for the collection of catalogs examined in this work, corresponding to our prior work [2]. We list the variables that have the most bearing on the established class divisions. The paper is organized as follows Sect. 1, comprises the Introduction, Sect. 2 literature review, and research gap, Sect. 3 methodology, and Sect. 4 results and discussion.

1.1 Dataset

In the current study, we look at data from five galaxy catalogs (that include features derived from spectroscopic and photometric observations), i.e., measurements of flux intensities in different wavelength bands from the Galaxy And Mass Assembly (GAMA) survey [3] for a sample of 1295 galaxies. Due to the catalogs' ability to give data for multiple groupings of galaxies, our data set consists of the set of galaxies for which a complete set of attributes may be accessed after balancing the relevant classes.

2 Literature Review and Research Gap Findings

In [4], the performance of the three machine learning techniques SVM, RF, and NB was evaluated. It was shown that RF outperformed SVM since the features were still not linearly separable even after being projected to higher dimensions using the RBF kernel. SVM, RF, and NB classifiers' classification accuracy was evaluated after they had been trained using training subsets in [4]. Weka Random Forest, Dai and Tong [5] NB, and LibSVM were employed. This contained the collection of 119 photos with category names from Zsolt Frei's galaxy database. Fuentes used automated morphological galaxy categorization in his experimental exploration of machine learning and age analysis in [2].

Locally weighted regression was produced using the feed-forward network included in the MATLAB neural network toolbox. We used cross-validation that was tenfold for each trial. They found that the ensembles and algorithms delivered superior results. Also, their analysis shows that local weighted regression using a homogenous ensemble outperformed prior literature-based work in terms of results. They also found that, in terms of accuracy and speed, local weighted regression outperforms artificial neural networks. According to Khalifa et al. [6], astronomers employed fuzzy differencing techniques and decision tree learning algorithms to classify the types of galaxies. They investigated the CART, C4.5, Random Forest, and fuzzy logic algorithms and created a trustworthy classifier to discriminate between spiral galaxies, elliptical galaxies, stars, and other unnamed galactic objects. It was shown that Random Forest delivered the best percentage of accuracy, especially when 50 trees were utilized, while comparing the output from the fuzzy inference system, the CART method, the C4.5 algorithm, and the Random Forest decision tree technique. A stellar galaxy classification framework is provided in [3] by applying Deep Convolutional Neural Networks (ConvNets) directly to the reduced, calibrated pixel values. Edward J. Kim and Robert J. Brunner used data from the Sloan Digital Sky Survey and the Canada–France–Hawaii Telescope Lansing Survey (CFHTLenS) to demonstrate that ConvNets can produce accurate and calibrated probabilistic classifications that are competitive with conventional machine learning techniques (SDSS). ConvNet model appeared to be more correctly calibrated, and network could categorize the CFHTLenS data set as accurately as Random Forest method, they found (TPC). The SDSS data set was also subjected to the same network architecture, which fell short of TPC but nonetheless generated probabilities that were somewhat better calibrated. Lal et al., Lin et al., Lukic and Brüggen [7–9] examined many well-known machine learning methods, including as Convolutional Neural Networks (CNN), K-nearest Neighbor, logistic regression, Random Forest (RF), Support Vector Machine (SVM), and Neural Networks, for identifying galaxies using data from the Dark Energy Survey (DES). James Annis' study indicates that the CNN strategy is the most successful one to yet. They identified 2800 galaxies in Galaxy Zoo 1, with a detection accuracy of 0.99 for spiral and elliptical galaxies. Also, they found that there was little chance that visually ventricular galaxies were spiral or elliptical galaxies, demonstrating how supervised learning may be used to refresh knowledge of how this form of galaxy differs from both of these types of galaxies. They also considered the effects of categorizing their data differently in light of the fact that their data included rotated photos with rotational degrees that varied in increments of 10°.

3 Methodology

It doesn't matter if it can be measured or not; a feature is a property of an object that is employed to distinguish (or classify) two unrelated items. To sort arriving fish on a conveyor depending on species, for instance, use optical detection. When an input sample is represented by its features, points in the feature space are utilized as a representation. In the case of a single feature, the feature space has just one dimension. As is well known, pattern recognition seeks to identify the most appropriate decision rule for categorizing the incoming data (classes). Points in one class and points in another are separated by the decision boundary. Decision regions are created by dividing the feature space into decision areas [10] (Fig. 2).

It is simple to formulate pattern classification issues in terms of discriminant functions since they are readily constructed from data. When the decision limits are specified directly, the procedure is commonly referred to as discriminative modelling. The decision to select a particular functional form with parameters adjusted by a training strategy or prior knowledge of the patterns to be classified may serve as the basis for the discriminant function selection. In the literature, the difficulty of several discriminant function types has been studied. These functions vary from straightforward linear discriminant functions—where g is a linear combination of the xi—to sophisticated multi-parameter functions.

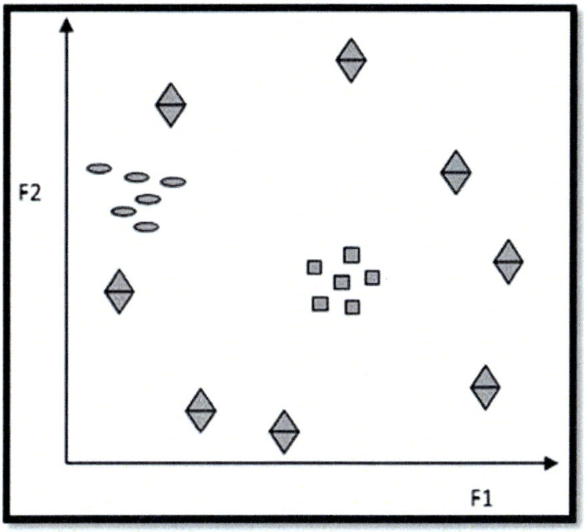

Fig. 2 Sample points in a 2-dimensional feature space

These nonlinear procedures include the multilayer perceptron. All model parameters, such as class priors and feature probability distributions, may be estimated using the relative frequencies of the training set. The probability is forecast with the highest level of accuracy [11–13]. The class probability from the training set can be calculated using the formula (prior for a specific class) = (number of samples in the class)/(total number of samples), or equiprobable classes can be assumed (i.e., priors = 1/(number of classes)). To estimate the parameters of the distribution for the attributes in the training set, either a distributional assumption or nonparametric models must be built. The assumption that each continuous value for a continuous variable is connected is a typical one when working with continuous data. Another well-liked approach for handling continuous variables is data binding and discretization. The distribution technique is frequently the better choice if the training data are sparse or the precise distribution of the data is known. The discretization algorithm often performs better in bigger training data sets than other techniques because of its versatility in responding to the distribution of the training data. When there is a lot of data available, the distribution approach is usually preferred over the discretization strategy, however naive Bayes is also frequently used in these circumstances (as more computationally intensive models can normally yield greater accuracy). In Fig. 3, the proposed model is given [14–16].

Deep Learning: As deep learning networks belong within the machine learning category and have three or more layers, they are also known as these networks. The artificial intelligence can "learn" from a tremendous quantity of data thanks to these neural networks' valiant but ultimately unsuccessful attempts to mimic the functioning of the human brain. In neural networks with just one layer, as demonstrated in Fig. 4, the inclusion of additional hidden layers can also assist to improve and increase accuracy. The ability to make approximations may still exist in neural networks with only one layer.

Logistic Regression: A machine learning technique called regression examines the outcomes of a variable that can only be answered yes or no. It belongs to the supervised learning methodology. As a result, it has a distinct and category value. The category of the variable that is more typical can be easily identified. Logistic regression and observations of various types of data are the two categories into which

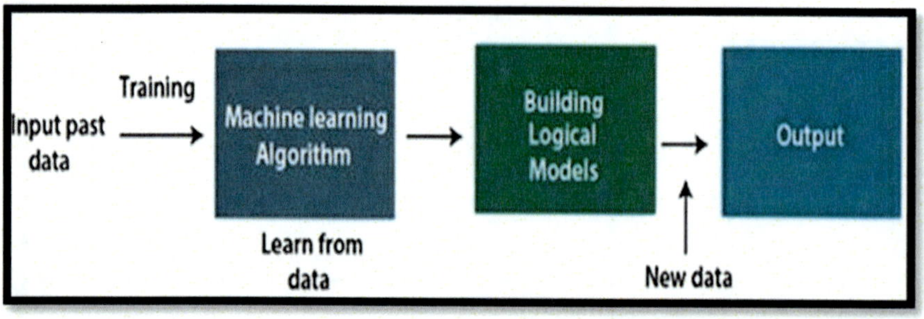

Fig. 3 Proposed model

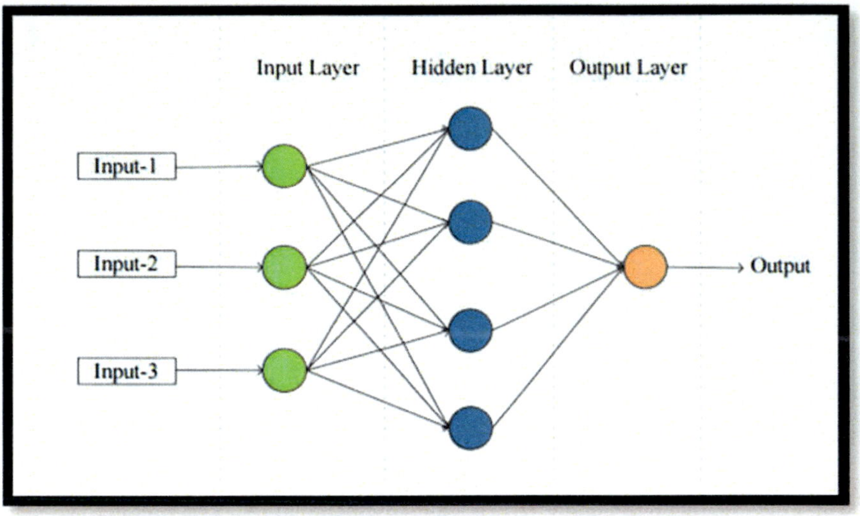

Fig. 4 Deep learning algorithms

it can be separated. Values from the memorial and binomial regression are used to define logistic functions.

Support Vector Machine: One of the most well-liked supervised learning methods is the SVM, which may be used to situations requiring classification, regression, or both at once. Yet, the majority of machine learning classification applications still have problems. The SVM technique tries to make future classification of new data points straightforward by specifying the decision boundary or ideal line that may categories the n-dimensional space. A hyper plane is the mathematical term for its optimal border.

Random Forest: It addresses classification and regression problems in machine learning (ML). This is based on the notion of ensemble methods, which is a technique for mixing several classifiers to handle difficult issues and improve the efficacy of the model. The Random Forest classifier, as implied by its name, averages several decision trees applied to different subsets of the provided dataset to increase the dataset's prediction accuracy [14].

Decision Tree: It is a supervised machine learning method that is generally used to address classification challenges, while it may also be applied to regression problems. It comprises of a classifier with a tree-like structure, with leaf nodes denoting classification results and branching denoting decision rules, respectively. A decision tree has two nodes: the Decision node and the Leaf node. A decision node has several branches, but a leaf node, which represents the outcome of the choice, only has one branch. As illustrated in Fig. 5, judgments or analyses are performed based on the features of the supplied dataset.

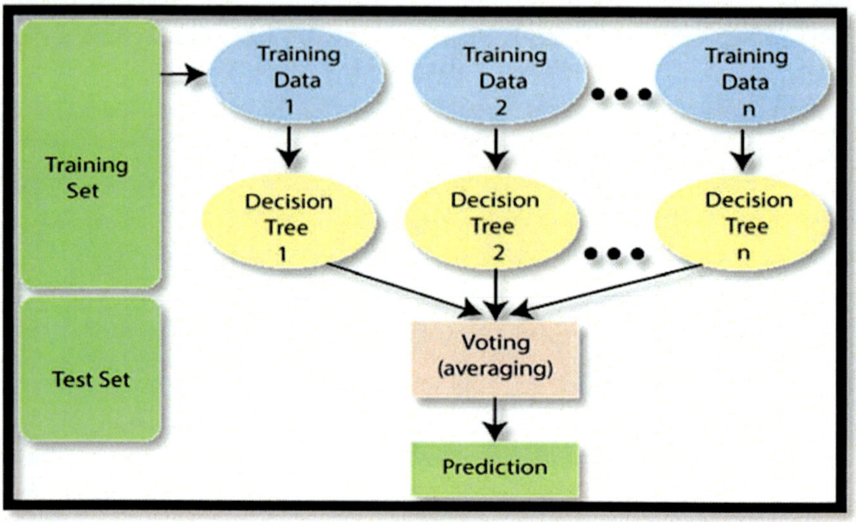

Fig. 5 Decision tree and random forest algorithm

4 Results

The voting classifier has the highest classification accuracy, with a rate of 86%, according to the confusion matrix for the classification based on the Gauss Fit Simple Catalogue. The accuracy in Class 3 is 47%. This is partially because there is more category overlap; 31% of class 1 samples are also categorised as class 3 samples, and 31% of class 3 samples are labeled as class 1 samples. Class 1 and class 3 overlap with LBS and class 7 overlap are both expanding, although they are still extremely tiny as seen in Fig. 6a. There is also increasing (12 cent) overlap between classes 1 and 5. In order to provide readers a better grasp of the nature of the used visual-based categorization technique, especially with regard to physical characteristics, particularly concerning physical variables, we have highlighted the relevance's of the catalog components for the analyzed class differences in Fig. 6b. In order to ensure appropriate interpretation, it is important to take into account the characteristics of the data sample, such as the correlations and classification performance shown in Fig. 6c, d. The development of a binary classification system was based on the area under the receiver operating characteristic curve (ROC). The True Positive Rate and False Positive Rate of the classifier are compared to produce the ROC curve. The score, which goes from 0 to 1, is inversely proportional to how accurate the prognosis was. By comparing this statistic to different methods, it essentially enables us to comprehend how the classifier performs when the threshold is changed, as shown in Table 1.

Fig. 6 **a** Sample data **b** morphological image **c** confusion matrix **d** ROC and AUC curve

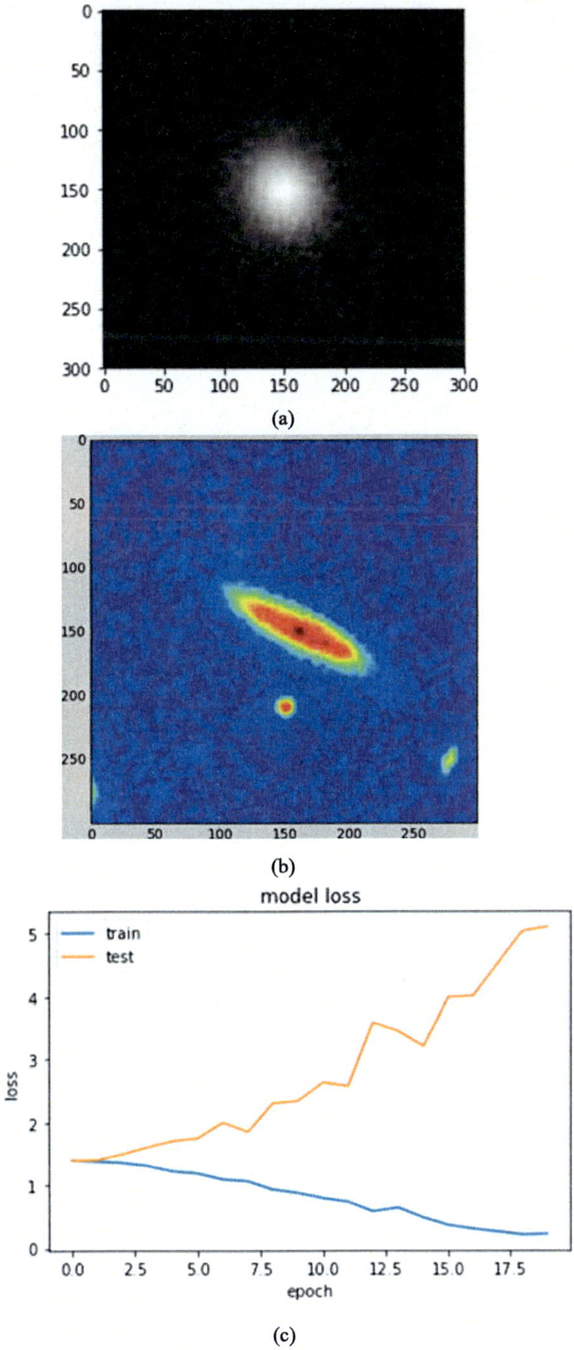

Fig. 6 (continued)

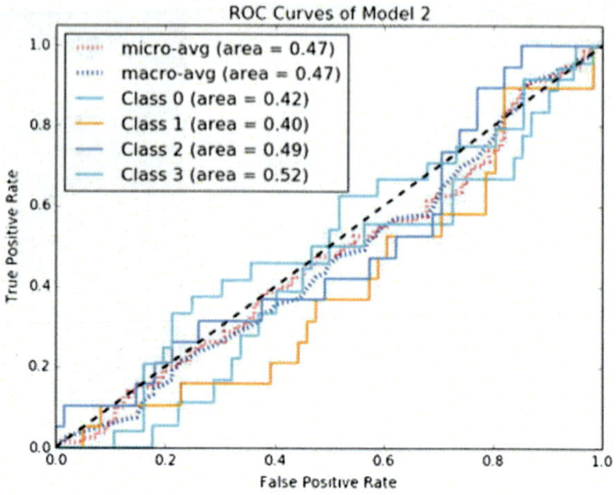

(d)

Table 1 Comparison of the existing method with proposed one

Paper no.	Data set	Methods	Accuracy (%)
[5]	GAMA	A Naive Bayes classifier, the C4.5 rule induction technique, and random forest were utilised	91.64
[7]	GAMA	Used weighted KNN along with Fischer score	90
[12]	GAMA	The algorithms CART, C4.5, random forest, and fuzzy were employed, with random forest (50 trees) proving to be the most accurate	91.1
Proposed one	GAMA	Morphological feature selection and reduction for polynomial time/computational complexity reduction and classification through ensemble method	93.55

5 Conclusion and Future Scope

The implemented system takes predefined constant image data set as input but it can be improved or updated to get a defined data set. The implemented system provides the result of multiple classified types of galaxies present in the given data set on which it has been trained on. After carrying out the implantation and contrasting the different Machine Learning algorithms, we can say that Random Forest is the best and has a high accuracy rate for categorizing galaxies into elliptical and spirals. The estimate of the accuracy of the morphological features is affected by several variables. The clump of stars and dust that ringed the galactic nucleus formed erroneous bounds, making it hard to distinguish the galaxy's edge. Future research will combine morphological traits with photometric information from additional spectral ranges to improve classification precision. The bar-to-bulge ratio and integrated curvature morphic features may also help to improve the accuracy of the seven-category classification. To get results that are reliable and acceptable, it is essential to use a competent data collection with the necessary features. The use of photo data sets of good quality will be rewarded.

References

1. Jones, E., Singal, J.: Analysis of a custom support vector machine for photometric redshift estimation and the inclusion of galaxy shape information. Astron. Astrophys. **600**, A113 (2017). https://doi.org/10.1051/0004-6361/201629558
2. Ibrahim, R.A., Elaziz, M.A., Ewees, A.A., Selim, I.M., Lu, S.: Galaxy images classification using hybrid brain storm optimization with moth flame optimization. J. Astron. Telesc. Instrum. Syst. **4**(03), 1 (2018). https://doi.org/10.1117/1.JATIS.4.3.038001
3. Kormendy, J., Bender, R.: A proposed revision of the Hubble sequence for elliptical galaxies. Astrophys. J. **464**(2), L119–L122 (1996). https://doi.org/10.1086/310095
4. Barchi, P., da Costa, F., Sautter, R., Rosa, R., Carvalho, R.: Improving galaxy morphology with machine learning. JCIS **7**(3) (2016). https://doi.org/10.6062/jcis.2016.07.03.0114
5. Dai, J.-M., Tong, J.: Galaxy morphology classification with deep convolutional neural networks. Astrophys. Space Sci. **364**(4), 55 (2019). https://doi.org/10.1007/s10509-019-3540-1
6. Khalifa, N.E.M., Taha M.H.N., Hassanien A.E., Selim, I.M.: Deep Galaxy: Classification of Galaxies Based on Deep Convolutional Neural Networks
7. Lal, D.V., et al.: Viewing classical radio galaxies with the upgraded GMRT and MeerKAT—a progress report. Galaxies **9**(4), 87 (2021). https://doi.org/10.3390/galaxies9040087
8. Lin, J.Y.-Y., Liao, S.-M., Huang, H.-J., Kuo, W.-T.: Galaxy Morphological Classification with Efficient Vision Transformer
9. Lukic, V., Brüggen, M.: Galaxy classifications with deep learning. Proc. IAU **12**(S325), 217–220 (2016). https://doi.org/10.1017/S1743921316012771
10. Nolte, L.W., Bilicki, M., Holwerda, B., Biehl, M.: Galaxy classification: a machine learning analysis of GAMA catalogue data. Neurocomputing **342**, 172–190 (2019). https://doi.org/10.1016/j.neucom.2018.12.076
11. O'Keefe, P.J., Gowanlock, M.G., McConnell, S.M., Patton, D.R.: Star-Galaxy Classification Using Data Mining Techniques with Considerations for Unbalanced Datasets
12. Ramírez, G., García, R.: Classification of Galaxies Using Machine Learning Methods
13. Mahalakshmi, G.S., Swadesh, B., Aswin, R.R.V., Sendhilkumar, S., Swaminathan, A., Surendran, S.: Classification and Feature Prediction of Star, Galaxies, Quasars, and Galaxy Morphologies Using Machine Learning. In: Review (2022). https://doi.org/10.21203/rs.3.rs-1885343/v1
14. Shi, H.: Galaxy Classification with Deep Convolutional Neural Networks
15. Whyte, L.F., Abraham, R.G., Merrifield, M.R., Eskridge, P.B., Frogel, J.A., Pogge, R.W.: Morphological classification of the OSU bright spiral galaxy survey. Mon. Not. R. Astron. Soc. **336**(4), 1281–1286 (2002). https://doi.org/10.1046/j.1365-8711.2002.05879.x
16. Awang Iskandar, D.N.F., et al.: Classification of planetary nebulae through deep transfer learning. Galaxies **8**(4), 88 (2020). https://doi.org/10.3390/galaxies8040088
17. Kennamer, N., Kirkby, D., Ihler, A., Sánchez, J.: ContextNet: Deep Learning for Star Galaxy Classification (2018)
18. Archana, P., Saketh, B.: Galaxies classification using deep learning algorithm in convolutional neural networks. **10**(5) (2020)

Evaluating the Performance of a Hybrid Model for Classification of Bicycle Crash Severity and Identification of Associated Risk Factors

Maruf Ahmed and Pradeep Hewage

Abstract This study conducted an exploratory analysis of bicycle crash data from Great Britain with the aim of identifying the key variables that influence the classification of such incidents. It also analysed data on a range of factors that may contribute to bicycle crashes, including the age of the cyclist, lighting conditions, weather conditions, road types, road conditions, and speed limits. Results indicated that these variables are among the most significant predictors of bicycle crashes, with road conditions, time of day, and lighting conditions being particularly vital factors. In addition, the study sought to compare the efficacy of different machine learning and deep learning models in predicting the severity of such incidents. Results indicated that these models demonstrated poor performance in predicting the severity of bicycle crashes. As a result, a hybrid model that combines the K-Nearest Neighbor and eXtreme Gradient Boosting algorithms was developed to improve accuracy. The hybrid model outperformed all other models, achieving an accuracy rate of 83.56%. The study, additionally, has put forward several recommendations, including the mandatory use of reflective clothing and the installation of Intelligent Transportation Systems (ITS) to enhance the safety of cyclists.

Keywords Bicycle crash · Exploratory analysis · Machine learning · Deep learning · Severity classification · K-nearest neighbor · eXtreme gradient boosting · Hybrid model

1 Introduction

Cycling has long been regarded as an optimal mode of transportation due to its numerous physical and environmental benefits. Among its physical advantages, cycling has been shown to improve bone strength [1], cognitive functioning [2], and decrease the risk of developing certain diseases such as breast cancer [3], obesity

M. Ahmed · P. Hewage (✉)
School of Creative Technologies, University of Bolton, Bolton BL3 5AB, UK
e-mail: p.hewage@bolton.ac.uk

C. Iwendi et al. (eds.), *Proceedings of ICACTCE'23—The International Conference on Advances in Communication Technology and Computer Engineering*, Lecture Notes in Networks and Systems 735, https://doi.org/10.1007/978-3-031-37164-6_44

[4], and diabetes [5]. Moreover, cycling contributes to environmental well-being by eliminating harmful emissions, reducing noise pollution, and promoting green spaces.

According to the British Broadcasting Corporation (BBC), the use of bicycles since Covid-19 has witnessed a dramatic rise, and there are estimates that the number of bicycle-users could go as high as five billion by 2050. Within the United Kingdom (UK) alone, several million people utilise bicycles for various purposes, making it the focal point of this study.

Nevertheless, despite the multitude of benefits associated with cycling, bicycles are often perceived as more dangerous than other forms of vehicles. This perception stems from the lack of external protective devices on bicycles, which increases the likelihood of rider injuries. Other factors contributing to the perception of danger include insufficient lane infrastructure and inadequate traffic laws.

To address these concerns, this study aims to employ multiple Machine Learning (ML) models to predict the severity of bicycle crashes based on a range of correlated variables. In doing so, the study seeks to identify the primary factors that contribute to the classification of crash severity levels, thereby providing valuable insights for enhancing bicycle safety.

1.1 Background

According to the data provided by the UK government, more than one hundred cyclists were killed, while a staggering 4,353 were severely injured and 11,994 were slightly injured only in 2021. Even though the number of casualties declined from 2014 to 2020, there was a rise in the number between 2020 and 2021. Machine Learning (ML), a branch of artificial intelligence, and its predictive methodologies are already being used by numerous sectors such as healthcare, finance, and marketing. They can be applied to increase efficiency, identify hidden patterns, improve accuracy etc., by analysing past data.

One of the main reasons why applying ML to cycling data can be beneficial is to improve road safety. When factors such as speed limit, road conditions, light conditions, and road types are taken into account while classifying the levels of severity, valuable insights, such as identifying high-risk conditions and age-group, can be extracted to decide whether there is a need to enhance either the infrastructure of the road or implement intervention strategies to make it safer for cyclists. In addition, the available resources can be properly utilised and allocated to areas where they are needed. This, in turn, can help authorities make sure that the resources are being used in an effective and efficient manner.

Another benefit of the implementation of ML models in cycling data is to grasp the associated risk factors accurately. These models can, for instance, assist in identifying the underlying factors—Such as the road type, road conditions, and the number of vehicles involved—to inform the decision-makers to work for the betterment of these factors and provide policies accordingly.

Although the statistical models—such as Random Forest, Decision Tree, and Logistic Regression—used in ML to perform classification are largely capable of making predictions quite accurately, they still have certain limitations like overfitting, lacking interpretability, and high error rate, also referred to as high variance. Mitigating these disadvantages often requires the development of either a hybrid model—where the models work together to create a new robust model—or an ensemble model—where the models work independently.

Therefore, it is crucial to analyse the risk factors and project the level of severity of bicycle accidents in Great Britain using an appropriate hybrid model in order to make cycling safer and improve the infrastructure and decrease the number of crash-related injuries and fatalities.

1.2 Aims and Objectives

The following are the propositional statements for the aim and objectives of this study:

Aim:

To enhance the accuracy of bicycle crash severity classification by developing a hybrid model integrating KNN and XGBoost models.

Objectives:

1. To develop a hybrid model that integrates KNN and XGBoost models for bicycle crash severity classification.
2. To compare the accuracy of the proposed hybrid model with other ML and deep learning models for bicycle crash severity classification.
3. To identify significant factors associated with bicycle crashes and determine their influence on the severity level.

2 Literature Review

Factors related to bicycle crashes and their influence on classifying the level of severity have been explored by a number of researchers using different methods. Although plenty of algorithms and models have been explored and implemented, Bayesian Network [6], Poisson Distribution [7], and eXtreme Gradient Boosting [8] are among the models that have been frequently implemented to analyse past data and make predictions.

One of the most comprehensive studies in identifying the risk factors and the severity of bicycle accidents using data mining techniques was carried out by Prati et al. [9]. They analysed the data provided by the Italian National Institute of Statistics using the Chi-square Automatic Interaction Detection (CHAID) decision tree technique and Bayesian Network analysis and discovered that according to the CHAID analysis, road type, crash type, and the age of the cyclist had been considered the most significant features while the Bayesian Network—in addition to road type and crash type—also marked the type of opponent vehicle as one of the substantial factors in determining the severity of bicycle crashes. Both techniques also demonstrated that rear-end collisions are more dangerous than angle collisions as they are more abrupt and unexpected, and also revealed that rural and urban roads impact crash severity in different manners due to the difference in their design and infrastructure. One of the interesting findings was that although in urban areas, the size of the opponent vehicle hugely affects the crash severity, that is, the bigger the vehicle, the more severe the outcome; it does not make much of a difference in rural areas. One of the suggestions put forward by the researchers was the addition of a line dedicated to cyclists and separated from motorised traffic. They also suggested that installing a detection system could be of assistance in warning motor vehicle drivers of cyclists' presence. Dash et al. [10] employed a different approach to identify the risk factors in which they applied multiple Machine Learning (ML) models—such as Logistic Regression, Decision Tree, and Random Forest. They mainly focused on the data collected from two of the largest cities in the USA—Nashville and Memphis—and found out that Random Forest performed better on the data as compared to the other implemented models. Santos et al., in their pursuit of comparing different ML models employed in recent studies, also came to the conclusion that Random Forest had a superior accuracy percentage as compared to other models in crash severity prediction. The critical variables in bicycle crashes, as per the analysis of Dash et al. [10] are insufficient lighting conditions, speed limits, and the number of lanes. Numerous measures were proposed by them based on the findings to discount the probability of bike accidents and ameliorate the safety of cyclists. The principal proposal was to install street lights as their result in line with the results of other researchers—such as Asgarzadeh et al. [11] and Kim et al. [12]—indicated that the absence of lights on in the streets substantially increases the risk of vehicle collisions. Their risk alleviation strategies, like by Prati et al. [9], also included the development of dedicated bike lanes; in addition, all urban roads with a speed limit above 30 miles per hour (mph) should be assessed on a regular basis as well as appropriate signage should be placed in vulnerable locations.

Single bicycle crashes (SBCs), defined as falls and obstacle collisions, are one of the common reasons why cyclists often get injured, and, as a consequence, a significant number of studies regarding bicycle accidents have focused on exploring and interpreting the associated data. Take the investigation carried out by Myhrmann et al. [13] in which they implemented a latent class-ordered probit model to identify the main factors that influence the injury severity of SBCs, for instance. The results they found indicated that the characteristics of the roads play perhaps the biggest role in determining the severity of injuries as it was found that crashes on road sections, as compared to bicycle lanes, considerably raise the probability of getting severely injured. Furthermore, lack of proper maintenance, poor lighting conditions, and the volume of traffic had also been found to be a few of the strongest determinants of injury severity. Errikson et al. [14] in a similar quest, revealed various insights related to SBCs in Sweden in which they performed the binary logistic regression on variables such as age-group, trip purpose, road state etc. Another feature to hugely influences the probability of SBCs is the lack of proper infrastructure, as shown by the study conducted in Finland by Utriainen [15]. One of the most influential factors in SBCs, according to their study, is the age of the cyclist as the results illustrated that the older the cyclist, the higher the risk of severe injury. The results also, axiomatically, showed that those with protective equipment—such as helmets—are less likely to be severely injured as compared to those without protective equipment. Additionally, crashes at intersections and road stretches, as opposed to pedestrian and cycle paths, escalate the probability of getting severely injured; this complements the findings of Myhrmann et al. [13].

Apart from the above factors, the propensity to take risks plays a crucial role in determining both crash and injury severity. This has been exhaustively shown in the study conducted by Prati et al. [16] in their pursuit of finding the differences between different genders among Italian cyclists. The data indicated that male cyclists are more likely (2.23 times higher) to be involved in serious crashes when compared to women cyclists; they, through the use of a series of chi-square tests, also found a number of differences between genders in various aspects of cycling such as road type, vehicle manoeuvre, and collision type. Many other research studies have also been conducted to check the differences between genders among cyclists and uncovered numerous fascinating insights. For instance, the study conducted in the Netherlands by Cobey et al. [17] on risk-taking behaviour exhibited that men are more prone to taking risks—such as not using bike lights and risky manoeuvres—than women.

Although there are millions of bicycle users in Great Britain (GB), the number of studies conducted on either classifying the risk factors associated with bicycle crashes or their severity level is comparatively lower than the studies carried out on accidents related to motor vehicles. Also, Most of the existing research studies include either a limited amount of data or data specific to only one region. Therefore, it is vital to investigate the risk factors and the level of crash severity involving a large number of variables and records to uncover relevant insights.

3 Research Methodology

This chapter provides an overview of the methods and models implemented in this study, including Naïve Bayes, Random Forest, Decision Tree, and the proposed hybrid model. The chapter further includes a comprehensive discussion of the dataset used in this study, its collection, cleaning, and preparation processes.

3.1 Proposed KNN-XGBoost Model

The proposed hybrid model developed for this study combines the K-nearest Neighbours (KNN) and eXtreme Gradient Boosting (XGBoost). KNN is an effective algorithm that works well for small to medium-sized datasets. It uses the distance between data points to determine the nearest neighbours and make predictions based on the majority class of the neighbours. It can be described as:

$$y_q = \text{mode}(y_i), \quad i = 1, \ldots, k \tag{1}$$

where:

- y_q is the predicted class label of the test instance
- y_i is the class label of the ith nearest neighbour to the test instance
- mode() is the function that returns the most frequent element in a list.

XGBoost is a powerful gradient-boosting algorithm that is effective in handling large datasets and high-dimensional feature spaces. It works by building an ensemble of decision trees and combining their predictions to make the final prediction. XGBoost has been shown to perform well in many machine learning tasks, including regression and classification, and can handle a wide range of data types. The XGBoost model can be expressed as follows:

$$h(x) = \sum_{m=1}^{M} f_m(x) \tag{2}$$

where:

- $h(x)$ is the predicted output for a given input x
- M is the number of trees in the ensemble
- $f_m(x)$ is the m-th tree in the ensemble and is defined as:

$$f_m(x) = w_{q(x)}^{(m)} \tag{3}$$

where:

- $q(x)$ is the index of the leaf node that input x falls into in the m-th tree
- w is the weight of the corresponding leaf node.

By combining KNN and XGBoost algorithms, the hybrid model benefits from the strengths of both algorithms. For example, the KNN algorithm can be used as a pre-processing step to identify the nearest neighbours and extract relevant features for XGBoost. This can help reduce the dimensionality of the dataset and improve the performance of XGBoost. Additionally, the hybrid model can be more robust to noisy and irrelevant features, as KNN can filter out such features during pre-processing.

A detailed explanation of the proposed hybrid model built for this study is given below:

Since we want to combine KNN and XGBoost to build a hybrid classification model, the predicted class of the KNN model can be represented as:

$$h_{\mathrm{KNN}}(x) = \{0, 1\} \tag{4}$$

where:

- x is the input data point.

Similarly, the predicted class of the XGBoost model can be represented as:

$$h_{\mathrm{XGBoost}}(x) = \{0, 1\} \tag{5}$$

For the final prediction, the soft voting method was used. For soft voting, we need to calculate the probability of the predicted class for each base model. Let's assume the predicted probability of class 1 for the KNN model is given by $p_{\mathrm{KNN}}(x)$, and the predicted probability of class 1 for the XGBoost model is given by $p_{\mathrm{XGBoost}}(x)$. Then, the final prediction for soft voting can be represented as:

$$H(x) = 1/2 * (p_{\mathrm{KNN}}(x) + p_{\mathrm{XGBoost}}(x)) \tag{6}$$

where:

- 1/2 is the normalisation factor since we have two base models.

While the combination of KNN and XGBoost models has been studied in other domains, such as recommendation filtering [18], its application in this context has not yet been investigated. Overall, the proposed hybrid model combining KNN and XGBoost is highly efficient on the provided dataset as it improves the accuracy and robustness of the predictions.

3.2 *Naïve Bayes*

The Naive Bayes model is a probabilistic classifier that uses Bayes' theorem to compute the probability of a class given a set of features or variables.

Bayes' theorem is used to calculate the posterior probability of a class given a set of features or variables. It is given by:

$$P(y|x) = P(x|y) * P(y)/P(x) \tag{7}$$

where:

- $P(y|x)$ is the posterior probability of class y given features x

- $P(x|y)$ is the likelihood of features x given class y
- $P(y)$ is the prior probability of class y
- $P(x)$ is the marginal probability of features x.

The Naive Bayes assumption is that the features or variables are conditionally independent given the class label. This assumption allows us to simplify the likelihood term in Bayes' theorem as follows:

$$P(x|y) = P(x1|y) * P(x2|y) * \ldots * P(xn|y) \tag{8}$$

where:

- $P(xi|y)$ is the probability of feature xi given class y.

Maximum a posteriori (MAP) estimation:

The parameters of the Naive Bayes model are estimated using maximum a posteriori (MAP) estimation. This involves finding the values of the model parameters that maximise the posterior probability of the model given the data. In the case of the Naive Bayes model, the MAP estimate of the parameters is given by:

$$\theta = \arg\max P(\theta|D) \tag{9}$$

where:

- θ is the set of model parameters (prior probabilities and feature probabilities)
- D is the training data.

Overall, the Naive Bayes model is an efficient classifier that can be trained quickly on large datasets and used in a wide variety of fields such as spam filtering [19], text classification [20], and sentiment analysis [21].

3.3 *Random Forest*

The Random Forest Classifier is based on decision trees, which recursively partition the feature space into regions that are homogeneous with respect to the target variable. The decision rule at each node of a decision tree can be expressed as:

$$\text{If } xj \leq t, \text{ go left} \tag{10}$$

$$\text{If } xj > t, \text{ go right} \tag{11}$$

where:

- xj is the value of the jth feature
- t is the threshold value for the jth feature

3.3.1 Bagging

Random Forest Classifier uses bagging to create multiple decision trees from bootstrap samples of the training data. Bagging involves randomly selecting subsets of the training data with replacement and training a decision tree on each subset.

3.3.2 Random Feature Subspace

Random Forest Classifier also uses random feature subspace sampling, where a random subset of features is selected at each node of the decision tree. This helps to decorrelate the trees and improve their diversity.

3.3.3 Prediction

To make a prediction, the Random Forest Classifier aggregates the predictions of all the decision trees. The most common way to do this is by taking the majority vote on the class labels predicted by the individual trees.

In this study, the hyperparameter n_estimators was set to 100 for the Random Forest model. This hyperparameter determines the number of decision trees that will be created and combined in the ensemble. The value for this hyperparameter can be increased to check if it provides a better accuracy percentage, but that will lead to an increased computational cost of training and prediction.

Another hyperparameter, random_state, was set to 42. Setting the random_state as 42 means that the algorithm will use the random seed 42. This value is arbitrary and can be any non-negative integer. By using the same random seed, it is ensured that the same sequence of random numbers is generated each time the algorithm is run, which can be useful for debugging and testing purposes.

Overall, the Random Forest Classifier is a powerful and versatile machine-learning algorithm that can be used for various classification tasks, such as classifying lymph disease [22] and detecting malware [23]. Its main advantage is its ability to handle high-dimensional and noisy data, as well as its ability to detect crucial features.

3.4 Decision Tree

The decision tree classifier is an intuitive classification model that uses a tree-like model of decisions and their possible consequences. At each internal node of the tree, a decision is made based on a specific feature, and the sample is split into two or more subsets.

Let X be a d-dimensional input feature vector and y be the corresponding class label. Let D be the training dataset containing N samples $\{(X1, y1), (X2, y2), \ldots, (XN, yN)\}$.

3.4.1 Splitting Criterion

At each node of the tree, a splitting criterion is chosen to determine the feature and threshold value that best separate the samples into different classes. The splitting criterion can be based on different measures of purity; in this case, entropy was chosen as the splitting criterion. Let j be the index of the splitting criterion, and let T be the threshold value. Then the decision rule at each node can be expressed as:

$$\text{If } Xj \le T, \text{ go left} \tag{12}$$

$$\text{If } Xj > T, \text{ go right} \tag{13}$$

where:

- $j = 1, 2, \ldots, d.$

3.4.2 Recursive Partitioning

Starting from the root node, the decision tree recursively partitions the feature space into regions that are homogeneous with respect to the target variable. The partitioning is done by selecting the splitting criterion that maximizes the information gain or minimizes the impurity at each node. The recursive partitioning continues until a stopping criterion is met, such as a maximum depth or minimum number of samples per leaf node.

The maximum depth hyperparameter for the implemented model was set to 7 to limit the decision tree to 7 levels. It controls the depth of the tree, which is the number of splits from the root node to the leaf nodes. By limiting the depth, the model is less likely to overfit the training data and can generalize better to new data.

The minimum number of samples per leaf node was set to 8, which means that each leaf node of the decision tree will contain at least 8 samples. This hyperparameter controls the minimum number of samples required to be in a leaf node, which can prevent the tree from being too specific to the noise or outliers in the data.

3.4.3 Prediction

To make a prediction for a new sample X, the decision tree starts at the root node and follows the decision rules at each node until it reaches a leaf node. The class label of the leaf node is then assigned as the prediction for the sample.

In summary, the decision tree classifier is a binary tree that recursively partitions the feature space into regions based on a chosen splitting criterion. It is a popular model in machine learning due to its simplicity, interpretability, and ability to handle nonlinear relationships between the features and target variables.

3.5 Feedforward Neural Network

Feedforward neural network (FFNN) is a type of artificial neural network where information flows in a single direction, from the input layer, through one or more hidden layers, and finally to the output layer.

Let X be the input vector of size n, and let Y be the output vector of size m. Let H be the hidden layer with k neurons, and let $f(x)$ be the activation function used for the hidden and output layers. Let W and b be the weight matrix and bias vector, respectively, for each layer.

3.5.1 Forward Propagation

The FFNN computes the output of each layer by matrix multiplication with the weight matrix, followed by the activation function:

$$a^1 = f\left(W^1 X + b^1\right) \tag{14}$$

$$a^2 = f\left(W^2 a^1 + b^2\right) \tag{15}$$

$$a^k = f\left(W^k a^{(k-1)} + b^k\right) \tag{16}$$

$$y = f\left(W^y a^k + b^y\right) \tag{17}$$

where:

- a^1 is the activation vector of layer 1
- W^1 is the weight matrix of layer 1
- b^1 is the bias vector of layer 1
- $f()$ is the activation function.

3.5.2 Backward Propagation

The FFNN uses backpropagation to update the weight and bias parameters of the network based on the error between the predicted output and the true output. The error is measured by a loss function, such as mean squared error or cross-entropy, and the weights are updated using the gradient descent algorithm:

$$\delta_y = (y - Y) * f'\left(W^y a^k + b^y\right) \tag{18}$$

$$\delta^k = \delta_y * W^y * f'\left(W^k a^{(k-1)} + b^k\right) \tag{19}$$

$$\delta^1 = \delta^2 * W^2 * f'\left(W^1 * X + b^1\right) \tag{20}$$

$$dW^y = \delta_y * a^k \tag{21}$$

$$db^y = \delta_y \tag{22}$$

$$dW^k = \delta^k * a^{(k-1)} \tag{23}$$

$$db^k = \delta^k \tag{24}$$

$$dW^1 = \delta^1 * X \tag{25}$$

$$db^1 = \delta^1 \tag{26}$$

where:

- δ^1 is the error vector of layer 1
- $f'()$ is the derivative of the activation function
- dW^1 and db^1 are the weight and bias gradients of layer 1.

3.5.3 Parameter Update

The FFNN updates the weight and bias parameters using the calculated gradients and a learning rate:

$$W^l = W^l - \text{learning\_ rate} * dW^l \tag{27}$$

$$b^l = b^l - \text{learning\_ rate} * db^l \tag{28}$$

where:

- learning_rate is a hyperparameter that controls the step size of the gradient descent algorithm.

This study proposes a feedforward neural network (FFNN) model for binary classification, where the first and second layers use the rectified linear unit (ReLU) activation function, and the third layer uses the sigmoid activation function. The model employs the binary cross-entropy loss function to measure the difference between predicted and actual outputs. The Adam optimiser is used to minimise the loss function and update the weights and biases in the network. The model's performance is evaluated using the accuracy metric, which measures the proportion of correct predictions. By leveraging the ReLU activation function, the FFNN model can efficiently handle the non-linear relationships between the input and output variables. The sigmoid activation function in the last layer ensures that the output is bounded between 0 and 1, which is ideal for binary classification tasks. The Adam optimiser's adaptive learning rate and momentum ensure the model can converge to the optimal solution quickly and robustly. Finally, the accuracy metric provides a reliable measure of the model's classification performance. Overall, the proposed FFNN model with ReLU and sigmoid activations, binary cross-entropy loss, Adam optimiser, and accuracy metric holds promise for effective binary classification tasks.

4 Data Collection and Description

The dataset used in this study was collected from Kaggle, a popular platform for hosting data science competitions and open datasets. The specific dataset used in this study, named 'Bicycle Accidents in Great Britain from 1979 to 2018', was uploaded to Kaggle by John Harshith [24]. The dataset was selected for its relevance to the research question.

Table 1 summarises the variables included in the dataset used for this study. The dataset contains 13 variables, including the accident index, number of vehicles, number of casualties, date, speed limit, road conditions, weather conditions, road type, day of the week, light conditions, gender, age group, and severity.

Table 1 Column names and their description

Column name	Column description
accident_index	A unique code is assigned to each crash
number_of_vehicles	The number of vehicles involved
number_of_casualties	The number of casualties
date	The date on which the accident occurred
speed_limit	The speed limit at the place of the crash
road_conditions	The condition of the road
weather_conditions	The condition of the weather
day	Day of the week
road_type	The type of the road
light_conditions	The condition of the light
gender	The gender of the rider
age_group	Age-group the rider belongs to
severity	Severity of the crash

4.1 Data Pre-processing

As part of the data preparation process, the dataset underwent various data pre-processing techniques, including the removal of missing data, elimination of duplicates, detection and verification of outliers, and feature engineering. This dataset had a limited amount of data on fatal crashes, resulting in the exclusion of such data from the prediction model. In addition, data from 2003 to 2018 were kept due to the fact that records before 2003 often had multiple variables missing. To facilitate the analysis, label encoding was employed to convert categorical data to numerical data. Furthermore, the dataset was partitioned into two sets, with 90% of the data used for training and the remaining 10% for testing purposes.

To balance the data in this study, Synthetic Minority Over-sampling Technique (SMOTE) was employed. SMOTE is a widely-used oversampling method that creates synthetic minority class samples by interpolating new samples between the existing minority samples. This technique effectively increases the number of minority samples in the dataset, thus mitigating the issue of class imbalance. The use of SMOTE in this study enabled a more accurate and reliable analysis of the data by reducing the impact of the class imbalance on the results.

5 Analysis and Discussion

This section includes the statistical analysis and discussion as well as results found from the implemented models.

5.1 Analysis

Both the number of crashes and the number of casualties witnessed fluctuations over the years, although the overall numbers increased marginally. The number of crashes went up to 16,783 in 2018 from 15,866 in 2003, and the number of casualties went up to 17,358 in 2018 from 16,484 in 2003 (Fig. 1). Among the months, June, July, and September had the highest number of incidents recorded, while December and February had the lowest (Fig. 2). Also, the lowest number of accidents was reported between 12:00 midnight and 5:00 in the morning, while the highest number was reported between 8:00 and 9:00 in the morning and between 4:00 and 7:00 in the evening. The age brackets that were involved in the highest number of crashes were between 26 and 45, whereas those involved in the lowest number of crashes were mostly between 6 and 10 and 56 and 75 (Fig. 3). However, riders aged between 46 and 75 were more likely to be involved in a serious crash, while those aged between 21 and 35 and 6 and 10 were least likely. Males were involved in a significantly higher number of crashes than females and were also more prone to be involved in

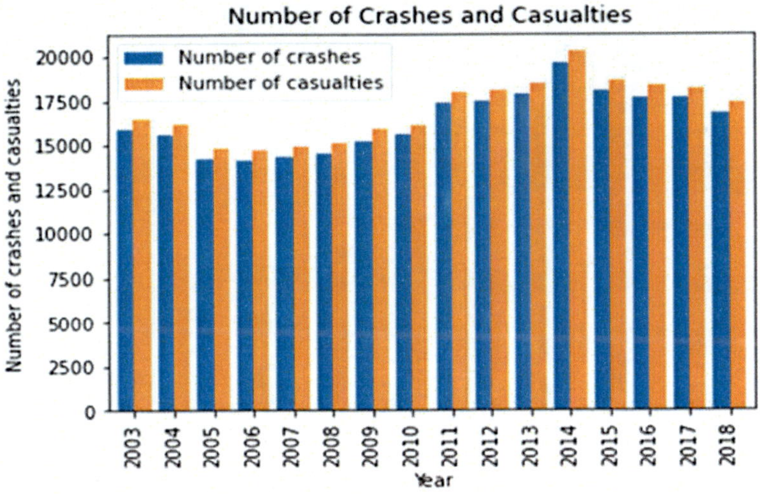

Fig. 1 Number of crashes and casualties each year

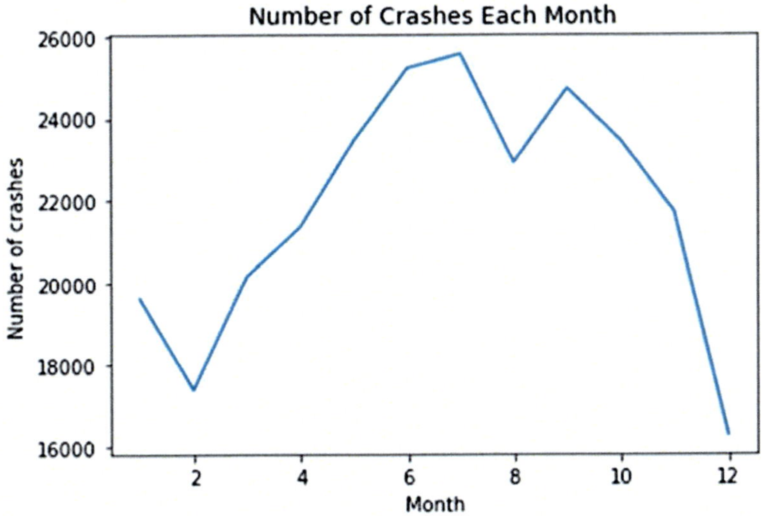

Fig. 2 Number of crashes each month

serious crashes. In fact, male riders accounted for approximately 80% of all crashes, while female riders accounted for only about 20% (Fig. 4).

Although the highest number of crashes took place in daylight and the lowest in places without lights, the probability of a crash being of serious severity increases substantially in places without lights. More than 80% of all crashes took place on single-carriageways, accounting for the highest number of accidents. One-way streets and slip roads accounted for the lowest number of crashes, with 2.02% and 0.25%,

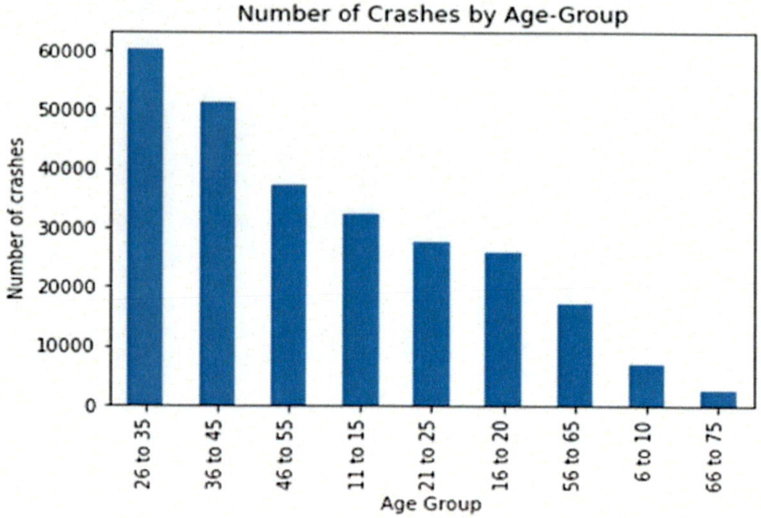

Fig. 3 Number of crashes by age group

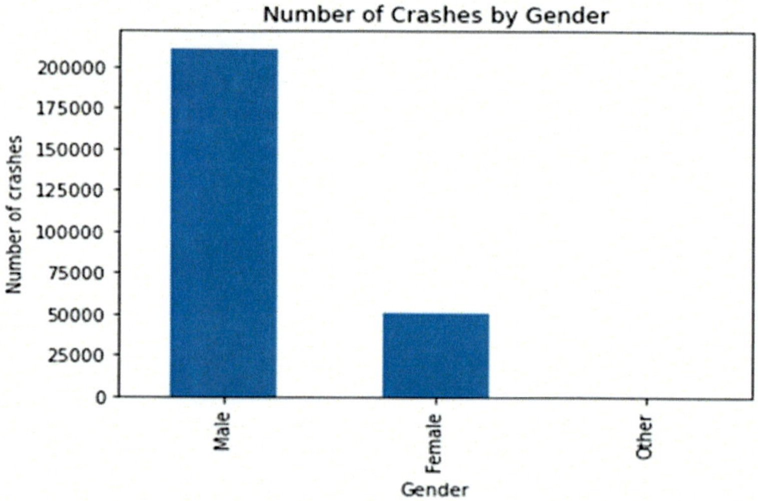

Fig. 4 Number of crashes by gender

respectively (Fig. 5). The type of road plays a vital role in determining the severity of the crash, as slip roads had the highest percentage (24.67%) of serious crashes, while roundabouts (16.15%) and one-way streets (16.09%) had the lowest. Upon analysing the data pertaining to the days of the week, it was discovered that there exists a notable disparity in the frequency of bicycle accidents between weekends and weekdays, with the former exhibiting significantly lower incident rates compared to the latter

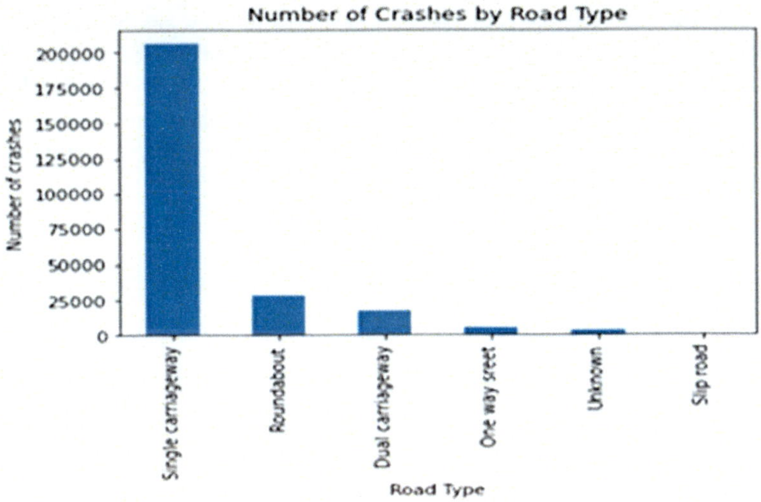

Fig. 5 Number of crashes by road type

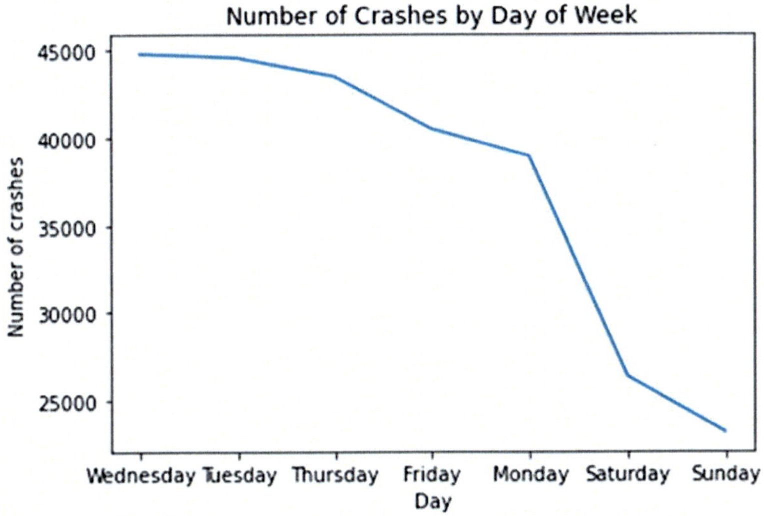

Fig. 6 Number of crashes by day of week

(Fig. 6). The weather conditions exhibited to be a significant determinant of the severity of crashes, wherein foggy weather appears to be associated with a markedly higher proportion (23.32%) of serious accidents. Conversely, clear (18.65%) and rainy (17.28%) weather conditions exhibit a relatively lower frequency of crashes that are classified as serious. When considering the likelihood of a crash being categorised as serious or slight, it can be inferred that the probability of a crash being serious was the highest when there is the presence of either frost or snow on the road, whereas the probability of a serious crash was at its lowest when the surface of the road was dry. Speed limit probably had the biggest impact on the severity of a crash since more than 65% of the crashes were classified as serious where the speed limit was 70 miles per hour (mph), 42.72% of the crashes were serious where the limit was 60 mph; the lowest 17.07% was where the limit was 20 mph.

5.2 *Discussion*

Even though the number of bicycle crashes as well as casualties in Great Britain has experienced an overall gradual upward trend, there has been a slight decline in these numbers over the past few years. The exploration of the data revealed, in line with the studies conducted by Liu et al. [25] and Prati et al. [9], that the number of crashes increases substantially in summer and at the beginning of autumn as compared to other seasons. One of the reasons for this can be the increasing number of cyclists on the roads in summer, which leads to more interactions between cyclists and other road users; also, motorised vehicle drivers might be less aware of this growth in the number of cyclists which, in turn, can lead to not taking necessary precautions to avoid accidents. Another finding, as opposed to the finding of the study carried out in a Swedish city by Dozza [26] suggests that cycling during weekdays is riskier than at weekends which could be a result of more people commuting to and from work during weekdays making it difficult for cyclists to navigate the roads safely. The data, in line with the investigation of Liu et al. [25], exhibits that cycling during the morning (07.00 a.m.–09.00 a.m.) and evening (03.00 p.m.–07.00 p.m.) hours is more dangerous than other periods of the day; these periods in the UK are considered to be rush hours as there is usually an increased volume of traffic on the roads during these times. One of the steps that cyclists can take in order to reduce the likelihood of collision is to be visible to other road users; although the Highway Code, a set of guidelines and rules provided by the authority for road users in the UK, advises cyclists to wear high-visibility clothing, it is not a legal requirement. This rule of wearing bright, reflective clothes can be made stricter for specific seasons and high-risk periods to lower the number of bicycle crashes. Additionally, widespread implementation of Intelligent Transportation Systems (ITS) has to be made as they come with the ability to provide real-time information about potential hazards and other vehicles on the road, which can make cycling smoother and more convenient.

One of the biggest features in determining the severity of bicycle crashes is the age of the rider, as the data demonstrates that riders aged 46–75 years are more likely to be in a serious crash compared to their younger counterparts; this has previously been substantiated by a number of researchers including Rodgers [27], Ekman et al. [28], Stone and Broughton [29], and Eluru et al. [30]. One reason for this disparity between older and younger cyclists may be that as people age, they experience difficulties in maintaining balance and coordination, as well as a decline in cognitive and physical abilities. Other possible reasons could include hearing loss, impaired vision, or reduced mobility. Encouraging older cyclists to switch to electric bicycles (e-bikes) can be a useful measure as they offer a less physically demanding alternative to conventional bikes. Along with promoting e-bikes, developing wider bike lanes, accessible parking spaces, and customised cycling training programs can be of assistance in discounting the probability of serious crashes among older cyclists. The condition of street lighting is another significant factor that can greatly affect the severity of bicycle crashes as the data, in line with the findings of Yan et al. [31] and Asgarzadeh et al. [11], shows that the probability of a crash being serious increases considerably when there are no street lights illuminated. Therefore, it is imperative to strategically install street lights in areas where they are needed to improve the safety of cyclists and reduce the likelihood of serious crashes. The type of road also plays a crucial role in bicycle crashes, as one of the findings suggests that accidents occurring on slip roads tend to be more severe compared to those on single or dual carriageways, one-way streets and roundabouts. Slips roads are deemed to be more dangerous for cyclists probably because they are designed for motorists to exit or enter a main road, usually at high speed, which can create a hazardous environment for cyclists, especially if the road is narrow and lacks proper infrastructure. Moreover, these roads often have limited visibility which makes it difficult for motor vehicle drivers to spot cyclists and react in a timely manner. To make cycling on slip roads safer, one solution could be to lower speed limits, giving motorists more time to react. Another solution would be to improve street lighting and visibility, ensuring that both cyclists and motorists are visible to each other.

In this study, weather conditions have been identified as a significant factor in determining the severity of bicycle accidents. The analysis reveals that foggy weather poses the highest risk for serious crashes, thus emphasising the need to consider weather conditions when assessing crash severity. The issue of conspicuity can be identified as a possible explanation for the elevated risk of serious crashes in foggy weather, particularly for bicycles which are typically less visible than motorised vehicles. This problem is further compounded by the reduced visibility caused by fog, exacerbating the issue of poor conspicuity. In addition to foggy weather conditions, the presence of frost and snow on the road has been identified as another factor that can increase the severity of bicycle crashes. To reiterate previous suggestions, improving the conspicuity of bicycles in foggy weather and reducing the risk of serious crashes in frosty and snowy conditions can be achieved through a combination of measures, including the development of adequate lighting systems, stricter regulations regarding the use of bright clothing, and the implementation of lower speed limits in adverse weather conditions. This study also indicates that males are slightly more susceptible to being involved in serious bicycle crashes than females, although the difference in risk is relatively small.

The findings of this study, consistent with the study carried out by Yan et al. [31], suggest that speed limit is one of the most significant factors influencing the severity of bicycle crashes, with a substantial increase in the proportion of crashes classified as serious as speed limits rise. Specifically, the data indicates that approximately 30% of total bicycle crashes are labelled as serious when the speed limit is 60 mph, while the proportion of serious crashes rises to around 40% when the limit is 70 mph. These results emphasise the critical role of calming traffic measures and speed reduction strategies in promoting road safety and preventing serious bicycle crashes, particularly in areas where bike lanes are present.

Among the various implemented and evaluated machine learning models, the FFNN model, although observed to be the worst-performing model with an overall accuracy rate of 69.24%, demonstrated a certain degree of efficiency in identifying and predicting crashes classified as serious as it outperformed several statistical models in projecting serious bicycle crashes. The Decision Tree classification model, although competent to an extent in accurately predicting slight crashes, had an overall accuracy rate of less than 70% (69.58%), which was the lowest among all models except for FFNN. The results of the Naïve Bayes classification model showed that it had the highest overall accuracy percentage (73.33%) compared to the FFNN, Decision Tree, or Random Forest models. However, the Random Forest model outperformed all other models in projecting crashes labelled as serious with a higher accuracy percentage, despite having an overall accuracy rate of only 72.84%. The proposed hybrid model built by amalgamating the KNN and XGBoost models proved to be the most effective model, with an overall accuracy rate of 83.56% (Fig. 7). It leveraged the strengths of both the KNN and XGBoost models to achieve enhanced accuracy in predicting the severity of bicycle crashes, especially in identifying and prognosticating slight crashes.

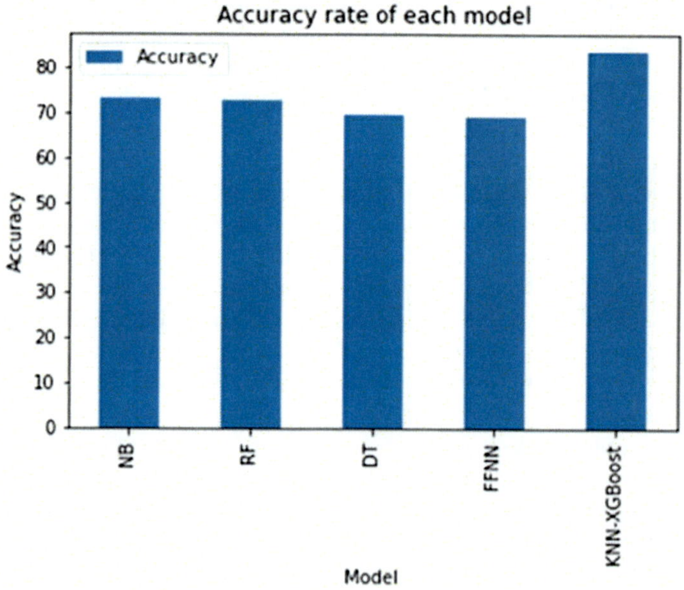

Fig. 7 Accuracy rate of each model

6 Conclusion and Future Research

6.1 Conclusion

The growing popularity of cycling as a means of transportation has led to an increase in the number of cyclists on the roadways, creating new challenges for ensuring the safety of these vulnerable road users. Cyclists are at a higher risk of being involved in crashes than other road users due to their increased vulnerability and exposure to hazards such as motor vehicles, weather conditions, and road infrastructure. As such, the need to prioritise cycling safety has become paramount in many areas around the world.

This study has sought to investigate the factors that contribute to the severity of bicycle crashes in Great Britain over the past few years. By analysing a range of variables such as season, time of day, age of the cyclist, lighting conditions, weather conditions, road types, road conditions, and speed limits, the study has identified the major factors that influence the severity of bicycle crashes.

The findings of the study highlight the complexity of the issue of bicycle safety, with multiple factors contributing to the severity of crashes. For instance, the data revealed that the time of day and season are significant predictors of the severity of bicycle crashes, with more serious crashes occurring during summer months and during early morning and evening hours. Moreover, the study also revealed that the age of the cyclist is a significant factor in determining the severity of a crash, with older cyclists more likely to be involved in serious crashes than younger ones. Additionally, lighting and weather conditions have been identified as key contributors to the severity of bicycle crashes, with poor lighting and adverse weather conditions increasing the likelihood of serious crashes. Furthermore, higher speed limits have been found to be associated with an increased likelihood of serious crashes.

To categorise the severity of bicycle crashes, various machine learning models have been applied, including singular machine learning models such as Naïve Bayes, Random Forest, and Decision Tree, as well as more complex deep learning models such as Feedforward Neural Networks. However, the results of these models have been limited, with none of them achieving an accuracy of 80% or higher. To address this issue, a hybrid machine learning model was developed, which combined the strengths of KNN and XGBoost models to improve accuracy. This approach allowed for a more robust and accurate classification of the severity of bicycle crashes. The results demonstrated that the proposed hybrid model outperformed all other models with an accuracy percentage of 83.56%. This represents a significant improvement over the previous models and highlights the potential of combining multiple machine-learning techniques to achieve better results.

Overall, this study provides valuable insights into the complex nature of bicycle safety and the multiple factors that contribute to the severity of crashes. The findings highlight the need for a multifaceted approach to promoting bicycle safety, incorporating measures to address environmental factors, cyclists' age, lighting and weather conditions, road infrastructure, and speed limits. Policymakers and transportation planners must consider these factors in their decision-making processes to create a safer and more accessible environment for cyclists. In addition, the success of the proposed hybrid model could have significant implications for improving the safety of cyclists on the road. By accurately categorising the severity of bicycle crashes, it could help inform the development of more effective safety measures and interventions to reduce the number of serious accidents.

6.2 Future Research

Future research can build on the findings of this study by combining other machine learning and deep learning models to further improve the accuracy of predicting the severity of bicycle crashes. For example, a hybrid model combining and utilising the strengths of supervised and unsupervised learning could be developed to capture the complex relationships between variables better and improve the accuracy of predictions.

Additionally, this study focused on a limited number of variables related to bicycle crashes, such as season, time of day, age of the cyclist, lighting conditions, weather conditions, road types, road conditions, and speed limits. Future research could explore additional variables, such as the presence of bike lanes, the type of bicycle involved in the crash, and the use of safety equipment such as helmets.

Furthermore, the data used in this study only included information up until 2018. Given the rapidly evolving landscape of transportation and safety technology, incorporating more recent data into the analysis could provide a more accurate picture of the current state of bicycle safety.

Overall, the potential for machine learning models to improve our understanding of bicycle crashes and inform the development of more effective safety interventions is vast. With continued research and refinement, these models could play a critical role in ensuring the safety of cyclists on the road, ultimately reducing the number of serious accidents and injuries.

References

1. Oja, P., et al.: Health benefits of cycling: a systematic review. Scand. J. Med. Sci. Sport. **21**(4), 496–509 (2011). https://doi.org/10.1111/j.1600-0838.2011.01299.x.
2. Leyland, L.-A., et al.: The effect of cycling on cognitive function and well-being in older adults. PLoS ONE **14**(2) (2019). https://doi.org/10.1371/journal.pone.0211779
3. Luo, J., et al.: Intentional weight loss and obesity-related cancer risk. JNCI Cancer Spectr. **3**(4) (2019). https://doi.org/10.1093/jncics/pkz054
4. Julian, V., et al.: Eccentric cycling is more efficient in reducing fat mass than concentric cycling in adolescents with obesity. Scand. J. Med. Sci. Sport. **29**(1), 4–15. https://doi.org/10.1111/sms.13301

5. Rasmussen, M.G., et al.: Associations between recreational and commuter cycling, changes in cycling, and type 2 diabetes risk: a cohort study of Danish men and women. PLOS Med. **13**(7) (2016). https://doi.org/10.1371/journal.pmed.1002076
6. Huang, H., et al.: County-level crash risk analysis in Florida: Bayesian spatial modeling. Transp. Res. Rec.: J. Transp. Res. Board **2148**(1), 27–37 (2010). https://doi.org/10.3141/2148-04
7. Oh, J., et al.: Assessing critical factors associated with bicycle collisions at urban signalized intersections. TRID (2007). https://trid.trb.org/view/847696. Accessed 13 Mar 2023
8. Wu, S., et al.: Analyzing accident injury severity via an extreme gradient boosting (XGBoost) model. J. Adv. Transp. **2021**, 1–11 (2021). https://doi.org/10.1155/2021/3771640
9. Prati, G., et al.: Using data mining techniques to predict the severity of bicycle crashes. Accid. Anal. Prev. **101**, 44–54 (2017). https://doi.org/10.1016/j.aap.2017.01.008
10. Dash, I., et al.: Factors impacting bike crash severity in urban areas. J. Saf. Res. **83**, 128–138 (2022). https://doi.org/10.1016/j.jsr.2022.08.010
11. Asgarzadeh, M., et al.: The impact of weather, road surface, time-of-day, and light conditions on the severity of bicycle-motor vehicle crash injuries. Am. J. Ind. Med. **61**(7), 556–565 (2018). https://doi.org/10.1002/ajim.22849
12. Kim, J.-K., et al.: Bicyclist injury severities in bicycle–motor vehicle accidents. Accid. Anal. Prev. **39**(2), 238–251 (2007). https://doi.org/10.1016/j.aap.2006.07.002
13. Myhrmann, M.S., et al.: Factors influencing the injury severity of single-bicycle crashes. Accid. Anal. Prev. **149**, 105875 (2021). https://doi.org/10.1016/j.aap.2020.105875
14. Eriksson, J., et al.: Injured cyclists with focus on single-bicycle crashes and differences in injury severity in Sweden. Accid. Anal. Prev. **165**, 106510 (2022). https://doi.org/10.1016/j.aap.2021.106510
15. Utriainen, R.: Characteristics of commuters' single-bicycle crashes in insurance data. Safety **6**(1), 13 (2020). https://doi.org/10.3390/safety6010013
16. Prati, G., et al.: Gender differences in cyclists' crashes: an analysis of routinely recorded crash data. Int. J. Inj. Control Saf. Promot. **26**(4), 391–398 (2019). https://doi.org/10.1080/17457300.2019.1653930
17. Cobey, K.D., et al.: Sex differences in risk taking behavior among Dutch cyclists. Evol. Psychol. **11**(2), 147470491301100 (2013). https://doi.org/10.1177/147470491301100206
18. Li, Y., et al.: Collaborative filtering recommendation algorithm based on KNN and XGBoost hybrid. J. Phys.: Conf. Ser. **1748**(3), 032041 (2021). https://doi.org/10.1088/1742-6596/1748/3/032041
19. Rusland, N.F., et al.: (2017) Analysis of naïve Bayes algorithm for email spam filtering across multiple datasets. In: IOP Conf. Ser.: Mater. Sci. Eng. **226**, 012091. https://doi.org/10.1088/1757-899x/226/1/012091
20. Wang, S., et al.: Adapting naive Bayes tree for text classification. Knowl. Inf. Syst. **44**(1), 77–89 (2014). https://doi.org/10.1007/s10115-014-0746-y
21. Troussas, C., et al.: Sentiment analysis of Facebook statuses using naive Bayes classifier for language learning. IISA **2013** (2013) [Preprint]. https://doi.org/10.1109/iisa.2013.6623713
22. Azar, A.T., et al.: A random forest classifier for lymph diseases. Comput. Methods Programs Biomed. **113**(2), 465–473 (2014). https://doi.org/10.1016/j.cmpb.2013.11.004
23. Alam, M.S., Vuong, S.T.: Random forest classification for detecting android malware. In: 2013 IEEE International Conference on Green Computing and Communications and IEEE Internet of Things and IEEE Cyber, Physical and Social Computing [Preprint] (2013). https://doi.org/10.1109/greencom-ithings-cpscom.2013.122
24. Harshith, J.: Bicycle Accidents in Great Britain (1979 to 2018), Kaggle (2021). https://www.kaggle.com/datasets/johnharshith/bicycle-accidents-in-great-britain-1979-to-2018. Accessed 24 Feb 2023
25. Liu, X., et al.: Analysis of bicycle accidents and recommended countermeasures in Beijing, China. Transp. Res. Rec. **1487**, 75–83 (1995)
26. Dozza, M.: Crash risk: how cycling flow can help explain crash data. Accid. Anal. Prev. **105**, 21–29 (2017). https://doi.org/10.1016/j.aap.2016.04.033

27. Rodgers, G.B.: Factors associated with the crash risk of adult bicyclists. J. Saf. Res. **28**(4), 233–241 (1997). https://doi.org/10.1016/s0022-4375(97)00009-1
28. Ekman, R., et al.: Bicycle-related injuries among the elderly—a new epidemic? Public Health **115**(1), 38–43 (2001). https://doi.org/10.1038/sj.ph.1900713
29. Stone, M., Broughton, J.: Getting off your bike: cycling accidents in Great Britain in 1990–1999. Accid. Anal. Prev. **35**(4), 549–556 (2003). https://doi.org/10.1016/s0001-4575(02)00032-5
30. Eluru, N., et al.: A mixed generalized ordered response model for examining pedestrian and bicyclist injury severity level in traffic crashes. Accid. Anal. Prev. **40**(3), 1033–1054. https://doi.org/10.1016/j.aap.2007.11.010
31. Yan, X., et al.: Motor vehicle–bicycle crashes in Beijing: irregular manoeuvres, crash patterns, and injury severity. Accid. Anal. Prev. **43**(5), 1751–1758 (2011). https://doi.org/10.1016/j.aap.2011.04.006

Digital Transformations in Medical Applications Using Audio and Virtual Reality Procedures

S. Shitharth, Hariprasath Manoharan, Omeoga Nkechi, and Celestine Iwendi

Abstract Numerous members of society struggle with health care issues, and despite the use of sensing technology, diseases in the body are still unable to be detected. The main cause of this identification process failure is the absence of any recognized virtual technology on the market. The majority of health care solicitations seek to create a specific application that simply delivers data on sensing values and ignores the virtual representation of those values. So, in order to detect the existence of viruses inside the body, this article offers an integration platform that links sensing devices with Virtual/Audio Reality (VR/AR) approaches. Additionally, a specific form of swarm intelligence algorithm known as Fruit Fly (FF) is used in the recognition process with a modified fitness function. The FF technique offers a lot of low layer awareness, which improves the output for efficient operation. The proposed AR/VR technique is used with biological sensors to analyze the real-time situations, and two case studies are divided. It is logical to conclude from the experimental results that all validated case studies offer excellent productivity and are adaptable to all environmental circumstances.

Keywords Digital techniques · Virtual reality · Audio reality · Medical applications · Biological sensors · Wireless applications

S. Shitharth (✉)
Department of Computer Science, Kebri Dehar University, Kebri Dehar, Ethiopia
e-mail: shitharth.it@gmail.com

H. Manoharan
Department of Electronics and Communication Engineering, Panimalar Engineering College, Poonamallee, Chennai 600123, Tamil Nadu, India

O. Nkechi
Manchester Metropolitan University, Manchester, UK

C. Iwendi
University of Bolton, Bolton, UK

C. Iwendi et al. (eds.), *Proceedings of ICACTCE'23—The International Conference on Advances in Communication Technology and Computer Engineering*, Lecture Notes in Networks and Systems 735, https://doi.org/10.1007/978-3-031-37164-6_45

1 Introduction

The process of monitoring all the body conditions [1] either it may be infection or some kind of disease it is necessary to visit nearby hospitals where the entire process is involved with scan type results. But scan-based results are much more time-consuming, and accurate results cannot be achieved. Therefore, with a communication network using signal transfer technology [2], it is much easier to monitor the condition of an individual. In addition, the same communication process can be handled using reality techniques that involve two types such as virtual and audio. A sensing unit can be directly implanted in the body using the abovementioned types by following the correct specification protocol. Even installing the sensing unit on the outer skin of the body is possible. Still, the sensing units can be installed directly inside the body to provide highly accurate monitoring results. Further, the sensing units are directly connected with VR/AR module [3], which provides monitoring depictions in necessary designs. Moreover, two individual platforms can be created using VR/AR modules in medical applications that enable specialists and individuals to be connected in the same communicating unit. This process of communication [2] involves monitoring segments, and diagnosis can also be processed in the virtual mode of operation at a remote location. The working principle of monitoring applications that are present at stage 1 provides a setup for AR/VR model Field [4, 5] with a high degree of freedom and acts as the first source of function, followed by synchronization of the sensing unit with VR/AR modules. Usually, the synchronization units check the compatibility nature and model of reality modules [6] as the arrangements are passed in the form of segments.

If synchronization is achieved, all values can be tracked and monitored using biomedical sensors, thus giving rise to capturing units in the system, referred to as pre-processing units. Once the images or videos are captured, data will be encoded in the cloud, thus providing possible interaction between modules and individuals using the signal transfer approach. Further, the health conditions [7] are examined using tracking systems where the values are decoded and transferred to the down method. After processing the data, system planning and other stages with two or three-dimensional visualization can be processed.

1.1 Major Contributions

The major contributions of proposed biomedical signal processing technique with audio and visual techniques are as follows.

The foremost novelty of projected model is to realize all signals using VR/AR techniques with specified equipment type which is designed using wireless modules.

To integrate wireless sensors that is highly flexible under distinct environmental conditions where all diseases are identified in an applicable way.

To enhance smooth operating conditions by implementing swarm optimization algorithm in order to fix proper scaling rate in detection process.

The ancillary novelty is also added in realization of signals that occurs due to sensing drive capability that minimizes distortion rate of multiple signal points.

1.2 Literature: A Comparative Analysis

Nevertheless, for such high standard electronic components, the Bluetooth module is used in real applications as it will lower the distance, and the data cannot pass to the receiver within an appropriate time. Hence there is a need to elucidate the range values of constituents, and measurable devices [8, 9], where all package ranging and installation controllers are provided. Then again, by using defined ranges, neither the system is tested in any real-time application platform, nor any simulation model is described, which adds confusion to the implementation strategy. Thus to avoid such confusion in the system, a regression analysis is used in the prediction model [4], which is directly used for estimating real-time sensor values at multiple events. Due to various circumstances, a hypothesis test is made by recognizing daily activities with training recognized models. As the training is given for all models, only the defined boundary ranges will be used to measure offset voltage. Still, the exercise requires more activities, requiring more sensors to be installed at a high-cost factor. Table 1 provides a detailed comparison on state-of-art models with distinct objective functions.

1.3 Problem Statement

All the existing models provides real time information related to medical applications only by implementing biological sensors where it is necessary to use device installation. In the existing models it is not possible to perform any physical stint representations that can be realized within short amount of time. However some of the approaches provides cognitive applications that provides outcomes with new representative pattern thus realizing all infections. Consequently if reality applications are introduced in medical related fields then it is much easier to realize the infections and serious cause of infections can also be avoided using audio and video patterns. Hence the projected model implements a physical stint representation system using AR/VR applications that provides complete information about different infections which in turn leads to much easy way of diagnosis.

Table 1 Comparison of existing literatures

References	Application platform	Type of algorithm/ methods	Objectives		
			Minimization of distortion	Minimization of turning force	Representation of full scale
[1]	Sensing of multi-model applications	Machine learning	✓	–	✓
[3]	Clinical education applications	–	–	✓	–
[2]	Health monitoring applications	Rehabilitation training	–	✓	✓
[6]	Glove based health care applications	Glove reality model	✓	–	–
[7]	Biomedical applications	Bio measurement system	✓	–	✓
[8]	Daily activities applications	Poisson regression	–	✓	✓
[9]	Biomedical applications	Extended processing technique	✓	✓	–
[4]	Motion sickness	Long short term memory	–	–	✓
[5]	Cognitive applications	Cognitive workload	✓	–	–
[10]	Trauma applications	Hypnosis test based	✓	✓	–
[11]	Motion tracking	Fusion bases sensors	✓	✓	–
[12]	Industrial applications	Extended reality model	–	–	✓

2 Mathematical Model

The verification process is usually prepared using performance evaluation metrics which is carried out by considering convergence and robustness characteristics of a particular algorithm. Therefore, in the proposed method, algorithmic determinations are made to prove that FFO can be applied as a replacement algorithm that provides high support to all biomedical applications in virtual reality applications. Further, the designed system is subject to complex nature where the approximation is much needed in virtual segments, which is termed as the order of representation for virtual representations.

2.1 Complexity Analysis

Moreover, the performance evaluation is also carried out by examining the complexity of algorithms in real-time biomedical applications; hence a big-O-notation is used for such complexity analysis where it can be represented in mathematical terms.

$$Complexity_i = \sum_{i=1}^{n} O\begin{pmatrix} C_1 \; C_2 \\ C_i \; C_n \end{pmatrix} \tag{1}$$

where,

$C_1 \ldots C_n$ indicates complex intervals in virtual segments.

2.2 Robustness Analysis

Equation (2), which is framed as a minimization problem where all uncertainties are observed and added with total virtual segments, can be used to mathematically frame the robustness characteristics of the proposed method as shown below,

$$robust_i = \min \sum_{i=1}^{n} VR_d(i) + UC_i \tag{2}$$

where,

VR_d represents depiction of virtual segments

UC_i describes the insecure AR/VR links.

2.3 Working Functionalities

Input: Initialize the measurement angle with supplied voltage and offset voltage values and full-scale representation values of VR/AR l_i;

Output: Optimized values for realizing different portions using VR/AR with best fitness function values at a good control rate;

Step 1: At first, the objective function is constructed with the non-zero scaling factor

Step 2: Initialize the scaling matrix factors in x and y co-ordinates that must be followed by two-dimensional boundary index with $0 \leq i \leq 1$, and its deviation values with the indication of calibration and fitting scales;

Step 3: Provide the parametric falsehood values with the representation of distance in a systematic way for computing the distortion in VR/AR images;

Verify the value of change in previous and current state filter values using biomedical sensor signals $dist_i$ for identifying the location change values;

If the location change is higher do

Modify the filter values using force vector that is having different rotation axis with $1 \leq i \leq N$ into N number of rotated matrix values;

// Fitness phase

Update the fitness function with the correlation matrix by generating the antibodies using a degree of matching;

// Modified fitness phase

Select the best and worst smell factors with measurement of random values;

Update the modified fitness function with proper best and worst values followed by the rationalized initial position mechanism, and compute the new position values;

The time path of the initial position in x and y axis co-ordinates is updated;

$N_{new} = N_{old} + 1;$

End;

Step 4: If $(N_{new} < N)$ //Existing solution is replaced with the new solution

$f_i = f_{modified};$

$I_{min} = N$; //Attain the most feasible solutions for determining the overall best solution;

Increment the count N_{new} by 1;

Return the best overall solution;

End;

3 Results and Discussions

To examine the complete effectiveness of the proposed method with combined Fruit Fly (FF), two different scenarios are considered as follows,

Scenario 1: Scale factor measurements

Scenario 2: Distortion less operation

Scenario 1

The VR/AR models used in biomedical applications are used concerning full-scale measurement values. There is a need to measure the scale factor in this case. Two factors are calculated for realizing the objects in real-time, such as calibration and fitting scale, where both factors must be set at proper proportions. In case if calibration scale is not placed in the appropriate position, then line errors will be maximized; thus, the full-scale representation will be reduced. Further, if virtual reality segments are fitted in two axis scales, measured concerning the x and y axis, then nearly half of the full-scale reading is neglected. The abovementioned scaling process is denoted as a numb computation method as no values are obtained at this stage. This method is avoided in the proposed way by providing proper scaling values. Two scaling factor values are kept at a minimum for the simulation case, and a simulation study is carried out, as indicated in Fig. 1 and Table 2.

From Fig. 1 and Table 2 it is perceived that calibrated scale at the initial position is set at 0.5, and it is varied in the same value terms. On the other hand, fitting values are set at 2 points at the initial state of measurement, and for each significant part

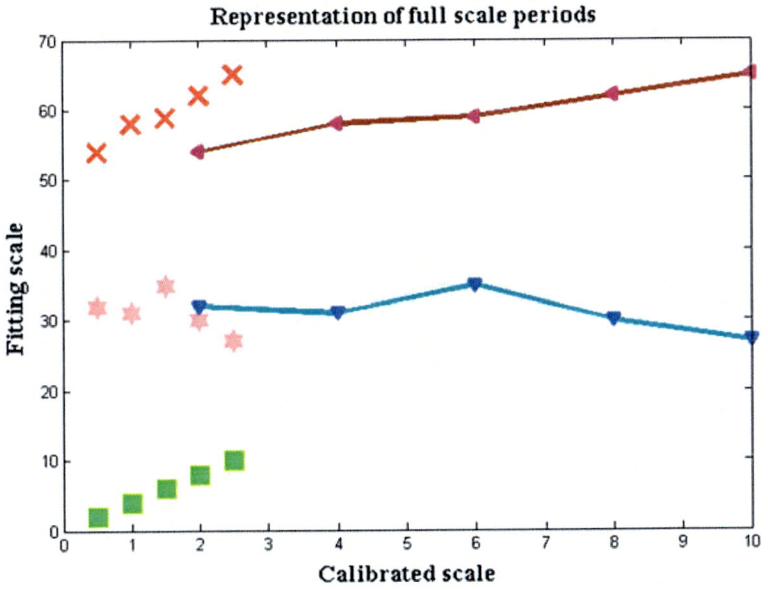

Fig. 1 Scale of representations

Table 2 Comparison of scale representations

Calibrated scale	Fitting scale	Percentage of full scale representation (existing)	Percentage of full scale representation (proposed)
0.5	2	32	54
1	4	31	58
1.5	6	35	59
2	8	30	62
2.5	10	27	65

is varied up to a factor of 10. Since all the values are provided in equal proportion, it is possible to offer a full-scale representation of marked segments. Moreover, a comparison analysis is made with the data set that is present in the existing method [7] by distributing the scale factors equally, and it is realized that much low full-scale VR/AR representation is achieved. This can be proved for the third scaling factor (i.e.) when calibration and fitting scales are set at 1.5 and 6 respectively. With the abovementioned scale factor the full scale representation of proposed method is 59% whereas existing method even with equal distribution provides only 35%.

Scenario 2

If VR/AR devices are used in medical applications, there is a need to provide distortion-less operation during the signal transfer stage. Thus, this case examines the amount of distortion, which is also expressed using a percentage of full-scale representations. Further, the distortion rate must always be less than 50% as medium illusion is provided in this value. But if the distortion falls above the additional 50% then the illusion of signals used for monitoring or diagnosing body conditions using biomedical signals will be lost entirely. Moreover, the illusion of VR/AR device models must be higher to achieve cognitive benefits, thus realizing all sensitive parts of the human body.

Additionally, if any sudden change in the behavior is present, it can also be recognized with high illumination capability, thus providing reasonable access to the monitoring devices. Figure 2 portrays the amount of distortion achieved with full scale factors. From Fig. 2 and Table 3, it is detected that the illusion rate concerning the number of segments in virtual reality pictures and videos is provided from 1000 pixels per second to 50,000 pixels per second. For each pixel value, the amount of distortion is observed, and the full-scale representations are directly taken from Scenario 1. With the changing illusion values, a comparison analysis is made, and it indicates that the proposed method's distortion is much lesser than the existing model [7]. Further, it is shown using a computer design model where varied pixels are directly provided in the image processing tool and then directed to the virtual reality toolbox. Using the provided direction flow, all the illusion rates in the proposed method provide a low distortion rate below 0.5. The same design model provides high distortion equal to 3 and then falls below a specific value. This proves that all problems can be easily identified at low distortion using the projected method.

4 Conclusions

Even though the data isn't actually stored anywhere, it can still be made accessible by reflective operations, which handle false variables after they've already been processed. An optimization technique using FF is provided with modified fitness function values that suitably turn the scale factor, offering a low distortion rate, for testing the proposed method of VR/AR applications. The factor of scale described above is included as one of the test cases, and the force vector approach is used to

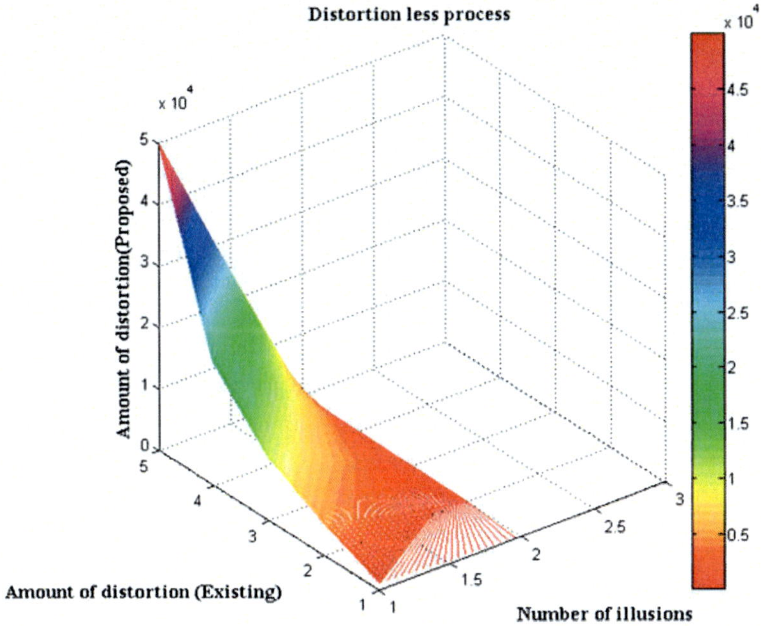

Fig. 2 Comparison of distortion with illusions

Table 3 Amount of observed distortions

Number of illusions	Amount of distortion (existing)	Amount of distortion (proposed)
1000	2.4	0.5
5000	2.6	0.4
10,000	2.9	0.5
20,000	3	0.3
50,000	2.8	0.3

guide the sensing apparatus in the right directions. The predicted approach using FF has been shown to be around 66% more efficient than the current process in all of the case studies that have been verified. In the future, a robotic model that uses machine learning techniques can be implemented in place of sensing devices, hence expanding the capabilities of the suggested approach.

References

1. Khadidos, A.O., Shitharth, S., Khadidos, A.O., Sangeetha, K., Alyoubi, K.H.: Healthcare data security using IoT sensors based on random hashing mechanism. J. Sens. **2022**, 17 (2022), Article ID 8457116. https://doi.org/10.1155/2022/8457116
2. D'Ippolito, F., Massaro, M.: A multi-sensor multi-rate algorithm for motor rehabilitation with augmented reality devices. In: 2016 6th IEEE International Conference on Biomedical Robotics and Biomechatronics (BioRob), pp. 759–765. IEEE (2016)
3. Bhattacharya, P., et al.: Coalition of 6G and blockchain in AR/VR space: challenges and future directions. IEEE Access **9**, 168455–168484 (2021)
4. Munirathinam, R., Ponnan, S., Chakraborty, C., Umathurai, S.: Improved performance on seizure detection in an automated electroencephalogram signal under evolution by extracting entropy feature. Multimed. Tools Appl. **81**(10), 13355–13370 (2022)
5. Salankar, N., Koundal, D., Chakraborty, C., Garg, L.: Automated attention deficit classification system from multimodal physiological signals. Multimed. Tools Appl. 1–16 (2022)
6. Almagrabi, H., Alshareef, A.M., Manoharan, H., Mujlid, H., Yafoz, A., Selvarajan, S.: Empirical Compression Features of Mobile Computing and Data Applications Using Deep Neural Networks, vol. 2022 (2022)
7. Putze, F., et al.: Editorial: brain-computer interfaces and augmented. Virtual Real. Front. Hum. Neurosci **14**, 144 (2020)
8. Gnacek, M., et al.: EmteqPRO-fully integrated biometric sensing array for non-invasive biomedical research in virtual reality. Front. Virtual Real. 3 (2022)
9. Teng, C.-C., Jensen, N., Smith, T., Forbush, T., Fletcher, K., Hoover, M.: Interactive augmented live virtual reality streaming: a health care application. In: Proceedings of the 2nd International Conference on Medical and Health Informatics, pp. 143–147 (2018)
10. Anitha, G., Manoharan, A., Manoharan, H., Ganesan, P.: A Survey of Security Issues in IoT and Fault Identification using Predictive Analysis in Industry 4.0, vol. 70, no. 12, pp. 99–108 (2022)
11. Gupta, A.K., Chakraborty, C., Gupta, B.: Secure transmission of EEG data using watermarking algorithm for the detection of Epileptical Seizures. Traitement du Signal **38**(2), 473–479 (2021)
12. Hasanin, T., Kshirsagar, P.R., Manoharan, H., Sengar, S.S., Selvarajan, S., Satapathy, S.C.: Exploration of Despair Eccentricities Based on Scale Metrics with Feature Sampling Using a Deep Learning Algorithm (2022)

Special Session 1: Digital Transformation of Contemporary Business Practices

A Review of Social Media Marketing on Digital Savvy Brand Shoppers

Sepideh Samadi and Imran Akhtar

Abstract Social Media (SM) has recently become a platform and a tool for both individuals and companies to communicate in a virtual platform. E-retails is another virtual platform where creates cybernetic communication between brands and customers. Companies create effective marketing strategy through both platforms to influence their customer behavior. On the other hand, digital savvy customers who prefer to take the advantage of saving time and have a convenience shopping experience, buying their desired products through e-shopping is another fact that makes digital marketers to implement new strategies to communicate with their customers. Thus, social media has become an effective tool to create an interaction to influence digital savvy shoppers. The purpose of this paper is to examine different aspects of social media as a tool that can develop marketing strategies and how they influence individuals who prefer to buy products through e-retails. This paper aims to provide a comprehensive review of the impact of social media marketing on digital savvy brand shoppers. The scope of this review encompasses various literature reviews that has been done to illustrate a framework design to represent the impacts that social media has on a specific target audience. The research methodology used for this review involves a systematic literature review of relevant academic and industry publications. The contribution of this paper lies in providing insights into the effectiveness of social media marketing strategies for engaging digital savvy brand shoppers and improving brand loyalty. The findings of this paper represent how social media can influence e-shoppers and have impact on their decision-making process. On the other hand, the findings suggest that social media marketing can enhance brand awareness and influence consumer purchase decision through advertising, user-generated content, and social media influencers. The review also identifies key challenges and limitations of social media marketing for digital savvy brand shoppers.

Keywords Social media (SM) · SM marketing · E-shoppers · E-retail supply chains · Decision making process

S. Samadi (✉) · I. Akhtar
Greater Manchester Business School, University of Bolton, Bolton, UK
e-mail: ss5res@bolton.ac.uk

C. Iwendi et al. (eds.), *Proceedings of ICACTCE'23—The International Conference on Advances in Communication Technology and Computer Engineering*, Lecture Notes in Networks and Systems 735, https://doi.org/10.1007/978-3-031-37164-6_46

1 Introduction

Traditionally, going to malls and do even a simple window shopping was a habit where people would like to take it as a hobby. With the raise of digital platforms such as Instagram, Facebook, Tiktok and other social medias, people started to spend most of their time, even at malls on surfing into their social media pages. By the rise of social media platforms, comes the digital retails and shops, where it gives the opportunity to customers to experience digital shopping rather than physical ones [1]. This trend is not only being limited to digital buyers, but it has been observed that even digital screen shopping is also taking place by customers, no matter if they add any product to their cart or not.

Social media as a strong tool has made its way in a fast phase to communicate with customers. On the other side, companies have also come to the reality of its effect on customers, especially when it comes for them to decide on buying a specific product. thus, other than creating a digital shop, a social media page by companies have also been created and a platform on digital shops, to communicate in a faster way with their customers, hence the communication leads them to influence their decision making. Examples of these types of digital shops can be Amazon, E-Bay, Noon, Alibaba, and others, where different types of trades, B2C, B2B, C2B and even C2C is also being seen in these platforms. Different brands that are selling their products through these platforms are not only limiting themselves in providing digital shopping to their customers, but they also create an awareness of their products being existent in these digital retailers through social media. Thus, the brand's social media has become a place to familiarized and aware customers of the product availability in certain digital shops [2].

1.1 Introduction to the Research

The presence of social media has changed customer's decision making in an outstanding way. Not only social media but surfing in digital shops and retails have also brought a considerable shift in customer's decision-making process. Traditionally, customers would realize the need of a certain product, though going to shopping malls or the retail shop, searching for their desired product, trying it, evaluating it with other shops, and finally leading to act of buying, makes customers to go through a long process of decision-making process [3]. Meanwhile, evaluation is somehow complex and complicated when physical shopping takes place as customers require to spend lots of time searching for the product, they want to be able to compare and evaluate variety of products with each other, thus making the final decision of buying to take place.

As a result of this, social media comes to the picture to make customers realize and find more products while saving time and at the same time customer's feedback and comments under each post would make it faster for them to evaluate the products through digital word of mouth. Digital shops and retails have also brought this process faster for customers as customer would be able to compare, evaluate and make their final purchase through digital shops in a specific category in this platform [4]. Hence, decision making process has faced a great shift and phase comparing to traditional buying in digital retails and shops and social media is a tool where customers can get more feedback and increase the speed of product evaluation through digital word of mouth in this platform. This paper will examine the impact that social media has on customers who are digital savvy shoppers' decision-making process and identify the influences that this digital platform can have on customer's purchase behavior in decision making phases.

1.2 Research Objectives

With rise of digital retails, comes the digital platforms for shopping and social media as another platform where can make customers decision making process easier, yet complex in buying a specific product. Thus, some questions would stand out to clarify how customers, who prefer online shopping over physical shopping decision making process would be influenced by social media. Below objectives are aimed to be reached by this research paper:

1. To identify the power of social media on customer's decision-making process.
2. To study the effectiveness of social media tools on customer's shopping behavior.
3. To determine the impact of social media on e-shoppers purchase behavior.

Social media, comparing to other traditional marketing communication tools implied different marketing strategies to communicate and reach their customers. Thus, this paper will examine the impacts that this communication tool has on digital savvy shopper's decision making in purchasing products through digital platforms.

1.3 Research Objectives

While there is extensive literature available on the impact of social media marketing on consumer behavior, there is a significant gap in the literature regarding the specific influence of social media marketing on digital savvy brand shoppers. Most studies have focused on the broader impact of social media marketing on consumer behavior, without specifically examining the behavior and preferences of digital-savvy consumers. This finding highlights the need for further research that specifically explores the relationship between social media marketing and digital-savvy brand shoppers, including their attitudes towards different social media platforms and the effectiveness of various marketing strategies in engaging this audience. Such research can provide valuable insights for marketers seeking to optimize their social media marketing efforts to better reach and engage with this tech-savvy consumer segment.

1.4 Research Contribution

From a theoretical perspective, this paper highlights the importance of understanding the behavior and preferences of digital savvy consumers in the context of social media marketing. The review identifies key factors that influence the effectiveness of social media marketing, including platform selection, content creation, and influencer marketing. These insights can inform future research in this area and help marketers better understand how to engage with this tech-savvy consumer segment. From a methodological perspective, this paper provides a systematic and comprehensive review of the available literature on social media marketing and digital savvy brand shoppers. The rigorous approach ensures that the review is comprehensive and reflective of the most up-to-date knowledge in the field. From a practical perspective, this paper provides valuable insights and recommendations for marketers seeking to optimize their social media marketing strategies to engage with digital savvy brand shoppers. The review identifies key challenges and opportunities in this area, and provides guidance on effective marketing strategies, such as user-generated content, influencer marketing, and targeted advertising. Overall, this paper's contribution highlights the importance of understanding the behavior and preferences of digital savvy brand shoppers in the context of social media marketing. By providing valuable insights and recommendations for marketers, this paper can help businesses better engage with this increasingly important consumer segment and ultimately improve their bottom line.

2 Literature Review

Talking about marketing strategy makes us think of a traditional marketing strategy that takes place as a first step in all businesses. Starting of finding the right target customer to create positioning strategy requires considerable attention. Before, traditional marketing strategy steps were taking place in a long period of time, whereas with the existence of digital world and Internet, the process has become faster, easier, and more accurate. Companies can use digital platforms to find their target market and practice a benchmarking by comparing what their competitors are doing as the same time. Thus, this is where digital marketing comes into picture to create a new strategic function and approaches, designing, planning, and implementing marketing strategies [5] (Fig. 1).

Conventional marketing is the first approach where being given to digital marketing as a definition that all its function and strategies takes place on the Internet. However, the tools and functions are suitable for digital world, but it has become

Fig. 1 Overview of digital marketing (DM) tools [6]

appropriate for marketing purposes to develop channels, languages and formats that leads companies to design their marketing strategies. Today, digital marketing has become a tool that brings customers, suppliers, and distributors all together and make the communication process faster, easier, and more effective. On the other hand, not only communication process has become a function in digital marketing, but all other marketing functions, such as pricing, branding, advertising, competition virility and other functions that are taking place in digital marketing. As the word, "Digital" implies all activities that is related to marketing purposes can happen through a virtual platform.

However, the system requires a technological updates and advances, it has become a trendy and vital tool for all marketing purposes. As a result of this, all organizations must ensure that their company and function is meeting all technological advances and updates to maintain this trendy, significant and vital tool, and platform for their business function. Other than organization's functions in all areas through virtual platform, communicating with customers has also become easier and now it is a necessity for all companies to create awareness, communicate, retail, and maintain their relationship with customers, though lacking it can bring a lot of inevitable results [7]. Meanwhile that digital marketing has become a platform where organizations can have all the necessary functions in relation to their customers, a digital marketing plan is a strategy where it requires an organization to take into consideration to implement the right digital marketing strategy to the right customer target segment. Social media platforms have become a place where all marketing strategy functions can easily take place. This platform has given the space and ease to marketers to design and plan the right strategy to reach out to their customers, through the right platform and in a right time. At the same time, this platform has made it different for customers as well. Digital marketing has impacted customer decision making process as well, while new digital marketing strategies is required to be implemented [8].

2.1 Social Media Marketing

Social medias are mainly known as a social networking platform, where individuals can create a profile, upload pictures, write blogs, podcasting and participate in different virtual social functions. The main purpose behind social medias were the networking, however, the platform now is being used for other purposes as well. Social media platforms were used by computers and laptops, and now all individuals can have access to it through their phone, and even have more than one account (Fig. 2).

On the other hand, not only an individual personal profile is being made by individuals, but people started to create pages and profiles where they can present their small business, talent, art works or any other capabilities they have, and this is where it comes the greatest social media bloggers who have millions and millions of followers [5]. As a result of this trendy increase, bloggers started to create marketing campaigns, dealing with brands to advertise them. This has become an income source and a kind of career for bloggers in a way that companies use them to advertise their products to a great mass number of audiences. Traditionally, a celebrity was being used in media advertisement, using a specific product to influence customers, whereas now social media bloggers and their activity in social media platform is replacing the traditional way of celebrity advertisement. With the growth of fast trending marketing activities in social media, companies started to create new social media marketing campaigns and strategies to influence their target audience, and this is where brands underway creating their own social media page and have social media events. This strategy has made customer's engagement to be more and visible with the respective brand, hence decision-making process of customers is also being impacted.

Fig. 2 Social media marketing practices 6

As mentioned earlier, the whole process of decision making in physical shopping were being divided into different stages. Now, with the existence of social media, the stages in decision making have also faced an incredible shift. The complexity of decision-making process where companies and brands must take into consideration is that now customers have lots of information, comparison, and evaluation substitute through their social media, and even negative or positive word of mouth can be created through this platform. As a result of this, we can say that social media is a strong and powerful tool where consumer behavior phases and elements has been significantly changed and impacted and this is where marketing strategy should implement major changes and strategic plans [8]. This is exactly where new Social Media Strategy comes into picture where companies and brands need to carefully apply it and not forget to update it with the trending activities taken place in social media due to the strong impact that social media has on customer's behavior, and it can easily lead them to make the final purchase or create a negative impact on their attitude.

2.2 *Impact of E-Commerce on E-Retail Supply Chain (E-RSC)*

A supply chain is defined as a three or more organizations that are directly linked with each other to drive the product manufacturing flow or service providing to customers. Managing the supply chain is essentially important as it is related to all the activities that takes place among the chain members and organization. These relationships vary from a single transaction between different elements to the complex one, as it can also be related to direct or indirect channel of distribution [7]. An indirect channel will lead customers to get the product or service needed through different channels such as physically visiting a retail shop and purchase the product, while the indirect channel is more of a customer and producers' relationship directly with each other and this is where so many channels are not participating in the transformation and delivering of the product or service to the end user. With the existence and revolution that technology has created in all aspects of the business, supply chain is another sector that has been influenced greatly by the industry 4.0 revolution. This is where the impact of e-retail and e-commerce comes into the picture and has shifted the supply chain and distribution channels function to face a great shift in their transformation of the product to delivering the final product to customers. Thus, this revolution requires the supply chain management to implement a new strategy, tools and techniques that can be unique [9] (Fig. 3).

One of the main changes that technology and 4.0th industrialized revolution has brought is the important role it has on reducing supply chain costs. If the manager creates an effective and efficient tool, using technology and e-commerce, it can lead the supply chain function to reduce lead time, increase rate of information availability, delivering the products to the right customers at the right time (Just-in-Time- JIT). This method and the usage of technology can also influence the inter-organizational

Fig. 3 E-commerce and E-retail supply chains (E-RSC) 6

communication and reduce time interims with further cooperation of the channels with other sectors. Due to the revolution and impacts that technology and 4.0th industrialized revolution has brought into the supply chain, this has affected the consumer's purchase behavior as well, thus, the location where customers were getting the final product from the producer is also transformed to be through online platforms [8]. This is where the entire process of producing a product to giving to final customers has faced lots of changes with the existence of industry 4.0 revolution.

2.3 E-Retails and E-Shops

4.0th Industrialized Revolution is characterized by new and innovative technologies that brings a shadow or better say blur the lines and connections between physical and digital worlds. E-Retailing is a subdivision of e-commerce where all business trading happens through the Internet. The place where selling a product through online shopping and an Internet is what E-Retail is being defined. Comparing E-retail or shops to traditional stores is that the customers will not get the tangible experience of touching the products when making their purchase, while on e-shopping customer can only evaluate the product by looking at their pictures but have all the information needed included underneath the product.

The main purpose behind all e-shops and e-retails were to save time, cost and space and make it more convenient for customers to order any product their want online. Problems were still there as customers might not trust the online platform as were afraid of phishing or identity thefts, thus it was not a very practical manner taken place by customers [10]. As technology started to dominate all business functions, customers underway practice using online platforms as well, and this is where social media practical behavior has come into the picture. So that is why we can say that the existence of social media made customers to be ready and be more aware of buying products through online platforms and online stores. Amazon and E-Bay were the first online e-shops that customers were aware of it and would trust, and then with the social media promotions, other e-retailers and e-shops started to penetrate customer's awareness and market to sell different products and brands. This is where other e-retails started to collaborate with brands and sell their products to customers online [11].

Meanwhile, as the customer's awareness regarding e-retails and e-shops increase, here it comes to brands to see the cost effectiveness of selling their products online, thus collaboration with online e-retails and e-shops started to boost. As a result of this, new marketing strategies by e-retailers requires to be implemented. Some e-retailers and e-shops would provide the brand an exclusive marketing strategy and branding to maximize their e-shop brand customer's engagement and the product or brands awareness. One of the examples of this strategy can be given to company named Noon in UAE. Noon started to collaborate with brands and created its online shopping application and e-store. After a while of being successful in getting customer's trust and become successful, Noon started to have its own brand selling in its e-shop and e-retail. E-shops and e-retails were not that much popular long a long and right now, due to the COVID-19 pandemic, the image and usage of e-retails and e-shops has been seen significantly. The global pandemic caused e-shops and e-retails to be more visible and this is where customers started to realize that online-shopping through e-retails and e-stores can also be a good, easy, cheap, and reasonable option. Thus, the global pandemic brought the new opportunity for e-retailers and new customer behavior and engagement with customers. On the other hand, comes customer decision making process, where the steps and phases in the process has faced changes through e-retails and e-shops. This is where traditional decision-making process of customers varies when a brand or a product is going to be purchased through e-shops. And again, this can be implied as a shift and change that digitalization brings into customer's behavior.

2.4 *Digital Savvy Shoppers*

Digital Shoppers are being defined as customers who are technologically savvy and most of their purchases are with interaction through online platforms with retailers. The whole purchase behavior process for e-shoppers, including awareness, choosing, transaction, delivery, and aftersales support take place on digital channels. Thus, the whole purchase journey for digital shoppers occurs through digital technologies. Since this segment is technologically savvy, their purchase behavior and even decision-making process is different from regular customers. This type of shoppers would get more information about the desired product as they know search engines and finalize all the reviews and comparison online. On the other hand, product features and characteristics or any other relevant information about the product is very important for them to be seen in the online platform [12]. This type of customers is known as a different target segment for companies and brands as approaching strategy would be of a difference, thus relevantly, the marketing strategy would be also diverse from regular shoppers. E-shoppers have also given a new and novel way to retailers to track their demographics, behavioristic and physiographic background without them noticing it. Hence, a great deal of customer data would be available for e-retailers and brands that are collaborating with them for future marketing purposes.

2.5 E-Shoppers Decision Making Process

As mentioned earlier, digital shoppers are in different purchase journey from regular shoppers. Such difference, make digital shoppers decision making process to be different from regular physical shoppers. Before explaining this, let us have a quick glance of the whole decision-making process comparing to regular physical shoppers and e-shoppers. Below figures compare the two types of decision-making process in which Fig. 4 illustrates the physical shopper's decision-making process and the e-shopper's decision-making process. When talking about decision making process, the first step is realizing or recognizing the need. This recognition can take place either in advance, as an individual realize that he needs a specific product or by surfing on the Internet or going to shopping malls, an impulse or real need of a product will be raised.

Information search is the second step-in decision-making process where a customer would search for the information relevant to the need or a product or service of his need. Different platforms are being in used for these purposes as now Internet has made it easier for customers to search for all alternatives that they are looking for. Once alternatives are being found, the customer would rather evaluate and compare them with each other, and this is where the step leads the customer to take the final step of purchasing. Being satisfied or not satisfy would direct the customer to have post-purchase evaluation and behavior. The whole decision-making process of a physical shoppers can happen during a specific period, and even some of the steps such as evaluating the alternative, and search for information would be different. This is where physical tangible comparison of product would take place [10]. In contrast with physical experience and decision-making process of buying products, comes the digital shoppers' decision-making process. The same steps would take place by digital savvy shoppers whoever the amount of time that a person would spend in each step might be different [11]. On the other hand, the method and information search and evaluation would be different. In this case, the digital savvy shoppers would

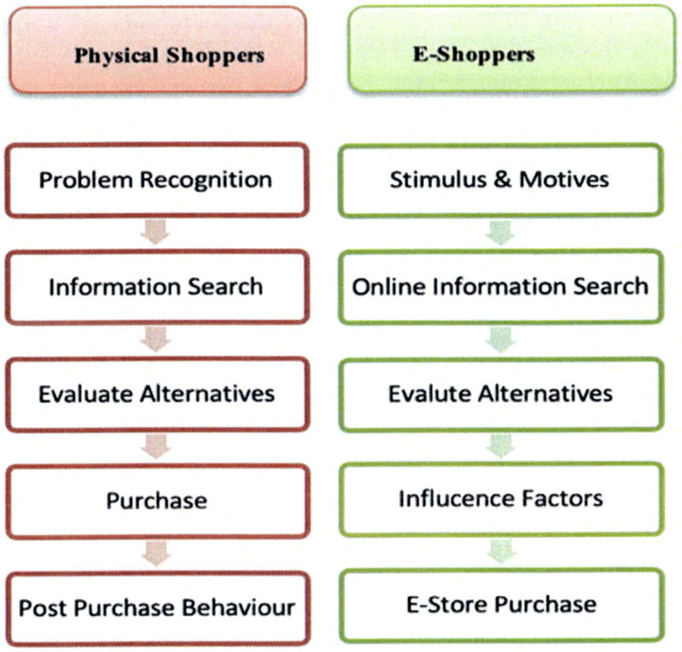

Fig. 4 Flow of decision-making processes 6

make all his information research through online platforms where users feedback and reviews would be also taken into consideration, while a physical shopper might not have this much of information and references as he would be able to get his surrounding people's and old users as a reference of feedback. As a result of this, evaluating products through online platforms would be easier while competitive for brands.

2.6 Social Media and E-Retails

Both social media and e-retails and e-shops have something in common with each other and that is the Internet platform. Both are tools that are communicating with people through online podiums and all individual interaction is happening in a virtual way. The difference that social media has with e-retails is that social media is not mainly focusing on purchasing products or services and it has other purposes as well, however e-retail's main and core purpose is to create an online shopping experience. Thus, customer's interaction and engagement through both platform is different from each other [9]. Since social media is an online platform for networking, brands have taken this into consideration and have taken advantage of this platform in a sense that having customer interaction and engagement has become easier and in different way. On the other hand, e-retails are also into the phase to create a social media page and make this interaction even more. The main core purpose of both platforms is to have as much as customer awareness, interaction and changing their attitude or creating a positive word of mouth among their users and followers. On the other side of the story, here comes the bloggers where recently, they have become a third party to create relation and interaction between brands and their targeted customers. E-retails have also used bloggers as a mean to increase purchase traffic through their platform where the bloggers would promote the e-retail shopping behavior [7].

Taking decision making process into consideration is also another factor that both social media and e-shops are in common, while the only difference is that sharing and reading reviews and feedback from a specific product would be in different methods. The decision-making process in social media consist of an influence (bloggers) while e-retailers are mainly a platform where people can buy and even evaluate different product with each other. The purchase behavior in e-shops is quite different from social media as social media is the main platform that can influence customer's information search, evaluation, and purchase action in different way while e-retail will lead customers to the final purchase action. As a result of this, we can conclude that both social media and e-shops can be a complementary platform for each other in a sense to enhance more customer interaction, engagement, and participation.

2.7 *COVID-19 Pandemic and Online Retails*

Since the hit of global pandemic of COVID-19 and its effects on the whole world, the plague brought a new trend and way of living in all sectors, such as business, airline and tourism industry, education, and consumer behavior. The pandemic force has shift everyone's life to adapt their behavior from having physical connection and behavior to have an online adjustment in all sectors. Purchase behavior is one of the areas that the pandemic has significantly influenced. People, disregard of their generation group had to adapt themselves with the usage of digital platforms. Social medias on the other hand, has emerged their existence more than before [12]. Since people had to stay at home, have less physical communication, they started to increase their communication manner and networking through social media, and this is where brands, companies and all sectors get the opportunity to develop a new strategy reaching to their target audiences. This is exactly where people started to get to know online shopping platforms more than before and now even social media's impact has become substantial more than before. As a result of this, we can conclude that with the pandemic phenomenon, the impact of social media on e-shoppers has reached its maximum.

3 Research Findings

Reading through excessive literature review on social media and online shopping behavior we can conclude that social media has become a significant tool and platform influencing customer's behavior, specifically the e-shoppers. Social media is now a complementary platform for e-retailers to reach, communicate and engage customer's behavior toward purchasing a specific product, especially its influence has become more outstanding during COVID-19 pandemic.

Moreover, customer's decision-making process has also impacted by social media and e-shops. Figure 5 illustrates the final literature review framework of this paper where it shows the relationship between social media and e-shops, and the decision-making process of customers through e-retails and e-shops. Summarizing the above framework can lead this paper for the bellow findings:

- Social media has brought more option of information search and evaluation progression in the decision-making process.
- Social media and e-retails are now in parallel use by e-shoppers.
- Social media and e-retails can significantly influence customer's decision-making process.
- Post purchase evaluation through social media and e-retails is being evaluated and influenced by online bloggers and other customer's feedback and comment.

4 Conclusion

For business, brands, and shops to be able to sell their products to their customers, they must know their potential customer behavior very well. Not only the knowledge of their behavior is enough, but decision-making process is an important element where brands and marketers can find ways to influence each step-in order to attract, retain and create customer loyalty. Due to the technological advancement, the creation of social media and e-shops has made it easier, yet complicated for companies to reach their customers, communicate with them, and influence their purchase behavior. Social media is an effective tool where online shoppers are using this platform as a first and main tool to buy products and get influenced through it. This paper makes a significant contribution to the field of social media marketing by providing a comprehensive review of the impact of social media marketing on digital savvy brand shoppers. The paper's theoretical contribution lies in synthesizing and analyzing relevant literature to identify key trends and insights related to the effectiveness of social media marketing for engaging digital savvy consumers. The paper's methodological contribution lies in the use of a systematic literature review approach, which ensures a rigorous and comprehensive analysis of the available literature. Finally, the paper's practical contribution lies in providing valuable insights and recommendations for marketers seeking to optimize their social media marketing strategies to engage with digital savvy brand shoppers.

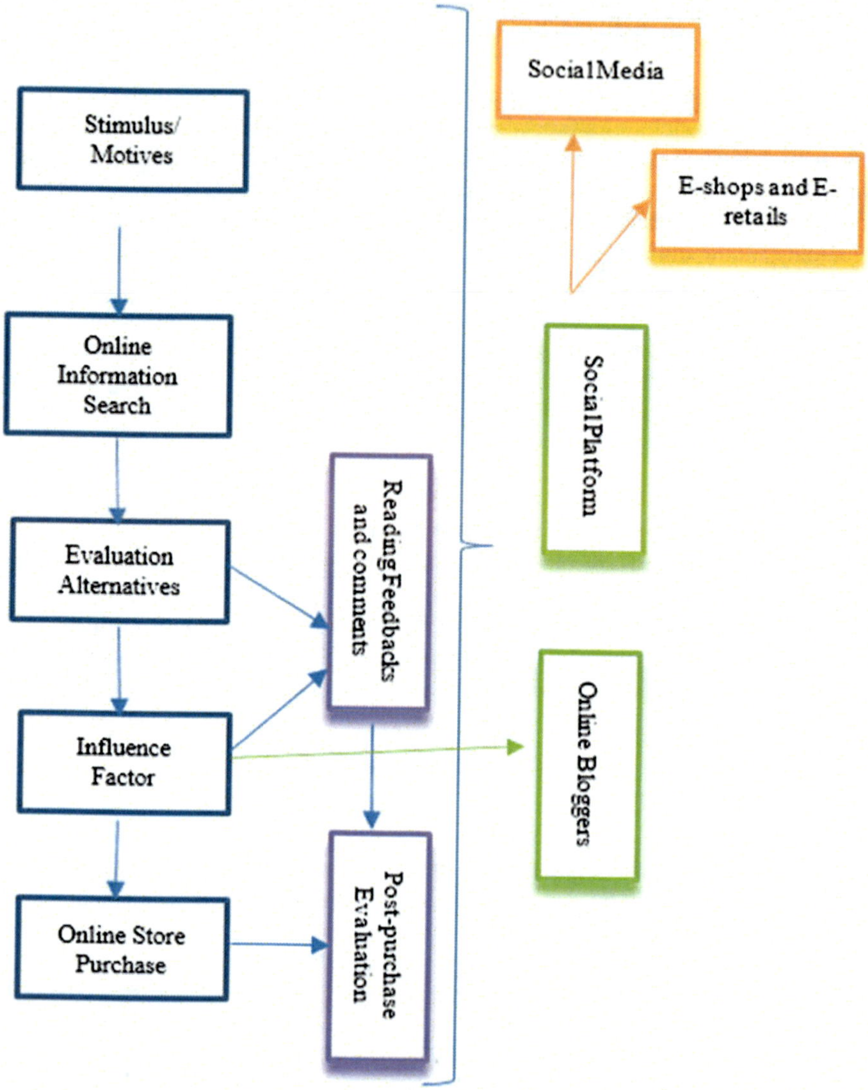

Fig. 5 Impact of social-media on E-shoppers decision making process

References

1. Laur, G.: Social media marketing. Asian J. Market. Commun. **22**(2), 189–214 (2019)
2. Weinberg, T.: The new community rules: Marketing on the social web. O'Reilly Media, Inc. (2019)
3. Goh, K.Y., Heng, C.S., Lin, Z.: Social media brand community and consumer behavior: quantifying the relative impact of user-and marketer-generated content. Inform. Syst. Res. **24**(1), 88–107

4. Kim, A.J., Ko, E.: Do social media marketing activities enhance customer equity? An empirical study of luxury fashion brand. J. Bus. Res. **65**(10), 1480–1486 (2019)
5. De Veirman, M., Cauberghe, V., Hudders, L.: Marketing through Instagram influencers: the impact of number of followers and product divergence on brand attitude. Int. J. Advert. **36**(5), 798–828 (2019). https://doi.org/10.1080/02650487.2017.1348035
6. Pushpalatha, M., Roshni, S., Ashwini V.: An Overview of Digital Marketing Trends. Int. J. Recent. Adv. Multi. Res. **6**, (2), 4659–4661 (2019)
7. Harsono, A.: The impact of E-commetce in supply chain management at Dell Inc. J. Multidiscipl. Eng. Sci. Technol. **1**(3)
8. Snyder, R.: E-commerce and supply chain manement. Am. Soc. Bus. Behav. Sci. **20**(1), 237–242 (2018)
9. UK Retail E-Commerce Sales to Reach £60 Billion.: http://www.emarketer.com/Article/UK-Retail-Ecommerce-Sales-Reach-60-BillionThis-Year/1012963. (2015). Accessed June 2021
10. Jeffrey, R.: Marketing in the Age of Subscribers, p. 2013. John Wiley & Sons Inc., Fans & Followers (2013)
11. Karve, S., Shinde, S.C.: Effectiveness of social networking sites. IBMRD J. Manag. Res. **2**(1), 2013 (2013)
12. Lee, C., Fung, S.: Online social networks and ecommerce. Student paper, TS085: Ethics and Law on the Electronic Frontier, MIT (2019)

Online Buying Behaviour of UK University Students

Farag Edghiem, Mohamed Shrief, and Siren De La Silva

Abstract Using a quantitative approach, the study is an attempt to comprehend the online buying behaviour of university located in northwest, Greater Manchester, UK students after the COVID-19 pandemic. 93 students were surveyed through questionnaires and seven in-depth interviews were conducted to collect data in relation to purchasing habits, behaviour and preferences of the students. The study revealed male students were more inclined towards online shopping than female students. In terms of frequency, it was noted that the number of times the respondents shopped and the time spent on shopping was more in the case of female respondents as compared to the male students. The study also identified reasons for shopping online and time-saving and convenience emerged as the most prominent reasons among the respondents. The study contributes to the existing literature on online buying behaviour of University students. It can assist marketers develop the strategies to increase their sales by enhancing the shopping experience of their prospective buyers. In addition, the study can serve as a basis to understand the buying behaviour through the aspects of economic, social, political, anthropological, etc. The use of quantitative method in the study enabled identifying the factors that have a significant impact on the online buying behaviour of consumers and the same can be used by strategists and companies to increase their profitability in the long run.

Keywords Buying behaviour · University students

F. Edghiem (✉)
Manchester Metropolitan University, Manchester, UK
e-mail: f.edghiem@mmu.ac.uk

M. Shrief
Arab Academy for Science, Technology and Maritime Transport, Giza, Egypt

S. De La Silva
Institute of Management Greater Manchester, Bolton, UK

C. Iwendi et al. (eds.), *Proceedings of ICACTCE'23—The International Conference on Advances in Communication Technology and Computer Engineering*, Lecture Notes in Networks and Systems 735, https://doi.org/10.1007/978-3-031-37164-6_47

1 Introduction

Online consumer buying behaviour is a study of customers in which marketers determine which individuals, groups, and organizations can be targeted and satisfied through the online delivery and presentation of goods, services, or experiences [1]. It was also explained as a set of activities that involve the purchase, and use of goods and services because of a customer's mental, and emotional behaviour [2]. Consumer behaviour is important to understand across any type of business, as consumers are continuously changing and varying. Hence, a failure to understand the preferences, needs, and wants of the customer can result in a loss of profits. However, understanding the behaviour of the consumer can be considered challenging [3]. Therefore, many marketers' decision is to focus on customer buying decisions instead. Consumer buying behaviour has changed dramatically due to the unfortunate event of the COVID-19 outbreak [4], as shoppers shifted most of their buying habits online. Moreover, many businesses took the path of stores closure. Thus, many stores found difficulty in meeting the customers' demand while they kept their online stores open [5]. According to [6], people tended to focus more on purchasing first-need products instead of luxury products as a reason of thinking of what can come after the pandemic. The consumer buying behaviour change was a result of the online transaction becoming a common practice [7]. Zaky et al. [8] believe that a panic buying phase distinguished between consumer behaviour before and after the COVID-19 pandemic.

The sector of higher education was one of the many sectors affected by the COVID-19 pandemic. Edghiem et al. [9] stated that the UK's higher education sector went through profound changes because of the COVID-19 pandemic. This raised the desire to study the university B students' buying behaviour online post-COVID-19 pandemic, their shopping habits, reasons for choosing to shop online rather than in-store, and the university students' opinions on shopping online. Hence, this paper aims to understand the online buying behaviour of University B students post-COVID-19.

2 Literature Review

In this section, we aim to explore the relationship between online consumer behaviour and its associated impacts. In doing so, an extensive literature review has been conducted to examine existing research on the topic, discuss key reports and studies, and determine the key elements and implications. The following literature review will provide an overview of the existing research surrounding online consumer behaviour and its associated implications.

2.1 Introduction the Concept of Consumer Buying Behaviour Online

The term consumer buying behaviour online was recently adopted [10]. It is considered a convenient type of shopping as consumers do their shopping in their comfort zone and chosen time. Online consumer behaviour is the process of how consumers interact with and experience products, services, and brands through digital channels. It's estimated that 86% of the world's population is expected to conduct their shopping online in 2022 [11]. Online shopping also allowed businesses to interact with their customers more and in a cost-effective manner without the need of bearing the costs of an actual store. Additionally, it allowed marketers to understand and stay up to date with trends in online consumer behaviour. With the aid of recent technology, such as artificial intelligence (AI) and machine learning, marketers can analyse data to gain insights into consumer needs, wants, and behaviours. Moreover, organizations use the data collected from their online buyers to improve their business creating more effective products, campaigns, and strategies that improve their return on investment (RoI). Sethna and Blythe [12] advocated that studying consumer buying behaviour online is fundamental for marketers to understand the reasons and the means of peoples' buying decisions, which is important in constructing a valuable marketing strategy.

Consumer behaviour is an activity that customers undertake when acquiring a product or a service [13]. Moreover, online consumers have accessibility, privacy, and convenience at any desirable time or place with unlimited options and an abundance of information; critics and reviews on each product or store that they decide to shop from. Online consumer behaviour takes into account the factors which influence the decisions and activities of consumers when using the internet for purchases. These factors include but are not limited to, purchase intention, attitude toward the product, trustworthiness, and security [13]. Through a detailed analysis of the browsing context, the effects of consumer behaviour, navigation, specific characteristics of the product, and user review scores, online retailers are able to understand the customer's behaviour [14]. As time progresses, online retailers have begun to use customer analytics, and customer relationship management in order to better understand who their customers are and what they value [15]. By accessing customer analytics, companies are able to develop effective online marketing strategies in order to reach their target customer base.

2.2 Buyer Behaviour Models

Research and analysis have continued to be conducted in order to effectively attract, engage, and retain customers. Businesses use buyer behaviour models to better understand and anticipate their customers' needs. Buyer behaviour models are important when it comes to marketing and understanding why people purchase products or services. For decades, marketers have been studying buyer behaviours through various models that help to explain why people perform certain purchases. Buyer behaviour models help to understand the decision-making process during customer purchases. Generally, such models involve the four stages; Awareness, Interest, Desire, and Action (AIDA) [16, 17]. In the Awareness stage, the customer recognizes a need and begins to research possible solutions. In the Interest stage, the customer starts to search for more information about some of the options and compares them. In the Desire stage, the customer puts a preference for one of the options. In the final stage, Action, the customer purchases the selected product or service [18].

Buyer behaviour models aid in analysing consumer decision-making in response to marketing variables such as product characteristics, price, advertising, and promotions. These models are useful to marketers and consumer researchers to understand consumer preferences and behaviour, as well as to generate insights on the best way to reach existing and potential customers. Moreover, they are used to identify and predict consumer behaviour. The purpose of creating such models is to understand how people come to make buying decisions and how different factors influence their decisions. These models take into account various aspects that may influence buyers, such as cultural values, geographical, economic, and other factors [19]. Consumer behaviour models can be separated into two major categories: traditional and contemporary [20]. A traditional model was introduced by [21], including the learning model, which is based on the belief that the customer is bounded by satisfying basic, and learned needs. The sociological model is based on the idea that someone's social status and influence in a societal group can affect their purchasing decisions. The psychoanalytic model argues that conscious, and subconscious influences are both involved in driving consumer behaviour. The economic model considers that consumers meet their needs at the least cost possible. Contemporary models focus more on the process of decision-making to purchase a product or a service [22].

Contemporary models include the Black Box model also named as Stimulus-Response model in which the customer responds to various stimuli and transforms these stimuli into responses on which customers base their purchasing decisions. Howard-Sheth's model also focuses on the concept of stimulus-response but goes beyond the Black Box model, discussing it in a more comprehensive manner, and in-depth exploration of the input and output variables, as well as the perceptual and learning constructs. The model suggests that consumer decision-making is a multi-step process including problem recognition, information search and evaluation, purchase decision, and post-purchase evaluation. Engel-Kollat-Blackwell's (EKB) model outlines five rational decision-making steps in the process of purchasing a product or service. Information collection and evaluation are utilized by consumers when making their purchases [23].

3 Methodology

This paper adopts a qualitative research design method based on a survey questionnaire to focus on studying consumer buying behaviour, targeting university students. Qualitative research based on a survey questionnaire is one effective way for researchers to garner information, and understand people's beliefs, attitudes, and behaviours. As a general approach, this type of research design allows researchers to present individuals with a set of questions, giving them the opportunity to provide feedback based on predetermined criteria [24]. These types of surveys are far-reaching, as they can be conducted both in-person and electronically, thereby increasing the potential response rate. Furthermore, since surveys collect data from a larger sample size, they can provide a more well-rounded view of the targeted population. Thus, the use of survey questionnaires as a qualitative research design method is an effective technique for data collection for the online consumer behaviour of university students. The questions posed on the survey are relevant to the researched topic in the study and based on previously published research and theory [25–27]. The survey is also designed to ensure response bias is avoided, and appropriate for the targeted population. Hence, the results concluded in this study can provide a wealth of data that can be further analysed, interpreted, and used as a building stone in future research.

A convenience sampling method was followed to collect 93 responses, sending online and offline questionnaire forms, as well as one-on-one interviews with seven university students. This type of sampling is considered a non-probability sampling design. As it depends on the available members of the population to form a sample size. Two common types of convenience sampling are opportunity sampling and quota sampling. Opportunity sampling is a type of convenience sampling in which participants are selected based solely on their availability and willingness to participate. Quota sampling is structured in a way to ensure the sample is representative of the population by selecting from pre- determined subgroups with predetermined amounts of individuals from each subgroup (Bryman, 2016). In this paper, we relied on the opportunity sampling method in collecting the sample size. One-on-one interviews were conducted with students and considered as depth interviews. Which are unstructured interviews to get a better understanding of the student's responses to the questionnaire. Zhang and Wildemuth [28] posit its effectiveness as it can be useful in unique problem situations as well as collecting detailed understanding around complex behaviour. The questionnaire responses and one-on-one interviews were conducted over a period of six months. The duration of the data collection ended in November 2022. The length of the interviews varied between 20 and 40 minutes, which was considered sufficient based on the questionnaire's number of questions. The software used for forming the questionnaire is Google forms, while the ones used for listing the responses and data analysis are Excel 365, and SPSS Statistics [29].

4 Demographic Characteristics

In this study, a total of 93 university students participated out of which twelve male students, and 15 female students fell in the age group of 18 to 24. 24 male students and 15 female students were in the age group of 25 to 35, and 18 male students and 9 female students were in the age group of 36 to 45. Making the percentage of female participants equivalent to 42%. The students were given four possible reasons about what most affects their buying behaviour. 8% of the participants chose ethnicity, while 23% chose religion, but the two choices that were most chosen were education and economic conditions with percentages 34 and 35% respectively.

5 Shopping Online and In-Store Preference and Frequency

Findings show that 72% of the male participants prefer shopping online to in-store, while a less percentage of 67% of the female participants prefer shopping online as shown in Figure 1. During the depth interviews, the differences were that the male participants discussed their opinion of shopping as a leisure activity and part of the household chores. Based on the conducted conversations, they seemed more concerned with reaching their goal of purchasing the intended items without taking the time to enjoy many of the shopping experiences that come along with it. They also clarified that online shopping simplified this process by making it easier and more convenient. The female participants had the opposite opinion as they focus more on the experience, using words like fun and interesting to describe it. They also added that they take advantage of online shopping in offers and discounts presented online, adding that they go online before shopping in-store, so they be able to compare between stores and know which one to go to. We concluded that they view shopping as not just a purchasing action, but a wholesome experience. The sample numbers show that a greater percentage of the male participants prefer to shop online while less percentage of female participants prefer online shopping, showing a difference of 5%. However, after computing a 90% confidence interval for the difference in proportions, it yields that it lies from approximately −10 to 20%. This implies that there is not enough evidence given the sample size, that can guarantee that the male percentage will always come higher than the female one, regarding preferring online shopping.

Comparing online and in-store shopping frequency of university students. We divided the number of shopping per month into three categories, the participant can choose any of which as shown in Figs. 2 and 3. The first category is shopping ten times or more per month. 10% male students chose that answer regarding online shopping, and 6% male students chose that answer regarding in-store shopping per month. On the other hand, 19% female students had that choice regarding online shopping, and a less percentage of 16% female students had that choice regarding in-store shopping per month. The second category is shopping five to nine times per month.

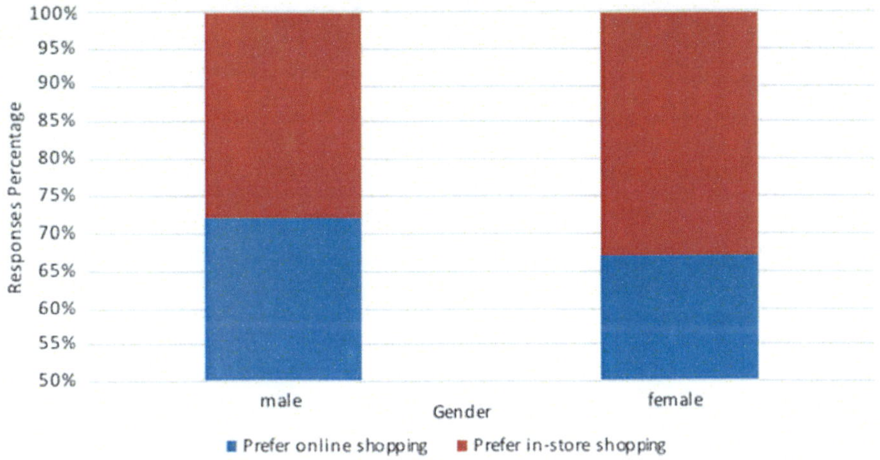

Fig. 1 Online and in-store shopping preferences (by respondent's gender)

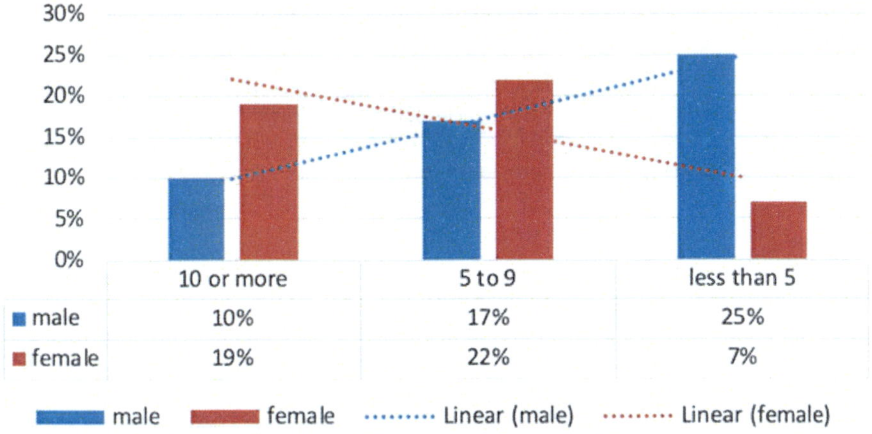

Fig. 2 Online and in-store shopping preferences (monthly online shopping by gender)

17% male students had the choice regarding online shopping, while only 10% male students had that choice regarding in-store shopping. 22% female students chose that category regarding online shopping, and a bigger percentage of 27% female students had that choice regarding in-store shopping. The third and last category is shopping less than five times per month. 25% male students chose that answer regarding online shopping, and 34% male students had that choice.

Regarding in-store shopping. 7% female students had that choice regarding online shopping, while approximately the same female student's percentage chose the same answer regarding in-store shopping. A study of the responses concluded that 80% of the male students did not break the barrier of shopping online seven times per month,

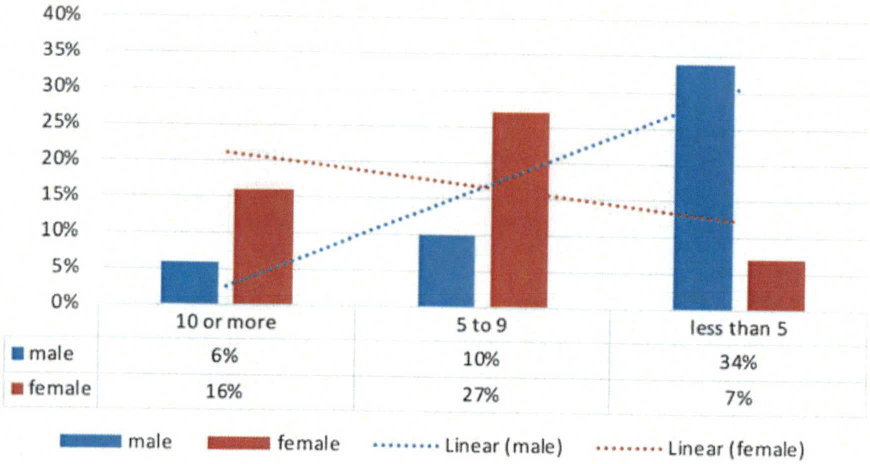

Fig. 3 Online and in-store shopping preferences (monthly in-store shopping by gender)

and even fewer did so regarding in-store shopping. However, the same percentage of female students exceeded six times per month in both categories.

Regarding the time spent shopping per week, the responses were classified into four categories. The first one was less than an hour. Approximately 17% male students had that response, while 6% female students chose the same. The second category was two to three hours of shopping per week. 20% male students had that choice, while 14% female students chose the same response. 8% male students and 20% female students chose the third answer which is four to five hours of shopping per week. The last category was six or more hours, 5% male students chose that answer, while 10% female students chose the same.

In the questionnaire the students were given a list of possible reasons that can be the motives of what drive them to prefer shopping online and were asked to rate each one choosing from strongly agree which is equivalent to five on the Likert scale down to strongly disagree, which is equivalent to one on the Likert scale.

Table 1 shows the results of a statistical analysis of responses to a questionnaire given to students about their reasons for preferring online shopping. The table includes the reasons for preferring online shopping, the lower limit (LL), the mean, the upper limit (UL), the sample standard deviation (S), and the margin of error.

Based on the results shown in Table 1, it appears that the majority of students surveyed strongly agree that online shopping saves time and allows them to shop anytime. Additionally, a wide variety of items and the ability to compare prices with more information available about each item were also highly rated reasons for preferring online shopping. On the other hand, "Can be considered risky" and "hesitation in providing personal or financial information" are the least preferred reasons for preferring online shopping, as they have the lowest mean. It can also be observed that the sample standard deviation (S) is relatively low for most reasons, indicating that the majority of the students have similar opinions. However, for "Can

Table 1 The reasons behind online shopping preferences

Reasons for preferring online shopping	LL	Mean	UL	Sample STD(S)	Margin of error
Time saving	3.93	4.11	4.29	1.06	0.18
Ability to shop anytime	3.85	4.05	4.26	1.19	0.21
Can be considered risky	2.51	2.77	3.03	1.52	0.26
Wider variety of items	4.62	4.79	4.95	0.98	0.17
Hesitation in providing personal or f financial formation	2.71	2.96	3.20	1.43	0.25
Privacy and convenience while shopping	4.34	4.53	4.71	1.07	0.18
Easier to compare prices with more information available about each item	4.63	4.77	4.91	0.81	0.14
Ability to check feedback and reviews	4.66	4.81	4.97	0.91	0.16
You can have more shopping experience in-store than online	4.23	4.39	4.55	0.93	0.16
You prefer physically sensing the products before purchasing them	4.45	4.61	4.78	0.94	0.16
In-store shopping takes more time than online shopping	4.70	4.93	5	1.34	0.23

be considered risky" and "hesitation in providing personal or financial information" the S is relatively high, indicating that there is more variation in opinions among the students. The margin of error is also relatively low for all reasons, indicating that the results are likely to be accurate.

6 Conclusion

Fast pacing improvements in technology have altered the nature of consumer buying behaviour [30]. Presently, online purchases tend to be immediate, and the simplicity of the platform adds to the consumer's experience [29]. The easier is the browsing through the online stores, lower will be the time and effort devoted by the consumer towards shopping [31].

The present study adopts a quantitative research approach to study the consumer buying behaviour amongst University B students and aimed to identify the reasons that determine the consumer's online buying behaviour. Adopting a non-probability sampling design, the study includes 93 responses, collected through online and off-line questionnaire forms and seven one-on one interviews with the respondents. Out of 93 respondents, 54 respondents were males while 39 respondents were females. The data was further divided on the basis of age group where, 12 males were in the age group of 18 to 24, 24 males in the age group of 25–35 and 18 males in the age group of 36 to 45. In the case of female respondents, 15 were in the age group of

18 to 24, 15 in the age group of 25–35 and 9 in the age group of 36 to 45. The data was collected over a span of six months, after which responses were analysed using Excel and SPSS.

According to the data, it was noted that higher proportion of males (72%) preferred to shop online as compared to females (67%). This was because males considered online shopping as a part of their household chores and online shopping made their task easier. In contrast to that, women considered it to be an experience and a leisure activity. The consumer behaviour is not determined by purchase decision alone, but also the experience associated with it [32]. Asa result, traditional shopping gives the consumer an immediate satisfaction which is lagging in the case of online shopping [33], especially when the delivery is delayed [34].

In terms of frequency of online shopping, it was higher in the case of women as compared to men. It was noted that 19 per cent of the female respondents shopped online for 10 times or more as compared to 10 per cent of the male respondents. On the other hand, a significant proportion of male respondents (25%) as compared to females (7%) shopped online for less than five times. The previous research have shown that shopping is mainly associated with women [35]. However, the recent research state that it would be inaccurate to stereotype the shopping behaviours and therefore, there is a need to identify the male shopping behaviour in more detail [36].

The time spent on shopping is an important determinant to analyse the consumer buying behaviour. From the data, it was noted that on an average, female respondents spent more time in shopping as compared to male respondents. 17 per cent of the male respondents spent less than one hour on shopping as compared to 6 per cent of the female respondents. On the other hand, 5 percent of the male respondents spent more than six hours on shopping and proportion was double for female respondents (10%). Higher amount of time can be linked to the buying pleasure developed because of the environment that is created at the time of shopping [37]. This can induce the consumer to purchase the product impulsively [38].

The respondents were asked about the reasons for preferring online shopping. Some of the most prominent responses were "time saving", "wider variety of items", and "Privacy and convenience while shopping". Madhavaram and Laverie [39], too, stated that online purchasing saves time and effort as it is easier to browse through several options available by not leaving the comfort of home. It is worth noting that according to the result, the standard deviation in case of reasons such as "can be considered risky" and "hesitation in providing personal or financial information" is high implying the presence of significant variation in opinion amongst the respondents regarding online shopping. Low margin of error in the case of all reasons reflect on accuracy of data.

With respect to consumer buying behaviour, majority of the research focus on the purchase in traditional stores with limited studies on the impact of growing online presence of stores on consumer's purchase decisions [40]. The present study opts towards addressing the gap by identifying the determinants that affect online buying behaviour amongst university students. Through the empirical evidence, the study adds to the methodology that could be used to comprehend the consumer's online buying behaviour by using a quantitative approach, rather than adopting the binaries in a quantitative study. This could help the researchers understand the purchase decisions through cultural, socio-economic, political, anthropological and several other lenses.

The study is also significant for the contemporary retailers, especially those using online platforms to enhance their sales. One of the prominent findings of the study was with respect to high amount of time spent on shopping by females. The results showed that marketers could adopt several strategies to encourage prospective buyers to purchase from the company. For instance, personalisation of website can help provide each consumer a shopping experience as per his/her preferences which can consequently enhance the sales made by the company [41]. Additionally, the visual appearance of the online store has a positive impact on the consumer's behaviour (Van Der Heijden, 2003). Therefore, marketers can focus on online display of the products to ensure maximum satisfaction to the consumers.

The study also makes a significant contribution in terms of the research methodology. It applied a quantitative approach to understand the consumer buying behaviour and identified several reasons that motivate the buyers to prefer online buying as compared to off-line shopping. [42] highlighted that due to absence of human contact, the risk in online shopping is high. The present study, too, highlighted that risk associated with personal and financial information can affect the online buying behaviour. However, the high variation in the sample reflects a difference of opinion amongst the respondents.

In conclusion, for the future, an elaborative research can be conducted using a larger sample size that could reveal different facets of online buying behaviour amongst the university students. Additionally, the web-based retailers have brought about a significant change in the retailing industry [43]. The current trend of online retailing has minimised the spatial–temporal restrictions that are faced by the consumers in the traditional retail industry [44]. Thus, considering the exceeding use of online marketing techniques used by companies to attract consumers to purchase their products, it would be beneficial to include the idea of impulse buying in detail in the research.

References

1. Kotler, P., Keller, K.: Marketing Management, 14th edn. Prentice Hall (2011)
2. Stallworth, P.: Consumer behaviour and marketing strategic. Int. J. Manag. Commerce Innov. **2**, 44–52 (2008)
3. Parment, A., Kotler, P., Armstrong, G.: Principles of Marketing Scandinavian Edition, 3rd edn. uPDF eBook, Pearson Higher Ed (2021)
4. Laato, S., Islam, A.N., Farooq, A., Dhir, A.: Unusual purchasing behavior during the early stages of the covid-19 pandemic: the stimulus- organism-response approach. J. Retail. Consum. Serv. **57**, 102224 (2020)
5. Kumar, A., Luthra, S., Mangla, S.K., Kazançog˘lu, Y.: Covid-19 impact on sustainable production and operations management. Sustain. Operat. Comp. **1**, 1–7 (2020).
6. Sheth, J.: Impact of covid-19 on consumer behavior: Will the old habits return or die? J. Bus. Res. **117**, 280–283 (2020)

7. Sharma, A., Jhamb, D.: Changing consumer behaviours towards online shopping-an impact of covid 19. Acad. Market. Stud. J. **24**, 1–10 (2020)
8. Zaky, A., Mohamed, H., Saxena, G.: Conceptualising the panic buying phenomenon during covid-19 as an affective assemblage. Eur. J. Market. (2022)
9. Edghiem, F., Abualqumboz, M., Mouzughi, Y.: Covid-19 transition, could twitter support uk universities? Knowl. Manag. Res. Pract. **19**, 439–444 (2021)
10. Yoo, B., Donthu, N.: Developing a scale to measure the perceived quality of an internet shopping site (sitequal). Q. J. Electron. Commer. **2**, 31–45 (2001)
11. Cheetah Digital.: Digital Consumer Trends Index 2022. https://www.cheetahdigital.com/report/digital-consumer-trends-index-2022/ (2022)
12. Sethna, Z., & Blythe, J. (2019). Consumer behaviour. Sage.
13. Pachauri, M.: Consumer behaviour: a literature review. Mark. Rev. **2**, 319–355 (2001)
14. Hwang, Y., Jeong, J.: Electronic commerce and online consumer behavior research: a literature review. Inf. Dev. **32**, 377–388 (2016)
15. Hirschowitz, A.: Closing the crm loop: the 21st century marketer's challenge: transforming customer insight into customer value. J. Target. Meas. Anal. Mark. **10**, 168–178 (2001)
16. Hassan, S., Nadzim, S.Z.A., Shiratuddin, N.: Strategic use of social media for small business based on the aida model. Procedia Soc. Behav. Sci. **172**, 262–269 (2015)
17. Nevo, A.: Empirical models of consumer behavior. Annu. Rev. Econ. **3**, 51–75 (2011)
18. Ghirvu, A.I.: The aid a model for advergames. USV Ann. Econ. Public Admin. **13**, 90–98 (2013)
19. Sheth, J.: New areas of research in marketing strategy, consumer behavior, and marketing analytics: the future is bright. J. Market. Theory Pract. **29**, 3–12 (2021)
20. Johnson, A.R., Stewart, D.W.: A reappraisal of the role of emotion in consumer behavior: traditional and contemporary approaches
21. McLeod, S.: Maslow's hierarchy of needs. Simply Psychol. **1** (2007)
22. Koufaris, M.: Applying the technology acceptance model and flow theory to online consumer behavior. Inf. Syst. Res. **13**, 205–223 (2002)
23. Sheth, J.N.: Models of buyer behavior: conceptual, quantitative, and empirical. Marketing Classics Press (2011)
24. Ricci, L., Lanfranchi, J.-B., Lemetayer, F., Rotonda, C., Guillemin, F., Coste, J., Spitz, E.: Qualitative methods used to generate questionnaire items: a systematic review. Qual. Health Res. **29**, 149–156 (2019)
25. Singh, A.K., Sailo, M.: Consumer behavior in online shopping: a study of aizawl. Int. J. Bus. Manag. Res. (2013)
26. Jadhav, V., Khanna, M.: Factors influencing online buying behavior of college students: a qualitative analysis. Qual. Rep. **21**, 1 (2016)
27. Lester, D.H., Forman, A.M., Loyd, D.: Internet shopping and buying behavior of college students. Serv. Mark. Q. **27**, 123–138 (2006)
28. Zhang, Y., Wildemuth, B.M.: Unstructured interviews. In: Applications of social research methods to questions in information and library science
29. Rezaei, S., Ali, F., Amin, M., Jayashree, S.: Online impulse buying of tourism products: the role of web site personality, utilitarian and hedonic web browsing. J. Hosp. Tour. Technol. **7**(1), 60–83 (2016)
30. Dawson, S., Kim, M.: Cues on apparel web sites that trigger impulse purchases. J. Fash. Market. Manag. Int. J. **14**(2), 230–246 (2010)
31. Castaneda, J.A., Frías, D.M., Rodríguez, M.A.: Antecedents of internet acceptance and use as an information source by tourists. Online Inf. Rev. **33**(3), 548–567 (2009)
32. Schiffman, L.G., Hansen, H., Kanuk, L.L.: Consumer Behaviour: An European Outlook. UK, Pearson Education, Cambridge (2008)
33. Jones, M.A., Reynolds, K.E., Weun, S., Beatty, S.E.: The product-specific nature of impulse buying tendency. J. Bus. Res. **56**(7), 505–511 (2003)
34. Verhagen, T., Dolen, W.: The influence of online store beliefs on consumer online impulse buying: a model and empirical application. Inform. Manag. **48**(8), 320–327 (2011)

35. Dholakia, R.R., Chiang, K.P.: Shoppers in cyberspace: Are they from venus or Mars and does it matter? J. Consum. Psychol. **13**, 171 (2003)
36. Otnes, C., McGrath, M.A.: Perceptions and realities of male shopping behaviour. J. Retail. **77**, 111–137 (2001)
37. Reynolds, K.E., Beatty, S.E.: A relationship customer typology. J. Retail. **75**(4), 509–523 (1999)
38. Vonkeman, C., Verhagen, T., Van Dolen, W.: Role of local presence in online impulse buying. Inform. Manag.
39. Madhavaram, S.R., Laverie, D.A.: Exploring impulse buying on the internet. Adv. Consum. Res. **31**(1), 59–66 (2004)
40. Zhou, L., Wong, A.: Consumer impulse buying and in-store stimuli in Chinese supermarkets. J. Int. Consum. Mark. **16**(2), 37–53 (2004)
41. Chakraborty, G., Lala, V., Warren, D.: What do customers consider important in B2b websites? J. Advert. Res. **43**(1), 50–61 (2003)
42. Hwang, Y., Kim, D.J.: Customer self-service systems: the effects of perceived web quality with service contents on enjoyment, anxiety, and E-trust. Decis. Support Syst. **43**(3), 746–760 (2007)
43. Hausman, A.: A multi-method investigation of consumer motivations in impulse buying behavior. J. Consum. Mark. **17**(5), 403–426 (2000)
44. Virvilaite, R., Saladiene, V., Bagdonaite, R.: Peculiarities of impulsive buying in the market of consumer goods. Inzinerine Ekonomika Eng. Econ. **2**(62), 101–108 (2009)

Factors Impacting a UK Based Software Business on International Expansion

Mohamed Shrief, Farag Edghiem, and Brittany Lea

Abstract This paper empirically examines the factors influencing a UK-based software businesses' expansion into the EU through a case study of Deontics, a UK-based software company. By conducting eight structured interviews with the company's top management team and analysing their responses using thematic analysis. The study identifies common challenges, including finance, marketing strategy, and leadership. Additionally, the study highlights the unique challenges posed by Brexit and COVID-19 pandemic, covering the gap in the body of literature regarding the expansion journey of UK-based software businesses, and evaluates how Deontics navigated these obstacles during their international expansion journey.

Keywords Software · International expansion

1 Introduction

Many of the UK-based businesses, especially the small and medium-sized enterprises (SMEs), aim to expand globally. One of the main reasons behind such common goal for SMEs, is to tap into a new ever-growing market, and impact the business growth [1]. Liñán et al. (2019) go beyond advocating that global expansion became a necessity for survival and sustainability of the business. However, high-tech firms often internationalize at a faster pace and utilize distinct market entry strategies compared to those in low-tech markets, but this expansion also comes with a range

M. Shrief (✉)
Manchester Metropolitan University, Manchester, UK
e-mail: mohamedhassansherif@student.aast.edu

F. Edghiem
Arab Academy for Science, Technology and Maritime Transport, Cairo, Egypt

B. Lea
Institute of Management Greater Manchester, Manchester, UK

C. Iwendi et al. (eds.), *Proceedings of ICACTCE'23—The International Conference on Advances in Communication Technology and Computer Engineering*, Lecture Notes in Networks and Systems 735, https://doi.org/10.1007/978-3-031-37164-6_48

of challenges that must be overcome to ensure not only the success, but also long-term sustainability [2].

These challenges can differ depending on the type of business that the company is conducting. For example, High tech businesses can face technology design and build challenges [3]. While other businesses such as retail can face challenges such as cultural adaptation or an increase in production costs (Liñán et al. 2019). Additionally, some factors can play a key role in impacting any UK based businesses while expanding to EU market such as Brexit and COVID19 pandemic [4]. This paper aims to explore the various factors that impact a UK-based software business on its international expansion. The research will draw on a case study performed on Deontics Company.

1.1 Background on Deontics

Deontics, founded in 2014, is a deep tech company which span out from Oxford University, UCL and Cancer Research UK. They solve some of the most complex healthcare challenges to date around personalized healthcare and precision medicine. The software was initially developed from the early work with the Nobel Laureate winner Herbert Simon who is known as the father of artificial intelligence (Al) and Prof John Fox, and their inspirations to apply AI in medicine. Deontics have built a powerful AI analytics and decision engine. It has multiple features, but most importantly it blends clinical process and reasoning, which is then powered by knowledge and data. The engine supports a suite of Deontics own tools and software applications. In addition, it also can power third party software with Application Programming Interfaces (APIs). The software has advanced interoperability and flexibility and is often integrated into third party electronic medical record vendor software. The core software is an AI based engine and is a class one medical device; it supports its users with advanced third generation clinical decision support as well as analytics. Deontics vision is to be the most clinically recommended AI analytics and decision engine which will touch the lives of 100 million patients every year in the next five years. The company has over 300 scientifically and clinical peer reviewed published papers.

They report that their software uses an intuitive reasoning framework which uses pros and cons when making decisions, which is similar to what clinicians do when they come to decision making. Many AI software solutions are known as "black box" solutions, this simply means the AI solution does not report why it reached a certain decision. The difference that Deontics advertise is their solution can explain why it came with a certain conclusion.

1.2 Challenges Facing SMEs When Expanding Globally

For SMEs that are based within the UK, it can be challenging to expand internationally due to multiple factors including Regulatory compliance, cultural differences, language barriers, competition, tariffs and trade agreements, data protection and privacy laws [5]. Expanding a business internationally can be a critical time for businesses and can increase the risk of financial and legal complications [6]. Further Research that was carried out by [7] has demonstrated that 29% of SMEs in the UK with the prospect of expanding globally failed due to financial difficulties before they even had the opportunity to expand globally. Financial difficulty is a key factor that is largely impacting businesses, causing cutbacks that affect essential areas such as staffing and investment in management, research, and growth initiatives. As a result, SMEs are less likely to achieve expansion success and maximize their business potential, whereas if they were adequately funded, the number of failed businesses would decrease [7].

Many companies resort to financial funds from Investors. While most investors provide funding to businesses in return for shares of the business and to make a return on their investments (ROI). A company may struggle to attract investors if it does not meet certain criteria such as having potential, a solid business plan, risk management strategy, traction in the targeted market, and a high return on investment. These strict criteria make it difficult for unproven companies to attract investors as they are considered high-risk [8]. In such case, SMEs expansion plans beyond the UK market is challenging due to the lack of funds, which is considered as a key barrier. There are also gaps in literature on how to implement strategic planning methods. Following and implementing a solid strategy can help to prevent poor performance and business failures as well as avoiding unnecessary risks [9]. This also helps when managing business expansion for example managing team members, keeping track of multiple processes, and being prepared for the steps that are to happen next [10].

One of the key factors that will be discussed regarding challenges that UK-based businesses face when expanding globally is the difference in laws and legislation between countries, especially after Brexit, as it is largely different for companies that are outside of Europe. Some of the factors that can impact a business are the rules and legislations regarding employees, tax, data protection [6]. This will impact each company differently depending on the nature of the business and the countries that they choose to expand into.

1.3 Issues Caused by Brexit on Business Expansion

UK voted to leave the EU on the 23rd of June 2018, but this did not officially come into place until the 31st of January 2020 [11]. Before Brexit, the UK was part of a single trading area meaning that trade was able to flow freely with no additional paperwork or implications, and goods were able to be imported and exported between the UK and the EU without any tariffs, documentation, and border checks [12]. Now there are border checks, different regulations, and documentation that have been implemented between the trading blocs [4]. This has caused challenges affecting all industries both large corporates and SMEs, especially issues with exporting and importing goods and services to and from the EU. As a result, businesses are facing difficulties expanding abroad, making it harder to operate in their own countries because of the new challenges in force. These range from new regulatory requirements, affected staffing including overseas workers, visa requirements leading to staffing shortages, legal changes, additional associated costs, and increased paperwork [12].

2 Methodology

The methodology employed in this paper combines both qualitative and quantitative research methods. A case study approach is performed to inform the literature review. Structured interviews were conducted with multiple individuals involved in the company, utilizing a set of eight pre-determined questions as a guide to gather relevant information while allowing for flexibility to expand and ask additional questions as needed during the interviews. The findings from the literature review will be utilized to critically evaluate the information obtained through the interviews, in order to compare the experiences of the studied SME that have recently undergone expansion, with relevant theoretical frameworks. The interview questions will facilitate an understanding of various perspectives within the company by examining the processes undertaken in comparison to established theory, in order to determine the extent to which these processes have been beneficial and to critically analyse the approach and levels of success achieved. The research and analysis will be guided by the main research question which is, what are the factors impacting a UK software businesses ability for global expansion? The participants involved will remain anonymous throughout this research and had the right to withdraw at any point throughout this research without any implication to all parties involved. This study followed theoretical approaches to help understand the research that is being carried out. This method can help hold and structure the research [13]. Interpretivist approach was preferred as it allows the researchers own deductions to be considered as an influencing factor towards the research, and that is more suited for research revolving around interviews [14], unlike positivism where the researcher has no personal input on the final interpretation. Meaning that the interpretivism approach is best suited when carrying out this type of research.

Despite the researcher's inclination towards interpretivism, it is acknowledged that the incorporation of positivist elements will provide a more comprehensive and valid justification for the study's conclusions. Specifically, that interpretive approach will be utilized to elicit the thoughts, perceptions, and feelings of the participants while conducting the interviews, allowing for a more nuanced understanding of their perspectives. However, the inclusion of positivist elements will serve to contextualize these observations within established theoretical frameworks, thereby increasing the overall validity of the study's findings.

3 Results and Discussion

In this chapter, we present the findings and results of our research, which was conducted through a literature review and structured interviews with five members of the top management team at Deontics company. To maintain anonymity, we have coded the participants with letters. Hence, the participants were identified as participant A, B, C, D, and E, each represents a different department within the company. The responses to the eight questions asked during the interviews were analysed using thematic analysis to identify key themes related to the challenges faced by the company. These themes were further evaluated to determine whether they were influenced by external factors, such as Brexit and the global pandemic, or internal factors, such as the company's desire to expand globally or other internal business considerations.

3.1 Issues Caused by Brexit on Business Expansion

Question 1—What Challenges Has the Business Faced Over the Last Couple of Years?

COVID-19 pandemic was identified as the primary external factor that contributed to the challenges encountered by the business in the last couple years. Through interviews with team members and performing thematic analysis of their responses, it became apparent that customer relations were significantly impacted. Additionally, the pandemic also affected the workplace environment, product development, and finance. The implementation of multiple lockdowns and the shift away from traditional in-person working methods presented challenges for the team, who had to quickly adapt in order to maintain operations while also complying with government guidelines. As participant A noted, "A number of the challenges that we have faced have been common to all companies during this period namely COVID-19 which has changed how we work as a company". Participant A also mentioned marketing was directly impacted by the pandemic and said, "Access to the key clinical decision makers was also more limited making new sales difficult". The responses also revealed that the business faced internal challenges as well, including those related to marketing strategy, workplace/environment, leadership, finance, and product development. These challenges were determined to be internal in nature and not directly attributed to external factors such as Brexit or the COVID-19 pandemic. As participant D emphasized, the internal factors that the business faced were primarily centred around, "Leadership of the business and finance". Participant E also supports that these internal factors contributed to challenges and said, there was a "Lack of strategy, Lack of focus, Lack of marketing expertise. General poor management". Participant A mentions finance as a challenge which then further impacted the final product, they said "There are also the challenges we face because of where we are in a company's lifecycle namely the challenge of balancing product development with revenue / funding demands".

Question one has given the opportunity to expand further in question two, where the internal and external factors along with the business challenges can be explored further.

Question 2—How Have the Challenges Impacted the Business?

The responses from question two highlighted that the biggest challenges facing the business were internal factors. Finance was the biggest challenge followed by marketing, and leadership both being equal. Participant E mentions the challenges of finance and strategy/ leadership being linked when they said, "Because of the lack of strategy & focus the investors were unwilling to invest more money in a business that lacked direction". Participant B also mentions finance impacting on the business as a general impact saying, "The business being under resourced has meant a need to focus on revenue-based work". When the responses from participant E and participant B were analysed together, it seemed there was a lack of investment into the business, because of the lack of strategy and leadership which meant the company was underfunded so finance has impacted on the product development. That was due to the switch to revenue generation instead of the original business plan and product development. This seems to be confirmed when participant D said the following, "for whatever reason the original executive team failed to execute on this plan. The board lost confidence in the executive team and by direction of the board". It seems the board recognised the challenges the business was facing, and they sought to resolve them by appointing a new leadership team. The second impact affecting the business was external, and that was COVID-19 pandemic. From analysing the responses, it seems the global pandemic created challenges in finance, workplace environment and customer relations. Participant A said, "A number of the challenges that we have faced have been common to all companies during this period namely COVID-19 which has changed how we work as a company". i.e., moving to being 100% remote has meant we have had to adopt new ways of interacting such as, slack, hangouts. Participant A highlighted the ways in which the business had to adapt. Participant D also talked about the impact and challenges they faced with the workforce and the role during a pandemic, they said, "the pandemic made the job tougher". They also said how hard it was running a team virtually and having to make decisions about this team and quoted "Having virtual interactions and working remotely at any time can be difficult, however due to the nature of my employment and my remit there was more pressure to get things right first time" participant D clearly articulates the impacts that the external factor of a global pandemic had on the workforce and leadership function. Participant A also highlighted the impacts of the workforce working remote when they said, "Working remotely is also a challenge for a number of employees".

From the interview responses to these questions, it was evident that the global pandemic impacted the business and caused challenges with customer relations. Most of their customers were NHS and Healthcare providers who were facing their own challenges. Participant B said, "Hospitals being overwhelmed has meant they haven't been able to prioritise implementing new technology and the business has been unable to sell the product". Another challenge with customer relations was the interaction with the customer base. Participant A mentioned, "It has also changed how we engage with our customer base, partners and suppliers so less, if any, face to face interactions in favour of Zoom, Teams etc....". The challenge of customer relations, in particular the NHS, and their focus and reprioritisation impacted finance, with sales not happening due to the NHS having a different focus was problematic for the business. In fact, participant D further expands on this they said, "Monies which were ring fenced by the NHS to invest in technology like ours were ploughed into Covid-19 spending", explaining that this caused cashflow issues.

Brexit was also mentioned as a major external factor that caused the business a challenge, this challenge was regarding regulatory requirements, participant D said that "Brexit was also a challenge, this was for several reasons, one our need to scale and no longer being part of the EU required legal and regulatory changes".

Question 3—Has the Marketing Plan Changed to Support the Launch of the Business Globally?

The objective of this question was to evaluate the validity of the literature review findings in a real-world business setting. Through the collection and analysis of responses, the research aimed to identify key themes that may have influenced changes in the marketing plan.

It was concluded that the main driver for changing the marketing plan was to support the launch of the business globally and every response sets out the rational reason. There was also reference to Brexit impacting part of the marketing plan. Participant C said, "The 2022 plan is to advance on global opportunities within the B2B market" They then went on to say that research was undertaken, and they would be sticking to English speaking countries to allow them to scale globally. Moreover, they added that marketing route had changed from that within the UK, as the focus was on direct sales. However, to scale internationally the focus became on the B2B market. Business plans seem to have aligned in the last two year for scaling globally. Many changes have already taken place, and the focus has changed from B2C to B2B. Participant C describes marketing B2C with the provider hospitals and then changing this to B2B. They said this switch was "to avoid the time lag and delays with individual end users" It also seemed that research has played a big part in marketing and that additional research was carried out to expand globally. Participant C said, "Research the key issues we can resolve within each country, align these with new messaging/campaigns, and then progress to testing and roll-out". Brexit and COVID-19 pandemic seemed to have an impact on marketing activities, participant C said, "Brexit and Covid-19 have forced us to change our strategy for all areas of the business". It seems the company had to adopt more of a digital strategy to expand globally, this was because the business could not attend in person events.

Question 4—Have Different Marketing Approaches Been Used or Considered When Looking at Expanding Globally Compared to the Approaches that are Taken for the UK Market?

Deontics did have a marketing strategy and had adopted different marketing approaches. Participant D said, "in our GTM we have defined our market segments and then route to market". They described that to expand globally they will focus on B2B customers and that the marketing will be around the joined added value of the product. This is a very different approach than the one was followed within the UK market. Participant D also confirms that the marketing channels for customer acquisition are very different when scaling globally, they said, "The advertising platforms are also different, our focus are global platforms" They also described the difference in the targeted segments. Within the UK it was clinicians and financial staff, while globally the target revolve around chief product officers and chief technology officers.

Question 5—Have any Theories or Marketing Models Been Used by the Business? If Yes, THEN How Were They Useful and Were There Any Limitations with THE Models and Theories?

It was concluded from responses that several theories were used such as, competitor analysis, Unique Selling Points (USPs), and a S.W.O.T analysis. Participant C described in detail how they use a S.W.O.T analysis, and how using this approach ensures they are in a strong position when they approach a client, "This enables our marketing strategies to be as efficient and effective as possible and means that by the time we approach a potential customer we are already in a strong position". They also described using a competitor review and using this proactively to target competitors. Participant D also mentioned using a S.W.O.T analysis and a competitor analysis, they also mentioned aligning their marketing to the business goals when they determined the geographical areas they wished to expand into, "this helped us to determine the geographical focus, we ensured these aligned with the business goals".

Question 6—Has Brexit Impacted This Business If Yes Then How So?

All the participants recognised that Brexit had a strong external impact on the business with the overwhelming challenge being regulatory and then finance. Participant A spoke at length around the challenges of Deontics being a registered medical device, and that since Brexit, the routes that need to be taken to have this medical device registration had changed. They described a stringent process since leaving the EU they said, "UK approval is no longer accepted by the EU, as it has adopted more stringent requirements" participants B, D, and E also expressed the challenges with regulation and medical device registration since Brexit. Participant A, B, and D also mentioned the higher costs and financial impact associated with the medical device registration post Brexit. The other area affecting the business was compliance with data laws, again confirming legal and regulatory was the biggest challenge of Brexit, participant D described how they have had to move data centres to comply with EU data laws.

Question 7—What Has Been Done to Overcome the Problems Caused By Brexit?

After looking and analysing the challenges caused by Brexit it led to asking the team what they did to overcome these issues within regulatory and finance. Participant A said, "We are engaging bespoke consultancies to advise". It would seem the business did not necessarily have the inhouse resources to overcome the challenges around regulations. Participant D also confirmed this and said they were "engaging with specialist consultants to help us with our medical device registration, we need to navigate the EU notified bodies". It seems that the financial challenges increased by the challenges caused by Brexit, and more money has had to be spent to address the issues. This is confirmed when participant D said, "there has been a cost implication to us". Meaning that while the business is trying to overcome the challenges caused by Brexit, uncertainty around medical device regulation since Brexit makes it difficult to achieve this.

Question 8—Looking Back, What Would Be Done Differently from the Start?

A lack of leadership seemed to have an overwhelming response and was the most common theme, next was strategy and finally marketing strategy. It seems the board had recognised the internal impacts and the challenges that were causing and had implemented changes. New leadership was implemented, which proved effective since the business became more focused, employees shown more motivation within their roles, and challenges are being addressed. Participant B said. "The business in its early years was very unfocussed which wasted time. A properly implemented strategy and associated specific marketing would have been very useful'. Participant D also recognised the gaps in strategy and marketing, as well as leadership, in fact they had been brought in to sort these issues, they said, "the business is now thriving, within two years we are delivering on our plans, even in the adversity of Brexit and a pandemic. What this demonstrates to me is, if you fail to plan, you plan to fail. I have implemented a strategy, clear company objectives a go to market plan and execution plan. I really believe that if the business had a different leadership from the start and had implemented such plans, they may have already achieved their exit or be in a very different position". These are extremely strong statements that participant B, and E all confirmed. A strong strategic plan should have been in place from the start and different leadership.

4 Evaluation of Key Findings

From the literature review and research conducted, common themes were identified for each question. Each of the common identified themes is reviewed against the interview responses to evaluate the relevance of key findings within the responses.

4.1 Common Challenges Faced When Expanding a Business

Finance was an overwhelming factor when conducting the interviews and reviewing the body of literature. There were many topics around finance that were described as common challenges faced when expanding a business. These were around businesses not having enough funding to do proper research to allow them to expand globally [7]. Finance was also linked to not having enough knowledge and resources to develop robust strategic plans, which would give confidence to investors to fund a business that allow them to scale. It seemed a vicious circle of finance and then resources, and good research, which translate to developing business strategies to give confidence to investors. Looking at the responses from the participants, it shows that finance has had a key role within the business to allow them to scale, in fact participant D mentions in question two, that investments were centred around scaling internationally, in which the business was having to focus on several objectives to help them scale which all had finance at the centre. These objectives include balancing the product development, expanding, and revenue generation to support cash flow.

Research and strategic planning were also a crucial finding in the body of literature, as key enablers for a business expanding [9]. After conducting a review of the responses from staff members, it was found that there was a consensus among staff members in support of the findings presented from the literature review. All staff members discussed the importance of leadership and strategy in their responses. Literature review findings suggest that investors often have strict criteria and policies for their investments and require evidence of potential and a solid business plan in order to proceed. The staff members' responses confirmed this, as they reported that the board of the company made significant changes, including a change in leadership, which led to the implementation of a new strategy and marketing plan. As a result, the company was able to secure investments from various investors. This supports the idea that having robust plans and strategies in place can instil confidence in potential investors, and highlights the importance of addressing this issue early on in the process. participant E confirms this when they were asked what could have been done differently from the start they said, "develop a strategy and products that investors could get behind which would have meant that enough cash was injected into the company" The responses support the literature review where it is described how hard it is for investors to come on board [8].

Another discovery made during the review and research of literature around common challenges of expanding a business were around legal requirements. From the research the major factor was around laws and legislation [6]. However, from analysing the literature and the interviewees responses it does not seem that legal and legislation were a challenge for business expansion on their own. It seems that legal and legislation challenges for this business were a direct result from Brexit.

4.2 Issues Caused by Brexit on Business Expansion

Legal and regulatory impacts were identified as key challenges when conducting the literature review. Visa changes leading to staffing shortages are also mentioned in the research and the body of literature [12]. From the responses there was no evidence that linked to the business facing challenges from visa changes and staffing shortages. However, there was strong evidence to suggest the business was and still is impacted from regulatory and legal issues arising from Brexit. This is because the businesses product is software which is a registered medical device. All staff members speak of the challenges faced post Brexit with regulations of medical devices in the EU. Participant D said, "prior to Brexit, approval as a medical device at a UK level allowed you to be approved at EU level". Most of the staff interviewed conveyed the challenges on the business post Brexit around regulatory and legal requirements. Therefore, the findings in the literature review around legal and regulatory challenges support the findings from the interviews. In addition there were specific challenges to software companies that host data that were discovered during the research. The business described how they had been impacted as a UK software business and had to move their data centre to comply with data cloud hosting.

4.3 Theories and Models Used

After reviewing the body of literature and analysis of the interviews, it was clear a business needed to adopt several theories and models to support the business in global expansion such as, SWOT Analysis, Competitor analysis and Marketing/ Strategic planning. From interviewing the staff, it was clear that the main cause of the business not being successful at the start, was due to a lack of understanding of the importance of using the correct strategies and theories, and the benefits they can offer when used correctly. It seems when the board recognised these challenges they brought in new leadership. Participant B mentions the challenges of not having appropriate marketing and a strategy in place when they said "the business in its early years was very unfocussed which wasted time. A properly implemented strategy and associated specific marketing would have been very useful—this is being done now but the previous approach has made the work harder". Therefore, the implementation of effective marketing theories and strategies can play a crucial role in the success of a business. In the case of Deontics, the utilization of a competitor analysis, SWOT analysis, and identification of unique selling points (USPs) in combination with the development of a five-year plan, specific strategies, and a comprehensive go-to-market plan, allowed the company to excel despite external factors such as Brexit and the global pandemic. This illustrates the importance of having a robust marketing strategy in place to evaluate potential risks and mitigate them effectively. Had these marketing theories, strategies, and plans been implemented from the outset, the business may have been further along in its goals of international expansion.

Additionally, the results from the interviews align with the literature review and highlight the significance of conducting a SWOT analysis [15]. Implementing these methods has allowed Deontics to create a value proposition that is unique and differentiate itself in the different markets [16]. Participant C talks in depth on this matter describing their use of a SWOT analysis and competitor review, so they have a clear idea within their marketing plan of where to start.

5 Conclusion

This research provided an in-depth exploration of the challenges faced by Deontics during its international expansion, compared with the ones revealed in the body of literature. The study found that finance, marketing and strategy, and leadership were the common challenges encountered by the company. The study revealed a correlation between leadership and strategy, which in turn impacted marketing. The findings also confirmed the importance of using appropriate theories and models to develop effective strategies and marketing plans for scaling internationally and achieving business goals. The success of Deontics after implementing new plans and strategies, even amidst the challenges posed by Brexit and the global pandemic, further underlines the significance of having a robust strategy in place. Furthermore, the study highlighted that internal and external factors play a crucial role in the success of a business, and that the right planning and strategies can help overcome these challenges, which aligns with the results found after reviewing the body of literature.

Acknowledgements We would like to express our gratitude to all those who have contributed to the success of this research project. First and foremost, we would like to thank Deontics, and to acknowledge the participants of the study for their time and willingness to contribute to the project. Without the contributions and support of all these individuals and organizations, this research would not have been possible.

References

1. Chung, J.Y., Yoon, W.: Technological capabilities and internationalization of high-tech ventures: the moderating role of strategic orientations. Manag. Decis. Econ. (2020). https://doi.org/10.1002/mde.3195
2. Crick, D., Spence, M.: The internationalisation of 'high performing' UK high-tech SMEs: a study of planned and unplanned strategies. Int. Bus. Rev. **14**(2), 167–185 (2005). https://doi.org/10.1016/j.ibusrev.2004.04.007
3. Dawson, P.: Establishing a high tech company in a global competitive market: a sociomaterial process perspective. Ind. Mark. Manag. (2019).https://doi.org/10.1016/j.indmarman.2019.12.002
4. Colwill, S.: Impact of Brexit on UK Businesses: Positive & Negative Effects. MSC Notaries (2021). https://www.mscnotaries.com/impact-brexit-uk-businesses/

5. Saint-Leger (2019) Risk involved in an international business. Bizfluent [online]. Available at: https://bizfluent.com/info-7766597-risk-involved-international-business.html
6. Nommick C (2018) Main legal issues in international business expansion. Soulier Avocats Strategic Lawyering [online]. Available at: https://www.soulier-avocats.com/en/main-legal-issues-in-international-business-expansion/
7. Kepka, A.: Business startup statistics UK (2020 update)|Fundsquire (2020). https://fundsquire.co.uk/
8. Kollenda, P.: Financial returns or social impact? What motivates impact investors' lending to firms in low-income countries. J. Bank. & Financ. 106224 (2021)
9. Raymond (2020) 9 Biggest business expansion problems faced by entrepreneurs – Ray's now [online]. Available at: https://www.raysnow.com/9-biggest-business-expansion-problems-faced-by-entrepreneurs/
10. Abdel-Maksoud A, Jabbour M, Abdel-Kader M (2021) Stakeholder pressure, eco-control systems, and firms' performance: Empirical evidence from UK manufacturers. Account Forum 45(1):30–57. https://doi.org/10.1080/01559982.2020.1827697
11. Islam F (2021) Brexit: one year on, the economic impact is starting to show. BBC News [online] 24 Dec. Available at: https://www.bbc.co.uk/news/business-59761292
12. Morgan K (2022) How has Brexit affected small business imports and exports [online]. Available at: https://www.unbiased.co.uk/life/small-business/how-has-brexit-affected-small-business-imports-and-exports
13. Ullrich C, Stürmlinger A, Wensing M, Krug K (2020) Qualitative research methods in medical dissertations: an observational methodological study on prevalence and reporting quality of dissertation abstracts in a German university. BMC Med Res Methodol 20(1). https://doi.org/10.1186/s12874-020-01186-6
14. Saunders M, Lewis P, Thornhill A, Bristow A (2012) "Research methods for business students" Chapter 4: Understanding research philosophy and approaches to theory development. Research Gate [online]. Available at: https://www.researchgate.net/publication/330760964_Research_Methods_for_Business_Students_Chapter_4_Understanding_research_philosophy_and_approaches_to_theory_development
15. Chereau, P., Meschi, P.-X.: Expanding Internationally. Strategic Consulting, pp.111–143 (2017). https://doi.org/10.1007/978-3-319-64422-6_5
16. Czepiel JA, Kerin RA (2012) Competitor Analysis [online]. Available at: http://pages.stern.nyu.edu/~jczepiel/Publications/CompetitorAnalysis.pdf

The Role of Blockchain Technology and Student Coin Case in Saudi Arabia's Higher Education Sector. The Challenge of Achieving 2030 Vision

Nasser Alshareef and Imran Akhtar

Abstract At present, researchers and practitioners are increasingly interested in blockchain technology. This is due to its unique characteristics, including safety, regionalisation, dependability, and data security. Even with the increased attention and engagement, few individuals are aware of the present state of knowledge and usage of Blockchain (BC) technology, more so in the context of higher education in the Kingdom of Saudi Arabis. This paper offers a comprehensive evaluation of prior research pertaining to blockchain-based educational applications. It focuses on three primary themes: (1) educational applications and its use created by using blockchain technology, (2) the advantages of blockchain technology in higher education and (3) the adoption area in higher education challenges and problems of blockchain technology. Detailed analysis of the results for each topic and an in-depth discussion was carried out based on the survey results. This study also sheds light on other crucial areas of education where blockchain technology might help.

Keywords Higher education · Blockchain application · Educational technology · Vision 2030

1 Introduction

As Qu et al. [39] illustrate blockchain to have been widely employed as the critical technology underpinning cryptocurrencies. Recently, it has influenced previously untapped businesses as a data storage possibility like manufacturing. As per [7, 28], healthcare, as well as education, amongst other sections, all have positively benefited

N. Alshareef (✉)
College of Business Administration (CBA) in Majmaah, Majmaah University, Al Majma'ah 11952, Kingdom of Saudi Arabia
e-mail: nm.alshareef@mu.edu.sa

I. Akhtar
Graeter Manchester Business School, University of Bolton, Bolton, UK

C. Iwendi et al. (eds.), *Proceedings of ICACTCE'23—The International Conference on Advances in Communication Technology and Computer Engineering*, Lecture Notes in Networks and Systems 735, https://doi.org/10.1007/978-3-031-37164-6_49

from such an application [36]. This research paper investigates the possible benefits, problems, and overall impact of using blockchain in the education sector. This study will look at two significant difficulties. To begin, how can blockchain technology increase educational institutions' performance and student learning? This question will look at three distinct areas where a blockchain solution may be beneficial: (1) Educational institutions (universities, start-ups, and non-governmental organisations) are searching for ways to enhance the efficiency and security of student data storage and administration; (2) Beginners (in this example, students) can benefit from a more appealing, dependable, and long-term method of accumulating, proving, and sharing the advantages of higher education and knowledge decoding; (3) Employers want a secure and trustworthy means of evaluating the legitimacy of student abilities and credentials. The motives, worries, and ultimate aims of these three groups will be discussed in this essay and blockchain's capacity to produce personal and community value through educational applications. The second issue that this research report attempts to address is the barriers to blockchain deployment in Saudi Arabia's higher education industry. This issue will focus on identifying and assessing the obstacles private, and public education organisations may encounter while developing or implementing blockchain solutions.

1.1 What is Blockchain?

As described by [45], the blockchain is a secure, regionalised database. A "block" holds transaction date, time, amount and participants' other information (competitors on the blockchain usually cannot identify individuals). Blockchains can be classified as public, private, or permissioned. The public blockchain network is open to anyone to join and contribute to [9]. Public blockchains are helpful in this approach because they enable completely decentralised, democratic, and permissionless procedures. Permissioned blockchains, unlike public blockchains, only enable verified parties, such as members of an organization, to participate in order to join the network after being invited and validated. The third form of blockchain is a private blockchain, similar to permissioned blockchains but is owned and controlled by a single entity.

Several processes ensure the blockchain's security. Each block on the blockchain, for example, stores the preceding block's hash value. The hash function takes variable-length input and outputs a fixed-length result. As a result, the hash inside the blockchain (i.e., the hash chain) makes changing the previous block extremely difficult, assuring immutability [24]. Miners that contribute blocks to the blockchain are also urged to protect the network's integrity by rejecting any fraudulent operations. Depending on the blockchain technology in use, the nature of such incentives differs. It incentivizes miners not to squander precious resources by approving fraudulent transactions/blocks by requiring them to "work" (i.e. expend computational power) to add blocks to the chain [42].

Another term to be familiar with in blockchain technology, is a smart contract that is a blockchain-based computer program that specifies the parameters of a transaction between buyers and sellers and may be enforced automatically by miners [14]. Smart contracts, as a result, force and facilitate contract negotiation and execution. Blockchain technology is still in its early stages of development. Today, there is a keen interest in Saudi Arabia's education and financial sectors. Various educational systems exist in Saudi Arabia. These strategies have been combined in order to promote student employment. To tackle the problem of individualized learning and secure the integrity of academic credentials, blockchain technology is required [20]. It is envisaged that blockchain will continue to dominate as a promising new technology in the Kingdom of Saudi Arabia (hereafter referred to as KSA).

1.2 The Study of Blockchain in the Educational Sector

It is important to note that blockchain's use in education is still in its infancy, which will influence the accessibility and quality of current research on the subject. As [5] pointed out, "although the number of documents on the application of blockchain in education has increased in the past few years, it is still fragmented, and the subject has not yet been systematically reviewed." Today's blockchain technology may not have advanced sufficiently to apply to all use cases [40]. Such a novel and a new idea is particularly problematic for educational network use cases (such as blockchain record-keeping or the adoption of electronic assets). Because blockchain research in the education industry is still in its early stages, many existing and prominent education blockchain efforts have yet to be extensively explored and documented to evaluate its usefulness in the higher education context [25].

2 Literature Review

The paper "Blockchain in Higher Education" by Grech and Camilleri [21] is the first study of its kind located in the higher education arena proposing the use of Blockchain As per the Scholar Google and Semantic Scholar (64 citations), this research is one of five notable publications related to blockchain in education (181 citations). Based on European Commission's Science and Knowledge Services Joint Research Center (JRC) has eloquently explained the practical application of blockchain in universities [20, 25]. The University of Nicosia (UNIC) in Cyprus was the first to adopt blockchain technology in higher education in 2014. When it began legally using the technology to store and authenticate its certificates, it was also the first to collect tuition fees in the form of Bitcoin. Since 2017, UNIC has used its development, the "block.co" blockchain platform, to publish all of its graduate diploma programs (Bachelor, Master, and Ph.D.) on the Blockchain [19]. Educational and medical institutions, government offices, transportation, and other organisations utilise the platform. Sony

Global Education established a blockchain platform for storing, safeguarding, and transmitting information on student performance and development in 2016, with IBM's help. According to the business, blockchain is an essential technology that will shape the future of education [22]. In 2017, the Massachusetts Institute of Technology (MIT) created a blockchain-based block certificate that simplifies the issuance of digital degrees and professional certifications [19].

Grech and Gamileri used qualitative research methodologies, including expert interviews and literature reviews in their study and in doing do outlined eight possibilities for how blockchain may be used in education establishments: (1) Permanent certificate protection; (2) Multi-stage recognition verification; (3) Automatic credit identification and transaction; (4) Training passport for life; (5) Track intellectual property rights and their reasonable use; (6) To accept the expenses from students via blockchain; (7) Using Blockchain (BC), provide financial help to students (issue vouchers to them) (8) Determine who the students are (Grech and Galimeri 2017). The proposed blockchain development scenarios in education, with the "excluding (7)" are medium and short-term in nature. The long-term future was not taken into consideration in this study. The two writers feel that the following unsolved issues are stumbling blocks to blockchain adoption: data verification in third-world nations is difficult, there is no centralised data administration, and there is no global standard system. The author examined only three of the six unique characteristics of blockchain.

Furthermore, all initiatives in this field are still in the pilot phase at writing. All of the evaluations we looked at were classified based on the following academic publications that looked into the limitations of Blockchain (BC): 1. The process of storing and processing information is decentralised [46]. Each network participant saves all recorded data. It allows for globally distributed networks, centralised databases, and backups [22] 2. Information that cannot be changed. The blockchain allows for verifying the complete chain of information at any moment [46] 3. Keep note of transactions. Any network participant can read the entire content of the transaction, but if there is no unique key, they cannot access and change the records of other participants [12, 22, 46] 4. Consensus mechanism: Participants are responsible for maintaining the network on their own. It refers to the maintenance of the blockchain network's nodes cooperatively. It aids in the resolution of issues such as data tampering hazards and does not necessitate the involvement of a third party (Grech and Galimeri 2017) [22, 46]. The fifth option is cryptocurrency. It is used to overcome challenges relating to the execution of proof of commitments; for example, smart contracts can provide incentives in the form of bitcoin [12, 22]. Additionally, smart contracts are based on the direct involvement of all parties. It is self-contained functional software that is transferred over the blockchain's nodes. It is a real-time proof of promise execution tool that assists in the reduction of commission and internal audit losses [12, 44]. Chen et al. [12] proposed four application trends in the educational process based on the blockchain's unique characteristics and benefits and limitations: (1) Identify the participants in the education process, (2) Legal rights and promises are digitally encoded of the parties; (3) Student performance evaluation; (4) As a source of revenue, education is essential. According to paper by [19] identified four (04) areas of Blockchain (BC) applications, including seventeen developments plans.

3 Methodology

This research article combines two primary ways to address the above-mentioned critical issues: a case-based method and a research-based method. This study looks at educational programs incorporating blockchain technology as part of the first approach. The projects addressed span small to large educational institutions, private to public, national (Saudi) to international. At the same time, the research paper reviews existing research to give a more comprehensive understanding of the benefits and drawbacks of applying blockchain in education. In general, the holistic approach mentioned in this article focuses on educational institutions' (knowledge providers') and learners' viewpoints (knowledge beneficiaries). In addition to the information gleaned from interviews with technical experts, entrepreneurs, policymakers and academic community who have used blockchain education solutions, this research is based on the comprehensive and substantive opportunities and challenges that arise in the development of blockchain applied in the field of higher education.

3.1 Higher Education and Blockchain

Before delving into the two primary study objectives listed above, it is vital to answer a question: Should blockchain technology be utilised in schools? As a result, this study was conducted. According to [38], the teaching process may be enhanced in several ways by using blockchain.

3.2 Empowering Learner Autonomy

Blockchain and data (for example, certificates, learned skills, etc.) are helpful for student identity, not a central management agency such as academia (varsity), but a student. The students must have the option of storing their lifetime book data collection (both within and outside the classroom), eventual ownership, and control over who gets access to it (for example, employers). This might lead to learners proving the accuracy of their qualifications on their resumes and giving them more control over what employers can see. Even though students profit from the blockchain "wallet," they may keep and share all learning data with anybody. Students are the exclusive proprietors of their identity-related data, but they also benefit from the professor's assistance, guaranteeing that they are not alone in their educational endeavours.

3.3 *Improved Safety and Efficiency of Higher Education Institutions, Enterprises and Learners*

The identity, privacy, and security of student data may also be ensured through blockchain. The blockchain-primarily ensures immutability through its hash chain, which guarantees security and efficacy. Students and learners, for example, cannot alter prior education certificates recorded on the blockchain, but they may readily alter paper records. Furthermore, the blockchain, not the storage of data, but the hash of the data, is the primary means of ensuring anonymity. Before being stored on the blockchain, the data might be encrypted if desired. By record-keeping, [40] mentioned several blockchain-driven effectiveness functions, including digital certificate and intellectual property management, expedited diploma verification, and rapid and secure student payments. Not only do these applications help higher education institutions save money and time, but they also help companies and individual higher education students.

3.4 *Trust Support and Transparency Integration*

Blockchain also assures that students and users/learners cannot alter their grades, degrees, or certificates, giving businesses confidence that job applicants possess the necessary abilities to thrive in the field. As a result, the blockchain has established itself as a "trust anchor for a single source of credential truth" Thayer et al. Furthermore, the anchor allows for a better match between job candidates and companies. Distributed ledger technology, in general, improve learning and safeguard academic records by adding trust and transparency to the talents trade and sharing process. They also strengthen linkages between colleges, universities, and companies, as well as their ties to society.

4 Impact of Blockchain (BC) Technology on Saudi Arabia's Education Sector

Bitcoin and other cryptocurrencies function because of blockchain, which is a distributed ledger. It is without a doubt the most widely used cryptocurrency technology. Because of the security and integrity of blockchain technology, it is looking for a means to infiltrate the Saudi Arabian education business. There are several advantages to implementing blockchain technology in educational institutions around the Kingdom today [27]. The many uses of Blockchain in Saudi education are examined in this research paper.

4.1 To Help in the Verification of Student Records and Certification in the Kingdom

The record-keeping of certificates and student credentials at educational institutions has been transformed by blockchain technology. There is no need for a middleman when using blockchain technology to validate degrees, certificates, diplomas, and other academic papers [3]. The use of blockchain technology in Saudi educational institutions can also aid with certification. Most nations find certifying and certifying most learning institutions problematic [11]. Most educational institutions' credentials and educational quality are simple to check using blockchain technology.

4.2 Reduce Fraud Cases in Educational Ministry

There have been several reported instances of job candidates falsifying academic papers. When applying for employment in Saudi Arabia, many applicants presented falsified accounts of their academic qualifications. In certain circumstances, educational employers or interviewers are unable to verify the applicant's abilities due to a lack of a score sheet. As a result, they are more prone to recruit inexperienced or untrained workers to finish the job. Education is one of the most vulnerable businesses to fraud and hacking assaults. Hackers might theoretically change and erase data in the educational system. They do this to politicians, producing false certifications. All educational scams can be avoided with the use of blockchain technology. It certifies that the ledger of all academic qualifications is consistent and transparent [3]. Anyone needs the permission of the network user to manipulate the information [11]. In addition, Saudi universities can use custom blockchain protocols to develop ledgers. This platform is essential for issuing student transcripts and certificates in the Kingdom.

4.3 Reduce Fraud Cases in Educational Ministry

There have been several reported instances of job candidates falsifying academic papers. When applying for employment in Saudi Arabia, many applicants presented falsified accounts of their academic qualifications. In certain circumstances, educational employers or interviewers are unable to verify the applicant's abilities due to a lack of a score sheet. As a result, they are more prone to recruit inexperienced or untrained workers to finish the job.

4.4 Decentralization of E-Learning

Online courses are now available for both domestic and foreign students at educational institutions. Students can enroll in online courses at their preferred university. Inconsistency is a problem while choosing online courses from several educational institutions. Different learning institutions have distinct characteristics and profiles from one another [8]. The majority of them tell their students what they should study. This is due to the lack of real-time data interchange. They supply pupils with pre-recorded lessons that they must watch while studying. Students have little opportunity to communicate with the tutor or even ask questions. It generally leads to pupils paying other online instructors and essay writing firms for assistance (such as essayservice.com). Blockchain technology may be used by educational institutions to enable decentralized online learning [37].

4.5 Create a Better Learning Platform in Saudi Arabia

Blockchain technology will help develop a better online learning platform. It's worth emphasizing that colleges have the ability to create user-friendly education systems and programs that link students and professors. Schools improve the acquisition and sharing of learning resources by employing educational principles. Clients (users) can also download learning materials and access other educational services in the Kingdom. When students on the education platform watch the tutorial and invite new users to join the system, they will receive tokens [41]. When people connect with their material, content providers will receive extra tokens. Learning becomes more convenient for everyone across the Kingdom as a result of the speedy interaction between students, instructors, and developers [37].

5 Blockchain's Application in Education Sector

Their comprehensive research [5] identify twelve types of blockchain-based educational applications in their comprehensive research. Despite their exhaustive list, the author does not detail each type of application. As a result, this essay will use their list as a starting point while also looking at the work of other experts and entrepreneurs to define the essential blockchain applications in education.

5.1 Create a Better Learning Platform in Saudi Arabia

According to Sanz-Cobena et al. [43] blockchain technology allows that the students' and learners' academic data will be made public and easily shared with corporations and institutions, allowing for more opportunities for personal growth. "An authorised educational schedule may be utilised to anticipate future potential based on individual students' learning history" in this way. The program will give proper tools for students/learners and employers who can accurately and honestly represent students' potential based on academic success and students who can track and convey their academic progress (trustworthy authentication). It allows students/learners to profit.

The Digital Credentials Alliance [43], which was established in 2018 and is headed by partnerships amongst significant institutions throughout the world, is a significant academic project to develop a worldwide infrastructure for digital credentials. Its goal is to create a secure platform for academic digital certificates. According to a McKinsey report, according to certificate alliance, partners intend to create a vast worldwide network of educational institutions and industry ecosystems (i.e., employers) utilising their specified standards in the next five years.

In this case, learners who use the Digital Certificate Alliance platform. It provides access to verified transcripts of lifelong learning to share with employers. It can earn certificates safely and securely without requiring or requiring their colleges to pay for copies of transcripts and curatorial certificates received from several colleges they provide. Higher education institutions can generate profit from a streamlined procedure for issuing many certificates to one learner source and keeping and exchanging student information cost-effectively and safely. Finally, as previously said, having simple access to confirmed academic qualifications of potential employees would assist companies. MIT, one of the Digital Certificate Alliance's sponsors, a university, has already developed several blockchain applications to aid with education accreditation. For example, certification developed by the Massachusetts Institute of Technology Media Lab and Learning Machine allows users top first download the app and generate a secret password to protect their ownership; next they add the certificate issuer to their applications. Such measure are a fantastic example of a (blockchain) certification project [26]. Where traditional academic degrees cannot guarantee topic skills. Furthermore, educational institutions implementing certificate and identity management blockchain applications include the University of Nicosia [26]. American students/study and recognized courses through verifiable course certificates on the blockchain. However, the matching system developed by the open-source university will assist firms in finding eligible applicants and will leverage users' degree credits to help them see prospective and suggested job alternatives. Furthermore, the platform includes teaching elements, and smart contracts are used to manage all fees paid to learn content providers.

If students acquire a new skill, collaborate on assignments, or manage colleagues at work, [38] said, "these abilities and experience may also be recognised on the academic transcript". As a result, certificate management blockchain technology in education extends beyond academic performance to learning outside of the classroom. In addition to higher education institutions implementing blockchain for learning certification, many private start-ups are doing so as well. The blockchain [6] is a well-known mechanism that "dematerialises and automates the issuance of certification diplomas and certificates". Many universities, large corporations, and start-ups are using blockchain technology to strengthen and simplify the process of giving learners with everlasting digital certificates in order to identify and verify success validity (e.g., degrees) in and out of the classroom (e.g., MOOC or large-scale Open online courses), as well as other online courses.

5.2 *Motivating and Enhancing Lifelong Learning*

Blockchain has several uses in teaching and learning more attractive and engaging is a part of the educational process. According to [1], students, learners and teachers are "frustrated with many common online learning technologies" that cannot effectively attract learners in this scenario. As a result, they looked into open-source blockchain technology as a viable tool for improving or enhancing the current state of online education and learning. Devine characterises blockchain learning as a decentralised learning contract focusing on students/learners/exchange, as defined by Melanie Swan's "Blockchain: Blueprint for a New Economy" [6].

"Tokenisation" is the economy's future, and everyone and companies have their tokens. Student Coin lets users create their bespoke coins for use in crowdfunding, trading, governance, voting, and payment. Tokens created using Student Coin are valued using the STC Token as the primary reference currency and may be exchanged and crowdfunded on Student Coin's platform STC Exchange [15]. For example, how do Students issue deFi tokens to decentralise student loans? DeFi tokens are "tokenised" using STC-based tokens as mentioned below Fig. 1. Students create their tokens to obtain decentralised student loans and pay tuition for all promising universities' STC exchange users may trade and finance a variety of personal tokens established using student currencies to earn recurring pay-outs.

The project's primary purpose is to bring tokens and blockchain into the mainstream-Student Coin is a virgin platform that allows large-scale tokenisation of individuals, organisations, companies, start-ups, and decentralised finance (De-Fi). Large-scale tokenisation is a process that requires an easy-to-use and universal ecosystem built by student currency. With the STC ecosystem, everyone can easily create any token and carry out long-term development. Student Coin will allow tokenisation of individuals, companies, and organisations, as well as management of token utilities, crowdfunding and easy listing of new projects [17].

Higher education universities are places to carry out and cultivate innovation, find new technological research and be the source of great research ideas. In addition, universities worldwide are now more connected than ever, and there is a lot of external value transfer. Finally, the global education market is one of the fastest-growing

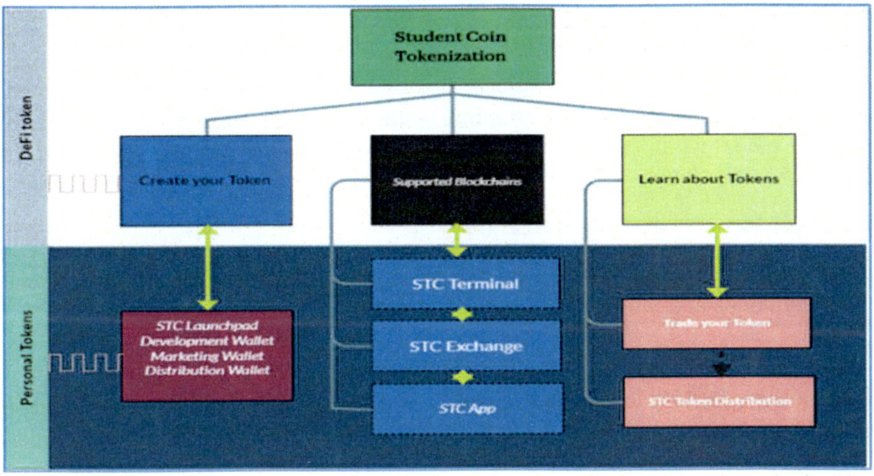

Fig. 1 Student coin tokenization

industries, and it creates a suitable environment for building disruptive blockchain-based solutions (such as student coins). These three factors make the university the perfect hotbed for creating the first large-scale tokenisation platform, similar to the origins of the internet and social media.

The Student Coin has a disruptive business model. It is focused on creating the educational ecosystem and platform for tokenisation with STC as the reference currency. Student Coin attempts to create an ecosystem and a platform for tokenisation, which will maximally benefit from the application of tokens. The project is based on the fundamentals and accrued value of the native, digital STC token and the natural, unique solutions provided by blockchain-based assets. The STC token will power the STC Terminal, grant access to the entire educational panel, and work as the primary reference currency at STC Exchange—where only STC-based personnel, De-Fi, NFT, start-up, and corporate tokens will be traded. The value will be taken from setting the STC as the only reference currency for STC-based tokens, which will cause its value to grow with the growth of STC-backed assets and further community growth. The primary mechanism of the value proposition can be compared to international trade, where local currencies appreciate the increased demand for locally produced goods and exports.

Moreover, the world's education is struggling with digital transformation. Education is a defined market that grows yearly and creates sustainable demand without big macroeconomic swings [39]. Because it entirely focuses on the long-term and stable development of the project and the flexibility of our STC ecosystem.

This study also aims to revolutionise the world of higher education by implementing blockchain technology and creating tokens for everyone. Cooperation with academic institutions is key to the broader use of tokens. Internet and social media went mainstream due to the involvement of communities from the top world universities, with a similar path currently being experienced with the Student Coin [45].

This study aims to achieve all our goals through the stable and deliberate growth of the project, consistent with the progress made in the last two years in the Kingdom. For the Student Coin, the launch of Launchpad ICO is the natural consequence of the sustainable development of the project.

Academic Reports and Identification

Universities have a tough time keeping track of their students' identities and crucial expert guidance and information.

Pedagogy in the Twenty-First Century

Every learner and student have their own set of learning abilities. It is critical to creating new educational frameworks that cater to the individual needs of each learner.

Students Loans

Education spending persuades students to choose their preferred career path. Learner incentive programs were created to acknowledge the abilities of the students.

A University or College of Higher Education

New higher education frameworks are being implemented to satisfy expected demands or to realize a desire. Purpose of this recommended approach is to provide new ideas to higher education shareholders in order to merge and join a group. After that, encourage the safe exchange of ideas. It significantly enhances the security of IoT (Internet of Things) and provides safe or real interaction between all users [18]. The goal of this study is to develop a communication framework for the educational system and to use blockchain technology to provide a safe, dependable, and fast link between connected users on the Internet of Things [4].

6 Why is Blockchain an Important Function

Data packets are mainly transmitted in the cloud through blockchain technology nodes. Every node is sometimes referred to as a block. Each block includes its hash code (a one-of-a-kind identifier, similar to a fingerprint), as well as the preceding block's hash code and contents. Each node has a link via which data may be exchanged. Hash codes and data for numerous blocks and their prior blocks are stored in each connection that transfers data. As a result, these blocks are linked to security via encryption technology. A point-to-point topology network distributes these blocks in a decentralised system [44]. Once a new block is formed, it will migrate to the network and check the authenticity of each linked node. Different experiments are used to assess the proposed system's performance. To begin, the author used open-source software to produce thousands of fixed-size blocks [44]. The authors co-created a cloud connected IoT network as well as fog and blockchain.

Meanwhile, the proposed structure covers the usage of fog computing and IoT devices in the education sector to interact, interchange, and then distribute information amongst IoT endpoints utilising blockchain technology at the network's edge [41]. The structure's transactions are proposed to be sent via a point-to-point system architecture [43]. If the transaction is confirmed, transform it into blocks, add it to the blockchain, or send it over the channel. Its miners are critical to the blockchain's ability to respond to the generation of new blocks. We employed a variety of criteria to identify the structure throughout the study process. A hash algorithm was introduced using Hyperledger's IROHA tool. The primary device is where Docker-compose is installed. Decentralised Hyperledger, proof of work (POW) methods, P2P (point-to-point) systems, and other services are among the Hyperledger IROHA technologies [42]. The Hyperledger Sumeragi IROHA technique was added to the blockchain. To interface with IoT (Internet of Things) endpoints, IROHA Android and iOS software packages may require blockchain facilities.

IoT (Internet of Things) nodes may have requested transactions, according to the Sumeragi algorithm, and they typically follow the following steps:

Stage 1: Transmission

Members can approve, arrange, or sign-up contracts and give data to the system.

Stage 2: Identity Verification and Signature

It verifies, authorizes, or certifies transactions, as well as broadcasting permitted IoT transactions to participant networks as explained below Figs. 2, 3 and 4.

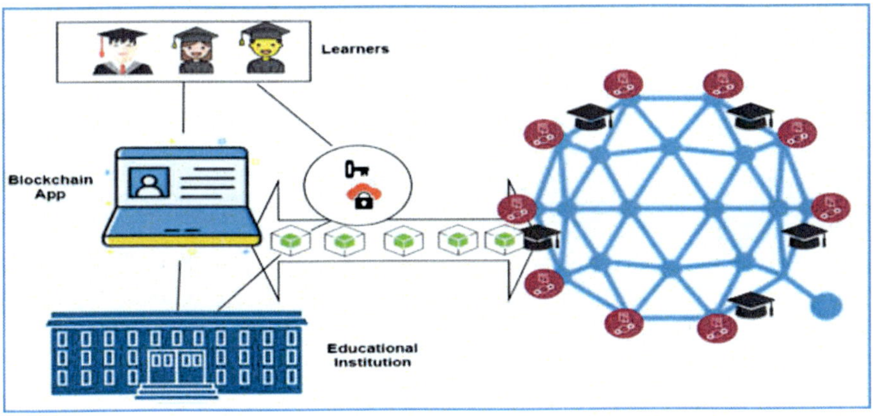

Fig. 2 Blockchain identification

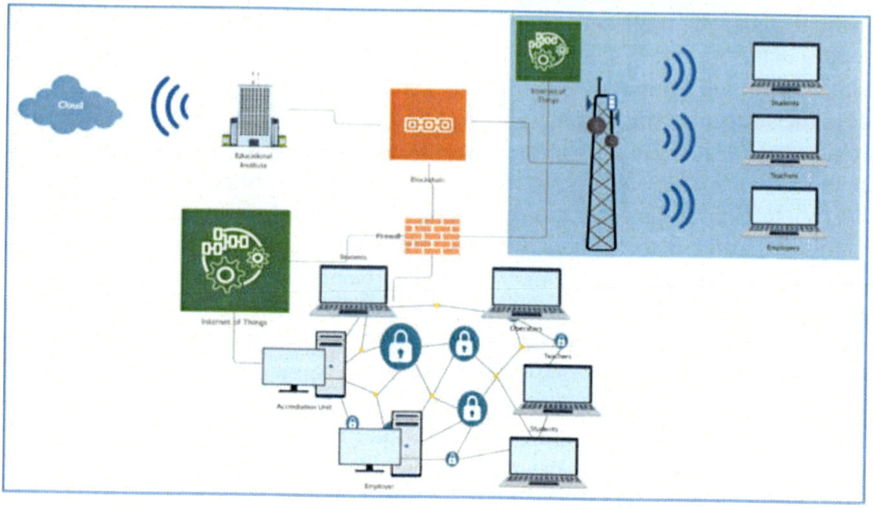

Fig. 3 Blockchain (BC)-based educational system

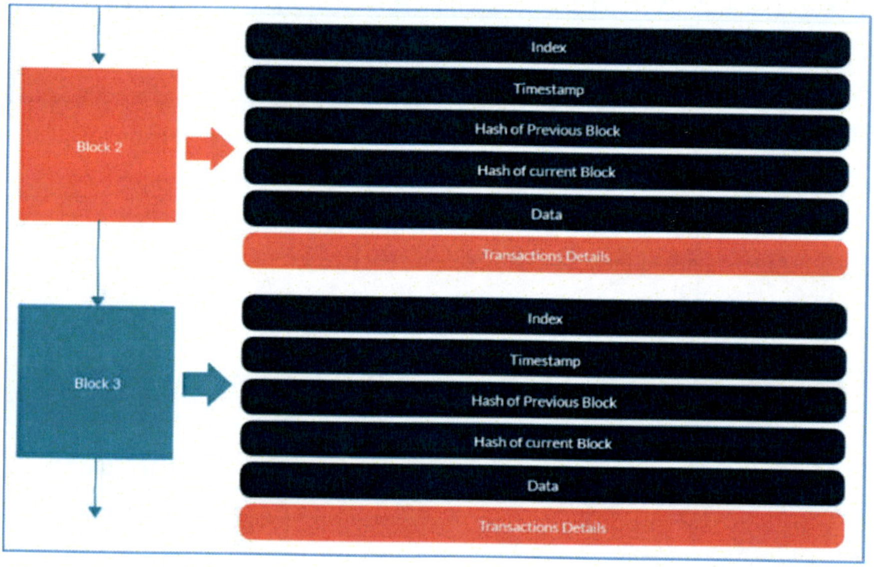

Fig. 4 Blockchain-IoT framework for the educational system

7 Limitations and Future Work

The education blockchain platform is currently in a prototype stage and can be expanded and optimised in different aspects. It impossible to anticipate how and whether blockchain would have a substantial and long-term influence on education

[13]. If major multinational corporations and governments in Saudi Arabia do not begin to utilize and appreciate digital certificates soon, some limitations on academic digital certificates may be lifted within five years. The notion of lifelong education, which entails constantly acquiring new abilities and updating old ones, will necessitate reliable and unalterable learning certification records. As a result, as lifelong education becomes more important in a society dominated by rapid technology breakthroughs, the need for blockchain-based certification may rise.

Blockchain education solution providers, such as the Digital Certificate Partnership, want to create an ecosystem of educational institutions that employ the standards they specify. However, they recognize the need and complexity of developing a long-term business model and market acceptance plan in order to make their vision a reality. Consequently, global world has moved its perception of blockchain from its use in bitcoin to its promise in new areas like healthcare and education. However, in other sectors, such as the supply chain, technology has shown to be successful [40]. The education industry's early "prototyping" stage [5]. To have a large-scale beneficial influence on blockchain solutions in education, the commercial sector (transnational corporation's enterprises) as well as the government sector (academic institutions, authorities) must work together to test, study, create, implement, and fund such innovations.

Furthermore, blockchain should not be viewed as a threat to or a replacement for educational institutions. Innovative technology, on the other hand, may provide value to a wide range of educational processes by making learning more appealing and efficient, lowering costs, enhancing trust, and improving security and privacy.

8 Discussion

A few years ago, "blockchain" was not widely used. Today, we have heard a lot of it and its application in education. Due to its nature and security as a decentralised ledger, blockchain can find applications in healthcare, land documents and registry and now education. The advantage of using blockchain in education lies in the security it provides for data, and it can protect data even if any node is compromised [2]. In addition, the General Data Protection Regulation (GDPR) was implemented in Europe, and other countries adopted similar guidelines. Including India (the Personal Data Protection Act, which was introduced in the Indian Parliament in 2019–20, currently to be determined) may cause concerns about such issues [10] implementing blockchain for student information. The main obstacle is whether student information is classified as personal data, especially stored on a distributed ledger. With the introduction of cryptocurrency, blockchain has begun to change how the financial sector operates. In education, the possibilities may be just the tip of the iceberg in terms of how blockchain will affect the future education system. The application of blockchain in education seems promising.

According to estimates by the Saudi communications sector regulator, the blockchain market in Saudi Arabia is expected to grow by 41% between 2021 and

2025. According to the Saudi Press Agency, the rapid growth of the blockchain market is part of the broader expected growth in the IT and emerging technology sectors. It will reach 26.63 billion USD by 2025, a compound annual growth rate of 10%, and today's market size is 17.31 billion USD across the KSA.

Technologies like blockchain can accelerate this process in KSA's education sector. Corporate governance and regulation are facilitated by blockchain technology [23]. As a preferred teaching approach, ledger technology, artificial intelligence, cell phones, and iPad are increasingly replacing cumbersome desktop computers and outmoded books. According to recent research by the University of Cincinnati, technological advancements benefit the classroom greatly, with 92 percent of teachers believing that technology has a substantial influence on how students learn and interact. Blockchain is one of the technologies that has a favourable influence on the learning process. Its ledger technology may help students and instructors manage accountability, transparency, and their entire educational experience [38]. Although still in its early stages, the blockchain in education technology has been hailed as the "right technology" that can completely transform a system that has been considered obsolete for many years. Below are examples, as mentioned in Table 1, of how blockchain can be implemented in the education sector.

9 Discussion

Student Coin is being touted as a game-changing blockchain-based credit platform for higher education. The suggested platform combines the advantages of blockchain and other technologies to produce a highly dependable and internationally trusted credit and score system for higher education [16]. We demonstrated a prototype version of the Student Coin platform that may be used as an open-source Ark blockchain platform as a proof of concept. The proposed student currency platform gives a global uniform perspective for students and organisations. Students have access to a single, transparent picture of their completed courses. Higher education institutions gain access to the most updated data information, independent of the student's educational background. Potential employers would also benefit from the proposed approach since they will be able to check the information given by the pupils immediately.

A distributed P2P (peer-to-peer) network system might be used to implement the suggested method. It aids in the transition of higher education scoring systems from existing conventional digital or physical records (such as databases) to a more efficient, accessible, omnipresent, and efficient blockchain (BC) technology. It is predicted that such a system would grow into a united, streamlined, universal credit and score system for higher education. All HEIs (higher education institutions) are highly encouraged to contact us about joining the Student Coin initiative. The author wants to test prototypes in real-world settings in the future, including institutions, students, and businesses. The suggested notion may be further validated in this way.

In addition, the author wants to use the student coin blockchain to assign each course a unique blockchain address and token pool. Instead of receiving tokens

Table 1 Blockchain education companies changing the nexus of education sector

Blockchain education company	Location	Usage of blockchain in education	Industry impact
Blockcerts	Cambridge, USA	Blockcerts, an open standard platform for producing, issuing, and validating blockchain support certificates, was created by Learning Machine in collaboration with the MIT Media Lab. Companies can examine the integrity of papers and uncover fraudulent information by establishing records (such as transcripts and certificates) on the blockchain	In 2018, almost 600 MIT alumni elected to get their diplomas digitally using Blockcerts' Blockchain. Students' academic records will be preserved indefinitely, and potential employers will be able to check them at any time
APPII	London, England	Users of APPII establish personal profiles and academic resumes, which include educational background and transcripts. Apple then utilizes the blockchain to validate the user's identity and save their data in its system	The Open University and APPII cooperated to develop a qualification platform for new student management. They're leaving alumni with irreversible academic records
Gilgamesh Platform	California, USA	Gilgamesh is an Ethereum smart contract-based knowledge-sharing network. Book readers, students, and authors can assemble to discuss books or other works, just like they do on social networking platforms. The most notable distinction is that people earn GIL tokens for sharing, discussing, and commenting about the information	The Gilgamesh app (iOS only) includes book recommendations, social media feeds, and a GIL token wallet for interacting with other knowledge seekers

(continued)

Table 1 (continued)

Blockchain education company	Location	Usage of blockchain in education	Industry impact
ODEM	Chiasso (Switzerland)	Professors and students may agree on a certain course using smart contracts, which will assist students advance their education and solidify their professional credentials. The ODEM ledger will then identify every course the student has taken or that the professor has taught, improving both of their reputations on the ODEM platform. ODEM tokens can also be used to pay for course expenses	For instructors and students to demonstrate expertise in certain disciplines, ODEM has established "skills badges." This is because more ODEM students will desire to take classes offered by professors who have improved their ODEM badges
Sony Global Education	Tokyo (Japan)	Sony Global Education teamed up with IBM to create a blockchain platform that allows numerous schools to upload personal academic achievements and other essential student information to the ledger, allowing students to transfer or continue their education with an unmistakable record	For the 2018 Global Math Challenge, Sony used their blockchain to give certificates to contestants. These certificates can be utilized as permanent score reports to aid participants in their future academic or professional endeavors
BitDegree	Lithuania, Kaunas	BitDegree's technology-focused online education platform integrates blockchain ledger and tokenization. "Cryptocurrency for Dummies," for example, is one of the company's online courses. The site also encourages students to learn by awarding tokenized scholarships for successfully completing courses	From cryptocurrency to blockchain implementation, the online portal provides both free and paid courses. To the world of gamified coding. From bitcoin to blockchain implementation to gamification coding, rm provides both free and paid courses

directly from the university, students will receive tokens from the course address after fulfilling the course duties. The course address will be the institution's and professor's multi-signature address. The platform presented is currently based on the ECTS scoring system. However, it may be developed and adapted to any existing credit or certification system to include other components of the higher education certification version.

10 Declaration

The authors of this review have no conflict of interest among them or anybody else regarding the conceptual content, financial matters, or otherwise.

Acknowledgements This research study was supported by ***the College of Business Administration (CBA), Majmaah University (MU)***, Kingdom of Saudi Arabia.

References

1. Aggarwal, S., Kumar, N.: Basics of blockchain. In: Advances in Computers, vol. 121, pp. 129–146. Elsevier (2021)
2. Alam, S.: A blockchain-based framework for secure educational credentials. Turk. J. Comput. Math. Educ. (TURCOMAT) **12**(10), 5157–5167 (2021)
3. Alam, S., et al.: Blockchain-based Initiatives: current state and challenges. Comput. Netw. **198**, 108395 (2021)
4. Alam, T., Benaida, M.: Blockchain and internet of things in higher education. Univ. J. Educ. Res. **8**, 2164–2174 (2020)
5. Alammary, A., Alhazmi, S., Almasri, M., Gillani, S.: Blockchain-based applications in education: a systematic review. Appl. Sci. **9**(12), 2400 (2019)
6. Almuqren, L., Cristea, A.: AraCust: a Saudi telecom tweets corpus for sentiment analysis. PeerJ Comput. Sci. **7**, e510 (2021). https://doi.org/10.7717/peerj-cs.510
7. Angrish, A., et al.: A case study for blockchain in manufacturing:"FabRec": a prototype for peer-to-peer network of manufacturing nodes. Procedia Manuf. **26**, 1180–1192 (2018)
8. Arcinas, M.M.: 2021 A blockchain based framework for securing students' educational data. Linguist. Antverp. **2**, 4475–4484 (2021)
9. Benisi, N.Z., Aminian, M., Javadi, B.: Blockchain-based decentralized storage networks: a survey. J. Netw. Comput. Appl. **162**, 102656 (2020)
10. Bhoi, B.K.: Digital Transformation in India's Payment and Settlement Systems: The Way Forward (2021)
11. Bucea-Manea-Ţoniş, R., Martins, O.M.D., Bucea-Manea-Ţoniş, R., Gheorghiţă, C., Kuleto, V., Ilić, M.P., Simion, V.-E.: Blockchain technology enhances sustainable higher education. Sustainability **13**(22), 12347 (2021). https://doi.org/10.3390/su132212347
12. Chen, G., et al.: Exploring blockchain technology and its potential applications for education. Smart Learn. Environ. **5**(1), 1–10 (2018)
13. Chiu, H.-Y., Lim, E.W.K.: Technology vs ideology: how far will artificial intelligence and distributed ledger technology transform corporate governance and business? Berkeley Bus. Law J. **18**(1), 1–63 (2021)

14. Cong, L.W., He, Z.: Blockchain disruption and smart contracts. Rev. Financ. Stud. **32**(5), 1754–1797 (2019)
15. Curchoe, C.L.: The blockchain and decentralized manipulation of confidential information: uses in medical healthcare and assisted reproduction. J. Assist. Reprod. Genet. 1–3 (2022)
16. de Castro, R.Q.: Blockchain applications for diploma verification in higher education institutions in Portugal-a mixed methods study, pp. 1–101 (2021)
17. Dorsala, M.R., Sastry, V., Chapram, S.: Blockchain-based solutions for cloud computing: a survey. J. Netw. Comput. Appl. **196**, 103246 (2021)
18. Dotan, M., et al.: Survey on blockchain networking: context, state-of-the-art, challenges. ACM Comput. Surv. (CSUR) **54**(5), 1–34 (2021)
19. Durant, E., Trachy, A.: Digital diploma debuts at MIT. Massachusetts Institute of Technology (2017)
20. Fedorova, E.P., Skobleva, E.I.: Application of blockchain technology in higher education. Eur. J. Contemp. Educ. **9**(3), 552–571 (2020)
21. Grech, A., Camilleri, A.F.: Blockchain in Education. Publications Office of the European Union, Luxembourg (2017)
22. Hameed, B., et al.: A review of blockchain based educational projects. Int. J. Adv. Comput. Sci. Appl. **10**(10), 1–9 (2019)
23. Jiang, S., et al.: Policy assessments for the carbon emission flows and sustainability of bitcoin blockchain operation in China. Nat. Commun. **12**(1), 1–10 (2021)
24. Kamal, Z.A., Fareed, R.: A Proposed hash algorithm to use for blockchain base transaction flow system. Period. Eng. Nat. Sci. **9**(4), 657–673 (2021)
25. Kamišalić, A., et al.: A preliminary review of blockchain-based solutions in higher education. In: International Workshop on Learning Technology for Education in Cloud. Springer (2019)
26. Karau, H., et al.: Learning Spark: Lightning-Fast Big Data Analysis. O'Reilly Media, Inc. (2015)
27. Kilfoil, W.R., et al.: Teaching and learning review. Department for Education Innovation, University of Pretoria (2019)
28. Pina, A.R.B., et al.: Blockchain en Educación: introducción y crítica al estado de la cuestión. Edutec. Revista Electrónica de Tecnología Educativa **61**, 363–363 (2017)
29. Pokhrel, S., Chhetri, R.: A literature review on impact of COVID-19 pandemic on teaching and learning. High. Educ. Future **8**(1), 133–141 (2021)
30. Pregowska, A., et al.: A worldwide journey through distance education—from the post office to virtual, augmented and mixed realities, and education during the COVID-19 pandemic. Educ. Sci. **11**(3), 1–26 (2021)
31. Qu, C., Tao, M., Zhang, J., Hong, X., Yuan, R.: Blockchain based credibility verification method for IoT entities. Secur. Commun. Netw. 1–11 (2018)
32. Razia, B., Awwad, B.: A comprehensive review of blockchain technology and its related aspects in higher education. In: Technologies, Artificial Intelligence and the Future of Learning Post-COVID-19, pp. 553–571 (2022)
33. Rivera, A.V., Refaey, A., Hossain, E.: A blockchain framework for secure task sharing in multi-access edge computing. IEEE Netw. **35**(3), 176–183 (2020)
34. Salman, T., et al.: Security services using blockchains: a state of the art survey. IEEE Commun. Surv. & Tutor. **21**(1), 858–880 (2018)
35. Sanz-Cobena, A., et al.: Strategies for greenhouse gas emissions mitigation in Mediterranean agriculture: a review. Agr. Ecosyst. Environ. **238**, 5–24 (2017)
36. Sharples, M., Domingue, J.: The Blockchain and Kudos: A Distributed System for Educational Record, Reputation and Reward. Springer International Publishing, Cham (2016)
37. Shuaib, M., et al.: Blockchain-based framework for secure and reliable land registry system. Telkomnika **18**(5), 2560–2571 (2020)
38. Shuaib, M., Hafizah Hassan, N., Usman, S., et al.: Identity model for blockchain-based land registry system: a comparison. Wirel. Commun. Mob. Comput. 1–17 (2022)
39. Siddiqui, S.T., et al.: Blockchain security threats, attacks and countermeasures. In: Ambient Communications and Computer Systems, pp. 51–62. Springer (2020)

40. Thayer, T.: Promising and ambitious blockchain initiatives for higher education. Gartner Research (2018)
41. Tian, Y., et al.: Asset tokenization: a blockchain solution to financing infrastructure in emerging markets and developing economies, pp.1–27. Institute of Global Finance Working Paper (2020)
42. Tribis, Y., El Bouchti, A., Bouayad, H.: Supply chain management based on blockchain: a systematic mapping study. In: MATEC Web of Conferences, pp. 1–8. EDP Sciences (2018)
43. Turkanović, M., et al.: EduCTX: A blockchain-based higher education credit platform. IEEE Access **6**, 5112–5127 (2018)
44. Yadav, D., et al.: Enhancing data security in cloud using blockchain. In: 4th International Conference on Intelligent Computing and Control Systems (ICICCS). IEEE (2020)
45. Yang, Z., et al.: A blockchain-based reputation system for data credibility assessment in vehicular networks. In: IEEE 28th Annual International Symposium on Personal, Indoor, and Mobile Radio Communications (PIMRC) (2017)
46. Yumna, H., et al.: Use of blockchain in education: a systematic literature review. In: Asian Conference on Intelligent Information and Database Systems. Springer (2019)

Special Session 2: Applications of Technology for Mental Health

Research Protocol for a Preliminary Evaluation of a Gamified Tool for Adult Wellbeing

Jacob Gavin and Steven Barnes

Abstract In this paper, the authors present the rationale and protocol regarding a preliminary assessment of an early-prototype serious game, designed in conjunction with a multidisciplinary team of academics, game-developers and potential end-users. The present study aims to provide a preliminary evaluation of the game in terms of its ability to improve mental-wellbeing in an adult sample. The proposed evaluation will consist of a quasi-experimental intervention utilising a pre-post design, with self-reported wellbeing data collected at each of the three timespoints. Additionally, engagement in ongoing and non-directed game-engagement in the period following post-intervention will be examined with analysis of the extent to which this may predict follow-up outcomes. Suggestions for ongoing assessment and potential directions for ongoing development are discussed.

Keywords Wellbeing town · Serious games · Gamification · The five ways to wellbeing · Mental wellbeing · Self-directed engagement

1 Introduction

1.1 Mental Wellbeing

Mental wellbeing refers to an individual being capable of dealing with life stressors, making positive contributions to their community, and realising their potential [1]. Positive mental wellbeing is associated with increases in psychological resilience [2],

Special Session on "Applications of Technology for Mental Health".

J. Gavin (✉)
University of Bolton, Bolton, UK
e-mail: jmg1eps@bolton.ac.uk

S. Barnes
Uniwersytet SWPS Humanistycznospołeczny, Katowice, Poland

C. Iwendi et al. (eds.), *Proceedings of ICACTCE'23—The International Conference on Advances in Communication Technology and Computer Engineering*, Lecture Notes in Networks and Systems 735, https://doi.org/10.1007/978-3-031-37164-6_50

better occupational functioning [3], reduced mortality [4], and improved self-esteem [5], and better relationships [6].

Some have argued that wellbeing is opposingly situated spectrally to mental illness and can therefore be considered to be opposed to poor mental health [7]. The negative implications of experience of mental illness are well established [8–12], as is the elevated risk of mental illness in the context of COVID-19 social restrictions [13]. Wellbeing is significant as a psychological construct in that it not only serves to help prevent psychological distress and experience of mental illness, but may also aid in the recovery process [14]. The increasing burden of poor mental health [15] and the associated increases in demand for services [16] emphasise the relevance and value of effective and accessible tools for improving wellbeing.

1.2 Five Ways to Wellbeing

The Five Ways to Wellbeing (FWTW) framework [17] comprises of evidence-based behaviours designed to improve wellbeing [18], including: (1), connecting with others; (2), being active; (3), taking notice (aware of surroundings); (4) keep learning; and (5), giving (volunteering, doing good deeds for others) [19]. The New Economics Forum [20] recently published a series of proposals for using the framework to help individuals with coping with social restrictions. The framework holds appeal in that it is readily adapted for use in daily life [21]. While FWTW does appear to produce improvements to wellbeing at even minor levels of use [22], encouraging engagement, as with many psychological interventions, remains a barrier to entry. Mediums capable of fostering intrinsic motivation may be valuable in improving engagement with this validated framework.

1.3 Serious Games and Applications to Psychological Intervention

Using games to deliver educational and serious content is, in itself, not a new phenomenon, with the term having existed for some decades [23]. Serious games have been developed for numerous applications, including corporate [24], healthcare [25], and education [26]. Their motivational value is largely attributed to the engaging capacity of games themselves [27], with a number of theoretical explanations available, such as Self-Determination Theory [28–31], to explain how aspects of game-design foster intrinsic motivation.

In terms of alleviating psychiatric symptoms, serious games have also made significant strides [32–39], with the benefits carried over into behaviour in the real-world [40]. Even minimal use appears valuable in improving outcomes and reducing drop-out [41, 42].

Videogames can also act as a means to overcome a significant number of barriers to psychological treatment associated with non-technological means of supporting mental health [43–47]. Authors in the field have called for more scientifically rigorous research to assist with developing a sound basis for confidence in the medium [48].

While the use of games for the purposes of psychological intervention is growing rapidly, and while games have the potential to engage users and improve wellbeing [40, 49, 50], applications of serious games for positive psychology interventions are less well-documented [51]. 'The Wellbeing Game', (Mental Health Foundation of New Zealand), is one example of a gamified tool shown to be effective in stress-reduction and in making users more aware of how they might improve their wellbeing [52, 53]. Unfortunately, the game is not publicly accessible.

1.4 Serious Games and Self-directed Engagement

As discussed above, existing literature frequently points to serious games as being a means to increase and maintain engagement [54–56]. Despite a number of studies pointing to the benefits of games in structured interventions and trials, comparatively little research exists to determine the extent to which games and engagement function in more ecologically valid environments. As serious games are likely to be used by outpatients in real-world settings, and as ongoing engagement with psychological interventions is required for the maintenance of positive outcomes [57], this represents an important area for examination [51]. Determining the extent to which players may choose to use a serious game without structured direction, a more valid representation of the games motivational power may be achieved, along with an inference of what 'dose' of the game may be required to elicit improvement and to prevent relapse.

1.5 "Wellbeing Town"

"Wellbeing Town" is a prototype videogame designed by a multidisciplinary team of potential players, game developers (both professional and student), and with practitioner support from a provider local to the area (Mental Health Independent Support Team (MhIST), Bolton, United Kingdom) for the Microsoft Windows platform [51]. Based around the FWTW framework, "Wellbeing Town" is designed as a role-playing game and adopts a non-linear, open-world design with a series of interactive non-playable characters (NPCs).

The game structure is designed to utilise mechanics aimed at fostering intrinsic motivation [58, 59]. NPCs issue players with tasks corresponding to aspects of FWTW (see Appendix), the accomplishment of which is recognised via a reward structure, and a growing affiliation with in-game NPCs [60], and can be concluded in a manner and order determined by the player. The design of the game has been carefully constructed to be accessible by a broad demographic of players with varying previous experience of videogaming, ensuring the game provides a fitting level of challenge, while also remaining enjoyable to play [61]. While adults are increasingly technologically active, lack of access, experience using technology, and poor electronic health literacy can still be a barrier [51, 62, 63].

1.6 The Present Study

This study aims to provide an initial evaluation of "Wellbeing Town" in respect of its present capacity to make improvements to self-reported wellbeing. As little research currently exists to assess the use of gamified tools in outpatient settings, self-directed engagement post-intervention and prior to follow-up will be examined, along with the extent to which such self-directed play may be predictive of follow-up outcomes.

It is hypothesized that:

H1: Engagement with the game will result in significant improvements in wellbeing scores across a structured intervention period.

H2: Users who choose to voluntarily use the game during the follow-up period will demonstrate better wellbeing outcomes than users who do not use the game following the end of the structured trial. Extent of use will be related to the degree of ongoing improvement.

2 Method

2.1 Research Design and Procedure

The present research represents a preliminary evaluation of the game and utilises a quasi-experimental, single condition, pre-post experimental design and has received full ethical approval from the Ethics Committee of the University of Bolton, UK. After participants have been provided with full information regarding the research and have consented to take part, all will be assessed for baseline wellbeing scores using the Warwick-Edinburgh Mental Wellbeing Scale (WEMWBS) [64], the psychometric properties of which are provided below.

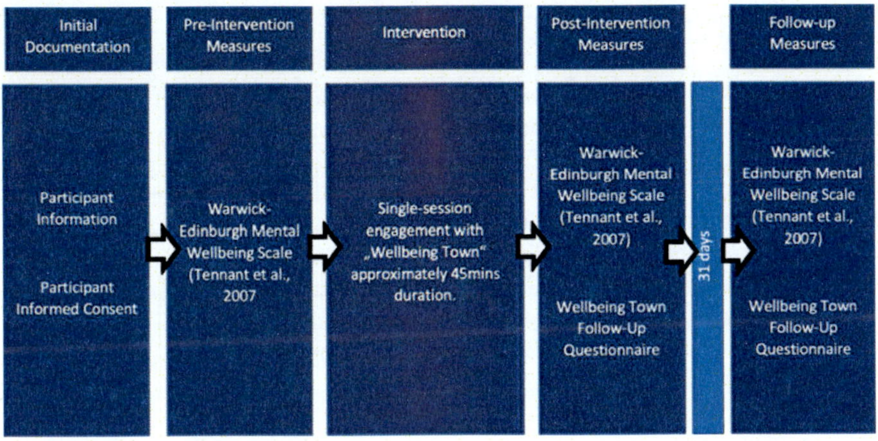

Fig. 1 Protocol for initial evaluation

As the present investigation represents an initial pilot, and as the game is currently undergoing development, participants will be asked to play "Wellbeing Town" for a single session of 45 minutes, which will be followed by a second instance of the WEMWBS. Following this designated period of use, participants will be advised that while they are no longer formally required to engage with the game, they may choose to do so as much or little as they wish over the forthcoming 31 days. Participants will not be provided with any further information concerning self-directed play, to avoid biasing ongoing use by creating an expectation effect regarding anticipated use of the game.

Finally, at the conclusion of the thirty-one day period following the structured trial period, WEMWBS data will again be obtained. At this point, an additional data regarding the game will be collected using the Wellbeing Town Follow-Up Questionnaire. This scale has been created to record ongoing voluntary use of the game following the structured intervention (see Materials).

A graphical representation of the study has been provided below (Fig. 1).

Should the results indicate that the game possesses the capability to improve well-being scores, and should there be no fundamental problems concerning to the game's design and function, it is anticipated that ongoing assessment will be conducted to more rigorously evaluate "Wellbeing Town" as a scalable tool.

2.2 Participants

Participants for this initial pilot will consist of adults aged 16 years and over and will initially be recruited via convenience sampling from the University of Bolton, an provider of higher education in England with a diverse cohort. The sampling frame will be broadened from this point via snowball sampling utilising word of mouth, to provide some further heterogeneity. As this study will not examine the game in the same manner as a more extensive clinical assessment, no selection criteria pertaining to pre-participation wellbeing scores has been determined, though it is anticipated that participants who report poorer wellbeing before the study commencing may be more likely to report more noticeable benefits of gameplay.

2.3 Materials

Warwick Edinburgh Mental Wellbeing Scale (WEMWBS). The WEMWBS [64] is a 14-item scale which examines positive functioning, positive affect, and fulfilling interpersonal relations. Responses are obtained via a 5-point Likert scale ranging from 1 ("none of the time") to 5 ("all of the time"). Responses to individual items are summed to produce an overall wellbeing score, with higher scores representing better psychological wellbeing. Previous examinations of the scale reveal it to have good internal consistency (α=.87 (95%CI [0.85; 0.88]) and good test- retest reliability (rtt=.66 (95%CI [0.59; 0.72], N=212) [65].

Wellbeing Town Follow-up Questionnaire. This scale comprises of five items concerning the extent to which users voluntarily used the game during the follow-up period (if at all) and any motivations for choosing to/not to engage with the game during this period. This questionnaire is intended to be used solely for this purpose and to allow for examination in relation to hypothesis 2.

At present, there is no suitable scale which exists to measure this kind of voluntary use of serious and therapeutic games, or its potential implications for wellbeing or clinical outcomes. Items address different features of self-directed play, and as such scores on individual items are examined individually. The authors also anticipate, given a suitable sample size, that the scale itself will be examined to determine its value as a measure for this purpose.

3 Results

3.1 Plan for Analysis

This research has two primary aims: firstly, to evaluate „Wellbeing Town in terms of self-reported wellbeing, in a preliminary manner using a controlled, structured trial. Secondly, this research will assess the extent of ongoing voluntary play following the structured trial and any potential implications for follow-up wellbeing scores.

It is expected at this stage that data analysis of wellbeing scores will comprise of repeated measures ANOVA. For participants engaging in gameplay over the followup period, dependent-samples t-test will be used to assess wellbeing scores, along with simple-linear regression analysis of time spent engaged in play and change in wellbeing score between the timepoints. Initial approximation suggests that a sample of 40–50 participants (accounting for attrition of 15–20%) will likely be required to ensure sufficient power in these analyses.

4 Discussion

4.1 Projected Significance

To our best knowledge, this study represents the first evaluation of a wellbeing game designed for adults, with the consultation and contribution of potential end-users. Additionally, this research will also provide the first assessment of voluntary play and its potential impact on ongoing outcomes. Depending on outcomes of the current evaluation, a path for further development will be determined for "Wellbeing Town".

4.2 Dissemination of Findings

In addition to the presentation of this protocol at the International Conference on Advances in Communication Technology and Computer Engineering, and the publication of the protocol in the journal Mental Health and Social Inclusion, data- collection is due to conclude by early-mid 2022. Once outcomes are established, it is intended that these also will be published in an appropriate, peer-reviewed outlet.

4.3 Anticipated Limitations

A number of limitations are anticipated at this stage due to the preliminary nature of the present evaluation. Firstly, attrition is expected during the course of the trial–it is hoped that initial recruitment will compensate for this to ensure initial analyses are not underpowered. Secondly, as the development of "Wellbeing Town" is not finalised, reaching definitive conclusions regarding its capabilities and efficacies will require further evaluation in controlled trial with randomisation and suitable control condition.

5 Conclusions

This paper has presented a protocol for preliminary assessment of a serious game designed to improve adult wellbeing. The paper has outlined the planned structure and nature of the research and route for data-analysis. Finally, this paper provides an overview of the anticipated significance and ongoing research plan concerning this project.

Acknowledgements Authors thank the development team for their work on the game:
Tanya Madeva
Callum Pogson
Robert Valache
Chloe Webb-Adam
Additionally, authors are grateful to Col Harding and Rebecca Mayhew of the Centre for Games Design at the University of Bolton for assisting in advising and guiding the development team during this process.

Funding Statement The presented research has not received funding or financial contribution from any funding agency in public, commercial, or not-for-profit sectors.

Publication Statement The present paper represents a summary of this work intended for publication in IEEE conference proceedings. A full version of this paper can be found in the journal 'Mental Health and Social Inclusion'.

Appendix: In-Game Screenshots

References

1. Burns, P., Van Der Meer, R.: Happy Hookers: findings from an international study exploring the effects of crochet on wellbeing. Perspect. Public Health **141**(3), 149–157 (2021)
2. Yıldırım, M., Özaslan, A., Arslan, G.: Perceived risk and parental coronavirus anxiety in healthcare workers: a moderated mediation role of coronavirus fear and mental well-being". Psychol. Health Med. **27**(5), 1095–1106 (2022)

3. Vainre, M., Galante, J., Watson, P., Dalgleish, T., Hitchcock, C.: Protocol for the work engagement and well-being study (SWELL): a randomised controlled feasibility trial evaluating the effects of mindfulness versus light physical exercise at work. BMJ Open **12**(4), e059051 (2022)
4. Hobbs, C., Jelbert, S., Santos, L.R., Hood, B.: Evaluation of a credit-bearing online administered happiness course on undergraduates' mental well-being during the COVID-19 pandemic. PLoS ONE **17**(2), e0263514 (2022)
5. Czenczek-Lewandowska, E., Wyszyńska, J., Leszczak, J., Baran, J., Weres, A., Mazur, A., Lewandowski, B.: Health behaviours of young adults during the outbreak of the Covid-19 pandemic–a longitudinal study. BMC Public Health **21**(1), 1–10 (2021)
6. Alanko, K., Lund, H.: Transgender youth and social support: a survey study on the effects of good relationships on well-being and mental health. Young **28**(2), 199–216 (2020)
7. Connor, C., De Valliere, N., Warwick, J., Stewart-Brown, S., Thompson, A.: The COV-ED Survey: exploring the impact of learning and teaching from home on parent/carers' and teachers' mental health and wellbeing during COVID-19 lockdown. BMC Public Health **22**(1), 1–15 (2022)
8. Kiely, K.M., Sutherland, G., Butterworth, P., Reavley, N.J.: Age and gender differences in the reciprocal relationship between social connectedness and mental health. Soc. Psychiatry Psychiatr. Epidemiol. **56**(6), 1069–1081 (2021)
9. Radicke, A., Sell, M., Adema, B., Daubmann, A., Kilian, R., Busmann, M., Winter, S.M., Lambert, M., Wiegand- Grefe, S.: Risk and protective factors associated with health-related quality of life of parents with mental illness. Front. Psychiatry **12** (2021)
10. Stevenson, J., Miller, C.L., Martin, K., Mohammadi, L., Lawn, S.: Investigating the reciprocal temporal relationships between tobacco consumption and psychological disorders for youth: an international review. BMJ Open **12**(6), e055499 (2022)
11. Ghossoub, E., Cherro, M., Akil, C., Gharzeddine, Y.: Mental illness and the risk of self- and other-directed aggression: results from the National Survey on Drug Use and Health. J. Psychiatr. Res. **132**, 161–166 (2021)
12. Launders, N., Hayes, J.F., Price, G., Osborn, D.P.: Clustering of physical health multimorbidity in people with severe mental illness: an accumulated prevalence analysis of United Kingdom primary care data. PLoS Med. **19**(4), e1003976 (2022)
13. Campion, J., Javed, A., Sartorius, N., Marmot, M.: Addressing the public mental health challenge of COVID-19. Lancet Psych **7**(8), 657–659 (2020)
14. Van Agteren, J., Bartholomaeus, J., Fassnacht, D.B., Iasiello, M., Ali, K., Lo, L., Kyrios, M.: Using internet-based psychological measurement to capture the deteriorating community mental health profile during COVID-19: observational study. JMIR Mental Health **7**(6), e20696 (2020)
15. Agboola, A.A., Esan, O.T., Afolabi, O.T., Soyinka, T.A., Oluwaranti, A.O., Adetayo, A.: Economic burden of the therapeutic management of mental illnesses and its effect on household purchasing power. PLoS ONE **13**(9), e0202396 (2018)
16. Tingley, J., Greenhalgh, I., Stallard, P.: Technology Matters: BlueIce–using a smartphone app to beat adolescent self-harm. Child Adolesc. Mental Health **25**(3), 192–194 (2020)
17. Rath, T., Harter, J.K., Harter J.: Wellbeing: The Five Essential Elements. Simon & Schuster (2010)
18. Abdallah, S., Main, G., People, L., Rees, G.: Ways to wellbeing: Exploring the links between children's activities and their subjective wellbeing. The Children's Society. Report. The Children's Society, London (2014)
19. Ng, S.S., Leung, T.K., Cheng, E.K., Chan, F.S., Chan, J.Y., Poon, D.F., Lo, A.W.: Efficacy of 'Five Ways to Well-Being Program'in promotion of mental wellbeing for persons admitted to acute psychiatric service. J. Psychosoc. Rehabil. Mental Health **2**(2), 143–151 (2015)
20. Stephens, L.: Five ways to wellbeing at a time of social distancing. New Economics Foundation. Retrieved from https://neweconomics.org/2020/03/five-ways-to-wellbeing-at-a-time-of-social-distancing (2020)
21. Phillips, R., Evans, B., Muirhead, S.: Curiosity, place and wellbeing: encouraging place-specific curiosity as a 'way to wellbeing. Environ. Plan. **47**(11), 2339–2354 (2015)

22. Mackay, L., Egli, V., Booker, L.J., Prendergast, K.: New Zealand's engagement with the five ways to wellbeing: evidence from a large cross-sectional survey, Kōtuitui: New Zealand J. Soc. Sci. Online **14**(2), 230–244 (2019)
23. Abt, C.: Serious Games. Viking Press (1970)
24. Larson, K.: Serious games and gamification in the corporate training environment: a literature review. TechTrends **64**(2), 319–328 (2020)
25. Maheu-Cadotte, M.A., Cossette, S., Dube, V., Fontaine, G., Lavallee, A., Lavoie, P., Mailhot, T., Deschenes, M.F.: Efficacy of serious games in healthcare professions education: a systematic review and meta-analysis. Simul. Healthcare **16**(3), 199–212 (2021)
26. Zhonggen, Y.: A meta-analysis of use of serious games in education over a decade. Int. J. Comp. Games Technol. **2019** (2019)
27. Sipiyaruk, K., Hatzipanagos, S., Reynolds, P.A., Gallagher, J.E.: Serious games and the COVID-19 pandemic in dental education: an integrative review of the literature. Computers **10**(4), 42 (2021)
28. Ghoman, S.K., Patel, S.D., Cutumisu, M., von Hauff, P., Jeffery, T., Brown, M.R., Schmölzer, G.M.: Serious games, a game changer in teaching neonatal resuscitation? A review. Arch. Dis. Childhood-Fetal Neonatal Ed. **105**(1), 98–107 (2020)
29. Guillaume, R.O., Kalkbrenner, M.T.: The utility of self-determination theory in faculty of color's successful pursuit of tenure and promotion to the rank of associate professor. Int. J. Educ. Res. **98**, 272–279 (2019)
30. Passalacqua, M., Léger, P.M., Nacke, L.E., Fredette, M., Labonté-Lemoyne, É., Lin, X., Caprioli, T., Sénécal, S.: Playing in the backstore: interface gamification increases warehousing workforce engagement. Ind. Manag. Data Syst. **120**(7), 1309–1330 (2020)
31. Ryan, R.M., Deci, E.L.: Self-determination theory and the facilitation of intrinsic motivation, social development, and well-being. Am. Psychol. **55**(1), 68–78 (2000)
32. Baranowski, T., Buday, R., Thompson, D., Lyons, E.J., Shirong-Lu, A., Baranowski, J.: Developing games for health behaviour change: getting started. Games for Health **2**(4), 183–190 (2013)
33. Giner-Bartolomé, C., Fagundo, A.B., Sánchez, I., Jiménez-Murcia, S., Santamaría, J.J., Ladouceur, R., Menchón, J.M., Fernández-Aranda, F.: Can an intervention based on a serious videogame prior to cognitive behavioural therapy be helpful in bulimia nervosa? A clinical case study. Front. Psychol. **6**, 982 (2015)
34. Tárrega, S., Castro-Carreras, L., Fernández-Aranda, F., Granero, R., Giner-Bartolomé, C., Aymamí, N., Gómez-Peña, M., Santamaría, J.J., Forcano, L., Steward, T., Menchón, J.M., Jiménez-Murcia, S.: A serious videogame as an additional therapy tool for training emotional regulation and impulsivity control in severe gambling disorder. Front. Psychol. **6**, 1721 (2015)
35. Wijnhoven, L.A., Creemers, D.H., Engels, R.C., Granic, I.: The effect of the video game Mindlight on anxiety symptoms in children with an autism spectrum disorder. BMC Psychiatry **15**(1), 1–9 (2015)
36. Rica, R.L., Shimojo, G.L., Gomes, M.C., Alonso, A.C., Pitta, R.M., Santa-Rosa, F.-A., Pontes Junior, F.L., Ceschini, F., Gobbo, S., Bergamin, M., Bocalini, D.S.: Effects of a Kinect-based physical training program on body composition, functional fitness and depression in institutionalized older adults. Epidemiol. Clin. Pract. Health **20**(3), 195–200 (2021)
37. Derks, S., Willemen, A.M., Sterkenburg, P.S.: Improving adaptive and cognitive skills of children with an intellectual disability and/or autism spectrum disorder: Meta-analysis of randomised controlled trials on the effects of serious games. Int. J. Child- Comp. Interact. 100488 (2022)
38. Doumas, I., Everard, G., Dehem, S., Lejeune, T.: Serious games for upper limb rehabilitation after stroke: a meta-analysis. J. Neuroeng. Rehabil. **18**(1), 1–16 (2021)
39. Min, A., Min, H., Kim, S.: Effectiveness of serious games in ånurse education: a systematic review. Nurse Educ. Today **108**, 105178 (2022)
40. Bindoff, I., de Salas, K., Peterson, G., Ling, T., Lewis, I., Wells, L., Gee, P., Ferguson, S.G.: Quittr: the design of a video game to support smoking cessation. JMIR Serious Games **4**(2), e19 (2016)

41. King, G., Currie, M., Petersen, P.: Child and parent engagement in the mental health intervention process: a motivational framework. Child Adolesc. Mental Health **19**(1), 2–8 (2012)
42. Scherer, E.A., Ben-Zeev, D., Li, Z., Kane, J.M.: Analyzing mHealth engagement: joint models for intensively collected user engagement data. JMIR Mhealth Uhealth **5**(1), e1 (2017)
43. Barnes, S., Prescott, J.: Empirical evidence for the outcomes of therapeutic video games for adolescents with anxiety disorders: systematic review. JMIR Serious Games **6**(1), e9530 (2018)
44. Ralston, A.L., Andrews, A.R., III., Hope, D.A.: Fulfilling the promise of mental health technology to reduce public health disparities: review and research agenda. Clin. Psychol. Sci. Pract. **26**(1), e12277 (2019)
45. Ong, J.G., Lim-Ashworth, N.S., Ooi, Y.P., Boon, J.S., Ang, R.P., Goh, D.H., Ong, S.H., Fung, D.S.: An interactive mobile app game to address aggression (RegnaTales): pilot quantitative study. JMIR Serious Games **7**(2), e13242 (2019)
46. Vallury, K.D., Jones, M., Oosterbroek, C.: Computerized cognitive behaviour therapy for anxiety and depression in rural areas: a systematic review. J. Med. Internet Res. **17**(6), e:139 (2015)
47. Wiljer, D., Shi, J., Lo, B., Sanches, M., Hollenberg, E., Johnson, A., Abi-Jaoudé, A., Chaim, G., Cleverley, K., Henderson, J., Isaranuwatchai, W., Levinson, A., Robb, J., Wong, H.W., Voineskos, A.: Effects of a mobile and web app (Thought Spot) on mental health help- seeking among college and university students: randomized controlled trial. J. Med. Internet Res. **22**(10), e20790 (2020)
48. Dewhirst, A., Laugharne, R., Shankar, R.: Therapeutic use of serious games in mental health: scoping review. BJPsych Open **8**(2) (2022)
49. Fleming, T. M., Bavin, L., Stasiak, K., Hermansson-Webb, E., Merry, S. N., Cheek, C., ... & Hetrick, S.: Serious games and gamification for mental health: current status and promising directions. Frontiers in psychiatry **7** 215 (2017). https://doi.org/10.3389/fpsyt.2016.00215
50. Anastasiadis, T., Lampropoulos, G., & Siakas, K.: Digital game-based learning and serious games in education. International Journal of Advances in Scientific Research and Engineering, **4**(12), 139–144 (2018). https://doi.org/10.31695/IJASRE.2018.33016
51. Barnes, S., Bradley, M., Williams, A.: "Piece of mind" and "Wellbeing town": engaging service users in the development of a wellbeing game. In: Prescott, J. (ed.) Digital Innovations for Mental Health Support, pp. 151–186. IGI Global (2022)
52. Green, J.: Evaluation of the 2012 Wellbeing Game. Canterbury District Health Board (2013)
53. Keeman, A., Näswall, K., Malinen, S., Kuntz, J.: Employee wellbeing: evaluating a well- being intervention in two settings. Front. Psychol. **8**, 505 (2017)
54. de Carvalho, C. V., & Coelho, A.: Game-Based Learning, Gamification in Education and Serious Games. Computers **11**(3) 36 (2022). https://doi.org/10.3390/computers11030036
55. Ijaz, A., Khan, M. Y., Ali, S. M., Qadir, J., & Boulos, M. N. K.: Serious games for healthcare professional training: A systematic review. European Journal of Biomedical Informatics **15**(1), (2019)
56. Naul, E., & Liu, M.: Why story matters: A review of narrative in serious games. Journal of Educational Computing Research **58**(3), 687–707 (2020). https://doi.org/10.1177/0735633119859904
57. O'Sullivan, S., Schmaal, L., D'Alfonso, S., Toenders, Y. J., Valentine, L., McEnery, C., ... & Alvarez-Jimenez, M.: Characterizing use of a multicomponent digital intervention to predict treatment outcomes in first-episode psychosis: Cluster analysis. JMIR Mental Health **9**(4), e29211 (2022). https://doi.org/10.2196/29211
58. Lawrence, J., Herrick, T.: Supporting wellbeing through scholarship of teaching and learning. J. Appl. Res. High. Educ. **12**(5), 871–881 (2020)
59. Cheater, S.: Promoting positive mental health. International Journal of Health Promotion and Education **57**(2), 112–114 (2019). https://doi.org/10.1080/14635240.2019.1579477
60. Krath, J., Schürmann, L., Von Korflesch, H.F.: Revealing the theoretical basis of gamification: a systematic review and analysis of theory in research on gamification, serious games and game-based learning. Comput. Hum. Behav. **125**, 106963 (2021)

61. Belchior, P., Marsiske, M., Sisco, S., Yam, A., Mann, W.: Older adults' engagement with a video game training program. Act. Adapt. Aging **36**(4), 269–279 (2012)
62. Czaja, S.L., Sharit, J., Lee, C.C., Nair, S.N., Hernández, M.A., Arana, N., Fu, S.H.: Factors influencing use of an e-health website in a community sample of older adults. J. Am. Med. Inform. Assoc. **20**(2), 277–284 (2013)
63. Jimison, H., Gorman, P., Woods, S., Nygren, P., Walker, M., Norris, S., Hersh, W.: Evidence Reports/Technology Assessment No. 175: Barriers and drivers of health information technology use for the elderly, chronically ill, and undeservedAgency for Healthcare Research and Quality, Rockville, MD. Retrieved from http://www.ncbi.nlm.nih.gov/books/NBK38653 (2008)
64. Tennant, R., Hiller, L., Fishwick, R., Platt, S., Joseph, S., Weich, S., Parkinson, J., Secker, J., Stewart-Brown, S.: The Warwick-Edinburgh mental well-being scale (WEMWBS): development and UK validation. Health Qual. Life Outcomes **5**(1), 1–13 (2007)
65. Clarke, A., Friede, T., Putz, R., Ashdown, J., Martin, S., Blake, A., Adi, Y., Parkinson, J., Flynn, P., Platt, S., Stewart-Brown, S.: Warwick-Edinburgh Mental Well-being Scale (WEMWBS): validated for teenage school students in England and Scotland. A mixed method assessment. BMC Public Health **11**, 487 (2011)
66. Abd-Alrazaq, A., Al-Jafar, E., Alajlani, M., Toro, C., Alhuwail, D., Ahmed, A., Reagu, S.M., Al-Shorbaji, N., Househ, M.: The effectiveness of serious games for alleviating depression: systematic review and meta-analysis. JMIR Serious Games **10**(1), e32331 (2022).

Development and Evaluation of "Perspective: Counselling Simulator": A Gamified Tool for Developing Risk-Assessment Skills in Trainee Counsellors

Charlotte Conn, Aashiya Patel, Jacob Gavin, Mishell Granda Salazar, Andrew Williams, and Steven Barnes

Abstract Self-efficacy is a key factor in reducing the disparity between theoretical knowledge of counselling techniques and their practical application [Akinlolu and Chukwudi in Int. J. Sci. Res. Publ. (IJSRP) 9:8785, 2019]. Assessing and managing risk are critical aspects of counselling training, with self-efficacy a central component of ethical practice. Gamification presents an opportunity to motivate trainees to engage with serious content. In this paper, two studies are presenting to chart the development and preliminary assessment of "Perspective: Counselling Simulator", a game-based tool for developing self-efficacy in risk-assessment skills in trainee counsellors, to BACP (United Kingdom) standards. Prototype evaluation yielded good system usability scores and encouraging feedback from users and identified areas for further development. Subsequently, a preliminary assessment of the game suggests that system usability was improved and that the game was able to make improvements to self-efficacy at a descriptive level. Users provided further feedback on the game to allow for further development and assessment. This work represents, to our knowledge, the first efforts concerning the development and evaluation of a game-based tool which can be used to teach and consolidate risk-assessment self-efficacy in UK counselling students to BACP standards.

Keywords Counselling · Risk-assessment · Training · Self-efficacy · Serious games · Games for education

---

C. Conn (✉) · A. Patel · J. Gavin · M. G. Salazar · A. Williams
University of Bolton, Bolton, UK
e-mail: C.Conn@bolton.ac.uk

S. Barnes
Uniwersytet SWPS Humanistycznospołeczny, Katowice, Poland

C. Iwendi et al. (eds.), *Proceedings of ICACTCE'23—The International Conference on Advances in Communication Technology and Computer Engineering*, Lecture Notes in Networks and Systems 735, https://doi.org/10.1007/978-3-031-37164-6_51

1 Introduction

1.1 Risk-Assessment in Counselling

Counselling settings can be unpredictable, with counsellors often encountering situations where clients may present in a highly emotional and challenging manner. This requires the practitioner to undertake an assessment of unmet demands and to develop and maintain meaningful relationships, while remaining calm and in control [23]. Consequently, practitioners are trained extensively to develop confidence in delivering safe and effective practice [21], with a strong sense of self-efficacy an essential component [2].

Self-efficacy, as applied to counselling settings, affords managing in-session activities and dealing with challenging situations and preventing avoidance and underperformance [16, 26]. It is therefore key in narrowing the gap between theoretical knowledge and practical application of effective counselling techniques. Practitioners must develop knowledge around indicators of risk across numerous dimensions (situational, relational, contextual, professional, personal) and structure session dialogue accordingly to minimise harm, or the risk thereof, to the client or others [1, 22, 34].

Training in the United Kingdom requires would-be practitioners to complete 400 h of work related to skill-development (100 h) to achieve BACP individual accreditation eligibility [8]. Trainees are prepared to identify numerous risk-factors and appropriate action, including when external agencies should be contacted. While counselling courses indicate why risk-training is valuable and important, [43] report that only half offered sufficient time for the development of such skills with appropriate clients, indicating that additional opportunities for practical skill development are required. Online educational tools represent an opportunity to provide such support to enhance learning [17].

1.2 Gamification

The use of games for serious applications is a rapidly expanding field [4, 38]. 'Serious games' apply strategies for entertainment but also to deliver formal and educational content [40], such as in healthcare [12], vocational training [41], mental health [30], and marketing [44]. The motivating capacity of games is leveraged to encourage users to engage with serious content whilst increasing commitment and academic performance [10, 28]. Research indicates that games actively elicit lasting changes to behaviour in short periods of time [7], which is likely a substantial factor in their increasing popularity as a tool for teaching and learning [20].

Previous research shows games can improve learner interest in a topic, deliver opportunities to practice intentional clinical reasoning [24] and improve outcomes in education [5, 31]. Games have emerged to encourage client participation in counselling and exposure programmes [27, 32], with some use of practitioners as co-designers [39]. Professional development enables counsellors to develop both theoretical knowledge and application [25], yet no game currently exists for skill-development in practitioners and may represent a meaningful contribution to the field [37].

1.3 Users in the Design and Development of Serious Games

As in games for entertainment, the requirements of a user and their experience of gaming are important factors when designing serious games [18]. For instance, research suggests adults may pürefer games using puzzle-style tasks [13] with enhanced colour, reduced visual complexity and a limited number of intermediary steps before completing a task [19]. While games must remain challenging and enjoyable to elicit flow [6, 14], a greater focus on accessibility may be needed for adults, to ensure a game does not become overwhelming or complicated [36]. The practice of learning using videogames may also be less common in adults [13], with use potentially dependent on numerous factors, e.g. context (using games a means to substitute for less-desirable real-world activity), and/or content (e.g. enjoying gameplay) [15].

1.4 "Perspective: Counselling Simulator"

"Perspective: Counselling Simulator" was built by a multidisciplinary team of practicing counsellors and providers of counselling training, professional and student game-designers, and academics. The game uses an isometric, third-person perspective (akin to "Theme Hospital" [11]), animating clients as non-playable characters (NPCs). Players take the role of clinician and are presented with clients serially in a text-format similar to "Papers, Please" 33, drawing on a repository of cases which can updated by training providers. Players select their responses from a series of pre-prepared options, with cases at times also requiring the user to identify the need for to call for emergency support. Players are provided with a facility to call for guidance to help them to identify the appropriate answer.

Feedback is provided for all responses, with correct responses rewarded with in-game currency, exchangeable for in-game items (such as upgrades to the in-game environment) or to hire extra clinic staff. While new clinical staff are rendered cosmetically, their presence does not allow for items in the case repository to be skipped, nor do they have any influence on the progression of the game, aside from accruing additional reputation and money for the clinic.

1.5 The Present Study

This paper reports on two studies concerning "Perspective: Counselling Simulator" [29]. Research was conducted on two samples of counselling trainees in a Higher Education setting. Firstly, we report on the development and initial usability and acceptability trial of an early prototype version. Secondly, following updates and ongoing development of the game utilsiing prior user-feedback, we report the outcomes of a second usability and acceptability study, as well as an initial assessment of the game's ability to make improvements to self-reported self-efficacy in risk-assessment knowledge and practice.

2 Method

2.1 Game Development

Initial development was conducted at the Centre for Games Design at the University of Bolton, UK. Students on an MA Games Development course were split into three development teams and given a design brief by the research team (using risk-assessment case studies taken from university training materials). One-week later, teams presented their initial designs and were provided with guidance from the multi-disciplinary research panel of professional game-development staff, practicing counsellors and counselling trainers, and academic input from serious game researchers. After further development, each team conceptualised their design and presented this to the research teaml, with the most established game ("Perspective: Counselling Simulator) "selected for further development and user-evaluation. A refined initial sketch for "Perspective" is provided in Fig. 1.

User-feedback collected during the first study was then utilized by the game-developers to make improvements. The resulting second version was then evaluated by a second group of participants, the outcomes of which are reported here as study two. Screenshots of the final version of "Perspective" are provided in Fig. 2.

2.2 Design

In study one, an initial version of the game was evaluated in a scheduled in-class session with first-year MSc Counselling and Positive Psychology students, who provided quantitative and qualitative feedback regarding acceptability, usability, and areas for ongoing development. Following further development of the game, a new sample of students on the same course trialled the game using a pre vs. post design, to again assess usability, as well as any benefits of gameplay for self-reported counselling self-efficacy.

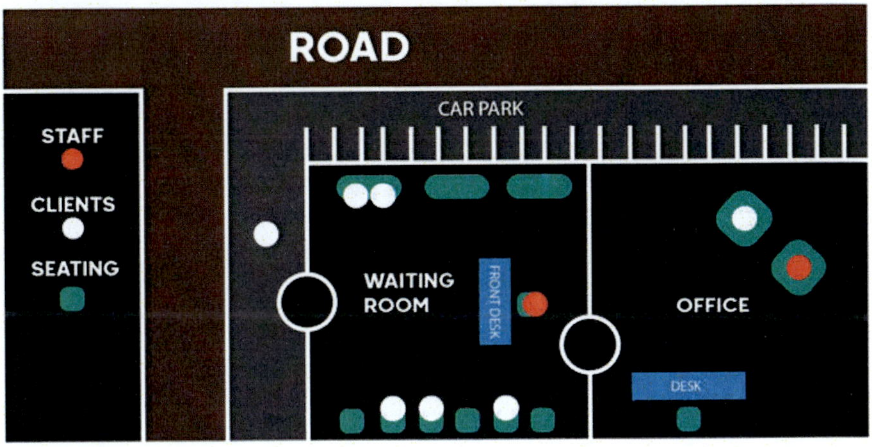

Fig. 1 Refined initial sketch for "Perspective: Counselling Simulator" (copyright: Daniel Mercer)

2.3 Participants

Study one consisted of a sample of participants recruited using convenience sampling (N = 24; mean age = 34.08, SD = 9.75; 20f, 4m) reporting an average of 4.71 (SD = 11.91) h of gameplay in the previous week. At this stage of their programme, no students had commenced any trainingrelating to assessment of risk, and thus were least unlikely to have been exposed to content presented in the game.

Study two consisted of a separate group of students sourced through convenience sampling and yet to commence risk-assessment training or clinical practice (N = 24; mean age = 36.79, SD = 12.50; 18f, 2m, 1non-binary). Participants reported an average of 3.38 (SD = 6.37) h of gameplay in the previous week. Participants in studies one and two did not differ significantly in age ($t(46) = 0.837$, $p = 0.407$), the average number of hours spent engaging in video-gaming per week ($t(46) = 0.227$, $p = 0.631$), or pre-play PCES scores ($t(46) = 1.374$, $p = 0.088$) and therefore were similar to participants in study one.

2.4 Materials

System Usability Scale (SUS). The usability of the game was trialled in both studies using the System Usability Scale (SUS) [9]. The SUS is a 10-item scale used to assess how useable, functional and learnable a product may be and has been used across a broad range of both hardware and software. Participants respond using a five-point Likert scale from 'Strongly agree' to 'Strongly disagree'.

Fig. 2 Screenshots of final version of "Perspective: Counselling Simulator" demonstrating the game environment; example of case study and response screen; and case study update screen

The SUS is noted for its ease of administration and is considered to be robust and versatile 3. Final scores are obtained by subtracting one from responses to odd-numbered items and by subtracting responses from five for even-numbered items, which scales resulting values from 0 to 4. These values can then be summed and multiplied by 2.5 to produce a final score out of 100.

Additional Feedback. To allow for ongoing game-development, additional feedback was collected comprising of the following: (1) Likert-scale questions used to determine whether the game was useful and helpful in teaching/practicing risk-assessment skills which could be applied to real-world practice, whether the game was entertaining and fun as a game, and whether participants believed they would continue to use the game to develop their risk-assessment skills. Qualitative feedback was also collected regarding instances of bugs or instances where players encountered game-behaviour that was confusing or unexpected. Demographic data was also obtained (age, gender, prior videogaming experience).

Psychologist and Counsellor Self-Efficacy Scale (PCES). To determine whether the game was able to develop risk-assessment self-efficacy, we utilised an amended version of the Psychologist and Counsellor Self-Efficacy Scale (PCES) [42]. In original guise, the PCES is a self-report scale consisting of 31-items and can be used to examine perceptions of competency across a range of areas of professional practice and national and international standards for psychologists and counsellors, including research, ethics, intervention, legal requirements, and assessment and measurement. Participants respond to the stem of "How confident are you that you have the ability to…", using a five-point Likert scale ('1' (not at all) to '5' (extremely)). Total scores are obtained by summing item scores.

As the PCES is a general means for measuring self-efficacy in psychologists and counsellors, a number of items were not appropriate or required for assessing risk-assessment. Therefore, a modified version of the scale was used to more appropriate assess risk-assessment as an isolated skill. A copy of the edited version of the PCES used to obtain data reported in this paper is provided in the appendices. An analysis of the reliability of the edited version of the scale yielded a Cronbach's alpha for pre-play scores of $\alpha = 0.953$, and for post-play scores of $\alpha = 0.937$, indicating a very high internal consistency, without the need to remove items. As the sample size for the present research was small, a more statistically complex validation of the scale using factorial analysis was not considered appropriate at this stage.

2.5 Procedure

In both studies, participants played the game as part of a pre-scheduled one-hour teaching session at the university campus, with each individual playing the game to completion (defined as answering the full selection of case studies once each). In study one, all data presented were collected following engagement with the game. In

Table 1 Participant feedback

Item	Agree/Strongly agree
Game was useful in teaching me about counselling risk-assessment skills	62.8
What I learned from the game will be useful in my real-world practice	70.9
I would play the game again in future to practice/develop my risk-assessment skills	66.6

study two, while most data were collected following gameplay, PCES scores were collected both prior to, and following gameplay. The study was ethically approved by the University of Bolton Ethical Committee.

3 Results

3.1 *Study One: Acceptability, Usability and Feedback*

Study one assessed an initial prototype version for acceptability, to identify bugs or aspects limiting usability, and provide potential users to provide feedback to aid development. Normality testing revealed some variables to be non-normal, with evidence of skew and kurtosis. Due to the small sample size, and to protect the real-world validity of the data collected, transformation and normalisation was not pursued, with non-normal variables analysed via non-parametrically. The game achieved an SUS score of 61.46 (SD = 16.25), with did not correlate with age or prior videogaming experience, and players did not report any game-breaking bugs or issues. Participants reported positively on the game as indicated in Table 1.

In addition, participants provided qualitative feedback. Again, due to the small number of participants, and as not all participants provided qualitative feedback, qualitative data were not subjected to formal structured protocols and were instead analysed with surface analysis. Players reported that the game would benefit from improvements to feedback on their performance as well as identifying further usability improvements. The game was subsequently developed to attend to this feedback.

3.2 *Study Two: Initial Evaluation of the Game as a Learning Tool*

Study two consisted of an initial evaluation of enhancements to self-reported risk-assessment self-efficacy, and updated usability assessment. As in study one

Table 2 Participant feedback

Item	Agree/Strongly agree
Game was useful in teaching me about counselling risk-assessment skills	100
What I learned from the game will be useful in my real-world practice	100
I would play the game again in future to practice/develop my risk-assessment skills	95.8

, some variables were not normally distributed and were not transformed/normalised. The game achieved an SUS score of 86.25 (SD = 12.25), which did not correlate with age or videogaming experience, and was a significant improvement on study one ($t(46) = 5.969$, $p < 0.001$). Again, participants did not report any game-breaking issues. Participants again reported positively on the game as indicated in Table 2.

While PCES scores improved descriptively between pre- and post-play, this was not statistically significant (pre: 52.62(SD = 8.64), post: 57.00(SD = 6.62), $Z = -1.826$, $p = 0.068$). Qualitative feedback was again examined using surface analysis, and identified further opportunities for improvements to usability and functionality, along with suggestions to enhance the game's appeal.

4 Discussion

4.1 Principal Findings

These two studies report on the development and initial evaluation of "Perspective: Counselling Simulator" as a learning and consolidation tool for counselling risk-assessment self-efficacy. Findings indicate the use of puzzle tasks in the game was well-received by participants [13], and SUS outcomes indicated significant improvement during development, supporting the involvement of users when designing games for serious contexts [4, 18].

Areas identified for further development reflected previous findings regarding clarity in design and ease of use [19, 36], and speak to a need for game-developers, clinicians/training providers, and end-users to communicate and collaborate on the development of serious games. As adults may be less likely to use games for learning, and therefore may not be as familiar with the construct [13], it is important reduce barriers to entry [15]. Gamification presents an opportunity to deliver serious content in an engaging manner [40]. However, the design of serious games must account for the complexities of combining serious and fun elements in a way that remains accessible to players.

4.2 Limitations

At this stage, PCES improvements did not reach statistical significance. As players only played the game for a relatively brief period of time, it may be that exposure to the learning materials was too brief to exert more effect. The small sample size and underpowered analyses ought also to be considered as factors influencing outcomes presently. While the game is a promising and well-received tool for developing risk-assessment self-efficacy, further research is warranted to establish the game as a scalable training tool.

Considerations for ongoing development are outlined below, however participant comments regarding some issues encountered when playing are salient in the discussion of outcomes. It may be that particular aspects of the game at present may have negatively affected its capacity to teach and train. While such comments were often minor, unexpected issues or lack of sufficient challenge can interrupt flow [14].

4.3 Further Game Development

Following on from the present studies, ongoing development work is planned for "Perspective: Counselling Simulator", following which additional investigation will be conducted. Following player feedback, we plan to enhance the challenge and difficulty [6, 14] of the game, including developing in-game levels where task-difficulty increases in a more gradual and linear manner. This should also help to alleviate some concerns regarding clarity and difficulty when learning the game. Further case studies will also be added to expand the repertoire and add to a sense of in-game variety. To enhance visualisation of in-game progress, improvements to feedback are planned to more clearly demarcate advancing in-game competency and improve intrinsic motivation [35].

Balancing these improvements against ensuring such games remain accessible and intuitive to use will be important in creating a challenging gaming experience which delivers educational value without compromising ease of use [19, 36]. Any subsequent improvement will again be subjected to user-assessment to ensure the game remains accessible and engaging.

4.4 Conclusions

"Perspective: Counselling Simulator" is a promising tool for developing self-efficacy for risk-assessment in trainee counsellors. Developing the game with end-user involvement at each step has been of value in ensuring the end-product is accessible, useable and appealing once completed. The game currently is presently able to make descriptive improvements in self-efficacy, which with further development, rigorous evaluation, and appropriate engagement time may become statistically significant. Developing an engaging tool for practicing this critical professional skill in line with BACP guidelines has both individual and scalable value for counselling training.

Acknowledgements Authors thank the developers of "Perspective: Counselling Simulator":

Daniel Mercer
Daniel Węgrzyn
Kayew Yang

Funding Statement A University of Bolton Jenkinson Award 2023 (£2,000) was used to pay developers (also in the case of developer DM for maintenance). A Ryley Internship Award (Reference: J-2312) (£2,400) was used to employ authors JG and MGS at a cost of £1,200 each. Authors CC, AP, SB and AW did not benefit personally from funding.

Publication Statement The present paper represents a summary of this work intended for publication in conference proceedings. A full version of this paper can be found in the journal 'Mental Health and Social Inclusion'.

Appendix 1: Amended Version of the Psychologist and Counsellor Self-Efficacy Scale (PCES) Used in the Present Study for the Assessment of Self-Efficacy in Risk-Assessment

Please score responses on a scale from 1 to 5, where 1 means 'not at all confident' and 5 means 'extremely confident'.

How confident are you that you have the ability to...

Ethics

E39 Manage informed consent with all your clients?
E49 Observe ethical guidelines?
E52 Refer clients whose needs are better met with someone else?

Legal matter

Understand legislation and best practice relating to...
E79a ...the management of anger, violence and aggression?
E79b ...self-harm?

Intervention

E6 Monitor clients' progress?
E9 Correctly respond to a range of presenting issues?
E32 Effectively end a working relationship with clients?
E35 Help clients to identify situations that increase their risk of getting worse?
E42 Communicate constructively with other professionals?

E46 Support clients and families/caregivers through the mental health system?
E72 Be alert to signs of client disengagement?
E74 Communicate information clearly to clients?

References

1. Andrew, R.: Working with suicidal clients in the counselling professions (2020). https://www.pacfa.org.au/common/Uploaded%20files/PCFA/Documents/Member%20Resources/bacp-working-with-suicidal-clients-fact-sheet-gpia042-jun20.pdf
2. Akinlolu, A.D., Chukwudi, A.R.: Counselling self-efficacy and professional commitment: the mediating role of emotional intelligence and gender identification. Int. J. Sci. Res. Publ. (IJSRP) **9**(3), 8785 (2019)
3. Bangor, A., Kortum, P.T., Miller, J.T.: An empirical evaluation of the system usability scale. Int. J. Hum.-Comput. Interact. **24**(6), 574–594 (2008)
4. Barnes, S., Prescott, J.: Therapeutic gaming for adolescent anxiety: development and evaluation of a mobile intervention. In: Prescott, J. (ed.) Digital Innovations for Mental Health Support, pp. 187–227 (2022)
5. Beemer, L.R., Ajibewa, T.A., DellaVecchia, G., Hasson, R.E.: A pilot intervention using gamification to enhance student participation in classroom activity breaks. Int. J. Environ. Res. Public Health **16**(21), 4082 (2019)
6. Belchior, P., Marsiske, M., Sisco, S., Yam, A., Mann, W.: Older adults' engagement with a video game training program. Act. Adapt. Aging **36**(4), 269–279 (2012)
7. Bindoff, I., de Salas, K., Peterson, G., Ling, T., Lewis, I., Wells, L., Gee, P., Ferguson, S.G.: Quittr: the design of a video game to support smoking cessation. JMIR Serious Games **4**(2), e19 (2016)
8. British Association for Counselling and Psychology, BACP Course Accreditation Scheme Eligibility Guide -Criteria A1 to A10 (2019). https://www.bacp.co.uk/media/1505/bacp-course-accreditation-eligibility-guide
9. Brooke, J.: System Usability Scale (SUS): A Quick-and-Dirty Method of System Evaluation User Information, pp. 1–7. Digital Equipment Co. Ltd, Reading, UK (1986)
10. Buckley, P., Doyle, E.: Gamification and student motivation. Interact. Learn. Environ. **24**(6), 1162–1175 (2016)
11. Bullfrog Productions. Theme hospital [PC]. Electronic Arts (1996)
12. Carlier, S., Coppens, D., De Backere, F., De Turck, F.: Investigating the influence of personalised gamification on mobile survey user experience. Sustainability **13**(18), 10434 (2021)
13. Charlier, N., Ott, M., Remmele, B., Whitton, N.: Not just for children: game-based learning for older adults, In: 6th European Conference on Game Based Learning (ECGBL 2012), pp. 102–108 (2012)
14. Csikzentmihalyi, M.: Beyond Boredom and Anxiety. Jossey-Bass (1985)
15. de Schutter, B., Malliet, S.: The older player of digital games: a classification based on perceived need satisfaction. Communications **39**(1), 67–87 (2014)
16. Delgadillo, J., Moreea, O., Outhwaite-Luke, H., Dace, T., Nicholls, B., Ramseyer, G., Dale, V.: Confidence in the face of risk: the Risk Assessment and Management Self-Efficacy Study (RAMSES). Psychiatr. Bull. **38**(2), 58–65 (2014)
17. Dumford, A.D., Miller, A.L.: Online learning in higher education: exploring advantages and disadvantages for engagement. J. Comput. High. Educ. **30**(3), 452–465 (2018)
18. Fitzgerald, M., Ratcliffe, G.: Serious games, gamification, and serious mental illness: a scoping review. Psychiatr. Serv. **71**(2), 170–183 (2020)
19. Gamberini, L., Alcaniz, M., Barresi, M., Fabregat, M., Ibanez, D., Prontu, L.: Cognition, technology and games for the elderly: an introduction to ELDERGAMES project. PsychNology **4**, 285.308 (2006)

20. Gentile, D.A., Gentile, J.R.: Violent video games as exemplary teachers: a conceptual analysis. J. Youth Adolesc. **37**(2), 137–141 (2008)
21. Grafanaki, S.: Counsellors in the making: Research on counselling training and formative experiences of trainee counsellors. Couns. Psychother. Res. **10**(2), 81–82 (2010)
22. Kirkbride, R.: Counselling Young People: A Practitioner Manual, pp 57–60. SAGE Publications Ltd., London (2017)
23. Kottler, J.A.: Growing a Therapist. Jossey-Bass (1995)
24. Krishnamurthy, K., Selvaraj, N., Gupta, P., Cyriac, B., Dhurairaj, P., Abdullah, A., Krishnapillai, A., Lugova, H., Haque, M., Xie, S., Ang, E.-T.: Benefits of gamification in medical education. Clin. Anat. **25**(6), 795–807 (2022)
25. Kumaria, S., Bhola, P., Orlinsky, D.E.: Influences that count: professional development of psychotherapists and counsellors in India. Asia Pacific J. Couns. Psychother. **9**(1), 86–106 (2018)
26. Lent, R.W., Hill, C.E., Hoffman, M.A.: Development and validation of the counselor activity self-efficacy scales. J. Couns. Psychol. **50**(1), 97–108 (2003)
27. Lindner, P., Rozental, A., Jurell, A., Reuterskӧld, L., Andersson, G., Hamilton, W., Miloff, A., Carlbring, P.: Experiences of gamified and automated virtual reality exposure therapy for spider phobia: qualitative study. JMIR Serious Games **8**(2), e17807 (2020)
28. Manzano-León, A., Camacho-Lazarraga, P., Guerrero, M.A., Guerrero-Puerta, L., Aguilar-Parra, J.M., Trigueros, R., Alias, A.: Between level up and game over: a systematic literature review of gamification in education. Sustainability **13**(4), 2247 (2021)
29. Mercer, D., Węgrzyn, D., Yang, K.: Perspective: counselling simulator [PC]. Centre forGames Development, University of Bolton, UK (2022)
30. Nicolaidou, I., Aristeidis, L., Lambrinos, L.: A gamified app for supporting undergraduate students' mental health: a feasibility and usability study. Digital Health **8**, 20552076221109060 (2022)
31. Ofosu-Ampong, K.: The shift to gamification in education: a review on dominant issues. J. Educ. Technol. Syst. **49**(1), 113–137 (2020)
32. Piao, S., Joo, J.: A behavioural strategy to nudge young people to adopt in-person counselling: gamification. In: Bruyns, G., Wei, H. (eds.) With Design: Reinventing Design Modes. IASDR 2021. Springer, Singapore (2022)
33. Pope: Papers, please [PC]. 3909 LLC (2013)
34. Reeves, A.: Working with Risk in Counselling and Psychotherapy. SAGE Publications Ltd., London, UK (2015)
35. Ryan, R.M., Deci, E.L.: Self-determination theory and the facilitation of intrinsic motivation. Soc. Dev. Well-Being Am. Psychol. **55**(1), 68–78 (2000)
36. Salmon, J.P., Dolan, S.M., Drake, R.S., Wilson, G.C., Klein, R.M., Eskes, G.A.: A survey of video game preferences in adults: building better games for older adults. Entertainment Comput. **21**, 45–64 (2017)
37. Sochor, R., Schenk, J., Fink, K., Berger, J.: Gamification in industrial shopfloor– development of a method for classification and selection of suitable game elements in diverse production and logistics environments. Procedia CIRP **100**, 157–162 (2021)
38. Toda, A.M., Klock, A.C., Oliveira, W., Palomino, P.T., Rodrigues, L., Shi, L., Cristea, A.I., et al.: Analysing gamification elements in educational environments using an existing Gamification taxonomy. Smart Learn. Environ. **6**(1), 1–14 (2019). https://doi.org/10.1186/s40561-019-0106-1
39. Tochilnikova, E., Patnaik, A., Alsebayel, G., Narayan, U., Coeytaux, A., Ramdin, V., Kim, M., Harteveld, C.: "Guilty of talking too much": how psychotherapists gamify therapy. In: CHI Conference on Human Factors in Computing Systems (CHI '22), New Orleans, LA (2022)
40. Treiblmaier, H., Putz, L.M.: Gamification as a moderator for the impact of intrinsic motivation: findings from a multigroup field experiment. Learn. Motiv. **71**, 101655 (2020)
41. Ulmer, J., Braun, S., Cheng, C.T., Dowey, S., Wollert, J.: Human-centred gamification framework for manufacturing systems. Procedia CIRP **93**, 670–695 (2020)

42. Watt, H.M.G., Ehrich, J., Stewart, S.E., Snell, T., Bucich, M., Jacobs, N., Furlonger, B., English, D.: Development of the psychologist and counsellor self-efficacy scale. High. Educ. Skills Work-Based Learn. **9**(3), 485–509 (2018)
43. Wheeler, S., Bowl, R., Reeves, A.: Assessing risk: confrontation or avoidance—what is taught on counsellor training courses. Br. J. Guid. Couns. **32**(2), 235–247 (2004)
44. Yıldırım, İ, Şen, S.: The effects of gamification on students' academic achievement: a meta-analysis study. Interact. Learn. Environ. **29**(8), 1301–1318 (2021)

Initial Evaluation of a Mobile Therapeutic Game for Adolescent Anxiety Disorders

Steven Barnes, Julie Prescott, and Joseph Adams

Abstract While validated treatment options exist, barriers to access or engagement to psychological therapies for anxious adolescents remain. Therapeutic games have emerged as a viable route to reducing the impact of these issues. The present study represents a preliminary assessment of a therapeutic game (utilizing elements of cognitive-behavioural therapy and attention-bias modification) for mobile devices, designed with end-users and intended to reduce anxiety in adolescents. The study aimed to assess the game over a structured intervention period, and to examine any existing relationship between self- directed play and ongoing clinical outcomes. Two groups of clinical (N = 16) and subclinical (N = 15) participants engaged with a within-groups assessment of self-reported anxiety and threat-detection bias at both pre- and post- two-week intervention, with two-month follow-up. It was observed that state and trait self-reported anxiety and threat-detection bias significantly reduced over the intervention period, with reductions maintained at two-months. Participants continued to use the game voluntarily, with the extent of engagement in the clinical groups significantly predicting follow-up outcomes. Considerations for ongoing research and further evaluation are provided.

Keywords Therapeutic games · Gaming · eHealth · mHealth · Adolescence · Anxiety disorders

---

S. Barnes (✉)
Uniwersytet SWPS Humanistycznospołeczny, Katowice, Poland
e-mail: sbarnes@swps.edu.pl

J. Prescott
University of Law, Leeds, UK

J. Adams
Independent Game Developer, Manchester, UK

C. Iwendi et al. (eds.), *Proceedings of ICACTCE'23—The International Conference on Advances in Communication Technology and Computer Engineering*, Lecture Notes in Networks and Systems 735, https://doi.org/10.1007/978-3-031-37164-6_52

1 Introduction

1.1 Anxiety Disorder in Adolescence

Anxiety disorders (AD) are the most commonly diagnosed psychiatric conditions in adolescence [62], with onset increasingly significantly during this life stage [4, 16, 30, 35, 63]. Adolescence has been characterised as a particularly vulnerable period [58], with biomaturational progress often insufficient to deal with a rapidly rising pervasiveness of anxiogenic agents, such as social and academic demands [27]. Experience of anxiety has both short-term [15, 45] and long-term negative implications [24, 28] with early detection of symptoms, and, more ideally, the prevention of escalation to clinical disorder, important in reducing lasting effects [10].

Cognitive-behavioural therapy (CBT) and attention-bias modification (ABM) have each been shown independently to reduce AD experience [8, 18, 23, 31, 32, 34, 65]. However, numerous barriers to treatment-seeking [50] mean as few as 25–50% of affected individuals may seek treatment [36, 51], with those who do often facing long-waiting times [26].

1.2 Therapeutic Games

Young people are often described as having a familiarity with technological tools [47, 49], and as being open to using such provisions for wellbeing [61]. Serious and therapeutic games (TGs) are an engaging means of delivering therapeutic content in a personalisable and adaptable environment affording exploration and practice of behaviours [11]. TGs for anxiety hold promise in alleviating symptoms [53, 54], though with some signs of early disengagement. To date, little is known of their use for anxiety in adolescence either when delivered using mobile devices, or in self-directed contexts [22].

TGs also offer opportunities for expanded outpatient care, with effective oversight an important consideration for practitioners [2]. The remote observation of health information can reassure clients [44] and positively influence outcomes [52], allowing also for observation of how TGs operate in self-directed circumstances.

Finally, as clinically anxious individuals often form the majority of the sampling frame in such research, comparatively little is understood of how TGs may act as a preventative tool for those deemed subclinical or 'at risk'—individuals who, due to the subclinical nature of their current presentation, often do not receive clinical attention.

1.3 "Rocket"

'Rocket' is a mobile TG designed for iOS and Android devices, developed by a multidisciplinary group encompassing researchers, practicing therapists, and developers of games for entertainment at the Centre for Games Development at the University of Bolton. An overview of the game's structure, development and user-feedback study has been published previously [2]. 'Rocket' adopts elements of CBT and ABM and using a level-based structure which gradually exposes players to new content and increasing challenge. For the purposes of this initial evaluation, a sideloaded APK of the Android version was used.

'Rocket' stores data relating to performance on in-game tasks, which are stored securely using Amazon Web Services (AWS). Ecological momentary assessment is employed, with participants prompted to enter current experienced anxiety symptoms pre- and post-play, allowing for evaluation of the effects of play to go beyond retrospective self-report and allowing examination of behavioural and emotional changes in real-time [56]. Example screenshots can be found in Fig. 1.

1.4 The Present Study

'Rocket' was evaluated over a two-week intervention period, with participants assessed again two months later as a follow-up. Reductions in self-reported anxiety and threat detection bias were assessed, along with the extent of self-directed play in the follow-up period and implications for outcomes.

The study aimed to:

(1) Examine the ability of 'Rocket' to reducing state and trait anxiety over a two-week structured trial.
(2) Examine whether anxiolytic effects were maintained after two-months at follow-up.
(3) To assess if participants who voluntarily engaged with the game following the end of the structured trial, and if so, whether the extent of this ongoing use had implications for their levels of anxiety and threat-detection at follow-up.

2 Method

2.1 Design

Two parallel conditions were examined (clinical anxiety, at-risk anxiety) across three timepoints (pre-intervention, post-intervention and two-month follow-up). The structured trial period comprised of two weeks of daily 30-min sessions playing 'Rocket'

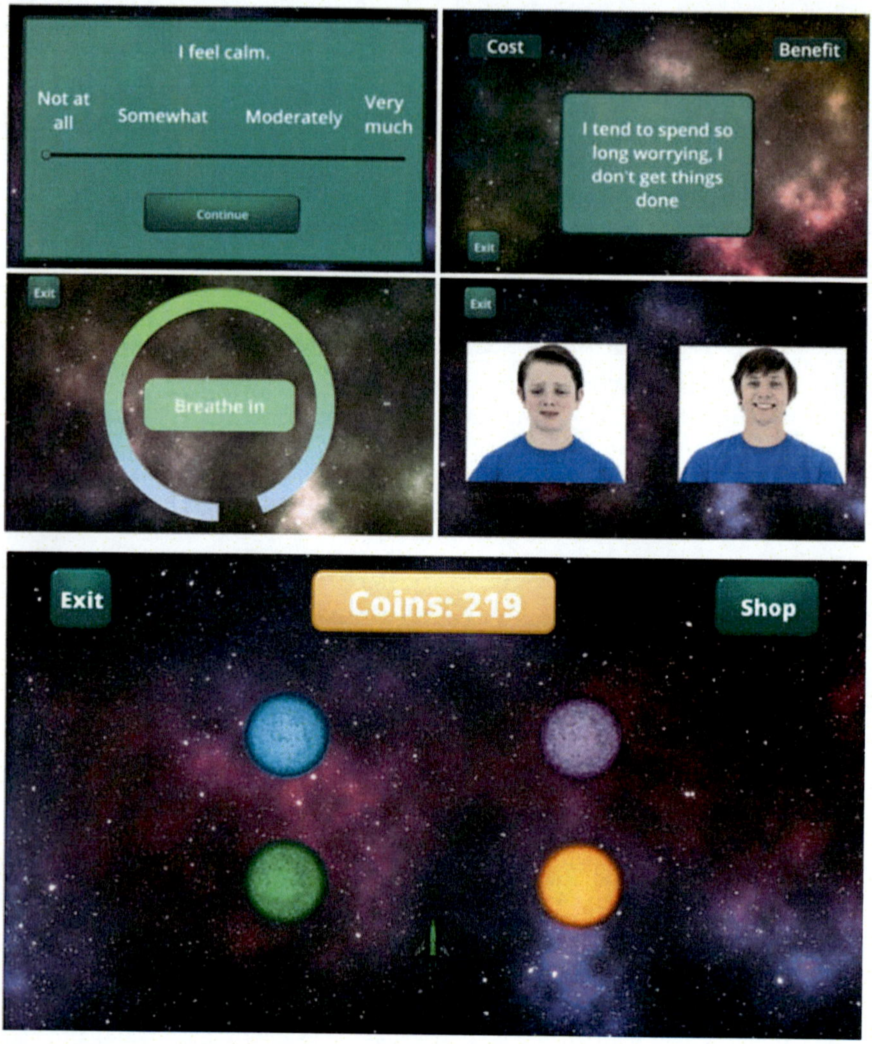

Fig. 1 Screenshots taken from "Rocket"

in addition to any currently received treatment. Following the intervention, participants were informed that they could now choose to use 'Rocket' (or not) on a frequency of their choosing before follow-up measures were collected.

Table 1 Participant demographics

Item	Data
Mean age	16.39 (SD = 1.94)
Male	8
Female	23
Mean *N* hours spent videogaming p/week	15.79 (SD = 13.15)
Diagnosis of GAD	5
Diagnosis of social anxiety	2
Receiving medication (unspecified)	1

2.2 *Participants*

G*Power 3 was employed to calculate an estimated *N* for the sample [25], assuming a small to medium effect of d = 0.36 [21], estimated correlation of r = 0.5 and non-sphericity correction of $\varepsilon = 1$. This analysis provided a target sample of N = 22.

Ongoing restrictions and lockdown measures associated with COVID-19 required the research to be conducted online in its entirety. After attrition and withdrawal for insufficient engagement (e.g. did not complete the requisite 30m per day), a group of 31 participants aged 11–19 years finished the trial (8m, 23f, mean age = 16.39, SD = 1.94). Further information regarding the sample is provided below (Table 1).

2.3 *Materials*

To capture the effects of both CBT and ABM, self-report inventories and threat-detection bias tasks were used.

Beck Anxiety Inventory (BAI) The BAI [57] comprises of 21 items to measure AD symptoms. Participants respond considering the degree of experienced emotion over the seven days prior to assessment using a four-point scale (from 0 = not at all, to 3 = severely). Responses are summed to create a final score, which can also be used to interpret symptom severity.

The BAI has high internal consistency ($\alpha = 0.92$) and test-retest reliability (rtt = 0.75) [46, 60], and is responsive to temporal change [7].

Symptom Checklist Anxiety Scale (SCL-A20). A 20-item subscale of the Symptom Checklist 90 Revised (SCL-90-R) [19], the SCL-A20 examines anxiety over the seven days prior to assessment using a five-point Likert scale (0 = not at all, to 4 = extremely). As with the BAI, individual responses can be summed to obtain an overall score. In addition to the BAI, scores ranging from 20 to 29 are classified as subclinical but 'at-risk', with scores of $\geq$30 classified as indicative of clinical levels of anxiety.

Spielberger State Trait Anxiety Inventory (STAI-Y6). The STAI-Y6 [42] is a six-item self-report inventory. Responses are based on present feelings using a four-point scale (from 1 = not at all, to 4 = very much). A total score can be obtained by summing individual items. A strong correlation has been observed between the Y6 and the full STAI (r = 0.95) [42]. As the STAI-Y6 contains a small number of items and is quick to complete, it was embedded in 'Rocket' as a pre- and post-play ecological momentary assessment of state-anxiety.

Threat Detection Bias (TDB). TDB was assessed via dot-probe consisting of detection of happy face (present/not present) and detection of unhappy face (present/not present). Before trials commenced, participants were provided with instructions and offered a small number of attempts to practice the procedure. Reaction times were collected and used to assess threat-detection bias. Stimuli for this assessment came from the EU-Emotion Stimulus Set (European Community's Seventh Framework Programme: ASC-Inclusion Project (FP7/2007–2013)) and were assessed previously in terms of expression-intensity to determine whether individual items were suitable for use [2]. An example can be observed in Fig. 2.

TDB measures were conducted using OpenSesame v3.2.8 [43] and stored using JATOS (Just Another Tool for Online Studies), an open-source, cross-platform distribution service using JavaScript [38].

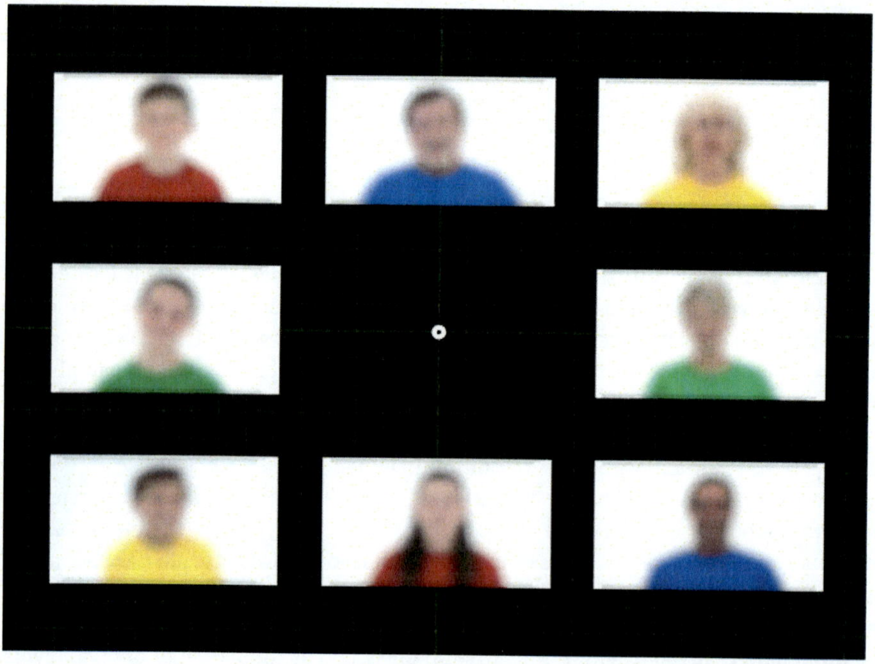

Fig. 2 Example of TDB task (detection of unhappy face)

2.4 *Procedure*

The study was ethically approved by the Ethics Committee at the University of Bolton, UK. Schools local to the Greater Manchester area provided participants. Additionally, participants volunteered online via Reddit and in response to an advertisement placed in local media. Snowball sampling was used to expand the sample where possible.

Participants were provided with full study information and provided their consent prior to participating (this was obtained from a parental/guardian for participants aged <16 years). All participants were assessed using self-report measures and dot-probe task to determine their for the trial. If eligible, SCL-A20 scores were used to allocate participants to the appropriate group (clinical, at-risk. All participating individuals were provided with an onboarding introduction to the trial consisting of an introduction to 'Rocket' and the various in-game tasks and features (including an additional reminder of which data the game would capture and store) to control for any existing knowledge and familiarity with videogames. Participants were informed that the game had been designed to alleviate user anxiety symptoms. Additional information and guidance was restricted at this point in order to reduce any potential expectation effects.

During the intervention period, participants used the game for 30-min per day. Ecological momentary assessment was employed, with participants completing the STAI-Y6 before and after each play session, with engagement verified via deposits in the AWS bucket. Following the two-week intervention, self-report and threat-detection bias measures were repeated. Participants were instructed at this point that they may now continue playing the game (or not) as they wished over the following two-months, and were reminded that, should they continue using the game, that it would continue to collect data as it had previously. Once more, direction to participants at this point was kept limited to essential information in order to limit expectation effects.

Two months after their participation in the structured intervention, anxiety assessments were repeated. Once the study was completed, participants were provided with a full debriefing.

3 Results

3.1 *Normality Testing*

While some measures (SCL-A20; BAI; TDB) were normally distributed, some measures (hours spent in self-directed play (at-risk group)) were non-normal. So as not to negatively affect the real-world validity of the data, no attempts at transformation or normalization were made.

A strong relationship was observed between self-report anxiety measures across the timepoints (baseline: $r = 0.904$, $p < 0.001$; post-intervention: $r = 0.922$, $p < 0.001$; follow-up: $r = 0.937$, $p < 0.001$), with all coefficients above 0.8, indicating a strong likelihood of multicollinearity and confirming that the two inventories had performed as expected in measuring the same construct comparably.

3.2 Examination of State- and Trait-Anxiety and TDB

Paired-samples t-test revealed significant reductions in STAI-Y6 scores for both clinical participants ($t(15) = 9.575$, $p < 0.001$, $g = 0.83$) and at-risk participants ($t(14) = 9.071$, $p < 0.001$, $g = 0.77$). An overview of RM ANOVA analyses of the game in terms of BAI, SCL-A20 and TDB measures can be found in Table 2.

Table 2 Table of outcomes

Item	Pre-intervention	Post-intervention	Follow-up	*p*	ηp^2
Clinical group					
BAI	26.13 (SD = 5.10)	25.63 (SD = 7.05)	26.19 (SD = 10.17)	<0.001	0.575
SCL-A20	39.63 (SD = 6.66)	27.63 (SD = 6.16)	28.38 (SD = 10.43)	<0.001	0.527
RT unhappy (ms)	1251.53 (SD = 359.16)	1429.90 (SD = 360.87)	1488.81 (SD = 526.74)	0.016	0.285
AC unhappy	15.81 (SD = 2.20)	14.31 (SD = 1.89)	14.56 (SD = 2.00)	0.002	0.348
RT happy (ms)	1830.24 (SD = 433.75)	1530.09 (SD = 379.47)	1569.75 (SD = 464.79)	0.004	0.365
AC happy	11.44 (SD = 2.13)	15.00 (SD = 1.71)	14.56 (SD = 3.22)	<0.001	0.577
At-risk group					
BAI	22.40 (SD = 3.64)	16.13 (SD = 5.83)	18.40 (SD = 4.61)	<0.001	0.480
SCL-A20	25.00 (SD = 2.67)	15.47 (SD = 5.33)	17.80 (SD = 5.54)	<0.001	0.694
RT unhappy (ms)	1544.16 (SD = 412.87)	1614.07 (ms = 365.71)	1593.20 (SD = 513.60)	0.498	0.033
AC unhappy	15.00 (SD = 1.25)	13.20 (SD = 1.86)	14.13 (SD = 1.60)	<0.001	0.429
RT happy (ms)	1698.13 (SD = 564.59)	1407.00 (SD = 507.79)	1464.33 (SD = 542.29)	0.004	0.379
AC happy	13.60 (SD = 1.18)	16.27 (SD = 1.39)	14.93 (SD = 2.15)	<0.001	0.552

Overall, significant reductions in anxiety (state and trait) were observed in self-report outcomes, and in several TDB outcomes.

3.3 *Implications of Self-Directed Play*

While a Gaussian distribution was observed for the time participants spent engaging in self-directed play for the clinical group, this was not the case for the at-risk group. SDP correlated negatively with changes in BAI ($rs = -0.641$, $p = 0.001$) and SCL-A20 ($rs = -0.659$, $p = 0.005$) scores between post-intervention and follow-up. SDP correlated positively with RT change for detecting unhappy faces ($rs = 0.702$, $p = 0.002$), and negatively with RT for detecting happy faces ($rs = -0.315$, $p = 0.042$). Assessment of accuracy in unhappy ($rs = -0.372$, $p = 0.156$) or happy face trials ($rs = 0.430$, $p = 0.097$) did not produce any significant correlations.

Speculative regression analysis revealed SDP significantly predicted changes in SCL-A20 outcomes over the follow-up period ($F(1, 14) = 8.187$, $p = 0.013$, $\beta = -0.607$, $R2 = 0.369$) with SDP accounting for 36.9% of variance SCL-A20 score changes ($Y' = 4.103 \pm 0.239(X)$). A significant model was also observed for BAI outcomes ($F(1, 14) = 10.087$, $p = 0.007$, $\beta = -0.647$, $R2 = 0.419$), with SDP accounting for 41.9% of the variance in BAI score changes ($Y' = 3.690 \pm 0.223(X)$).

Simple linear-regression confirmed that SDP significantly predicted changes in RT for detecting unhappy faces ($F(1, 14) = 9.297$, $p = 0.009$, $\beta = -0.632$, $R2 = 0.399$), with SDP accounting for 39.9% of the variance in RT change ($Y' = -87.466 \pm 10.433(X)$). SDP did not however predict changes in RT when participants were asked to detect happy faces ($F(1, 14) = 3.679$, $p = 0.076$, $\beta = -0.456$, $R2 = 0.208$).

A normal distribution was observed for SDP time and most TDB measures in at-risk participants. SDP did correlate significantly with changes in self-reported anxiety (SCL-A20: $r = -0.458$, $p = 0.086$; BAI: $r = -0.280$, $p = 0.312$). Furthermore, no significant correlations emerged between SDP time and either changes in RT (unhappy: $r = 0.017$, $p = 0.953$; happy: $rs = -0.463$, $p = 0.082$) or accuracy (unhappy: $r = -0.266$, $p = 0.338$; happy: $r = 0.299$, $p = 0.278$). As a result, no modelling analysis was undertaken.

4 Discussion

4.1 *Principal Findings*

Anxiety disorders are associated with significant deficits and emotional disturbance [1, 3]. Despite this, the increased risk of anxiety in adolescence [3, 5], the risk of long-term mental health burden following adolescent anxiety [1, 28, 37, 64], the comparatively low levels of treatment-seeking in anxious adolescents [14, 36], and previous work suggesting digital games may help to overcome such issues, [17, 39, 53–55], few validated tools exist publicly using mobile devices to address these problems.

Digital interventions can overcome barriers to treatment accessibility and assist in overcoming additional barriers associated with pathological factors such as avoidance or low motivation, even when symptoms are chronic in nature [12, 20]. Additionally, digital tools can assist in nurturing the development of skills applicable to non-game environments [6]. Based on these initial findings, 'Rocket' appears to hold value as a valid therapeutic tool, capable of producing enduring positive change [6, 29]. While the third hypothesis was partially confirmed, participants did engage in SDP, reinforcing the use of TGs as a motivating tool capable of elciting repeated engagement [33]. That the game proved to demonstrate continued acceptability also points to the value of engaging end-users in the design of TGs [2].

SDP, however, did not predict all TDB outcomes, nor did it predict measures in at-risk participants. It may be the case that clinical participants were more likely to benefit from ongoing intervention, as the scope for clinical improvement was greater. Additional data concerning voluntary use of serious and therapeutic games will be vital in determining how valuable TGs may be as an outpatient tool.

The findings also provide encouraging signs for using ecological momentary assessment to remotely collect user data in TGs. All systems functioned as expected, and while ethical considerations remain important [59], this represents a valuable addition to TGs as a potential outpatient service.

4.2 Limitations

A number of potential limitations must be considered. Firstly, COVID-19 social restrictions necessitated the study be conducted purely online. While data provide valuable insight into 'Rocket' as an anxiolytic tool, participants were recruited via self- selection and are subsequently unlikely to represent the wider anxious demographic, where avoidance is a common pathological barrier to treatment. Of note however, a number of participants scoring above clinical thresholds did not hold a clinical diagnosis, suggesting that TGs may attract an audience who, perhaps due to avoidance, are otherwise unknown or unattended to by traditional care systems [40].

Attrition was high, sometimes due to participants experiencing technical problems, usually associated with game-installation, as the game was installed using sideloading procedures. Due to COVID restrictions, providing help was at times challenging and, while not an insurmountable issue under more normal circumstances, this reinforces the need for TGs to be accessible. A large number of participants were withdrawn during the follow-up period, largely for loss of communication, suggesting that ongoing clinical input, reminder systems, and caution in discharging to outpatient status may be needed when TGs are used for outpatients.

Finally, while the order of TDB stimulus presentation was randomised, repetitive use may have produced improvements in scores related to memory and learning rather than reduction of detection biases. In anxiety disorders, learning is a major driver of the condition [48]. Aversive associative learning outlines the persistence and resurgence of anxiety and avoidance strategies. While attentional biases could be seen as simply a heightened degree of attention to 'threatening' stimuli [41], difficulties with disengaging may, over time, play an important role [13]. For individuals with successful TDB outcomes, this may explain why some self-reported anxiety still remained. Additionally, familiar faces are often detected with superior accuracy [9]. In games utilizing TDB using face-stimuli, it may be that a significantly larger library of stimuli is required to avoid this issue.

4.3 Conclusions

This study presents a preliminary assessment of a mobile TG in clinical and sub-clinical samples, and, to our best knowledge, the first evaluation of SDP and its potential effects ongoing anxiety presentation. While findings must be considered in terms of the somewhat limited window of clinical severity covered by this sample, engaging TGs which provide accessibility and appeal as a means for practicing maintenance behaviours represent a means of reducing risk of relapse. Ecological momentary assessment also represents developers of therapeutic games with an tool for assessment of SDP and ongoing clinical status.

As adolescents with anxiety have typically proven to be challenging to engage in the therapeutic process, often demonstrazing unfavourable outcomes, ways of engaging with therapeutic interventions using more impersonal and asynchronous may be valuable in providing some degree of intervention to an underserved population.

Acknowledgements Authors thank all of the participants involved in evaluating the game for their engagement with the project.

Funding Statement This research did not receive any funding or financial contribution from the public, commercial, or not-for-profit sectors.

Publication Statement The present paper represents a summary of this work intended for publication in conference proceedings. A full version of this paper can be found in the journal 'Mental Health and Social Inclusion'.

References

1. Bandelow, B., Michaelis, S.: Epidemiology of anxiety disorders in the 21st century. Dialogues Clin. Neurosci. **17**(3), 327–335 (2015)
2. Barnes, S., Prescott, J.: Therapeutic gaming for adolescent anxiety: development and evaluation of a mobile intervention. In: Prescott, J. (ed.) Digital Innovations for Mental Health Support, pp. 187–227. IGI Global (2022)
3. Baxter, A.J., Scott, K.M., Vos, T., Whiteford, H.A.: Global prevalence of anxiety disorders: a systematic review and meta-regression. Psychol. Med. **43**(5), 897–910 (2013)
4. Beesdo, K., Bittner, A., Pine, D.S., Stein, M.B., Höfler, M., Lieb, R., Wittchen, H.U.: Incidence of social anxiety disorder and the consistent risk for secondary depression in the first three decades of life. Arch. Gen. Psychiatr. **64**(8), 903–912 (2007)
5. Bhatia, M.S., Goyal, A.: Anxiety disorders in children and adolescents: need for early detection. J. Postgrad. Med. **64**(2), 75–76 (2018)
6. Bindoff, I., de Salas, K., Peterson, G., Ling, T., Lewis, I., Wells, L., Gee, P., Ferguson, S.G: Quittr: the design of a video game to support smoking cessation. JMIR Serious Games **4**(2), e:19 (2016)
7. Brown, G.K., Beck, A.T., Newman, C.F., Beck, J.S., Tran, G.Q.: A comparison of focused and standard cognitive therapy for panic disorder. J. Anxiety Disord. **11**(3), 329–345 (1997)
8. Brown, J.S., Elliott, S.A., Boardman, J., Andiappan, M., Landau, S., Howay, E.: Can the effects of a 1-day CBT psychoeducational workshop on self-confidence be maintained after 2 years? A naturalistic study. Depress. Anxiety **25**(7), 632–640 (2008)
9. Bruce, V.: Changing faces: visual and non-visual coding processes in face recognition. Br. J. Psychol. **73**(1), 105–116 (1982)
10. Cabral, M.D., Patel, D.R.: Risk factors and prevention strategies for anxiety disorders in childhood and adolescence. In: Kim, Y.K. (ed.) Anxiety Disorders: Advances in Experimental Medicine and Biology, vol. 1191, Springer, Singapore (2020)
11. Cannon-Bowers, J.A., Bowers, C., Procci, K.: Using video games as educational tools in healthcare. In: Tobias, S., Fletcher, J.D. (eds.) Computer Games and Instruction, pp. 47–72. Information Age Publishing, Charlotte, NC (2011)
12. Charlier, N., Zupancic, N., Fieuws, S., Denhaerynck, S., Zaman, K., Moons, P.: Serious games for improving knowledge and self-management in people with chronic conditions: a systematic review and meta-analysis. J. Am. Med. Inf. Assoc. **23**(1), 230–239 (2016)
13. Cisler, J.M., Koster, E.H.: Mechanisms of attentional biases towards threat in anxiety disorders: an integrative review. Clin. Psychol. Rev. **30**, 203–216 (2010)
14. Collins, K.A., Westra, H.A., Dozois, D.J.A., Burns, D.D.: Gaps in accessing treatment for anxiety and depression: challenges for the delivery of care. Clin. Psychol. Rev. **24**, 583–616 (2004)
15. Copeland, W.E., Angold, A., Shanahan, L., Costello, E.J.: Longitudinal patterns of anxiety from childhood to adulthood: the great smoky mountains study. J. Am. Acad. Child Adolesc. Psychiatr. **53**(1), 21–33 (2014)
16. Costello, E.J., Angold, A.: Epidemiology. In: March, J.S. (ed.) Anxiety Disorders in Children and Adolescents, pp. 109–124. Guilford Press, New York (1995)
17. Coyle, D., McGlade, N., Doherty, G., O'Reilly, G.: Exploratory evaluations of a computer game supporting cognitive behavioural therapy for adolescents. In: Proceedings of the SIGCHI Conference on Human Factors in Computing Systems (CHI' 11), pp. 2937–2946 (2011)
18. Davis, T.E., May, A., Whiting, S.E.: Evidence-based treatment of anxiety and phobia in children and adolescents: current status and effects on the emotional response. Clin. Psychol. Rev. **31**(4), 592–602 (2011)
19. Derogatis, L.R., Unger, R.: Symptom checklist-90 revised. In: Weiner, I.B., Craighead, E. (eds.) The Corsini Encyclopedia of Psychology (2010)
20. de Smet, A., van Ryckeghem, D., Compernolle, S., Baranowski, T., Thompson, D., Crombez, G., Poels, K., van Lippevelde, W., Bastiaensens, S., van Cleemput, K., Vandebosch, H., de

Bourdeauhuji, I.: A meta-analysis of serious digital games for healthy lifestyle promotion. Prev. Med. **69**, 95–107 (2014)
21. de Swart, J.J.W., van den Broek, H., Stams, G.J.J.M., Asscher, J.J., van der Laan, P.H., Holsbrink-Engels, G.A., van der Helm, G.H.P.: The effectiveness of institutional youth care over the past three decades: a meta-analysis. Child Youth Serv. Rev. **34**, 1818–1824 (2012)
22. Dewhirst, A., Laugharne, R., Shankar, R.: Therapeutic use of serious games in mental health: scoping review. BJPsychOpen **8**, e37 (2022)
23. Eldar, S., Apter, A., Lotan, D., Edgar, K.P., Naim, R., Fox, N.A., Pine, D.S., Bar-Haim, Y.: Attention bias modification treatment for pediatric anxiety disorders: a randomized controlled trial. Am. J. Psychiatr. **169**(2), 213–220 (2012)
24. Essau, C.A., Conradt, J., Petermann, F.: Course and outcome of anxiety disorders in adolescents". J. Anxiety Disord. **16**(1), 67–81 (2002)
25. Faul, F., Erdfelder, E., Buchner, A., Lang, A-G.: Statistical power analyses using G*Power 3.1: tests for correlation and regression analyses. Behav. Res. Methods **41**, 1149–1160 (2007)
26. Fehm, L., Pélissolo, A., Furmark, T., Wittchen, H.U.: Size and burden of social phobia in Europe. Eur. Neuropsychopharmacol. **15**, 453–462 (2005)
27. Giedd, J., Blumenthal, J., Jeffries, N.O., Castellanos, F.X., Liu, H., Zijdenbos, A., Paus, T., Evans, A.C., Rapoport, J.L.: Brain development during childhood and adolescence: a longitudinal MRI study. Nat. Neurosci. **2**(10), 861–863 (1999)
28. Ginsburg, G.S., Becker, E.M., Keeton, C.P., Sakolsky, D., Piacentini, J., Albano, A.M., Compton, S.N., Iyengar, S., Sullivan, K., Caporino, N., Peris, T., Birmaher, B., Rynn, M., March, J., Kendall, P.C.: Naturalistic follow-up of youths treated for pediatric anxiety disorders. JAMA Psychiatr. **71**(3), 310–318 (2014)
29. Granic, I., Lobel, A., Engels, R.C.: The benefits of playing video games. Am. Psychol. **69**(1), 66–78 (2014)
30. Grant, D.M.: Anxiety in adolescence. In: W.T. Donohue, L. Benuto, L. Woodward-Tolle (eds.) Handbook of Adolescent Health, pp. 507–519. Springer, New York (2013)
31. Hakamata, Y., Lissek, S., Bar-Haim, Y., Britton, J.C., Leibeinluft, E., Ernst, M., Pine, D.S.: Attention bias modification treatment: a meta-analysis toward the establishment of novel treatment for anxiety. Biol. Psychiatr. **68**(11), 982–990 (2010)
32. Henin, A., Micco, J.A., Schoeller, M., Boudreaux, A., Hirshfeld-Becker, D.: Cognitive behavioral therapy with children and adolescents. In: Massachusetts General Hospital Handbook of Cognitive Behavioral Therapy, pp. 259–275 (2016)
33. Ibrahim, E.N.M., Jamali, N., Suhaimi, A.I.H.: Exploring gamification design elements for mental health support. Int. J. Adv. Technol. Eng. Explor. **8**(74), 114 (2021)
34. Kazdin, A.E., Rabbitt, S.M.: Novel models for delivering mental health services and reducing the burdens of mental illness. Clin. Psychol. Sci. **1**, 170–191 (2013)
35. Kessler, R.C., Berglund, P., Demler, O., Jin, R., Merikangas, K.R., Walters, E.E.: Lifetime prevalence and age-of-onset distributions of DSM-IV disorders in the National Comorbidity Survey Replication. Arch. Gen. Psychiatr. **62**, 593–602 (2005)
36. Kessler, R.C., Wang, P.S.: The descriptive epidemiology of commonly occurring mental disorders in the United States. Annu. Rev. Public Health **29**, 115–129 (2008)
37. Kessler, R.C., Ruscio, A.M., Shear, K., Wittchen, H.U.: Epidemiology of anxiety disorders. Curr. Top. Behav. Neurosci **2**, 21–35 (2010)
38. Lange, K., Kühn, S., Filevich, E.: Correction: "Just Another Tool for Online Studies" (JATOS): an easy solution for setup and management of web servers supporting online studies. PLoS ONE **10**(7), e0134073 (2015)
39. Li, W.H., Chung, J.O., Ho, E.K.: The effectiveness of therapeutic play, using virtual reality computer games in promoting the psychological well-being of children hospitalised with cancer. J. Clin. Nurs. **20**, 2135–2143 (2011)
40. Lorian, C.N., Grisham, J.R.: Clinical implications of risk aversion: an online study of risk-avoidance and treatment utilization in pathological anxiety. J. Anxiety Disord. **25**(6), 840–848 (2011)

41. MacLeod, C., Mathews, A., Tata, P.: Attentional bias in emotional disorders. J. Abnorm. Psychol. **95**(1), 15 (1986)
42. Marteau, T.M., Bekker, H.: The development of a six-item short- form of the state scale of the Spielberger State-Trait Anxiety Inventory (STAI). Br. J. Clin. Psychol. **31**(3), 301–306 (1992)
43. Mathôt, S., Schreij, D., Theeuwes, J.: OpenSesame: An open- source, graphical experiment builder for the social sciences. Behav. Res. Methods **44**(2), 314–324 (2012)
44. McCann, L., Maguire, R., Miller, M., Kearney, N.: Patients' perceptions and experiences of using a mobile phone-based advanced symptom management system (ASyMS©) to monitor and manage chemotherapy related toxicity. Eur. J. Cancer Care **18**(2), 156–164 (2009)
45. Office for National Statistics: Mental health of children and young people in Great Britain (2004). Office for National Statistics, London. http://www.hscic.gov.uk/catalogue/PUB06116/ment-heal-chil-youn-peop-gb-2004-rep1.pdf
46. Osman, A., Hoffman, J., Barrios, F.X., Kopper, B.A., Breitenstein, J.L., Hahn, S.K.: Factor structure, reliability, and validity of the Beck Anxiety Inventory in adolescent psychiatric inpatients. J. Clin. Psychol. **58**(4), 443–456 (2002)
47. Papastergiou, M.: Exploring the potential of computer and video games for health and physical education: a literature review. Comput. Educ. **53**(3), 603–622 (2009)
48. Pittig, A., Treanor, M., LeBeau, R.T., Craske, M.G.: The role of associative fear and avoidance learning in anxiety disorders: Gaps and directions for future research. Neurosci. Biobehav. Rev. **88**, 117–140 (2018)
49. Prensky, M.: Digital natives, digital immigrants. On Horizon **9**(5), 1–6 (2001)
50. Radez, J., Reardon, T., Creswell, C., Lawrence, P.J., Evdoka-Burton, G., Waite, P.: Why do children and adolescents (not) seek and access professional help for their mental health problems? A systematic review of quantitative and qualitative studies. Eur. Child & Adolesc. Psychiatr. **30**, 183–211 (2020)
51. Sawyer, M.G., Patton, G.C.: Unmet need in mental health service delivery: children and adolescents. In: Andrews, G., Henderson, S. (eds.) Unmet Need in Psychiatry: Problems, Resources, Responses, pp. 330–344. Cambridge University Press, New York (2000)
52. Scherr, D., Kastner, P., Kollmann, A., Hallas, A., Auer, J., Krappinger, H., Schuchlenz, H., Stark, G., Grander, W., Jakl, G., Schreier, G., Fruhwald, F.M.: Effect of home-based telemonitoring using mobile phone technology on the outcome of heart failure patients after an episode of acute decompensation: randomized controlled trial. J. Med. Internet Res. **11**(3), e1252 (2009)
53. Scholten, H., Malmberg, M., Lobel, A., Engels, R.C.M.E., Granic, I.: A randomized controlled trial to test the effectiveness of an immersive 3D video game for anxiety prevention among adolescents. PLoS ONE **11**(1), e0147763 (2016)
54. Schoneveld, E.A., Malmberg, M., Lichtwarck-Aschoff, A., Verheijen, G.P., Engels, R.C.M.E., Granic, I.: A neurofeedback video game (MindLight) to prevent anxiety in children: a randomized controlled trial. Comput. Hum. Behav. **63**, 321–333 (2016)
55. Schuurmans, A.A., Nijhof, K.S., Vermaes, I.P., Engels, R.C.M.E., Granic, I.: A pilot study evaluating "Dojo", a videogame intervention for youths with externalizing and anxiety problems. Games Health **4**(5), 401–408 (2015)
56. Shiffman, S., Stona, A.A., Hufford, M.R.: Ecological momentary assessment. Annu. Rev. Clin. Psychol. **4**, 1–32 (2008)
57. Steer, R.A., Beck, A.T.: Beck anxiety inventory. In: Zalaquett, C.P., Wood, R.J. (eds.) Evaluating Stress: A Book of Resources, pp. 23–40. Scarecrow Education (1997)
58. Steinberg, L.: Cognitive and affective development in adolescence. Trends Cogn. Sci. **9**(2), 69–74 (2005)
59. Sundram, F., Hawken, S.J., Lucassen, M., Greenwood, A., Merry, S.N.: Tips and traps: lessons from codesigning a clinician e-monitoring tool for computerized cognitive behavioural therapy. JMIR Mental Health **4**(1), e3 (2017)
60. Tabachnick, B.G., Fidell, L.S.: Using Multivariate Statistics, 2nd edn. HarperCollins, New York (1989)
61. Torok, M., Han, J., McGillivray, L., Wong, Q., Werner-Seidler, A., O'Dea, B., Calear, A., Christensen, H.: The effect of a therapeutic smartphone application on suicidal ideation in

young adults: findings from a randomized controlled trial in Australia. PLoS Med. **19**(5), e1003978 (2022)
62. Vizard, T., Sadler, K., Ford, T., Maschesli, F., Pearce, N., Mandalia, D., David, J., Brodie, E., Forbes, N., Goodman, A., Goodman, R., McManus, S., Collison, D.: Mental health of children and young people in England, 2017 summary of key findings (2017). https://files.digital.nhs.uk/A6/EA7D58/MHCYP%202017%20Summary.pdf
63. Wittchen, H.U., Fehm, L.: Epidemiology and natural course of social fears and social phobia. Acta Psychiatr. Scand. **417**, 4–18 (2003)
64. Woodward, L.J., Fergusson, D.M.: Life course outcomes for young people with anxiety disorders in adolescence. J. Am. Acad. Child Adolesc. Psychiatr. **40**(9), 1086–1093 (2001)
65. Zhang, A., Borhneimer, L., Weaver, A., Franklin, C., Hang Hai, A., Guz, S., Shen, L.: Cognitive behavioral therapy for primary care depression and anxiety: a secondary meta-analytic review using robust variance estimation in meta-regression. J. Behav. Med. **42**(6), 1117–1141 (2019)

New Technologies and Interventions to Improve the Mental Health of People with Diabetes

Ayesha Adeel, Chathurika Kannangara, Harni Bharaj, Ambar Basu, and Barrie Green

Abstract The present paper represents an initial research protocol for the development and evaluation of the Positive Psychology and Artificial Intelligence based Chatbot; Artificial Intelligence Diabetes Assistant—Mental Health (AIDA-MH) and an integrated mobile app designed to enhance mental health outcomes of people with diabetes. The rationale for the proposed research, planned methodology and analysis are discussed in this paper briefly. This proposed project will have quantitative and qualitative data and will be using pre and post intervention with randomized controlled trials, follow-ups and semi-structured interviews at the later stage. This paper is a proposed protocol for future evaluation, therefore, no data is reported and analysed at this stage. The findings and conclusions from this project will be a foundation ground for policy implementation in diabetes services, to integrate digital technology to improve mental health outcomes alongside diabetes outcomes.

Keywords Digital Mental Health Interventions (DMHIs) · Diabetes Mellitus (DM) · Positive Psychology Interventions (PPIs)

1 Introduction

According to World Health Organization (WHO, 2002), mental health (MH) is more than the absence of mental disorders. It is a state of mental well-being that helps an individual to overcome life stressors, realize abilities, and contribute to the community. MH conditions are not only limited to mental disorders and psychological

A. Adeel (✉) · C. Kannangara
University of Bolton, Bolton, UK
e-mail: aa60mpo@bolton.ac.uk

H. Bharaj · A. Basu
Bolton NHS Foundation Trust, Bolton, UK

B. Green
Forensic Nursing Consultant, Bolton, UK

C. Iwendi et al. (eds.), *Proceedings of ICACTCE'23—The International Conference on Advances in Communication Technology and Computer Engineering*, Lecture Notes in Networks and Systems 735, https://doi.org/10.1007/978-3-031-37164-6_53

disabilities but also include all mental conditions related with substantial distress, impaired functioning, or threat of self-harm. Multiple factors contribute to the mental health conditions including individual, social, and structural determinants. One such factor is being diagnosed with a chronic health condition [1].

Diabetes mellitus (DM) is a chronic condition with high levels of blood glucose level. This happens if the body does not produce insulin, then the person is diagnosed with type 1 diabetes or if the produced insulin is insufficient or ineffective, this is type 2 diabetes. The other type of diabetes is gestational diabetes, which is diagnosed in pregnant women usually in the third trimester. Over a period, poorly managed diabetes has serious complications including damage to the heart, eyes, feet, and kidneys [2]. Living with diabetes does not only contribute to physical complications but also to mental health conditions. Depression is 2–3 times more likely in people with diabetes than people without. Similarly, anxiety is 20% more common in people with diabetes than otherwise. Managing diabetes itself is stressful for many people, where they feel discouraged, worried, frustrated, or tired of dealing with diabetes. This overwhelming stress is called diabetes distress. Some 30% to 50% of diabetic people develop diabetes distress in any 18-month period [3].

Diabetes distress is the emotional reaction to living with diabetes, trouble of daily self-management along with long-term complications of diabetes. Social and financial implications also likely to impact diabetes distress including stigma, discrimination, insurance and treatment costs. Emotional symptoms involved in diabetes distress often overlaps with several other mental health conditions, among which depression is common. Despite their similarities, they are different constructs where diabetes distress does not assume the psychopathology like major depression. The '7As' model of diabetes distress describes seven steps (Aware, Ask, Assess, Advise, Assist, Assign, and Arrange) to be clinically applied across two stages: identify diabetes distress and support the person who experiences diabetes distress. An increase in diabetes distress is linked with adverse physiological and psychological consequences in people with diabetes including poor self-management, higher HbA1c, severe hypoglycaemia and compromised quality of life [4].

Positive psychology interventions (PPIs) increase positivity in individuals to help them deal with their negative life events and moods. PPIs include enhancing happiness, well-being, gratitude, empathy, kindness, optimism, meaning in life and strengths [5, 6]. A systematic review identified that several wellbeing interventions including positive psychology interventions, mindfulness-based interventions, Acceptance and Commitment Therapy interventions, and resilience-based interventions, were promising in improving psychological and physical health in people with type 1 and 2 diabetes [7]. Similarly, several studies have found that digital health interventions improved diabetes outcomes [8]. However, the literature provides limited data about the efficacy and effectiveness of digital mental health interventions for improving psychological outcomes in people with diabetes. Therefore, there is a need for more thorough, controlled, and quality studies to develop and evaluate appropriate digital mental health interventions for people with diabetes [9]. Therefore, the aim of this project is to develop and evaluate a user-led clinically approved transition chatbot based on Positive Psychology theory, using Artificial Intelligence (AI) and integrated Mobile App, for people with diabetes to improve their wellbeing. The proposed hypotheses for the study will be:

H1: AIDA-MH will significantly improve the psychological outcomes (diabetes distress, depression, anxiety, and PERMA profile) of people with diabetes.

H2: AIDA-MH will also significantly impact on the diabetes related physiological outcomes (blood sugar levels, HbA1c, Oxytocin levels, blood pressure and weight) of people with diabetes.

2 Method

2.1 *Research Design and Procedure*

A user-led clinically approved transition chatbot using artificial intelligence will be developed and evaluated along with an integrated mobile application. This chatbot will be based on positive psychology theory and will help people with diabetes to improve their psychological wellbeing. The AI Chatbot, AIDA-MH (Artificial Intelligence Diabetes Assistant- Mental Health), will be developed and evaluated over a series of stages, where Public Patient Involvement and Engagement (PPIE) will lead to the design and development of the AIDA-MH Chatbot and a Mobile App. The developed components will be tested in a pilot project with a clinical sample of people with diabetes, to evaluate their effectiveness for mental health outcomes and user friendliness. This will then be followed by rigorous randomized controlled trials of the final prototypes to assess their effectiveness, in comparison with an alternative Cognitive Behaviour Therapy digital mental health intervention, not focused on people with diabetes. The study will be facilitated by a steering group of researchers, consultants, patients with lived experience of diabetes, technical staff with expertise in the algorithms of Artificial Intelligence and Machine Learning, and psychologists with expertise in Positive Psychology Interventions. This steering group will have regularly scheduled meeting and will further contribute to the Patient and Public Involvement and Engagement (PPIE) at several stages of the study to guide it and improve the quality and relevance of the work (Fig. 1).

2.2 *Participants*

This study will mainly be carried out in the Bolton Diabetes Centre, Bolton NHS Foundation Trust and participants will be recruited using purposive sampling. Consultants from the Diabetes Centre will refer diabetes patients suitable for the study which will improve the sample size. No exclusion or inclusion criteria have been applied as part of the recruitment strategy at this stage of the study, however, it will be added as the study will progress further. A heterogenous sample of people

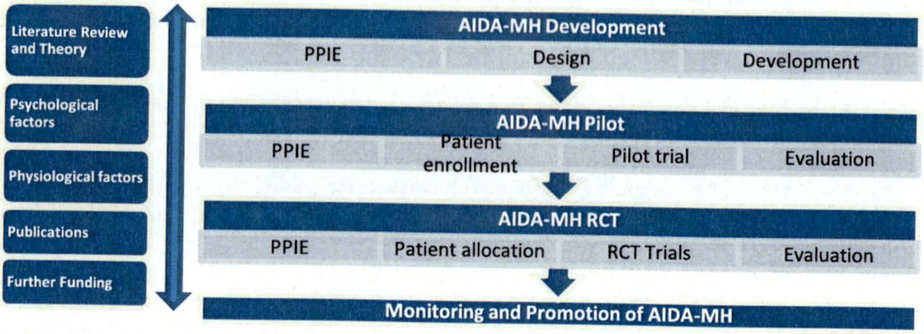

Fig. 1 AIDA-MH design, development, and implementation framework

with diabetes who will have mental health concerns and may be the potential users of the AIDA-MH. It is anticipated that this study will be likely to have more evident outcomes in participants with lower wellbeing scores at the initial stage of the study.

2.3 *Materials*

A range of psychological and physiological measures will be used during different stages of the project. Psychological measures will include the Diabetes Distress Scale (DDS) [10] to assess diabetes distress, Generalized Anxiety Disorder Assessment (GAD) [11] to assess anxiety levels, The Patient Health Questionnaire (PHQ-9) [12] to assess depressive symptoms, The Warwick-Edinburgh Mental Wellbeing (WEBMWBS) [13] to assess overall psychological wellbeing and the PERMA questionnaire [14] to assess flourishing. Physiological measures will include recording blood sugar levels, HBA1c, Oxytocin levels, blood pressure and weight of the people with diabetes.

Diabetes Distress Scale (DDS)

The DDS is widely employed as a clinical instrument to measure diabetes-related emotional distress and for planning clinical interventions for adults with type 1 and type 2 diabetes. This scale covers four areas: emotional burden subscale (EB), physician-related distress subscale (PD), regimen-related distress subscale (RD) and diabetes related interpersonal distress (ID). The 17 items of the scale cover possible problem areas that diabetic people may suffer, ranging from daily hassles to major life events. Respondents are asked to indicate which each of these 17 problems may have upset them over the month and circle the suitable number. The 6-point scale varies from 1 meaning "Not a problem" to 6 meaning "A very serious problem". DDS average score reflects overall diabetes distress, and each subscale score reflects the significance of the items in that scale. EB measures feelings like being exhausted, or fearful about managing the challenges of diabetes, whereas RD measures feelings like failing by not controlling diabetes well e.g. meal plan etc. ID measures feelings like not receiving sufficient support for diabetes from friends and family and PD measures worries about healthcare and acquiring adequate support from health care providers. The average score less than 2.0 reflects little or no distress, score between 2.0 and 2.9 reflects average distress whereas score more than 3.0 reflect severe distress. The 17-item scale has a very high correlation with the 28-item original scale ($r = 0.99$) indicating the efficacy of the 17-item scale with 40% fewer items. DDS has good internal reliability and validity across four different clinical sites. The internal consistency of the 17-item scale ($\alpha = 0.93$) and subscales (EB = 0.88, PD = 0.88, RD = 0.90 and ID = 0.88) is adequate. Elevated DDS scores are related to being younger and depressed, being on insulin and having elevated lipid levels with poor self-care [10, 15].

Generalized Anxiety Disorder Assessment (GAD-7)

The GAD-7 is a short self-report measure to screen cases of Generalized Anxiety Disorder (GAD). GAD-7 evaluates the incidence and severity of GAD in practice. This 7-item anxiety scale assesses the symptom criteria for GAD from the Diagnostic and Statistical Manual of Mental Disorders, Fourth Edition (DSM-IV-TR). The participants are asked to specify how frequently they have been worried by the given problems over the last 2 weeks. The 4-point Likert scale ranges from 0 meaning "Not at all" to 3 meaning "Nearly every day". A score of 10 or more signifies the cut point for detecting cases of GAD whereas a cut-off points of 5 is mild level of anxiety, 10 represents moderate levels and 15 represents severe levels of anxiety. The internal consistency ($\alpha = 0.92$) and test–retest reliability ($r = 0.83$) of the scale are adequate. Good convergent validity of the scale was determined by its correlation with two existing anxiety scales: the Beck Anxiety Inventory ($r = 0.72$) and the anxiety subscale of the Symptom Checklist-90 ($r = 0.74$) [11].

Patient Health Questionnaire (PHQ-9)

The PHQ-9 is a short, self-administered scale to establish a depressive disorder diagnosis and depression symptom severity. This 9-item scale is based on the 9 criteria of DSM-IV major depressive disorder diagnosis. The 4-point Likert scale ranges from 0 meaning "not at all" to 3 meaning "nearly every day". PHQ-9 scores range from 0 to 27, where the cut-off points of 5, 10, 15 and 20 represent mild, moderate, moderately severe, and severe depression, respectively. PHQ-9 has excellent internal consistency ($r > 0.8$) and good test–retest reliability ($r = 0.84$) [12].

Warwick-Edinburgh Mental Wellbeing Scale (WEBMWBS)

This scale measures mental wellbeing in the general population. The 14-item measure blends eudemonic and hedonic components of mental health, namely positive functioning (i.e., self-acceptance, autonomy, competence, and personal development), positive affect (i.e., happiness, calmness, and optimism), and fulfilling interpersonal relations. This scale has 14 statements on feelings and thoughts and participants are asked to select the number that best describes their experience over the last 2 weeks. The scale uses a 5-point scale ranging from 1 meaning "None of the time" to 5 meaning "All of the time". The 5 categories are added together to provide a single score ranging from 14 to 70, with higher score indicative of higher degree of wellbeing. It has strong internal consistency ($\alpha = 0.87$) and acceptable test–retest reliability ($r = 0.66$) [13, 16].

PERMA-Profiler

The PERMA-profiler measures flourishing based on Seligman's wellbeing theory of five pillars, consisting of **p**ositive emotion, **e**ngagement, **r**elationships, **m**eaning and **a**ccomplishment and negative emotion and health. This 23-item scale consists of 15 items for PERMA domains, three items for each and 8 additional filler items to determine overall wellbeing, negative emotions, loneliness, and physical health. The 11-point scale ranges from 0 to 10 with different response anchors for eight different blocks. Average scores are calculated for the items within each factor. Scores are reported visually using a bar chart to provide a person's complete PERMA profile. The PERMA-profiler demonstrates acceptable internal consistency and validity [14].

3 Results

This project will generate both qualitative and quantitative data to be able to evaluate the validity, reliability and the effectiveness of AIDA-MH and integrated Mobile App for people with diabetes. Pre- and post- comparisons will include descriptive and inferential statistics and the semi-structured interviews will be analysed using thematic analysis to understand the clinical impact and benefits to patients with diabetes.

4 Discussion

4.1 Project Significance

The proposed project will be the first, to our knowledge, to develop and evaluate the effectiveness of a positive psychology based digital mental health intervention for people with diabetes. Secondly, AIDA-MH and the mobile app will be a foundation ground for policy implementation in diabetes services to integrate digital technology to tackle mental health outcomes, along with diabetes related outcomes. This project will also be significant to be carried out in the UK, because more than 4.9 million people have diabetes in the UK, 13.6 million individuals are at risk of type 2 diabetes and if nothing changes, 5.5 million individuals will have the diagnosis by 2030 in the UK. According to Diabetes UK, the NHS uses at least 10 billion a year on diabetes and almost 80% of this amount is spend on treating diabetes complications [17]. It is anticipated that findings of this project will feed into further clinical evaluations.

4.2 *Dissemination of Findings*

Data for this project will be collected at different intervals from mid-2022 to mid-2025 and will be analysed using the above-mentioned protocol. We anticipate a minimum of 2 articles from the pilot study and 3 articles from the final project to be published in appropriate peer-reviewed journals along with poster and oral presentations in national and international conferences.

4.3 *Anticipated Limitations*

It is anticipated that attrition may have an impact on the analysis and may introduce bias which could be due to the loss of participants to follow-up, and/or participants may have missing data at one or more points of the study.

5 Conclusion

This paper proposed a protocol for the development and evaluation of the effectiveness of a chat-based chatbot AIDA-MH and an integrated Mobile App for people with diabetes. The proposed methodology and analytic approach will be presented with the projected significance of the project, its limitations, and implications.

Funding Statement The authors are grateful to Health Innovation Manchester (HInM) for funding the pilot study. It is hoped that the pilot study will lead to further funding bids.

Collaborators The project will be carried out in collaboration with Consultants from the Bolton NHS Trust and the Diabetes Centre, Bolton (Dr Harni Bharaj and Professor Ambar Basu), and long-term insulin dependent diabetic patient with lived experience (Dr Barrie Green). Additionally, this project has full support from Diabetes UK, Policy, Campaigns and Improvement provision and Research Provision.

References

1. World Health Organization (WHO): Mental health: strengthening our response. https://www.who.int/. Accessed 15 Jan 2023
2. Diabetes UK: Diabetes: the basics. https://www.diabetes.org.uk/. Accessed 15 Jan 2023
3. Centres for Disease Control and Prevention (CDC): Mental health. https://www.cdc.gov/diabetes/managing/mental-health.html. Accessed 15 Jan 2023
4. Diabetes UK: Diabetes distress (Chap. 3). https://www.diabetes.org.uk/. Accessed 15 Jan 2023
5. Parks, A., Titova, M.: Positive psychology interventions: an overview. In: Wood, A.M., Johnson, J. (eds.) The Wiley Handbook of Positive Clinical Psychology. https://www.researchgate.net/publication/304930527_Positive_Psychology_Interventions_An_Overview (2016)

6. Bolier, L., Haverman, M., Westerhof, G.J., Riper, H., Smit, F., Bohlmeijer, E.: Positive psychology interventions: a meta-analysis of randomized controlled studies. BMC Public Health **13**, 119 (2013). https://doi.org/10.1186/1471-2458-13-119
7. Massey, C.N., Feig, E.H., Duque-Serrano, L., Wexler, D., Moskowitz, J.T., Huffman, J.C.: Well-being interventions for individuals with diabetes: a systematic review. Diab. Res. Clin. Pract. **147**, 118–133 (2019)
8. Maiga, B., Bagayoko, C., Ali, A.A.M., Anne, A., Gagnon, M., Traoré, S.A., Landrier, J.F., Geissbuhler, A.: Effectiveness of digital health interventions for diabetes: systematic review of systematic reviews (2021). https://doi.org/10.21203/rs.3.rs-722330/v1
9. Adeel, A., Kannangara, C., Bharaj, H., Basu, A., Green, B., Ogilvie, L., Carson, J.: Importance of digital mental health interventions to support people with diabetes related psychological outcomes: a narrative review. Mental Health Soc. Incl. (2023). (In Press)
10. Polonsky, W.H., Fisher, L., Earles, J., Dudl, R.J., Lees, J., Mullan, J., Jackson, R.A.: Assessing psychological distress in diabetes. Diab. Care **28**(3), 626–631 (2005)
11. Spitzer, R.L., Kroenke, K., Williams, J.B., Lowe, B.: A brief measure for assessing generalised anxiety disorder. Arch. Intern. Med. **166**(10), 1092 (2006)
12. Kroenke, K., Spitzer, R.L., Williams, J.B.: The PHQ-9. J. Gen. Intern. Med. **16**(9), 606–613 (2001)
13. Stewart-Brown, S., Tennat, A., Tennat, R., Platt, S., Parkinson, J., Weich, S.: Internal construct validity of the Warwick-Edinburgh mental well-being scale (WEMWBS): a Rasch analysis using data from the Scottish Health Education Population Survey. Health Qual. Life Outcomes **7**(1) (2009). https://doi.org/10.1186/1477-7525-7-15
14. Butler, J., Kern, M.L.: The PERMA-Profiler: a brief multidimensional measure of flourishing. Int. J. Wellbeing **6**(3), 1–48 (2016)
15. Diabetes Distress Assessment and Resource Center (DD): The original, 17-item DDS assess diabetes distress for adults with type 1 and type 2 diabetes. https://diabetesdistress.org/dd-assess-score-3. Accessed 06 Feb 2023
16. Clarke, A., Friede, T., Putz, R., Ashdown, J., Martin, S., Blake, A., Adi, Y., Parkinson, J., Flynn, P., Platt, S., Stewart-Brown, S.: Warwick-Edinburgh Mental Well-being Scale (WEMWBS): validated for teenage school students in England and Scotland. A mixed methods assessment. BMC Public Health (2011). https://doi.org/10.1186/1471-2458-11-487
17. Diabetes UK: Diabetes statistics. https://www.diabetes.org.uk/. Accessed 15 Jan 2023

Help over Harm: Practical and Ethical Considerations for the Evaluation and Deployment of Therapeutic Games

Steven Barnes, Tara Chandler, and Mishell Granda Salazar

Abstract This paper collates and discusses key issues pertaining to therapeutic game development, deployment and monitoring. In this paper, we present four areas of challenge, concerning the effective evaluation of therapeutic games; data-security and management; ensuring effective game-design; and barriers to therapeutic game uptake and engagement. The implications of these issues are discussed and authors provide suggestions for ongoing research and towards establishing consensus for standards of practice.

Keywords Therapeutic games · Ethics · Gaming · e-health · m-health · Mental health

1 Introduction

Mental health conditions (MHC) are highly prevalent [5], with reports that 1 in 4 people can be anticipated to have an experience of poor mental health at some point in life [19]. Despite numerous scientifically validated therapies existing for the treatment of MHCs, accessibility, availability and stigma remain problematic [7, 33].

Poor availability, and the infeasibility of treating MHCs solely via in-vivo care has resulted in a rise in interest in digital innovations for mental health (DIMH) [32], tools for MHC intervention deliverable across a multitude of platforms including smartphones, virtual reality, gaming, and via the Internet. DIMHs offer a number of benefits for improving access to psychological treatment, cost-efficiency, and stigma.

Therapeutic games (TGs) utilise ubiquitous mediums (e.g. PCs, smartphones) to deliver treatment via engaging gamification, using an adaptable environment with instant feedback [9]. Research suggests TGs are capable of making significant

S. Barnes (✉)
Uniwersytet SWPS Humanistycznospołeczny, Katowice, Poland
e-mail: sbarnes@swps.edu.pl

T. Chandler · M. G. Salazar
University of Bolton, Bolton, UK

C. Iwendi et al. (eds.), *Proceedings of ICACTCE'23—The International Conference on Advances in Communication Technology and Computer Engineering*, Lecture Notes in Networks and Systems 735, https://doi.org/10.1007/978-3-031-37164-6_54

improvements to clinical presentations and in eliciting lasting behavioural changes [3], including in groups typically challenging to and underserved by traditional treatments, across a number of conditions [4, 5, 12, 24], Tárrega et al. (2015). While implemented across ages, TGs are considered to be highly suited to younger cohorts [8, 10].

While initial research is promising, several important considerations remain to ensure effective TG evaluation and trust in the medium as a means of intervention. Torous et al. [31] recently published a valuable insight paper regarding the use of smartphone apps for health purposes, noting the lack of agreed standards of practice in app evaluation and subsequent implications for users and healthcare professionals in evaluating app quality. This paper follows in a similar vein, in that an urgent need exists for moving towards agreed principles and guidelines in TGs.

2 Areas for Consideration

2.1 *Ensuring and Communicating Effectiveness*

While a number of TGs have been successfully subjected to scientific evaluation, a number remain publicly accessible without such empirical support. A recent study found as few as 3.41% of apps for anxiety and depression, available in the Apple AppStore and Google Play store, were unable to provide evidence to support claims of their effectiveness [17]. Furthermore, despite advertising themselves as anxiety reducing, a number of games available in major app-stores do not reduce anxiety [5]. For users and clinicians, determining appropriate support tools is challenging, with implications for trust. At present, there is no overseeing regulatory body covering the development, evaluation and dissemination of TGs, allowing for market saturation with unproven, ineffective and potentially damaging products.

The lack of consensus regarding TG evaluation, and 'early stage' nature of TG research has led some reviewers to call for more rigorous methodological designs with more extensive follow-up periods, to better evaluate intervention-dose and TG suitability among differing cohort/condition combinations [11, 13, 28] (Kowal et al. 2021). Others point to some evaluations as at risk of bias, the lack of meta-analyses, lack of research into clinically severe populations, and overreliance on non-clinical participants in TG evaluation [1].

TGs can reach groups often underrepresented in in-vivo treatment. Consequently, a degree of TG engagement is likely to be achieved in outpatient contexts by individuals not currently known to formal healthcare systems. Despite this, a large proportion of TG research has been conducted in structured and monitored intervention trials, which while of great value, may not help in understanding TG engagement and efficacy in self-directed, unsupervised environments. As outpatient engagement will depend largely on the user's intrinsic motivation—a challenging requirement in some MHCs—further research into TG use in outpatient settings is warranted.

2.2 Data Security and Management

TGs afford a variety of opportunities for expanding use of eHealth measures, including remote collection and observation of client data. Remote monitoring has numerous advantages for detecting worsening symptom presentation and allowing for earlier intervention, reducing need for future hospitalisation [27], and can be reassuring to users [18]. As videogames often collect numerous metrics indicative of progress and learning (e.g. experience points (XP)), they are fertile ground for embedding such systems.

These systems however collect and share sensitive data with a range of stakeholders (including a range of practitioners and systems for data-storage). User-data allowances ought to be considered, with TGs perhaps storing data locally to the device and only importing/exporting data at fixed intervals [26]. Also, security implications can be of concern to users [30], particularly as a number of health apps which collect and transmit user health data do so despite not having disclosed this in their application privacy policy [14]. Clear and transparent regulations for practice will be valuable in assuaging such concerns. It is also likely that clinicians may need some degree of training in accessing and processing data, particularly if extensive. A number of standards for practice are already in operation for handling sensitive client data in the context of healthcare, meaning a path exists for developing similar standards for TGs.

2.3 Ensuring Effective Therapeutic Game Design and Acceptability/Usability

While TGs exist to support a variety of MHCs, user-involvement in their design is inconsistent, with studies concerning user-acceptance and ease of use often conducted long after a game has been conceptualized, structured, and developed, thereby limiting the potential for meaningful change. As suggested by [25], a player is central to the experience of a game. A client's expectations of the outcome of psychological treatment, and their motivation to engage with the process, are also both key in determining whether engagement will be sustained and whether outcomes will be positive [16, 29]. User-centred design is of significant value in ensuring an acceptable and engaging product.

Effective TG design requires the successful amalgamation of the sometimes-contrasting elements of serious content and entertaining gamification. Collaborative and multidisciplinary teams of practitioners, academics, end-users and designers of games for entertainment are likely to yield the most engaging and therapeutically effective end result [11, 13].

2.4 *Barriers to Therapeutic Game Uptake and Engagement*

While TGs can reduce barriers to access and engagement with psychological therapy, they may also create them. While young people have long been described as being 'digitally native' [23], this is unlikely to always be accurate. Initial acceptability studies also find poorer electronic health literacy in older audiences [6], suggesting an ongoing need for clinician support in TG use [15].

TG use necessitates access to technology and sufficient technological competency, with the potential to alienate those without these means. While smartphone access in the UK is high and growing [22], this is not universally true [2]. Digital healthcare protocols which do not sufficiently account for these issues risk widening existing socioeconomic disparities in access to mental healthcare [20, 21].

3 Conclusions

TGs hold significant potential in expanding reach and accessibility to validated psychological treatments. To ensure trust, practitioners, researchers, government and health policy, end-users and other parties with an interest in the field (e.g. application stores and game-stores) to coalesce around a set of rigorous standards to ensure quality.

The National Health Service in the UK had previously obliged itself to open an online library for the recommendation of health applications, which was unfortunately decommissioned in December 2021. While a small number of NHS trusts have operated similar systems independently (e.g. Central and North-West London NHS Foundation Trust (see Appendix)), there currently is no cohesive and unified system to recommend proven and useful digital tools. The British Standards Institute has, however, published standards and guidelines for health applications, which could potentially be developed further to apply to therapeutic games.

This paper presented a number of areas for discussion and considering regarding TG development, evaluation and dissemination. Authors have outlined the potential implications of the issues raised and aim to prompt discussion to move the field towards establishing standards of practice and to protect the potential value of TGs as a tool for improving mental healthcare.

Funding Statement This research has not received, and is not currently due to receive, funding or financial contribution from any funding agency in the public, commercial, or not-for-profit sectors.

Publication Statement The present paper represents a summary of this work intended for publication in IEEE conference proceedings. A full version of this paper can be found in the journal 'Mental Health and Social Inclusion'.

Appendix

Central and North-West London NHS Foundation Trust (https://www.cnwl.nhs.uk/news/new-local-health-apps-library-help-you-manage-your-health-and-wellbeing).

References

1. Abd-alrazaq, A., Alajlani, M., Alhuwail, D., Schneider, J., Akhu-Zaheya, L., Ahmed, A., Househ, M.: The effectiveness of serious games in alleviating anxiety: systematic review and meta-analysis. JMIR Serious Games **10**(1), e29137 (2022)
2. Ancis, J.R.: The age of cyberpsychology: an overview. Technol. Mind Behav. **1**(1) (2020)
3. Baranowski, T., Buday, R., Thompson, D., Lyons, E.J., Shirong-Lu, A., Baranowski, J.: Developing games for health behaviour change: getting started. Games Health **2**(4), 183–190 (2013)
4. Barnes, S., Prescott, J.: Empirical evidence for the outcomes of therapeutic video games for adolescents with anxiety disorders: systematic review. JMIR Serious Games **6**(1), e3 (2018)
5. Barnes, S., Prescott, J.: Therapeutic gaming for adolescent anxiety: Development and evaluation of a mobile intervention, In: Prescott, J. (ed.) Digital Innovations for Mental Health Support, pp. 187–227. IGI Global (2022)
6. Barnes, S., Bradley, M., Williams, A.: "Piece of mind" and "wellbeing town": engaging service users in the development of a wellbeing game. In: Prescott, J. (ed.) Digital Innovations for Mental Health Support, pp. 151–186. IGI Global (2022)
7. Bradbury, A.: Mental health stigma: the impact of age and gender on attitudes. Community Ment. Health J. **56**(5), 933–938 (2020)
8. Ceranoglu, T.A.: Video games in psychotherapy. Rev. Gen. Psychol. **14**(2), 141–146 (2010)
9. Charlier, N., Zupancic, N., Fieuws, S., Denhaerynck, S., Zaman, K., Moons, P.: Serious games for improving knowledge and self-management in people with chronic conditions: a systematic review and meta-analysis. J. Am. Med. Inform. Assoc. **23**(1), 230–239 (2016)
10. David, O.A., Predatu, R., Cardoş, R.A.I.: Effectiveness of the REThink therapeutic online video game in promoting mental health in children and adolescents. Internet Interv. **25**, 100391 (2021)
11. Dewhirst, A., Laugharne, R., Shankar, R.: Therapeutic use of serious games in mental health: scoping review. BJPsychOpen **8**, e37 (2022)
12. Giner-Bartolomé, C., Fagundo, A.B., Sánchez, I., Jiménez-Murcia, S., Santamaría, J.J., Ladouceur, R., Menchón, J.M., Fernández-Aranda, F.: Can an intervention based on a serious videogame prior to cognitive behavioural therapy be helpful in bulimia nervosa? A clinical case study. Front. Psychol. **6**, 982 (2015)
13. Halldorsson, B., Hill, C., Waite, P., Partridge, K., Freeman, D., Creswell, C.: Annual research review: immersive virtual reality and digital applied gaming interventions for the treatment of mental health problems in children and young people: the need for rigorous treatment development and clinical evaluation. J. Child Psychol. Psychiatr. **62**(5), 584–605 (2021)
14. Huckvale, K., Torous, J., Larsen, M.E.: Assessment of the data sharing and privacy practices of smartphone apps for depression and smoking cessation. JAMA Netw. Open **2**(4), e192542 (2019)
15. LaFleur, L.B., Herbert, Z.B., Dupuy, A.S.: Leveling up your game: the use of video games as a therapeutic modality. J. Creat. Ment. Health **13**(1), 58–67 (2018)
16. Lewis, C.C., Simons, A.D., Kim, H.K.: The role of early symptom trajectories and pretreatment variables in predicting treatment response to cognitive behavioral therapy. J. Consult. Clin. Psychol. **80**(4), 525–534 (2012)

17. Marshall, J.M., Dunstan, D.A., Bartik, W.: The digital psychiatrist: in search of evidence- based apps for anxiety and depression. Front. Psychiatr. **10**, 831 (2019)
18. McCann, L., Maguire, R., Miller, M., Kearney, N.: Patients' perceptions and experiences of using a mobile phone-based advanced symptom management system (ASyMS©) to monitor and manage chemotherapy related toxicity. Eur. J. Cancer Care **18**(2), 156–164 (2009)
19. Mind UK: How common are mental health problems? (2020). https://www.mind.org.uk/information-support/types-of-mental-health-problems/statistics-and-facts-about-mental-health/how-common-are-mental-health-problems/
20. Neuhauser, L., Kreps, G.: Rethinking communication in the e-health era. J. Health Psychol. **8**(1), 7–23 (2003)
21. Payne, L., Flannery, H., Kambakara Gedara, C., Daniilidi, X., Hitchcock, M., Lambert, D., Taylor, C., Christie, D.: Business as usual? Psychological support at a distance. Clin. Child Psychol. Psychiatr. **25**(3), 672–686 (2020)
22. Poushter, J., Stewart, R.: Smartphone ownership and internet usage continues to climb in emerging economies: but advanced economies still have higher rates of technology use. Pew Research Centre (2016). https://www.diapoimansi.gr/PDF/pew_research%201.pdf
23. Prensky, M.: Digital natives, digital immigrants. On Horizon **5**, 1–6 (2001)
24. Ruiz, M., Moreno, M., Girela-Serrano, B., Diaz-Olivan, I., Muñoz, L.J., Gonzalez-Garrido, C., Porras-Segovia, A.: Winning the game against depression: a systematic review of video games for the treatment of depressive disorders. Curr. Psychiatr. Rep. **24**, 1–13 (2022)
25. Salen, K., Zimmerman, E.: Rules of Play: Game Design Fundamentals. The MIT Press, Cambridge (MA) (2004)
26. Scherer, E.A., Ben-Zeev, D., Li, Z., Kane, J.M.: Analyzing mHealth engagement: joint models for intensively collected user engagement data. JMIR Mhealth Uhealth **5**(1), e1 (2017)
27. Scherr, D., Kastner, P., Kollmann, A., Hallas, A., Auer, J., Krappinger, H., Schuchlenz, H., Stark, G., Grander, W., Jakl, G., Schreier, G., Fruhwald, F.M.: Effect of home-based telemonitoring using mobile phone technology on the outcome of heart failure patients after an episode of acute decompensation: randomized controlled trial. J. Med. Internet Res. **11**(3), e1252 (2009)
28. Schoneveld, E.A., Wols, A., Lichtwarck-Aschoff, A., Otten, R., Granic, I.: Mental health outcomes of an applied game for children with elevated anxiety symptoms: a randomized controlled non-inferiority trial. J. Child Fam. Stud. **29**, 2169–2185 (2019)
29. Stice, E., Rohde, P., Seeley, J.R., Gau, J.M.: Testing mediators of intervention effects in randomized controlled trials: an evaluation of three depression prevention programs. J. Consult. Clin. Psychol. **78**(2), 273–280 (2010)
30. Sundram, F., Hawken, S.J., Lucassen, M., Greenwood, A., Merry, S.N.: Tips and traps: lessons from codesigning a clinician e-monitoring tool for computerized cognitive behavioural therapy. JMIR Mental Health **4**(1), e3 (2017)
31. Torous, J., Andersson, G., Bertagnoli, A., Christensen, H., Cuijpers, P., Firth, J., Haim, A., Hsin, H., Hollis, C., Lewis, S., Mohr, D.C., Pratap, A., Roux, S., Sherrill, J., Arean, P.A.: Towards a consensus around standards for smartphone apps and digital mental health. World Psychiatr. **18**(1), 97–98 (2019)
32. Torous, J., Bucci, S., Bell, I.H., Kessing, L.V., Faurholt-Jepson, M., Whelan, P., Carvalho, A.F., Keshavan, M., Linardon, J., Firth, J.: The growing field of digital psychiatry: current evidence and the future of apps, social media, chatbots, and virtual reality. World Psychiatr. **20**(3), 318–335 (2021)
33. World Health Organization: Mental health (2021). https://www.who.int/health-topics/mental-health#tab=tab_1